Internal Rotation
in Molecules

WILEY MONOGRAPHS IN CHEMICAL PHYSICS

Editor
John B. Birks, *Reader in Physics, University of Manchester*

Internal Rotation in Molecules

Edited by

W. J. Orville-Thomas

*Department of Chemistry and Applied Chemistry,
University of Salford*

Assistant Editor: **Mavis Redshaw**

A Wiley–Interscience Publication

JOHN WILEY & SONS

London · New York · Sydney · Toronto

Library of Congress Catalog card No 73-2791

ISBN 0 471 65707 7

Printed in Great Britain by
J. W. Arrowsmith Ltd., Bristol, England

Preface

In the first half of this century chemical methods proved a powerful tool in the elucidation of molecular configuration and conformation. Their importance then decreased as they gradually became superseded by physical methods of structural determination.

The general position up to 1954 is excellently summarized in the book *Structure of Molecules and Internal Rotation* by S. Mizushima, in which accounts are given of the use in structural determination of infrared and Raman spectroscopy, of dipole moments and electron diffraction data. The current position is that the emphasis has tilted even further in favour of the use of a greater range of physical methods and a new factor has emerged to the extent that theoretical methods are becoming of increasing importance and power in the elucidation of conformational problems. In addition, the theoretical approach is adding to the sparse knowledge available on the nature of potential barriers to rotation.

It is the aim of this book to emphasize the relatively greater importance nowadays of the use of physical and theoretical techniques in conformational analysis. As is appropriate, greater space has been devoted to the newer approaches which will be less familiar to the general reader.

Finally, it is with great pleasure that the editor thanks the contributors for devoting so much valuable time to this project, and Dr. G. J. Szasz for bringing to his attention the work of early pioneers. Thanks are also due to Ann Foster, Peter Walkden and John Bignall for assistance in preparation of the final version.

August 1972   W. J. ORVILLE-THOMAS

Contributing Authors

ABRAHAM, R. J. *Department of Organic Chemistry, The Robert Robinson Laboratories, University of Liverpool, Liverpool, U.K.*

ALLEN, G. *Department of Chemistry, University of Manchester, Manchester, U.K.*

BRETSCHNEIDER, E. *Department of Organic Chemistry, University of Liverpool, Liverpool, U.K.*

CLARK, A. H. *Biophysics Division, Unilever Research Laboratories, Sharnbrook, Bedfordshire, U.K.*

CUNLIFFE, A. V. *Ministry of Aviation Explosives Research and Development Establishment, Waltham Abbey, Essex, U.K.*

FEWSTER, S. *Department of Electronics and Electrical Engineering, The University, Glasgow, U.K.*

GITTINS, MISS V. M. *Department of Chemistry and Applied Chemistry, University of Salford, Salford, U.K.*

LADD, J. A. *Department of Chemistry and Applied Chemistry, University of Salford, Salford, U.K.*

ORVILLE-THOMAS, W. J. *Department of Chemistry and Applied Chemistry, University of Salford, Salford, U.K.*

OWEN, N. L. *School of Physical and Molecular Sciences, University College of North Wales, Bangor, Caerns, U.K.*

PARK, P. J. D. *B.P. Research Centre, Sunbury-on-Thames, Middlesex, U.K.*

PETHRICK, R. A. — *Department of Chemistry, University of Strathclyde, Glasgow, U.K.*

RIDDELL, F. G. — *Department of Chemistry, University of Stirling, Stirling, U.K.*

SMYTH, C. P. — *Department of Chemistry, Princeton University, Princeton, New Jersey, U.S.A.*

THOMAS, B. H. — *Department of Chemistry, City of London Polytechnic, London, U.K.*

VEILLARD, A. — *Institut de Chimie, Strasbourg, France.*

WALKER, S. M. — *Donnan Laboratories, University of Liverpool, Liverpool, U.K.*

WARDALE, H. W. — *Department of Chemistry and Applied Chemistry, University of Salford, Salford, U.K.*

WHITE, R. F. — *Department of Chemistry, City of London Polytechnic, London, U.K.*

WYN-JONES, E. — *Department of Chemistry and Applied Chemistry, University of Salford, Salford, U.K.*

Contents

1. Internal rotation in molecules

W. J. ORVILLE-THOMAS

1.1 Introduction	1
1.2 Development of stereochemical theory	1
1.2.1 Theories of Van't Hoff and Le Bel	1
1.2.2 Free rotation	2
1.2.3 Restricted rotation	2
1.3 Conformation and conformational analysis	7
1.3.1 Conformation	7
1.3.2 Conformational analysis	7
1.3.3 Simple hydrocarbons and their derivatives	7
1.3.4 Cyclic compounds	10
1.4 Methods of study	12
1.4.1 Introduction	12
1.4.2 Chemical methods	12
1.4.3 Physical methods	13
1.4.4 Theoretical methods	15
1.5 Results	17
1.6 References	17

2. Studies of conformational equilibria in cyclic compounds by chemical methods

F. G. RIDDELL

2.1 Introduction	19
2.2 Equilibration methods	19
2.3 Kinetic methods	24
2.4 Acknowledgment	26
2.5 References	27

3. Dipole moment, dielectric loss and intramolecular motion

C. P. SMYTH

3.1 Dipole moment	29
3.2 Dielectric loss and relaxation	30

3.3   Molecules containing one axis of dipole rotation .   .   .   31
3.4   Effects of internal field   34
3.5   Molecular relaxation times of substituted ethanes   .   .   37
3.6   Molecules containing more than one axis of dipole rotation .   40
3.7   More rigorous analyses of long molecules   .   .   .   43
3.8   High polymers   44
3.9   Intramolecular relaxation times.   45
3.10 Recent advances (1971–1973 and related papers) .   .   .   51
3.11 References   54

4. Infrared and Raman band intensities and conformational change

P. A. PARK, R. J. D. PETHRICK AND B. H. THOMAS

4.1   Introduction   57
4.2   Concept of rotational isomerism   57
4.3   Classical description of infrared and Raman activity   .   .   58
4.4   Correlation theory   63
      4.4.1   Absorption of radiation   64
      4.4.2   Raman scattering   68
4.5   Thermodynamic description of rotational isomerism   .   .   70
4.6   Assignment of vibrational frequencies   74
      4.6.1   Isotopic effects   74
      4.6.2   Comparison of the spectra of similar compounds   .   .   74
      4.6.3   Vibrational analysis   74
      4.6.4   Intensities   75
      4.6.5   Gas-phase band contours   75
      4.6.6   Polarized infrared studies   75
4.7   Quantitative absolute band intensity determinations   .   .   76
      4.7.1   Gases   77
              4.7.1.1   Method of Wilson and Wells.   .   .   .   77
      4.7.2   Liquids   78
              4.7.2.1   Ramsay method   78
              4.7.2.2   Dispersion methods   79
              4.7.2.3   Attenuated total reflection   80
              4.7.2.4   Interferometry   81
              4.7.2.5   Raman spectroscopy   82
4.8   Discussion of data   84
      4.8.1   Ethane-like molecules (sp^3–sp^3)   85
      4.8.2   Acid chlorides, aldehydes and related molecules (sp^3–sp^2)   90
      4.8.3   Unsaturated aldehydes, ethers and anisoles (sp^2–sp^2)   .   92
      4.8.4   Miscellaneous studies on small molecules   .   .   .   94
      4.8.5   Studies on macromolecules.   95

Contentsxi

4.9   Conclusions 100
4.10 Recent advances (1971–1973) 107
4.11 References 108

5. NMR and ESR studies on simple rotamers

J. A. LADD AND H. W. WARDALE

5.1   Nuclear magnetic resonance 115
      5.1.1   Completely averaged spectra 116
              5.1.1.1   Temperature dependence studies . . . 117
              5.1.1.2   Solvent dependence studies 119
      5.1.2   Dynamic equilibria and line-shape analysis . . 120
              5.1.2.1   Classical line-shape theory . . . 121
              5.1.2.2   The Bloch formulation 123
              5.1.2.3   The Bloch equations for rotating axes . . 124
              5.1.2.4   The Bloch equations modified to account for dynamic
                        exchange processes 125
              5.1.2.5   Quantum mechanical theory of exchange . 127
      5.1.3   Transient phenomena 136
              5.1.3.1   The spin-echo technique 136
              5.1.3.2   Double-resonance techniques . . . 139
5.2   Electron spin resonance 141
      5.2.1   Isotropic hyperfine interactions 142
      5.2.2   Modification of spectra by intramolecular processes . 144
      5.2.3   Nitroaromatic radical anions 150
      5.2.4   Ring inversion in cyclic nitroxides . . . 151
5.3   References 154

6. Studies of internal rotation by microwave spectroscopy

N. L. OWEN

6.1   Introduction 157
6.2   Microwave spectroscopy and the structure of molecules . 158
      6.2.1   Shapes of molecules 158
      6.2.2   Accurate molecular structure 158
      6.2.3   Low-frequency molecular vibrations . . . 159
      6.2.4   Dipole moments 159
      6.2.5   Quadrupole moments 160
      6.2.6   Unstable species 160
      6.2.7   Chemical reactions 160
      6.2.8   Energy transfer and relaxation processes . . 161
      6.2.9   Magnetic measurements 161

6.3   The microwave experiment 161
  6.3.1   Sources of microwave power 161
  6.3.2   Microwave absorption cells 164
  6.3.3   Detection of microwaves 164
  6.3.4   Stark modulation 165
  6.3.5   Frequency measurement 165
  6.3.6   Commercial spectrometers 166
6.4   Rotational spectra of molecules 168
  6.4.1   Spherical top molecules 168
  6.4.2   Linear molecules 169
  6.4.3   Symmetric top molecules 169
  6.4.4   Asymmetric top molecules 170
6.5   The effect of molecular vibrations on rotational spectra . 174
6.6   Microwave methods of studying internal rotation . . 177
6.7   The potential energy associated with internal rotation . . 178
6.8   Solution of the internal rotation Hamiltonian . . . 183
  6.8.1   Principal axis method 184
  6.8.2   Internal axis method 187
6.9   Calculation of potential barriers—an example . . . 188
6.10   Higher order terms in the potential energy . . . 190
6.11   Molecules with low internal rotation barriers . . . 191
6.12   Molecules with more than one internally rotating group . 193
6.13   Potential barriers and energy differences from relative intensity
    measurements 197
6.14   Inversion and ring puckering 199
6.15   Asymmetric rotating groups and rotational isomerism . . 206
6.16   Conclusion 211
6.17   References 211

7. The calculation of barriers to internal rotation from torsional frequencies

A. V. CUNLIFFE

7.1   Introduction 217
7.2   Molecules with a single top with N-fold symmetry . . 218
  7.2.1   Kinetic energy 219
  7.2.2   Derivation of the kinetic energy 221
  7.2.3   The Harmonic oscillator approximation . . 226
  7.2.4 V_{2N} and higher terms 229
  7.2.5   Selection rules 230
7.3   Single-top molecules for which neither the top nor framework
    has a rotational symmetry axis 235
  7.3.1   High barriers—harmonic oscillation approximation . 238

7.3.2   Complete treatment   240
7.3.3   Symmetric well   241
7.3.4   Unsymmetric well   243
7.3.5   Mixing of the torsion with other vibrations   244
7.4   Molecules with more than one top   245
7.4.1   Molecules with two CX_3 tops   245
7.4.2   Selection rules   248
7.4.3   Complete treatment   249
7.5   References   252

8. Torsional vibrations and rotational isomerism

G. ALLEN AND S. FEWSTER

8.1   Introduction   255
8.1.1   Molecules possessing symmetrical tops   255
8.1.2   Molecules possessing asymmetric tops   258
8.1.3   Effect of medium   260
8.2   Spectroscopic techniques   260
8.2.1   Far infrared interferometers   261
8.2.2   Neutron incoherent inelastic scattering   263
8.3   Spectroscopic measurements   267
8.3.1   Molecules possessing symmetrical tops   267
8.3.1.1   Chloroethanes   267
8.3.1.2   Fluoroethanes   269
8.3.2   Molecules possessing two symmetrical tops   270
8.3.2.1   Molecules with two methyl tops   271
8.3.2.2   Molecules with two non-identical tops   274
8.3.3   Molecules possessing two coupled asymmetric rotors   278
8.4   Limitations in the use of torsional frequencies   281
8.4.1   Theoretical assumptions   281
8.4.2   Structural parameters   281
8.4.3   Potential function   281
8.5   Acknowledgment   282
8.6   References   282

9. Molecular acoustics and conformational behaviour

S. M. WALKER

9.1   Introduction   285
9.2   Relaxation theory   287
9.3   Kinetics   292
9.4   Thermodynamic properties   294

9.5   Apparatus   298
    9.5.1   Reverberation   299
    9.5.2   Streaming   300
    9.5.3   Optical methods   301
    9.5.4   Resonance   302
    9.5.5   Interferometry   303
    9.5.6   Pulse methods   304
9.6   Experimental results   308
    9.6.1   Alkanes   308
    9.6.2   Esters   314
    9.6.3   Aldehydes and ketones   317
    9.6.4   Amines   319
    9.6.5   Cyclic molecules   320
9.7   Bibliography and references   322

10. Electron diffraction studies and rotational isomerism

A. H. CLARK

10.1   Introduction   325
10.2   The history of gas-phase electron diffraction   .   .   .   325
10.3   The electron diffraction experiment   326
10.4   The theoretical treatment of electron scattering by semi-rigid
        molecules   329
    10.4.1   Preliminary definitions and relationships   .   .   .   329
    10.4.2   Theoretical expressions for the scattered intensity.   .   331
    10.4.3   Subtraction of the atomic background   .   .   .   334
    10.4.4   Modification of the molecular intensity   .   .   .   335
    10.4.5   The radial distribution curve   335
    10.4.6   Least-squares refinement   337
10.5   Electron diffraction and internal rotation   338
    10.5.1   Introduction   338
    10.5.2   Molecular intensity and radial distribution functions for
              a molecular executing hindered or free internal rotation   339
10.6   Electron diffraction studies of ethane derivatives and related
        molecules   342
    10.6.1   Compounds with only one staggered conformation   .   342
    10.6.2   Compounds with two possible staggered conformations   348
    10.6.3   Compounds with three possible staggered conformations   353
10.7   Electron diffraction studies of sandwich compounds   .   .   354
    10.7.1   Ferrocene, ruthenocene and nickelocene   .   .   .   354
    10.7.2   Biscyclopentadienylmanganese and -beryllium   .   .   356
    10.7.3   Biscyclopentadienyltin and -lead   .   .   .   356

10.8   Electron diffraction studies of compounds containing vinyl,   357
       carbonyl and phenyl groups   358
       10.8.1   Results for vinyl derivatives   360
       10.8.2   Results for carbonyl derivatives   361
       10.8.3   Results for benzene derivatives . . . . .
10.9   Electron diffraction studies of dienes, polyenes and related   362
       compounds   362
       10.9.1   Dienes   364
       10.9.2   Polyenes.   366
       10.9.3   Ene-ones and di-ones   366
       10.9.4   Biphenyls   369
       10.9.5   Diboron tetrachloride . . . .
10.10 Electron diffraction studies of non-cyclic compounds with two
       or more torsional degrees of freedom   370
       10.10.1 The n-alkanes   371
       10.10.2 Compounds with two equivalent rotors . . .   372
       10.10.3 Compounds with three equivalent rotors . .   374
       10.10.4 Compounds with four equivalent rotors . .   375
10.11 Electron diffraction studies of cyclic molecules . . .   376
       10.11.1 Four-membered rings   376
       10.11.2 Five-membered rings   377
       10.11.3 Six-membered rings   378
       10.11.4 Cyclooctane   378
       10.11.5 *cis,cis*-Cyclodeca-1,6-diene   379
       10.11.6 Cyclotetradeca-1,8-diyne   379
10.12 Conclusions   380
10.13 References   380

11.   *Ab initio* calculations of barrier heights

A. VEILLARD

11.1   Introduction   385
11.2   Outline of the theoretical and computational methods .   386
       11.2.1   Approximate solutions of the Schrödinger equation—the
                LCAO–MO–SCF method.   386
       11.2.2   Choice of the expansion basis set—Slater and Gaussian
                functions   388
       11.2.3   The bond-orbital approach   391
       11.2.4   Correlated wave-functions and energies . . .   392
11.3   Discussion of the factors influencing the computed barrier .   393
11.4   Survey of the computed barrier heights   396
       11.4.1   Near Hartree–Fock calculations with geometry optimi-
                zation   396

11.4.2 Calculations with a limited basis set 399
11.4.3 Shape of the barrier 402
11.5 The origin of rotation barriers from *ab initio* calculations . 403
11.5.1 Energy component analysis of the rotation barriers . 404
11.5.2 Analysis of rotational barriers in terms of wave-function
and electron density 409
11.6 Conclusion 411
11.7 Preface to Table 11.5 412
11.8 Recent developments 420
11.9 References 421

12. Ring inversion in some six-membered heterocyclic compounds

MISS V. M. GITTINS, E. WYN-JONES AND R. F. M. WHITE

12.1 Introduction 425
12.2 General concepts of ring inversion 425
12.2.1 Cyclohexanes 425
12.3 Conformational analysis of 1,3-dioxans 429
12.3.1 Introduction 429
12.3.2 Equilibrium studies 432
12.3.2.1 Steric effects 432
12.3.2.2 Anomeric effect 448
12.3.2.3 Internal rotation in exocyclic groups . . 450
12.3.3 Kinetic studies 452
12.3.3.1 Methods of study 452
12.3.3.2 Entropy of activation 454
12.3.3.3 Discussion of experimental results . . 458
12.3.4 Twist-boat conformation 461
12.4 General review of the conformational analyses of cyclic
sulphites 466
12.5 References 478

13. Medium effects on rotational and conformational equilibria

R. J. ABRAHAM AND E. BRETSCHNEIDER

13.1 Introduction 481
13.2 General principles and methods of measurement . . 484
13.2.1 Static methods 485
13.2.2 Dynamic methods 493
13.2.3 Other methods 497
13.3 Theory of solvent effects 498

13.3.1   The reaction-field equation 499
13.3.2   Dipole–dipole interactions 502
13.3.3   Generalized polar interactions 504
13.3.4   Assumptions and limitations of the theory . . . 505
13.3.5   The standard model used 507
13.4   The solvent dependence of rotamer energies . . . 511
13.4.1   Basic compounds, chloro and bromoethanes . . 512
13.4.2   Halogenated alkanes 526
13.4.3   Aldehydes and ketones 537
13.4.4   Other acyclic compounds 545
13.5   The solvent dependence of conformer energies . . . 553
13.5.1   Halocyclohexanes 554
13.5.2   Furfuraldehyde 564
13.5.3   5-Substituted 1,3-dioxans 568
13.5.4   Cyclic ketones 572
13.6   References 579

Author Index 585

Subject Index 602

1 *Internal rotation in molecules*

W. J. Orville-Thomas

1.1 Introduction

The way in which information on the molecular structures of organic compounds is obtained has changed radically with the passage of time. The classical approach consisted of a chemical analysis leading to an empirical formula. This was followed by determining experimentally the patterns of behaviour of a particular substance with reference to those of other compounds. This procedure led to information on the various structural features present in the compound. From the second half of the nineteenth century onwards, thanks to the theories of Kekule and others, this experimental approach aided by inductive reasoning enabled a constitution to be assigned to each new organic substance as it was isolated or synthesized. The constitution of a molecule consists of information on the valence bonding: i.e. which atoms are bonded to which in the molecule, e.g. ethane (I) and formic acid (II). The constitutional formula obtained in this way is usually unambiguous but provides no information on the stereochemistry or the shape

$$
\underset{\text{(I)}}{\text{H}-\overset{\displaystyle \text{H}}{\underset{\displaystyle \text{H}}{\text{C}}}-\overset{\displaystyle \text{H}}{\underset{\displaystyle \text{H}}{\text{C}}}-\text{H}}
\qquad\qquad
\underset{\text{(II)}}{\text{H}-\text{C}\overset{\displaystyle \text{O}}{\underset{\displaystyle \text{OH}}{\big\langle}}}
$$

of the molecule. The overriding importance of the stereochemical aspect was thrust into prominence, however, when two or more substances with distinctly different chemical and physical properties were found to have the same constitution. It thus became obvious that the stereochemistry of molecules had to be elucidated.

1.2 Development of stereochemical theory

1.2.1 *Theories of Van't Hoff and Le Bel*

The first concepts of stereochemistry were put forward by Van't Hoff and Le Bel in the latter part of the nineteenth century. Van't Hoff postulated a tetrahedral model for the bonds associated with a saturated carbon atom. The tetrahedral theory of Van't Hoff was experimentally confirmed by Fischer in 1914, who demonstrated that by interchanging two substituents it was possible to transform an optically active compound into its antipode, which had an equal but opposite optical activity. He also showed that optical activity disappears when two identical substituents are placed on a central carbon atom. The chemical proofs for the tetrahedral distribution of valence bonds around a saturated carbon atom have since been confirmed many times by means of physical methods. These experimental studies have been reinforced by wave-mechanical calculations, which show that the energy of the ethane molecule is at its lowest when the carbon atom uses four sp^3 hybrid atomic orbitals for bond forming.

1.2.2 *Free rotation*

Van't Hoff made one further great contribution to the theory of stereochemistry. In his view, free rotation could occur around any carbon–carbon single bond, but rotation around a double bond was restricted. By rotation about one or more single bonds it is possible to obtain arrangements of the atoms in a molecule which are non-identical with each other. If one considers simple examples, such as 1,2-dichloroethane, this means that the chlorine atoms can assume an infinite number of relative positions by rotation of one CH_2Cl group with respect to the other. It appears, therefore, that a very large number of stereoisomers are possible. Since a large number of isomers for such compounds had never been isolated, it was concluded that carbon atoms bound by a single bond had free rotation around this bond.

1.2.3 *Restricted rotation*

As time passed, the concept of free rotation around single bonds became suspect, as indications were obtained that in many compounds free rotation was restricted and that the molecule existed to a greater or lesser extent in a number of preferred stereoisomers. These non-identical arrangements are known as conformations and the determination of the preferred forms of a particular molecule, i.e. those most favoured energetically, is known as conformational analysis. Clearly, only rotation about bonds between atoms carrying at least one other substituent can lead to conformational differences. Examples of simple molecules in which different conformations are possible are: hydrogen peroxide (III), hydrazine (IV) and ethane (V).

(III)   (IV)   (V)

Structural studies nowadays depend as much upon highly sophisticated equipment as they do upon human intuition. It is proper, however, to acknowledge the great debt owed by modern workers to the classical work done in the field of organic stereochemistry. The advances made by Van't Hoff and Le Bel need no emphasis, but scant regard is given to other pioneers, such as Wislicenus and, in particular, Bischoff (1855–1908). Bischoff was born on April 8th, 1855 in Wurzburg. He went to Wurzburg University in 1873 to study medicine. He gradually lost interest in medical studies and became involved in chemical problems. In order to further his chemical knowledge, Bischoff left Wurzburg for Wiesbaden and worked in the laboratory of Fresenius. From there he proceeded to Heidelberg and worked with Bunsen. Returning to Wurzburg, he carried out research under the general direction of Wislicenus on acetopropionic acid and received his doctorate in 1879. In 1885 Wislicenus succeeded Kolbe at Leipzig University and took Bischoff with him as a lecturer. In 1887 Ostwald left the University of Riga and was succeeded, on the recommendation of Wislicenus, by Bischoff. Here Bischoff remained until April 1908, when severe illness caused him to resign. Bischoff was concerned almost entirely with stereochemical problems arising from his work in the organic synthetic field. He synthesized many hundreds of new materials during his studies in the stereochemical field and his contributions are of lasting importance. He produced a wealth of new material, which supported the ideas of Van't Hoff and Le Bel, based on the idea of asymmetric carbon atoms. So far as I am aware, however, Bischoff's great contributions, which led directly to the important field of conformational analysis, have not received the recognition they deserve.

The first indications that rotation around a single bond was not always free but could be restricted were provided by Bischoff in 1891. He studied the constitution of a large number of organic substances and in a number of key papers[1] he suggested for the first time: (a) that ethane in its equilibrium position had a *staggered conformation*, and (b) that *restricted rotation* occurred in multiple-substituted ethanes, such as the isomeric disubstituted succinic acids, which he correctly formulated as (VI) and (VII). When it is

$$
\begin{array}{ccc}
& CH_3 & \\
H_3C & \diagdown \!\!\!\diagup & COOH \\
& \ominus & \\
H & \diagup \!\!\!\diagdown & COOH \\
& H & \\
& (VI) &
\end{array}
\qquad\qquad
\begin{array}{ccc}
& CH_3 & \\
H_3C & \diagdown \!\!\!\diagup & COOH \\
& \ominus & \\
COOH & \diagup \!\!\!\diagdown & H \\
& H & \\
& (VII) &
\end{array}
$$

remembered that Bischoff also introduced the term 'dynamic isomerism' it becomes clear that, although not hitherto generally recognized, he did anticipate rotational isomerism and, therefore, had great claims to be considered as one of the most important progenitors of conformational analysis. Between Bischoff's work in the 1890's and 1930 it became very clear that rotation around single bonds was not always free and, as a consequence, some conformations are more stable than others.

The first experimental proof that restricted rotation occurred about single bonds was produced in 1922 when Christie and Kenner resolved 2,2'-dinitrodiphenyl-6,6-dicarboxylic acid (VIII) into optically-active forms.[2]

$$O_2N \qquad COOH$$

$$HOOC \qquad NO_2$$

(VIII)

Resolution is possible in this case because of the presence of two bulky groups on each ring in close proximity to each other. These prevent rotation about the central carbon–carbon single bond. In this and similar cases, the barrier to rotation is high. At this time organic chemists were not particularly concerned with the possibility that in certain molecules the energy barrier to rotation could be low. Chemical physicists, however, were becoming more and more intrigued by the increasing indications that comparatively small barriers to rotation existed. Studies on systems in which low barriers were suspected were difficult, however, at this time, owing to the absence of any technique which could demonstrate the phenomenon experimentally. From 1930 onwards, evidence began to rapidly accumulate that the phenomenon of restricted rotation about single bonds was very widespread. One of the first observations involved ethane.[3] It was shown that the observed and calculated entropy were not equal. The most sensible explanation was that a barrier to free rotation of the two methyl groups with respect to each other existed in ethane. At the time, it was not possible to decide unequivocally whether the interaction was attractive, leading to an eclipsed form, or repulsive, leading to a form for the molecule in which the methyl groups were staggered with respect to each other. The existence of such a barrier to free rotation in ethane indicated quite clearly that such barriers must exist for all aliphatic and alicyclic groups in general.

Concurrently, similar evidence was being accumulated on a second important group of compounds. These studies have contributed greatly to the understanding of the phenomenon of restricted rotation. This group consists of compounds containing condensed-ring systems, such as cyclic hydrocarbons, steroids etc. The same ideas of stereochemical theory which hold for aliphatic hydrocarbons and their derivatives, which lead to the idea that certain conformers are energetically preferred to others, can also be applied to cyclic systems. Following Van't Hoff's proposal, in 1874, that the valence bonds around a saturated carbon atom were tetrahedrally disposed, Baeyer, some ten years later, studied ring compounds.[4] He pointed out that the bond angles in cyclopropane and cyclobutane were far from the tetrahedral value. He therefore attempted to explain the properties of ring compounds in terms of a 'strain' theory. According to Baeyer's theory, the strain in cycloalkanes should decrease from that in cyclopropanes to a minimum in cyclopentane and then increase again for larger rings. These predictions were found to be only partially in accord with experimental measurements based on heats of combustion.

This discrepancy arose because of Baeyer's error in assuming that all the ring carbon atoms in cyclic compounds were coplanar. The problem was solved in 1890 by Sachse[5] who abandoned the arbitrary hypothesis of Baeyer that all carbon atoms in cyclic compounds were in the same plane and returned to the simple Van't Hoff picture of a tetrahedral arrangement of valence bonds. As a result, he produced models for cyclic compounds which are strain free. The three carbon atoms in cyclopropane are necessarily coplanar, but in cyclobutane and cyclopentane less strain is introduced into the molecule if all carbon atoms are not coplanar. For six-membered rings and upwards, the construction of rings free of strain is only possible by using puckered structures. For example, the cyclohexane ring can occur in a relatively strain-free form either as a rigid chair form (IX) or a second flexible structure, the boat form (X) (see however § 1.3.4). For a long time, these essentially correct concepts of the spatial structure of cyclohexane put

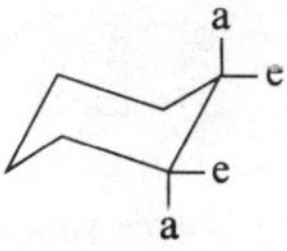

a   e
e
a

(IX) 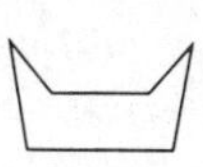   (X)

forward by Sachse were rejected, since it was argued that no evidence for the existence of two cyclohexanes existed. The essential correctness of Sachse's ideas was, however, confirmed in 1918 by Mohr,[6] who showed that the chair and boat forms of cyclohexane were easily interconvertible with the expenditure of a relatively small amount of energy and hence it was unreasonable to expect to be able to isolate the separate isomers.

Sachse made one further important contribution in that he was the first to point out that there could be two monosubstitution products of the chair form, namely, a form in which the substituent was equatorial (XII) and the other in which the substituent was axial (XI) and that these two forms could be interconverted by a process of chair inversion, that is, they would be expected to be in dynamic equilibrium:

$$\text{(XI)} \quad \rightleftharpoons \quad \rightleftharpoons \quad \text{(XII)}$$

As a result of this chair inversion all equatorial bonds become axial and all axial bonds become equatorial. In cyclohexane itself this process occurs very rapidly, since the barrier height to chair inversion is comparatively low.

The final piece of evidence was provided in 1943 by Hassel,[7] who carried out an electron-diffraction study and proved conclusively that cyclohexane was non-planar and existed in the chair conformation only. In the chair conformation the hydrogen atoms are as far apart as possible, corresponding to the staggered form of ethane and other aliphatic compounds in general. This proves that the barrier to rotation is repulsive rather than attractive in character.

It appears, therefore, that Sachse and Bischoff have considerable claims to be considered the founders of conformational analysis. Inexplicably, however, although the basic concepts had already been enunciated as early as the 1890's, their complete acceptance did not come about until 60 years later, when Barton, in a classic paper,[8] emphasized in detail the many chemical consequences of the difference between equatorial and axial substituents. His ideas were taken up enthusiastically, particularly by workers interested in natural products and those concerned with mechanistic studies. Since the appearance of Barton's paper in 1950, the subject has developed rapidly. The experimental work of Hassel and others was soon confirmed by semi-quantitative calculations of non-bonded interactions by Dostrovsky, Hughes and Ingold[9] and by Westheimer and Mayer.[10] These calculations were extended by Barton, who simplified the approach by constructing special models which quickly became an important tool for working out the principles of conformational analysis.[11] The conformational method as developed by Barton, Eliel and others developed into a very powerful tool which, purely by chemical procedures alone, reduced the possible choice between the various configurations.[12] Nowadays, physical methods, based mainly on spectroscopy and diffraction techniques and chemical relaxation methods, have largely superseded the chemical method. In addition, conformational preferences, thanks to the general availability of large computers, can now be calculated by semi-empirical methods.

1.3 Conformation and conformational analysis

1.3.1 *Conformation*

In a compound in which two carbon atoms are joined by a single bond, if free rotation occurs, the substituents attached to the carbon atoms can assume an infinite number of positions relative to each other. The various relative positions of the substituents to each other, brought about by rotation around the carbon–carbon link, are not, however, all of the same energy. Depending upon the nature of the substituents, one or more of the possible arrangements correspond to a minimum in the potential energy of the system. The various shapes which a molecule can assume by means of free rotation around a single bond are known as conformations of the molecule. The expression 'conformation' was first introduced in 1929 by Haworth in his work on the constitution of sugars.[13] Several definitions of conformation have been put forward, including: (a) the different possible arrangements in space of atoms corresponding to a single classical configuration, and (b) those arrangements in space of the atoms of a molecule that arise from the rotation or twisting of bonds and which are not superposable.

1.3.2 *Conformational analysis*

Conformational analysis is concerned with the detailed arrangement in space of the atoms comprising a molecule, i.e. with the three-dimensional structures of molecules. The concept of conformational analysis was introduced by Barton in his fundamental investigations.[8,12,14] The basic premise underlying conformational analysis is that the chemical and physical properties of compounds are closely related to preferred conformations. This is true not only for systems in the ground state, but also for those in transition states and excited states.

1.3.3 *Simple hydrocarbons and their derivatives*

The existence of isomerism in biphenyls was recognized in 1922. Apart from this, the first experimental suggestion based on physical methods that rotation about single bonds could be restricted came in 1930, when it was shown that the *meso* and *dl* forms of stilbene dichlorides differed appreciably in dipole moment.[15] This conclusion was supported by many other investigations of physical properties and this body of data led to speculation about the energy barrier to rotation about carbon–carbon single bonds. This information is admirably summarized in Reference 15, Chapters 1–3.

The next important step forward occurred in 1936, when Kemp and Pitzer[3] made the valuable suggestion that even in ethane a potential barrier to rotation of approximately 3 kcal/mole existed. This assumption made possible the reconciliation of experimental data on heat capacity and entropy of ethane with the values calculated theoretically on the basis of statistical mechanics.

This breakthrough highlighted the fact that as one methyl group in ethane rotates with respect to the other the potential energy of the molecule changes as illustrated in Figure 1.1, in which the potential energy is plotted vertically,

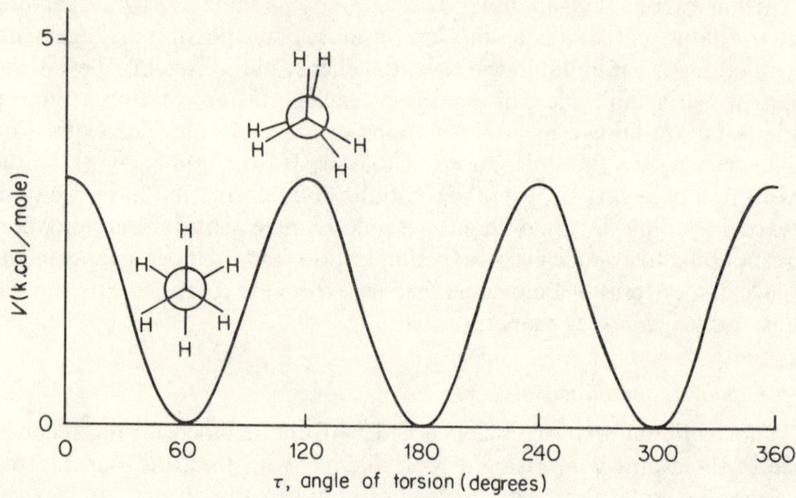

Figure 1.1   Potential energy curve for ethane

whilst the abscissa represents the angle of torsion or azimuthal angle. The figure represents the change of potential energy during the course of one complete rotation about the carbon–carbon single bond. Included in the figure are common representations of the staggered and eclipsed conformers in terms of the Newman projection formulae. In these formulae, for the sake of clarity, the eclipsed conformations are drawn with an angle of torsion of about 10° instead of the proper value, which is 0°. The general form of Figure 1.1 is closely expressed by the expression

$$V(\tau) = \tfrac{1}{2}V_0(1 + \cos 3\tau)$$

where V_0 is the height of the potential barrier.

Figure 1.1 shows graphically that an energy barrier (repeated three times) has to be overcome by the rotating methyl group. In order to cross this barrier an activation energy must be supplied and as a consequence the molecule will tend to adopt one of the relatively stable arrangements corresponding to a minimum point in the potential energy curve. That is, there will be preferred conformations for the molecule. Since the energy difference between individual rotamers or conformers is quite small, the individual isomers cannot, in general, be isolated. However, it should be emphasized that definite physical properties can be attributed to the various conformational isomers and these are usually determined by spectroscopic methods.

Specific heat and spectroscopic data indicate that the value of this energy barrier lies between 2·8 and 3·04 kcal/mole. This value indicates that completely free rotation around the carbon–carbon bond does not exist. An energy barrier of $E \ll kT$ ($RT = 0.6$ cal at room temperature) corresponds to free rotation around the bond, since the molecules in general would possess more than this amount of energy from thermal sources. If $E \gg kT$, then rotation around the carbon–carbon bond does not occur and is replaced by a torsional vibration. These considerations show that ethane is therefore a hindered rotator.

As in the case of ethane, n-butane can exist in a number of possible staggered conformations. Because the substituents are now different, the potential energies of the various staggered and eclipsed conformers differ. Since the stable conformations of ethane are identical, the energy minima have the same value. In butane, however, which can be considered as ethane in which two of the hydrogens are replaced by methyl groups, the potential energy diagram is as shown in Figure 1.2; there are two different minima corresponding to two energetically-feasible conformations. The Newman projections of these conformers are also shown. Examination of the potential energy curve for n-butane shows that the most stable form consists of a staggered

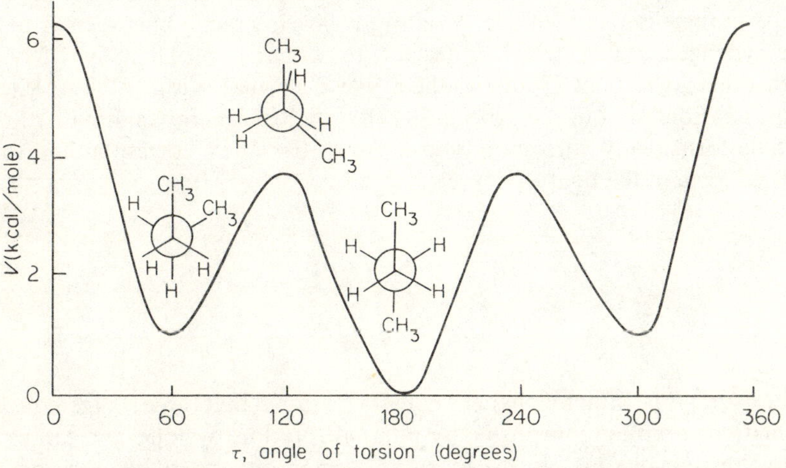

Figure 1.2 Potential energy curve for n-butane

conformation in which the two methyl groups are as far apart as possible; this is known as the *anti*-conformation. Two other conformations which are mirror images of each other are also energetically feasible. They correspond to conformations in which the two methyl groups are adjacent but not eclipsed. These two enantiomorphic forms are called *gauche* conformations. The three possible eclipsed conformations are unstable. In butane, since the *gauche* and *anti* forms differ in potential energy by 0·8 kcal/mole (in favour of the *anti* form), there will be roughly twice as many molecules in the *anti* form as there are in the *gauche* form at room temperature. A simple calculation shows that at room temperature a barrier height of about 20 kcal/mole would be necessary to permit the separation of two conformers.

1.3.4 *Cyclic compounds*

A convenient example is cyclohexane, of which two non-planar forms can be constructed free of strain, as Sachse and Mohr pointed out.[5,6] The two forms consist of a rigid chair form (IX) and a flexible form (XIII) which was for a long time erroneously thought to be a boat form (X) (see page 425). In the chair form, as pointed out by Sachse, the CH bonds are either axial (a in IX) or equatorial (e in IX). A model of the chair form shows that it is free of angle and torsional strain and also free of Van der Waals strain, since no pair of non-bonded atoms approach to within the sum of their Van der Waals radii of each other. These factors are sufficient to ensure that cyclohexane and most of its derivatives exist preferentially in the chair form. This statement is supported by a great deal of physical evidence obtained from X-ray diffraction, electron diffraction, infrared and Raman spectroscopy and from thermodynamic data. It has been pointed out that the boat form, although free of angle strain, does have a certain amount of torsional strain and that there is also a Van der Waals repulsive interaction between some of the hydrogen atoms, which approach to within approximately 1·8 Å of each other, whereas the sum of their Van der Waals radii is 2·4 Å. These considerations lead to the expectation that the true energy minimum of the flexible form really corresponds to a skew boat form, corresponding to a twisted form of the boat (XIII).

(XIII)

In a similar fashion to ethane, where it was found that three staggered conformations are interconvertible by internal rotation, cyclohexane can exist in two distinct chair conformations. In passing from one to the other, a particular chair conformation is first converted to the boat form, which is then converted to the alternative chair conformation, as shown below. This is a simplified representation since, clearly, other forms, such as the twist-boat (XIII), must be involved somewhere along the inversion path (Figure 1.3).

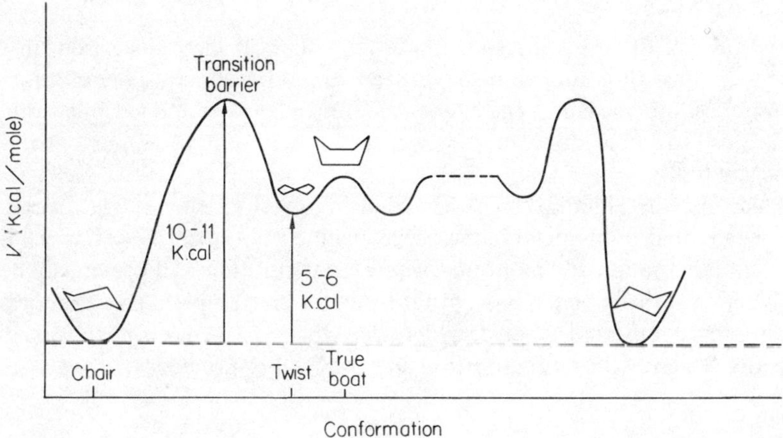

The chair form can be converted into the boat form by the flipping up of one carbon atom or vice-versa. During this process, the valence bond angles change to a small extent and the energy of the system increases. The potential energy curve for the chair inversion process is therefore as shown in Figure 1.3. The height of the energy barrier is about 10 kcal/mole and the chair form

Figure 1.3   Potential energy curve for the chair inversion process in cyclohexane

corresponds to the most stable conformer. In this form it is found that all the hydrogen atoms of the adjacent methylene groups have a staggered conformation relative to each other. The considerations given above indicate that the boat form is unfavourable from the energy point of view. Besides the boat form, there are many other flexible forms possible and it is accepted that the twist-boat conformation has a lower energy than the boat form. The energy difference between the chair and the twist-boat form is nowadays taken to be about 5 kcal/mole and this means that at 25°C the ratio of concentrations of the twist to chair forms would be less than $0 \cdot 001 : 1$, which effectively means that the flexible twist forms can be neglected.

1.4 Methods of study

1.4.1 *Introduction*

In the first half of this century the main method of determining the configuration of a molecule was based on chemical methods. To a great extent these have now been superseded by physical methods of structure determination. Another big change has been the increasingly successful use of theoretical approaches to the determination of molecular stereochemistry.

1.4.2 *Chemical methods*

In principle, a very direct method of determining conformational structures is by the use of calorimetric measurements, involving the determination of the enthalpy and entropy of a substance. This method has been used to demonstrate that the five carbon atoms in cyclopentane are non-planar and that cyclohexane prefers the chair conformation to the boat. This approach has not been used extensively because of the difficult and painstaking experimental work necessary.

Because of the difficulties encountered, chemical methods based on the determination of equilibrium have been developed. In these equilibration techniques, one measures the energy difference between the various isomers and this can be done quite accurately without too much trouble experimentally.

Basically, the chemical methods used involved intelligent guesswork to the extent that configurations were determined not absolutely but relative to a similar 'standard' compound whose configuration had previously been assigned arbitrarily on the basis of agreement. For example, a most important 'standard' compound is glyceraldehyde, whose $(+)$-enantiomer was universally assigned the configuration, using a Fischer projection, of

$$
\begin{array}{c}
CHO \\
| \\
H\text{—}{\longleftarrow}\text{—}OH \\
| \\
CH_2OH
\end{array}
$$

(XIV)

This approach and particularly the symbolism of Fischer projections do not provide much information, however, on the most probable conformer of the molecule.

Using chemical methods conformations can be studied by two main means[12,14]: (a) a study of the thermodynamics of an equilibrium process similar to the conformational equilibrium of interest leads to stereoisomeric information in suitable cases, (b) kinetic methods can be used to compare the relative reactivities of conformers. Following this second approach, chemical reactions are carried out from which predictions of the spatial structure of the molecule can be made. These reactions were usually carried out on model compounds.

Unfortunately, chemical methods do not always produce unequivocal results, because the relation between molecular structure and, for example, rates of chemical reaction are not at all simple. These methods are critically reviewed by Riddell (Chapter 2), who discusses their scope and limitations. In subsequent chapters, particularly that by Gittins, Wyn-Jones and White (Chapter 12), many of the results obtained by chemical methods are discussed and compared with those obtained for similar systems by physical and theoretical methods.

1.4.3 *Physical methods*

Nowadays, physical methods of structural, including conformational, determination are predominant. Amongst the methods used are diffraction techniques (both X-ray and electron), a variety of spectroscopic methods, including microwave, infrared, Raman, NMR, ESR, NQR and ultraviolet; in addition, valuable information has been obtained from dipole moments, the Kerr effect and optical rotatory dispersion.

1.4.3.1 *Diffraction methods*   In principle, an X-ray crystallographic study can be made to yield highly accurate structural details of any crystalline compound. The structural details include, of course, information on preferred conformation. This method is extremely powerful and can cope with very complicated molecules such as vitamin B_{12}. The single greatest limitation of the method, which can be made to yield very accurate bond lengths and bond angles, lies in the fact that only compounds in the crystalline state can be studied. Since the conformation of a particular molecule is not necessarily the same in a crystal as in solution and chemists are more interested in the liquid than in the solid phase, information provided by X-ray methods is sometimes insufficient.

Electron diffraction led to the first historic structural breakthrough in the conformational field. Hassel's lead has been followed by many workers who have obtained a wealth of accurate data. Nowadays, not only accurate geometry can be obtained, but rough values for barrier heights and some information on the form of the potential energy function, the topic dealt with in Chapter 7. The electron diffraction method is most suitable for the study of substances in the gas phase. A limiting factor here is that only comparatively small molecules can be studied, since in a complicated organic compound there are so many different carbon–carbon bond lengths that the radial distribution curve forms almost a continuum and, hence, a large number of suggested structures might very well fit the experimental data obtained. Clark (Chapter 9) describes the various refinements which are currently being made to the standard electron diffraction technique and then shows how this method can be applied to internal rotation.

1.4.3.2 *Spectroscopic methods*   The various spectroscopic techniques yield information which differs in value. For example, Owen (Chapter 6) shows how microwave spectroscopy, where it is applicable, yields very accurate structural information. Bond lengths are determined to within a few thousandths of an angstrom and bond angles to a fraction of a degree. In addition, molecular dipole moments and rotational energy barriers can also be determined from microwave spectra.

Vibrational spectroscopy was one of the earliest physical methods to be used for conformational studies. It is still widely used, as is demonstrated in Chapter 4 by Park, Pethrick and Thomas. In the case of simple molecules containing elements of symmetry, vibrational spectroscopy has been able to establish a complete and exact structure. The long series of papers by Mizushima and his coworkers provides elegant examples of how spectroscopic investigations lead to the determination of conformation.[15] In addition to this method of attack, which involves precise knowledge of group theory, a simpler, less sophisticated approach has been found to be most useful, based on the concept of definite chemical groupings, such as the carbonyl group, giving rise to characteristic frequencies, whose position in the spectrum remains comparatively invariant to molecular environment. These approaches depend upon the measurement of band frequencies. Nowadays band intensities and correlation theory are extensively used and accounts of both these developments are given.

The use of NMR and ESR to yield information on conformation is described by Ladd and Wardale (Chapter 5). NMR spectroscopy can be used in one of two ways. The first is based on the use of chemical shifts, whereas the second uses coupling constants. Up to 1965, virtually all conformational studies carried out by NMR were based on proton resonances but the position changed radically in the subsequent period and now the literature contains a large number of studies involving the fluorine and ^{13}C nuclei.

During the last ten years, ESR techniques have been developed which enable the change in free energy for conformational equilibria involving radicals or radical ions to be obtained. This work has important implications for reaction kinetic studies.

When the barrier height becomes appreciable, free rotation is replaced by restricted rotation in which a torsional vibration occurs around the bond. Clearly, the torsional frequencies are related to the potential barrier restricting rotation. A theoretical analysis of torsional vibrations is given by Cunliffe (Chapter 7) and this forms the basis of examples analysed by Allen and Fewster (Chapter 8). The latter account also describes the methods available for the experimental determination of torsional frequencies. The use of conventional far infrared and laser Raman spectrometers is well known and hence not described fully. The newer methods of Michelson interferometry and neutron incoherent inelastic scattering are, however, dealt with in greater detail.

1.4.3.3 *Relaxation methods* Because of the low energy barriers between conformational isomers these substances exist as an equilibrium mixture at normal temperatures. Again owing to the energetic factors involved, the systems correspond to very fast chemical equilibria and as such cannot be studied by conventional chemical means. Experimental methods to study very fast reactions have developed enormously since 1950. Since conformational equilibria correspond to very fast reactions, these techniques have been widely used in order to determine the thermodynamic data and kinetic parameters involved. The position up to 1954 is summarized in the Faraday Discussion Volume, **17**, 9–234 (1954).

The passage of a sound wave through a liquid produces periodic temperature changes which accompany the passage of the wave. If the liquid consists of an equilibrium mixture of conformers, the fluctuating temperature changes will perturb the conformational equilibrium. The raising of the temperature of the liquid increases the translational energy and part of this extra energy will be transferred via a coupling mechanism and transformed into conformational energy. This process effectively disturbs the equilibrium and increases the fraction of molecules in the higher energy state, i.e. the concentration of molecules having the less energetically favoured conformation increases. This process is reversed when the temperature decreases in the next half-cycle, when the energy begins to return to the sound wave. Under certain conditions, depending on the frequency of the sound wave, there is a net loss of sound energy. This behaviour corresponds to acoustic relaxation in the liquid and measurement of absorption of sound energy as a function of frequency leads to the possibility of determining the energies associated with internal rotation.[16] The application of sound waves to study liquid systems represents one of the most powerful tools which have recently become available to study conformational equilibria. The mode of procedure and some typical results are given by Walker (Chapter 9).

Dipole moments have long been used in conformational studies. Nowadays, this classical approach is powerfully augmented by the measurement of relaxation times which can be related to the rotation of dipolar molecules and dipolar groups within molecules. The resulting analysis, as shown by Smyth (Chapter 3), leads to valuable information on intramolecular motion.

1.4.4 *Theoretical methods*

The exact conformation taken up by a substance has a critical effect on its reactions and properties. Nowadays conformational arrangements are to a very large extent determined by physical methods but it should be stated that in many cases, where an experimental study is not possible, theoretical techniques have provided very useful information on the preferred conformation or conformations of particular substances. The advent of large computers has enabled this third line of attack to be widely exploited. In general, the methods of theoretical chemistry are used to calculate the energy of a molecule as a function of a torsional angle and in this way it is possible to predict the most stable conformer or conformers.

The theoretical approach also throws some light on the important problem concerned with the nature of the repulsive and other forces which exist between the substituents and the bonds of a molecule capable of internal rotation. The most obvious way to explain the appearance of a potential barrier during internal rotation is to carry out a complete wave-mechanical calculation of the total energy of the molecule as a function of the angle of twist; an examination of the magnitudes of the various energy terms will then provide an insight into the nature of the forces involved. Such calculations are extremely complicated, however, and the various approximations lead to grossly inaccurate results, since the energy term due to rotation is a very small factor in the total picture. The first serious calculation was that of Eyring, who in 1932 calculated the interactions of the hydrogen atoms of the methyl groups.[17] This calculation led to a potential barrier of 0·3 kcal/mole, a value not in good accord with the present accepted barrier of 3·0 kcal/mole. Later wave-mechanical calculations[18] have introduced assumptions such as a lack of cylindrical symmetry of the carbon–carbon σ-bond and attempts have been made to explain hindrance to rotation on the basis of polar bonds present in the molecule and, also, steric repulsion forces.[15,19] Undoubtedly, dipole–dipole and higher multipole interactions are present but they are not sufficient to account for the experimentally measured barrier heights.[20] In the same way, steric repulsion forces seem to be inadequate. An attempt to relate the potential barrier in part to the properties of the carbon–carbon σ-bond with a non-cylindrical electron distribution was also made by Pauling. He suggested that the potential barrier arises from exchange interactions of the hybrid bond orbitals, which are predominantly of the sp^3 type but which were also postulated to possess a small amount of d and f character.[21] An essentially different theory[22] was developed by Eyring and his coworkers in an attempt to explain the potential barrier in ethane and other rotamers. They enunciated the following three main principles for chemical bonds: (a) an electron seeks to lower its potential energy by moving close to the nucleus, (b) an electron seeks to lower its kinetic energy by maximizing its path-length by moving over several nuclei, (c) an electron seeks to lower its kinetic energy by avoiding orbital bending. According to Eyring, in the staggered form of ethane three sets of four nuclei lie in a plane, thereby providing a smooth zig-zag path for the movement of electrons. In the eclipsed form, only semi-circular paths are possible and such paths are not considered to be as favourable as those provided in the staggered form. The idea is that the straighter, smoother paths in the staggered conformation of ethane account for the greater stability of the staggered conformation and illustrate the idea that electrons hate to go around corners.

In larger molecules, such as n-butane, steric factors are clearly of much more importance than in ethane in deciding the relative stability of the various conformers. The present state of the art is indicated by Veillard (Chapter 11), who describes *ab initio* calculations of barrier heights.

1.5 Results

Many experimental data are presented and analysed in terms of particular models in Chapters 2–11. In addition, two topics of particular importance are dealt with separately.

Most of the early work on cyclic compounds dealt with molecules with all-carbon ring systems. Since then, considerable attention has been paid to heterocyclic compounds. Many studies have been made on the intriguing problem concerned with the ring-inversion process and this topic is covered by Gittins, Wyn-Jones and White in Chapter 12.

The determination of the preferred conformations adopted by molecules in solution is obviously important, for example, in the case of cyclic poly-peptides and depsipeptides. This is so since many of these compounds possess biological activity and it is reasonable to expect some relationship between their biological activity and their solution conformers. These molecules are relatively complex, however, and a definitive answer to the problems raised by these studies[23] must necessarily await the production of a coherent theory on the effects of the medium on conformational equilibria of simpler mole-cules. This topic is dealt with by Abraham and Bretschneider in Chapter 13.

1.6 References

1. C. A. Bischoff, *Chem. Ber.*, **23**, 620 (1890); **24**, 1074, 1086 (1891); **26**, 1452 (1891).
2. G. H. Christie and J. Kenner, *J. Chem. Soc.*, **LXXI**, 614 (1922).
3. J. D. Kemp and K. S. Pitzer, *J. Chem. Phys.*, **4**, 749 (1936); *J. Am. Chem. Soc.*, **59**, 276 (1937).
4. A. Baeyer, *Chem. Ber.*, **18**, 2269 (1885).
5. H. Sachse, *Chem. Ber.*, **23**, 1363 (1890); *Z. Physik. Chem.* (*Leipzig*), **10**, 203 (1892).
6. E. Mohr, *J. Prakt. Chem.*, [2] **98**, 315 (1918); *Chem. Ber.*, **55**, 230 (1922).
7. O. Hassel, *Tidsskr. Kjemi Bergvesen Met.*, **3**, 32 (1943). [English transl.: *Topics in Stereochemistry*, **6**, 11 (1971).]
8. D. H. R. Barton, *Experientia*, **6**, 316 (1950).
9. I. Dostrovsky, E. D. Hughes and C. K. Ingold, *J. Chem. Soc.*, 173 (1946).
10. F. H. Westheimer and J. E. Mayer, *J. Chem. Phys.*, **14**, 733 (1946).
11. D. H. R. Barton, *J. Chem. Soc.*, 340 (1948); 1027 (1953).

12. E. L. Eliel, N. L. Allinger, S. J. Angyal and G. A. Morrison, *Conformational Analysis* (London: Interscience, 1965).
13. W. N. Haworth (Ed.), *The Constitution of the Sugars* (London: Arnold, 1929).
14. M. Hanack, *Conformational Theory* (London: Academic Press, 1965).
15. S. Mizushima, *Structure of Molecules and Internal Rotation* (New York: Academic Press, 1954).
16. W. J. Orville-Thomas and E. Wyn-Jones, *Transfer and Storage of Energy by Molecules*, Vol. 2, Ed. G. M. Burnett and A. M. North (London: Interscience, 1969), p. 265.
17. H. Eyring, *J. Am. Chem. Soc.*, **54**, 3191 (1932).
18. E. Bright-Wilson, Jr., *Advan. Chem. Phys.*, **2**, 367 (1959).
19. E. A. Mason and M. M. Kreevoy, *J. Am. Chem. Soc.*, **77**, 5808 (1955).
20. E. N. Lasettre and L. B. Dean, Jr., *J. Chem. Phys.* **17**, 317 (1949).
21. L. Pauling, *Proc. Nat. Acad. Sci. U.S.*, **44**, 211 (1958).
22. H. Eyring, G. H. Stewart and R. P. Smith, *Proc. Nat. Acad. Sci. U.S.*, **44**, 259 (1958).
23. F. A. Bovey, A. L. Brewster, D. J. Patel, A. E. Tonelli and D. A. Torchia, *Acc. Chem. Research*, **5**, 193 (1972).

2 Studies of conformational equilibria in cyclic compounds by chemical methods

F. G. Riddell

2.1 Introduction

In a book devoted largely to physical methods for the determination of molecular structure and conformation, a chapter on chemical methods for conformational studies may seem somewhat out of place. However, a large amount of conformational information has come from chemical methods alone. It is not hard to see why this should be. The mainspring of the enormous interest in conformational analysis in the 1950's and 60's arose from earlier chemical work on steroids and terpenoids which was elegantly interpreted in a pioneering paper by Barton.[1] A few years later Barton and Cookson[2] proposed, as the fundamental tenet of conformational analysis, that the chemical and physical properties of organic molecules depend not only on their gross structure and stereochemistry, but also on the conformations that they prefer to adopt. It follows from this principle that chemical methods can yield useful conformational information.

Basically we can study conformations by two main chemical means. These are: (a) by studying the thermodynamics of a chemical equilibrium corresponding closely to the conformational equilibrium of interest, and (b) by using kinetic methods either to compare reactivities of conformations or to trap a slow conformational interchange by a much more rapid chemical reaction. The purpose of this chapter is to review these methods and to discuss their scope and limitations, rather than to detail results. A comprehensive compilation of conformational data from chemical and other methods may be found in Reference 3.

2.2 Equilibration methods

If a chemical equilibrium, which resembles a conformational equilibrium, can be established, it may be used as a model system. Certain assumptions are involved in this method, and certain precautions must be taken. These are best made apparent in the context of some actual examples.

One of the most widely studied conformational equilibria is that in cyclohexanol (I) $\rightleftharpoons$ (II) (Figure 2.1a). The chemical equilibrium (III) $\rightleftharpoons$ (IV) imitates this conformational equilibrium.[4]

The 4-t-butyl group in (III) and (IV) has such a large axial equatorial energy difference that conformations with axial t-butyl groups can be ruled out. (Other conformation 'locking' groups have been employed notably methyl group in a 1,3-diequatorial relationship or a second six-membered ring fused to the substrate ring with *trans* stereochemistry.) The energy difference between (III) and (IV) therefore arises solely from the change in orientation of the hydroxyl group.

If we make the following assumptions then (III) $\rightleftharpoons$ (IV) is a good model for (I) $\rightleftharpoons$ (II): (a) the cyclohexane ring is in no way deformed by the t-butyl group, (b) the t-butyl group has no electronic effect on the ring or its substituents, and (c) if either assumption (a) or (b) above is not true then these effects are still negligible.

We must also be sure that the equilibrium (III) $\rightleftharpoons$ (IV) is the actual equilibrium established under the experimental conditions. For Raney nickel this is probably true as it serves as a hydrogenation–dehydrogenation catalyst and a probable intermediate is the ketone (V). The equilibrium may be established by other catalysts, for instance, aluminium isopropoxide.[5] In the case of this catalyst it is likely that a fair proportion of the alcohols (III) and (IV) are complexed with the aluminium as an alkoxide and so the equilibrium established is as shown in Figure 2.1(b), the energy difference measured

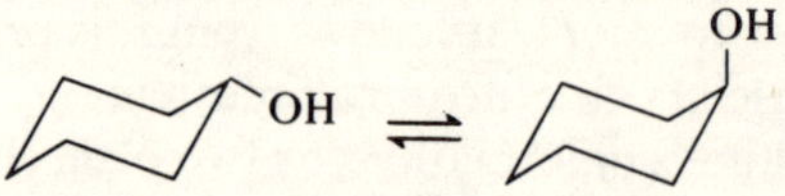

(I)   (II)

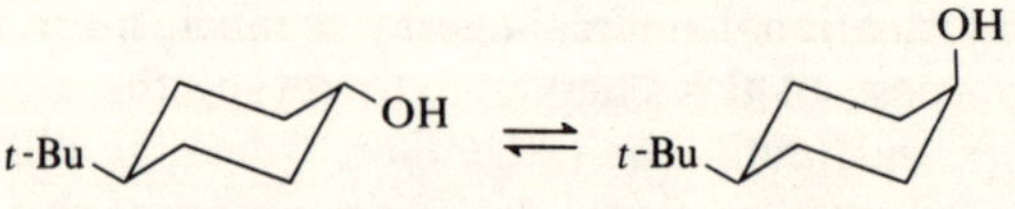

(III)   (IV)

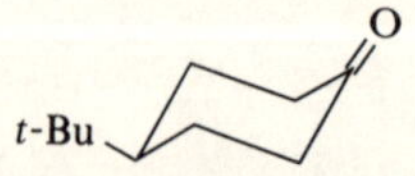

(V)

Figure 2.1(a)

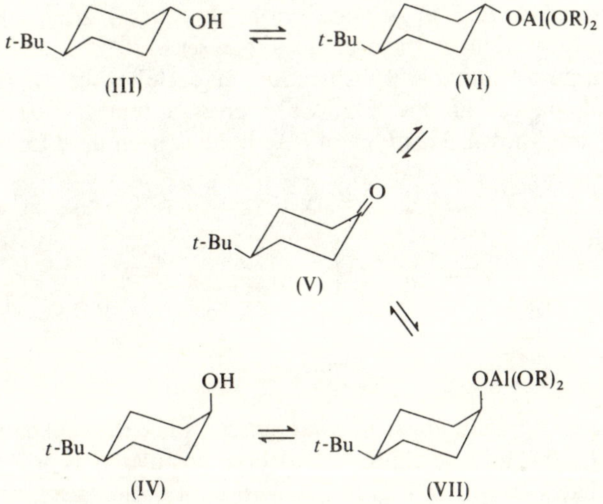

Figure 2.1(b)

corresponding to $(III + VI) \rightleftharpoons (VII + IV)$ which is not what is required.[6] Fortuitously the results with both catalysts agree fairly well. Care must therefore be exercised in choosing the correct catalytic conditions.

When the equilibrium has been established, and this is usually checked by approaching equilibrium from both sides, a sensitive means of analysis is required. This is normally gas–liquid chromatography, although a variety of other techniques, such as NMR and IR, have been used.

Any chemical reaction which allows an equilibrium to be established cleanly and in a reasonable period of time may be used. Hydrogenation–dehydrogenation has proved a useful and versatile reaction and has been used extensively. Raney nickel is the most common catalyst for cyclohexanols[4,7–9] and palladium on charcoal has been widely used for hydrocarbons.[10–16] Rhodium has been tried and rejected for methyl cyclohexanes[17] although other catalysts have been successfully employed.[17–19]

The reversible formation of carbanions in basic media is perhaps the most widely used reaction for equilibration purposes, its importance in this field

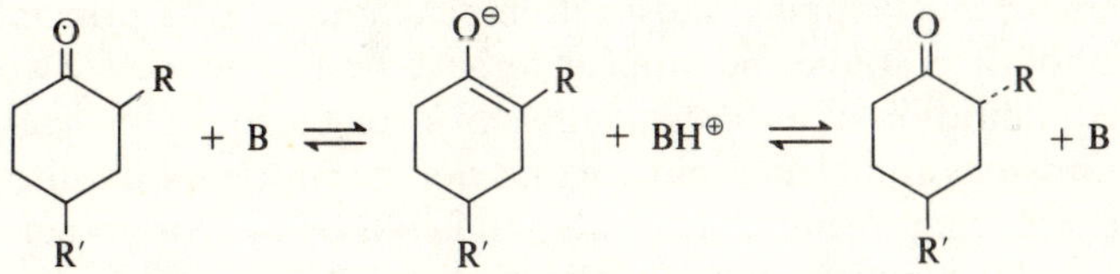

Figure 2.2

matching its importance in organic synthesis. Several groups have used carbanion equilibrations in the cyclohexanone series.[20–24]

Epimerization of suitable substituents on cyclohexane rings may be accomplished via carbanions. Figure 2.3 gives a typical example,[25] but many others are known. This type of reaction has been used for carboxylic

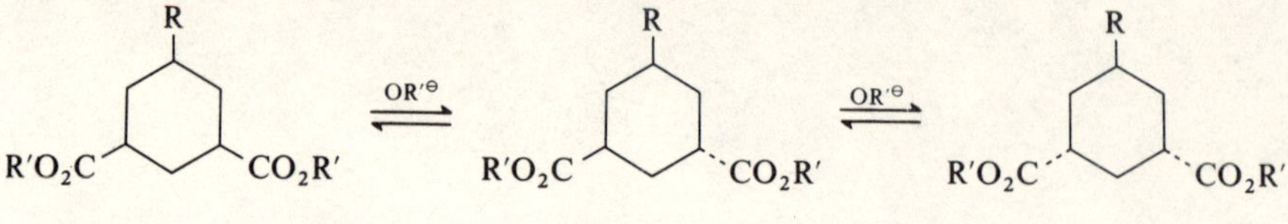

Figure 2.3

acid derivatives[26–35] and for nitro compounds.[36] The thermal equilibration of acids, acid salts and acid chlorides has been postulated as proceeding via the intermediates shown in Figure 2.4, whose formation may involve carbanions.[37]

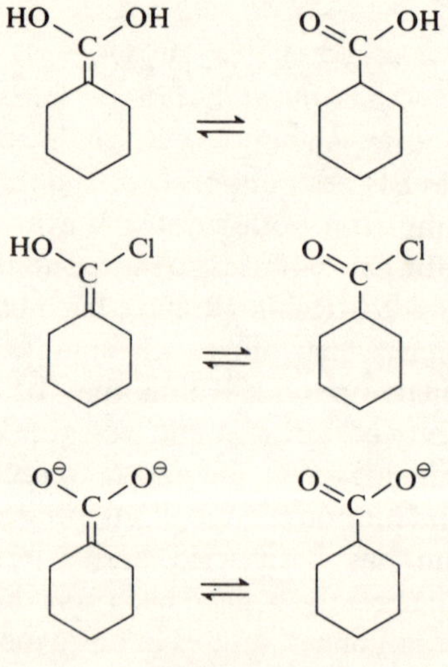

Figure 2.4

Reversible nucleophilic substitution reactions are in principle suitable for equilibration studies but appear to have been little used. Two examples of the equilibration of tertiary alcohols in strong acid[38,39] probably involve carbonium ions and olefins and are thus not true nucleophilic displacements.

Reversible ring-opening of acetals and ketals by anhydrous acids has been a very useful reaction in both alicyclic and heterocyclic conformational studies. The mechanism of the reaction is probably that shown in Figure 2.5.

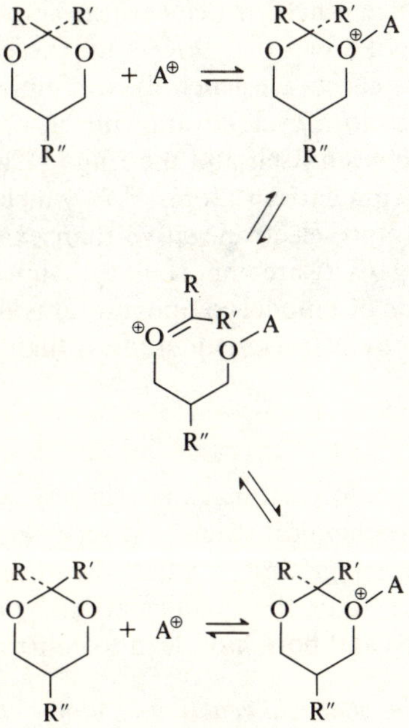

Figure 2.5

In the 1,3-dioxan series BF_3,[40–43] trifluoracetic acid[44] and HCl[45] have been used as catalysts. Complete exchange of the groups at the 2-position of a 1,3-dioxan ring between two molecules setting up a four-component equilibrium has been reported.[46] This technique has rendered possible the study of some previously inaccessible conformational equilibria. Acid-catalysed equilibrations have also been used for thioxolanes[47,48] and dithianes,[49] and have facilitated the study of the anomeric effect.[50,51]

Eliel and Rerick studied the epimerization of cyclohexanols with lithium aluminium hydride and aluminium trichloride.[52] They noted that the more stable isomer was almost exclusively formed in all cases, suggesting that the epimerizing group attached to oxygen was bulky and a suitable conformation-locking agent. This further suggested that $LiAlH_4/AlCl_3$ should be a suitable equilibrating system for conformational studies, and further work using this reagent[53] has given results which agree well with other values.

Cyclohexane mercurials are equilibrated by benzoyl peroxide in pyridine,[54] which indicates a possible radical mechanism for the equilibration. Equilibria in sulphoxides have been established by epimerization at the sulphur atom.[55,56] These equilibria may be accomplished in a variety of ways.[57]

Fundamental to the whole of the preceding discussion of the use of equilibration techniques for conformational equilibria is the assumption that the conformation-locking agent neither deforms the ring sterically nor affects it electronically. It is now known that this is not the case,[58–63] although for practical purposes these effects are generally negligible. A t-butyl group gives a slight flattening effect to a cyclohexane ring in order to minimize non-bonded interactions between itself and the ring.[58] This flattening increases the s character of the ring carbon atoms,[59,60] which consequently behave as if they are slightly more electronegative than expected.[64] Where slight differences in electronegativity are important the t-butyl group is not suitable to lock the conformation of a model compound. This effect is more important in studies involving reactivity (i.e. kinetic studies), than in studies of equilibria.

2.3 Kinetic methods

Kinetic methods for conformational analysis may be split into two distinct categories: (a) where the chemical reaction is very much slower than the rate of conformational interchange, and (b) where the opposite is the case and the reaction is much faster than the conformational process. Both types of methods have been used and both have lead to controversy in the literature.

2.3.1 *Methods where the chemical reaction is slower than the conformational process*

Typical of this type of reaction is the solvolysis of cyclohexyl tosylate (Figure 2.6). This reaction was originally studied by Winstein and Holness.[65]

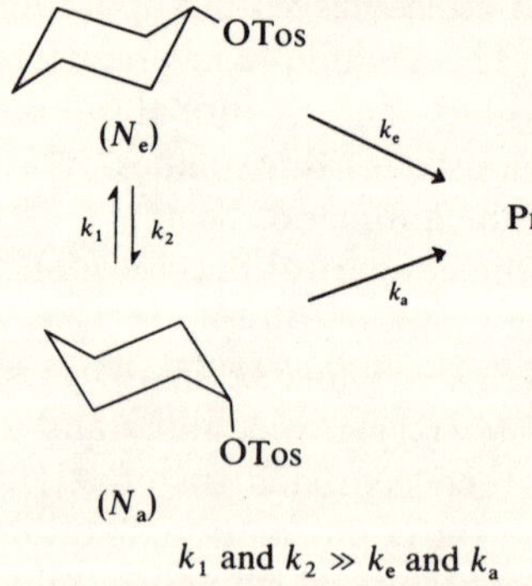

Figure 2.6

They suggested that both the axial and equatorial conformations of cyclohexyl tosylate reacted through separate transition states. As models for these transition states they studied the solvolyses of the *cis*- and *trans*-4-t-butyl compounds, which have axial and equatorial tosylate substituents respectively. The t-butyl group was said to be '... a compelling but remote control of conformation', the first time this group had been used for conformational control.

If both conformations react via distinct transition states the overall rate of reaction (k) is controlled by the mole fractions (N_e, N_a) and the rates of reaction of each conformation (k_e, k_a)

$$k = N_e k_e + N_a k_a$$

In general $k = \sum_1^n N_n k_n$ where there are n distinct transition states corresponding to n different conformations. The model compounds with 4-t-butyl groups provide values of k_e and k_a and so, if all else is correct in the theory, one may find N_e and N_a, enabling the position of the equilibrium to be found.

The ideas put forward by Winstein and Holness were rapidly used by other groups to measure the conformational equilibria in a variety of systems, using many different types of reaction.[66]

Considerable doubt has now been cast upon the validity of the assumptions of this kinetic method. Firstly, evidence has been produced for a *single* transition state in the solvolysis of cyclohexyl tosylate.[67] Secondly, several groups have reached the conclusion that twisted boat-like transition states are probably involved in solvolytic displacement reactions.[68–71] Thirdly, it is now known, as discussed earlier, that 4-t-butyl groups although 'remote' can none the less exert some electronic influence on a ring substituent. All this evidence strikes at the heart of the assumptions necessary for the kinetic method. In view of this mounting criticism the kinetic method of conformational analysis must now be considered dead, and unless some compelling counter-evidence is forthcoming its resurrection is most unlikely.

2.3.2 *Methods where the conformational process is slower than the chemical reaction*

If the rate of a chemical reaction is very much more rapid than that of a conformational interchange the reaction may be used to trap and identify the conformations. This idea is elegantly illustrated by some recent work from Grant's laboratory (Figure 2.7).[72]

The equilibrium between the *cis* and *trans* phosphonites is established slowly (6 hours at 40°). To identify the predominant conformation the mixture was oxidized to the phosphonates at 0°C with N_2O_4, a reaction which is 'instantaneous', and in all probability involves retention of configuration at phosphorous. The major product of the oxidation is the isomer with t-butyl and *cis*-methyl, whose structure was already known from X-ray crystallography. This identifies the more stable component in the phosphonite equilibrium. The argument is made more convincing by extra evidence being produced for retention of configuration during the oxidation reaction.[72]

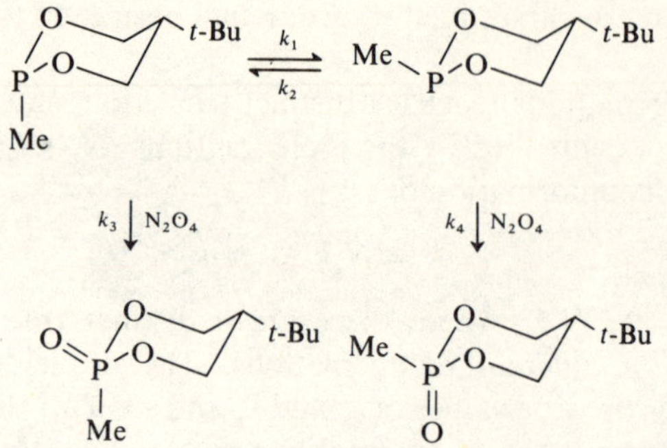

$$k_1 \text{ and } k_2 \ll k_3 \text{ and } k_4$$

Figure 2.7

A more controversial use of this method is in the measurement of con-formational equilibria at nitrogen by rapid protonation or deuterionation. Booth suggested that piperidines reacted so quickly with deuteriotrifluor-acetic acid that the N—H axial–equatorial equilibrium could be trapped.[73] Although there is little doubt that reaction of the base with D^+ is much faster than nitrogen inversion this approach has been criticized by the McKennas.[74] They pointed out that the mixing process of acid with amine is likely to be long in relation to the half-lives of these very fast reactions no matter how efficient the mechanics of mixing may be. Hence there will be local volume elements formed during the mixing where the acidity is not high enough to prevent partial or complete equilibration of the diastereo-isomers. The final ratio of diastereoisomeric cations is therefore likely to be some way between the genuine kinetically-trapped products and the thermo-dynamically-equilibrated products. The same criticism has also been voiced by Robinson who treated dimethylamine with deuterated acid.[75] The product contained dimethyl ammonium ions with two, one and no deu-terium atoms, a clear indication of equilibration during mixing.

Although it may prove possible to use these very rapid reaction methods for the measurement of conformational equilibria it is now apparent that there are limitations set by speed of mixing and speed of conformational inter-change. This is an area which should prove fruitful in the future if these difficulties can be overcome.

2.4 Acknowledgment

It is a pleasure to acknowledge some helpful discussion with Dr. H. Maskill during the preparation of this chapter.

2.5 References

1. D. H. R. Barton, *Experientia*, **6**, 316 (1950).
2. D. H. R. Barton and R. C. Cookson, *Quart. Rev. (London)*, **10**, 44 (1956).
3. J. A. Hirsch in *Topics in Stereochemistry*, Vol. 1, Ed. Allinger and Eliel (New York: Interscience, 1967).
4. E. L. Eliel and S. H. Schroeter, *J. Am. Chem. Soc.*, **87**, 5031 (1965).
5. E. L. Eliel and R. S. Ro, *J. Am. Chem. Soc.*, **79**, 5992 (1959).
6. G. Chuirdoglu, H. Gonze and W. Masschelein, *Bull. Soc. Chim. Belges*, **71**, 484 (1962).
7. E. L. Eliel and E. C. Gilbert, *J. Am. Chem. Soc.*, **91**, 5487 (1969).
8. E. L. Eliel, S. H. Schroeter, T. J. Brett, F. G. Biros and J. C. Richer, *J. Am. Chem. Soc.*, **88**, 3327 (1966).
9. D. J. Pasto and R. D. Rao, *J. Am. Chem. Soc.*, **91**, 2792 (1969).
10. N. L. Allinger and J. L. Coke, *J. Am. Chem. Soc.*, **81**, 4080 (1959).
11. N. L. Allinger and J. L. Coke, *J. Am. Chem. Soc.*, **82**, 2553 (1960).
12. N. L. Allinger and L. A. Freiberg, *J. Am. Chem. Soc.*, **82**, 2393 (1960).
13. N. L. Allinger and S. E. Hu, J. Am. Chem. Soc., **84**, 370 (1962).
14. N. L. Allinger and S. E. Hu, *J. Org. Chem.*, **27**, 3417 (1962).
15. N. L. Allinger and L. A. Freiberg, *J. Am. Chem. Soc.*, **84**, 2201 (1962).
16. N. L. Allinger, W. Szkrybalo and F. A. Van Catledge, *J. Org. Chem.*, **33**, 784 (1968).
17. C. J. Egan and W. C. Bus, *J. Phys. Chem.*, **63**, 1887 (1959).
18. N. D. Zelinsky and E. I. Margolis, *Chem. Ber.*, **65B**, 1613 (1932).
19. A. K. Roebuck and B. L. Evering, *J. Am. Chem. Soc.*, **75**, 1631 (1953).
20. N. L. Allinger and H. M. Blatter, *J. Am. Chem. Soc.*, **83**, 994 (1961).
21. B. Rickbourn, *J. Am. Chem. Soc.*, **84**, 2414 (1962).
22. W. D. Cotterill and M. J. T. Robinson, *Tetrahedron*, **20**, 765, 777 (1964).
23. H. O. House and G. H. Rasmusson, *J. Org. Chem.*, **28**, 31 (1963).
24. J. L. Coke and M. C. Mourning, *J. Org. Chem.*, **32**, 4063 (1967).
25. B. J. Armitage, G. W. Kenner and M. J. T. Robinson, *Tetrahedron*, **20**, 740 (1964).
26. N. L. Allinger and R. J. Curby, *J. Org. Chem.*, **26**, 933 (1961).
27. E. L. Eliel, H. Haubenstock and R. V. Achyra, *J. Am. Chem. Soc.*, **83**, 2351 (1961).
28. N. L. Allinger, L. A. Freiberg and S. E. Hu, *J. Am. Chem. Soc.*, **84**, 2836 (1962).
29. G. J. Fonken and S. Shiengthong, *J. Org. Chem.*, **28**, 3435 (1963).
30. N. L. Allinger and L. A. Tushaus, *J. Org. Chem.*, **30**, 1945 (1965).
31. N. L. Allinger and L. A. Freiberg, *J. Org. Chem.*, **31**, 894 (1966).
32. C. B. Anderson and D. T. Sepp, *J. Org. Chem.*, **33**, 3272 (1968).
33. N. L. Allinger and W. Szkrybalo, *J. Org. Chem.*, **27**, 4601 (1962).
34. B. Rickbourn and F. R. Jensen, *J. Org. Chem.*, **27**, 4606 (1962).
35. K. Brown, A. R. Katritzky and A. J. Waring, *Proc. Chem. Soc.*, 257 (1964).
36. R. J. Ouellette and G. E. Booth, *J. Org. Chem.*, **30**, 423 (1965).
37. E. L. Eliel and M. C. Reese, *J. Am. Chem. Soc.*, **90**, 1560 (1968).
38. J. J. Uebel and H. W. Goodwin, *J. Org. Chem.*, **33**, 3317 (1968).
39. N. L. Allinger and C. D. Liang, *J. Org. Chem.*, **33**, 3319 (1968).
40. E. L. Eliel and M. C. Knoeber, *J. Am. Chem. Soc.*, **88**, 5347 (1966).
41. E. L. Eliel and M. C. Knoeber, *J. Am. Chem. Soc.*, **90**, 3444 (1968).
42. E. L. Eliel and D. I. C. Raileanu, *Chem. Commun.*, 291 (1970).
43. E. L. Eliel and M. K. Kaloustian, *Chem. Commun.*, 290 (1970).
44. F. G. Riddell and M. J. T. Robinson, *Tetrahedron*, **23**, 3417 (1967).
45. B. J. Hutchinson, R. A. Y. Jones, A. R. Katritzky, K. F. Record and D. J. Brignell, *J. Chem. Soc. (B)*, 1224 (1970).

46. E. L. Eliel, Symposium on Conformational Analysis, Brussels, 1969.
47. E. L. Eliel, L. A. Pilato and V. G. Badding, *J. Am. Chem. Soc.*, **84**, 2377 (1962).
48. M. P. Mertes, H. K. Lee and R. L. Schowen, *J. Org. Chem.*, **34**, 2080 (1969).
49. E. L. Eliel and R. O. Hutchins, *J. Am. Chem. Soc.*, **91**, 2703 (1969).
50. C. B. Anderson and D. T. Sepp, *Chem. Ind.* (*London*), 2054 (1964).
51. E. L. Eliel and C. A. Giza, *J. Org. Chem.*, **33**, 3754 (1968).
52. E. L. Eliel and M. N. Rerick, *J. Am. Chem. Soc.*, **82**, 1367 (1960).
53. E. L. Eliel and T. J. Brett, *J. Am. Chem. Soc.*, **87**, 5039 (1965).
54. F. R. Jensen and L. H. Gale, *J. Am. Chem. Soc.*, **81**, 6337 (1959).
55. C. R. Johnson and D. McCants, *J. Am. Chem. Soc.*, **86**, 2935 (1964).
56. C. R. Johnson and W. O. Siegl, *J. Am. Chem. Soc.*, **91**, 2796 (1969).
57. K. Mislow, *Record Chem. Prog. Kresge–Hooker Sci. Lib.*, **28**, 217 (1967).
58. G. Berti, B. Macchia, F. Macchia, S. Merlino and U. Muccini, *Tetrahedron Letters*, 3205 (1971).
59. F. Shah-Malak and J. P. H. Utley, *Chem. Commun.*, 69 (1967).
60. G. E. Hawkes and J. P. H. Utley, *Chem. Commun.*, 1033 (1969).
61. H. Booth and P. R. Thornburrow, *J. Chem. Soc.* (*B*), 1051 (1971).
62. D. J. Pasto and F. M. Klein, *J. Org. Chem.*, **33**, 1468 (1968).
63. M. Pánková, J. Sicher, M. Tichý and M. C. Whiting, *J. Chem. Soc.* (*B*), 365 (1968).
64. H. A. Bent, *Chem. Rev.*, 275 (1961).
65. S. Winstein and N. J. Holness, *J. Am. Chem. Soc.*, **77**, 5562 (1955).
66. E. L. Eliel and R. S. Ro, *Chem. Ind.* (*London*), 251 (1956); E. L. Eliel and C. A. Lukach, *J. Am. Chem. Soc.*, **79**, 5986 (1957); N. Mori, *Bull. Chem. Soc. Japan*, **35**, 1755 (1962); N. Mori and F. Sudi, *Bull. Chem. Soc. Japan*, **36**, 227 (1963); E. L. Eliel, H. Haubenstock and R. V. Achyra, *J. Am. Chem. Soc.*, **83**, 2351 (1961); J. E. Nordlander, J. M. Blank and S. P. Jindal, *Tetrahedron Letters*, 3477 (1969).
67. J. L. Mateos, C. Perez and H. Kwart, *Chem. Commun.*, 125 (1967).
68. H. Kwart and T. Takeshita, *J. Am. Chem. Soc.*, **86**, 1161 (1964).
69. V. J. Shiner, Jr. and J. G. Jewett, *J. Am. Chem. Soc.*, **87**, 1382, 1383 (1965).
70. W. H. Saunders and K. T. Finley, *J. Am. Chem. Soc.*, **87**, 1384 (1965).
71. N. C. G. Campbell, D. M. Muir, R. R. Hill, J. H. Parish, R. M. Southam and M. C. Whiting, *J. Chem. Soc.* (*B*), 355 (1968).
72. W. G. Bentrude, K. C. Yee, R. D. Bertrand and D. M. Grant, *J. Am. Chem. Soc.*, **93**, 797 (1971).
73. H. Booth, *Chem. Commun.*, 802 (1968).
74. J. McKenna and J. M. McKenna, *J. Chem. Soc.* (*B*), 644 (1969).
75. M. J. T. Robinson, International Symposium on Conformational Analysis, Brussels, 1969.

3 Dipole moment, dielectric loss and intramolecular rotation

Charles P. Smyth

3.1 Dipole moment

The electric dipole moment of the molecule was one of the first tools used in the study of intramolecular rotation and the establishment of the existence of rotational isomerism. An electric dipole is a pair of electric charges, equal in size but opposite in sign and very close together. The dipole moment is the product of one of the two charges and the distance between them. The system of negative electronic charges and positive nuclear charges formed by a molecule may be treated as forming a single dipole, the moment of which is the molecular dipole moment, or it may be treated as a system of dipoles lying in molecular axes, or bonds or groups. The vector sum of these dipole moments is the molecular dipole moment. When an electric field acts upon a molecule, the charges shift so as to form an induced dipole, usually much smaller than the permanent dipole arising from structural asymmetry of the molecule. The order of magnitude of a molecular dipole moment is that of the product of an electronic charge, 4.80×10^{-10} electro-static units, times an atomic radius, 10^{-8} cm, that is, 4.80×10^{-18} e.s.u. cm. Dipole moment values are expressed in 10^{-18} e.s.u. cm frequently called a 'debye' and abbreviated to 'D'.

The relation of the low-frequency or so-called static dielectric constant or permittivity ε_0 to dipole moment μ is given by the equation developed by Debye:[1]

$$\text{polarization, } P = \frac{\varepsilon_0 - 1}{\varepsilon_0 + 2}\frac{M}{d} = \frac{4\pi N}{3}\left(\alpha_0 + \frac{\mu^2}{3kT}\right) \tag{3.1}$$

in which M is the molecular weight of the substance in question, d is the density, N is the Avogadro number, the number of molecules per mole, α_0 is the molecular polarizability, i.e. the dipole moment induced in the molecule by unit electric field, k is the Boltzmann constant and T is the absolute temperature. A frequently used approximation is to substitute for the first term in equation (3.1) the expression

$$\frac{4\pi N}{3}\alpha_0 = \frac{n_D^2 - 1}{n_D^2 + 2}\left(\frac{M}{d}\right) \tag{3.2}$$

where n_D is the refractive index for the sodium D line. As dielectric constant or permittivity is usually determined by measuring the capacitance of a condenser first empty and then filled with the material in question, the molecules orienting in the alternating electric field used in the measurements must be able to attain equilibrium with the field in order that equation (3.1) may hold. Consequently, dipole moments are normally measured most accurately in the vapour state, but most conveniently in the liquid state, dilute solutions in non-polar solvents being used to reduce the errors due to the effects of the internal field in the liquid. The methods of measurement and of calculation are described in detail elsewhere.[2–4] The Onsager equation[5] makes possible the calculation of dipole moment from the dielectric constant or permittivity of a pure liquid, sometimes with an error no greater than that in measurements on dilute solutions.[2] The equation may be written

$$\frac{(\varepsilon - \varepsilon_\infty)(2\varepsilon + \varepsilon_\infty)}{\varepsilon(\varepsilon_\infty + 2)^2} = \frac{4\pi Nd}{3M}\left(\frac{\mu^2}{3kT}\right) \tag{3.3}$$

where ε_∞ has been substituted[2] for the square of Onsager's 'internal refraction index' n; ε_∞, variously called, is the infinite frequency or optical dielectric constant.

3.2 Dielectric loss and relaxation

At very high frequencies or high viscosities, the dipolar molecules do not attain equilibrium with the applied field, an absorption of energy occurs, the dielectric constant or permittivity decreases and the ratio of the loss current to the charging current is the loss tangent $\varepsilon''/\varepsilon'$, where ε'' is the dielectric loss and ε' is the measured dielectric constant.[2] The relations of these quantities to ε_0, ε_∞ and the angular frequency ω ($2\pi \times$ frequency in c/s) are given by the equations of Debye[1]

$$\varepsilon' = \varepsilon_\infty + \frac{\varepsilon_0 - \varepsilon_\infty}{1 + \omega^2\tau^2} \tag{3.4}$$

$$\varepsilon'' = \frac{(\varepsilon_0 - \varepsilon_\infty)\omega\tau}{1 + \omega^2\tau^2} \tag{3.5}$$

in which τ is the dielectric relaxation time. Dielectric relaxation is the exponential decay with time of the polarization in a dielectric when an externally applied field is removed; τ is the time in which the polarization is reduced to $1/e$ times its original value, where e is the natural logarithmic base. Examination of equation (3.5) shows that ε'' approaches zero both for small and for large values of $\omega\tau$, while it reaches a maximum value ε_m when $\omega\tau = 1$. The corresponding value of the angular frequency ω_m is evidently the reciprocal of τ and thus provides an experimental definition of τ as $\tau = 1/\omega_m$. The various methods of using these relationships and equations derived from them to obtain τ are described in detail elsewhere.[2]

When two mutually independent relaxation processes exist in a system, having relaxation times τ_1 and τ_2, equations (3.4) and (3.5) become[6,7]

$$\frac{\varepsilon' - \varepsilon_\infty}{\varepsilon_0 - \varepsilon_\infty} = C_1\left(\frac{1}{1 + (\omega\tau_1)^2}\right) + C_2\left(\frac{1}{1 + (\omega\tau_2)^2}\right) \tag{3.6}$$

$$\frac{\varepsilon''}{\varepsilon_0 - \varepsilon_\infty} = C_1\left(\frac{\omega\tau_1}{1 + (\omega\tau_1)^2}\right) + C_2\left(\frac{\omega\tau_2}{(1 + \omega\tau_2)^2}\right) \tag{3.7}$$

where C_1 and C_2 are the relative weights of each relaxation term and $C_1 + C_2 = 1$. These equations may be used when the loss process involves dipole orientation by molecular rotation and by an intramolecular process as well.

3.3 Molecules containing one axis of dipole rotation

In very early work on dipole moments approximate values were obtained for the molecules of 1,3-dichloroethane, 1,2-dibromoethane and 1,1,2,2-tetrabromoethane and it was pointed out that rotation of one half of the molecule relative to the other around the C—C bond could change the resultant molecular moment from zero to a maximum value.[8] It was further indicated as probable that these substances consisted of molecules, the two halves of which, -CH_2X or -CHX_2, occupied every possible stage between the positions for minimum and for maximum moment, although it would not follow that all positions were equally probable. It thus appeared that the moment value calculated from experimental data represented an average of values running from zero to a certain maximum value. The molecules of glycol and glycerol were similarly regarded as having moments which were the resultants of those of movable dipoles. An equation was developed later for calculating the moment of the substituted-ethane type of molecule without regard for differences in the probabilities of the rotational positions,[9] as were equations for several other molecular types,[10] which later became special cases of a general equation[11] for the mean-square sum of n dipole moments, freedom of rotation around the bonds being assumed:

$$\mu_a^2 = \sum_{j=1}^{n} m_j^2 + 2\sum_{j=1}^{n}\sum_{s<j}\prod_{k=j}^{s+1} \cos\theta_k m_j m_s \tag{3.8}$$

The meaning of the various quantities will be evident from the statement, sufficient in itself for the calculation, that for free rotation about connecting lines the mean-square sum of n vectors is equal to the sum of the squares of the lengths of the separate vectors, plus twice all the products of the lengths of two vectors multiplied by the product of the cosines of the angles θ_k made by the directed lines connecting the pair; θ_k is zero when the directed lines are continuations of each other.

The product $m_j m_s$ of the lengths of the vectors, that is, the values of the dipole moments, is positive if they point in the same direction in passing along a chain and negative if they point in opposite directions. If a portion of a molecule is stiff, the dipoles existing in that portion may be resolved into a single resultant vector passing through the nearest point of rotation. If rotation around any bond is hindered, it must receive special consideration.*

So little was known about interatomic forces within molecules that the variation of intramolecular potential energy with rotation around the C—C bond was, at first, calculated in terms of dipole–dipole interaction energies, simplified so as to give maximum repulsive energy for the *cis* position and minimum for the *anti*.[12,13] A calculation of the potential energy of the ethane molecule as a function of the azimuthal angle of rotation of one methyl group relative to the other around the C—C bond was carried out in terms of the energies between the three hydrogens in one methyl group and those in the other.[14] Of course, maxima occurred at each of the three equivalent positions, 120° apart, at which the three hydrogens of one methyl group were as close as possible to the three of the other methyl, a position often called 'eclipsed', and minima occurred at each of the three 'staggered' positions, 60° from the eclipsed, in which the three hydrogens of one methyl were as far as possible from those of the other. The height of the maxima above the minima, the rotational barrier, was in good agreement with available experimental information and so low as to indicate rotational freedom. However, subsequent experimental information and calculations showed the potential barrier to rotation to be about ten times as high as that indicated by this pioneering calculation.

More rigorous calculations,[13] taking into account dipole–dipole forces and Van der Waals attractive and repulsive forces, gave an improved but still approximate potential energy curve for 1,2-dichloroethane. In general, molecules of this type showed an increase of dipole moment with rising temperature because the *anti* form tended to have minimum or zero dipole moment and minimum potential energy and the *cis* form maximum moment and potential energy.

* From *Dielectric Behavior and Structure* by Charles P. Smyth. Copyright 1955 by the McGraw-Hill Book Company.

When 1,1,2,2-tetrachloroethane was measured in the vapour state its dipole moment was found[15] to vary by no more than the experimental error 0·01 from the value 1·36 D over a range of 35°C, while the similar value of 1,2-dichloroethane increased by 0·06 over the same range of temperature. A very approximate potential energy curve was calculated for the molecule as a function of the azimuthal angle of rotation around the C—C bond, both dipole–dipole forces and steric repulsive forces being taken into account. This curve contained three rather deep minima, of which the lowest corresponded to the *anti* conformation and the other two, which were equivalent, corresponded to staggered conformations of the chlorines and were only 0·36 kcal/mole higher than the *anti* minimum. The molecules were regarded as distributed among the three potential energy troughs having conformations which oscillated around those for the minima. The distribution of the two molecular conformations among the three troughs was calculated by means of the Boltzmann equation

$$n_1/n_2 = 2\,\mathrm{e}^{-E/kT} \tag{3.9}$$

where the factor 2 takes account of the existence of two equivalent troughs for the staggered conformation and E is the energy difference 0·36 kcal/mole between the minimum for each of the two staggered conformations and that for the *anti*. The calculated ratio of the total number of molecules in the two troughs for the staggered conformation to the number in the trough for the *anti* conformation comes out at 1·27:1 at 401°K and 1·32:1 at 436°K. From these relative numbers of molecules and dipole moment values estimated for each of the two molecular conformations, the dipole moment of 1,1,2,2-tetrachloroethane was calculated as 1·46 D at 401°K and 1·47 D at 436°K as compared with the observed value 1·36 D at both temperatures. The difference of 0·1 D between the observed and estimated moments is within the uncertainty of the latter and the variation or absence of variation of moment with temperature is within the experimental error. However years later,[16] the constancy of the moment was observed over a much wider range of temperature and used to estimate an energy difference of $0 \pm 0·2$ kcal/mole between the two conformations in satisfactory agreement with the earlier and less accurate result.

In 1932 Mizushima[17] published the first of many papers (77 from 1932–1959) reporting the extensive work of his group on internal rotation. Further dipole moment measurements were made and studies of Raman spectra added greatly to the understanding of the phenomena involved. The description of the different molecular conformations existing in different potential minima as 'rotational isomers' gave realism to the molecular pictures. The results of infrared, electron diffraction and thermal measurements were also interpreted. An admirable account of this work and other investigations in this field has been given by Mizushima in his book,[18] and a survey of the dielectric aspects of the field was published a year later.[13] In summary, rotational isomers generally exist in *anti* and skew or *gauche* forms, separated by rotational potential energy barriers normally too low to permit of separation of the forms. The most frequently studied rotational isomers or rotamers are those formed as the result of the three potential wells encountered in rotation around a carbon–carbon single bond (C—C). If the carbon–carbon bond is double (C=C), there are two wells, separated by a potential barrier so high that it cannot be readily crossed and stereo-isomerism occurs. The three potential minima around the C—C bond may be of equal energies, as in the ethane or hexachloroethane molecule, or there may be two of equal energies, the skew or *gauche* form, and one of lower energy, the *anti* form, as in 1,2-dichloroethane. Other possibilities obviously exist.

As the dipole moments of the *anti* and *gauche* forms of the rotational isomers can be calculated from bond or group moment values,[10] the relative numbers of the two rotational isomers, n_t and n_g, of 1,2-dichloroethane can be calculated from the measured dipole moment of 1,2-dichloroethane, which is the mean square moment[19] (equation 3.10) in which m_t and m_g are the

$$\mu^2 = \frac{n_t m_t^2 + n_g m_g^2}{n_t + n_g} = \frac{m_t^2 + m_g^2 n_g/n_t}{1 + n_g/n_t} \tag{3.10}$$

calculated moments of the *anti* and *gauche* forms. The energy difference ΔE between the two forms is obtained from the Boltzmann equation

$$\frac{n_g}{n_t} = \frac{2f_g}{f_t} e^{-\Delta E/kT} \tag{3.11}$$

in which f_g and f_t are the partition functions of the two forms and the factor 2 is present because of the presence of the two equivalent *gauche* forms. If the ratio f_g/f_t is approximated as 1, equation (3.11) becomes identical with equation (3.9). Mizushima and his coworkers calculated the ratio f_g/f_t as 0.95 at 298°K and 0.90 at 500°K and obtained ΔE as 1·21 kcal/mole. With rising temperature, the proportion of the more polar molecules (*gauche*), and hence, the observed moment, must increase. For pure liquid 1,2-dichloro-ethane, the isomer ratio was obtained from dipole moment and Raman line intensity measurements, the two methods giving reasonably good agreement.[19,20]

3.4 Effect of internal field

The value of ΔE obtained for the pure dipolar liquid was 1·0–1·2 kcal lower than that for the vapour, which was in good agreement with the lowering of energy of 1·0 kcal for the *gauche* form by electrostatic interaction calculated by Mizushima and his coworkers.[20-22] This stabilization of the *gauche* or polar form in the liquid state results in a higher mean square dipole moment than that observed for the vapour state. However, in the solid state. Mizushima[18] has found that the *anti* form is stabilized almost to the complete exclusion of the *gauche* form in 11 out of 13 molecules with a central C—C bond. In ClH_2C—$CH_2\overset{.}{O}H$ the stable form in the solid is the *gauche* because of hydrogen bonding between the two rotating groups. In Cl_3C—CCl_3 the choice is between a staggered and an eclipsed form, and the staggered form is naturally the more stable.

In view of the demonstrated effect of internal field upon the concentration ratios of the rotational isomers, one would expect solvent to exert an important effect, but, although the moment values found for 1,2-dichloroethane in solution were much higher than those found for the vapour,[23] the solvent dielectric constant was frequently not the determining factor. Benzene and toluene as solvents raised the apparent dipole moment of 1,2-dichloroethane to an abnormal extent. It was suggested that this might result from the formation of an easily dissociated solute–solvent complex, which, however, was not observed in freezing-point measurements.[24]

The energy difference between the *gauche* and *anti* forms is less in solution than in the vapour, with consequent increase in the amount of *gauche* form and decrease in that of *anti*. This is because the highly polar *gauche* form is stabilized in solution by the reaction field which it induces in the surrounding medium. Wada[25] calculated the energy difference in terms of Onsager's field as

$$\Delta E_{gas} - \Delta E_{soln.} = \frac{\varepsilon - 1}{2\varepsilon + 2}\left(\frac{\mu^2}{a^3}\right) \tag{3.12}$$

in which μ is the dipole moment of the *gauche* form in the gaseous state, and a the radius of the spherical cavity occupied by the molecule. Although good agreement with observed results was obtained with equation (3.12), when ΔE_{gas}–$\Delta E_{soln.}$ was plotted against $\varepsilon - 1$ for solutions of 1,2-dichloroethane and 1,2-dibromoethane in various solvents, Wada pointed out that the moment to be used in the equation was not the gas value but the solution value. Unfortunately, when this change was made the agreement between calculated and observed results was unsatisfactory. The model, upon which the calculation was based, was therefore changed from a spherical cavity with a dipole at its centre to a prolate spheroid with two dipoles at the supposed contact points of the carbon and chlorine atoms. With this new model fair agreement was obtained with measurements in hexane, heptane, carbon tetrachloride, carbon disulphide, and even methanol. However, it was pointed out[25,26] that for solutions in benzene and acetone the stabilization energy could not be explained in terms of such a simple electrostatic model and that a kind of short-range force must be considered in these cases.

More measurements (Table 3.1) on 1,2-dichloroethane and 1,2-dibromoethane in mixtures with cyclohexane and benzene[27] gave dipole moment values for 1,2-dichloroethane in good agreement with the results of earlier measurements.[26]

Table 3.1 Dipole moments at 20°C of 1,2-dichloroethane and 1,2-dibromoethane in mixtures with cyclohexane and benzene (x = mole fraction of solute)[27]

x	μ(D) 1,2-$C_2H_4Cl_2$		x	μ(D) 1,2-$C_2H_4Br_2$	
	C_6H_{12}	C_6H_6		C_6H_{12}	C_6H_6
0	1·37	1.76	0	0·94	1·28
0·10	1·36	1·76	0·20	0·90	1·22
0·19	1·41	—	0·40	0·98	1·18
0·27	1·49	1·80	0·60	1·05	1·18
0·53	1·69	1·85	0·80	1·12	1·19
1·00	2·10	2·10	0·90	1·15	1·19
			1·00	1·20	1·20

These dipole moment values were used[27] to calculate the concentrations of molecules in the *gauche* form, the energies of transformation of the *anti* into the *gauche* form, and the energies of stabilization of the *gauche* form. It would appear from Table 3.1 that the *gauche* form, with its relatively large dipole moment, is more strongly stabilized in benzene than in cyclohexane. For 1,2-dichloroethane–cyclohexane mixtures, the calculated mole fraction x_g of a *gauche* form ($x_t + x_g = 1$) increases from 0·289 to 0·660 as x_a, the mole fraction of 1,2-dichloroethane in the mixture, increases from 0 to 1 and the calculated transformation energy ΔE decreases from 0·897 to $-0·012$ kcal/mole. For 1,2-dichloroethane–benzene mixtures, x_g appears to increase from 0·476 to 0·660, while ΔE decreases from 0·430 to $-0·012$. For 1,2-dibromoethane–cyclohexane mixtures, x_g increases from 0·220 to 0·360 as x_a increases from 0 to 1 and ΔE decreases from 1·140 to 0·730 kcal/mole. For 1,2-dibromoethane–benzene mixtures, x_g appears to decrease only from 0·412 to 0·360, while ΔE increases only from 0·610 to 0·738 kcal/mole.

It will be apparent in the next section that solute–solvent interaction may change the dipole moment by a small amount and thus increase the error in the calculation of isomer concentration from dipole moment. Neckel and Volk[27] have taken from the literature values of the transformation energy ΔE_{gas} of the *anti* into the *gauche* form in the gas phase as 1·210 kcal/mole for 1,2-dichloroethane and 1·450 kcal/mole for 1,2-dibromoethane. The difference ΔE between the potential minimum for the *gauche* form and that for the *anti* form is then obtained as $\Delta E_{gas} - \Delta E_{liq.} = \Delta E$. The maximum difference in energy between the *gauche* and *anti* potential minima is 1·222 kcal/mole for pure liquid 1,2-dichloroethane and the minimum difference 0·251 kcal/mole for a 1,2-dibromoethane–cyclohexane solution ($x_a = 0.195$). The strong effect of the solvent benzene upon the observed moment is attributed to a proton donor–acceptor exchange effect between benzene and the *gauche* form, which will be discussed again later.

For three solutions of 1,2-dichloroethane, the excess enthalpies arising from the internal rotation were calculated[28] from a theory derived from the Onsager model, with not very good agreement with experiment being obtained. A similar attack upon the dielectric constants of mixtures of 1,2-dichloroethane[29] yielded better agreement with the measured values.

Piekara and Chelkowski[30] have found that, while the dielectric constants or permittivities of most polar liquids decrease slightly when very high voltages are applied, because of the so-called dielectric saturation effect, the reverse is observed for solutions of 1,2-dichloro- and 1,2-dibromoethane in carbon tetrachloride for voltages of $30–100 \, kV \, cm^{-1}$. The strong field tends to stabilize the polar *gauche* form and increase its concentration at the expense of the practically non-polar *anti* form, with a consequent small increase in dielectric constant or permittivity.

3.5 Molecular relaxation times of substituted ethanes

When the measurement of dielectric relaxation time was first extensively used as a rather rough tool for the investigation of molecular structure and intermolecular forces, the relaxation times of 1,2-dichloroethane and 1,2-dibromoethane, determined in the pure liquid state, showed nothing to differentiate them markedly from the alkyl monohalides.[31] Chitoku and Higasi[32] determined relaxation times for 1,2-dichloroethane in dilute solution in several solvents at 20°C, obtaining the following values in 10^{-12} s (ps) in the solvents listed in order of increasing viscosity: 1·21 in hexane, 4·53 in *p*-xylene, 3·69 in benzene, 2·15 in carbon tetrachloride and 2·16 in cyclohexane. No serious attempt was made to separate the observed absorption into two peaks. Actually, one would expect the relaxation time for possible rotational orientation of the CH_2Cl group to be too close to that of the molecule as a whole to permit of analytical separation in terms of the available data. Chitoku and Higasi suggested a possible approximate picture of *gauche* molecules rotating in a viscous medium with additional hindrance caused by loose complexing by means of hydrogen bonding with solvent molecules. Dipole moment values were obtained in good agreement with those in the literature. Using $\mu_g = 2.57 \, D$ and $\mu_t = 0$, values of x_g and ΔE were calculated in satisfactory agreement with previous values.[27] It had been suggested much earlier[33] that the higher moment of 1,2-dichloroethane in benzene solution resulted from pulling of the two chlorines into the *cis* position by hydrogen bonds formed by them with two adjacent benzene hydrogens. This structure looked attractive on paper, but involved an unstable, eclipsed structure of the 1,2-dichloroethane molecule and was little used. However, Chitoku and Higasi concluded that the hydrogens of the 1,2-dichloroethane were more important than the chlorines in solute–solvent bonding, thus seeming to eliminate the rather ingenious structure proposed by Müller.

Crossley and Walker[34] extended the dielectric absorption measurements of several haloethanes in cyclohexane solution to a higher frequency of 70 Gc/s (4 mm) and calculated mean relaxation times and apparent dipole moments. Pentachloro-, 1,1,1,2-tetrachloro- and 1,1,2,2-tetrabromoethane, which had high potential barriers to internal rotation (6–9 kcal/mole), were found to show no distribution of relaxation times, the dielectric relaxation being characterized by a single value for the time, while 1,1-dichloro-, 1,1-dibromo-, 1,2-dichloro- and 1,1,2-trichloroethane, which have potential energy barriers of about 3 kcal/mole, have sufficiently large distribution coefficients to suggest a possibility of contribution from an intramolecular process. 1,1,2,2-Tetrachloroethane has a distribution coefficient of 0·07, identical with that for 1,2-dichloroethane, but seemingly not enough to evidence intramolecular relaxation, since molecular interaction should be greater in this case. Indeed, only for 1,1-dichloroethane and 1,1-dibromoethane are the distribution coefficients, 0·18 and 0·14 respectively, large enough to suggest, in the absence of other evidence, any probability of intramolecular relaxation. It will be noted that, in several of these molecules, one half of the molecule is symmetrical around the C—C axis so that rotation around this axis will not change the molecular dipole moment, but may change its direction and so contribute to dielectric relaxation.

In continuing their work on the relaxation processes of solutions of halo-ethanes Crossley and Walker[35] cite their previous work[34] in which the relaxation times of dilute cyclohexane solutions of haloethanes plotted against molecular volume and molar volume of the solute gave approximately straight lines showing that the interaction of these compounds with cyclohexane as solvent is constant. They further point out that solvent molecules such as benzene and p-xylene, which possess π-electron bonds, may act as donors with acceptor molecules such as 1,2-dichloroethane in donor–acceptor interactions,[32,36] and cite the relaxation times of chloroform at $20°: 3·2 \times 10^{-12}$ s in cyclohexane, $5·0 \times 10^{-12}$ s in carbon tetrachloride and $7·1 \times 10^{-12}$ s in benzene,[37] which showed that the relaxation time of the chloroform was sensitive to the capacity of the solvent for donating electrons to the protonic hydrogen of the solute. When they measured in p-xylene solution the compounds which they had previously measured in cyclohexane, they found longer relaxation times, which gave free energies of activation for molecular reorientation 0·2–0·5 kcal/mole higher than those for the process in cyclohexane, presumably because of weak hydrogen bonding between the hydrogen (acceptor) of the substituted ethane and the π-electron cloud of the aromatic ring (donor). The difference observed between the solute dipole moments in the two solvents may be dependent upon the volume of the haloethane molecule and the protonic character of its hydrogens.

The chloroethanes studied by Crossley and Walker have been further investigated in somewhat more dilute solution in cyclohexane, benzene, mesitylene and dioxane, the measurements being extended to 2 mm, which should increase the accuracy of the values for short relaxation times.[38] For each chloroethane the dielectric relaxation time was found to increase with increased solvent basicity, an approximately linear relationship existing between the solute relaxation time and the ionization potential of the hydrocarbon solvent. The weak solute–solvent interaction, presumably hydrogen bonding of solute C—H to solvent π-electron cloud in the aromatic hydrocarbons and to oxygen in dioxane, hinders the rotation of the solute molecule and lengthens the molecular relaxation. The solute–solvent interaction in excess of that in cyclohexane, where it is primarily Van der Waals interaction, was calculated[35] as the free energy of activation difference $\Delta\Delta G_0^{\pm}$ for molecular reorientation in two solvents, B and cyclohexane, as obtained from

$$\tau_{\text{B}}/\tau_{\text{cyc.}} = \exp\left(\Delta\Delta G_0^{\pm}/RT\right) \tag{3.13}$$

The values for the different molecules in different solvents at 25°C and 55°C varied from 0·140 kcal/mole for 1,1,1-trichloroethane in benzene at 25° to 0·850 kcal/mole for pentachloroethane in p-dioxane at 55°.

The relaxation times of the chloroethanes in dilute solution increase with increasing basicity and ionization potential of a hydrocarbon solvent because of weak solute–solvent interaction, presumably hydrogen bonding of haloethane hydrogen to the solvent molecule, and the molecular dipole moment shows a parallel increase of value. This increase seems to occur more or less independently of the presence or absence of possible rotation of a dipole around an axis with which it makes an angle as in 1,2-dichloroethane. The increase in moment can be attributed to shift of charge produced by the formation of the hydrogen bond.[2] Thus the increase in dipole moment does not necessarily give evidence of an increase in the proportion of the polar *gauche* form in a mixture of *gauche* and *anti* forms. There would seem, therefore, to be serious question of the accuracy of a calculation of isomer concentration from dipole moments measured in a solution where even weak hydrogen bonding can occur. Increase in dipole moment due to weak hydrogen bonding between solute and solvent would increase the value calculated for the apparent concentration of the *gauche* form, even if no such increase occurred. However, since the *gauche* form has a large dipole moment, the addition of the small moment caused by hydrogen bonding would not invalidate the calculation of an approximate value for the *gauche* form concentration, but would tend to obscure the effect of solvent upon isomer concentration.

3.6 Molecules containing more than one axis of dipole rotation

For molecules containing a single axis of dipole rotation, it has been shown that the two groups rotating around the axis are distributed among a small number of potential energy troughs encountered as the groups rotate. For an equal distribution among an infinite number of such troughs, equation (3.8) makes possible the calculation of the molecular dipole moment. For a simple class of molecules represented by the formula $X(CH_2)_nX$, equation (3.8) gives a dipole moment

$$\mu = m[2 - 2(-\cos\theta)^n]^{1/2} \tag{3.14}$$

where m is the moment in the direction of each C—X bond and θ is the carbon valence angle $= 110°$. When n is 2, the equation applies to the disubstituted ethanes, which have been considered in some detail in the previous sections. In general, equation (3.8) may best be formulated for a specific case by following its statement in words given after the equation. In case of doubt, it may be wise to check the method against one or two of the equations previously developed for several different molecular types.[10]

If we put a suitable value for m (the C—Br moment) and $n = 2$ in equation (3.14), we obtain $\mu = 1·99$ D for BrH_2C—CH_2Br, as compared to an observed value increasing from 0·75 in heptane solution at $-30°$ to 0·97 at $50°$ and a value in benzene solution at $20°$ to $50°$ of 1·19. The reasons for this large difference between the observed and the calculated values for heptane solution and for the solvent effect are apparent from the previous discussion. When n is 3 in this series, that is, for 1,3-dibromopropane, the calculated moment is 2·36 D as compared to 2·03 observed in heptane solution and 1·98 in benzene solution without significant change with temperature in either solvent.

It should be mentioned that, in order to allow for environmental effects, a value of 1·5 D was assigned to m for 1,2-dibromoethane, 1·7 for 1,3-dibromopropane and 1·9 for the molecules with larger n. It would appear that the calculated values of the molecular moment could have an absolute error of 0·1–0·2 D due to uncertainty in m. The moments observed for 1,4-dibromobutane are 2·02 D in benzene and, in heptane, 1·96 at $25°$ and 2·01 at $50°$, as compared to 2·67 calculated. With m unchanged, the calculated value increases only from 2·67 through 2·68 to 2·69 D, as n increases from 4 through 5, 6 and 9 to 10, while the observed moment changes from 2·02 to 2·27 to 2·40 to 2·57 to 2·55, with little effect of solvent or temperature. The differences between the values observed for the 9- and 10-carbon-chain molecules and that calculated for free rotation are within the uncertainty of the latter. Analogous results were obtained for the dicyanoalkanes[39] (X = CN), the difference between the observed and the calculated moments decreasing with increasing chain-length until it was 0·50 for $n = 8$ and 0·05 for $n = 10$.

When X is NH_2, m makes an angle of about 100° with the direction of the C—N bond.[10] The diamines, $H_2N(CH_2)_nNH_2$, consequently, have two more angles in the effective molecular chain than the dihalo- and dicyano-alkanes having the same n, and also four dipoles, the two N—H dipoles in each NH_2 group being resolved into one instead of two. The calculated moment values increase from 1·88D for $n = 2$ to 1·95 for $n = 5$, 6 and 8, while the moment values observed in benzene solution vary somewhat randomly from 1·92 to 1·99, the difference between the calculated values and the observed being negligible. The diols, $HO(CH_2)_nOH$, resemble the diamines in having two more angles and two more dipoles than the corresponding dihaloalkanes. Their dipole moments in dioxane solution increased only from 2·29 D for $n = 2$ to 2·51 for $n = 3$, 2·48 for $n = 6$ and 2·53 for $n = 10$, as compared to 2·1 D calculated for all of them. The diol moments are almost certainly raised by hydrogen bonding to the oxygens of the dioxane molecules, while the diamine moments are probably raised a little by weak hydrogen bonding to the π-electron bonds of the solvent benzene molecules. Agreement between observed moment values for some molecules and those calculated on the assumption of free rotation does not mean that these molecules actually have freedom of rotation around the bonds in their chains, but that there are so many conformations that the mean moment is indistinguishable from that for free rotation. For the dihalo- and dicyanoalkanes, this follows from equation (3.14), which shows that as n increases μ approaches $2^{1/2} m$ as a limit, which corresponds to completely random orientation of the two dipoles relative to each other.

α,α'-Dichloro-p-xylene and α,α'-dibromo-p-xylene are of particular interest because the two CH_2X groups can rotate around a common axis as in 1,2-dichloro- and 1,2-dibromoethane, but the forces between the two groups are replaced by those between the groups and the intervening benzene ring. These molecules will also be found of interest in connection with the interpretation of relaxation times (§ 3.9). The moments calculated for these molecules on the basis of free rotation around the common axis through the ring are, of course, equal to those calculated for 1,2-dichloro- and 1,2-dibromoethane, i.e. 2·0 D, if the value 1·5 D is used for m. If, however, the moment of the benzyl halide,[40] 1·85 D, is used for m, the resultant calculated molecular moment is 2·46 D, as compared to 2·20 observed for the chloro compound and 2·04 for the bromo. Evidently, the lower value for m is more nearly correct, which is as it should be, since the inductive effects, which are presumably responsible for the increase in moment of the benzyl halide, should partially cancel each other in the *para*-disubstituted compounds. There is a possibility that some double-bond character in the bond between the CH_2X group and the ring would tend to stabilize the groups in *cis* and *anti* positions coplanar with the ring. The dipole moments tell us nothing as to this because the same resultant moment values would be calculated, if the energies of the *cis* and *anti* positions differed little, as seems probable. However, the single dielectric relaxation time found for α,α'-dichloro-*p*-xylene[41] is so small, 3.7×10^{-12} s at 20°, as to show little or no contribution from a polar *cis* molecule, group rotation being the only evident mechanism for relaxation. Rotation of the CH_2Cl group is virtually eliminated by steric hindrance by the four methyl groups in bis-(chloromethyl)-durene, which has a relaxation time of 43.5×10^{-12} s at 20°, evidently that of the molecule as a whole.[41] For α,α'-dibromo-*o*-xylene,[10] the observed moments of 2.02 D in benzene at 25° and 2.08 at 50° and 1.79 in carbon tetrachloride at $-15°$ up to 1.92 at 45° are lower than the calculated value of 2.54, suggesting that repulsion between the two bromine atoms may considerably reduce the stability of the conformations of maximum moment, in which the molecular models show the bromine to be touching.

Some similar effects are evident in the dipole moments of the dimethoxy-benzenes.[13,42] o-$C_6H_4(OCH_3)_2$ has a moment of 1.18 D in benzene at 10° rising to 1.32 at 40° as compared to a calculated value of 1.93, presumably because, as shown by the molecular model, steric repulsion makes certain conformations of maximum moment unstable or impossible. The moments of the *meta* and *para* compounds, in which there is no appreciable steric repulsion between the two methoxy groups, show no dependence on temperature and fair agreement between observed and calculated values. In 1,4-di-*t*-butyl-2,5-dimethoxybenzene,[42] steric repulsion between the *t*-butyl and the methoxy groups reduces the stabilities of some conformations of high moment and consequently reduces the resultant moment below the calculated value. With hydroxyl instead of methoxy groups, the steric repulsion is reduced, but the dipole moment is still reduced below the calculated value.[42] Evidence of the smallness of steric repulsion between the hydroxyl groups is given by the agreement between the observed and calculated moments of *o*-dihydroxybenzene.[42] In *p*-dimethoxybenzene, there may be enough double-bond character in the bonds between the oxygens and the ring to give some stability to coplanar *cis* and *anti* structures of almost equal energies. The mean square dipole moment value for the consequent equal distribution between these two conformations would be the same as that for a distribution based on equal probability of orientation around the common axis of rotation. In 1,4-di-*t*-butyl-2,5-dimethoxybenzene, the *t*-butyl group would block the *cis* position and possibly create a displaced, shallower minimum, less occupied than the supposed *anti* minimum, and therefore resulting in a lower and temperature-dependent mean-square moment.

The dipole moment method may be applied to many other molecules, such as alkyl ethers, alkyl orthoborates, organic phosphites, phosphates and thiophosphates, and dicarboxylic acid esters.

3.7 More rigorous analyses of long molecules

More rigorous analyses of the dipole moments of the α,ω-dihaloalkanes have been carried out, among which those of Smith and coworkers[43–45] may be mentioned. Additional measurements of the dipole moments of the α,ω-dibromoalkanes in benzene solution[46,47] agreed well with the results of the earlier measurements on the same substances and, in the case of the second set,[47] attained such precision that the moment values lay well on either of two curves, one calculated for μ as a function of the odd values of n, and the other as a function of the even values, the latter being the lower of the two, although the difference decreased to a hardly detectable amount at $n = 10$. The calculation of the moment of the dibromoalkane $Br(CH_2)_nBr$ was carried out by means of an equation developed previously by Hayman

$$\mu_n^2 = 2\mu_{pr}^2 - [2/(\lambda_1 - \lambda_2)][P(\lambda_1)\lambda_1^{n-3} - P(\lambda_2)\lambda_2^{n-3}] \qquad (3.15)$$

and Eliezer,[48] (equation 3.15) where μ_{pr} is the moment of n-propyl bromide, the function $P(\lambda)$ is defined by

$$P(\lambda) = \mu\xi(1 - \alpha^2)\{\mu\xi + 2\mu'' + \lambda[2\mu' + \alpha\mu(2 - \xi)]\}$$
$$+ (1 + \alpha\eta/\lambda)[\mu'' + \lambda(\mu' + \alpha\mu)]^2 \qquad (3.16)$$

and λ_1 and λ_2 are the roots of the equation

$$\lambda^2 - \alpha(1 - \eta)\lambda - \eta = 0 \qquad (3.17)$$

such that $\lambda_1 > \lambda_2$, α is the cosine of the supplement of the valence angle of carbon and is therefore $1/3$, and μ, μ' and μ'' are the primary and the induced bond moments along the Br—C bond and the first and second C—C bonds, respectively, at each end of the molecule. These bond moments were calculated by means of the theory of the inductive effect developed by Smith and coworkers.[49] The parameters ξ and η measure the degree of freedom of rotation around the first and second C—C bonds, respectively, at each end of the molecule. They are the average value of the cosine of the angle of rotation around these bonds. The analogous parameters for the remaining C—C bonds in the molecule were taken as equal to η. Hayman and Eliezer point out that equation (3.15) is based on three simplifying assumptions: (a) the electrostatic interactions between the dipoles at the two ends of the chain are negligible, (b) no correlation exists between the rotations around the various C—C bonds in the molecule and (c) no correlation exists between the rotational partition function of a conformation and the value of the dipole moment of the conformation. They conclude that, although their experimental results are in excellent agreement with the simple picture of a hydrocarbon chain with a flexibility due to independent restricted rotations about the various C—C bonds, they are not sufficient to give any detailed information about the form of the potential barrier restricting these rotations.

After a long series of papers (which are referred to in Reference 50) treating the conformations of polymeric chain molecules in the rotational–isomeric state approximation, Leonard, Jernigan and Flory[50] examined the α,ω-dibromoalkanes by exact methods applicable to a linear sequence of bonds each constrained to choice among several discrete rotational states, due account being taken of neighbour dependence in assignment of statistical weights to the various conformations. One *anti* and two *gauche* states ($\pm 120°$) were used for each bond, an energy of 500 cal/mole being assigned to *gauche* relative to *anti* in keeping with spectroscopic evidence for n-alkanes and with successful calculations on polymethylene chains. The fact that successive *gauche* states of opposite sign are excluded by steric overlap was taken into account in the calculations. Non-bonded interactions involving a bromine atom were assigned statistical weights consistent with Van der Waals radii and with evidence from Raman and infrared spectra. The values of the mean-square dipole moments averaged over all conformations were found to be appreciably affected by the dipole–dipole interaction energy for $n < 7$, but not for longer chains.

Leonard, Jernigan and Flory point to errors in previous treatments, such as incorrect weighting of permissible conformations through erroneous introduction of symmetry numbers.[50] They take account of the previously neglected effect of the dipole–dipole energy, but ignore inductive effects on the ground that, except for the lowest members of the series, they can be included in the group moments. This very rigorous and thorough treatment gives calculated moment values for the α,ω-dibromoalkanes in fair agreement with the observed, but the agreement is actually not as good as that obtained by Hayman and Eliezer.[48] A difference in the values of some of the quantities used in the calculation may well affect the result more than a minor error in theory or neglect of minor influences.

3.8 High polymers

The structural relationships and dielectric behaviour which have been discussed in detail for small molecules and chains having as many as 10 carbon atoms may extend to much larger molecules, those of high polymers. Measurement and analysis of the dielectric relaxation in undiluted liquid atactic polypropylene oxides by Baur and Stockmayer[51] may serve as an illustration of polymer behaviour. Here we are concerned with the effect of C—O dipoles in the chains of the molecules $H(-O-CHR-CH_2)_x-O(CH_2-CHR-O)_yH$, where $10 \gtrsim x \cong y \gtrsim 35$ and R is CH_3 for the samples investigated. The principal dielectric absorption has a temperature-dependent maximum at about 1 MHz and is obviously dependent on the motion of the C—O dipoles through change in the conformation of the chain backbone. A secondary dispersion, which accounts for only about 4 or 5 % of the total polarization, is observed at lower frequencies of 1–10 KHz. Because of the methyl side-group, the dipole in each monomer unit does not exactly bisect the C—O—C angle, so that each dipole has a small component parallel to the contour of the chain. In each of the two arms of the chain, consisting of x or y segments, the parallel components of the unit dipoles in each arm add vectorially to give a net cumulative moment in the direction of, and proportional to, a vector drawn from end to end of the arm. The secondary dispersion results from relaxation of the cumulative dipole moment, found to be about 0·18D per monomer unit, and is strongly dependent on molecular weight, while the principal dispersion occurs at essentially the same frequency for all molecular weights at a given temperature. These polyether polymers with their dipoles in the chains or backbones of the molecules contrast with those of vinyl acetate and methyl methacrylate, which have mobile side-groups,[52] somewhat analogous to the CH_3O and CH_2Cl groups in the substituted benzenes which have been considered.

A full treatment of the dielectric behaviour and intramolecular motion of polymers is far beyond the scope of this chapter. General information may be obtained from some text-books on polymers, while detailed treatments of some aspects of the subject are available.[53–56]

3.9 Intramolecular relaxation times

In the preceding sections the dielectric relaxation time has been used occasionally to obtain information concerning molecular interactions possibly affecting the relative concentration and behaviour of rotational isomers. The present section will be devoted to a survey of the information given by these relaxation times concerning intramolecular motions.

The dielectric relaxation time τ can be obtained from measurements of dielectric loss ε'' as the reciprocal of the angular frequency ω_m for which ε'' is a maximum (equation 3.5). Equations (3.4) and (3.5) are derived by separation of real and imaginary parts in the Debye equation

$$\varepsilon^* = \varepsilon_\infty + \frac{\varepsilon_0 - \varepsilon_\infty}{1 + i\omega\tau} \tag{3.18}$$

in which ε^*, the complex dielectric constant, is defined as

$$\varepsilon^* = \varepsilon' - i\varepsilon'' \tag{3.19}$$

Cole and Cole[57] showed that equations (3.4) and (3.5) could be combined in the form of the equation for a circle

$$\left[\varepsilon' - \left(\frac{\varepsilon_0 + \varepsilon_\infty}{2}\right)\right]^2 + \varepsilon''^2 = \left(\frac{\varepsilon_0 - \varepsilon_\infty}{2}\right)^2 \tag{3.20}$$

Since all values must be positive, this gives a semi-circular plot of ε'' against ε', and when ε'' is plotted as ordinate against ε' as abscissa, the diameter of the semi-circle lies on the abscissa axis. When the measured dielectric constants or permittivities and losses of a material give such a semi-circle, they conform to the Debye theory, and the material is often said to show Debye behaviour. Many materials give a semi-circular arc intersecting the abscissa axis at the values of ε_∞ and ε_0 and having one end of its diameter at ε_∞. For such materials, equation (3.18) was modified[57] by the introduction of an empirical constant α to give

$$\varepsilon^* = \varepsilon_\infty + \left[\frac{\varepsilon_0 - \varepsilon_\infty}{1 + (\mathrm{i}\omega\tau_0)^{1-\alpha}}\right] \tag{3.21}$$

where τ_0 is the most probable relaxation time, the reciprocal of the angular frequency at which ε'' has its maximum value, and α is an empirical constant with a value between 0 and 1, a measure of the distribution of the relaxation times. The centre of the circle of which the arc is a part lies below the abscissa axis and the diameter drawn through the centre from the ε_∞ point makes an angle $\alpha\pi/2$ with the abscissa axis; α and τ_0 may be obtained from the Cole–Cole plot.[58]

Instead of the usual symmetrical arc plot, a skewed-arc plot is sometimes obtained, representable by the Cole–Davidson empirical equation[59]

$$\varepsilon^* = \varepsilon_\infty + \frac{\varepsilon_0 - \varepsilon_\infty}{(1 + \mathrm{i}\omega\tau_0)^\beta} \tag{3.22}$$

where β is an empirical constant with a value between 0 and 1 and τ_0 is a characteristic relaxation time. When $\alpha = 0$ in equation (3.21) or $\beta = 1$ in equation (3.22), the equation becomes identical with equation (3.18), which represents Debye behaviour.[60] When two distinct relaxation processes occur, equations (3.6) and (3.7) should be applicable, but unless the relaxation times τ_1 and τ_2 are considerably separated, the absorption regions for the two processes overlap to such an extent that the plot of the dielectric loss against log frequency may merely show a slightly flattened and widened maximum and the arc plot may appear symmetrical with an appreciable value of α, suggesting a distribution of relaxation times around a most probable value.

If the intensities of the two absorption processes differ considerably, the ε''-log-frequency curve will be unsymmetrical and the arc plot may be skewed. These are possibilities which may well be borne in mind in the interpretation of measurements where both a molecular and an intramolecular rotation may occur. It has already been noted (§ 3.5) that the molecular relaxation time and the probable relaxation time for CH_2Cl rotation in 1,2-dichloroethane are probably too close together to be distinguished.

The actual molecular relaxation time τ_μ is smaller than the directly determined macroscopic relaxation time $\tau_m = 1/\omega_m$, as shown by the expression[61–63]

$$\tau_\mu = \left(\frac{2\varepsilon_0 + \varepsilon_\infty}{3\varepsilon_0}\right)_{\tau_m} \tag{3.23}$$

but, for dilute solutions, ε_0 is so close to ε_∞ that the difference between τ_μ and τ_m can be neglected as a fair approximation, as it is unimportant in its effect on the value obtained for the intramolecular relaxation time τ_2. From the approximate nature of the calculation when two not widely separated relaxation times appear to exist, it is evident that caution must be exercised in recognizing the existence of a second relaxation process which makes only a small contribution to the dielectric loss. Otherwise, it may merely replace α of equation (3.21) or β of equation (3.22) as an adjustable parameter to fit the data. When only a few experimental points are available, a considerable value found for α may be indicative of the existence of two relaxation times rather than a distribution of relaxation times around a single most probable value (equation 3.21).

The evidence of intramolecular motion, usually group rotation, given by relaxation times is illustrated in the following tables and discussion. When evidence indicates a relaxation time to be that of the molecule as a whole, the time is listed as τ_1, while a relaxation evidently associated with group motion is listed as τ_2. Some of the molecules in Table 3.2 have already been discussed, primarily from the point of view of their dipole moments. The fact that α,α'-dichloro-p-xylene has a single relaxation time only half as large as that of the considerably smaller and rigid chlorobenzene molecule shows that the relaxation time observed must be that for rotational orientation of the CH_2Cl groups, the resultant dipole moment in the axis of rotation of the two groups, which would be the fixed moment of the molecule, being zero. This conclusion receives confirmation from the large single relaxation time of bis-(chloromethyl)-durene, in which adjacent methyl groups block the rotation of the CH_2Cl groups, leaving a fixed molecular moment, which causes the molecule to relax by rotational orientation as a whole. The molecules with CH_3O and CH_2CN instead of CH_2Cl show similar short relaxation times as the result of group rotation.

Table 3.2 Relaxation times of substituted benzenes and naphthalenes in benzene solution at 20°[64]

1,4-Substituents, benzenes			Times (10^{-12} s)	
1		4	τ_1	τ_2
Cl		—	7·5	—
CH_2Cl		CH_2Cl	—	3·7
CH_2Cl	2,3,5,6-$(CH_3)_4$	CH_2Cl	43	—
CH_2O		CH_3O	—	5·6
CH_2CN		CH_2CN	—	8·9
CH_3CO		CH_3CO	28	7·7
OH	2,6-$(CH_3)_2$	—	13·7	3·4

1,2-Substituents, naphthalenes		Times (10^{-12} s)	
1	2	τ_1	τ_2
OH	—	15	3·4
—	OH	24	3·2
CH_2Cl	—	26	—
CH_2CN	—	31	—
CH_3CO	—	25	—
—	CH_3CO	33	7·3

If the acetyl groups were capable of free rotation, 1,4-diacetylbenzene would be expected to give relaxation times characteristic of internal rotation only, since the fixed dipole-moment components in the molecular axis are equal and oppositely directed.[65] However, the relaxation time obtained for this compound is much larger than would be expected for a purely internal relaxation mechanism, indicating molecular rotation as the major relaxation mechanism. The potential-energy barriers hindering rotation of the acetyl groups may cause the molecules to exist in unstable *cis* and *anti* conformations, the *cis* conformation contributing to dielectric relaxation by molecular rotation. It appears that the potential-energy barriers hindering rotation of the acetyl groups are larger than those in the case of the methoxy groups, and that because of this, group rotation is much less probable. Since Stuart–Briegleb models indicate that steric hindrance to rotation of the acetyl groups should be low, it appears probable that the energy of resonance of the acetyl groups with the benzene ring is the main source of the hindering potential barriers. The large relaxation time τ_1, presumably due to the overall molecular rotation, is consistent with the size of the molecule, which is roughly the same as that of *p*-phenylphenol. The small relaxation time τ_2 for the acetyl group rotation is indistinguishable from those found for the acetyl group in *p*-

phenylacetophenone, 2-acetonaphthone and 4-acetyl-o-terphenyl.[66] The short relaxation time τ_2 for 2,6-dimethylphenol of $3{\cdot}4 \times 10^{-12}$ s, which is close to the values of τ_2 for p-phenylphenol[65] and 1- and 2-naphthol[65,67] in benzene solution, and to the value $3{\cdot}74 \times 10^{-12}$ found by Davies and Meakins[68] for the OH in 2,4,6-tri-t-butylphenol in decalin solution, is clearly the value for OH rotation. The 2-substituted naphthalenes in Table 3.2 have longer relaxation times than the corresponding 1-substituted compounds, because, although the molecular volumes are nearly the same, the molecular axis in the direction of the larger dipole moment component is longer. The relaxation times for the small OH group (Table 3.2) are virtually the same in the two positions, but the absence of a measurable contribution from the three larger groups (Table 3.2) shows that the steric repulsion of the hydrogen in the 8-position blocks the rotation of the CH_3CO, CH_2Cl and CH_2CN groups, confirming the indications given by the Stuart–Briegleb models of the molecules. A relaxation similar to that attributed to OH rotation here has been ascribed to a similar mechanism in the aliphatic alcohols.

The very small values of τ_2 found for pure aniline, $0{\cdot}9 \times 10^{-12}$ s, and pure dimethylaniline, $1{\cdot}5 \times 10^{-12}$ s, as compared to the larger values for group rotation in Table 3.2, suggest the possibility of an inversion analogous to that occurring in the ammonia molecule.[64] The surprisingly low relaxation time, 2×10^{-12} s found for the large molecule of triphenylamine[69] may best be attributed to inversion. The very small dipole moment of the molecule indicates that the three nitrogen valencies are so little out of one plane that the barrier to inversion should be small and the relaxation time correspondingly short. The single relaxation time found for the large molecule of phenoxathiin[70] is sufficiently low to show the existence of an intramolecular relaxation mechanism. As there can be no group rotation in this three-ring molecule a bending or inversion of the slightly bent molecule is indicated as the intramolecular relaxation process. The single low relaxation time found for diphenyl ether,[71] 4 in benzene and 6×10^{-12} in Nujol, shows intramolecular motion and illustrates the smallness of the effect of liquid viscosity upon an intramolecular relaxation process, a 250-fold increase in viscosity from benzene to Nujol increasing the relaxation time by only 50%.[64]*

Equation (3.8) gives the mean-square sum of the n dipole moments in a molecule taking account of their direction as well as their size, and this is the square of the dipole moment calculated from the measured dielectric constant on the assumption that the molecules are free to orient as a whole in the field applied in the measurement. One would therefore expect that the

---

* From 'Dielectric Relaxation Times' by Charles P. Smyth in *Molecular Relaxation Processes* (1966). Reproduced with the permission of the Chemical Society.

squares of the dipole moments in Table 3.3 would increase approximately linearly with the number of oxygen atoms, which is $x + 1$. An exact linear dependence is found[73] for μ^2 as a function of x. An exact linear plot would also be obtained if all of the n dipoles in the molecule were free to orient independently of one another.

Table 3.3 Dipole moments and dielectric relaxation times of pure liquid polymethylpolysiloxanes, $(CH_3)_3Si[OSi(CH_3)_2]_xOSi(CH_3)_3$ at $25°$ [72,73]

x	μ	τ_1	τ_2
0	0·38	—	0·3
1	0·64	2·97	0·4
2	0·79	3·85	0·47
3	0·93	4·77	0·51
5	1·17	6·85	0·67

The larger of the two relaxation times, τ_1, increases linearly with the number of oxygen atoms, $x + 1$, that is, with the volume of the molecule, as required by the approximate relationship of Debye: $\tau = 4\pi\eta a^3/kT$, in which η is the internal friction coefficient, a is the radius of the molecule treated as if it were spherical, k is the Boltzmann constant and T is the absolute temperature. However, the values of τ_1 and their rate of increase with increase in molecular size are somewhat small for a mechanism consisting wholly of dipole orientation by molecular rotation, suggesting some contribution from intramolecular motion to the dipole orientation process. The values of τ_2 are so small as to indicate a wholly intramolecular mechanism for this second relaxation process, which the values of C_2 show to be responsible for 73–82% of the relaxation. The size of τ_2 is smaller than that generally found for rotation of a small polar group inside a molecule. The small increase in the value of τ_2 with increase in molecular size is like those usually observed for intramolecular rotation and attributed to a small effect of increase in viscosity, much smaller than the effect of viscosity on molecular rotation.

Measurements[74] on hexamethyldisiloxane at $2°$ and $-20°$ indicate the existence of two relaxation processes, $\tau_1 = 2·4$ and $\tau_2 = 0·6$ at $-20°$. At temperatures of $20°$ and $40°$, the two processes presumably exist, but with relaxation times so near each other as to be indistinguishable in the measurements. For the larger hexaethyldisiloxane molecule,[72] which has the same dipole moment, the much longer relaxation time, $\tau_1 = 11·4$ at $25°$, presumably associated with rotational orientation of the whole molecule, is easily distinguished from the short intramolecular relaxation time, $\tau_2 = 0·6$.

The apparent molecular dipole moment obtained for each substance could result from the average of the moment contributions from the orientation of molecules having different conformations around their Si—O bonds, or it could result from orientation of the group by bending of the Si—O—Si bonds and by their rotation around adjacent Si—O bonds, or it could result from both orientational processes. Any of the three alternatives would give the observed linear dependence of μ^2 on chain-length. That both orientational processes occur, the last of the three alternatives, is shown by the two relaxation times. The first of these, τ_1, is associated primarily with molecular rotation, while the second, τ_2 (contributing 73–80% of the total relaxation), is associated primarily with group dipole orientation. The resolution of the dielectric relaxation of these liquids into two processes provides a good representation of the observed dielectric data, but may well be an over-simplification.

The general problem of the interpretation of the relaxation time has been discussed before. These experimental data can be well represented in terms of a most probable relaxation time τ_0 and an empirical constant α, which gives a symmetrical distribution of relaxation times around the most probable value τ_0. The values of τ_0 are so low as to indicate the presence of intramolecular relaxation, which involves a time or times lower than that or those for relaxation by overall molecular rotation. A symmetrical distribution of relaxation times around a single most probable value seems, therefore, to be improbable. The experimental data can be equally well represented in terms of two relaxation times τ_1 and τ_2 which have values of such magnitudes as to be interpretable in terms of molecular and intramolecular orientation. The experimental data are not sufficiently detailed to establish this as the only possible interpretation. Indeed, it is probable that there are more than two relaxation times. However, the values of τ_1 and τ_2 obtained are sufficient to indicate the general nature of the relaxation processes and the approximate frequencies of two of the principal molecular motions.[73]

The stereochemistry of the poly-(dimethylsiloxane) chain in very large molecules has been extensively investigated by Flory and coworkers.[56] It has been concluded that the chain has a preferred conformation that is a rather flat helix of six residues per turn, resulting from the inequality of the bond angles at silicon (110°) and oxygen (143°), so that the all-*trans* conformation is not a straight zig-zag.

3.10 Recent advances (1971–1973 and related papers)

Most of the very recent dielectric investigations of intramolecular motion have employed high-frequency measurements of dielectric constant and loss and the calculation of dielectric relaxation times. Following upon the theoretical work of Provder and Vaughan[75] on dielectric relaxation and hindered internal rotation, Klug, Kranbuehl and Vaughan[76] presented a model calculation of the dielectric behaviour of a liquid whose molecules had conformation-dependent dipole-moment components, measured the dielectric constants and losses of dilute solutions of 4-bromobiphenyl, 2,2′-dibromobiphenyl and 2,2′-bipyridine at wavelengths of 14,990 and 3·0–1·2 cm, and interpreted the results in the light of the model calculation.

Higasi, Koga and Nakamura[77] have developed a method of calculating two relaxation times from measurements of dielectric constant and loss at a single frequency at several temperatures. Chitoku and Higasi[78] have applied the method to their measurements at 100 GHz (0·3 cm) of several substances in which internal rotation is possible.

Mark and Sutton[79] have calculated dipole moments for all the members of the chloroethane series, using charge distributions based on the semi-empirical method of R. P. Smith, H. Eyring and their coworkers, and obtaining satisfactory agreement with the measured values in the literature. For 1,2-dichloroethane, 1,1,2-trichloroethane and 1,1,2,2-tetrachloroethane, the dipole moments were averaged over all conformations of the molecules, but, for chloroethane, 1,1-dichloroethane, 1,1,1-trichloroethane, 1,1,1,2-tetra-chloroethane, pentachloroethane and hexachloroethane, they are of course independent of internal rotation. Hasan, Pal and Ghatak[80] have calculated a single molecular relaxation time for 1,2-dichloroethane and for 1,2-dibromoethane from measurements at low frequency and at 1·62 and 3·20 cm wavelength over a range of temperature and have used the slope of the plot of log τ against the reciprocal of the absolute temperature to calculate the activation energy for the dipole orientation process. They point out that the energy value 1·29 kcal/mole thus obtained for 1,2-dichloroethane is in good agreement with the value 1·22 obtained by Wada[25] from the calculated electrostatic self-energies of the polar isomers, while there is a more notice-able difference between their value of 1·09 for 1,2-dibromoethane and the value of 0·800 calculated by Wada. Chandra and Prakash[81] have measured dielectric constants and losses at several frequencies from 1·8 to 36·8 GHz (16·7–0·82 cm) for 1,1-dichloro-1-nitroethane and 1,1,2,2-tetrabromoethane, as pure liquids and in benzene solution, and have obtained Cole–Cole arc plots giving single relaxation times with no distribution ($\alpha = 0$), from which they conclude that the relaxation process is indistinguishable from that of a rigid spherical molecule.

Measurements of the dielectric constants and losses[82] at wavelengths of 1·25 cm, 3·22 cm and 575 m on three isomeric dibromobutanes and 1,4-dibromopentane as pure liquids at 20°, 40° and 60° gave single relaxation times with virtually no distribution for the three isomers and practically indistinguishable from one another. The somewhat larger molecule of 1,4-dibromopentane showed a correspondingly larger relaxation time. The dipole moments of the molecules showed marked differences because of differences in molecular geometry and rotamer concentrations. Rates of the rotation around C—C bonds or transformation of one rotamer to another were not such as to lower the relaxation times below those to be expected for rotation of the entire molecules.

The measurements of Chandra and Prakash[83] at 450 kHz and 1·8–36·8 GHz (16·7–0·82 cm) on four dibromoalkanes $Br(CH_2)_{3-6}Br$ gave Davidson–Cole skewed arcs, which they attributed to the presence of intramolecular cooperative phenomena. The dielectric constants and losses at 0·22, 0·43, 1·25 and 3·22 cm wavelengths have been measured for four pure liquid α, ω-dibromoalkanes, $Br(CH_2)_nBr$, where $n = 4, 6, 8$ and 10, and combined with earlier measurements at lower frequencies to calculate dielectric relaxation times.[84] The empirical Cole–Cole symmetrical arc plot failed to fit the very high frequency data and the equation[85] based on a distribution of relaxation times between two limits proved inadequate. However, the empirical Davidson–Cole equation represented the data well, as found by Chandra and Prakash[83] for four α,ω-dibromoalkanes having $n = 3, 4, 5$ and 6 at wavelengths down to 8 mm. The extension of the measurements[84] down to 4 and 2 mm gave an even more rigorous test of the applicability of the Davidson–Cole equation than the measurements which stopped at 8 mm. Analysis[84] in terms of two superimposed, non-interacting Debye-type absorptions (equations 3.6 and 3.7) proved as successful as the Davidson–Cole equation. The values obtained for one relaxation time τ_1 were of the magnitude and temperature dependence to correspond to overall molecular rotation, while the closeness of the values of the other relaxation time τ_2 to the value $3·7 \times 10^{-12}$ s attributed to the rotation of the CH_2Cl group around its bond to the ring in α,α'-dichloro-p-xylene suggested that τ_2 may be associated with rotational orientation of the two terminal CH_2Br groups in the dibromoalkanes. These two terminal groups might be expected to rotate more readily around their C—C bonds than would the larger molecular segments, which should have longer relaxation times, possibly not analytically separable from τ_1.

Cumper and Wood[86] have measured seventeen pyrroles in solution in benzene and in dioxane, using one radio frequency and two microwave frequencies, 9·8 and 24·2 GHz (3·06 and 1·24 cm wavelengths). They were able to calculate two very approximate relaxation times, 12 and 1×10^{-12} s, for each of several of these substituted pyrroles, attributing the small value to intramolecular rotation, but concluding that this was absent in the 2-acylpyrroles. Cumper and Rossiter[87] have made similar measurements on symmetrically substituted benzophenones and have obtained extremely small relaxation times, some as low as $0·05 \times 10^{-12}$ s, which they plausibly interpreted as corresponding to the tail-end of a far infrared absorption region due to bending. In view of the apparent short relaxation times ($\sim 1 \times 10^{-12}$ s) obtained for several non-polar liquids[88,89] and attributed to collision absorption, the explanation of this absorption may be somewhat uncertain. Cumper and Rossiter[87] also measured a series of compounds of the formula $(CH_3)_{4-n}M(OCH_3)_n$, in which M = C or Si and $n = 1, 2, 3$ or 4, and attributed the very short relaxation times which they obtained, 0.8×10^{-12} s for both $C(OCH_3)_4$ and $Si(OCH_3)_4$, to rotation of the methoxy groups around their M—O bonds, hindered in the carbon compounds (M = C) and free in the silicon (M = Si) where M—O distance is greater.

Klages and Kraus[90] have made extensive measurements from 0.3 to 135 GHz (100–0.22 cm) on solutions of mono- and disubstituted benzenes with —CH_2Cl, —OCH_3, and —$COCH_3$ as the substituent groups. Dilute solutions in mesitylene were measured from $-30°$ to 70°C, while the solutions in benzene and carbon disulphide were measured at 20° only. They found additional absorption at millimetre wave frequencies which could be approximated by a Debye term, although they attributed it to the influence of far infrared bands. The correction for this small additional absorption raises somewhat the values calculated for τ_2, but they are still of the same order of magnitude as those previously observed. They found only one relaxation time for p-diacetylbenzene, acetophenone and p-chloroacetophenone, and for p-methoxyacetophenone, a molecular relaxation time and a second relaxation time corresponding to —OCH_3 rotation, and therefore concluded that the —$COCH_3$ group was rigidly attached to the ring. This result differs but little from the result in Table 3.2, where the supposed intramolecular process makes only a 15% contribution.[65] Similar apparent contributions from intramolecular rotation were 14% for 2-acetonaphthone[66] in dilute benzene solution at 20°, and 12% for 3-acetyl-o-terphenyl and p-phenylacetophenone[65] in benzene at 20°. The 8% correction for far infrared absorption would account for any differences between the results of the investigators in excess of the probable experimental errors.

3.11 References

1. P. Debye, *Polar Molecules* (New York: Chemical Catalog Co., 1929).
2. C. P. Smyth, *Dielectric Behavior and Structure* (New York: McGraw-Hill, 1955).
3. W. E. Vaughan, C. P. Smyth and J. G. Powles, *Physical Methods of Chemistry*, Vol. 1, Ed. A. Weissberger and B. W. Rossiter (New York: Interscience, 1972), Part IV, Chap. V; C. P. Smyth, *Physical Methods of Chemistry*, Vol. 1, Ed. A. Weissberger and B. W. Rossiter (New York: Interscience, 1972), Part IV, Chap. VI.
4. N. E. Hill, W. E. Vaughan, A. H. Price and M. Davies, *Dielectric Properties and Molecular Behaviour* (London: Van Nostrand Reinhold, 1969).
5. L. Onsager, *J. Am. Chem. Soc.*, **58**, 1486 (1936).
6. K. Bergmann, *Doctoral Dissertation*, Freiburg/Breisgau, West Germany (1957).
7. K. Bergmann, D. M. Roberti and C. P. Smyth, *J. Phys. Chem.*, **64**, 665 (1960).
8. C. P. Smyth, *J. Am. Chem. Soc.*, **46**, 2151 (1924).
9. J. W. Williams, *Z. Physik. Chem.*, **138**, 75 (1928).
10. C. P. Smyth, *Dielectric Behavior and Structure* (New York: McGraw-Hill, 1955), Chap. VIII.

11. H. Eyring, *Phys. Rev.*, **39**, 746 (1932).
12. C. P. Smyth, R. W. Dornte and E. B. Wilson, Jr., *J. Am. Chem. Soc.*, **53**, 4242 (1931).
13. C. P. Smyth, *Dielectric Behavior and Structure* (New York: McGraw-Hill, 1955), Chap. XI.
14. H. Eyring, *J. Am. Chem. Soc.*, **54**, 3191 (1932).
15. C. P. Smyth and K. B. McAlpine, *J. Am. Chem. Soc.*, **57**, 979 (1935).
16. J. R. Thomas and W. D. Gwinn, *J. Am. Chem. Soc.*, **71**, 2785 (1949).
17. S. Mizushima and K. Higasi, *Proc. Imp. Acad. Tokyo*, **8**, 482 (1932).
18. S. Mizushima, *Structure of Molecules and Internal Rotation* (New York: Academic Press, 1954).
19. I. Watanabe, S. Mizushima and Y. Morino, *Sci. Papers Inst. Phys. Chem. Res. (Tokyo)*, **39**, 401 (1942).
20. I. Watanabe, S. Mizushima and Y. Masiko, *Sci. Papers Inst. Phys. Chem. Res. (Tokyo)*, **40**, 425 (1943).
21. S. Mizushima, Y. Morino, I. Watanabe, T. Simanouti and S. Yamaguchi, *J. Chem. Phys.*, **17**, 591 (1949).
22. Y. Morino, S. Mizushima, K. Kuratani and M. Katayama, *J. Chem. Phys.*, **18**, 754 (1950).
23. A. E. Stearn and C. P. Smyth, *J. Am. Chem. Soc.*, **56**, 1667 (1934).
24. H. Huettig, Jr. and C. P. Smyth, *J. Am. Chem. Soc.*, **57**, 1523 (1935).
25. A. Wada, *J. Chem. Phys.*, **22**, 198 (1954).
26. A. Wada and Y. Morino, *J. Chem. Phys.*, **22**, 1276 (1954).
27. A. Neckel and H. Volk, *Z. Elektrochem.*, **62**, 1104 (1958).
28. K. Amaya and R. Fujishiro, *Bull. Chem. Soc. Japan*, **31**, 90 (1958).
29. R. Fujishiro and K. Kimura, *Bull. Chem. Soc. Japan*, **32**, 1237 (1959).
30. A. Piekara and A. Chelkowski, *J. Chem. Phys.*, **25**, 795 (1956).
31. E. J. Hennelly, W. M. Heston, Jr. and C. P. Smyth, *J. Am. Chem. Soc.*, **70**, 4102 (1948).
32. K. Chitoku and K. Higasi, *Bull. Chem. Soc. Japan*, **40**, 773 (1967).
33. H. Müller, *Z. Physik*, **34**, 689 (1933).
34. J. Crossley and S. Walker, *J. Chem. Phys.*, **45**, 4733 (1966).
35. J. Crossley and S. Walker, *J. Chem. Phys.*, **48**, 4742 (1968).
36. M. Tamies, *J. Am. Chem. Soc.*, **86**, 152 (1964).
37. A. A. Antony and C. P. Smyth, *J. Am. Chem. Soc.*, **86**, 152 (1964).
38. J. Crossley and C. P. Smyth, *J. Am. Chem. Soc.*, **91**, 2482 (1969).
39. P. Trunel, *Ann. Chim.*, **12**, 93 (1939).
40. C. P. Smyth, *Dielectric Behavior and Structure* (New York: McGraw-Hill, 1955), Chap. X.
41. W. P. Purcell, K. Fish and C. P. Smyth, *J. Am. Chem. Soc.*, **82**, 6299 (1960).
42. P. F. Oesper, C. P. Smyth and M. S. Kharasch, *J. Am. Chem. Soc.*, **64**, 937 (1942).
43. R. P. Smith and E. M. Mortensen, *J. Am. Chem. Soc.*, **78**, 3932 (1956).
44. R. P. Smith and E. M. Mortensen, *J. Chem. Phys.*, **32**, 508 (1960).
45. R. P. Smith and J. J. Rasmussen, *J. Am. Chem. Soc.*, **83**, 3785 (1961).
46. J. A. A. Ketelaar and N. van Meurs, *Rec. Trav. Chim.*, **76**, 437 (1957); N. van Meurs, Thesis, Amsterdam (1956).
47. H. J. G. Hayman and I. Eliezer, *J. Chem. Phys.*, **35**, 644 (1961).
48. H. J. G. Hayman and I. Eliezer, *J. Chem. Phys.*, **28**, 890 (1958).
49. R. P. Smith, T. Ree, J. L. Magee and H. Eyring, *J. Am. Chem. Soc.*, **73**, 2263 (1951).
50. W. J. Leonard, Jr., R. L. Jernigan and P. J. Flory, *J. Chem. Phys.*, **43**, 2256 (1965).
51. M. E. Baur and W. H. Stockmayer, *J. Chem. Phys.*, **43**, 4319 (1965).
52. W. Peticolas, *Rubber Chem. Technol.*, **36**, 1422 (1963).

53. M. V. Volkenstein, *Configurational Statistics of Polymeric Chains* (New York: Interscience, 1963).
54. T. M. Birshtein and O. B. Ptitsyn, *Conformations of Macromolecules* (New York: Interscience, 1966).
55. N. G. McCrum, B. E. Read and G. Williams, *Anelastic and Dielectric Effects in Polymeric Solids* (London: Wiley, 1967).
56. P. J. Flory, *Statistical Mechanics of Chain Molecules* (New York: Interscience, 1969).
57. K. S. Cole and R. H. Cole, *J. Chem. Phys.*, **9**, 341 (1941).
58. C. P. Smyth, *Dielectric Behavior and Structure* (New York: McGraw-Hill, 1955), Chap. II.
59. D. W. Davidson and R. H. Cole, *J. Chem. Phys.*, **18**, 1417 (1951); **19**, 1484 (1951).
60. N. E. Hill, W. E. Vaughan, A. H. Price and M. Davies, *Dielectric Properties and Molecular Behaviour* (London: Van Nostrand Reinhold, 1969), Chap.1 and 5.
61. J. G. Powles, *J. Chem. Phys.*, **21**, 633 (1953).
62. S. H. Glarum, *J. Chem. Phys.*, **33**, 639 (1960).
63. R. C. Miller and C. P. Smyth, *J. Am. Chem. Soc.*, **79**, 3310 (1957).
64. C. P. Smyth, *Molecular Relaxation Processes*, Chemical Society Special Publication No. 20 (London: Academic Press, 1966), pp. 1–12.
65. F. K. Fong and C. P. Smyth, *J. Am. Chem. Soc.*, **85**, 1565 (1963).
66. F. K. Fong and C. P. Smyth, *J. Am. Chem. Soc.*, **85**, 548 (1963).
67. F. K. Fong and C. P. Smyth, *J. Phys. Chem.*, **67**, 226 (1963).
68. M. Davies and R. J. Meakins, *J. Chem. Phys.*, **26**, 1584 (1957).
69. E. N. di Carlo and C. P. Smyth, *J. Am. Chem. Soc.*, **84**, 3638 (1962).
70. J. E. Anderson and C. P. Smyth, *J. Chem. Phys.*, **42**, 473 (1965).
71. D. M. Roberti, O. F. Kalman and C. P. Smyth, *J. Am. Chem. Soc.*, **82**, 3523 (1960).
72. S. Dasgupta, S. K. Garg and C. P. Smyth, *J. Am. Chem. Soc.*, **89**, 2243 (1967).
73. S. Dasgupta and C. P. Smyth, *J. Chem. Phys.*, **47**, 2911 (1967).
74. S. Dasgupta and C. P. Smyth, *J. Chem. Phys.*, **54**, 4648 (1971).
75. T. Provder and W. E. Vaughan, *J. Chem. Phys.*, **46**, 848 (1967).
76. D. D. Klug, D. E. Kranbuehl and W. E. Vaughan, *J. Chem. Phys.* **53**, 4187 (1970).
77. K. Higasi, Y. Koga and M. Nakamura, *Bull. Chem. Soc. Japan*, **44**, 988 (1971).
78. K. Chitoku and K. Higasi, *Bull. Chem. Soc. Japan*, **44**, 992 (1971).
79. J. E. Mark and C. Sutton, *J. Am. Chem. Soc.*, **94**, 1083 (1972).
80. A. Hasan, A. Pal and A. Ghatak, *Bull. Chem. Soc. Japan*, **44**, 322 (1971).
81. S. Chandra and J. Prakash, *J. Chim. Phys. Physicochim. Biol.*, **68**, 1128 (1971).
82. J. E. Anderson and C. P. Smyth, *J. Phys. Chem.*, **77**, 230 (1973).
83. S. Chandra and J. Prakash, *J. Phys. Chem.*, **75**, 2616 (1971); **54**, 5366 (1971).
84. S. K. Garg, W. S. Lovell, C. J. Clemett and C. P. Smyth, *J. Phys. Chem.*, **77**, 232 (1973).
85. K. Higasi, K. Bergmann and C. P. Smyth, *J. Phys. Chem.*, **64**, 880 (1960).
86. C. W. N. Cumper and J. W. M. Wood, *J. Chem. Soc.* (*B*), 1811 (1971).
87. C. W. N. Cumper and R. F. Rossiter, *J. Phys. Chem.*, **76**, 525 (1972).
88. S. K. Garg, J. E. Bertie, H. Kilp and C. P. Smyth, *J. Chem. Phys.*, **49**, 2551 (1968).
89. G. W. F. Pardoe, *Trans. Faraday Soc.*, **66**, 2699 (1970).
90. G. Klages and G. Kraus, *Z. Naturforsch.* (*A*), **26**, 1272 (1971).

4 Infrared and Raman band intensities and conformational change

P. J. D. Park, R. A. Pethrick and B. H. Thomas

4.1. Introduction

The concept of rotational isomerism was first proposed by Kohlrausch[1] and his coworkers to explain their observation of more lines in the Raman spectrum of molecules of the type C_3H_7X and C_4H_9X, where X is a halogen, than would be allowed by the selection rules for a single spatial structure. The excess bands could not be assigned to overtones or combination modes and could only be satisfactorily explained on the basis of these molecules possessing more than one geometric structure. A study of the infrared spectra of dihalogen-substituted ethanes confirmed this proposition, as the spectra of these molecules were temperature sensitive, certain lines becoming decidedly weaker when the temperature of the sample was lowered. This observation may be rationalized in terms of a depopulation of the thermodynamically less stable isomeric structure. The appearance of automatic recording infrared spectrometers gave a stimulus to this work and the progress in this field of research up until about 1959 has been reviewed by Mizushima[2] and Sheppard.[3]

4.2 Concept of rotational isomerism

The rotation of one part of a molecule with respect to another about an internal chemical bond is usually accompanied by a change in the potential energy of the molecule. In some molecules internal rotation of this kind gives rise to the phenomenon of rotational isomerism, where the molecules exist as an equilibrium mixture of two or more conformers. One of the classical examples of a molecule exhibiting rotational isomerism is 1,2-dibromoethane, where the variation of potential energy with azimuthal angle is shown in Figure 4.1. Also included in this diagram are the conformations of the molecule corresponding to the various maxima and minima of the potential energy curve. The minima in these potential energy curves correspond to the staggered conformations of the molecules which are

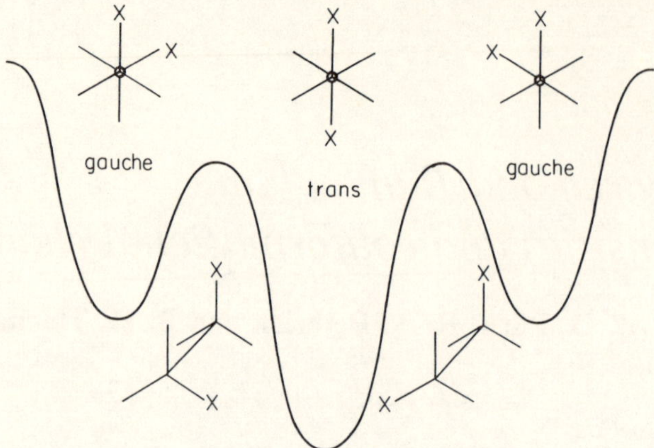

Figure 4.1   Rotational isomerism in 1,2-dibromoethane

the so-called 'rotational isomers'. The difference in energy between the minima is known as the enthalpy difference between the isomers, and the difference between neighbouring maxima and minima is termed the potential barrier hindering internal rotation. This chapter is concerned with a description of the use of Raman and infrared studies in the investigation of the energy differences between rotational isomers. The study of torsional frequencies in the evaluation of the activation energies is dealt with in Chapters 7 and 8.

4.3 Classical description of infrared and Raman activity

The total energy of a molecule is the sum of its translational, electronic, vibrational and rotational energy components. In the gas phase, the last three of these energy forms may be considered to be quantized, and any allowed transition between two energy states in the molecule will be associated with an absorption or emission of radiation of wavelength $\lambda = (E'_f - E''_i)/hc$, where E'_f and E''_i are the two energy states involved in the transition, c is the velocity of light and h is Planck's constant. To a first approximation the internal energy components of the molecule may be considered to be independent of each other. Thus energy changes between electronic states occur at wavelengths in the visible or ultraviolet region, while transitions between vibrational energy states give rise to bands in the infrared region. The spectrum of absorbed or emitted radiation by a molecule is known as a band spectrum to distinguish it from atomic spectra where the spectral lines are narrow. The reason for the relative width of molecular vibrational lines is the coupling of the atomic displacements with rotation of the molecule. It is possible in the gas phase, with a high resolution spectrometer, to separate a molecular band into a set of closely spaced vibrational–rotational peaks.

A diatomic molecule has just one degree of vibrational freedom. Its vibration can be approximately treated as that of a classical simple harmonic oscillator. The energy (eigenvalues) of any vibrational state will therefore be equal to $E_\omega = (v + \frac{1}{2})hc\omega$, where v is the vibrational quantum number which equals zero or an integer and ω is the vibrational frequency of the oscillator expressed in wave numbers (m^{-1}). Since each vibrational state has associated with it the rotational states of the molecule, a more correct expression of the energy of a molecular vibration would be $E_\omega = (v + \frac{1}{2})hc\omega + h^2 J(J + 1)/8\pi^2 I$, where J is the rotational quantum number and I the moment of inertia. It is easy to picture the vibrational states of a diatomic molecule as being associated with the energy states of a quantized simple harmonic oscillator motion for the two atoms. For a polyatomic molecule, the picture becomes more complex since in any vibration all the atoms of the molecule will be in motion. By definition, each atom may be represented by a set of coordinates (q), which define its position relative to its equilibrium position. It is possible to show that the kinetic energy of the molecule, $T = \frac{1}{2}\sum_{i \neq j} a_{ij} q_i q_j$, where a_{ij} is a function of the mass of the ijth atom, and the potential energy $V = \frac{1}{2}\sum_{i \neq j} b_{ij} q_i q_j$, where $b_{ij} = (\partial^2 V/\partial q_i . \partial q_j)_0$. Inserting these terms in the Lagrange equations of motion $\partial/\partial t(\partial L/\partial q_k) - \partial L/\partial q_k = 0$ (where L is the Lagrange function equal to $T - V$) and manipulating the results, one can show that a series of s equations of simple harmonic motion can be derived, where $s = 3n - 6$, the number of vibrational degrees of freedom for a molecule with n atoms. Transforming these coordinates, q, to normal coordinates, Q, the Lagrange equation becomes

$$\partial/\partial t(\partial T/\partial Q_L) + (\partial V/\partial Q_L) = 0 \quad \text{where } L = 1, 2, 3, 4 \ldots s, \tag{4.1}$$

and it is possible to show that all the nuclei undergo simple harmonic motion with different amplitudes. The amplitude of each atom is found to be proportional to its mass. Vibrations of this type are known as the normal vibrational modes of the molecule and their energy terms summed together give the total vibrational energy

$$E_\omega = (v_1 + \tfrac{1}{2})hc\omega_1 + (v_2 + \tfrac{1}{2})hc\omega_2 + \cdots (v_s + \tfrac{1}{2})hc\omega_s \tag{4.2}$$

Electromagnetic energy interacts with a molecule in two ways to change the vibrational energy.

Radiation in the infrared region can be absorbed by the molecule if its frequency is equivalent to the quantized energy jump of the vibrational state and the electric vector of the radiation finds a suitable electric field within the molecule with which to interact. A suitable electric field is caused by an oscillating dipole moment which is associated with these normal modes of vibration in which the electron density changes with movement of the atoms. The probability of transfer between the two energy states m and n can be determined from quantum mechanics as being equal to

$$P_{mn} = \int \chi_m^* \mu \chi_n \, \partial \tau \tag{4.3}$$

where χ_m and χ_n are the eigenfunctions of the two states (the asterisk indicating the complex conjugate) and μ is the component of the electric dipole moment in the direction of the applied field. For vibrational–rotational spectra it is found that even if there is a change in dipole moment this probability differs from zero only if the change in vibrational quantum number (as given in equation 4.1) of the two states taking part in the transition is equal to unity, i.e. $\Delta v = +1$. A diatomic molecule treated as a harmonic oscillator absorbs radiation of the same frequency as its own classical vibrational frequency ω. A molecule containing n atoms will therefore have a spectrum containing $(3n - 6)$ fundamental bands. For a linear molecule the number of vibrational bands will be reduced to $(3n - 5)$ bands. For infrared activity the vibration must be asymmetric in its vibration with respect to charge density. The above theory assumes that all vibrations are pure and harmonic in nature. This is an approximation. Due to the anharmonicity of vibrations the observed frequency will differ slightly from that calculated for harmonic motion. Because of anharmonicity overtone and combination bands also occur, but these are much weaker than the normal bands.

In this classical description of the origin of infrared spectra, the molecule is assumed to be a distinct entity whose position in space and configuration are time independent. The process of interaction of an electromagnetic wave with a molecule requires a finite time, this, in the case of an infrared transition, will vary between 1×10^{-14} to 1×10^{-11} s. A molecule the size of carbon tetrabromide is able to execute approximately a quarter of a rotation in the period of 10^{-12} s. It is obvious, therefore, that the orientation of the dipole relative to the direction of the electric vector of the radiation alters during the vibration period, leading to time-dependent contributions being involved in the description of such interactions.

Apart from the direct interaction of the electric vector with a particular dipole in the molecule a second order interaction can take place and give rise to energy exchange. The field associated with the radiation may interact with the molecule leading to an induced dipole. The induced polarization μ is a vector whose direction, in general, differs from that of the applied field $\mathbf{F}$. The proportionality constant α is termed the polarizability and is a tensor. Since the electric field associated with the electromagnetic wave oscillates in a simple harmonic nature with respect to time, then so will the induced dipole moment. Thus

$$\boldsymbol{\mu} = \alpha \mathbf{F}_0 \sin 2\pi v t \tag{4.4}$$

where v is the frequency of the radiation and $\mathbf{F}_0$ is the equilibrium value of the field. According to classical theory this oscillating dipole will emit radiation which is equal in frequency to that of the exciting radiation. This is known as the Rayleigh scattering. It is found that if the molecule involved in the scattering is small, the intensity of the scattered radiation is independent of direction. On the other hand if the dimensions of the molecule are of the order of the wavelength, interference can occur between the elements arising from various parts of the molecule and the resulting intensity distribution with direction becomes complex.

By considering the internal vibrations or rotations of a molecule it is easily realized that these motions will affect the induced dipole moment and hence they will influence the scattered radiation. For a diatomic molecule, for example, where there is a small displacement from equilibrium along the axis, the polarizability α then becomes

$$\alpha = \alpha_0 + \beta x/A \tag{4.5}$$

where α_0 is the equilibrium polarizability, β is the rate of variation of polarizability with displacement and A is the vibrational amplitude. Since the oscillations within the molecule are considered to be simple harmonic, the variation of displacement x with time may be represented by $x = A \sin 2\pi v_v t$, where v_v is the vibrational frequency of the oscillation in the molecule. It follows therefore that $\alpha = \alpha_0 + \beta \sin 2\pi v_v t$, and that

$$\mu = \alpha_0 \mathbf{F}_0 \sin 2\pi v t + \beta \mathbf{F}_0 \sin 2\pi v t \sin 2\pi v_v t \tag{4.6}$$

This may be written as

$$\mu = \alpha_0 \mathbf{F}_0 \sin 2\pi v t + \tfrac{1}{2}\beta \mathbf{F}_0[\cos 2\pi(v - v_v)t - \cos 2\pi(v + v_v)t] \tag{4.7}$$

Thus the scattered radiation is of frequency v and $v - v_v$ and $v + v_v$ where the shift in frequency from v is equal to a normal vibrational mode frequency of the molecule. The scattered radiation of frequency $v - v_v$ and $v + v_v$ are known as the Stokes and anti-Stokes lines of the Raman effect.[4,5]

According to quantum theory, Rayleigh scattering is the result of the absorption of incident radiation which causes a rise in the energy of the molecule to a higher energy state and which later returns to the ground state. The light emitted is therefore of the same frequency as that of the absorbed radiation. Raman scattering is caused by the molecule returning not to the original state but to a higher or lower vibrational and/or rotational state. The probability of transition in the Raman effect in the x direction is

$$P_{mn}(x) = \int \chi_m^*(\alpha_{xx}\mathbf{F}_x)\chi_n \, \partial \tau \tag{4.8}$$

where α_{xx} is the polarizability in the x direction when the field $\mathbf{F}_x$ acts in the same direction. Assuming α_{xx} to vary in the course of an oscillation such that

$$\alpha_{xx} = \alpha_{xx}^0 + \beta_{xx}(x/A) \tag{4.9}$$

then

$$P_{mn}(x) = \mathbf{F}_x\alpha_{xx}^0 \int \chi_m^*\chi_n \, \partial\tau + (\mathbf{F}_x/A)\beta_{xx} \int \chi_m^*\chi_n \, \partial\tau \tag{4.10}$$

Since the vibrational eigenfunctions are orthogonal, $\mathbf{F}_x\alpha_{xx}^0 \int \chi_m^*\chi_n \, \partial\tau$ is zero except when m and n are equal, that is, there is no vibrational transition and therefore the radiation emitted corresponds to Rayleigh scattering. If m and n are different then the above equation will be zero unless the polarizability varies during the course of the molecular vibration, i.e. $\beta_{xx} \neq 0$. For vibrational Raman spectra, the second term of equation (4.10) differs from zero only when $\Delta v = \pm 1$; that is, the change in the vibrational quantum number is unity. The above discussion has assumed the polarizability change to be along the axis of the molecular vibration. The polarizability of a molecule may be considered as resolvable into three mutually orthogonal components. These give the dimensions of the so-called polarization ellipsoid.

The three components can be represented by

$$\mu_x = \alpha_{xx}\mathbf{F}_x + \alpha_{xy}\mathbf{F}_y + \alpha_{xz}\mathbf{F}_z$$

$$\mu_y = \alpha_{yx}\mathbf{F}_x + \alpha_{yy}\mathbf{F}_y + \alpha_{yz}\mathbf{F}_z \tag{4.11}$$

$$\mu_z = \alpha_{zx}\mathbf{F}_x + \alpha_{zy}\mathbf{F}_y + \alpha_{zz}\mathbf{F}_z$$

Since it can be shown that $\alpha_{xy} = \alpha_{yx}$, $\alpha_{xz} = \alpha_{zx}$ and $\alpha_{yz} = \alpha_{zy}$, the number of uniquely defined components of the polarizability tensor are reduced to six. These are used as coefficients in the equation of the ellipsoid.

$$\alpha_{xx}x^2 + \alpha_{yy}y^2 + \alpha_{zz}z^2 + 2\alpha_{xy}xy + 2\alpha_{yz}yz + 2\alpha_{zx}zx = 1 \tag{4.12}$$

If any of the three components of the polarization ellipsoid change during the course of a molecular vibration then that mode of vibration will interact with radiation to give a Raman effect.

Observation of the polarization of the scattered radiation at right angles to the incident radiation shows that the ratio of the intensity of the light polarized perpendicular to the plane of observation $I_\perp$ and that parallel to the plane of observation $I_\parallel$ differs from unity. The ratio is known as the depolarization ratio $p_n = I_\perp/I_\parallel$. By averaging over all orientations of the polarizability ellipsoid, Born[6] has shown that for unpolarized incident light the depolarization ratio is $p_n = I_\perp/I_\parallel = 6\gamma^2/[45(\alpha')^2 + 7\gamma^2]$, where $\alpha' = \frac{1}{3}(\alpha_{xx} + \alpha_{yy} + \alpha_{zz})$ and is termed the spherical part of the polarizability and γ is the completely anisotropic part:

$$\gamma^2 = \tfrac{1}{2}[(\alpha_{xx} - \alpha_{yy})^2 + (\alpha_{yy} - \alpha_{zz})^2 + (\alpha_{zz} - \alpha_{xx})^2$$
$$+ 6(\alpha_{xy}^2 + \alpha_{yz}^2 + \alpha_{zx}^2)] \tag{4.13}$$

This applies for Rayleigh scattering. For Raman scattering the components α_{xx}, α_{yy} etc. are replaced by the change in polarizability during the vibration $(\partial\alpha_{xx}/\partial x)$. The maximum value for the depolarization ratio is 6/7 when $\alpha' = 0$. A Raman line with such a value is depolarized. If the depolarization ratio is less than 6/7 then the line is polarized. Only totally symmetrical vibrations yield polarized Raman lines.

As in the case of infrared activity, the classical theory does not include any consideration of the possibility of dynamic changes of molecular structure and environment. Such factors can alter the description of a Raman line. For instance, the fact that the molecule is rotating leads to the effects of changes in environment being included in the terms describing the polarizability tensor. The time dependence may be introduced into the description of infrared and Raman theory by using the concepts of correlation theory.

4.4 Correlation theory

Correlation theory was first introduced into the general field of spectroscopic investigation by its use in the analysis of the decay of polarization in nuclear magnetic resonance experiments.[7] Qualitatively, a correlation function describes how long some property of a system persists in the presence of microscopic motions which tend to average the system. Mathematically, the autocorrelation function $C(t)$ of a quantity A is given by the ensemble average

$$C(t) = \langle A(0) \,.\, A(t)\rangle_0 \tag{4.14}$$

In this system A can be any dynamic function of the variables of the system, such as the momentum of a certain atom. The time dependence indicated in A is that produced by the natural motion of the system. The average $\langle - \rangle_0$ is over an ensemblage of systems at the reference time 0. Ordinarily, this ensemblage is the canonical Boltzmann distribution appropriate to systems in thermal equilibrium. The quantities of interest, $A(t)$, are usually defined in such a way that the ensemblage average of A is zero. A system in which the time average is the same as averages taken over the initial ensemblage is termed ergodic and the correlation function of A will approach zero as the time becomes large $(t \rightarrow \infty)$.

Correlation functions are particularly useful whenever one has two physically weakly coupled systems, such as radiation weakly interacting with matter. In such a situation one need only follow the free motion of the separate systems in the absence of the weak coupling between them. In particular, it is only necessary to know how the free motion of the separate systems affects the weak coupling between them. The information is contained in the correlation function of the coupling Hamiltonian describing the interaction between the two systems. To illustrate the application of correlation theory, the absorption of light due to its weak coupling with matter will be considered. This is particularly pertinent in that it allows postulates to be made as to band shapes of various rotational isomers.

4.4.1 *Absorption of radiation*[8]

Electromagnetic radiation of frequency ω, interacting with a group of molecules in a quantum state described by the vector $|i\rangle$ may induce transitions to another state $|f\rangle$. For this to occur the frequency of the light must match the energy gap between two quantum states of a molecule

$$\omega_{if} = (E_f - E_i)/h \tag{4.15}$$

It can be shown that the probability per unit time that a transition takes place is given by time-dependent perturbation theory

$$\mathbf{P}_{f \leftarrow i}(\omega) = \frac{\pi}{2\hbar}\{\langle f|\mathbf{E} \cdot \boldsymbol{\mu}|i\rangle\}^2\{\delta(\omega_{fi} - \omega) + \delta(\omega_{fi} + \omega)\} \tag{4.16}$$

where $\mathbf{E}$ is the amplitude of the electric field associated with the light wave and $\boldsymbol{\mu}$ is the total electric dipole moment operator for the molecules in the system. Equation (4.16) indicates that the interaction between radiation and matter has been approximated by the electric dipole interaction. The Dirac δ function appears at both $\omega_{fi} - \omega$ and $\omega_{fi} + \omega$ because the electric field vector $\mathbf{E}(t)$ is periodic and may be written as a sum of

$$\mathbf{E}(t) = \mathbf{E}_0 \cos \omega t = \frac{\mathbf{E}_0}{2}(e^{i\omega t} + e^{-i\omega t}) \tag{4.17}$$

The usual infrared experiment involved measurement of the rate of energy absorption $-\mathbf{E}_{\text{rad}}$ from the radiation, rather than the transition rate $\mathbf{P}_{f \leftarrow i}(\omega)$. The probability as described by equation (4.16) is related to the energy loss by

$$-\mathbf{E}_{\text{rad}} = \sum_f \hbar\omega_{fi}\mathbf{P}_{f \leftarrow i}$$

$$= \frac{\pi}{2\hbar}\sum_f \sum_i \omega_{fi}(\rho_i - \rho_f)\{\langle f|\mathbf{E}_0 \cdot \boldsymbol{\mu}|i\rangle\}^2 \, \delta(\omega_{fi} - \omega) \tag{4.18–4.19}$$

using $\omega_{fi} = -\omega_{fi}$. The initial ensemble of states may be chosen to be that of a Boltzmann distribution, in which case

$$\rho_f = \rho_i \exp(-\hbar\omega_{fi}/kT) \tag{4.20}$$

The energy loss can then be expressed as

$$-\mathbf{E}_{\text{rad}} = \frac{\pi}{2\hbar}\left(1 - \exp(-\hbar\omega/kT)\sum_i \sum_f \rho_i\{\langle f|\mathbf{E}_0 \cdot \boldsymbol{\mu}|i\rangle\}^2\delta(\omega_{fi} - \omega)\right) \tag{4.21}$$

where ω_{fi} is replaced by ω because of the energy conservation of the δ function. To obtain the absorption (imaginary) part of the dielectric constant, $\varepsilon''(\omega)$, the angular frequency ω and the average energy density in the radiation field $\bar{E}_{rad}$, are required.

$$\varepsilon''(\omega) = -E_{rad}/\omega\bar{E}_{rad} \tag{4.22}$$

where $\bar{E}_{rad} = E_0^2/8\pi$

$$\varepsilon''(\omega) = (4\pi^2/\hbar)[1 - \exp(-\hbar\omega/kT]\sum_f\sum_i\{\langle f|\hat{\varepsilon}.\boldsymbol{\mu}|i\rangle\}^2\delta(\omega_{fi} - \omega) \tag{4.23}$$

where $\hat{\varepsilon}$ is a unit vector along the electric field of the radiation. The form of the relationship can be simplified if the distribution factor is included in the definition of the shape of an absorption band $I(\omega)$:

$$I(\omega) = 3\hbar\varepsilon''(\omega)/4\pi^2[1 - \exp(-\hbar\omega/kT)]$$

$$= 3\sum_i\sum_f\rho_i\{\langle f|\hat{\varepsilon}.\boldsymbol{\mu}|i\rangle\}^2\delta(\omega_{fi} - \omega) \tag{4.24}$$

This formula represents the Bohr–Schrödinger view of spectroscopy as transitions between Bohr stationary states, represented by the time-dependent Schrödinger states $|i\rangle$ and $|f\rangle$. This form is similar to the classical description except that the latter relationship contains the time dependence of the system. The time dependence of equation (4.24) may be introduced in the form of the Fourier expansion of the Dirac δ function

$$\delta(\omega) = 1/2\pi\int_{-\infty}^{+\infty} e^{i\omega t}\,dt \tag{4.25}$$

giving

$$I(\omega) = 3/2\pi\sum_{if}\rho_i\langle i|\hat{\varepsilon}.\boldsymbol{\mu}|f\rangle\langle f|\hat{\varepsilon}.\boldsymbol{\mu}|i\rangle\int_{-\infty}^{+\infty}dt\,\exp[i\{(E_f - E_i)\hbar - \omega\}t] \tag{4.26}$$

The energy eigenvalues E_f and E_i may be represented using the Hamiltonian of matter, giving

$$I(\omega) = 3/2\pi\int_{-\infty}^{+\infty}dt\,e^{-i\omega t}\sum_{if}\rho_i\langle i|\hat{\varepsilon}.\boldsymbol{\mu}(0)|f\rangle\langle f|\hat{\varepsilon}.\boldsymbol{\mu}(t)|i\rangle \tag{4.27}$$

where the time-dependent Heisenberg operator $\boldsymbol{\mu}(t)$ for the dipole at time t, is defined as

$$\boldsymbol{\mu}(t) = e^{iHt/\hbar}\boldsymbol{\mu}(0)\,e^{-iHt/\hbar} \tag{4.28}$$

It should be remembered that the sum over the set of states $|f\rangle$ must, in this case, equal unity, that is

$$\sum_f|f\rangle\langle f| = 1 \tag{4.29}$$

giving

$$\mathbf{I}(\omega) = 3/2\pi \int_{-\infty}^{+\infty} dt\, e^{-i\omega t} \sum_i \rho_i \langle \mathbf{i}|\hat{\boldsymbol{\varepsilon}} \cdot \boldsymbol{\mu}(0) \cdot \hat{\boldsymbol{\varepsilon}} \cdot \boldsymbol{\mu}(t)|\mathbf{i}\rangle \qquad (4.30)$$

The sum over initial states i, weighted by ρ_i, is simply an equilibrium average, which will be denoted by $\langle - \rangle_0$:

$$\mathbf{I}(\omega) = 3/2\pi \int_{-\infty}^{+\infty} dt\, e^{-i\omega t} \langle \hat{\boldsymbol{\varepsilon}}\boldsymbol{\mu}(0) \cdot \hat{\boldsymbol{\varepsilon}}\boldsymbol{\mu}(t)\rangle_0 \qquad (4.31)$$

For an isotropic system of absorbing molecules, such as in the gas or ideal liquid phase, the same result is obtained for any direction of polarization and equation (4.31) then has the form

$$\mathbf{I}(\omega) = 1/2\pi \int_{-\infty}^{+\infty} dt\, e^{-i\omega t} \langle \boldsymbol{\mu}(0) \cdot \boldsymbol{\mu}(t)\rangle_0 \qquad (4.32)$$

Thus the Heisenberg-type expression for a spectrum has the form of a Fourier transform of a correlation function of the dipole moment operator for the absorbing molecule. The above treatment is due to Gordan[8,9] and its relevance to infrared studies on rotational isomeric equilibria is that it allows for a time dependence to be introduced into the description of the system. It should be pointed out that in general the time required for the inter-conversion is typically much longer than the characteristic time for an infrared transition. The observed spectrum is thus the superposition of those of the separate isomeric states. Complications such as are found in nuclear magnetic resonance measurements, when the time of interconversion be-comes comparable with that of the interaction, can be neglected to a first approximation in infrared studies. The time dependence of the above equations arises purely from the effects of random collisions (Brownian motion) in the system.

No assumption has yet been made about the random motion of the dipoles. Their movement as a function of time will be described by the Hamiltonian for the system. A correlation function describes the decay of our knowledge about a system as it approaches equilibrium. The direction in which a dipole points at a certain time, which is taken as a reference zero, may be defined uniquely in terms of a set of coordinates. After a time (t), the dipole's direction will cease to be uniquely defined, having undergone a number of collisions thus losing correlation with its initial direction. In general, one must consider the dipole moment to be that of a whole group of interacting molecules. The product of $\boldsymbol{\mu}(0) \cdot \boldsymbol{\mu}(t)$ will contain cross-terms arising from dipole moments of different molecules $\boldsymbol{\mu}_i(0) \cdot \boldsymbol{\mu}_j(t)$ provided that the total dipole moment is simply the sum of the individual molecular dipoles. The justification of this assump-tion requires a specific model for liquid structure to be chosen and a detailed discussion is beyond the scope of this chapter. Assuming that this simplification is justified it is possible to consider the implications which the above formulation has on the description of molecular vibrations. However, these cross-terms cannot be simply interpreted in terms of reorientation of a single molecule when the conçentration of dipolar molecules is high. This implies that the above formulation is only strictly applicable for the gas phase. The treatment of absorption due to rotational motion of the molecular dipoles will not be considered.

The interpretation of the dipole correlation function is often simpler in the case of rotation–vibration bands in the near infrared. The vibration of a molecule is localized within that molecule and so the frequency in the term $\exp(-i\omega t)$ of equation (4.32) is a measure of the way in which the direction of the vibrational transition dipoles is changed by rotation of the molecule. A correlation function of this form describes the average projection of a molecule's vibrational transition dipole moment at a time (t) onto the direction that the dipole moment has at the reference time taken as zero. This therefore gives a quantitative picture of the way in which a molecule's orientation is randomized by the thermal molecular motion of the surrounding medium.

Measurement of the shape of a rotation–vibration band in the near infrared and subsequent Fourier analysis enables the dipole correlation function to be obtained as the integral over the frequency spectrum:

$$\langle \boldsymbol{\mu}(0) \cdot \boldsymbol{\mu}(t) \rangle = \int_{\text{Band}} \mathrm{d}\omega \, \mathrm{e}^{i\omega t} \mathbf{I}(\omega) \tag{4.33}$$

It has been found that in the case of light diatomic molecules in the gas phase the correlation function possesses negative values after a certain time lapse. The interpretation of this observation being that after a time lapse, t', it is highly probable that a molecule has swung round to a point in the opposite direction from that which it had at $t = 0$. So correlated, easy reorientation of a molecule is not permitted in a liquid due to the fact that neighbouring molecules exhibit a torque and the curves retain a positive correlation with the original direction. The correlation function is normally normalized to its initial value at $t = 0$.

$$\langle \boldsymbol{\mu}^2(0) \rangle = \int_{\text{Band}} \mathrm{d}\omega \, \mathbf{I}(\omega) \tag{4.34}$$

If the correlation function of a unit vector $\boldsymbol{\mu}(t)$ along the direction of the transition dipole moment is defined by

$$\langle \boldsymbol{\mu}(0) \cdot \boldsymbol{\mu}(t) \rangle = \langle \boldsymbol{\mu}(0) \cdot \boldsymbol{\mu}(t) \rangle / \langle \boldsymbol{\mu}(0)^2 \rangle \tag{4.35}$$

the normalized intensity distribution may then be written as

$$\bar{\mathbf{I}}(\omega) = \mathbf{I}(\omega) \bigg/ \int_{\text{Band}} \mathbf{I}(\omega)\, d\omega \tag{4.36}$$

Detailed consideration of the autocorrelation function indicates that it may be further factorized into two functions, one describing the translational motion and the other the rotational motion of the molecule. Empirically it can be shown that factors such as change of dipole moment or moment of inertia should profoundly influence the form of the autocorrelation function describing a particular band. In the case of the classical system 1,2-dibromoethane, the net dipole of the *trans* structure is zero, whereas that of the *gauche* isomer is the vectorial addition of the dipole components of the two carbon–bromine bonds. The principal moments of inertia will change markedly with the interconversion of states from the *trans* to *gauche* structures. The implication is that conformational changes through the modification of the autocorrelation function will lead to different band shapes for the carbon–halogen stretching vibrations associated with a particular isomeric structure. The time for the average interchange between states is very long on the time scale of the autocorrelation function and as a result does not enter into the above considerations, since as far as an absorption transition is concerned the distribution between *gauche* and *trans* states is stationary. In general, molecules which are highly polar and have high moments of inertia for rotational motion will possess motion which is highly correlated. On the other hand, a small dipole and a low moment of inertia illustrates the situation of a low correlation of the molecular dipole transition with its initial direction. Inspection of the band contour and calculation of the appropriate correlation functions may be used to gain further information on the structure of various conformational isomers. The use of autocorrelation functions in the interpretation of band shapes has so far been restricted to diatomic molecules which are rigid and in which conformational changes are not possible. Study of different vibrational bands of the same molecules enables the correlation of different polarization directions of the vector $\boldsymbol{\mu}$ to be investigated. In a linear molecule, $\boldsymbol{\mu}$ can be either parallel or perpendicular to the molecular axis in different vibrational bands, and the shape of these two types of band measure somewhat different aspects of the reorientation process.

4.4.2 *Raman scattering*[10]

Electromagnetic radiation passing through a medium when its frequency is well above that for absorption by a molecular vibration may be scattered as described by classical theory. The quantum-mechanical description of the scattering process in terms of the differential scattering cross-section, for scattering of radiation with a frequency range $d\omega$ and of solid angle $d\Omega$, is given by

$$\frac{d^2\sigma}{d\omega\,d\Omega} = \lambda_S^{-4} \sum_{fi} \{\langle i|\hat{\varepsilon}_I \boldsymbol{\alpha} \hat{\varepsilon}_S|f\rangle\}^2 \rho_i\, \delta(\omega - \omega_{fi}) \tag{4.37}$$

where $2\pi\lambda_S$ is the scattered wavelength, $\hat{\varepsilon}_I$ and $\hat{\varepsilon}_S$ are unit vectors along the direction of the electric vectors in the incident and scattered directions respectively, and $\boldsymbol{\alpha}$ is the polarizability tensor of a group of interacting molecules. The appropriate correlation function has the form[10]

$$\lambda_S^4 \frac{d^2\sigma}{d\omega\,d\Omega} = 1/2\pi \int_{-\infty}^{+\infty} dt\, e^{-i\omega t} \langle(\hat{\varepsilon}_I\boldsymbol{\alpha}(0)\hat{\varepsilon}_S)(\hat{\varepsilon}_I\boldsymbol{\alpha}(t)\hat{\varepsilon}_S)\rangle \tag{4.38}$$

where $\omega = \omega_f - \omega_i = \omega_I - \omega_S$ by conservation of energy.

It is convenient to divide the scattered intensity into two parts, the polarized and depolarized components. They are distinguished experimentally by the different ways in which they depend on the angle between $\hat{\varepsilon}_I$ and $\hat{\varepsilon}_S$. For $\hat{\varepsilon}_I$ and $\hat{\varepsilon}_S$ perpendicular, only the depolarized component appears, whereas for the $\hat{\varepsilon}_I$ parallel to $\hat{\varepsilon}_S$, both components appear:

$$\frac{d^2\sigma_{\text{pol.}}}{d\omega\,d\Omega} = \left(\frac{d^2\sigma}{d\omega\,d\Omega}\right)_{\parallel} - \frac{4}{3}\left(\frac{d^2\sigma}{d\omega\,d\Omega}\right)_{\perp} \tag{4.39}$$

and

$$\frac{d^2\sigma_{\text{depol.}}}{d\omega\,d\Omega} = \frac{1}{10}\left(\frac{d^2\sigma}{d\omega\,d\Omega}\right)_{\perp} \tag{4.40}$$

It is also possible to make a corresponding division of the polarizability $\boldsymbol{\alpha}$ into its average

$$\bar{\boldsymbol{\alpha}} = \text{Tr}\,\boldsymbol{\alpha}/3 \tag{4.41}$$

and its traceless anisotropy

$$\boldsymbol{\beta} = \boldsymbol{\alpha} - \bar{\boldsymbol{\alpha}} \tag{4.42}$$

The choice of these definitions is made so that the polarized intensity is determined by the correlation function of the average polarizability:

$$\lambda_S^4\left(\frac{d^2\sigma_{\text{pol.}}}{d\omega\,d\Omega}\right) = 1/2\pi \int_{-\infty}^{+\infty} dt\, e^{-i\omega t} \langle\bar{\boldsymbol{\alpha}}(0)\,.\,\bar{\boldsymbol{\alpha}}(t)\rangle \tag{4.43}$$

Similarly, the depolarized scattering is determined by the correlation function of the anisotropy of the polarizability:

$$\lambda_S^4\left(\frac{d^2\sigma_{\text{depol.}}}{d\omega\,d\Omega}\right) = 1/2\pi \int_{-\infty}^{+\infty} dt\, e^{-i\omega t} \langle\text{Tr}\,\boldsymbol{\beta}(0)\,.\,\boldsymbol{\beta}(t)\rangle \tag{4.44}$$

If there is no term in the scattering due to vibrational motion of the molecule then this polarized term corresponds to the Rayleigh scattering and the depolarized scattering is the conventional rotational scattering. If the scattering process excites a molecular vibration, then the polarized scattering corresponds to the sum of the isotropic part and the depolarized or anisotropic part of that transition. The frequency of a vibrational band is normally measured with reference to the average frequency of the polarized part of the band. The vibrational frequency may also be obtained from the average frequency of the depolarized part of the band, by subtracting a correction for the asymmetry, or first moment of the depolarized band centre.

When a molecule has some element of symmetry, the form of the polarizability anisotropy is completely determined, as was described in the earlier section. Carbon tetrabromide possesses a three-fold axis of symmetry and the correlation function for the depolarized scattering from a totally symmetric vibration has the form

$$\langle \mathrm{Tr}\, \boldsymbol{\beta}(0) \cdot \boldsymbol{\beta}(t)\rangle \alpha^{\frac{1}{2}}\{[3\boldsymbol{\mu}(0) \cdot \boldsymbol{\mu}(t)]^2 - 1\} = P_2 \langle \boldsymbol{\mu}(0) \cdot \boldsymbol{\mu}(t)\rangle \tag{4.45}$$

where $P_2(x)$ is the second Legendre polynomial. As in the case of the infrared transition, these correlation functions decay to zero after a long time, the rotational motion having become randomized by the molecular motion.

The time dependence of the correlation function may be obtained from the Fourier transformation of the frequency distribution of the Raman band using

$$\langle \mathrm{Tr}\, \boldsymbol{\beta}(0) \cdot \boldsymbol{\beta}(t)\rangle = \int_{-\infty}^{+\infty} d\omega\, e^{i\omega t} \lambda_S^4 \left(\frac{d^2 \sigma_{\mathrm{depol.}}(\omega)}{d\Omega\, d\omega} \right) \tag{4.46}$$

Such an analysis of bands of a system capable of conformational isomerism should lead to a more detailed understanding of the rotational motion and change of structure during isomerism.

Correlation theory has had little impact, so far, on the study of rotational isomerism. It is hoped that its use in future studies may assist in the assignment of vibrational modes to particular structures and also provide a detailed understanding of the geometric and dipole moment changes involved in the isomeric process.

4.5 Thermodynamic description of rotational isomerism

Normally, in the vapour or liquid states, molecules exist as an equilibrium mixture of isomers. The recorded spectra in these phases will contain the vibrational frequencies of all isomers, provided their populations are in excess of about 4%. In practice, the vibrational modes of the individual isomers may take place in similar environments and hence their frequencies may be nearly coincident. It is, however, possible that the spectra of the two isomers may be quite different even though only a very small energy difference exists between states, but this does not happen often. In addition many isomers may have vibrational modes whose frequencies are accidentally degenerate. Thus the spectra of the mixture of isomers is found to contain less bands than is first expected.

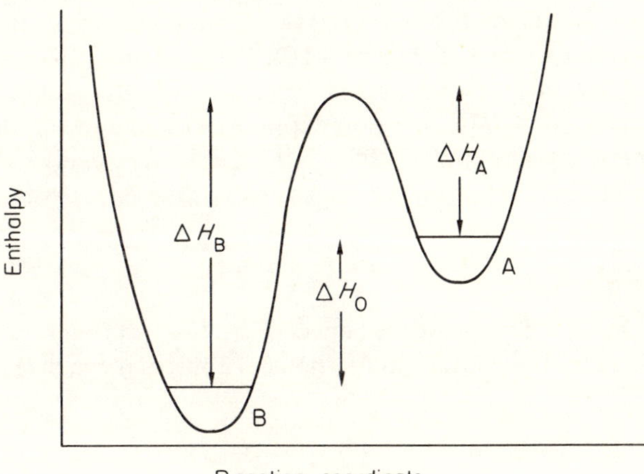

Figure 4.2   Rotational isomerism in a two-state system

In the two state equilibrium (Figure 4.2)

$$A \rightleftharpoons B \tag{4.47}$$

the populations of the two isomeric states A and B will be given by the relationship

$$C_B/C_A = \exp(-\Delta G_0/RT) \tag{4.48}$$

where C_A and C_B are the isomeric populations and ΔG_0 is the Gibbs free energy difference between the isomers. According to the above, the population of isomer B will increase as the temperature of the liquid is lowered. At the boiling and melting points, discontinuities will arise due to changes of phase. At the melting point, which corresponds to the liquid–crystalline phase transition, all the molecules in the liquid with conformation A will be converted into the more stable form B. A number of examples exist in which molecules in both forms exist in the solid state. The changes of conformation to one specific low-energy form is typical of the behaviour observed in many 1,2-disubstituted ethanes. It should be noted that due to balance between intra- and intermolecular forces in the determination of the energy difference, the conformation of the 'stable' form may change with phase. The normal thermodynamic distribution is usually heavily biased and may be assisted by the stabilizing influences of the intermolecular forces of crystal packing. The vibrational spectrum is usually considerably simplified in the solid and corresponds to that of one isomeric form. This fact can be used to assist in the assignment of the observed bands to specific vibrational modes.

For a system of two rotational isomeric states at equilibrium the ratio of C_B/C_A equals K, the equilibrium constant for the system. In the case of 1,2-dichloroethane (X = Cl in Figure 4.1), C_A and C_B refer to the relative populations of the *gauche* and *anti* isomers respectively. Actually, $C_A = C_A' + C_A''$, where C_A' ($= C_A''$) is the population of any one of the individual optically isomeric gauche forms; ΔG_0 is the Gibbs free energy difference related to the enthalpy (ΔH_0) and the entropy (ΔS_0) differences by the relation

$$\Delta G_0 = \Delta H_0 - T\Delta S_0 \tag{4.49}$$

The distribution in the population between the two rotationally isomeric states may be treated by a statistical mechanical approach which leads to the relationship

$$C_B/C_A = \prod_{fB} \Big/ 2 \prod_{fA} [\exp(-\Delta E_0/RT)] \tag{4.50}$$

where ΔE_0 is the internal energy difference between the isomers and $\prod_{fB}$ and $\prod_{fA}$ are the usual products of the partition functions for the *anti* and *gauche* isomers. The factor 2 accounts for the statistical weight of the *gauche* isomer. To a first approximation $\Delta E_0 \approx \Delta H_0$; this assumes that the pressure and volume dependence of the distribution is negligible. This has only been shown to be true in a very limited number of cases. It follows that

$$\Delta S_0 = R \ln \left(\prod_{fB} \Big/ \prod_{fA} \right) - R \ln 2 \tag{4.51}$$

The translational partition functions for the two isomers are assumed to be equal in magnitude and it is also assumed that the partition functions for the overall rotation and vibration are approximately the same, so that $\Delta S_0 = -R \ln 2$. The justification can be seen for this approximation when it is realized that rotation about an internal bond may lead to only 5–15% change in the moments of inertia of the forms involved. Such a relatively small change in moments of inertia will not significantly alter the position of vibrational–rotational bands but may significantly modify their shape. If the sum of the vibrational frequencies associated with each of the isomeric forms is obtained, it is usually found that they differ by less than 5%. When these limits are introduced into the consideration of the partition functions, it can be seen that the above approximation holds to approximately 5%.

In the studies of the infrared and Raman spectra of a mixture of conformational isomers, many of the vibrational frequencies will be different. Since the intensity of a vibrational band is proportional to the population of that state, it follows that by choosing two corresponding bands, thermodynamic parameters may be calculated. The band intensities for each isomer may be given by

$$I_A = \alpha_A C_A l \quad \text{and} \quad I_B = \alpha_B C_B l \tag{4.52}$$

where I_A and I_B are the band areas proportional to the band intensities of isomers A and B, α_A and α_B are the integrated absorption coefficients and l is the cell length. The expression for the equilibrium constant K can be written as

$$K = C_B/C_A = I_B\alpha_A/I_A\alpha_B \tag{4.53}$$

and using the Van't Hoff isochore

$$\ln(I_B/I_A) = -\Delta H_0/RT + \ln(\alpha_B/\alpha_A) + \Delta S_0/R \tag{4.54}$$

By measuring intensities at different temperatures, ΔH_0 can be derived from the slope of a plot of $\ln(I_B/I_A)$ against reciprocal temperature. This treatment, which is widely used, assumes that ΔH_0, ΔS_0 and the quantity $\ln(\alpha_B/\alpha_A)$ are independent of temperature. In most of the work carried out in this field optical densities have been used rather than integrated intensities.

If C is the total concentration, then $C = C_A + C_B$ at all temperatures, and from equation (4.52):

$$I_B/\alpha_B l + I_A/\alpha_A l = C \tag{4.55}$$

and thus

$$I_B = (-\alpha_B/\alpha_A)I_A + \alpha_B lC \tag{4.56}$$

Assuming α_A and α_B are independent of temperature, if I_B is plotted against I_A at different temperatures, the quantity $-\alpha_B/\alpha_A$ will be the slope of the resulting straight line. The entropy difference, ΔS_0 can then be found from the previous equations. This procedure can only be applied when the integrated intensities of the absorption bands are known. Alternatively, the entropy can be obtained by comparing the optical densities or intensities of three bands, one belonging to each of the isomers and the third being an absorption band whose frequency is common to both isomers.

Before proceeding to discuss the results of such studies, the question of frequency assignment and experimental error will be discussed.

4.6 Assignment of vibrational frequencies

The first step in the assignment of the vibrational bands of a molecule capable of rotational isomerism, is to obtain the temperature dependence of the vibrational spectrum. This will give a clue as to the origin of each of the vibrational bands. Next, it is usual to assign bands to the fundamental modes. Group theory enables one to predict the number of modes and their activity in the infrared and Raman spectra and their depolarization ratios if any. The assignment of the experimentally observed bands can then be attempted. A good starting point for this is the use of the known characteristic group frequencies. This is not always satisfactory as the required group frequencies sometimes overlap. Detailed assignments are difficult to make with confidence, but a number of techniques are available.

4.6.1 *Isotopic effects*

If we assume, as is generally possible, that the shift in the observed frequencies on isotopic substitution can be attributed principally to mass effects, then only the frequencies of the particular band containing the isotope will be appreciably altered, provided its motion hardly interacts with that of the other modes. A general qualitative relation, which involves the isotopic effect, is the Teller–Redlich product rule. This states that the product of the zero-order frequencies for all the vibrations of a given symmetry type is independent of the potential constants and depends only on the relative masses of the atoms and the geometric structure of the molecule.

4.6.2 *Comparison of the spectra of similar compounds*

This is the most widely used method for organic molecules. By comparing the spectra of a large series of compounds of similar chemical structure, assignment of fundamental vibrational frequencies to specific geometries can be made. Shifts in the frequencies of particular vibrational modes with chemical structure may also be used to distinguish closely related spatial configurations.

4.6.3 *Vibrational analysis*

By fitting force constants which have been obtained for similar molecules into an assumed 'model' it is possible to calculate the experimental frequencies. Many attempts have been made to use this approach.[11] In most cases computer programmes which allow small ranges in force constant variation to be investigated are used. The reliability of the normal modes obtained is limited by the quality of the force field and the limitation that force constants are never strictly transferable. Since the number of frequencies is far smaller than the number of interactions, a 'generalized force field' cannot be used and either a simple valence force field or a Urey–

Bradley force field must be assumed.[12,13] The latter assumes that the repulsive energy between non-bonded atoms is proportional to $1/r^9$ and hence gives force constants a value of $\frac{1}{10}$ of the stretching force constants. The description of the force field in simple molecules has been treated in some detail elsewhere and will not receive further comment here.[4]

4.6.4 *Intensities*

The intensity of an infrared band is proportional to the change in the dipole moment during the vibration. The magnitude of this change bears some relation to spatial array but is most sensitive to the type of chemical bond. It is found in practice that certain bands are always strong. Thus the C—Cl stretching and bending vibrations are usually stronger than any other bands in their region of the spectrum.

The intensity of Raman bands is dependent on many factors, and may be influenced by sampling methods as well as instrumental and molecular effects. According to Placzek's theory[15] of vibrational fundamentals, the Raman intensity for a given mode is governed by the derived polarizability tensor of the molecule as a whole. Wolkenstein[16] and Long[17] derived an expression which shows that the derived polarization tensor of the whole molecule is made up of contributions resulting from the stretching and changes of orientation of the bonds appropriate to the particular normal vibrational modes in question. Thus when a band is stretched, the polarizability associated with it changes. As the polarizability is concerned with the movement of electrons it is reasonable to assume that the change, when stretching occurs, is related to the degree of covalent character of the bond. Thus the ratio of the intensities of the C—C, C=C and C≡C bonds is in the order $1:2:3$.

It is therefore possible to distinguish some modes by the strength of their infrared and Raman bands. Where dual activity is allowed, the symmetric modes tend to be more prominent in the Raman spectrum, the asymmetric in the infrared spectrum.

4.6.5 *Gas-phase band contours*

In the vapour phase, the contours of the vibrational–rotational bands may be studied and are dependent upon the relative magnitudes of the principal moments of inertia relative to the orientation of the vibration being studied. If an assumed structure is used in the calculation of the principal moments it is possible from inspection of the contours to distinguish between rotational isomeric states.[18,19]

4.6.6 *Polarized infrared studies*

A study of the effect of polarized infrared on an orientated solid can reveal information about the direction of the change in dipole moment of the unit cell as a result of a molecular vibration. If the molecular orientations in the solid sample are known, the polarized infrared spectrum can frequently be of help in making band assignments.

Use of the above methods give reasonable assignments, as long as certain points are remembered. Since all the atoms of the molecule are in motion during a vibration it is not difficult to see that one vibration could interfere or 'couple' with another provided they are of the same symmetry. The force constants are normally assumed to be constant but in actual fact, due to interactions, other modes cause a periodic variation in them. The extent of this coupling is usually unknown, but is found to be greater when a common atom is involved in both modes, or when the angle between modes is large, i.e. 180°, and they thus have similar frequencies. In vibrational analysis the extent of interaction is given by the off-diagonal term in the irreducible representation of the force field matrix. Not only do interactions occur between two stretching frequencies, but there are also bending–bending, bending–stretching interactions between these modes, and overtones or combinations may occur as well. For all organic molecules, the most difficult region to assign is that between 1200 and 700 m^{-1}. Here one finds the C—C stretching modes, the bending, wagging and other deformations, and rocking modes. Since many molecules have little symmetry, appreciable coupling may occur and a unique assignment is almost impossible. A number of specialized texts have appeared dealing in depth with these problems[20,21] to which the reader is referred for further discussion.

4.7 Quantitative absolute band intensity determinations

Of primary importance in the discussion of rotational isomerism is the method of measurement of the band intensity which is intimately related to the population of a particular isomeric state. The true integrated absorption intensity, A, of an infrared band is defined as

$$A = \int_{\text{Band}} \alpha_\omega \, d\omega = 1/Cl \int_{\text{Band}} \ln (I_0/I_\omega) \, d\omega \qquad (4.57)$$

where α_ω is the absorption coefficient for frequency ω, C is the concentration of the solute in mole/l. and l is the cell length. The intensities I_0 and I_ω are respectively those of the incident and transmitted beam.

In practice, the resolving power of the spectrometer is an important factor in intensity determination. The use of finite slit-widths means that the radiation is not monochromatic, but depends on a slit function, so that the quantity measured is the apparent integrated absorption intensity, B, defined as

$$B = 1/Cl \int_{\text{Band}} \ln\left(T_0/T_\omega\right) d\omega = 1/Cl \int_{\text{Band}} \ln\left\{\frac{I_0(\omega)g(\omega, \omega')\,d\omega}{I_\omega(\omega)g(\omega, \omega')\,d\omega}\right\} d\omega' \qquad (4.58)$$

where T_0 and T_ω are apparent intensities when the spectrometer is set at the frequency ω. The slit function $g(\omega, \omega')$ describes that fraction of light of frequency ω which passes through the slit when the instrument is set at a frequency ω'. It will be non-zero only when ω is very close to ω'. The integration is carried out over all values of ω for which $g(\omega, \omega') \neq 0$. The width of this range of ω will be of the order of the resolving power of the spectrometer. The apparent intensities are measured from the spectrum, and when the $\ln\left(T_0/T_\omega\right)$ values are plotted versus ω, BCl, the area under the absorption band may be determined. A slow scanning rate for the spectrometer and a fast response of the recorder are important for accurate intensity measurements. The slit should be manually set so that the optical slit-width divided by the band-width at half peak height is less than 0·3. In addition to having good signal to noise ratios, the linearity of the comb in the reference beam of the spectrometer and of the pen displacement should be checked with rotating sector standards. Difficulties arise from wings of the bands, which in theory, for the absolute intensity to be determined, should extend to infinity. In practice, the area of the band is only measured over a finite frequency range on either side of the band centre, as the absorption falls eventually to the same order as the experimental error in the absorption measurement. A correction has to be made for the residual area lying outside this interval.

Overlapping bands, which often occur in infrared spectra, may be separated arbitrarily into approximately symmetrical bands by inspection, or by computer techniques.[22] Such methods are not very suitable for the evaluation of thermodynamic data as the error is not strictly temperature insensitive. The use of cells with built-in heaters, thermocouples and chambers through which a refrigerant, such as liquid nitrogen, may be passed, enables spectra of samples over a wide range of temperature to be obtained.

4.7.1 Gases

4.7.1.1 *Method of Wilson and Wells*[23] This method assumes that, firstly, the incident intensity is constant over the slit-width and, secondly, that the resolving power is constant over the width of the band. Then $\lim_{pl\to 0} B = A$, where p is the pressure of the absorbing gas. Measurement of B at a series of values of pl enables extrapolation to zero pl to be carried out and the limiting value of B which equals A to be found. This overcomes the problem of finite slit-widths, but does rely on values of B obtained at low absorption, when instrumental errors are greatest.

If I_0 and I_ω are constant over the range when $g(\omega, \omega') \neq 0$, then equation (4.57) becomes

$$B = (1/Cl) \int_{\text{Band}} \ln (I_0/I_\omega) \, d\omega' = A \qquad (4.59)$$

so that A may be determined directly, provided that no other absorbing gas is present in the spectrometer and I_0 is constant. To avoid complications arising from the presence of rotational fine structure, pressure broadening is obtained by using an inert non-absorbing foreign gas such as argon or nitrogen, thus producing a band with a smooth contour. If the spectrometer has sufficiently high resolving power A can then be obtained directly; otherwise, the extrapolation procedure must be used. It is important that the pressure used is sufficient for complete broadening of the absorption band; this will be so, provided the Beer's law plot is a straight line through the origin.

The experiment consists of measuring the transmission curves for the pressurized sample (T) and for the pressurizing gas only (T_0), in a high-pressure cell. No satisfactory wing correction is known for gaseous band shapes and the graphical integration is usually restricted to the main part of the band as far out as the absorption is measurable.

4.7.2 *Liquids*

4.7.2.1 *Ramsay method*[24]   This is an extension of the extrapolation procedure used by Wilson and Wells.[23] The true integrated absorption intensity of a band is obtained by extrapolation to zero Cl of the apparent integrated absorption intensity at a series of concentrations and/or pathlengths. Ramsay showed that the percentage wing correction to the measured intensity for a Lorentz curve is a function of the ratio of the integration interval to the true half intensity widths. Computation of the effect of mathematically scanning a Lorentz curve with a triangular slit function which represents approximately the spectral energy distribution across the exit slit of a spectrometer, has enabled tables of corrections of various band shapes to be calculated. On applying these corrections the basic assumption of the applicability of the Lorentz curve function must be kept in mind. For isolated liquid bands the agreement is close.

The choice of base line is important as it defines the value of T_0. For pure liquids the base line can be obtained by running a trace with no sample in the beam and using this to connect the points of levelling-off of the band. For solution studies independent runs are taken on the same thickness of solvent and solution. A figure of 13% error in intensity has been reported as a result of a 1% absorption unit error in base-line position.[25]

A simpler method of direct integration was also put forward by Ramsay.[24] If the shape of the absorption band follows a Lorentz curve, then the band intensity can be obtained from the peak absorbance of the experimental curve and its half band-width; then

$$A = \frac{K[\ln (T_0/T)_{\omega\,\max}\,\Delta\omega^a_{\frac{1}{2}}]}{Cl} \tag{4.60}$$

where

$$K = \frac{\pi[\ln (I_0/I)_{\omega\,\max}][\Delta\omega^t_{\frac{1}{2}}/\Delta\omega^a_{\frac{1}{2}}]}{2 \ln (T_0/T)_{\omega\,\max}} \tag{4.61}$$

Tables of K and relations between apparent and true absorbances and half band-widths assuming a triangular slit function are given in Ramsay's paper.

This method is less accurate than the Wilson–Wells method. It is based only on two experimental values, the Lorentz curve may not be a good representation of the experimental band, and the triangular slit function may be an oversimplification. An improvement to this method[25,26] divides the observed bands into segments and integrates for each segment with a suitable width parameter. This gives a better fit to the experimental curve and can be applied to asymmetric bands.

Band intensity measurements are carried out using the usual liquid-phase sample cells of fixed or variable thickness with NaCl, KBr, CsI, ThBr/I or other windows. The cell length is determined by the standard interference methods of counting fringes given by the empty cell.

It can be said that whilst data may be obtained and it is usually used for thermodynamic studies from peak height measurements, this method is subject to a number of errors. Even when the total integrated intensity is calculated, it can be seen from what has been said above that the errors are by no means small. Often little consideration appears to have been given to the precise magnitude of the error in the intensity measurement and this topic may in future receive more critical review.

4.7.2.2 *Dispersion methods* It is possible to examine bands both in absorption and dispersion.

The electric moment change during a molecular vibration makes a contribution to the refractive index, n, given by the Kramers–Heisenberg formula as

$$\Delta_i(n-1) = \frac{N}{6\pi^2 C}\left[\left(\frac{\partial\mu}{\partial Q_i}\right)^2 \Big/ (\omega_i^2 - \omega^2)\right] \tag{4.62}$$

where N is the number of molecules per cm^3, μ the molecular dipole moment, Q_i the ith normal coordinate, ω_i the band centre of the corresponding vibration in cm^{-1} and ω the frequency of the incident light. The vibrational contribution to the refractive index is, therefore, intimately related to the infrared absorption intensity, so that the latter may be deduced from dispersion studies (Figure 4.3). Several relations connecting absolute integrated

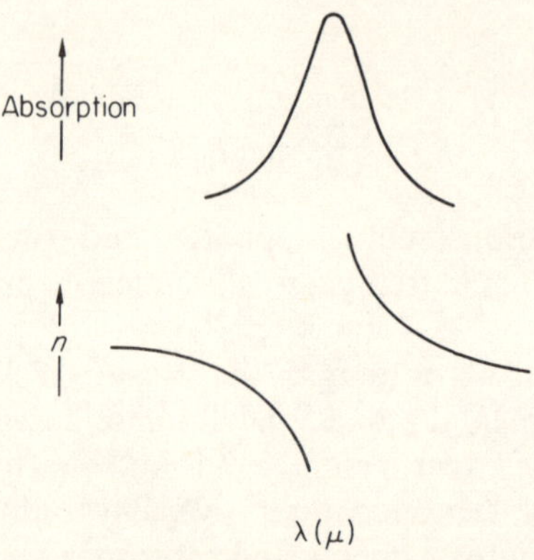

Figure 4.3   A typical dispersion curve

band intensities and the refractive index of an isotropic substance have been derived by Vincent–Geisse,[27] Kagarise[28] and Schatz.[29] Schatz's relation for an isolated band is

$$n(\omega) - n_e(\omega) = \frac{1}{2\pi^2}[A_i/(\omega_i^2 - \omega^2)] \tag{4.63}$$

where $n(\omega)$ is the refractive index at a frequency ω far from the absorption band which is centred at ω_i, $n_e(\omega)$ is the almost constant contribution to the refractive index from the region of electronic transitions. A plot of $n(\omega)$ versus $(\omega_i^2 - \omega^2)^{-1}$ is a straight line of slope $(A_i/2\pi^2)$, $n(\omega)$ can be obtained from fringe systems obtained with the sample in a germanium or silicon Fabry–Perot etalon. The characteristics of the fringe system from a parallel cells are given by $2nt = k\lambda$, where n is the refractive index of the medium between the plates, t is the cell thickness, λ is the wavelength of the fringe maxima, and k is the order of the interference (fringe number), which can be determined by a variable thickness method.[30]

4.7.2.3 *Attenuated total reflection*   Many solids and films are too strongly absorbing in the infrared to be studied by transmission techniques. For such samples a total internal reflection technique was developed by Fahrenfort.[31] A beam of radiation traverses a transparent prism or hemi-cylinder of a suitably high refractive index so that the beam is totally reflected from the back face and an absorbing substance is placed in very close, optical contact with this reflecting surface. A portion of the energy of the beam escapes from the prism face and is then returned into the prism and under suitable conditions the energy that escapes temporarily from the prism is selectively absorbed. The spectrum obtained is similar to that obtained from absorption measurements. This eliminates the very short path-lengths required for the measurement of highly absorbing samples.

The intensities of reflected radiation polarized in and perpendicular to the plane of incidence are given by[30,31]

$$R_p = \frac{(a - \cos \theta)^2 + b^2}{(a + \cos \theta)^2 + b^2} \quad \text{and} \quad R_i = R_p \frac{(a - \sin \theta \tan \theta)^2 + b^2}{(a - \sin \theta \tan \theta)^2 + b^2} \quad (4.64)$$

where

$$a^2 + b^2 = \{[n^2(1 - K^2) - \sin^2 \theta]^2 + 4n^4 K^2\}^{\frac{1}{2}}$$

$$a = \left(\tfrac{1}{2}[n^2(1 - K^2) - \sin^2 \theta] + \tfrac{1}{2}\{[n^2(1 - K^2) - \sin^2 \theta]^2 + 4n^4 K^2\}^{\frac{1}{2}}\right)^{\frac{1}{2}}$$

where θ is the angle of incidence and n and K are the optical constants of the absorbing medium, namely the refractive index and absorption index (which is related to the absorbance of the medium).

The reflected intensities at oblique incidence, then, are functions of n, K and θ. The value of θ is known from experiment so that n and K can be found from two measurements of R_p and/or R_i at θ_1 and/or θ_2, or R_i/R_p at θ_1 and θ_2, or from the reflected intensity of non-polarized radiation at θ_1 and θ_2.

For a particular θ value, the intensities of the radiation reflected by an absorbing medium, R_i and R_p, are much more sensitive to small changes in K when $n < 1$ than when $n > 1$. Thus, in practice, the reflection is measured at the interface between a transparent high refractive index material such as KRS-5, AgCl, Si or Ge. The relative refractive index at the interface will then be smaller than unity. The reflected radiation will be attenuated near the resonant frequencies so that the spectrum obtained is similar to a transmission spectrum.

The optical constants are usually obtained by measuring R_p or R_i at two angles of incidence θ_1 and θ_2. The bare reflecting surface of the high refractive index material serves as the reference mirror. Suitable prereflection optical layouts and graphical solution procedures have been given by Simon[32] Only a particular n, K combination will give the experimentally obtained reflecting power at a particular angle of incidence, so that the two optical constants can be obtained by consulting curves of reflecting power for various n, K values.

4.7.2.4 *Interferometry* Fourier transform infrared spectroscopy, based on the Michelson interferometer, was developed by Gebbie.[33] Radiation from a mercury discharge tube is divided by a beam splitter, which is a very thin sheet of polyethylene tetraphthalate, and it then passes on to two plane mirrors, one of which is moveable (Figure 4.4). The radiation emerging from the interferometer passes through the sample or reference cell and is focussed onto a Golay detector. If the interferometer is illuminated with

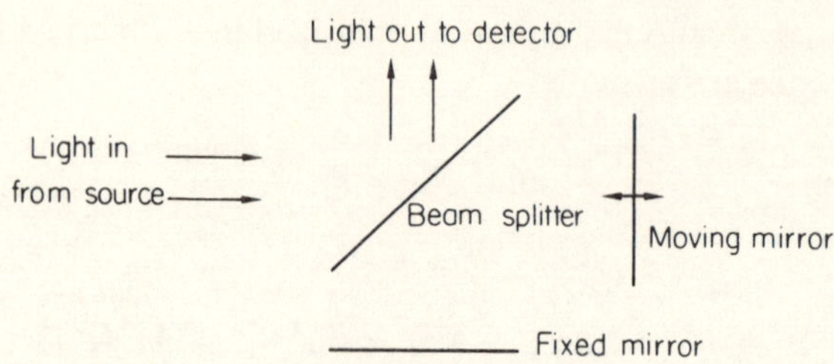

Figure 4.4   Schematic representation of a Michelson
interferometer

monochromatic radiation of wavelength λ, and the mirror is moved with a velocity B, then the signal from the detector has a frequency $f\lambda = 2B/\lambda$ and a plot of signal versus mirror distance is a pure cosine wave. With polychromatic radiation, as is the case in practice, the output signal is the sum of all the cosine waves, which form the Fourier transform of the spectrum. Thus, a Fourier analysis of the signal by a digital computer gives the desired spectrum. The computer usually controls the instrument operations and may signal average the results to improve the signal to noise ratio.

For double-beam operation the Fourier transform computations are done separately for the sample and reference beams and the spectrum plotted as wave number versus energy (single beam operation) or per cent transmittance, absorbance or log absorbance (double beam operation).

The interferometer spectrometer has certain advantages over conventional infrared spectrometers. Since the detector sees intensities of all wavelengths simultaneously, the measurement time is greatly reduced (up to 6 spectra per second may be obtained); also a very large entrance aperture instead of the slits of a grating spectrometer allows about 50 times more light to reach the detector. The combination of a large aperture and a more efficient use of scan time results in an overall advantage of about 10^3 in sensitivity. With a high-pressure mercury vapour lamp source and a Golay cell detector resolutions of $0\cdot1\,\mathrm{cm}^{-1}$ are easily achieved over a range 10–$500\,\mathrm{cm}^{-1}$.

4.7.2.5 *Raman spectroscopy*   In a Raman spectrometer the sample under examination is subject to a source of monochromatic radiation in the visible region and the spectrum is observed by using a highly discriminating grating double-monochromator, a detector and recorder. Due to the extreme weakness of the effect, the source must be intense and the spectrograph should be of high luminosity. Gone are the old mercury discharge tubes (Toronto arcs) and photographic recording devices. Raman spectrometers nowadays use laser excitation (e.g. He–Ne, krypton-ion, Ruby) and detect photoelectrically. The polarization, intensity and direction of the laser are well controlled; very accurate depolarization ratios can be obtained by measuring the intensity of a Raman emission line polarized parallel and perpendicular to the polarization of the laser beam.

The experimental determination of absolute Raman intensities is difficult. Many corrections and calibrations are necessary, especially with mercury discharge sources, before intensity values are obtained that can be used to evaluate derived bond properties. The errors are mainly due to sample geometry, refractive index and reflection losses, optical absorption and the limits of spectrometer collection efficiency.[34–37]

An approximate correction for these errors can be made by using an internal or external standard. However, with an external standard it is difficult to obtain geometrical accuracy even when using the same cell, and with internal standards, the intensity of the standard will vary with environmental changes. Relative intensities can, of course, be measured.

Some results have been obtained using laser excitation, when the corrections are much simplified.[38–42] From measurements of the angular dependence of the intensity of the Raman scattered radiation and depolarization ratios, absolute scattering cross-sections can be obtained.

Cells are available for liquid, solid and gas phase study. For liquids, capillary cells are much used. Multiple reflections of the Raman light from the walls, however, while giving high efficiency, tend to depolarize the laser beam. Before measuring depolarization ratios, then, a capillary tube should be precalibrated with CCl_4. A 1 cm^2 standard ultraviolet cell with one side plane-mirrored has been used by Allkins and Lippincott.[43] The cell is placed flat against the hemispherical lens of the spectrometer sample optical system and the intensity of the scattered radiation is increased by the back reflection through the sample. Good polarization results have been obtained.

In conclusion of this section on the experimental aspects it should be pointed out that the above methods are strictly measurements of the absolute intensities of bands and that more approximate techniques are usually used for relative intensities. It is obvious that methods such as optical density measurements do not allow for the change of band shape with amplitude or for corrections for wing intensities. These are, however, the most widely used for thermodynamic studies and the question of how reliable such measurements of energy differences are in practice will obviously be raised.

4.8 Discussion of data

Before reviewing the available data on energy differences between the rotational isomers of molecules, one must consider the effect that the environment has on the energy differences. Not long after the first observation of rotational isomerism, it was realized that the values obtained from enthalpy differences depended on the phase in which the molecule was studied. These differences could not be explained until they were rationalized in terms of the influence of intermolecular forces. These forces are of dipolar or multipolar origin, and their effect is to stabilize or destabilize the rotational isomers which exist in equilibrium. Differences in the enthalpies obtained in the gas and liquid phases for 1,2-dichloro- and 1,2-dibromoethane have been satisfactorily explained by application of the Onsager model for the description of the molecular electric field.[34] After considering the dipolar effects in the molecules Morino and Mizushima arrived at the following relationship:

$$\Delta H_v - \Delta H_1 = xh \tag{4.65}$$

where ΔH is the enthalpy difference, the subscripts v and 1 refer to the vapour and liquid phases respectively, $x = (\varepsilon - 1)/(2\varepsilon + 1)$ where ε is the dielectric constant of the liquid, and $h = (\mu_A^2/a_A^3) - (\mu_B^2/a_B^3)$, μ_A and μ_B being the dipole moments and a_A and a_B and being the molecular radii of isomers A and B respectively. Although the above equation appears to explain the data obtained on relatively simple molecules, modifications were required to account for the more complex fields found in larger molecules. Alternative approaches to this problem have been made by Wada[44] and Volkenstein and Brevdo.[45] These workers again based their work on the simple Onsager model. Recently Abrahams and his coworkers[46,47] reviewed the solvent problem in connection with their NMR studies (see also Chapter 13). They considered that the enthalpy difference was dependent on quadrupole electric fields produced by the environment as well as the dipolar fields. By using the Onsager model and taking into account the quadrupole fields as well as the dipolar they derived the following equation:

$$\Delta H_v - \Delta H_1 = [kx/(1 - lx)] + 3hx/(5 - x) \tag{4.66}$$

where $l = 2(\eta_d^2 - 1)/(\eta_d^2 + 2)$, η_d being the refractive index, and $h = (q_B^2 - q_A^2)/a^5$, q_A and q_B being the quadrupole moments of the isomers. This theory gave satisfactory explanations for the differences obtained in various solvents by NMR methods. It also gave satisfactory explanations to many of the values found by infrared methods.

It has been observed, in a study of a number of molecules, that the stable isomer in the gas phase is often the least stable one in the liquid. This change in the relative stabilities of the rotational isomers is usually associated with the conflict between the dipolar and quadrupolar stabilization forces, which become effective in the liquid phase, and the intramolecular forces which determine the most stable form in the gas phase. As mentioned in the discussion on the activation energy to internal rotation, the forces governing the energy of a particular isomer are not completely understood. A great number of questions remain to be answered. It is, however, possible to generalize and say that a structure in which steric interactions are minimized will be the most stable form in the gas phase. Also it is reasonable to assume that dipolar and quadrupolar interactions are non-significant in the gas phase, but are decisive factors in determining the most stable form in the liquid. An extensive study[46] of the variations of enthalpy differences in 1,2-dichloro- and 1,2-dibromoethane has shown good agreement with the results expected when using reaction field theory as a description of the forces determining the most stable isomer in the liquid phase. This simple theory does not, however, explain the changes observed when benzene is used as the solvent. This failure in the theory has been attributed to the existence of specific solvent interactions. These interactions are thought to lead to complex structure formation between the solvent and solute. It is worth pointing out that theoretically calculated values should only be compared with gas phase data. That complexes are formed between polar solvents and aromatic molecules is well known from NMR studies. The energy of interaction is calculated as being 4–6 kJ/mole for the benzene complexes and is believed to arise from the polarization of the benzene molecule by the polar molecule. The preferential interaction of the *gauche* isomer with the solvent in 1,2-dihaloethanes, to the effect of 1·6 kJ/mole, is consistent with the observed enthalpy differences in the solvent.

4.8.1 *Ethane-like molecules* ($sp^3 - sp^3$)

Studies of simple 1,2-dihaloethanes[1,2] indicate that in all cases the *trans* structure (Figure 4.1) is the more stable in the gas phase. The magnitude of the enthalpy difference is found to increase regularly with the sum of the Van der Waals radii of the two halogen atoms. Thus the value for 1-bromo-2-chloroethane[48] lies between those for the 1,2-dichloro-[49] and 1,2-dibromoethane.[50] The enthalpy difference in 1,2-difluoroethane[51] changes sign in passing from the gas phase to the liquid; that is, the *cis* configuration of the molecule is the more stable in the liquid phase.[52] In general, values of the enthalpy difference in the liquid and in solution are lower than in the gas phase. Solution values, as expected, are dependent on the polarity of the solvent. The effect of steric repulsive forces, as well as electrostatic (dipolar–dipolar) repulsive forces, on the enthalpy difference, may be illustrated by considering the populations in the vapour phase of the isomeric states of 1,2-dichloro- and 1,2-dibromoethane and n-butane. The greater enthalpy difference in the dibromo compound may be attributed to the larger non-bonded steric repulsive forces arising from the bulkier bromine atoms. Since non-bounded steric repulsive forces are proportional to the Van der Waals' radii of the substituents, the steric forces in 1,2-dibromoethane and n-butane should be very similar. However, the enthalpy difference in 1,2-dibromo-ethane is in fact found to be greater than that in n-butane. This difference has been attributed to the dipole–dipole repulsive forces in the more polar compound. The addition of non-polar flexible groups, as in the case of n-pentane and n-hexane, leads to a reduction in the observed energy difference from that found in n-butane. Light-scattering measurements[53] have shown that in the liquid phase, the higher members of the normal hydrocarbon series tend to associate with each other. Thus the observed values of the energy difference for the paraffins might be expected to show some effects from this interaction. Studies of 1,2-dichloro-,[54] 1,2-dibromo-[55] and 1,2-difluorotetrafluoroethane[56] indicate that the more stable form is the *trans* (C_{2h}) structure. The energy differences obtained from these studies are in good agreement with those calculated theoretically using a Scott–Scheraga potential function.[56] It must be remembered that this simple type of calculation does not take into account the possibility of bond distortion, either angular or along the bond direction, when one end of the molecule is rotated relative to the other end.

Substituted pentanes and butanes may in certain cases be considered to be 'ethane'-like and show a similar dependence of stability on steric interactions. It does however appear, that the presence of a methyl rotor as one of the substituents of the possible isomeric structures leads to complications when attempts are made to predict the relative stabilities of the possible states. An electrostatic attractive force between the alkyl group and the electronegative halogen atoms has been invoked to explain the enthalpy differences in the n-propyl and several other monoalkyl halides.[57] This interaction is largest in the *gauche* isomer and opposite in sign to the steric repulsive forces. The consequence of this interaction is an overall decrease in the energy of the *gauche* form which leads to a smaller enthalpy difference than those found in other substituted ethanes.

Enthalpy differences between rotational isomers of both the *meso* and ($\pm$) structures of 2,3-dichloro- and 2,3-dibromobutane have been derived[58] from temperature studies of the optical densities of the spectral bands. In the vapour phase there is a definite pattern in the enthalpy differences of both the ($\pm$) isomers, whereas in the liquid phase the trends are irregular. The vapour phase data on the *meso* isomer shows that the *trans* isomer is energetically more stable than the *gauche* by 5·0–6·3 kJ/mole. The magnitude of the experimental values compare well with those of gaseous 1,2-dihaloethanes and the 1,2-dihalogeno-2-methylpropanes.[59] In the gaseous state the relative energies for the ($\pm$) forms of the 2,3-dihalobutanes (Figure 4.5) are as

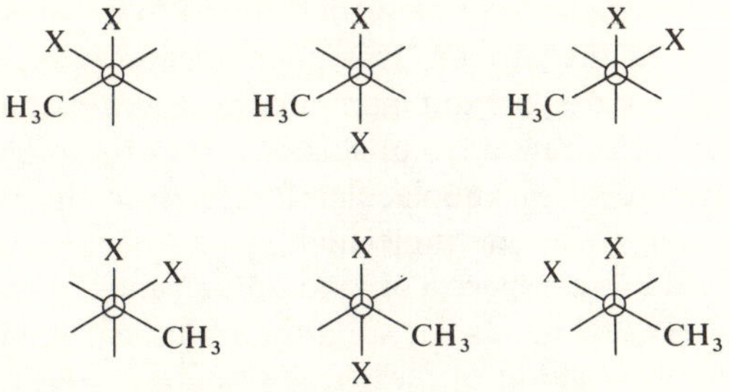

Figure 4.5　2,3-dihalobutanes

follows:

$$S_{HH} \approx S_{HHal} > S_{HCH_3}$$

Both S_{HH} and S_{HHal} structures will have four large substituents in the *gauche* position and consequently one should expect a great deal of steric repulsion. On the other hand, there is far less crowding in the S_{HCH_3} form. It is noticeable that in the ($\pm$) form the S_{HCH_3} isomer is the more stable in the gas phase, whereas the S_{HHal} form persists in the solid. In the liquid phase the complex array of intermolecular forces obviously causes the enthalpy differences to be modified considerably. An interesting observation is that for the solid 2,3-dibromo-2,3-dimethylbutane, extra bands are observed in the spectrum of the pressed disc.[60] These bands have been attributed to the presence of a small amount of *gauche* isomer which is formed during the preparation of the disc. Thus internal rotation must take place with some molecules going from the *trans* to the *gauche* form. A similar observation can be seen when liquids of rotational isomers are supercooled to form glassy solids.[61] In this state the molecules exist in both the *trans* and *gauche* forms. Slight heating and cooling of the glassy solid causes it to take up a more energetically stable form.

Infrared and Raman spectra of β-chloro-, β-fluoro-, β-bromo- and β-iodopropionitrile (XCH_2CH_2CN) show that each of these compounds exists as an equilibrium mixture of *gauche* and *trans* conformers.[62] In all cases, the *gauche* is the more stable. As with succinonitrile[63] and 1,2-difluoroethane,[64] β-fluoropropionitrile exists as a mixture of isomers in the solid state until temperatures below $-44°C$ are reached. Reversal in stability of the isomeric states when passing through the phase change has been observed in the case of 1-fluoro-2-halogenoethanes.[65] The *trans* structure is found to be the more stable in the vapour phase. The enthalpy differences obtained for the β-halogenopropionitriles follow a regular pattern with the *gauche* isomer becoming increasingly more stable as the size of the halogen is increased.

The ratio of the *gauche* to *trans* isomers is found to be considerably greater in the liquid than in the vapour. This phenomenon is ascribed to electrostatic effects being more predominant in the *gauche* isomer than in the *trans*. It is difficult to draw any conclusions from these observations with regard to the nature of intramolecular forces since the dominant effects in solution are likely to be intermolecular.

The infrared and Raman spectra of solid 2,3-dicyano-2,3-dimethylbutane[66] consist of relatively few bands. In solution additional bands are observed assignable to the production of the *gauche* isomeric state. The assignment of the skeletal modes for this molecule was made on the basis of the *trans* conformer having C_{2h} symmetry so that the mutual exclusion principle held for the activities of the fundamental vibrational modes. The relatively weak intensities of the bands due to the *gauche* rotamer indicate that in solution the proportion of this form is considerably smaller than that of the *trans* rotamer. This is consistent with the apparent intermolecular energy difference between the rotational isomers of 5·4 ± 0·42 kJ/mole,[67] obtained from a study of the temperature dependence of the dipole moments in carbon tetrachloride. This study indicated that ΔE for the vapour phase would have a value of 6·4 kJ/mole. Such a value should be compared with values of 4·2 kJ/mole for the vapour and −0·25 kJ/mole for the liquid phase of 1,2-dicyanoethane.[68] Studies have been reported for 2,3-dimethyl-2,3-diphenyl, and fluoro-, chloro- and bromo-substituted 2,3-dimethyl-2,3-di-*p*-halogeno-phenylbutanes and 1,2-diphenyl tetrachloroethane.[69] As for the previous compounds the simplicity of the Raman spectra suggests that only *trans* rotamer exists in the solid phase. Unlike 2,3-dibromo-2,3-dimethylbutane, preparation of the pressed disc samples does not cause a conversion of some of the *trans* molecules into the *gauche* conformation. It has previously been shown[68] that substitution of two phenyl groups for the hydrogen atoms in *sym*-tetrabromoethane has little effect on the relative stabilities of the rotational isomers in solution. On the other hand, substitution of *p*-halogenophenyl groups for two chloro- or bromo-atoms in 2,3-dihalogeno-2,3-dimethylphenyl (a) reduces the rotational isomeric energy difference almost to zero (in solution), thereby altering the *gauche–trans* isomeric ratio, and (b) reduces the dihedral angle from 70° to 65°. Studies of the relative stability of the halogen-substituted compounds suggests that the *gauche* rotamer will be the predominant one in solution. This conclusion is consistent with the analysis of the data in terms of 67–68 % of the *gauche* form.

When studying rotational isomerism the problem of band assignment is one of prime importance. A number of techniques have been developed to assist with this process. One of these[70] is to study changes in the relative spectral intensities with changes in the polarity of the solvent. The more polar conformer will be favoured in solvents of high polarity.

The molecules described above may be compared with those of the (2-haloethyl) benzenes (X = H, Cl, Br and I). In the spectra of each of these compounds two bands are assigned to the carbon–halogen stretching vibration, one to the *trans* and one to the *gauche* isomeric form.[71] For the iodo compound in the crystalline state, the band assigned to the *gauche* form is not detected. Of the halogens, iodine has the largest Van der Waals radius and the lowest electronegativity, which is approximately equal to that of carbon. Both these factors tend to favour the stability of the *trans* form. Any electrostatic repulsion between the benzene ring and the halogen atom would be greatest in the chloro compound and if such a repulsion were present it would be the driving force in making the isomeric forms of unequal stability. Since the spectra indicate the existence of both forms in the liquid and crystalline solid, it is proposed that the steric effects are not sufficiently large, even in the chloro analogue, to be the deciding factor.

Assignment of conformations in the (3-halopropyl) benzenes is less straight forward. In the 3-chloro and 3-bromo compounds, three carbon–halogen frequencies have been identified. If it is assumed that in the solid state steric interactions are minimized in the *trans* structure and that this property is the dominant factor in the lattice formation, then the predominant halogen stretching frequency may be assigned to the so-called *TT* conformation. The less intense bands which remain may be assigned to the *TG* and *GT* conformations which have only partially *trans*-type structures. If, alternatively, the steric interactions, due to the larger separations between bulky substituents involved, are the least important, then a more stable array may be obtained by the more compact *GG* conformation. This assignment is in agreement with that reported for the trimethylene halides.[72] The least stable forms must then be assigned to *TT* and *TG* isomeric structures. This *GG* conformation appears to be more stable than the totally *trans TT* form in the chloro compound and least stable in the bromo.

The importance of 1,3-interactions between bulky groups in chloropropanes has been the subject of a recent investigation (Table 4.1).[73–75] In a study of 17 chloropropanes, no instance was found where an isomer with a 'parallel (1,3) interaction' exists in a concentration which could be detected by infrared, provided there was another staggered isomer possible without this interaction. For those molecules where no isomer is possible without parallel (1,3) chlorine–chlorine interaction (CCl_3—CH_2—$CHCl_2$, CCl_3—$CClH$—$CHCl_2$, and CCl_3—CCl_2—$CHCl_2$) only one isomeric species was observed to be present. It is probable that there are considerable distortions involved in the formation of the stable isomeric state and obviously its stability is determined by the magnitude of the 1,3-interactions. The importance of this interaction is well known in the case of substituted hydrocarbons[76] and in polymer chemistry.[77,78]

Table 4.1   Internal rotational isomers in chlorinated propanes

	No. of possible isomers	No. excluding 1,3 interactions	No. isomers observed
$CH_2Cl—CH_2Me$	2	2	2
$CH_2Cl—CHCLMe$	3	3	3
$CH_2Cl—CH_2—CH_2Cl$	4	3	3
$CH_2Cl_2—CHClMe$	3	3	2 or 3
$CH_2Cl—CCl_2Me$	2	2	2
$CH_2Cl—CHCl—CH_2Cl$	6	4	2 or 3
$CCl_3—CH_2—CH_2Cl$	2	1	1 or 2
$CH_2Cl\text{-}CHCl\text{-}CH_2Cl$	9	5	2 or 4
$CH_2Cl_2—CH_2—CHCl_2$	4	1	1
$CHCl_2—CCl_2Me$	2	2	2
$CH_2Cl—CCl_2—CH_2Cl$	4	3	2
$CHCl_2—CHCl—CHCl_2$	6	1	1 or 2
$CCl_3—CHCl—CH_2Cl$	3	1	1
$CCl_3—CH_2—CHCl_2$	2	0	1
$CHCl_2—CCl_2—CH_2Cl$	5	3	1
$CCl_3—CHCl—CHCl_2$	3	0	1
$CCl_3—CCl_2—CHCl_2$	2	0	1

Spectroscopic studies of the 2-haloethanols indicate that the *gauche* isomer is the more stable form, presumably due to the influence of the internal hydrogen bond.[79,80] This interaction is strongest in the case of the 2-fluoroethanol, for which a single hydroxyl stretching frequency is observed in dilute solution in carbon tetrachloride. Apparent discrepancies in enthalpy differences estimated from the O—H and C—X stretching frequencies have been resolved.[81] These complications arose because no account was taken of the fact that in the less stable *gauche* isomer, the OH group is not engaged in intramolecular hydrogen bonding. The spectra of the solid show that 2-chloroethanol can exist in two different crystal phases, a stable one consisting of *gauche* molecules and a metastable one consisting of both isomers. The detailed interpretation of the spectra is a very complex problem and may be subject to reinterpretation in the future.

4.8.2   *Acid chlorides, aldehydes and related molecules* $(sp^3–sp^2)$

In the previous section, the molecules could take up only three well-defined rotational isomeric structures and this was often simplified as two of these had essentially equivalent energy. With acid chlorides, aldehydes and related systems, the nature of the stable configurations involved in the description of the rotational isomeric processes is often not well defined. A number of acid chlorides and simple aldehydes have been studied by either electron diffraction or microwave spectroscopy, enabling their structures to be characterized.[82] A large number of simple substituted aldehydes have been studied and shown to exhibit rotational isomerism; only a few of these have been studied sufficiently to provide enthalpy difference data.[83] Similarly, infrared studies of the rotational isomerism in 2,3-dihalopropanes have not been studied sufficiently to provide enthalpy data.[84] The measurements do, however, indicate that in this latter system there are two rotational isomers present. The effect of intermolecular interactions is assumed to be responsible for the reversal in relative stabilities of the isomers of 2,3-dichloropropane when going from vapour to liquid. It has also been suggested that significant vibrational coupling occurs in these molecules and that the stable conformations have a non-planar configuration. The existence of two stable rotational isomers has been confirmed by infrared studies on allyl halides.[85]

The acid chlorides and related substituted molecules have been extensively investigated during the last few years.[86–89] For instance, chloroacetyl fluoride, chloride and bromide have been studied by infrared spectroscopy to yield values for the enthalpy differences between the *cis* and *trans* states in the liquid and gas phases (Figure 4.6). NMR studies of these molecules

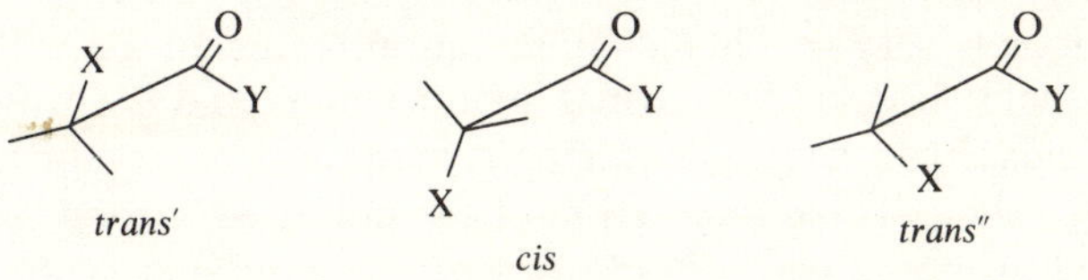

Figure 4.6 Conformations of haloacetyl halides

indicate a small negative enthalpy difference in the case of dichloroacetyl fluoride which suggests that the *gauche* conformation (chlorine eclipsing oxygen) is slightly more stable in the liquid phase than the *trans* conformation (hydrogen eclipsing oxygen).

The chloro-, bromo- and phenylacetyl fluorides have all shown positive *gauche/trans* and *cis/trans* enthalpy differences. Explanation of this stability of the *trans* structure may be based on arguments concerning resonance-delocalized structures. However, this explanation is not consistent with the observed shifts in fundamental vibrational frequencies when the electronegativities of the halogen substituents change. Symmetrically substituted difluoroacetone exists in at least two stable forms.[90] The less polar form is the more stable in the vapour phase, but the more polar form is the more stable in the liquid and the only form in the solid.

An interesting study of substituted butenes[91] has indicated how factors governing the stability of a particular rotamer are precisely those which determine the properties of polymeric materials. Variable temperature studies in both the liquid and vapour indicate that the *cis* form is more stable than the skew form. The *cis* form is almost as stable as the skew form in the gaseous state of 3-methyl-*trans*-2-pentene, and $1 \cdot 1 \pm 0 \cdot 4$ kJ/mole more stable in the liquid state. In 3-methyl-*cis*-2-pentene only one form is stable in the solid, this being ascribed to the skew form on the basis of force-constant calculations. The energy difference between the skew and *cis* forms is expected to be in excess of $12 \cdot 5$ kJ/mole. In the *cis* form the adjacent methyl groups would approach to within $2 \cdot 4$ Å of one another making this isomeric state far less stable than the skew one. Similarly 2-methyl-2-pentene is assumed to be almost completely in the skew form, steric hindrance between the two methyl groups making the *cis* form highly unstable. Comparison of these spectra with those of the previously studied isoprenes indicates that the *trans* form of the compound does not exist. It is probable that the barrier height between the skew form and its mirror image may be low, but the existence of a minimum at the *trans* form seems unlikely from this study. Four rotational isomers were observed to exist for 2-methyl-1-pentene in both the gas and liquid phases, while only one, the *cis–trans* form existed in the solid. It is of interest to note that, when the conformation about the first bond is the skew form, the energy difference between the *trans* and *gauche* configurations is very small. This observation is in contrast with values of $2 \cdot 1$–$3 \cdot 5$ kJ/mole observed for normal hydrocarbons, and is attributed to the absence of close hydrogen–hydrogen pairs in the SG′ form of this molecule.[92] There is also the possible interaction of the polarized carbon centre of the unsaturated grouping with a hydrogen on the saturated centre; this may also influence the relative stabilities of the conformers involved. Studies on the butenes may be used as models for the prediction of conformational stability in *trans*-1,4-polyisoprene. X-ray diffraction studies of the β-crystalline form show that the polymer adopts a $(\text{-S}\text{—}\text{S}'\text{—}\text{T-})_n$ structure which is consistent with the predictions from the above study on pentenes.[93–95] The α-crystalline form has been suggested to have either a $(\text{-SSTS}'\text{S}'\text{T-})_n$ or $(\text{-CSTCS}'\text{T-})_n$ structure.[96] The model studies described above favour the second of these forms as being the more probable structure for the β-form.

In this group of molecules the energy differences are determined by a complex array of interactions. Such interactions include those used in describing the ethanes, and those involved in resonance delocalization of the sp^2 centre with the sp^3 centre and even higher multipole interactions.

4.8.3 *Unsaturated aldehydes, ethers and anisoles* (sp^2–sp^2)

When considering (sp^2–sp^2) bonds, the effects of delocalization of electron density between the two π-orbitals should be observed. Measurements of the rotational barriers in a series of closely related α,β-unsaturated aldehydes[97] indicate that the activation energy does in fact show an approximate correlation with the degree of delocalization of the electron density between the ethylene and carbonyl bonds. Unfortunately gas-phase measurements of the energy difference are not at present available for an extensive enough range of molecules to indicate the precise influence delocalization has on the energy difference. A few general remarks may however be made about the expected behaviour in such systems. If one considers the general formula for the α,β-unsaturated aldehydes (Figure 4.7), then it is seen that the steric size

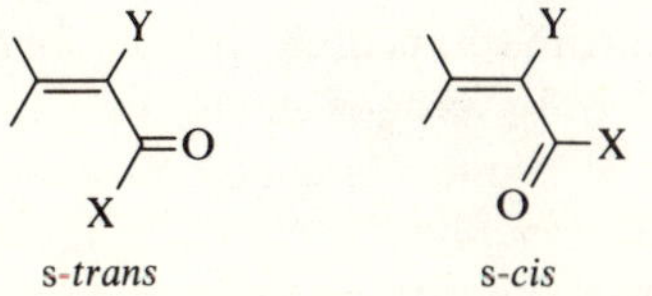

s-*trans*   s-*cis*

Figure 4.7 α,β-Unsaturated aldehydes

of X and Y will have significant effects on the relative stabilities of rotameric states. In the closely related systems 1,1-dibromo-3-fluorobutadiene, 1,1-dichloro-3-fluorobutadiene (Figure 4.8)[98] and the α,β-unsaturated acyl halides, interpretation[99] of the stable isomeric structures in terms of the

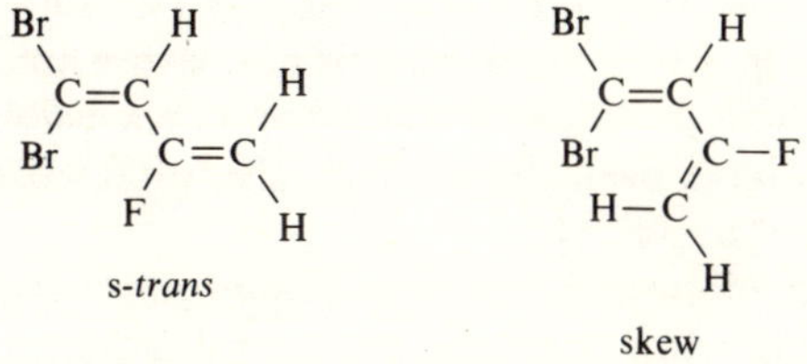

s-*trans*   skew

Figure 4.8   Substituted butadienes

magnitude of the 2,3-interactions has led to the suggestion that the equilibrium is between *trans* and *cis* in the acryl halides. It is, however, unfortunate that entropy data is not available for the latter system to support this assignment of the equilibria. In the case of the butadiene it is probably the 2,3-steric interactions which prohibit the molecule forming the delocalized *cis* structure, and consequently, the energy difference will be related to the magnitude of both steric and delocalized interactions. The presence of a highly polar bond as in the case of the α,β-unsaturated aldehydes may have two effects. Firstly, it changes the geometry around the carbon atom, and secondly, it causes an increase in the delocalization energy and hence in the stability of the *cis* structure. The assignment of a skew structure to one of the stable isomeric states is in accord with the conclusions made from the analysis of the low-temperature NMR studies of the butadiene molecules. Infrared measurements of acryl(propyl) chloride and bromide[101] indicate the existence of a mixture of two isomers for both of these molecules, both forms being present in appreciable proportions. In contrast, although evidence for the presence of two isomers in α,β-unsaturated aldehydes,[101,102] aromatic aldehydes,[102,103] and diazoketones[104] has been obtained, the energy difference between the two isomers is so great (5·0 kJ/mole) that only very small amounts of the second isomer are present at normal temperatures. Of the four possible conformations of the related formyl chloride,[99] observed skeletal vibrations are only assignable to the *trans–trans*, *cis–trans* and *cis–cis* isomers. The C—C band assigned to the *cis–trans* isomer has been found to disappear on crystallization of the liquid. Comparison of band intensities indicates that the energy differences between the *trans–trans* and *cis–trans* modifications is 0·25 kJ/mole, and between the *trans–trans* and *cis–cis* modifications, 1·25 kJ/mole. The effect of 2,3-interactions has also been dramatically observed in the infrared studies of liquid and solid methacryl and crotonyl chlorides.[105,106] In both cases the *trans* structure is observed to be the more stable form. In methacryl chloride at room temperature the second isomer does not exist in sufficient concentrations to be readily detectable. Apparently, the methyl group tends to destabilize the *cis* isomer to a considerable degree. In contrast, the α,β-methyl group in crotonyl chloride has very little effect on the *cis–trans* equilibrium. The enthalpy difference in the crotonyl chloride has been estimated to be approximately 1·25 kJ/mole in the liquid phase. Similar conclusions on the influence of 2,3-interactions have been drawn from electronic absorption studies of α,β-unsaturated ketones.[106]

Studies[107] of the vibrational spectra of *ortho-*, *meta-* and *para*-substituted fluoro-, chloro- and bromo-derivatives of anisoles indicate the existence of restricted rotation about the phenyl–oxygen bond. These studies did not attempt to investigate the magnitude of the energy differences, but rather concentrated on the evaluation of the size of the barrier heights restricting internal rotation. As might be expected, the activation energy to internal rotation appears to be consistently higher in the *meta*-derivatives than in both the *ortho-* and *para*-derivatives, the only exceptions being for the *ortho*-chloro- and bromoanisoles. Such an observation is indicative of the influence of halogen perturbation in the *meta*-position modifying the π-bond character of the phenyl–oxygen linkage. The exceptions presumably result from additional steric factors influencing the relative forms. The energy difference in *m*-fluoroanisole was found to be 2·4 $\pm$ 0·6 kJ/mole in the liquid phase.[107]

4.8.4 *Miscellaneous studies on small molecules*

The existence of rotational isomerism has been reported for simple molecules in a large number of papers on infrared and Raman spectra. Although it is impossible to attempt to list all of these, it is worth mentioning the studies on the organophosphorous compounds and trimethyl phosphite[108,109] which indicate that in O-methyl and O-methyl-d_3-phosphorodichlorodithionate and O-methyl phosphorodichloridate rotational isomerism is observed, though the nature of the structures involved is at present open to discussion.

In the gas phase, piperidine and morpholine[110] both exhibit two resolvable bands in the first N—H overtone region. These have been assigned (by comparison of their band contours with those predicted by theory) to the N—H equatorial and the N—H axial conformers. A temperature dependence study of the intensity ratios indicates that the N—H equatorial conformer predominates for both compounds with $\Delta H = 2 \cdot 1 \pm 0 \cdot 4$ kJ/mole. Examination of piperidine, morpholine and cis-2,6-dimethylpiperidine in carbon tetrachloride gives $\Delta H = 2 \cdot 5 \pm 0 \cdot 8$ kJ/mole, the N—H equatorial structure being the more stable in all cases.

4.8.5 *Studies on macromolecules*

A large number of studies have been carried out on macromolecules[111–113] and these have led to the identification of characteristic absorption regions which arise from particular elements of polymer structure. In this section, only data which provide either thermodynamic measurements or a new look at understanding the nature of the forces involved in macromolecular structure will be considered. Spectra of macromolecules are often used semi-empirically as a rapid means of identification of a particular polymeric species or merely as a test of purity. They may also be used quantitatively to provide detailed information on the dynamic nature of the polymer and for the evaluation of its structure and associated force field. A number of collections of polymer spectra have been made[112–114] and many more spectra are listed with the preparation of various stereospecific polymer species.[115] For brevity, only quantitative data, which have been obtained from detailed study of the spectra will be considered here.

The normal modes of vibration of the chemical bonds of simple molecules may, as indicated in the introduction, be represented using a simple force-field description.[116] The position of the constituent atoms of a molecule may be portrayed by a matrix representation (G matrix), and, similarly, the valence force field by the F matrix. The G matrix is generated from internal coordinates and atomic masses, while the F matrix possesses elements describing the forces holding the constituent chemical entities together. The product of the G and F matrices is related by a similarity transformation to a matrix whose elements depend only on the vibrational frequencies. In theory, calculation of the F and G elements should provide an accurate characterization of the valence force field for the molecule. In essence, if the force constants are known for a simple monomer, it should be possible to use the same force field to represent the polymer. The transferability of force constants from monomer to polymer has been questioned[117] and it appears that in general, such approximations are appropriate. An alternative force-field representation using the same basic ideas is due to Urey–Bradley. This differs from that mentioned previously[118] in that force constants representing non-bonded interactions are added to the off-diagonal valence-force constants. The 1,3 non-bonded interactions which exist between the hydrogen atoms in methane are introduced implicitly in this treatment, whereas in the former valence force-field description they are introduced in the form of the angle-deformation constants. The difference between the two approaches rests primarily in the way in which this non-bonded interaction is estimated. Unless the Urey–Bradley treatment is carried out to a high degree of refinement, it may lead to inaccurate results. This is also true, to a lesser extent, for the valence force-field approach. Force constant calculations, even on relatively simple molecules, require a large number of interactions to be considered for the results to be reliable. In the case of polymers, an even larger number of factors must be appraised. However, use of high speed computers has enabled refined calculations to be carried out on a number of molecules. Normally polymers are treated in terms of their repeating units which are assumed to be continuous and infinite. These calculations can, if necessary, be modified to allow for the influence of end-group effects. Since spectroscopic transitions are normally influenced by short-range interactions, the resulting representation is often found to be an accurate one. The force field for the polymer is usually constructed in the first instance by taking the values obtained from calculations on similar but smaller molecular structures. Overlay calculations[119] have been shown to give meaningful results when Coriolis constants, centrifugal distortion constants and vibrational mean-square amplitudes are not available for more exact least-square refinements.[120–122] The prediction of the infrared and Raman spectra relies on the application of selection rules to the transformation matrix.

What information can be obtained about the polymer from such calculations? The data can be represented in the form of separable G and F matrix representations. If agreement is to be obtained between calculation and experiment, the correct form of the G matrix, which generates the internal coordinates and atomic masses, must be chosen. Obviously, the matrix for a planar form of a polymer will differ from that of a helical structure and it is possible, using refined calculations, to investigate the effect of variations in the nature of closely related configurations on the predicted spectra. The F matrix, since it is modified during the calculation to give closer agreement between theory and experiment, is a probe of the true valence force-field of the polymer. Hence, the elements of the F matrix are indicative of the sizes of various types of interactions within the molecule and from their magnitude it is possible to investigate which of them are important in the determination of the stability of a given structure. It is of interest that the normal coordinate calculations on syndiotatic-1,2-polybutadiene[123] indicate that the four torsional coordinates about the carbon–carbon bond of the main skeleton appear to have little or no effect on the predicted spectra. However, these calculations indicate appreciable coupling between the modes localized in the vinyl group and the chain stretching and bending modes.[123] Calculations on isotatic polypropylene[124] indicate that it adopts a helix structure in the solid state and the observed changes on melting are consistent with the loss of this helical structure.[125] A detailed analysis, however, indicates that it possesses a regular helical structure even in the liquid phase,[126] the segments consisting of approximately 4–7 propylene units. The normal coordinate calculations show that within the $CH(CH_3)$—CH_2 monomeric unit, coupling of motions of the CH_3—C group with adjacent CH_2 and CH groups occurs. Studies of crystalline samples of polyvinylidene fluoride[125] show that this macromolecule may exist in at least two structural modifications. One form has a zig-zag chain structure, the other a glide plane. It may be noted that strong coupling is observed between some of the CH_2 modes, the magnitude of the interaction being sensitive to the conformation of the polymer chain. Polytetrafluoroethylene[127] also prefers the helical structure in the solid. Detailed calculations[128] show that appreciable coupling occurs between the chain stretch and the CF stretching modes. This is in contrast with the observation that in polyethylene[129] the majority of the motions are almost pure. Those which are not pure are the CH_2 bending modes, which mix slightly with, and have a large component of, CCC bending character. These observations would appear to explain in part the differences in the physical properties of these essentially similar polymers. Study of the infrared spectra indicates that the structure undergoes first- and second-order transitions at 90°C and 130°C respectively and this correlates with similar observations from X-ray studies.[129] Similar calculations have also been performed on polyethylene,[127,130] polybutene 1,[131] polypropylene[127] and poly-isobutylene.[127]

By using a simple model which is characteristic of the conformation of the polymer chain, it is, in principle, possible to calculate the appropriate dispersion relations required for the comparison of inelastic neutron scattering and infrared and Raman spectroscopy. The two may be related through the frequency distribution relationships.[132] These calculations involve a slightly different approach to that used in the *GF* valence force-field representation, mentioned previously. Although the *G* and *F* matrices are themselves symmetric, the product is asymmetric, requiring that both the *G* and *F* matrices be diagonalized separately. An alternative procedure, which yields a symmetric secular equation is to transform the potential energy matrix (*F*) rather than the kinetic energy matrix as is usually done. A transformation matrix (*B*) is introduced between the internal coordinates and the mass-adjusted

Cartesian coordinates. The secular determinant then has the form

$$|B'FB - \lambda E| = 0 \tag{4.67}$$

where B' is the inverse of the transformation matrix B and E represents the eigensolutions of the determinant. When the molecule is composed of an infinite number of atoms, the order of the band F matrices will be infinite. However, the translational symmetry of an infinite helix requires that the B and F matrices be a repetition of sub-matrices of order $3p$, where p is the number of atoms in the chemical repeat unit. The transformation involves the introduction of two constants n and θ;[133] n is an integer designating the unit and θ is the phase angle between the displacements of equivalent atoms in adjacent units. For any given phase difference θ (other than 0 and π) the Cartesian symmetry coordinates are complex, so that the secular equation is complex. Solutions of the secular equation will depend on the value chosen for θ. It is, however, possible to obtain the complete set of frequencies $\lambda(\theta)$ of the molecule by evaluation of equation (4.67) as a function of θ. The variation of frequency as a function of θ is called the dispersion curve and is useful in several ways. The dispersion curves themselves may be used in neutron-scattering calculations and the values of the frequencies, as a function of θ, can be compared with those observed from Raman and infrared spectra. This enables predictions to be made regarding the nature of the polymer structure from which the spectra were obtained. Such an approach to the investigation of polymer structure may in future become popular and lead to an understanding of the forces governing the detailed dynamics of these systems. Dispersion curves have been reported for polytetrafluoro-ethylene,[128] isotatic polypropylene,[134] polyethylene glycol,[135] polymethylene[136] and polyoxymethylene.[137] Model calculations have also been presented for a single chain model[137] and a random polymer.[138]

Study of the low-frequency Raman scattering spectra of polymethylene chains[136] shows a series of bands whose frequencies vary inversely and continuously as a function of chain-length. These spectral transitions are due to

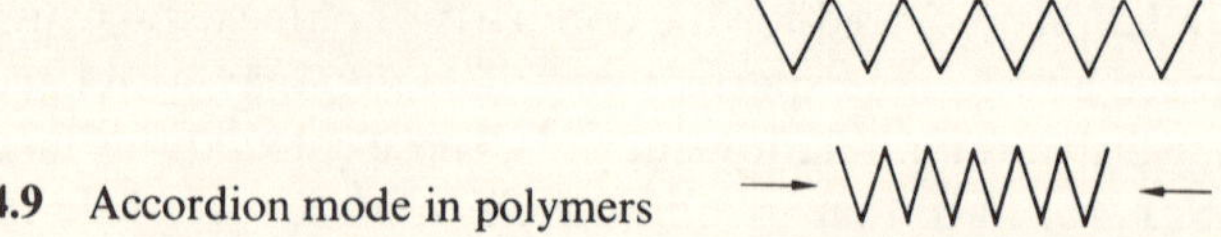

Figure 4.9   Accordion mode in polymers

the longitudinal acoustic 'accordion' modes (Figure 4.9). For this mode of vibration to be active in the Raman, when a change in polarizability occurs, it must be a longitudinal fundamental mode or an odd harmonic of the fundamental. The frequency corresponding to this longitudinal acoustic mode is obtained from the classical equation for a continuous elastic rod,

$$\omega = (m/2L)(E/\rho)^{\frac{1}{2}} \qquad (4.68)$$

where E is the Young's elastic modulus, ρ is the density, m is the vibration order and L is the rod length. To put things in perspective, the frequency for a polymer of 18 carbon atoms length would be 131 cm^{-1}, for one of 94 carbon atoms length it would be 26 cm^{-1} and for 10^5 carbon atoms it would be around 10^{-2} cm^{-1}. Breaks in the regularity of chain sequences may well prevent the observation of the mode corresponding to the overall accordion mode for these long polymers. In practice, the spectrum is made up of a number of lines which may be associated with the existence of a distribution of longitudinal acoustic modes, each characteristic of a particular length and abundance in the polymer. Clearly, this is a most exciting field, since analysis can be applied to helical as well as straight chains.

Infrared and Raman band intensities can also be useful in the study of the structural changes which occur in the polymer with changing temperature. For instance, polystyrene, polypropylene and poly-p-chlorostyrene in their isotatic modification exist in either helical or random-coil structures. The spectra observed at a given temperature will be the weighted summation of the spectra of the individual states. The weighting factor (p) will itself be related to the enthalpy and entropy difference between the two stable states by the usual thermodynamic relationships. For the simplest case in which the state of a monomer residue is not influenced by the states of the neighbouring residues, the total fraction $F(m)$, defined as the total fraction of monomer units which exist in the regular sequence of length $\geqslant m$, will be given by

$$F(m) = p^{m-1}[m - (m - 1)p] \qquad (4.69)$$

Also σ is defined as

$$\sigma = (1 - p)p \qquad (4.70)$$

$$= \exp(-\Delta G^0/RT) \qquad (4.71)$$

and $\Delta G^0 = G_r - G_h$, where G_r is the Gibbs free energy for the random coil and G_h that of the helix. Using $\Delta G^0 = \Delta H^0 - T\Delta S^0$ it can be seen that the observed transitions of $F(m)$ against $-\log \sigma$ ($=0.4343 \Delta G^0/RT$) agrees with that obtained experimentally only when the correct value of m is chosen. The value of ΔH^0 may then be obtained from a plot which should be linear of $\log \sigma$ against reciprocal temperature. Since these studies are often carried out in a solvent, it is important to appreciate that the thermodynamic functions refer to the entire system including the solvent used.[139] Studies of isotatic polystyrene indicate that m lies between 5 and 16 and $\Delta H^0 = 13.0$

kJ/mole and $\Delta S^0 = 9.9$ e.u./mole for the helical to random coil transition. Isotatic polypropylene on the other hand has a value of m equal to 10, $\Delta H^0 = 6.2$ kJ/mole and $\Delta S^0 = 2.2$ e.u./mole for the same process.[139] Isotatic polystyrene has also been shown to exhibit a variation in the number of segments constituting the helix structure; calculations indicate an average value of six for the sequence structure.[140] Infrared has also been used to study polymorphism involved between a four-fold (unstable form) and a stable three-fold helix conformation in polybutene-1.

Infrared measurements on linear polyethylene have enabled an estimate to be made of the degree of chain folding in the solid.[141] Assignment of the observed spectrum indicates that bands at 1350 and 1304 cm^{-1} are characteristic of the occurrence of *gauche* methylene units, which are associated with the disorder caused by folding in the structure of polyethylene. It should be noted that infrared absorption studies are sensitive only to greater than 5 % of the folded structure. It appears that there is little or no difference between solution-crystallized polyethylene and slow-cooled (atmospheric pressure) bulk polyethylene.

Examination of orientated samples using polarized infrared radiation enable dichroism and birefringence of polymers to be studied and thus allow structural changes to be investigated.[142] The dichroism and birefringence of a segment of a polymer molecule may be related to the length and stiffness of the segment.[143] Polymer films adopt different structures depending upon the method of preparation used. For instance, structural changes have been studied during the orientation process of cold-drawn samples of polyethylene[142] and polyvinyl chloride.[144] The origin of dichroism can be easily understood by consideration of the way in which polarized radiation interacts with an orientated polymer sample. When the electric vector of the source and the vector defining the movement of the atoms in the normal mode are parallel, the intensity of the observed transition will be a maximum. Rotation of the plane of polarization will lead to a variation of the band intensity. This is true for both infrared and Raman studies. The use of depolarization data in the assignment of infrared and Raman spectra of polymers has been recently discussed in the cases of polystyrene[145] and polyethylene.[146,147]

4.9 Conclusions

Although infrared and Raman studies of conformational changes have been in existence for some 25 years, new developments in the use of correlation theory and force-constant analysis promise an exciting future for this type of study. A collection of data on some of the molecules in which energy difference data has been reported are presented in Table 4.2.

Table 4.2 Enthalpy differences as determined by infrared and Raman measurements

Name	Enthalpy difference ΔH^0 (kJ/mole)	Phase	Method	Reference
1,2-Difluoroethane	0·84	Gas	IR	51
	3·76	Liquid	IR, Raman	51, 64
1,1,2,2-Tetrafluorcethane	4·85 $\pm$ 0·4	Gas	IR, Raman	64
1,2-Dichloroethane	4·88 $\pm$ 0·4	Gas	IR	50
	0·00	Liquid	IR, Raman	50
	1·67	Soln. in hexane	IR	148
1,1,2-Trichloroethane	10·9($\pm$1·2)	Gas	IR	149
	-0·8($\pm$0·12)	Liquid	IR	148
	-1·6	(1:2) Soln. in CCl_4	IR	148
	0·0	(1:2) Soln. in CH_3CN	IR	148
1,1,2,2-Tetrachloroethane	0($\pm$0·8)	Gas	IR	48
	-4·6 $\pm$ 0·4	Liquid	IR, Raman	148, 150
	-3·76	Soln. n-heptane	IR	148
	-5·86	Soln. CH_3NO_2	IR	148
1,2-Dibromoethane	7·4 $\pm$ 0·63	Gas	IR	50
	3·1 $\pm$ 0·5	Liquid	Raman, IR	150, 151
	3·95	Soln. n-hexane	IR	148
	1·67	Soln. CH_3NO_2	IR	148
1,1,2,2-Tetrabromoethane	2·85	Gas	IR	152
	-3·2	Liquid	IR, Raman	150
1,2-Dichloro-1,1-cifluoroethane	$\pm$1·8 $\pm$ 0·4	Gas, liquid	IR, Raman	153
1,1-Dichloro-2,2-cifluoroethane	-1·47	Liquid	IR	154
1,2-Dichlorotetrafluoroethane	2·1 $\pm$ 0·8	Gas	IR	55
1,1,2-Trichlorotrifluoroethane	1·47 $\pm$ 0·63	Gas	IR, Raman	155
1-Fluoro-1,1,2,2-tetrachloroethane	3·35 $\pm$ 0·4	Gas	IR	55
	-1·60	Liquid	IR	155

Table 4.2 Continued

Name	Enthalpy difference ΔH^0 (kJ/mole)	Phase	Method	Reference
1-Bromo-2-chloroethane	6.0 ± 0.4	Gas	IR	48
	2.05 ± 0.4	Liquid	Raman	48, 156
1,2-Dibromo-1,1-difluoroethane	4.30 ± 0.4	Gas	IR, Raman	157
1,2-Dibromotetrafluoroethane	3.92 ± 0.2	Gas	IR, Raman	158
	3.85 ± 0.2	Liquid	IR, Raman	158
	3.90 ± 0.3	Liquid	IR, Raman	159
1-Bromo-2-chlorotetrafluoroethane	2.5 ± 0.8	Gas	IR	55
	2.18 ± 0.8	Liquid	IR	55
1,2-Diiodotetrafluoroethane	7.70 ± 0.4	Gas	IR	56
1-Fluoro-2-chloroethane	0.84 ± 0.32	Gas	IR	65
	0.25 ± 0.17	Gas	IR	160
	2.0 ± 0.4	Gas	IR	161
1-Bromo-2-fluoroethane	1.25 ± 0.32	Gas	IR	65
	3.4 ± 0.8	Liquid	IR	65
1-Iodo-2-fluoroethane	6.30	Gas	IR	65
	2.5	Liquid	IR	65
Succinonitrile	-1.5 ± 0.20	Liquid	IR	63
β-Fluoropropionitrile	0.0	Liquid	IR	62
β-Chloropropionitrile	-1.85 ± 0.18	Liquid	IR	62
β-Bromopropionitrile	-2.26 ± 0.22	Liquid	IR	62
β-Iodopropionitrile	-1.59 ± 0.15	Liquid	IR	162
	-3.65 ± 0.4	Liquid	IR	62
	-2.26 ± 0.22	Liquid	IR	162
n-Butane	3.18 ± 0.4	Liquid	Raman	163
n-Pentane	1.88 ± 0.25	Liquid	Raman	163
n-Hexane	2.1 ± 0.4	Liquid	Raman	163
2-Fluoroethanol	8.7 ± 2.2	Soln. in CCl_4	IR	164

2-Chloroethanol	4.0 ± 0.38	Gas	IR, Raman	165
	5.0 ± 0.80	Soln. in CCl_4	IR	164
2-Bromoethanol	5.1 ± 0.32	Soln. in CCl_4	IR	164
2-Iodoethanol	3.4 ± 0.34	Soln. in CCl_4	IR	164
1-Chloropropane	-2.1	Gas	IR	166
	$+0.21 \pm 0.24$	Liquid	Raman, IR	166
	-1.25 ± 0.62	Liquid	Raman, IR	50, 167
1-Bromopropane	-1.18 ± 0.4	Gas	IR, Raman	50, 166
	1.85 ± 0.4	Liquid	IR, Raman	50, 167
2-Chloro-2-methylbutane	4.6	Gas	IR	58
	1.5	Liquid	IR	58
2-Bromo-2-methylbutane	5.88	Gas	IR	58
	1.6	Liquid	IR	58
2-Iodo-2-methylbutane	2.9	Liquid	IR	58
1-Iodo-2-methylpropane	-1.5	Liquid	IR	168
Ethane-1,2-dithiol	2.68	Gas	IR	169
1,2-Dimethylthioethane	4.60	Gas	IR	170
	0.40	Liquid	IR	170
2-Chloroethyl mercaptan	2.18	Gas	IR	171
	0.67	Liquid	IR	171
2-Bromoethyl mercaptan	4.00	Gas	IR	171
	2.60	Liquid	IR	171
Isopropyl mercaptan	3.54	Gas	IR	172
Isobutyl chloride	0.96 ± 0.08	Gas	IR	59
	1.55 ± 0.6	Liquid	IR	59
Isobutyl bromide	1.25 ± 0.12	Gas	IR	59
	1.10 ± 0.50	Liquid	IR	59
2,3-Dichloro-2,3-dimethylbutane	6.3	Liquid	IR	173
1,3-Dibromopropane	1.25	Liquid	Raman	167
1,2-Dichloro-2-methylpropane	4.06	Gas	IR	174
	0.00	Liquid	IR, Raman	174
1,4-Dibromobutane	-2.1 ± 0.4	Liquid	Raman	173

Table 4.2 Continued

Name	Enthalpy difference ΔH^0 (kJ/mole)	Phase	Method	Reference
meso-2,3-Dichlorobutane	5.4 ± 1.2	Gas	IR	175
	2.7 ± 0.4	Liquid	IR	175
meso-2,3-Dibromobutane	6.0 ± 0.88	Gas	IR	175
	1.82 ± 0.28	Liquid	IR	175
$(\pm)$2,3-Dichlorobutane S_{HCl}—S_{HH}	0.18	Gas	IR	176
	-0.29 ± 0.7	Liquid	IR	176
S_{HH}—S_{HCH_3}	6.7 ± 0.8	Gas	IR	176
	-1.63 ± 0.36	Liquid	IR	176
S_{HCl}—S_{HCH_3}	6.9 ± 0.6	Gas	IR	176
	1.75 ± 0.24	Liquid	IR	176
$(\pm)$2,3-Dibromobutane				
S_{HH}—S_{HCH_3}	6.6 ± 0.5	Gas	IR	176
	-1.8 ± 0.5	Liquid	IR	176
S_{HBr}—S_{HCH_3}	7.5 ± 1.2	Gas	IR	176
	1.9 ± 0.24	Liquid	IR	176
S_{HBr}—S_{HH}	0.75 ± 1.5	Gas	IR	176
	3.1 ± 0.50	Liquid	IR	176
2,3-Dicyano-2,3-dimethylbutane	6.4 ± 0.6	Gas	IR	66
	5.5 ± 0.4	Soln. CCl_4	IR	66
2-Chlorobutane				
$S_{H'H'}$—$S_{HH'}$	2.72 ± 1.2	Liquid	IR	177
$S_{H'C}$—$S_{H'H'}$	2.10 ± 0.8	Liquid	IR	177
$S_{H'H'}$—$S_{H'C}$	0.40 ± 0.40	Liquid	IR	177
2-Chloropentane				
$S_{CH'}$—$S_{H'H'}$	3.27 ± 0.12	Gas	IR	178
2-Bromopentane				
$S_{CH'}$—$S_{H'H'}$	3.94 ± 0.28	Gas	IR	178

3-Chloropentane				
$S_{CH'}-S_{H'H'}$	2.35 ± 0.16	Gas	IR	179
$S_{H'C'}-S_{CH'}$	0.55 ± 0.24	Liquid	IR	179
$S_{H'H}-S_{CC}$	3.80 ± 0.7	Liquid	IR	179
$S_{H'C}-S_{CC}$	1.10 ± 0.08	Liquid	IR	179
$S_{CH'}-S_{CC}$	0.54 ± 0.01	Liquid	IR	179
3-Bromopentane				
$S_{CH'}-S_{H'H'}$	3.65 ± 0.12	Gas	IR	178
2-Bromohexane				
$S_{CH'}-S_{H'H'}$	3.85 ± 0.12	Gas	IR	178
	1.85 ± 0.40	Liquid	IR	178
3-Bromohexane				
$S_{CH'}-S_{H'H'}$	3.10 ± 0.34	Gas	IR	178
	1.50 ± 0.63	Liquid	IR	178
Dichloroacetyl chloride	0.8	Gas	IR, Raman	180
	0.0	Liquid	IR, Raman	180
Bromoacetyl chloride	4.2 ± 0.4	Gas	IR, Raman	87
Bromoacetyl bromide	7.9 ± 1.2	Gas	IR, Raman	87
Fluoroacetyl chloride	3.18	Gas	IR	88
	8.25	Liquid	IR	88
Fluoroacetyl bromide	8.60 ± 1.0	Gas	IR	181
3-Methyl-*trans*-2-pentene	1.1 ± 0.4	Liquid	IR	91
3-Methyl-*cis*-2-pentene	12.5	Liquid	IR	91
1,1-Dichloro-3-fluorobutadiene	3.54 ± 0.4	Liquid	IR	98
1,1-Dibromo-3-fluorobutadiene	2.95 ± 0.4	Liquid	IR	98
N-Benzyl formamide	2.60 ± 0.26	Soln. in CCl_4	IR	182
N-Benzyl thioacetamide	0.92 ± 0.09	Soln. in CCl_4	IR	182
N-Benzyl thioformamide	-5.85 ± 1.2	Soln. in CCl_4	IR	182
Methyl vinyl ether	-4.80 ± 1.05	Gas	IR	183, 184
	-2.75 ± 0.84	Soln. in $CHCl_2-CH_2Cl$	IR	183, 184
Nitrous acid	1.63 ± 4.0	Gas	IR	185
	2.10	Gas	IR	186

Table 4.2   Continued

Name	Enthalpy difference ΔH^o (kJ/mole)	Phase	Method	Reference
Iodophenol	-4.0	Soln. in CCl_4	IR	187
m-Fluoroanisole	2.4 ± 0.6	Liquid	IR	107
Piperidine	2.2 ± 0.4	Gas	IR	110
	2.5 ± 0.8	Soln. in CCl_4	IR	110
Morpholine	1.97 ± 0.4	Gas	IR	110
	2.5 ± 0.8	Soln. in CCl_4	IR	110
cis-2,6-Dimethylpiperidine	2.5 ± 0.8	Soln. in CCl_4	IR	110
Methyl nitrite	2.5 ± 0.8	Gas	IR	185
2-Methyl-1-pentene				
CG—CT	0.8	Liquid	IR	91
ST—CT	3.2	Liquid	IR	91
SG′—CT	2.85	Liquid	IR	91
SG′—ST	0.32 ± 0.45	Liquid	IR	91
2-Methyl-1-butene	2.24 ± 0.33	Liquid	IR	91
Methyl acrylate	1.32 ± 0.19	Gas	IR	188
	1.67 ± 0.40	Soln. in CS_2	IR	188
Methyl-*trans*-crotonate	2.18 ± 0.10	Gas	IR	188
Acrylyl fluoride	1.25 ± 0	Gas	IR	189
Acrylyl chloride	2.50	Gas	IR	99
Crotonyl chloride	1.25	Gas	IR	99
Isoprene	6.30	Gas	IR	190
Oxalyl chloride	9.20	Gas	IR, Raman	191

4.10 Recent advances (1971–1973)

The significance of correlation theory in the discussion of infrared and Raman transitions was outlined at the beginning of this chapter. It is of interest to note that in the period following the initial preparation of this article a considerable increase in the number of studies of band contours has occurred. These studies have been primarily concerned with the investigation of simple molecules as a probe for the microscopic motions of particles in the gas and liquid phases. The implications outlined at the beginning of this chapter with regard to the effects of internal rotation on the band shapes have yet to be investigated. It is apparent that band-shape analysis will be a region of increasing research activity in the next few years and will hopefully provide valuable information on the short time motions of molecular species. It is not appropriate to expand this discussion further, but it should be noted that caution must be used when applying correlation theory to bands whose 'normal mode type' is ill-defined.

In the more classical field of infrared and Raman studies, a large number of papers have appeared on the rotational isomeric systems.[192–214] The majority of these studies are concerned with the detail of assignment of the vibrational spectra and hence the determination of the stable species present. A few of these studies report the temperature dependence of the spectra and derive energy differences. The systems studied range from the simple normal alkanes,[192] through substituted ethanes,[193–197] propanes,[198–203] propenes,[204–206] butanes[194,207] and butenes[208] to ketones[209–212] and amino compounds.[213–215] In the simple halogenated derivatives of ethane,[193–195] non-bonded intramolecular interactions are primarily responsible for the determination of the stable isomeric states. The stabilization of the *gauche* state through methyl–halogen interactions, discussed earlier in this chapter, have been identified in the studies of substituted propanes and butanes.[202,203,207] Intramolecular hydrogen bonds are similarly observed to stabilize the *gauche* conformer in dihydroxy,[216] halogeno and pseudo-halogeno hydroxy compounds.[196,197,199] Such studies further illustrate the effects of both these interactions upon the energy profile accompanying internal rotation, outlined earlier in this chapter. A detailed understanding of the interactions responsible for the determination of the stable structures in the α-halo ketonic[209–211] and allylic compounds[204–206,208] is yet to be obtained. It is however clear, that subtle effects associated with the changes in electron density distribution accompanying internal rotation play a significant role in the determination of the potential energy profile.[204–207,209–211,217] Certain of these molecular systems are small enough for detailed *ab initio* calculations to be performed in the not too distant future and these latter studies may be anticipated to provide a valuable insight into the mechanism of internal rotation in these electronically complex systems.

4.11 References

1. K. W. Kohlrausch, *Z. Physik. Chem.* (*Leipzig*), **18B**, 61 (1932).
2. S. Mizushima, *Structure of Molecules and Internal Rotation* (New York: Academic Press, 1954).
3. N. Sheppard, *Advan. Spectry.*, **1**, 288 (1959).
4. C. V. Raman, *Indian J. Phys.*, **2**, 387 (1928).
5. C. V. Raman and K. S. Krishnan, *Nature*, **121**, 501 (1928); *Proc. Roy. Soc.* (*London*), **112A**, 23 (1929); *Indian J. Phys.*, **2**, 339 (1928).
6. M. Born, *Optik* (Ann. Arbor, Michigan: Edwards Brothers, 1943).
7. J. W. Emsley, J. Feeney and L. H. Sutcliffe, *High Resolution NMR Spectroscopy*, Vol. 1 (Oxford: Pergamon Press, 1965).
8. R. G. Gordan, *J. Chem. Phys.*, **37**, 2587 (1962).
9. R. G. Gordan, *J. Chem. Phys.*, **42**, 3658 (1965).
10. R. G. Gordan, *J. Chem. Phys.*, **40**, 1973 (1964).
11. S. Mizushima and T. Shimanouchi, *Infrared Absorption and Raman Effect* (Tokyo: Kyorilsu, 1958).
12. H. C. Urey and C. A. Bradley, *Phys. Rev.*, **38**, 1969 (1931).
13. T. Shimanouchi, *J. Chem. Phys.*, **17**, 245, 734, 848 (1949).
14. I. M. Mills, *Infrared Spectroscopy and Molecular Structure*, Ed. M. Davis (Amsterdam: Elsevier, 1963), p. 166.
15. G. Placzek, 'Rayleigh Scattering and Raman Effect', in *Handbuch der Radiologie*, Vol. 6, Pt. 2 (Leipsig: Akademische Verlagegesellschaft, 1934), p. 405.
16. H. Wolkenstein, *Compt. Rend. Acad. Sci. USSR*, **30**, 791 (1941).
17. D. A. Long, *Proc. Roy. Soc.* (*London*), **A217**, 203 (1953).
18. R. H. Badger and L. R. Zumwalt, *J. Chem. Phys.*, **6**, 711 (1938).
19. H. J. Tsunetake and B. L. Crawford, *Appl. Spectr.*, **24**, 9 (1970).
20. E. B. Wilson, J. C. Decius and P. C. Cross, *Molecular Vibrations* (New York:
23. McGraw-Hill, 1955).
21. L. J. Bellamy, *The Infrared Spectra of Complex Molecules* (London: Methuen, 1958).
22. J. Pitha and R. N. Jones, *Can. J. Chem.*, **44**, 3031 (1966).
23. E. B. Wilson, Jr. and A. J. Wells, *J. Chem. Phys.*, **14**, 578 (1946).
24. D. A. Ramsay, *J. Am. Chem. Soc.*, **74**, 72 (1952).
25. G. J. Boobyer, *Spectrochim. Acta*, **23A**, 335 (1967).
26. A. Cabana and C. Sandorfy, *Spectrochim. Acta*, **16**, 335 (1960).
27. J. Vincent-Geisse and J. A. Ladd, *Spectrochim. Acta*, **17**, 627 (1961).
28. R. E. Kagarise and E. E. Ferguson, *J. Opt. Soc. Am.*, **48**, 430 (1958); *J. Chem. Phys.*, **30**, 1059 (1959); **31**, 236 (1959); **31**, 1258 (1959).
29. P. N. Schatz and S. Maeda, *J. Chem. Phys.*, **31**, 1146 (1959); **32**, 894 (1960); **36**, 571 (1962).
30. J. Vincent-Geisse and J. Lecomte, *Compt. Rend.*, **244**, 2152 (1957).
31. J. Fahrenfort, *Spectrochim. Acta*, **17**, 698 (1961); J. Fahrenfort and W. M. Visser, *Spectrochim. Acta*, **18**, 1103 (1962).
32. I. Simon, *J. Opt. Soc. Am.*, **41**, 336 (1951).
33. H. A. Gebbie, Proceedings of the Symposium on Interferometry, NPL June 1959 (London: HMSO, 1961); *Advances in Quantum Electronics*, Vol. II, Ed. J. R. Singer (Columbia: University Press, 1961), p. 155; *Nature*, **178**, 432 (1956); NPL Report Basic Physics Group (1962).
34. H. A. Szymanski (Ed.), *Raman Spectroscopy*, Vols. I and II (Plenum Press, 1965).
35. H. J. Bernstein and G. Allen, *J. Opt. Soc. Am.*, **45**, 237 (1955).

36. D. G. Rea, *J. Opt. Soc. Am.*, **49**, 90 (1959).
37. Yu. I. Naberukhin, *Opt. Spectry.*, **13**, 278 (1962).
38. J. G. Skinner and W. G. Nilsen, *J. Opt. Soc. Am.*, **58**, 113 (1968).
39. F. J. McClug and D. Weiner, *J. Opt. Soc. Am.*, **54**, 641 (1964); D. Weiner, S. E. Schwartz and F. J. McClug, *J. Appl. Phys.*, **36**, 2395 (1965).
40. G. Bret, *Compt. Rend.*, **260**, 6323 (1965).
41. Y. Kato and H. Takuma, *J. Opt. Soc. Am.*, **61**, 347 (1971).
42. S. P. S. Porto, *J. Opt. Soc. Am.*, **56**, 1585 (1966).
43. J. R. Allkins and E. R. Lippincott, *Spectrochim. Acta*, **25A**, 761 (1969).
44. A. Wada, *J. Chem. Phys.*, **22,** 198 (1954).
45. M. V. Volkenstein and V. I. Brevdo, *Zh. Fiz. Khim.*, **28**, 1313 (1954).
46. R. J. Abraham, L. Cavalle and K. G. Pachler, *Mol. Phys.*, **11**, 471 (1966).
47. R. J. Abraham, K. G. Pachler and P. L. Wessels, *Z. Phys. Chem.*, **58**, 257 (1968).
48. J. Powling and H. J. Bernstein, *J. Am. Chem. Soc.*, **73**, 1815 (1951).
49. S. Mizushima, Y. Morina, I. Wanutabe, T. Simanouti and S. Yamaguchi, *J. Chem. Phys.*, **17**, 591 (1949).
50. Y. A. Pentin and V. M. Tatetevskii, *Dokl. Akad. Nauk SSSR*, **108**, 290 (1956).
51. P. Klaboe and J. R. Neilson, *J. Chem. Phys.*, **33**, 1764 (1960).
52. J. P. Lowe, *Progress in Physical Organic Chemistry*, Vol. 6, Ed. A. Streitweiser and R. W. Taft (New York: Interscience, 1968).
53. P. Bothorel, *J. Colloid Sci.*, **27**, 529 (1968).
54. R. E. Kagarise and L. W. Daasch, *J. Chem. Phys.*, **23**, 130 (1955).
55. R. E. Kagarise, *J. Chem. Phys.*, **26**, 380 (1957).
56. C. Serbali and B. Minasso, *Spectrochim. Acta*, **24A**, 1813 (1968).
57. G. J. Szasz, *J. Chem. Phys.*, **23**, 2449 (1955).
58. P. J. D. Park and E. Wyn-Jones, *J. Chem. Soc. (A)*, 2944 (1968).
59. E. Wyn-Jones and W. J. Orville-Thomas, *Trans. Faraday Soc.*, **64**, 2907 (1968).
60. P. J. D. Park and E. Wyn-Jones, *Chem. Commun.*, 557 (1966).
61. J. K. Brown and N. Sheppard, *Discussions Faraday Soc.*, **10**, 144 (1950).
62. M. F. El. Bermani and N. Jonathan, *J. Chem. Soc. (A)*, 1711 (1968).
63. W. E. Fitzgerald and J. G. Janz, *J. Mol. Spectry.*, **1**, 49 (1957).
64. P. Klaboe and J. R. Neilsen, *J. Chem. Phys.*, **32**, 899 (1960).
65. M. F. El Bermani and N. Jonathan, *J. Chem. Phys.*, **49**, 340 (1968).
66. L. H. L. Chia, H. H. Huang and N. Sheppard, *J. Chem. Soc. (B)*, 359 (1970).
67. L. H. L. Chia, H. H. Huang and P. K. K. Lim, *J. Chem. Soc. (B)*, 608 (1971).
68. L. H. L. Chia and H. H. Huang, *J. Chem. Soc. (B)*, 1695 (1970).
69. L. H. L. Chia, K. K. Chia and H. H. Huang, *J. Chem. Soc. (B)*, 1117 (1969).
70. S. Mizushima, T. Shimanouchi, T. Miyazawa, I. Ichishima, K. Karatani, I. Nakagawa and N. Shido, *J. Chem. Phys.*, **21**, 815 (1953).
71. J. E. Saunders, J. J. Lucier and J. N. Willis, *Spectrochim. Acta*, **24**, 2023 (1968).
72. A. B. Dempster, K. Price and N. Sheppard, *Chem. Comm.*, **22**, 1457 (1968).
73. I. J. Gazzard and N. Sheppard, *Mol. Phys.*, **21,** 169 (1971).
74. A. B. Dempster, K. Price and N. Sheppard, *Spectrochim. Acta*, **A25**, 1381 (1969).
75. K. Price, Ph.D. Thesis, University of East Anglia (1969).
76. H. G. Harring, *Raman Analysis of Polychloro Compounds*, Thesis, Amsterdam (1955).
77. K. S. Pitzer, *Chem. Rev.*, **27**, 39 (1940).
78. T. Shimanouchi and M. Tasumi, *Spectrochim. Acta*, **17**, 755 (1961).
79. P. J. Krueger and H. D. Metec, *Can. J. Chem.*, **42**, 326 (1964).
80. E. Wyn-Jones and W. J. Orville-Thomas, *J. Mol. Struct.*, **1**, 29 (1967).
81. P. Buckley, S. Michel and P. A. Giguero, *Can. J. Chem.*, **47**, 901 (1969).
82. W. H. Flygare, *Adv. Phys. Chem.*, 331 (1967).

83. R. J. Abraham and J. A. Pople, *Mol. Phys.*, **3**, 609 (1960).

84. G. A. Crowder, *J. Mol. Spectry.*, **23**, 1 (1967).

85. R. D. Mihachlan and R. A. Nyquist, *Spectrochim. Acta*, **24A**, 103 (1968).

86. J. E. F. Jenkins and J. A. Ladd, *J. Chem. Soc.* (*B*), 1237 (1968).

87. I. Nakagawa, I. Ichishima, K. Kuratani, T. Miyazawa, T. Shimanouchi and S. Mizushima, *J. Chem. Phys.*, **20**, 1720 (1952).

88. V. I. P. Jones and J. A. Ladd, *J. Chem. Soc.* (*B*), 1719 (1970).

89. A. Y. Khan and N. Jonathan, *J. Chem. Phys.*, **50**, 1801 (1969).

90. G. A. Crowder and R. R. Cook, *J. Mol. Spectry.*, **25**, 133 (1968).

91. T. Shimanouchi and Y. Abe, *J. Polymer Sci.*, **A26**, 1419 (1968).

92. J. E. Mark, *J. Am. Chem. Soc.*, **88**, 4354 (1966).

93. C. W. Bunn, *Proc. Roy. Soc.* (*London*), **A180**, 40 (1942).

94. S. C. Nybury, *Acta Cryst.*, **7**, 385 (1954).

95. G. A. Jeffrey, *Trans. Faraday Soc.*, **40**, 517 (1944).

96. G. Natta, P. Corradini and L. Parri, *Alt. Acad. Maz. Lincei. Rad. Cl. Sci. Fir. Mat. Nat.*, **20**, 728 (1956).

97. R. A. Pethrick and E. Wyn-Jones, *Trans. Faraday Soc.*, **66**, 2483 (1970).

98. K. O. H. Hartmann, G. L. Carlson, R. E. Witkowski and W. G. Fateley, *Spectrochim. Acta*, **24A**, 157 (1968).

99. J. E. Katon and W. R. Fairheller, Jr., *J. Chem. Phys.*, **47**, 1248 (1967).

100. J. E. Katon and W. R. Fairheller, Jr., *J. Chem. Phys.*, **45**, 750 (1966).

101. M. S. deGroot and J. Lamb, *Proc. Roy. Soc.* (*London*), **A242**, 36 (1957).

102. F. A. Miller, W. G. Fateley and R. E. Witkowski, *Spectrochim. Acta*, **23A**, 891 (1967).

103. F. A. L. Anet and M. Ahmad, *J. Am. Chem. Soc.*, **86**, 119 (1964).

104. F. Kaplan and G. K. Meloy, *J. Am. Chem. Soc.*, **88**, 950 (1966).

105. W. P. Hayes and C. J. Timmons, *Spectrochim. Acta*, **24A**, 323 (1968).

106. F. H. Cotter, B. P. Staughan, C. J. Timmons, W. F. Forbes and R. Shelton, *J. Chem. Soc.* (*B*), 1146 (1967).

107. N. L. Owen and R. E. Hester, *Spectrochim. Acta*, **25A**, 343 (1969).

108. R. A. Nyquist and W. W. Maelder, *Spectrochim. Acta*, **22**, 1563 (1966).

109. R. A. Nyquist, *Spectrochim. Acta*, **22**, 1315 (1966).

110. R. A. Baldock and A. R. Katrizky, *J. Chem. Soc.*, 1470 (1968).

111. M. Davis, *Infrared Spectroscopy and Molecular Structure* (Amsterdam: Elsevier, 1963); G. Herzberg, *Spectra of Polyatomic Molecules* (London: Van Nostrand, 1948).

112. M. Tryan and E. Horowitz, *Analytical Chemistry of Polymers* (New York: Interscience, 1962).

113. G. Natta and G. Zerbi, *J. Polymer Sci.*, **C**, 1 (1963).

114. R. Zbinden, *Infrared Spectroscopy of High Polymers* (New York: Academic Press, 1964).

115. G. Natta and F. Danusso, *Stereoregular Polymers and Stereospecific Polymerizations*, Vol. I and II (Oxford: Pergamon Press, 1967).

116. E. B. Wilson, Jr., J. C. Decius and P. C. Cross, *Molecular Vibrations* (New York: McGraw-Hill, 1955), Chap. 4 and 8.

117. D. Lee, *Spectrochim. Acta*, **23A**, 453 (1967).

118. I. M. Mills, *Infrared Spectroscopy and Molecular Structure* (Amsterdam: Elsevier, 1963).

119. F. Overend, Ohio State Symposium, Molecular Structure and Spectroscopy, June (1961).

120. J. H. Schachtschneider and R. G. Snyder, *Spectrochim. Acta*, **19**, 177 (1963).

121. R. G. Snyder and J. H. Schachtschneider, *Spectrochim. Acta*, **21**, 169 (1965).

122. G. Zerbi and M. Gussoni, *J. Chem. Phys.*, **41**, 456 (1964).

123. G. Zerbi and M. Gussoni, *Spectrochim. Acta*, **22**, 2111 (1966).

124. H. Tabokoro, M. Kobayashi, M. Ukita, K. Yasufuka, S. Mibrahaski and T. Torii, *J. Chem. Phys.*, **42**, 1432 (1965).

125. G. Cortilli and G. Zerbi, *Spectrochim. Acta*, **23A**, 285 (1967).

126. G. Zerbi, M. Gussoni and F. Ciampelli, *Spectrochim. Acta*, **23**, 301 (1966).

127. P. DeSantis, E. Giglio, A. M. Liquori and A. Ripamonti, *J. Polymer Sci.*, **A1**, 1383 (1963).

128. M. J. Hannon, F. J. Boerio and J. L. Koenig, *J. Chem. Phys.*, **50**, 2829 (1969).

129. R. G. Snyder, *J. Chem. Phys.*, **47**, 1316 (1967).

130. M. Tasumi and S. Krimm, *J. Chem. Phys.*, **46**, 755 (1967).

131. J. P. Luongo and R. Salovey, *J. Polymer Sci.*, **A2**, 997; **B3**, 513 (1965).

132. G. Jannink, *J. Polymer Sci.*, **A2**, 529 (1968).

133. P. Higgs, *Proc. Roy. Soc. (London)*, **A220**, 472 (1953).

134. G. Zerbi and L. Piseri, *J. Chem. Phys.*, **47**, 3840 (1968).

135. H. Matsuura and T. Miyazawa, *J. Chem. Phys.*, **50**, 915 (1969).

136. R. F. Schaufele and T. Shimanouchi, *J. Chem. Phys.*, **47**, 3605 (1967).

137. L. Pisers and G. Zerbi, *J. Chem. Phys.*, **48**, 3561 (1968).

138. M. Tasumi and G. Zerbi, *J. Chem. Phys.*, **48**, 3813 (1968).

139. M. Koboyashi, K. Tsumura and T. Tadokora, *J. Polymer Sci.*, **A2**, 6, 1493 (1968).

140. M. Koboyashi, K. Akita and T. Tadokora, *Makromol. Chem.*, **118**, 324 (1968).

141. J. L. Koenig and D. E. Witenhofer, *Makromol. Chem.*, **99**, 193 (1966).

142. Y. Shindo, B. E. Rhead and R. S. Stein, *Makromol. Chem.*, **118**, 272 (1968).

143. C. R. Desper, *J. Polymer Sci.*, **A2**, 6, 1203 (1968).

144. T. Okada and L. Manderkern, *J. Polymer Sci.*, **A2**, 4, 239 (1967).

145. S. W. Cornell and J. L. Koenig, *J. Appl. Phys.*, **39**, 4883 (1968).

146. V. B. Carter, *J. Mol. Spectry.*, **34**, 356 (1970).

147. M. L. Peraldo and M. Combini, *Spectrochim. Acta*, **20**, 1509 (1964).

148. N. Sheppard and J. J. Turner, *Proc. Roy. Soc. (London)*, **A25**, 2506 (1959).

149. K. Kuratani and S. Mizushima, *J. Chem. Phys.*, **22**, 1403 (1954).

150. R. E. Kagarise and D. H. Rank, *Trans. Faraday Soc.*, **48**, 394 (1952).

151. S. Mizushima, Y. Morino, I. Wanutabe, T. Simaneruti and S. Yamaguchi, *J. Chem. Phys.*, **17**, 591 (1949).

152. G. L. Calson, W. G. Fateley and J. Hiraishi, *J. Mol. Struct.*, **6**, 101 (1970).

153. H. P. Bucker and J. R. Nielson, *J. Mol. Spectry.*, **11**, 47 (1963).

154. H. S. Gutowsley, G. G. Belford and P. E. McMahon, *J. Chem. Phys.*, **36**, 3353 (1962).

155. P. Klaboe and J. R. Nielson, *J. Mol. Spectry.*, **6**, 379 (1961).

156. J. K. Wilmshurst and H. J. Bernstein, *Can. J. Chem.*, **35**, 734 (1957).

157. R. E. Kagarise, *J. Chem. Phys.*, **24**, 1264 (1956).

158. R. E. Kagarise and L. W. Daasch, *J. Chem. Phys.*, **23**, 113 (1955).

159. P. J. D. Park, K. R. Crook and E. Wyn-Jones, *J. Chem. Soc. (A)*, 2910 (1969).

160. P. A. Bwahulin and L. P. Osipova, *Opt. Spectry., USSR (English Transl.)*, **6**, 406 (1959).

161. A. D. Giaiomer and C. P. Smyth, *J. Am. Chem. Soc.*, **77**, 1361 (1955).

162. E. Wyn-Jones and W. J. Orville-Thomas, *J. Chem. Soc. (A)*, 101 (1966); 5855 (1964).

163. N. Sheppard and G. J. Szusz, *J. Chem. Phys.*, **17**, 86 (1949).

164. P. J. Krueger and H. D. Methec, *Can. J. Chem.*, **42**, 326 (1964).

165. S. Mizushima, T. Shimanouchi, T. Miyazawa, K. Abe and M. Yasume, *J. Chem. Phys.*, **19**, 1477 (1951).

166. C. Komaki, I. Ichishima, K. Kuratani, T. Miyazawa, T. Shimanouchi and S. Mizushima, *Bull. Chem. Soc. Japan*, **28**, 330 (1955).

167. J. Goubeau and H. Pujenkamp, *Acta Phys. Austricia*, **3**, 282 (1949).

168. A. Houeix, G. Martin and M. Queneudec, *J. Mol. Struct.*, **2**, 369 (1968).

169. H. Hayashi, Y. Shiro, I. Oshima and K. Murata, *Bull. Chem. Soc. Japan*, **38**, 1734 (1965).

170. H. Hayashi, Y. Shiro, I. Oshima and K. Murata, *Bull. Chem. Soc. Japan*, **39**, 118 (1966).

171. M. Muirakami, Y. Shiro, I. Oshima and K. Murata, *Bull. Chem. Soc. Japan*, **38**, 1740 (1965).

172. G. A. Crowder and D. W. Scott, *J. Mol. Spectry.*, **16**, 122 (1965).

173. T. Koide, T. Oda, K. Ezumi, K. Iwatoni and T. Kubota, *Bull. Chem. Soc. Japan*, **41**, 307 (1968).

174. M. Hayushi, I. Ichishima, T. Shimanouchi and S. Mizushima, *Spectrochim. Acta*, **10**, 1 (1957).

175. P. J. D. Park and E. Wyn-Jones, *J. Chem. Soc.* (*A*), 2944 (1968).

176. P. J. D. Park and E. Wyn-Jones, *J. Chem. Soc.* (**A**), 422 (1969).

177. Yu. A. Pentin, L. P. Melikhova and O. D. Ul'Yanov, *Zh. Strukt. Khim.*, **4**, 535 (1963).

178. T. H. Thomas, E. Wyn-Jones and W. J. Orville-Thomas, *Trans. Faraday Soc.*, **65**, 974 (1969).

179. A. Caraculacu, J. Stokr and B. Schneider, *Collection Czech. Chem. Commun.*, **29**, 2783 (1964).

180. A. Miyake, I. Nakagawa, T. Miyazawa, I. Ichishima, T. Shimanouchi and S. Mizushima, *Spectrochim. Acta*, **13**, 161 (1958).

181. A. J. Khan and N. Jonathan, *J. Chem. Phys.*, **52**, 147 (1970).

182. I. Suzuki, M. Tsuboi, T. Shimanouchi and S. Mizushima, *Spectrochim. Acta*, **16**, 471 (1960).

183. N. L. Owen and N. Sheppard, *Proc. Chem. Soc.*, 264 (1963).

184. N. L. Owen and N. Sheppard, *Trans. Faraday Soc.*, **60**, 634 (1964).

185. G. E. McGraw, D. L. Burnitt and I. C. Hisatsune, *J. Chem. Phys.*, **45**, 1392 (1966).

186. L. Jones, R. Badger and G. Moore, *J. Chem. Phys.*, **19**, 1599 (1951).

187. A. W. Baker and A. T. Shulgin, *Spectrochim. Acta*, **22**, 95 (1966).

188. A. J. Bowles, W. O. George and D. B. Cunliffe-Jones, *J. Chem. Soc.* (*B*), 1070 (1970).

189. D. F. Foster, *J. Am. Chem. Soc.*, **88**, 5067 (1966).

190. S. Dzhassati, A. R. Kyazimova, V. I. Tyulin and Yu. A. Pentin, *Vest. Mosk. Univ. Ser. II: Khim.*, **23**, 19 (1968).

191. J. R. Durig and S. E. Hannum, *J. Chem. Phys.*, **52**, 6089 (1970).

192. J. D. Barnes and B. M. Fanconi, *J. Chem. Phys.*, **56**, 5190 (1972).

193. K. Kumar, *J. Mol. Struct.*, **12**, 19 (1972).

194. K. K. Chia, H. H. Huang and L. H. L. Chia, *J. Chem. Soc.* (*Perkin*), 268 (1972).

195. K. Tanabe, *Spectrochim. Acta*, **A28**, 407 (1972).

196. P. A. Gigiere and M. Schneider, *Can. J. Chem.*, **50**, 152 (1972).

197. M. Hayashi, K. Hamo, K. Ohno and H. Murata, *Bull. Chem. Soc. Japan*, **45**, 949 (1972).

198. J. Thorbjørnsrud, O. H. Ellestad, P. Klaboe and T. Torgrimsen, *J. Mol. Struct.*, **15**, 45 (1973).

199. S. K. Nandy, D. K. Mukharji, S. B. Roy and G. S. Kastha, *J. Phys. Chem.*, **77**, 469 (1973).

200. M. Carles Lorjou, A. Gourset Leroy, H. Bodat and R. Gaufres, *Spectrochim. Acta*, **A29**, 329 (1973).
201. A. B. Dempster, K. Price and N. Sheppard, *Spectrochim. Acta*, **A27**, 1579 (1971).
202. A. B. Dempster, K. Price and N. Sheppard, *Spectrochim. Acta*, **A27**, 1563 (1971).
203. R. Gaufes and C. Roulph, *J. Mol. Struct.*, **9**, 107 (1971).
204. G. A. Crowder, *J. Mol. Struct.*, **10**, 294 (1971).
205. G. A. Crowder, *J. Mol. Struct.*, **10**, 290 (1971).
206. B. Cadioli and U. Pincelli, *J. Chem. Soc.*, **68**, 991 (1972).
207. E. Benedetti and P. Cecili, *Spectrochim. Acta*, **A28**, 1007 (1972).
208. G. A. Crowder, *J. Mol. Struct.*, **12**, 302 (1972).
209. K. Tanabe and S. Sauki, *Spectrochim. Acta*, **A28**, 1083 (1972).
210. S. K. Nanby, D. K. Nukherji, S. B. Roy and G. S. Kastha, *J. Chem. Soc.* (*Faraday I*), **68**, 1312 (1972).
211. J. Goubeau and M. Addhelm, *Spectrochim. Acta*, **A28**, 2471 (1972).
212. G. A. Crowder and P. Pruettiangkura, *J. Mol. Struct.*, **15**, 161 (1973).
213. A. L. Verma, *J. Mol. Spectry.*, **40**, 554 (1971).
214. A. L. Verma, *Spectrochim. Acta*, **A27**, 2433 (1971).
215. A. L. Verma, *J. Mol. Struct.*, **11**, 241 (1972).
216. H. Matsuura, M. Hiraishi and T. Miyazawa, *Spectrochim. Acta*, **28A**, 2299 (1972).
217. C. Sourisseau and B. Pasquier, *J. Mol. Struct.*, **12**, 1 (1972).

5 NMR and ESR studies on simple rotamers

J. A. Ladd and H. W. Wardale

5.1 Nuclear magnetic resonance

Nuclear magnetic resonance spectroscopy is often described as the technique by which a nucleus is used as a sensitive probe into the structural and bonding properties of molecules. A typical high resolution spectrum is characterized by three types of parameters: chemical shifts, spin–spin coupling constants and the intensities of the resonance absorption signals. The latter are directly related to the numbers of magnetically equivalent nuclei giving rise to the resonance absorption and this one-to-one correspondence is a unique feature of this type of spectroscopy.

If a molecule can exist as rotational isomers, the stable forms may possess conformations in which the various nuclei find themselves in different magnetic environments, giving rise to different chemical shifts, and where the spin–spin coupling constants also differ as a result of changes in bond angles and bond lengths. Whether these isomers can be detected experimentally by the observation of their separate NMR spectra depends first of all on their abundances at a particular temperature and secondly on the lifetime of each species. The first factor is merely an instrumental limitation but the second is immutable in that it is connected with the inherent time-scale of the spectroscopic technique. The fact that radio frequencies are being employed ensures that only those species which have lifetimes of the order of 0·1 s or longer can be observed. This corresponds to activation energies of 20–25 kJ/mole, indicating that the rotation must be highly hindered by bulky substituents, by the existence of partial double-bond character in the torsional axis or by preferential intramolecular hydrogen bonding in one conformation. Most generally, none of these factors is present, and what one observes is a spectrum in which both the chemical shifts and spin–spin coupling constants are statistically weighted averages of these parameters for the possible rotamers. Nevertheless, if the rotamer populations can be varied by changes in temperature or in the permittivity of the solvent, then in principle it is possible to obtain values for the free energy differences ΔG_0 of the rotamers as well as for their individual chemical shifts and coupling constants.

Apart from these two extreme situations, there is the intermediate region where the transition from a single averaged spectrum to the superimposed separate spectra of the individual rotamers occurs, and it is here that NMR comes to the fore, particularly in cases where the rotamers are of equal energy and where their populations are therefore fixed. In the latter situation, even though the chemical conformations are identical, a particular nucleus exchanges position with another in a dynamic equilibrium. The uncertainty in establishing its magnetic environment on the NMR time-scale results in a 'blurring' of its resonance signal, and studies of the line-shapes in this situation can provide information concerning the rate processes involved and hence the free energy of activation $\Delta G^{\ddagger}$ of the process.

Over the last decade, refinements of both the classical and quantum-mechanical theory have been rapid and have now reached the stage where the line-shapes of quite complex systems can be analysed. New experimental techniques involving nuclear relaxation time measurements have also become available, and it will be the purpose of this section to provide a resumé of the various NMR methods of studying rotational isomerism, together with a discussion of their limitations and sources of error. The chemical literature in this field has recently been exhaustively reviewed by Sutherland.[1]

5.1.1 *Completely averaged spectra*

Consider the rotational isomerism about a single bond where there are two stable conformations. The free-energy profile is schematically shown in

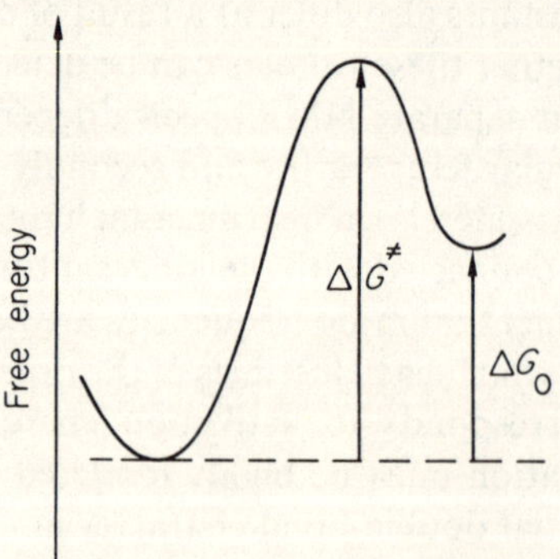

Figure 5.1   Free-energy profile for
two stable conformations

Figure 5.1. If the interconversion of the isomers is very rapid, then the NMR spectrum at ambient temperatures yields only the statistically averaged vicinal coupling constant $\langle J \rangle$ which is related to the mole fraction p_i and the characteristic spin–spin coupling constants J_i of the individual rotamers by equation (5.1).

$$\langle J \rangle = \sum_i p_i J_i \tag{5.1}$$

A similar equation can be written for the statistically averaged chemical shift $\langle \delta \rangle$, but the sensitivity of chemical shifts to macroscopic environmental changes[2] usually precludes their use in the type of analysis to be described here.

Assuming that the two energy forms of the molecule have degeneracies n_1 and n_2 (subscript 1 refers to the high-energy form and 2 refers to the low-energy form), then the equilibrium constant is given by equation (5.2).

$$K_{\text{equil}} = \frac{p_1}{p_2} = \exp\left(\frac{-\Delta G_0}{RT}\right)$$

$$= \frac{n_1}{n_2} \exp\left(\frac{\Delta S_0}{R}\right) \exp\left(\frac{-\Delta H_0}{RT}\right) \tag{5.2}$$

Since $p_1 + p_2 = 1$, then equations (5.1) and (5.2) yield

$$\langle J \rangle = p_1 J_1 + p_2 J_2$$
$$= (J_1 K + J_2)/(1 + K) \tag{5.3}$$

The observed value of $\langle J \rangle$ is influenced by both the temperature and the permittivity of the medium so that two methods of analysis have been developed.

5.1.1.1 *Temperature dependence studies* The determination of the values of the spin–spin coupling constants J_i and the free energy difference ΔG_0 from the temperature dependence of $\langle J \rangle$ was first advocated by Gutowsky, Belford and McMahon.[3] They accomplished this by obtaining the best least-squares fit of equation (5.3) to the observed data. Apart from the statistical entropy factor of $\ln(n_1/n_2)$ which has been treated separately in equation (5.2), it is usually a good approximation that $\Delta S_0 = 0$ for rotational isomers so that the analysis yields a value for the enthalpy difference ΔH_0 directly. If the J_i so determined refer to the vicinal (or 3-bond) coupling between two nuclear spins, then the problem is solved completely. However, quite often the J_i are themselves an average vicinal coupling for each rotamer as is illustrated by $CHX_2CF_2X(\mathbf{I})$ (Figure 5.2). $\mathbf{I_a}$ and $\mathbf{I_b}$ are optical isomers so that the coupling constants are the same but, in $\mathbf{I_c}$, $J_g = J_{g''}$ because of symmetry considerations, while $J_g \neq J_{g'}$ since the rotamers $\mathbf{I_a}$ and $\mathbf{I_c}$ possess different bond lengths and dihedral angles. Thus the individual coupling constants can be estimated only by assuming that such changes are slight. In the case of $CF_2BrCFBr_2$, however, Newmark and Sederholm[4] determined from the spectra of the 'frozen out' rotamers that $J_g = -18.4$ Hz and

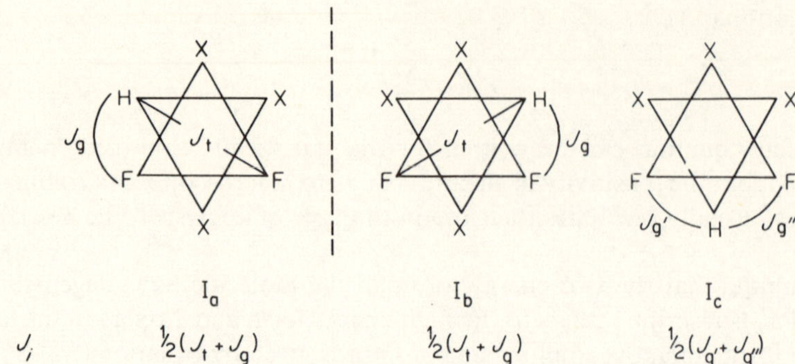

Figure 5.2 J_i for the rotamers cf CHX_2CF_2X

$J_{g'} = -21.5$ Hz, so that the assumption $J_g = J_{g'}$ may lead to quite erroneous results.

It is interesting to note that, in a situation such as CH_2XCH_2X or CF_2XCF_2X, the magnetic nuclei become chemically equivalent under conditions of rapid internal rotation and the NMR spectrum then consists of a single line. Nevertheless, information about the spin–spin coupling constants can be obtained by examining the spectra of those molecules containing one ^{13}C atom which are present in $\sim 2\%$ natural abundance.[5]

The major limitations of such temperature dependence studies are in the inherent assumption of the temperature invariance of the J_i. That this is not the case is convincingly demonstrated by the cases of CH_3CHO[6], CH_3COF[7], CH_3CHCl_2,[8] CF_3CHCl_2[8] and CF_3CFCl_2[8] where small changes in $\langle J \rangle$ with temperature have been observed. Since all the rotamers are identical, the relative populations must remain fixed and thus $\langle J \rangle$ should be constant. It is likely that the observed temperature dependence results from the fact that the potential minima are not sharp enough to make contributions to the coupling constants from torsional oscillations insignificant although Schug, McMahon and Gutowsky[9] have shown that this is limited to <0.2 Hz per $100°$ in the case of CH_3CHXY molecules.

The best confirmation of the values of the spin–spin coupling constants J_g and J_t obtained from such an analysis is by observing these parameters directly at low temperatures where the spectra of the individual rotamers are 'frozen out'. Thus, while Gutowsky, Belford and McMahon claim a high accuracy in this method, their results for $CF_2ClCFCl_2$ gave a value for ΔH_0 (4.1–11.3 kJ/mole) which was an order of magnitude different from that determined by vibrational spectroscopy (1.44 $\pm$ 0.62 kJ/mole). Furthermore, Newmark and Sederholm[10] showed that Gutowsky's values for J_g and J_t did not agree with those of the 'frozen out' rotamers. In the absence of such information from low-temperature studies reliance has therefore to be placed on a comparison of the J_g and J_t values derived with known values in similar compounds. The limitations of the method have been discussed at greater length by Govil and Bernstein[8] and by Gutowsky and coworkers.[11]

Variation of the temperature may also alter the permittivity of the medium and this, in turn, can alter the value of ΔG_0. This may well be an important consideration since it is known that, on going from the vapour to the liquid phase, the value of ΔG_0 changes markedly in some cases.[12]

5.1.1.2 *Solvent dependence studies*   The knowledge that the permittivity of the medium influences the rotamer populations suggests an alternative approach to the study of rotational isomerism and this method has been developed by Abraham, Cavalli and Pachler.[13] Each rotamer has an associated electric field due to the polarization of the surrounding medium by the solute and the calculation of the energy of these fields is based on the classical theory of dielectrics. Here the potential of any system of charges is represented as a power series (charge, dipole, quadrupole, etc.) so that, for a molecule of dipole moment μ, the dipole is the first term in the series. The details of this theory are dealt with in Chapter 13 but it is instructive to consider some of the implications here.

To the quadrupole approximation, the energy difference between the *trans* and *gauche* rotamers of a 1,2-disubstituted ethane in a solvent of permittivity ε (ΔE^s) compared with that in the vapour phase (ΔE^v) is explicitly given by

$$\Delta E^s = \Delta E^v - \frac{x\mu_g^2}{a^3 - 2\alpha x} + \frac{3(h_t - h_g)x}{5 - x} \tag{5.4}$$

where a is the radius of the solvent cavity (assumed spherical), α is the molecular polarizability, μ_g is the molecular dipole moment of the *gauche* rotamer ($\mu_t = 0$ by symmetry), and h_t and h_g are factors which depend only on the dipole moments of the C—X bonds and the molecular dimensions of the *trans* and *gauche* rotamers, respectively.

It is readily apparent from equation (5.4) that the assumption of a temperature-independent energy difference between the rotational isomers is invalid because x (a function of the permittivity) and both α and a (functions of the density) are temperature dependent. Indeed, since the temperature dependence of these terms can readily be measured it is possible to take proper account of this in the temperature dependence studies discussed earlier.

The limitations and sources of error of this method appear to lie in the fact that a knowledge of the molecular dimensions of each rotamer needs to be available. Although these can be roughly compounded from other molecules where the structural parameters are accurately known, the dihedral angle in the *gauche* conformation of ethane molecules, for example, is not necessarily 120°, so that the quadrupole contribution, which is often substantial, may be in error. Furthermore, there may well be an inherent solvent dependence of the coupling constants in some instances, since in CH_3COF and CH_3CF_3 the observed $\langle J \rangle$ value varies significantly with the permittivity of the medium although all the rotamers have the same energy.

5.1.2 *Dynamic equilibria and line-shape analysis*

If a magnetic nucleus can undergo exchange between two different positions in a molecule as a result of internal rotation, it often experiences a different effective magnetic field in the two sites and thus exhibits different chemical shifts. This means that, in an assembly of molecules where the exchange rate is slow, the residence time of nuclei at each site is long, compared with the reciprocal chemical shift difference (Hz^{-1}), and two different NMR signals will be observed. If the rate of exchange is increased by raising the temperature, eventually a single resonance signal will be observed, whose position will be intermediate between those of the previously separate signals. In this limit, the exchanging nuclei experience an effective field which is a weighted average of the fields at the different sites. At intermediate rates, an analysis of the change in line-shape as the signals coalesce can therefore yield the rate constants and activation energy of the process. One of the earliest applications of NMR in this field was the observation that restricted internal rotation existed about the C—N bond of amides.[14] In *N,N*-dimethylacetamide, for example, the room temperature ^1H spectrum of the liquid exhibits resonances associated with three chemically-shifted methyl groups, but, on raising the temperature, two of these broaden and gradually coalesce into a single sharp resonance signal. The explanation of this phenomenon is that the time-averaged environments of the two methyl groups attached to nitrogen only become equal at high temperatures where the rotation about the C—N bond is virtually free. At lower temperatures, however, these methyl groups find themselves in different environments, one being in closer proximity to the carbonyl group than the other. The reason for this is that the C—N bond acquires partial double bond character as a result of π-electron delocalization over the planar heavy atom skeleton and sufficient thermal energy has therefore to be supplied to surmount this barrier to free rotation. Similar observations have been made for nitrosamines by Phillips.[15]

In order to arrive at a quantitative measure of the activation energy of these processes, it is necessary to study the change in appearance of the spectrum as a function of temperature and the complete line-shape must therefore be calculated. Gutowsky and his coworkers[16–18] were the pioneers in this field and they showed how the phenomenological equations of Bloch[19] could be modified to account for the exchange of magnetism from one site to another during the interconversion of the isomers. Such a classical approach is completely valid when the exchanging groups are not involved in spin–spin coupling but a quantum-mechanical basis is required if coupling is involved.

5.1.2.1 *Classical line-shape theory* A single magnetic moment $\boldsymbol{\mu}$ in a uniform magnetic field $\mathbf{B}$ experiences a couple (or torque) given by $\boldsymbol{\mu}\,.\,\mathbf{B}$. According to Newton's second law of motion, force $=\dot{p}$ where p is the momentum, so that the equation of motion of the magnetic moment under these conditions is

$$\dot{\mathbf{J}} = \boldsymbol{\mu}\,.\,\mathbf{B} \tag{5.5}$$

where $\mathbf{J}$ is now the total spin angular momentum. Since $\boldsymbol{\mu} = \gamma\mathbf{J}$,

$$\dot{\boldsymbol{\mu}} = \gamma\dot{\mathbf{J}} = \gamma(\boldsymbol{\mu}\,.\,\mathbf{B})$$

or

$$\dot{\boldsymbol{\mu}} = -\gamma(\mathbf{B}\,.\,\boldsymbol{\mu}) \tag{5.6}$$

From the theory of infinitesimal rotations, if $\boldsymbol{\omega}$ is the angular velocity, then $\dot{\boldsymbol{\mu}} = \boldsymbol{\omega}\,.\,\boldsymbol{\mu}$.

$$\therefore \quad \boldsymbol{\omega} = -\gamma\mathbf{B} \tag{5.7}$$

Thus, if $\mathbf{B}$ is a constant field $\mathbf{B}_0$, the magnetic moment will precess about the direction of $\mathbf{B}_0$ in a negative sense (i.e. anticlockwise when viewed from the origin) and the magnitude of its angular velocity will be $\omega_0 = \gamma\mathbf{B}_0$. In the case of a nuclear magnetic moment this is referred to as Larmor precession and the movement of $\boldsymbol{\mu}$ traces out a cone. The case where the spin quantum number $I = \frac{1}{2}$ is shown in Figure 5.3; the energy of the magnetic moment in

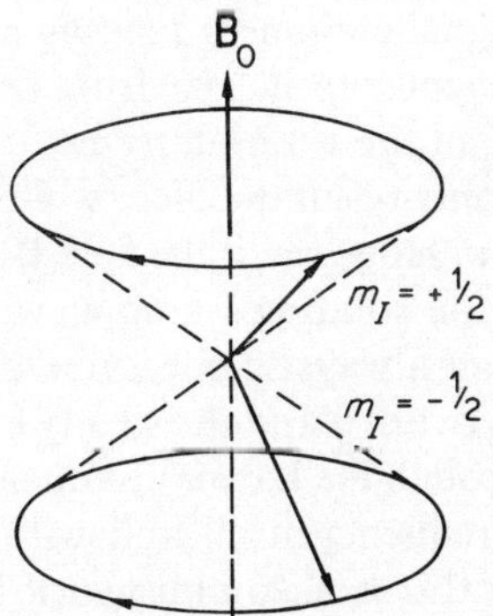

Figure 5.3 Precession cones in an applied field $\mathbf{B}_0$ where the spin quantum number $I = \frac{1}{2}$

the applied field $\mathbf{B}_0$ is quantized so there are two possible precession cones corresponding to $m_I = +\frac{1}{2}$ and $m_I = -\frac{1}{2}$.

If the system is now referred to another coordinate system which is rotating in the same sense and with the same angular frequency about the direction of $\mathbf{B}_0$, then the magnetic moment remains stationary in the new frame. In other words, the static magnetic field $\mathbf{B}_0$ is effectively reduced to zero. In the high-resolution experiment we are interested in inducing a transition of the magnetic moment from one energy state to another (i.e. from one cone to the other) and this is accomplished by applying a second weak magnetic field $\mathbf{B}_1$ which is perpendicular to the original field $\mathbf{B}_0$ and is rotating about that direction (Figure 5.4). If the angular velocity of

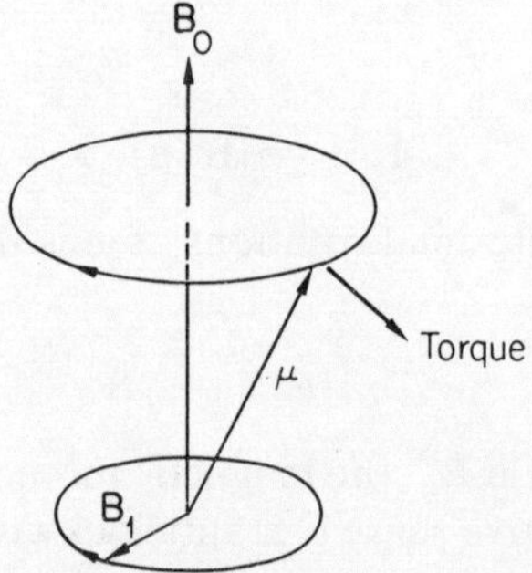

Figure 5.4   Transition of the magnetic moment from one
energy state to another

the second field $\mathbf{B}_1$ is different from that of Larmor precession, $\mathbf{B}_1$ will also be rotating in the rotating coordinate system and its effect will be to exert a torque $\mathbf{\mu} \cdot \mathbf{B}_1$ on the magnetic moment tending to tip the nuclear moment toward the plane perpendicular to $\mathbf{B}_0$, i.e. from one cone to another. If the direction of $\mathbf{B}_1$ is moving in the rotating frame, the direction of the torque will vary rapidly and the only resultant effect will be a slight wobbling of the steady precessional motion. However, if the field $\mathbf{B}_1$ is rotating at the Larmor frequency itself, then, in the rotating frame, it will behave like a constant field and the torque, being always in the same direction, will cause large oscillations in the angle between $\mathbf{\mu}$ and the steady field $\mathbf{B}_0$. Thus, if the rate of rotation of $\mathbf{B}_1$ is varied through the Larmor frequency, the oscillations will be greatest at the Larmor frequency itself and will show up as a resonance phenomenon. In practice this is accomplished if $\mathbf{B}_1$ is in a fixed direction but varies sinusoidally with a frequency in the vicinity of the Larmor value because such a variation can be regarded as a superposition of two equal fields rotating with equal angular velocities in opposite directions (the phases being appropriate). The component rotating in the direction opposite to the Larmor precession has very little effect on the resonance.

5.1.2.2. *The Bloch formulation*   Let $\mathbf{M}$ be the resultant magnetic moment of an assembly of identical nuclei with magnetogyric ratio γ. By analogy with the equation of motion of a single magnetic moment in a magnetic field $\mathbf{B}$, the macroscopic moment $\mathbf{M}$ satisfies the equation

$$\dot{\mathbf{M}} = \gamma(\mathbf{M}\,.\,\mathbf{B}) \tag{5.8}$$

and therefore precesses in the negative sense with an angular velocity $\omega_0 = \gamma\mathbf{B}_0$ about a steady magnetic field $\mathbf{B}_0$ along the z-direction. The three components of $\mathbf{M}$ are then

$$\dot{M}_x = \gamma[M_y B_z - M_z B_y] = \gamma M_y B_0 = \omega_0 M_y$$
$$\dot{M}_y = -\gamma[M_x B_z - M_z B_x] = -\gamma M_x B_0 = -\omega_0 M_x \tag{5.9}$$
$$\dot{M}_z = \gamma[M_x B_y - M_y B_x] = 0$$

The equations in this form do not, however, allow for relaxation effects. In the xy plane, the resultant macroscopic magnetization decays to zero because the individual moments lose phase with each other. This decay is characterized by the transverse relaxation time T_2. Along the z-direction the magnetization does not vanish due to the Boltzmann distribution of magnetic moments between the Zeemann levels so that M_z approaches its equilibrium value $\mathbf{M}_0$ at a rate characterized by the longitudinal relaxation time T_1. The appropriately modified equations are then

$$\dot{M}_x = \omega_0 M_y - \frac{M_x}{T_2}$$

$$\dot{M}_y = -\omega_0 M_x - \frac{M_y}{T_2} \tag{5.10}$$

$$\dot{M}_z = -\frac{M_z - M_0}{T_1}$$

The Bloch equations follow if the components of $\gamma(\mathbf{M}\,.\,\mathbf{B}_1)$ are added where $\mathbf{B}_1$ is the field perpendicular to $\mathbf{B}_0$ rotating with angular frequency $\omega = \gamma\mathbf{B}_1$ (in the same sense as the Larmor precession) with components

$$B_{1x} = B_1 \cos \omega t$$
$$B_{1y} = -B_1 \sin \omega t \tag{5.11}$$

Thus the complete equations are

$$\dot{M}_x = \omega_0 M_y - \gamma M_z B_{1y} - \frac{M_x}{T_2}$$

$$= \omega_0 M_y + \omega M_z \sin \omega t - \frac{M_x}{T_2}$$

$$\dot{M}_y = -\omega_0 M_x + \gamma M_z B_{1x} - \frac{M_y}{T_2}$$

$$= -\omega_0 M_x + \omega M_z \cos \omega t - \frac{M_y}{T_2} \qquad (5.12)$$

$$\dot{M}_z = \gamma M_x B_{1y} - \gamma M_y B_{1x} - \frac{M_z - M_0}{T_1}$$

$$= -\omega(M_x \sin \omega t + M_y \cos \omega t) - \frac{M_z - M_0}{T_1}$$

5.1.2.3 *The Bloch equations for rotating axes* The Bloch equations take a simpler form if, instead of being referred to a set of fixed axes (x, y, z), they are referred to a reference frame rotating about the z-axis with the same sense and angular velocity as the applied rf field $\mathbf{B}_1$ (Figure 5.5). The trans-

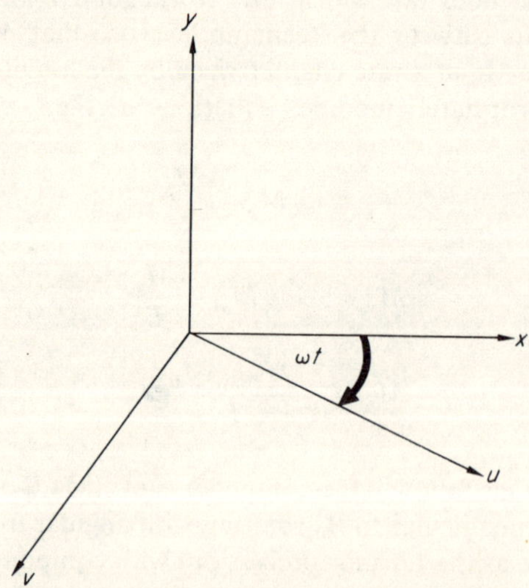

Figure 5.5 Rotating axes u, v in relation to the fixed frame x, y (the z-axis is perpendicular to the plane of the paper)

formation to these axes is:

$$u = x \cos \omega t - y \sin \omega t$$

$$v = -y \cos \omega t - x \sin \omega t$$

(5.13)

under which the Bloch equations become

$$\dot{u} + \frac{u}{T_2} + (\omega_0 - \omega)v = 0$$

$$\dot{v} + \frac{v}{T_2} - (\omega_0 - \omega)u + \gamma B_1 M_z = 0$$

(5.14)

$$\dot{M}_z + \frac{M_z - M_0}{T_1} - \gamma B_1 v = 0$$

5.1.2.4 *The Bloch equations modified to account for dynamic exchange processes* For a multiple-site system, the Bloch magnetizations can be written as the sum of the contributions from each site:

$$u = u_A + u_B + u_C + \cdots = \sum_i u_i$$

$$v = v_A + v_B + v_C + \cdots = \sum_i v_i$$

(5.15)

$$M_z = M_z^A + M_z^B + M_z^C + \cdots = \sum_i M_z^i$$

where u, v and M_z denote the components of the nuclear magnetization which are in-phase with the effective rotating component of the rf field, out-of-phase with this rotating rf field, and in the direction of the large stationary field, respectively.

McConnell[20] has shown how the Bloch equations can be modified to take account of chemical exchange by the inclusion of other terms representing the rate at which magnetization is lost and regained by the individual sites. These modifications are most readily understood in terms of the forward and reverse rate constants of the particular exchange system and this is particularly true for multiple site systems.

For a two site system

$$A \underset{r_{BA}}{\overset{r_{AB}}{\rightleftharpoons}} B$$

the modified equations for nuclei at site A are

$$\dot{u}_A + \frac{u_A}{T_{2A}} + \Delta\omega_A v_A = -r_{AB}u_A + r_{BA}u_B$$

$$\dot{v}_A + \frac{v_A}{T_{2A}} - \Delta\omega_A u_A = -r_{AB}v_A + r_{AB}v_B \qquad (5.16)$$

$$\dot{M}_z^A + \frac{M_z^A - M_0^A}{T_{1A}} - \gamma B_1 v_A = -r_{AB}M_z^A + r_{BA}M_z^B$$

where the first term on the right hand side represents the rate of loss of magnetization from site A to site B while the second represents the rate of gain of magnetization at site A from site B as a result of the exchange. A similar set of equations can be written for those nuclei at site B. Under equilibrium conditions (which normally prevail in the high resolution NMR spectrum at a particular temperature) we have

$$\dot{u}_i = \dot{v}_i = \dot{M}_z^i = 0 \qquad (5.17)$$

The six equations therefore become a complete set of ordinary simultaneous linear equations in six unknowns, which are the two components of u, v and M_z. Only $v = \Sigma_i\, v_i$ is required since this corresponds to the absorption mode in which the high resolution NMR spectrum is displayed.

There are two further points which need to be clarified before the equations developed above can be applied in practice. Firstly, while v is the required ordinate, the abscissa is required to be $\Delta\omega$. By reference to equation (5.14) it is evident that

$$\Delta\omega_A = \omega_A - \omega \quad \text{and} \quad \Delta\omega_B = \omega_B - \omega$$

A convenient origin in this instance is the average of the two resonance frequencies ω_A and ω_B. Let $\delta\omega_0 = \omega_A - \omega_B$, then

$$\Delta\omega_A = \frac{\delta\omega_0}{2} + \left(\frac{\delta\omega_0}{2} + \omega_B - \omega\right) = \frac{\delta\omega_0}{2} + \Delta\omega$$

$$\Delta\omega_B = -\frac{\delta\omega_0}{2} + \left(\frac{\delta\omega_0}{2} + \omega_B - \omega\right) = \Delta\omega - \frac{\delta\omega_0}{2} \qquad (5.18)$$

Secondly, values are required for ω_1 and M_z at each site. However, if the experimental conditions are such that there is no rf saturation (i.e. $\omega_1 \ll \gamma M_0$) then $M_z \approx M_0$. Since the fractional populations of nuclei at each site p_A and p_B must be known at any particular temperature, then $M_z^A \simeq M_0^A$ and $M_z^B \simeq M_0^B$ and it is unnecessary to solve the complete set of McConnell equations (5.16) to determine $\mathbf{v} = \Sigma_i\, \mathbf{v}_i$. Numerical tests have shown that $M_0^A/\omega_1 \geqslant 10^4$ is required for this approximation to be valid. Finally, to eliminate the dependence of v on the chosen value of ω_1, the calculated line-shape is normalized by dividing each ordinate by $v(\text{max})$.

5.1.2.5 Quantum-mechanical theory of exchange A quantum-mechanical theory is required to describe cases where the exchanging nuclei interact by strong spin–spin coupling. The state of the system before exchange may be described by a wave-function ψ which is usually expanded in a complete set of orthogonal and normalized spin functions $\{\phi_1, \phi_2, \ldots, \phi_n\}$ so that

$$\psi = c_1\phi_1 + c_2\phi_2 + \cdots + c_n\phi_n \tag{5.19}$$

The state of the system after exchange may similarly be characterized by another wave-function ψ'. However, if the same basis set can be used to describe the system before and after exchange, then the state of the system can be characterized either by the vector c of the coefficients c_i or alternatively by all their products $c_i c_j^*$. The latter comprise the elements ρ_{ij} of a matrix ρ called the density matrix; it is this formalism which has been successfully developed for the analysis of dynamic processes in NMR and ESR.[21–25]

Each state k of the exchanging system contributes to the total transverse magnetization $G = \Sigma_k p_k G_k$, and each G_k corresponds to the expectation value $\bar{I}^\pm$ of the spin operator $\hat{I}^\pm = \hat{I}_x \pm i\hat{I}_y$. In terms of the average density matrix $\bar{\rho}$, the expectation value $\bar{O}$ of an operator $\hat{O}$ is given by

$$\bar{O} = \sum_k \sum_{i,j} O_{ij}\rho_{ij} = N\,\text{Tr}\,(\bar{\rho}\mathcal{O}) \tag{5.20}$$

Where $\mathcal{O}$ is the matrix representation of $\hat{O}$. This expression is independent of the basis set used for the operators $\hat{O}$ and $\hat{\rho}$. Thus, for nuclei undergoing exchange,

$$G = \sum_k p_k G_k = \sum_k p_k\,\text{Tr}\left[\bar{\rho}_k \sum_l (\mathcal{I}_k^+)_l\right] \tag{5.21}$$

where the l's denote individual spins and $\mathcal{I}^\pm$ is the matrix representation of $\hat{I}^\pm$.

If the spin product wave-functions are chosen as the basis set then the elements $\langle\phi_i|\hat{I}^\pm|\phi_j\rangle$ of the matrix $\mathcal{I}^\pm$ are either 0 or 1. For example, considering an AB system with basis set $\{\alpha\alpha, \alpha\beta, \beta\alpha, \beta\beta\}$

$$\mathcal{I}^\pm = \begin{pmatrix} \langle\alpha\alpha|\hat{I}^\pm|\alpha\alpha\rangle & \langle\alpha\alpha|\hat{I}^\pm|\alpha\beta\rangle & \langle\alpha\alpha|\hat{I}^\pm|\beta\alpha\rangle & \langle\alpha\alpha|\hat{I}^\pm|\beta\beta\rangle \\ \langle\alpha\beta|\hat{I}^\pm|\alpha\alpha\rangle & \langle\alpha\beta|\hat{I}^\pm|\alpha\beta\rangle & \langle\alpha\beta|\hat{I}^\pm|\beta\alpha\rangle & \langle\alpha\beta|\hat{I}^\pm|\beta\beta\rangle \\ \langle\beta\alpha|\hat{I}^\pm|\alpha\alpha\rangle & \langle\beta\alpha|\hat{I}^\pm|\alpha\beta\rangle & \langle\beta\alpha|\hat{I}^\pm|\beta\alpha\rangle & \langle\beta\alpha|\hat{I}^\pm|\beta\beta\rangle \\ \langle\beta\beta|\hat{I}^\pm|\alpha\alpha\rangle & \langle\beta\beta|\hat{I}^\pm|\alpha\beta\rangle & \langle\beta\beta|\hat{I}^\pm|\beta\alpha\rangle & \langle\beta\beta|\hat{I}^\pm|\beta\beta\rangle \end{pmatrix} \tag{5.22}$$

leading to the results

$$\mathcal{I}^+ = \begin{pmatrix} 0 & 1 & 1 & 0 \\ 0 & 0 & 0 & 1 \\ 0 & 0 & 0 & 1 \\ 0 & 0 & 0 & 0 \end{pmatrix} \qquad \mathcal{I}^- = \begin{pmatrix} 0 & 0 & 0 & 0 \\ 1 & 0 & 0 & 0 \\ 1 & 0 & 0 & 0 \\ 0 & 1 & 1 & 0 \end{pmatrix} \qquad (5.23)$$

It remains then to compute those elements ρ_{kl} of the matrix ρ_k which constitute the trace in equation (5.21). These are $\rho_{12}, \rho_{13}, \rho_{24}, \rho_{34}$ if $\hat{I}^-$ is used or $\rho_{21}, \rho_{31}, \rho_{42}, \rho_{43}$ if $\hat{I}^+$ is used, and they are obtained from the equation of motion of ρ. In the absence of exchange and relaxation effects this is given by[26]

$$\frac{d\hat{\rho}}{dt} = -2\pi i[\hat{H}, \hat{\rho}] = -2\pi i[\hat{H}\hat{\rho} - \hat{\rho}\hat{H}] \qquad (5.24)$$

where $\hat{H}$ is the Hamiltonian operator for the system and the energy is in units of s^{-1}. The form of $\hat{H}$ is identical to that employed in high resolution NMR and is expressed in the rotating frame in terms of three components (a) the Zeeman part

$$\hat{H}_0 = \sum_k (v_k - v)\hat{I}_{z_k}$$

where the v_k are the chemical shifts (Hz) of the nuclei in the different environments k, (b) the spin-coupling part

$$\hat{H}_1 = +\sum_{j<l}\sum J_{jl}\hat{I}_j . \hat{I}_l$$

where J_{jl} are the coupling constants (Hz), and (c) the 'driving' part

$$\hat{H}_2 = \gamma B_1 \sum_j \hat{I}_{xj}$$

which corresponds to $\omega_1 M_z$ in the Bloch equations. The second term can be rewritten in terms of the raising and lowering operators $\hat{I}^+$ and $\hat{I}^-$ as

$$\hat{H}_1 = +\sum_{j<l}\sum J_{jl}\hat{I}_j . \hat{I}_l = +\sum_{j<l}\sum J_{jl}[\hat{I}_{z_j}\hat{I}_{z_l} + \tfrac{1}{2}(\hat{I}_j^+\hat{I}_l^- + \hat{I}_j^-\hat{I}_l^+)]$$

so that a consideration of the effects of $\hat{I}_x, \hat{I}_z, \hat{I}^+$ and $\hat{I}^-$ on the basis functions

$$\hat{I}_x|\alpha\rangle = \tfrac{1}{2}|\beta\rangle \qquad \hat{I}_z|\alpha\rangle = \tfrac{1}{2}|\alpha\rangle \qquad \hat{I}^+|\alpha\rangle = 0 \qquad \hat{I}^-|\alpha\rangle = |\beta\rangle$$

$$\hat{I}_x|\beta\rangle = \tfrac{1}{2}|\alpha\rangle \qquad \hat{I}_z|\beta\rangle = -\tfrac{1}{2}|\beta\rangle \qquad \hat{I}^+|\beta\rangle = |\alpha\rangle \qquad \hat{I}^-|\beta\rangle = 0$$

reduces the elements $H_{ij} = \langle \phi_i | \hat{H} | \phi_j \rangle$ of the Hamiltonian matrix $\mathcal{H}$ arising from $\hat{H}_0$ and $\hat{H}_1$ to expressions involving $(v_k - v)$, J_{jl} and multiplicative constants $\pm\frac{1}{2}$, while those arising from $\hat{H}_2$ appear as $\frac{1}{2}\gamma B_1$.*

The effects of exchange and relaxation are introduced into equation (5.24) by the addition of other terms in the same way that the Bloch equations were modified in the classical case:

$$\frac{d\rho}{dt} = -2\pi i[\mathcal{H}, \rho] + [(\mathcal{X}\rho\mathcal{X} - \rho)/\tau] - \rho T_2 \tag{5.25}$$

where ρ now denotes the state of the density matrix before exchange, whilst $\mathcal{X}\rho\mathcal{X}$, involving the exchange matrix $\mathcal{X}$, denotes the transformed ρ which results from the exchange; τ is the mean lifetime of each of the exchanging states, while T_2 is the value of the spin–spin relaxation time associated with that state.

For an AB system, with the same basis set as before, *mutual* exchange of nuclei causes the following changes in the basis functions:

$$
\begin{array}{ccccccc}
& \text{A} & \text{B} & & \text{A} & \text{B} & \\
\phi_1 & \alpha & \alpha & \leftrightarrow & \alpha & \alpha & \phi_1 \\
\phi_2 & \alpha & \beta & \leftrightarrow & \beta & \alpha & \phi_3 \\
\phi_3 & \beta & \alpha & \leftrightarrow & \alpha & \beta & \phi_2 \\
\phi_4 & \beta & \beta & \leftrightarrow & \beta & \beta & \phi_4 \\
\end{array}
$$

so that the exchange matrix takes the form

$$
\mathcal{X} = \begin{pmatrix}
1 & 0 & 0 & 0 \\
0 & 0 & 1 & 0 \\
0 & 1 & 0 & 0 \\
0 & 0 & 0 & 1
\end{pmatrix}
$$

Such a system is exemplified by ring inversion in 5,5-disubstituted-1,3-dioxans (Figure 5.6) where both the protons and the Me groups mutually

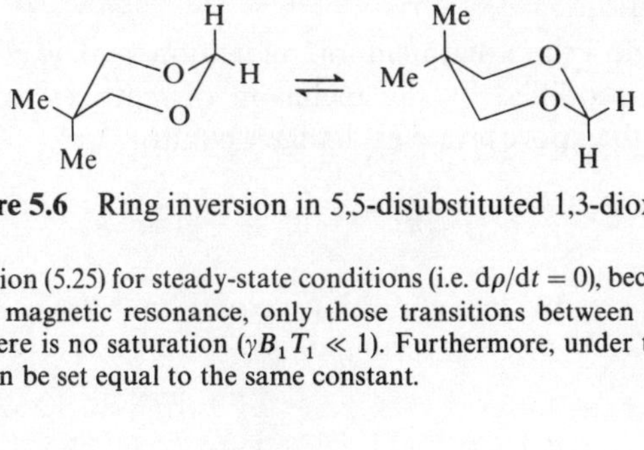

Figure 5.6 Ring inversion in 5,5-disubstituted 1,3-dioxans

* In solving equation (5.25) for steady-state conditions (i.e. $d\rho/dt = 0$), because of the selection rule $\Delta M = \pm 1$ for magnetic resonance, only those transitions between k and l need to be considered when there is no saturation ($\gamma B_1 T_1 \ll 1$). Furthermore, under these conditions, all the driving terms can be set equal to the same constant.

exchange positions. The form of $\mathscr{X}$ is similar in the case of intermolecular exchange, but now the wave-functions of different molecules are involved. For hindered internal rotation, however, exchange matrices which involve switching between different Hamiltonians must be constructed. The necessity for this becomes obvious in the ethane-type molecule CH_2XCXY_2 (Figure

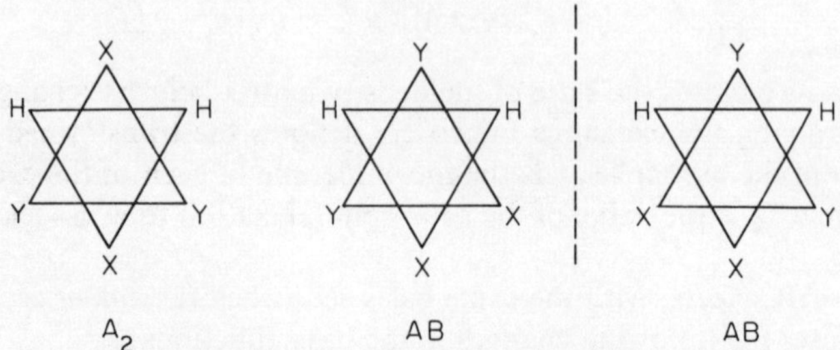

Figure 5.7   Isomers of CH_2XCXY_2

5.7) where the protons constitute an A_2 pair in one isomer but an AB pair in the other two isomers.

The modification of the density matrix formalism in such cases as these has been suggested by Johnson.[23] The commutator of equation (5.24) is considered to be an equation of transformation of $\hat{\rho}$ which can equally well be represented by a transformation operator $\hat{L}$ called the Liouville operator:

$$\hat{L}\hat{\rho} = [\hat{H}, \hat{\rho}] = \hat{H}\hat{\rho} - \hat{\rho}\hat{H}$$

The elements of $\mathscr{L}$, the matrix representation of $\hat{L}$, are then

$$L_{ij,kl} = H_{ik}\delta_{jl} - H_{lj}\delta_{ki}$$

where δ is the kroneker delta ($\delta_{\alpha\beta}$ is 1 if $\alpha = \beta$ or 0 if $\alpha \neq \beta$). The equation of motion is now written in matrix form as

$$\frac{d\rho}{dt} = -2\pi i\mathscr{L}\rho$$

and, as before, for the treatment of relaxation and exchange problems, this equation is modified by the inclusion of matrices for the relaxation operator $\hat{R}$ and the appropriate exchange operator $\hat{X}$:

$$\frac{d\rho}{dt} = (-2\pi i\mathscr{L} + \mathscr{R} + \mathscr{X})\rho \equiv \mathscr{M}\rho \qquad (5.26)$$

For unsaturated steady-state conditions, equation (5.26) is simply

$$\mathscr{M}\rho = 0$$

which can be written in the more convenient form

$$\mathcal{M}_0\rho = -i\sigma$$

with the definition that $\hat{L}^2\rho = -\hat{\sigma}$, where $\hat{L}^2$ corresponds to the driving part of the Hamiltonian. The complex magnetization G in the xy plane then becomes

$$G = \mathrm{Tr}(\mathcal{I}^{\pm}\rho) = \mathrm{Tr}(-i\mathcal{I}^{\pm}\mathcal{M}_0^{-1}\sigma)$$

from which the absorption line-shape and the dispersion line-shape are obtained as the imaginary and real parts respectively. This extended density matrix treatment of exchange is called the unified theory of exchange effects on NMR line-shapes and has been incorporated by Binsch[27] into a computer programme called DNMR.

The parameters which are required for the computation of a line-shape are the rates of exchange, the populations p_i and the spin–spin relaxation times T_{2i} of nuclei at each site, and finally their relative chemical shifts. It is the rate of exchange which one requires to establish by comparison of the calculated and observed line-shapes so that potential sources of errors in the other parameters need to be considered.

The site populations are fixed in an ideal case (N,N-dimethyl amides) and it is for this reason that many such systems have been chosen for study. Nonetheless, systems in which the rotamers do not have equal energy may be amenable for study if their energy difference (and thus the difference in their Boltzmann populations) is not appreciable. The temperature dependence of the populations in these cases can be roughly established by spectral integration of the signals from the separate rotamers at a range of low temperatures or, more satisfactorily, over the whole of the temperature range by IR intensity measurements.

The spin–spin relaxation time T_2 is related to the line-width Δ of a resonance signal by $T_2^{-1} = \pi\Delta$ in the absence of exchange, so that values of T_{2i} for nuclei at each site are obtained directly at low temperatures where the exchange is effectively stopped. At higher temperatures reliance is usually made on interpolation from the line-width of a sharp reference signal in the spectrum. The error associated with T_2 becomes most pronounced at the high and low temperature limits of the line-shape study since it is here that the lines are sharpest. This is illustrated in Figure 5.8 where the logarithm of the sum of the squares of the differences between two line-shapes which were computed with the same basic parameters apart from the T_2 values has been plotted versus the logarithm of the exchange rate τ. The lower curve results from the comparison of a 'true' line-shape where $T_2 = 0\cdot1$ s with an 'observed' line-shape where $T_2 = 0\cdot2$ s; for the upper curve the 'observed' line-shape had $T_2 = 0\cdot3$ s. In each case the fit is very good in the region of the coalescence point but is very poor in the extremes. This can have serious

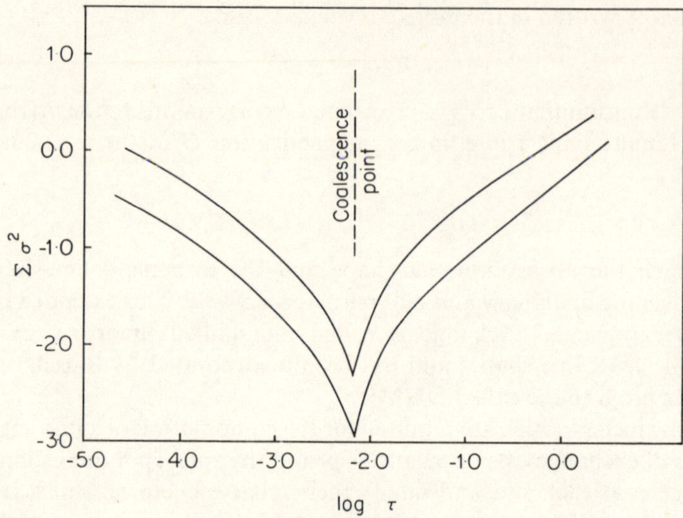

Figure 5.8

consequences in that finally the activation energies are derived from Arrhenius or Eyring plots. It is particularly important therefore that in-homogeneity broadening of the resonance signals is minimized at the high and low temperature extrema, since it is these rates of exchange which govern the slope of the subsequent Arrhenius or Eyring plot.

This source of error is compounded in many simple cases by the fact that the temperature range over which the line-shape changes from separate sharp signals to a single sharp signal is usually limited to 50–60°. Extrapolation over a much larger range is needed to obtain a reliable value of the entropy of activation from the intercept with the ln k axis in an Eyring plot. As is shown in Table 5.1, operation at higher field strength improves the position only marginally, but nevertheless this procedure is to be recommended. A knowledge of the relative chemical shift $\delta\omega_0$ in the absence of exchange is

Table 5.1

T_c(K)	$\Delta G^{\ddagger}$ (kJ/mole)	$\delta\omega_0$ (Hz)	
160·65	33·44 (8·0 kcal/mole)	18	
162·58	33·86 (8·1 kcal/mole)	18	at 60 MHz
163·85	33·44	30	
165·82	33·86	30	at 100 MHz
169·04	33·44	66	
171·07	33·86	66	at 220 MHz

also required for the calculation of a line-shape. Quite often problems of solubility are encountered when recording spectra at very low temperatures, while in some cases instrumental limitations do not allow sufficiently low temperatures to be attained. Since the coalescence temperature T_c of the spectrum is related to the relative chemical shift $\delta\omega_0$ for two-site exchange by the equation

$$\delta\omega_0 = \frac{\sqrt{2\kappa k_B T_c}}{\pi h} \cdot \exp\left(\frac{-\Delta G^\ddagger}{RT_c}\right)$$

it is apparent that the problem may be overcome by conducting the experiment at higher field strength (Table 5.1).

Most often the value of $\delta\omega_0$ is well established at low temperatures, but the possible temperature dependence of this parameter cannot be ignored. A simple check in the case of exchange between two equally populated sites is to ensure that the centre of the lineshape does not alter in chemical shift relative to an internal reference such as TMS over the whole temperature range. The temperature dependence of $\delta\omega_0$ is particularly marked in amides, probably as a result of molecular association. The position is considerably improved in these situations by the use of dilute solutions in an inert solvent such as CCl_4 or CS_2, although the difficulty is not entirely eliminated.

Ideally then the requirements for accurate line-shape analysis are the following: (a) An extremely homogenous magnetic field over the dimensions of the sample which also has the highest possible flux. (b) Very detailed comparisons of the calculated and observed line-shapes so that it is possible to correct for the temperature dependence of $\delta\omega_0$, for the temperature dependence of the p_i in cases where the rotamers are of different energies, and to improve the accuracy at the high and low temperature extremes where small differences in sharp signals are being examined. For this to be possible the line-shapes must be compared at as many points as possible, so that the limiting factor lies in the digitization of the observed spectrum. An analogue to digital converter with a storage capacity of 4K is probably the optimum and such a system has been employed successfully by Reeves.[28]

Although it is the absorption spectrum which is being compared, such a detailed comparison as advocated here allows further corrections to the base-line to be introduced to compensate for slight errors in the phase adjustment of the spectrometer. This is readily achieved by the introduction of a small contribution from the dispersion mode, rather than by computing the absorption mode alone.

Finally, it must be noted that while many of the systems chosen for dynamic studies in the past have often been deliberately simplified by isotopic substitution, it seems likely that much more accurate parameters can be derived from the observation of the numerous changes of appearance of complex spectra with changing rate (Figure 5.9).

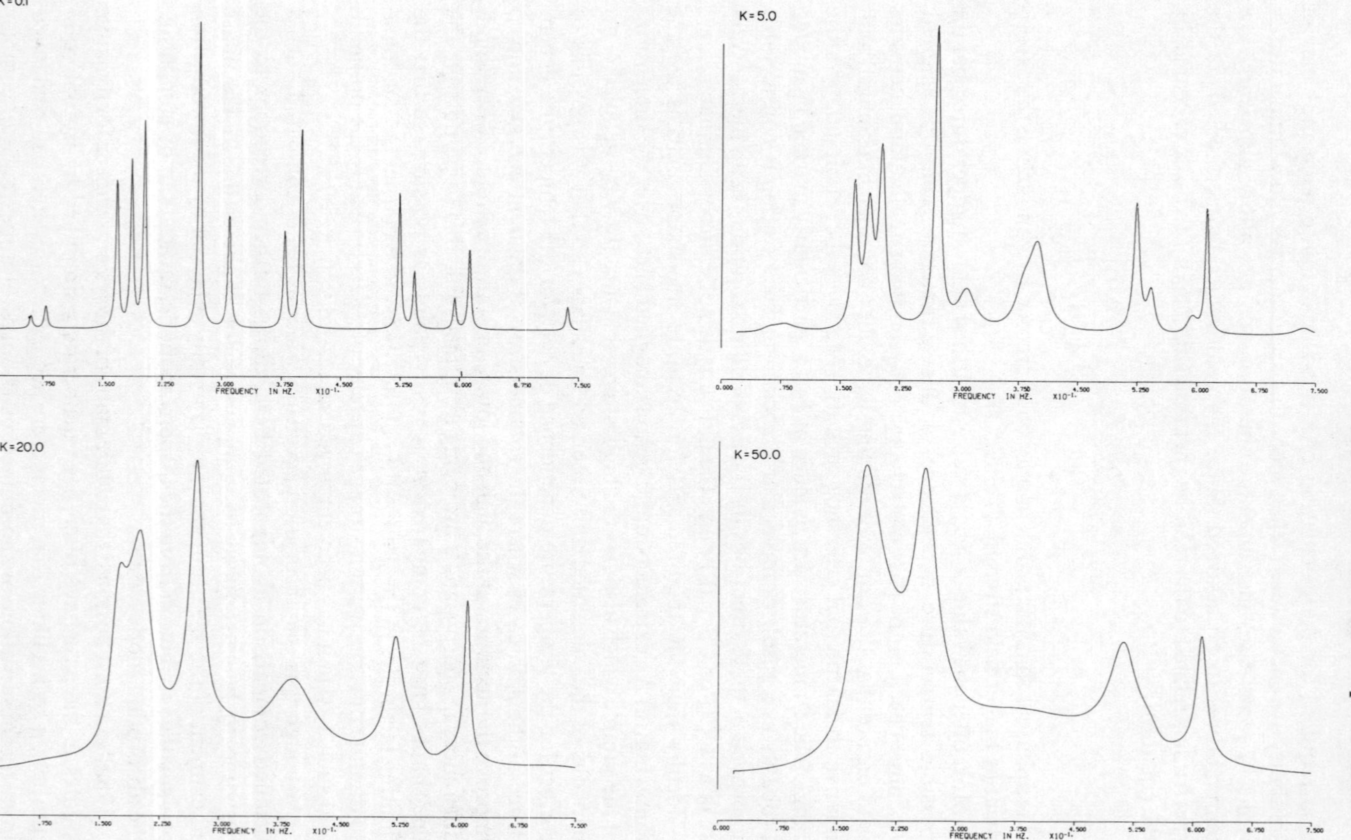
K = 0.1
K = 5.0
K = 20.0
K = 50.0
FREQUENCY IN HZ.   X10⁻¹.

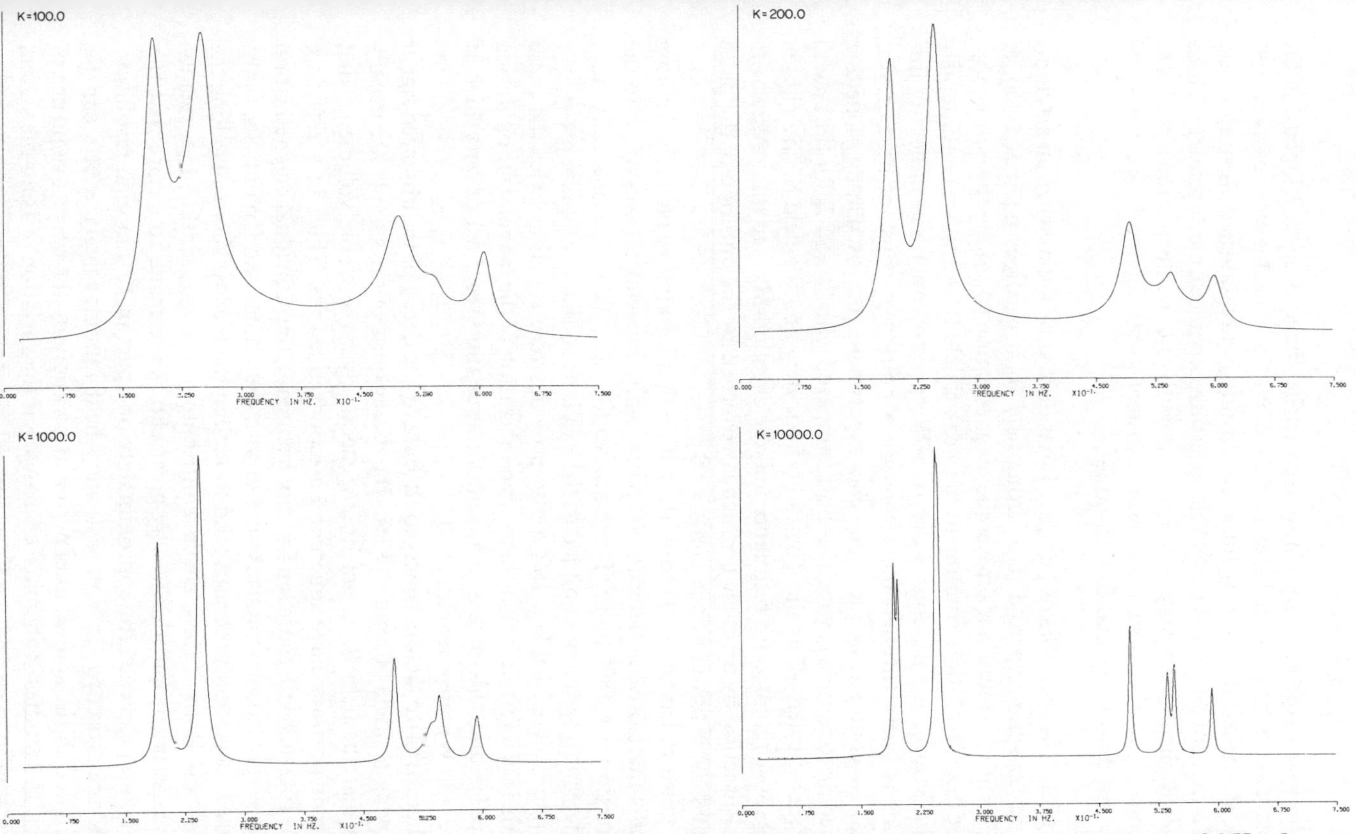

Figure 5.9 NMR line-shapes for an exchanging ABC system with $\mu_A = 19\cdot6$ Hz, $\mu_B = 22\cdot9$ Hz, $\mu_C = 52\cdot7$ Hz, $J_{AB} = 12\cdot3$ Hz, $J_{AC} = 7\cdot3$ Hz, $J_{BC} = -17\cdot7$ Hz and $T_2 = 0\cdot8$s. Nuclei A and B undergo mutual exchange characterized by rate constants K. From G. Binsch, *J. Am. Chem. Soc.*, **91**, 1304 (1969). Reproduced by permission of the American Chemical Society

5.1.3 *Transient phenomena*

The line-shape equations discussed earlier have been developed for the special case of unsaturated steady-state conditions, i.e. for slow passage and weak $\mathbf{B}_1$ fields. If either of these conditions is not satisfied then equation (5.17) does not apply and the NMR spectrum becomes time dependent. These transient phenomena may also be exploited to obtain information about rate processes and the most important transient NMR methods are the spin-echo and double-resonance techniques.

5.1.3.1 *The spin-echo technique*

It has been shown that when an assembly of nuclear spins is placed in a strong field $\mathbf{B}_0$ it becomes polarized and, at equilibrium, exhibits a macroscopic magnetization $\mathbf{M}_0$ along the z-direction. In the high resolution experiment the weak rf field $\mathbf{B}_1$ is applied continuously to the sample (continuous wave or CW spectroscopy) and either its frequency is swept through the resonance condition at constant field or vice versa. In contrast, in the spin-echo experiment, the oscillating rf field is turned on for a time t_p which is short compared with T_1 and T_2 for the system and then turned off again. The radio-frequency and the field $\mathbf{B}_0$ are chosen to be those required for resonance and the effect on $\mathbf{M}_0$ is to tilt it away from the z-direction by an angle θ which is governed by the duration of the pulse (t_p) and the rf power applied ($\omega_1 = \gamma\mathbf{B}_1$).

A pulse which takes $\mathbf{M}_0$ from the z-direction into the xy plane (in rotating axes) is therefore known as a 90° pulse, while one which takes $\mathbf{M}_0$ into the z-direction is a 180° pulse. These are readily identified experimentally since the detection system is confined to the xy plane so that a 90° pulse provides a maximum signal while a 180° pulse provides zero signal at twice the pulse length. It is important that there is no relaxation of the magnetization during the pulse and this means that short pulses (milliseconds) of very high rf power (20 W) must be employed.

There are two factors which contribute to the decay of the observed signal following a single 90° pulse. First, the dimensions of the sample are usually such that the field $\mathbf{B}_0$ is not homogeneous over the whole volume so that different portions have different Larmor frequencies. Thus, in a reference frame rotating with the mean Larmor frequency, the individual magnetization vectors (spin isochromats) lose phase with each other and fan out. Secondly, the true relaxation processes tend to realign the nuclear moments along the z-direction. Both of these effects contribute to the line-width Δ in a steady-state experiment and this is why great efforts are made to ensure that the field is homogeneous (to approximately one part in 10^9) in high resolution NMR spectroscopy. In contrast, such field inhomogeneity effects can be eliminated in the spin-echo experiment by applying the correct sequence of pulses. For studies of chemical exchange it is the variation of T_2 for the system which is of interest and the Carr–Purcell sequence of pulses[29] is required.

Following an initial 90° pulse, applied (say) along the x-axis in the rotating frame, the macroscopic magnetization vector lies along the y-axis and produces a maximum signal in the receiver (Figure 5.10a). This decays away as a result of the fanning-out of the spin isochromats and also due to spin–spin relaxation (Figure 5.10b). If the 90° pulse is followed by a 180° pulse at a time

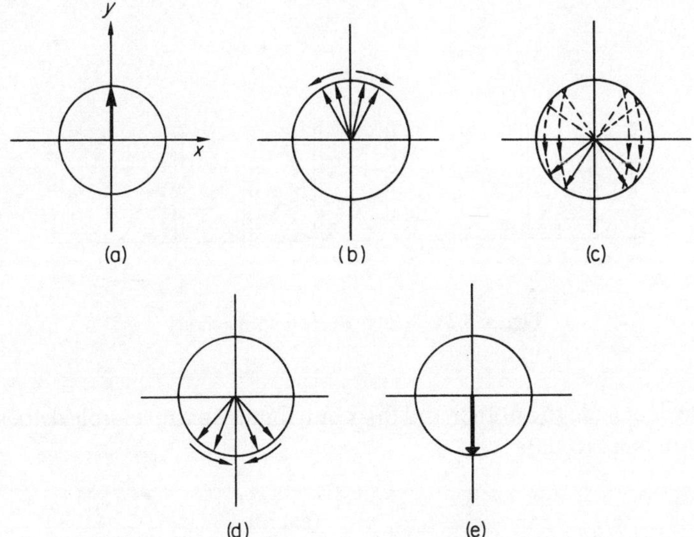

Figure 5.10   Spin isochromats following various pulses

t which is short compared to T_2, then the spin isochromats are inverted but preserve their sense and rate of precession (Figure 5.10c). The result is that they now refocus (Figure 5.10d) to provide another maximum signal (a spin-echo) in the receiver at a time $2t$ (Figure 5.10e). The process is repeated by the application of a train of 180° pulses at further times $3t$, $5t$, $7t$ etc., and the result is that spin-echoes are observed at times $4t$, $6t$, $8t$ etc. (Figure 5.11). The refocussing effect of the 180° pulses eliminates the effects of field inhomogeneity but they do not influence the spin–spin relaxation mechanisms of the system. The amplitudes of successive spin-echoes therefore decrease according to the relation

$$M(t) = M(0)\exp(-t/T_2)$$

In systems where there is a rapid interconversion of isomeric forms, a nucleus may alter its Larmor frequency by exchanging from one magnetic

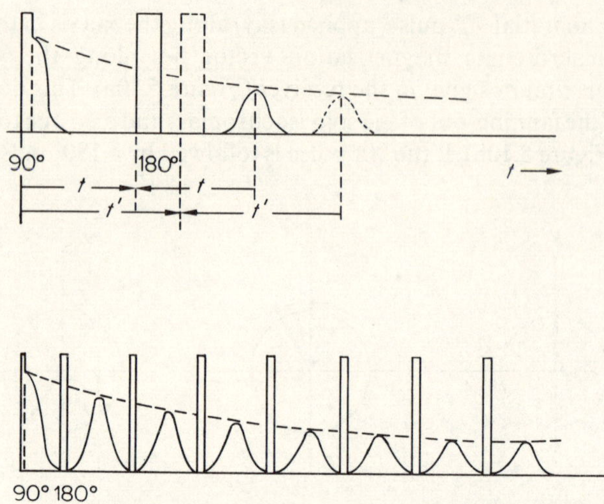

Figure 5.11   Carr–Purcell spin echoes

site in the molecule to another and this contributes an irreversible defocussing of the spin isochromats

$$\frac{1}{T_2} = \frac{1}{T_2^0} + \frac{1}{T_2}\,(\text{exch.})$$

A plot of the observed value of T_2^{-1} against the pulse repetition frequency t_{cp} under these circumstances produces a characteristic S-type curve which is described in the case of the exchange between two sites by the equation[30]

$$\frac{1}{T_2} = \frac{1}{T_2^0} + P_A P_B (\delta\omega_0)^2 \tau \left[1 - \left(\frac{2\tau}{t_{cp}}\right) \tanh\left(\frac{t_{cp}}{2\tau}\right) \right]$$

where τ^{-1} is the rate constant of the exchange process. It is evident that a plot of T_2^{-1} versus the function enclosed in brackets should provide a straight line graph of gradient $P_A P_B (\delta\omega_0)^2 \tau$ and intercept $(T_2^0)^{-1}$ on the ordinate. The value of τ at a particular temperature can therefore be extracted from the dependence of T_2 on t_{cp} by successive approximation to give the best least-squares fit to this straight line. If P_A and P_B are known, a value for $\delta\omega_0$ is provided by the analysis and any temperature dependence of this term is automatically accounted for.

The spin-echo technique is applicable to the whole range of rates which can be studied by the high resolution line-shape method and furthermore it can be extended to rates which are an order of magnitude larger. This increased temperature range immediately improves the reliability of the activation parameters obtained from Eyring or Arrhenius plots as compared to line-shape studies and the spin-echo technique is unrivalled in this respect.

A modulation of the amplitudes of the spin-echoes is introduced if homonuclear spin–spin coupling exists in the molecules under study and for this reason spin-echo studies have mainly been limited to rather simple uncoupled two-site exchange processes such as restricted internal rotation in amides. Apart from this limitation, this technique is the one of choice for exchange studies.

It is apparent that any temperature dependence of the site populations (P_i) or their relative chemical shift $(\delta\omega_0)$ does not influence the value of τ extracted in the analysis provided that the temperature remains constant throughout the determination of the dependence of T_2 on the Carr–Purcell pulse separation t_{cp}. The maintenance of this constant temperature is the most difficult aspect of the experiment, however, and therefore provides the most serious source of error.

A pulse of rf energy (20 W) applied for 5 ms will lead to a temperature rise of $\sim 0.05°C$ in 1 ml of a typical organic liquid assuming that all this energy is immediately converted into heat. The application of a train of pulses thus leads to a significant rise in the temperature of the sample unless the thermostatting is extremely efficient so that it is the amplitudes of the first few echoes which define T_2 best. It is also essential that the sample is allowed to equilibrate between pulse trains. The heating effect of the pulses is more likely to be apparent for short values of t_{cp} so that this must be allowed for in the analysis of the S-type plot of $(T_2)^{-1}$ vs. t_{cp}.

Another source of error lies in the neglect of diffusion effects. If, following the initial 90° pulse, a molecule diffuses from one part of the inhomogeneous field $\mathbf{B}_0$ to another, then its Larmor frequency also changes and this provides a further irreversible defocussing mechanism for the spin-isochromats similar to the effects of chemical exchange. One then has

$$\frac{1}{T_2} = \frac{1}{T_2^0} + \frac{1}{T_2}(\text{exch.}) + \frac{1}{T_2}(\text{diff.})$$

and the influence of the latter term becomes more pronounced at higher temperatures. However, quite large field gradients across the volume of the sample are usually required before this effect becomes significant so that generally this aspect may be safely disregarded.

5.1.3.2 *Double resonance techniques* Whereas the spin-echo technique permits an extension to faster rates of exchange using rather specialized equipment, Forsén and Hoffman[31] have devised a double-resonance method which is applicable to slow rates of exchange and which may be carried out with a standard high-resolution spectrometer equipped with an additional oscillator and a fast-writing recorder.

Experimentally one records the unsaturated slow-passage spectrum of the magnetization at one site while double-irradiating the other sites. Under these conditions the signal intensity is directly proportional to M_z and a decrease from $M_z(0)$ at $t = 0$ to $M_z(t)$ at time t is observed. Forsén and Hoffman used a multiple sweep device but it is easier to 'sit' on resonance if the spectrometer has a field-frequency lock system.

The underlying principles of the method are best understood in a discussion of exchange between two uncoupled sites A and B for which the Bloch equations have already been developed. Since the interest is confined to the z-component of the magnetization the relevant equations are:

$$\dot{M}_z^A = -\frac{M_z^A - M_0^A}{T_{1A}} - r_{AB}M_z^A + r_{BA}M_z^B + \gamma v_A B_1$$

$$\dot{M}_z^B = -\frac{M_z^B - M_0^B}{T_{1B}} - r_{BA}M_z^B + r_{AB}M_z^A + \gamma v_B B_1$$

If those nuclei at site B are now strongly irradiated at their resonant frequency then the magnetization at this site is destroyed through saturation and the time dependence of the z-magnetization of those nuclei at site A is then described by:

$$\dot{M}_z^A = -\frac{M_z^A - M_0^A}{T_{1A}} - r_{AB}M_z^A + \gamma v_A B_1$$

$$= -M_z^A\left(\frac{1}{T_{1A}} + \frac{1}{\tau_{AB}}\right) + \frac{M_0^A}{T_{1A}} + \gamma v_A B_1 \tag{5.27}$$

where $\tau_{AB} = r_{AB}^{-1}$.

Furthermore, if $\mathbf{B}_1$ is sufficiently weak to cause negligible saturation, then the last term may be dropped and the observed signal depends on M_z^A. Integration of equation (5.27) then yields

$$M_z^A(t) = \frac{M_0^A}{T_{1A} + \tau_A}\{T_{1A}\exp\left[-t(T_{1A} + \tau_A)/T_{1A}\tau_A\right] + \tau_A\}$$

and the new equilibrium state of the system obtained at $t \to \infty$ will be given by:

$$M_z^A(t \to \infty) = M_0^A\tau_A/(T_{1A} + \tau_A) \tag{5.28}$$

Thus a plot of $\ln\left[M_z^A(t) - M_z^A(t \to \infty)\right]$ versus t will provide a straight line of slope $(1/T_{1A} + 1/\tau_{AB})$ while equation (5.28) provides sufficient additional information for both T_{1A} and r_{AB} to be evaluated.

The observation of changes in magnetization at site B while nuclei at site A are being irradiated likewise yields values for T_{1B} and the reverse rate r_{BA}.

It is evident that there must be no overlapping of resonance lines for this technique to be accomplished and it is therefore most conveniently applied when there are no spin–spin coupling complications. The chemical shifts between the various sites need to be large to avoid beats between the two radio-frequencies being employed and it is for this reason that its application is very limited.

5.2 Electron spin resonance

The technique of electron spin resonance spectroscopy is only applicable to the study of radicals and radical ions because the spin of an unpaired electron is used to probe the molecular environment. However, a typical high resolution electron spin resonance spectrum contains the same kind of information as a nuclear magnetic resonance spectrum. The position of the resonance is measured by the g value. This gives a general indication of the electronic environment of the unpaired electron and is comparable to the chemical shift parameter. The hyperfine structure results from the interaction of the unpaired electron with its nuclear magnetic environment and is comparable to nuclear spin–spin coupling. The size of this interaction is strongly dependent on the structural relationship between the unpaired electron and the nucleus while the relative intensities of the hyperfine components indicate the degree of magnetic equivalence of the nuclei with respect to the unpaired electron. Thermodynamic and kinetic information can be obtained from the widths of the lines.

The relatively large magnetic moment of the electron makes ESR a more sensitive technique than NMR in comparable laboratory conditions and for high resolution work radical concentrations in the range 10^{-6}–10^{-3} M are normal. Higher concentrations should be avoided as radical–radical interactions then make an appreciable contribution to the line-width. In general relaxation processes are much more efficient in liquid phase ESR than in NMR. A typical relaxation time for the electron in these circumstances is 10^{-6} s compared with 10 s for a nucleus. Consequently, line-widths in ESR are a true reflection of the relaxation processes while the widths of the much narrower NMR lines are usually determined by inhomogeneities in the applied field. Because of this large difference in relaxation times the inherent time-scale of ESR is much shorter than that of NMR. Contributions to the line-width from uncertainty broadening are usually negligible for species having lifetimes as short as 10^{-6} s and the complete coalescence of hyperfine structure is an indication of a lifetime of about 10^{-11} s. Such short lifetimes imply that processes having relatively low activation energy barriers should be amenable to study by ESR spectroscopy.

5.2.1 *Isotropic hyperfine interactions*

The effect of the interaction of an unpaired electron with a neighbouring magnetic nucleus of spin I is to split the resonance into $2I + 1$ equally spaced lines of equal intensity. The splitting between adjacent lines, a_I, is called the hyperfine coupling constant. A consequence of the efficient relaxation is that, unlike NMR, interactions with nuclei having $I > \frac{1}{2}$ are observable and the usual rules for obtaining the relative intensities of groups of equivalent nuclei must be extended. The interaction of an unpaired electron with n inequivalent nuclei of spins $H, I, J, K \ldots$ produces a spectrum containing $(2H + 1)(2I + 1)(2J + 1)(2K + 1) \ldots$ lines of equal intensity with n different coupling constants a_H, a_I, a_J, a_K, etc. The interaction of an unpaired electron with n equivalent nuclei of spin I produces a spectrum containing $(2nI + 1)$ equally spaced lines with relative intensities given by the coefficients of the expansion $(x + x^2 + x^3 \ldots x^{2I + 1})^n$.

The hyperfine coupling observed from radicals in solution is normally due only to isotropic contributions to the coupling. Anisotropic contributions, which may be observed in the solid state or in highly viscous solutions, are averaged to zero by the rapid random tumbling of the radical through all possible orientations with respect to the field direction. The magnitude of the isotropic coupling is directly proportional to the square of the amplitude of the wave-function describing the unpaired electron at the interacting magnetic nucleus. All orbitals other than s-orbitals have nodes through the nucleus so the isotropic coupling is a measure of the s-character of the wave-function of the unpaired electron on that nucleus. Since the isotropic coupling to the proton, a_H, in the trapped hydrogen atom is 50·8 mT, the experimental values of a_H in other radicals can be interpreted in terms of a spin density, p_H, at that hydrogen by the relation $\rho_H = a_H/50\cdot8$. For other nuclei the s-character may be estimated by comparing the observed coupling with a value calculated for an unpaired electron entirely localized in the appropriate s-orbital or that nucleus.

Since the wave-function for an unpaired electron in a π-radical, such as an aromatic radical ion, has a nodal plane that corresponds with the molecular plane, it is at first sight surprising that isotropic coupling to magnetic nuclei in this plane can be observed. Couplings of this type are the result of polarization of the σ-bonding electrons by the unpaired π-electron (Figure 5.12) and the quantitative relationship between the coupling to a ring proton and the spin density at the adjacent carbon is given by the McConnell equation

$$a_{\mathrm{CH}}^{\mathrm{H}} = Q_{\mathrm{CH}}^{\mathrm{H}} \cdot \rho_{\mathrm{C}} \tag{5.29}$$

where ρ_{C} is the spin density on the carbon adjacent to the proton and $Q_{\mathrm{CH}}^{\mathrm{H}}$ is a constant having a value of $-2\cdot3$ mT. Detailed experimental evidence shows that $Q_{\mathrm{CH}}^{\mathrm{H}}$ is not strictly constant and a number of more complex

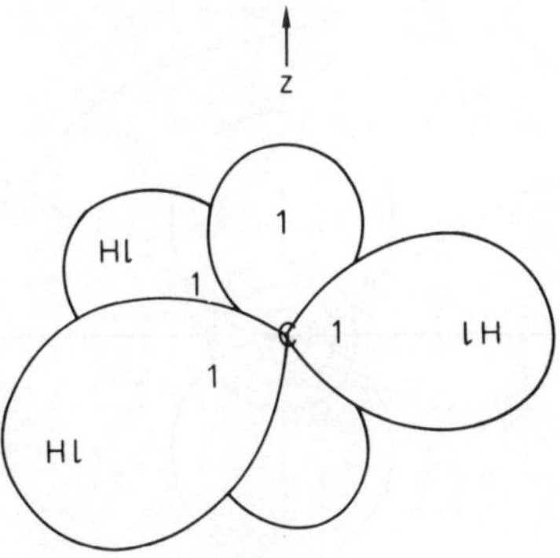

Figure 5.12 Polarization of bonding electrons in the methyl radical

expressions for the relationship between a_{CH}^{H} and ρ_C have been derived, but the simple expression is still a useful guide. The negative sign for the coupling constant implied by the negative value of Q_{CH}^{H} indicates that the spin of the polarized bonding electron that is directly responsible for the proton coupling is antiparallel to that of the unpaired π-electron.

Protons two bonds removed from the unpaired spin density generally have larger hyperfine couplings to the unpaired electron than α-protons because they are not normally confined to the nodal plane. The isotropic proton coupling is related to the spin density by

$$a_{CCH}^{H} = Q_{CCH}^{H} \cdot \rho_C \tag{5.30}$$

When rotation about the C—C bond is rapid on the ESR time-scale, Q_{CCH}^{H} is constant with a value of $+2{\cdot}93$ mT, but at slower rates of rotation this Q value has an angular dependence that is given by

$$Q_{CCH}^{H} = Q_n + Q_0 \cos^2 \theta \tag{5.31}$$

where θ is the dihedral angle between the plane containing the axis of the p orbital occupied by the unpaired electron and the carbon–carbon axis, and the plane containing the carbon–carbon axis and the carbon–hydrogen axis (Figure 5.13). The coupling is at a maximum when $\theta = 0$ and the proton is eclipsed with the unpaired electron orbital. The splitting is

$$a_{CCH}^{H} = (Q_n + Q_0)\rho_C \tag{5.32}$$

When $\theta = \pi/2$ the nucleus is once more in the nodal plane and the coupling is a minimum

$$a_{CCH}^{H} = Q_n\rho_C \tag{5.33}$$

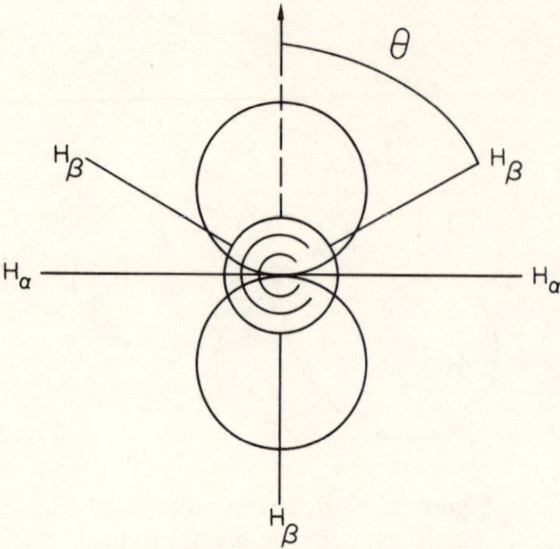

Figure 5.13   Definition of the dihedral angle for
coupling to β-protons

The value of Q_n is very small, less than 0·5 mT. The value of Q_0 is about 5·5 mT
so that with unit spin density on carbon the β-proton coupling varies from
about 6·0 mT to almost zero as the dihedral angle alters. This angular variation of the isotropic coupling to β-protons has been vigorously exploited in
the determination of conformational preferences in many radicals.[32,33]
Rapid rotation about the C—C bond results in the averaging of the β-proton
coupling. The average value of $\cos^2\theta$ is $\frac{1}{2}$ so that

$$Q_{\mathrm{CCH}}^{\mathrm{H}} = Q_n + \tfrac{1}{2}Q_0 = 2\cdot93 \tag{5.34}$$

In some circumstances the coupling to more remote nuclei is greater than
the natural line-width and may be resolved. Such circumstances are favoured
when the molecular geometry is rigid and restrains the remote nuclei in conformations that produce relatively small electron–nuclear distances and
small dihedral angles. Even then these remote couplings rarely exceed 0·1 mT
but they can still convey useful conformational information.[32]

5.2.2 *Modification of spectra by intramolecular processes*

How does intramolecular motion alter the ESR spectrum of a radical?
The observed effects are similar to those encountered in NMR and the
methods of line-shape analysis developed for NMR are applicable to the
ESR situation. If the lifetime of each conformation is long on the ESR
time-scale then the spectrum contains lines associated with the individual
conformations. When the magnetic environment of the unpaired electron
is identical in all conformations then the spectrum appears to represent only
one species. A hypothetical example of this situation would be the equilibrium

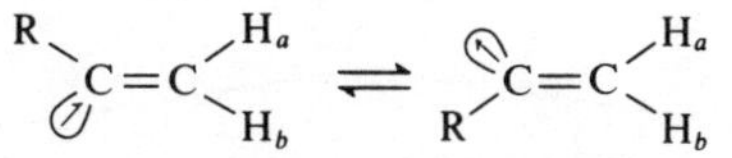

Figure 5.14

between two conformers of a substituted vinyl radical (Figure 5.14). If the coupling to the proton *trans* to the unpaired electron is x mT and that to the *cis* proton is y mT then the spectrum of the conformer (I) consists of a doublet splitting of x mT from H_a split into y mT doublets by H_b while that of (II) is a doublet splitting of x mT from H_b split into y mT doublets by H_a. The g values of the conformers are identical so the two spectra are precisely superposed and in slow exchange conditions a simple doublet of doublets is observed. In the terephthalaldehyde radical anion (Figure 5.15) the isomerization alters the magnetic environment of the unpaired electron and in

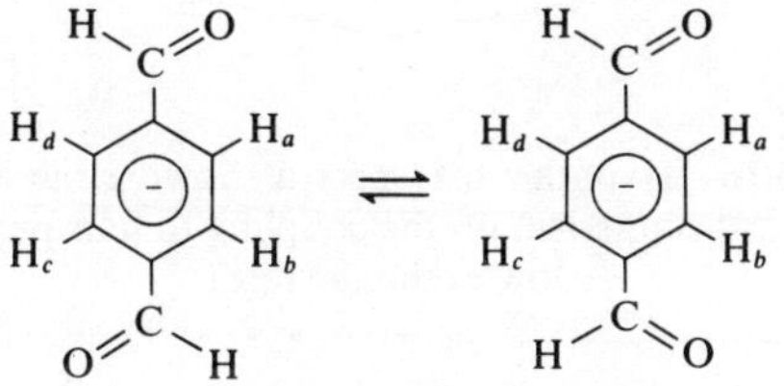

Figure 5.15

slow exchange conditions the separate spectra of the *cis* and *trans* forms are observable. Let us consider only the coupling of the unpaired electron to the four ring protons. In the *cis* isomer protons a and b are equivalent and protons c and d are equivalent. This would produce a nine-line spectrum, a 1:2:1 triplet split into 1:2:1 triplets. In the *trans* isomer protons a and c are equivalent and so are protons b and d. Again the spectrum would be a triplet of triplets but it would be fortuitous if the individual couplings were identical so that in the slow exchange region the spectrum should consist of two superposed triplet of triplets groups. Isomerization should not alter the relative distribution of the unpaired electron between the ring and the substituent, so that the sum of the ring proton splittings in the *cis* isomer should equal the sum of the ring proton splittings in the *trans* isomers. This means that the overall widths of the two spectra should be identical and since isomerization should not alter the g value the outermost pair of lines from each isomer should be coincident (Figure 5.16). Spectra from these two forms of the

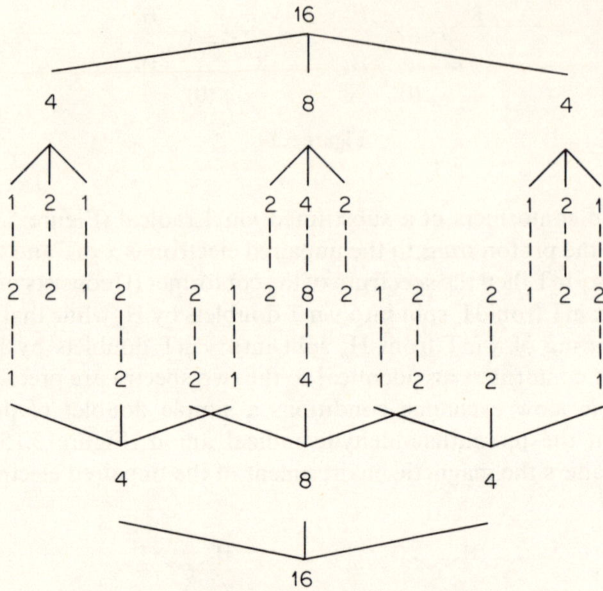

Figure 5.16 Hypothetical spectra showing the effect of geometrical isomerism on the coupling to four protons at slow exchange rates

terephthalaldehyde radical anion have been interpreted by Maki.[34] The two forms are not present in equal concentrations, the ratio being $[III]/[IV] = 1\cdot40$. Selective deuteration at positions 2 and 5 resulted in the assignment of (III) as the *trans* form of the radical. A minimum lifetime of $4\cdot7 \times 10^{-7}$ s was estimated from the line-width.

When the rate of exchange is comparable to the line-width then the operation of the Uncertainty Principle results in line broadening. The appropriate expression of the Uncertainty Principle is

$$\Delta v \Delta t \geqslant 1/2\pi \qquad (5.35)$$

where Δv is the uncertainty in frequency and Δt is the uncertainty in time. We can equate Δt with the lifetime of the species, τ, and the uncertainty of the resonance frequency is manifest by the line-width when measured in frequency units. The shape of the ESR lines of radicals in solution is normally Lorentzian and the spectrum is usually presented as the first derivative of the absorption so that the most convenient line-width parameter to measure is the peak width on the derivative, ΔB, and this is related to the half-width at half-height, Γ, by

$$2\Gamma = \sqrt{3}\Delta B \qquad (5.36)$$

and (5.35) may be rewritten

$$2\Gamma\tau = 1/2\pi \tag{5.37}$$

The relationship between the line-width parameter in field units and frequency units is

$$\Gamma(s^{-1}) = \Gamma(T)g\beta/h = \Gamma(T)\gamma_e/2\pi \tag{5.38}$$

where γ_e is the magnetogyric ratio of the electron. Hence the line broadening associated with a lifetime, τ, is

$$\Gamma(T) = (2\tau\gamma_e)^{-1} \tag{5.39}$$

and the overall width in the limit of slow exchange is

$$\Gamma = \Gamma_0 + (2\tau\gamma_e)^{-1} \tag{5.40}$$

where Γ_0 is the line-width in the absence of exchange. When the spin–spin relaxation time, T_2, is very much shorter than the spin-lattice relaxation time, T_1, it is related to the line-width of a Lorentzian line by

$$(T_2)^{-1} = \gamma_e\Gamma \tag{5.41}$$

At faster rates of exchange the rate begins to approach the frequency difference for the lines associated with the interconverting species. The lines are further broadened and begin to approach each other. This shift is related to the lifetime by

$$(\delta B_0^2 - \delta B_e^2)^{1/2} = \sqrt{2}/\tau\gamma_e \tag{5.42}$$

where δB is the line separation in the absence of exchange and δB_e is the line separation during exchange. In situations where the overall width of the spectrum is unaltered by the rate of exchange only certain transitions are broadened at intermediate rates giving rise to the alternating line-width effect.[35] A hypothetical situation is illustrated in Figure 5.17. Here the effect of an increasing exchange rate on two sets of *completely* equivalent pairs of protons is shown. The term completely equivalent is applied when hyperfine couplings are identical *at any instant*. At intermediate rates, components whose positions are either determined by the sum of the two couplings or independent of the hyperfine coupling are not broadened. Other transitions are broadened to varying extents depending on the nuclear spin quantum numbers involved. At fast rates of exchange the spectrum collapses to a quintet as all four protons show time-averaged equivalence. This coalescence and narrowing is a further consequence of the Uncertainty Principle. If the frequency uncertainty becomes greater than the frequency difference between two states in the absence of exchange, then τ becomes less

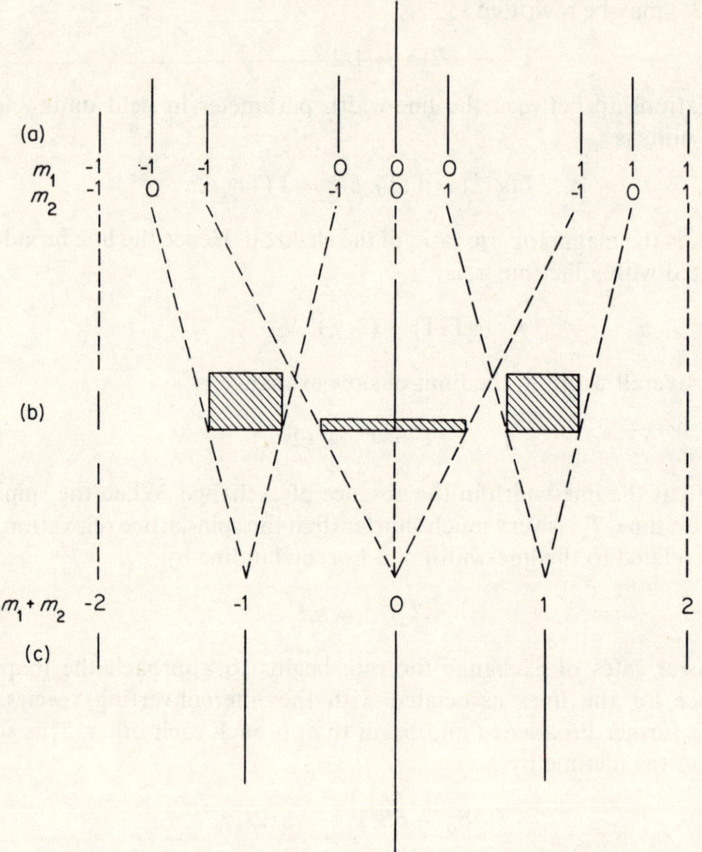

Figure 5.17 Hypothetical spectra for four equivalent protons comprised of two completely equivalent sets at (a) slow, (b) intermediate and (c) fast exchange rates

than Δt and the two states are no longer distinguishable and a single time average line is observed. The width is given by

$$\Gamma = \Gamma_0 + \gamma_e \tau \frac{\langle (\delta B_0)^2 \rangle}{4} \tag{5.43}$$

and at the fast exchange limit the coalesced line has narrowed to limiting width Γ_0. If the equilibrium constant for the interconversion is not 1 and the probabilities of the two forms are p_A and p_B then the position of the coalesced line is given by the weighted mean

$$\langle B \rangle = \frac{p_A B_A + p_B B_B}{p_A + p_B} \tag{5.44}$$

and the width by

$$\Gamma = \Gamma_0 + \gamma_e \tau p_A p_B \langle (\delta B_0)^2 \rangle \tag{5.45}$$

For a radical having a single nuclear spin I and conformations involving splittings of a_1 and a_2 the line-width shows dependence on the nuclear spin quantum number m_I and is given by

$$\Gamma = \Gamma_0 + \gamma_e \tau p_A p_B (a_1 - a_2)^2 m_I^2 \tag{5.46}$$

In very favourable cases with a simple two or three jump conformation model and a relatively simple spectrum it is possible to obtain rate constants directly from measurements of line-width and splitting from slow, intermediate and fast conversion spectra over a range of temperature. The Arrhenius or Eyring equations may then be used to obtain information about the size of the barrier. It is generally more satisfactory to obtain such information in more complex situations by matching experimental spectra from a range of conditions with spectra computed using either the modified Bloch equations or the relaxation matrix method.

The relaxation matrix method was developed by Redfield[36] and others[37] as a general theory of spin relaxation. It is an extension of the density matrix treatment described in the NMR section. Detailed development and application of the method to the ESR spectra of radicals undergoing dynamic processes in solution has been achieved by Freed and Fraenkel.[38,39] The total Hamiltonian, H, is considered to consist of two parts: a time-independent Hamiltonian, H_0', which determines the sharp-line spectrum, and the time-dependent Hamiltonian, $H'(t)$, which has zero time average. Relaxation is the result of fluctuations of $H'(t)$ and Redfield showed that for rapid fluctuations the elements of the density matrix obey a set of linear differential equations of the form

$$\frac{d\rho_{\alpha\alpha'}^*}{dt} = \sum_{\beta\beta'}{}' R_{\alpha\alpha'\beta\beta'} \rho_{\beta\beta'}^* \tag{5.47}$$

The prime on the summation indicates that it is only necessary to retain terms for which

$$\alpha - \alpha' = \beta - \beta' \tag{5.48}$$

where α, β, etc. are the eigenstates of H_0'. The asterisk indicates that the density matrix elements are defined in such a way that when $H'(t) = 0$, ρ^* is a constant. The relaxation matrix, R, plays a role analogous to the reciprocal of the relaxation times in the Bloch equations and has the important property that its eigenvalues when multiplied by -1 are the widths of the Lorentzian resonances at $\omega_{\alpha\alpha'}$

$$-R_{\alpha\alpha'\alpha\alpha'} = T_2^{-1} \tag{5.49}$$

Redfield's equation (5.47) is applicable only when the elements of the relaxation matrix are much less than the inverse of the correlation time. This condition implies that the treatment is valid only in the fast-exchange limit. The time dependence of hyperfine coupling is included in the Hamiltonian, $H'(t)$, in the form of a correlation function. For a single nucleus this function has the form

$$g(t') = \gamma_e^2 \langle [a(t) - \bar{a}][a(t + t') - \bar{a}] \rangle \tag{5.50}$$

which is a time average of the product of deviations in the hyperfine splitting at times t and $t + t'$. The line-widths are determined from the Fourier cosine transform

$$j(\omega) = \int_0^\infty g(t') \cos \omega t' \, dt' \tag{5.51}$$

Although they may be included, non-secular effects are normally neglected, because the frequency of the hyperfine fluctuations is usually much less than the microwave resonant frequency so that the approximation

$$j(\omega) = j(0) \tag{5.52}$$

can be made. These j factors are called spectral densities and for the transition with nuclear-spin quantum number m_I the line-width is given by

$$T_2^{-1}(m_I) = T_2^{-1}(0) + j(0)m_I^2 \tag{5.53}$$

or

$$\Gamma(m_I) = \Gamma_0 + j(0)m_I^2\gamma_e^{-1} \tag{5.54}$$

Extension of these expressions to more complex sets of nuclei and detailed working of the broadening calculations is given by Fraenkel.[39] The limitation of the method to the fast-exchange region means that it is less useful than the modified Bloch equations for the quantitative treatment of data, but it is generally more useful when a qualitative understanding is required.[40]

5.2.3 *Nitroaromatic radical anions*

The study of nitroaromatic radical anions has proved to be a particularly fruitful field for ESR spectroscopists. Many of these studies have been concerned with the dynamics of ion pairing but a substantial number are concerned with the steric hindrance of the nitro group by adjacent substituent groups. A number of methyl substituted nitrobenzene radical anions were examined by Geske and Ragle[41,42] who found that the nitrogen coupling constants for the radical anions of 2,6-dimethylnitrobenzene (1·78 mT) and 2,3,5,6-tetramethylnitrobenzene (2·04 mT) are much greater than the 1·03 mT coupling observed in the unsubstituted radical anion. The couplings to the ring protons in the methylated radicals are considerably smaller than in the unsubstituted radical. Since the inductive effect of the methyl group has been shown to have very little effect on the spin density in the ring and a single methyl group in the *ortho* position does not appreciably alter the nitrogen coupling the above effect is interpreted as being the result of the localization of the unpaired electron in the π-electron system of the nitro group which is sterically prevented from conjugating with the π-electron system of the aromatic ring by the presence of the two methyl groups *ortho* to the nitro group. Subsequently a large variety of alkyl-substituted nitrobenzene radical anions have been examined and this has enabled a correlation between the nitrogen coupling and the twist angle of the nitro group with respect to the plane of the ring to be made.[43]

As part of this study of steric hindrance McKinney and Geske[44] examined the spectrum of the radical anion of 2,3,5,6-tetraisopropylnitrobenzene. Severe twisting of the nitro group was to be expected but these workers were surprised to discover that non-identical conformational isomers exist. The spectrum consists of two overlapping $1:1:1$ triplets resulting from the nitrogen hyperfine coupling. The major conformer (B) has the smaller splitting of 2·20 mT. In the minor conformer (A) the splitting is increased to 2·36 mT. Both conformers have identical g values, so the central lines of each triplet are completely superposed, but the assignment has been confirmed by examining the spectrum of the radical after isotopic enrichment with nitrogen-15 $(I = \frac{1}{2})$. The equilibrium constant for the interconversion of these conformers $(A \rightleftharpoons B)$ can be determined from the relative intensities of the two signals. This equilibrium constant shows some solvent dependency ranging from 2·3 in acetonitrile to 6·0 in a water–dimethyl formamide mixture. The lower value corresponds to $\Delta G_{298}^{0} = -2·1$ kJ/mole. Proton couplings in both conformers are poorly resolved and have not been assigned but it is apparent that such coupling is larger in the less stable conformer. Correlation of the nitrogen couplings with the degree of twist suggests that in A the twist is about 80° and in B about 75°. Molecular models suggest that a highly coupled cog and gear interaction between the isopropyl groups and the nitro group restricts the nitro group to these twist angles. The two *meta* isopropyl groups play a vital buttressing role in producing this conformational isomerism, for the effect is not observed in the spectrum of the radical anion of 2,4,6-triisopropylnitrobenzene and the nitrogen splitting is reduced to 2·06 mT. The line-width of species A is 0·015 mT so that the minimum life-time of A is $3·8 \times 10^{-7}$ s and the corresponding rate of interconversion must be less than $2·6 \times 10^{6}$ s^{-1}.

5.2.4 *Ring inversion in cyclic nitroxides*

In the past few years there have been numerous studies of nitroxide radicals.[45] Of particular interest in this context have been the studies of piperidine and substituted piperidine nitroxide as these compounds have only one more electron than the corresponding cyclohexanone derivatives.[46–48] A number of these radicals have been examined over a wide temperature range and the ring inversion examined through slow, intermediate and fast rates. It is generally assumed that the inversion involves the conversion of one chair conformation into another and for symmetrical unsubstituted radicals such as piperidine nitroxide and morpholine nitroxide, the two conformations are identical and the populations equal so that $k = 1$ and $\Delta G_0 = 0$. Thus in the slow-exchange region the spectra of the two forms will be precisely super-posed. In nitroxide radicals the unpaired electron occupies a π-orbital on the N—O group so that there is considerable spin density on both nitrogen and oxygen atoms. Consequently the main hyperfine interaction is with the nitrogen and this produces a $1:1:1$ triplet with a splitting of about 1.7 mT. At fast inversion rates the four protons of the methylene groups attached to the nitrogen appear equivalent so that each component of the nitrogen triplet is split into a $1:4:6:4:1$ quintet with a splitting of about 1.3 mT. At 380°K the inversion rate is fast enough to produce a coalesced spectrum of this type but not fast enough to produce sufficient narrowing to allow coupling to more remote nuclei to be resolvable. At slow inversion rates the conformations appear locked. The two axial protons are equivalent and give rise to a relatively large $1:2:1$ triplet with a splitting of about 2.6 mT. The two equatorial protons are also equivalent and give rise to a small $1:2:1$ triplet with a splitting of about 0.4 mT. At 170°K the inversion of piperidine nitroxide produced photochemically from 1-hydroxypiperidine in methylene chloride solution is sufficiently slow that a coupling of 0.06 mT to three equivalent remote protons is resolved. The spectra of morpholine nitroxide are very similar to those of piperidine nitroxide but in the slow inversion region the coupling to only two remote protons is resolved. This suggests that this coupling is to two of the γ-protons that are three bonds removed from the nitroxide group, while in the piperidine nitroxide these two protons are fortuitously equivalent to one of the δ-protons. Rolfe, Sales and Utley[48] obtained activation parameters for piperidine nitroxide inversion by simula-tion of the spectra using the modified Bloch equations and a two-jump model followed by plotting the Eyring equation. They obtain $\Delta G^{\ddagger}_{298} = 22 \pm 2$ kJ/mole, $\Delta H^{\ddagger}_{298} = 24 \pm 1$ kJ/mole and $\Delta S^{\ddagger} = 9 \pm 3$ J/mole K^{-1}. Using the less sophisticated technique of measuring the peak separations in the intermediate range to obtain the rate constant Windle, Kuhnle and Beck[47] calculated $\Delta G^{\ddagger}_{253} = 21 \pm 0.5$ kJ/mole, $\Delta H^{\ddagger}_{253} = 22.7$ kJ/mole and $\Delta S^{\ddagger} = 6$ J/mole K^{-1}. Very similar values were obtained for morpholine nitroxide at 285°K. These values are very similar to the comparable parameters in cyclohexene.[49]

By replacement of one of the δ-protons with alkyl or aryl groups, deriva-tives of piperidine nitroxide with a strong conformational preference can be prepared.[46–48] At low temperatures the spectra of the separate conformations should be observable if the equilibrium constant is close to unity but in all these cases the conformational preference of the bulky substituent for the equatorial position results in only the spectrum of this conformer being observed. At high temperature the radical is rapidly interconverting but the spectrum is essentially the same as at low temperature except for the disappearance of the γ-proton coupling. Detailed examination of the β-proton hyperfine coupling constants shows that a small change has occurred. Under fast inversion conditions the mean value of the axial and the equatorial coupling is given by

$$\langle a_{ax.} \rangle = P_a a_{ax.} + P_b a_{eq.} \tag{5.55}$$

$$\langle a_{eq.} \rangle = P_a a_{eq.} + P_b a_{ax.} \tag{5.56}$$

where P_a and P_b are the relative populations of the two conformers a and b, and $a_{ax.}$ and $a_{eq.}$ are the axial and equatorial couplings in the absence of exchange. In the unsubstituted nitroxides $P_a = P_b = \frac{1}{2}$ and consequently $\langle a_{ax.} \rangle = \langle a_{eq.} \rangle = \langle a \rangle$, but in the substituted radicals $P_a \gg P_b$ so that the average values differ only slightly from the low-temperature values and the two spectra are of similar appearance. The above equations can be used to determine the relative populations of the two conformations if it is assumed that there is no marked temperature variation of $a_{ax.}$ and $a_{eq.}$. The free energy difference between the two conformations can be determined in the usual way. Values for 4-methyl, 4-isopropyl, and 4-phenyl piperidine nitroxide are -7, -9 and -10.5 kJ/mole respectively. Activation parameters have been determined either by simulation or calculation (Table 5.2).

Table 5.2

R	$\Delta G^\ddagger$	$\Delta H^\ddagger$	$\Delta S^\ddagger$
Me	26	32	18
	28	36	26
Ph	37	—	—

By analogy with the β-coupling in hydrocarbon radicals the β-proton hyperfine coupling in nitroxides is assumed to be proportional to the unpaired spin density on the adjacent nitrogen atom

$$a_{NCH}^H = (Q_n + Q_0 \cos^2 \theta)\rho_N \tag{5.57}$$

Windle, Kuhnle and Beck[47] estimated the dihedral angle θ by neglecting the contribution from $Q_n \rho_N$

$$a_{NCH}^H = Q_0 \cos^2 \theta \rho_N \tag{5.58}$$

In dimethyl nitroxide the methyl groups are rapidly rotating so that observed average β-proton coupling of 1.34 mT is associated with $\cos^2 \theta = \frac{1}{2}$. Thus $Q_0 \rho_N = 2.68$ mT. Substituting this value in (5.58) yields

$$a_{NCH}^H = 2.68 \cos^2 \theta \qquad (5.59)$$

This allows θ to be determined from the experimental splittings measured from the slow inversion spectra. In all of these radicals θ is close to 8° for the axial proton and 112° for the equatorial proton.

Low-temperature spectra of the substituted nitroxide radicals show interaction with two equivalent remote protons that is similar in size to the coupling to three remote protons in the unsubstituted piperidine nitroxide. The absence of coupling to a third remote proton in the substituted nitroxide is strong evidence that this third proton is the δ-proton in the equatorial position.

In the two preceding sections two classes of radicals have illustrated the potential of ESR spectroscopy in obtaining quantitative information about structure and conformation. A number of reviews[31,32,35,39,50–52] discuss and interpret both qualitative and quantitative aspects of the application of ESR spectroscopy to many other classes of radicals.

5.3 References

1. I. O. Sutherland, *Annual Review of NMR Spectroscopy*, Vol. 4, Ed. E. F. Mooney (London and New York: Academic Press, 1971).
2. J. W. Emsley, J. Feeney and L. H. Sutcliffe, *High Resolution Nuclear Magnetic Resonance* (Oxford: Pergamon Press, 1965).
3. H. S. Gutowsky, G. G. Belford and P. E. McMahon, *J. Chem. Phys.*, **36**, 3353 (1962).
4. R. A. Newmark and C. H. Sederholm, *J. Chem. Phys.*, **43**, 602 (1965).
5. A. D. Cohen, N. Sheppard and J. J. Turner, *Proc. Chem. Soc.*, 118 (1958).
6. J. G. Powles and J. H. Strange, *Mol. Phys.*, **5**, 329 (1962).
7. V. I. P. Jones and J. A. Ladd, *J. Chem. Soc.* (*B*), 1719 (1970).
8. G. Govil and H. J. Bernstein, *J. Chem. Phys.*, **47**, 2818 (1967).
9. J. C. Schug, P. E. McMahon and H. S. Gutowsky, *J. Chem. Phys.*, **33**, 843 (1960).
10. R. A. Newmark and C. H. Sederholm, *J. Chem. Phys.*, **39**, 3131 (1963).
11. H. S. Gutowsky, J. Jonas, Fu-ming Chen and R. Meinzer, *J. Chem. Phys.*, **42**, 2625 (1965).
12. J. P. Lowe, *Progress in Physical Organic Chemistry*, Vol. 6, Ed. A. Streitwieser, Jr. and R. W. Taft (London and New York: Interscience, 1968).
13. R. J. Abraham, L. Cavalli and K. G. R. Pachler, *Mol. Phys.*, **11**, 471 (1966).
14. W. D. Phillips, *J. Chem. Phys.*, **23**, 1363 (1955).
15. C. E. Looney, W. D. Phillips and E. L. Reilly, *J. Am. Chem. Soc.*, **79**, 6136 (1957).
16. H. S. Gutowsky, D. W. McCall and C. P. Slichter, *J. Chem. Phys.*, **21**, 279 (1953).
17. H. S. Gutowsky and A. Saika, *J. Chem. Phys.*, **21**, 1688 (1953).
18. H. S. Gutowsky and C. H. Holm, *J. Chem. Phys.*, **25**, 1228 (1956).
19. F. Bloch, *Phys. Rev.*, **70**, 460 (1946).
20. H. M. McConnell, *J. Chem. Phys.*, **28**, 430 (1958).

21. J. I. Kaplan, *J. Chem. Phys.*, **28**, 218 (1958); **29**, 462 (1958).
22. S. Alexander, *J. Chem. Phys.*, **37**, 967, 974 (1962); **38**, 1787 (1963); **40**, 2741 (1964).
23. C. S. Johnson, *J. Chem. Phys.*, **41**, 3277 (1964).
24. C. S. Johnson, *Advances in Magnetic Resonance*, Vol. 1, Ed. J. S. Waugh (New York and London: Academic Press, 1965).
25. R. M. Lynden-Bell, *Progress in NMR Spectroscopy*, Vol. 2, Ed. J. W. Emsley, J. Feeney and L. H. Sutcliffe (Oxford: Pergamon Press, 1967).
26. I. I. Rabi, N. F. Ramsey and J. Schwinger, *Rev. Mod. Phys.*, **26**, 167 (1954).
27. G. Binsch, *J. Am. Chem. Soc.*, **91**, 1304 (1969).
28. L. W. Reeves and K. N. Shaw, *Can. J. Chem.*, **49**, 3671 (1971).
29. H. Y. Carr and E. M. Purcell, *Phys. Rev.*, **88**, 415 (1952); **94**, 630 (1954).
30. A. Allerhand and H. S. Gutowsky, *J. Chem. Phys.*, **41**, 2115 (1964); **42**, 1587 (1965).
31. S. Forsén and R. A. Hoffman, *J. Chem. Phys.*, **39**, 2892 (1963); **40**, 1189 (1964).
32. G. A. Russell, *Determination of Organic Structures by Physical Methods*, Volume 3, Ed. F. C. Nachod and J. J. Zuckerman (New York and London: Academic Press, 1971).
33. D. H. Geske, *Progress in Physical Organic Chemistry*, Volume 4, Ed. A. Streitweiser, Jr. and R. W. Taft (New York: Interscience, 1967).
34. A. H. Maki, *J. Chem. Phys.*, **35**, 761 (1961).
35. P. D. Sullivan and J. R. Bolton, *Advances in Magnetic Resonance*, Volume 4, Ed. J. S. Waugh (New York and London: Academic Press, 1970).
36. A. G. Redfield, *IBM J. Res. Develop.*, **1**, 19 (1957); *Advances in Magnetic Resonance*, Volume 1, Ed. J. S. Waugh (New York and London: Academic Press, 1965).
37. R. K. Wangsness and F. Bloch, *Phys. Rev.*, **89**, 728 (1953); F. Bloch, *Phys. Rev.*, **102**, 104 (1956).
38. J. H. Freed and G. K. Fraenkel, *J. Chem. Phys.*, **39**, 326 (1963).
39. G. K. Fraenkel, *J. Phys. Chem.*, **71**, 139 (1967).
40. A. B. Barabas, W. F. Forbes and P. D. Sullivan, *Can. J. Chem.*, **45**, 267 (1967).
41. D. H. Geske and J. L. Ragle, *J. Am. Chem. Soc.*, **83**, 3532 (1961).
42. D. H. Geske, J. L. Ragle, M. A. Bambenek and A. L. Balch, *J. Am. Chem. Soc.*, **86**, 987 (1964).
43. P. H. Rieger and G. K. Fraenkel, *J. Chem. Phys.*, **39**, 609 (1963).
44. T. M. McKinney and D. H. Geske, *J. Chem. Phys.*, **44**, 2277 (1966).
45. A. Rassat, *Molecular Spectroscopy*, Ed. P. Hepple (London: The Institute of Petroleum, 1968).
46. A. Hudson and H. A. Hussain, *J. Chem. Soc.* (*B*), 251 (1968); 953 (1968).
47. J. J. Windle, J. A. Kuhnle and B. H. Beck, *J. Chem. Phys.*, **50**, 2630 (1969).
48. R. E. Rolfe, K. D. Sales and J. H. P. Utley, *Chem. Comm.*, 540 (1970).
49. F. A. L. Anet and M. Z. Haq, *J. Am. Chem. Soc.*, **87**, 3147 (1965).
50. A. Hudson and G. R. Luckhurst, *Chem. Rev.*, **69**, 191 (1969).
51. G. A. Russell, E. T. Strom, E. R. Talaty, K. Y. Chang, R. D. Stephens and M. C. Young, *Record Chem. Progr.* (*Kresge-Hooker Sci. Lib.*), **27**, 3 (1966).
52. G. A. Russell, *Radical Ions*, Ed. E. T. Kaiser and L. Kevan (New York: Interscience, 1968).

6 Studies of internal rotation by microwave spectroscopy

N. L. Owen

6.1 Introduction

In today's chemical literature, there is to be found an impressive compilation of experimentally determined potential barriers hindering rotation about single bonds, especially about carbon–carbon bonds. Most of these values have been determined from analysis of the pure rotational spectra of molecules by microwave spectroscopy. The microwave spectral region lies beyond the far infrared, spanning three decades of frequency (on a logarithmic scale) from about 10^9 Hz to 10^{12} Hz. In wave number units, such a gap represents only about 30 cm^{-1} (from 0·03 cm^{-1} to 33 cm^{-1}), but the quantity (and quality) of information which can be gleaned from this spectral region by microwave spectroscopy is impressive by any standard. The most useful region for absorption spectroscopy is found between 5 and 50 GHz (10^9 Hz), since many of the quantum transitions between different rotational states of molecules fall within these frequency limits. Microwave spectroscopy is a non-destructive technique which requires the sample to be gaseous, or to have a vapour pressure greater than about 10^{-4} mm Hg at the temperature of the experiment. Information relating to molecular structure derived from microwave spectroscopy refers to the freely rotating molecule in the vapour phase, and in general no intermolecular effects are involved.

Several specialized books on the subject are available[1–5] and there are numerous reviews on both general and specific aspects of microwave spectroscopy to be found in the literature.[6–15] This chapter will deal mainly with the applicability and usefulness of microwave spectroscopy for studying internal rotation in molecules, but will include a brief summary of the basic experimental and theoretical aspects of the subject. For more detailed and comprehensive accounts of the role of microwave spectroscopy in dealing with internal rotation, the reader is directed to articles by Lin and Swalen,[16] Wollrab,[17] Dreizler[18] and Gordy and Cook.[19]

6.2 Microwave spectroscopy and the structure of molecules

The rotational energies of free molecules are quantized, and radiation is absorbed or emitted by molecules at frequencies corresponding to the differences in energies of adjacent molecular eigenstates. Thus, pure rotation spectra reflect the rotational behaviour of molecules, which in turn is a function of their size, shape and mass. By careful study, sometimes under conditions of high resolving power, many interesting molecular structure details can be obtained from rotational spectra. The following summary includes some of the more important parameters which may, under certain circumstances, be obtained.

6.2.1 *Shapes of molecules*

The complicated rotational motion of molecules may be considered to be composed of three independent components rotating about axes at right angles to each other. Thus, the key parameters governing the rotational motion are the three mutually perpendicular moments of inertia, I_x, I_y and I_z, where $I = \sum_i m_i r_i^2$ (here m_i represents the mass of atom i and r the distance of the ith atom from the centre of gravity of the molecule). Where the reference axes are chosen so that the inertial tensor is diagonal (i.e. one of the moments of inertia has the largest possible value, while another has the lowest possible value), the three moments are called the *principal* moments of inertia, and are labelled I_A, I_B and I_C. A successful analysis of a microwave spectrum implies that the values of the three principal moments of inertia are known, which in turn often helps to establish the gross molecular shape and conformation. Molecules which are capable of existing in two or more isomeric forms can often be identified very clearly, since different molecular configurations generally have quite different moments of inertia, and hence very different spectral patterns. In the special instance of planar molecules, the largest principal moment of inertia is equal to the sum of the other two (in the molecular plane).

$$I_C = I_A + I_B$$

This expression holds exactly, provided a small term known as the inertial defect of the molecule is included (see §6.5), and thus the planarity (or otherwise) of a molecule can be established very conveniently and quickly.

6.2.2 *Accurate molecular structure*

This has been, and probably still remains, one of the most important single features of microwave spectroscopy—the possibility of establishing the precise geometry of a molecular species. With the exception of diatomic molecules, a single moment of inertia is not sufficient to describe the molecular structure completely, and so several different isotopically substituted molecular species have to be studied. The assumption is made that exchanging an isotope changes the mass, but not the internuclear distance. Calculations based on observed ground vibrational state rotational constants lead to internuclear values known as r_0 parameters. Unfortunately, these values have little physical significance, since different isotopic species have different vibrational ground states, and the subsequent calculation of the more meaningful equilibrium (r_e) structures from r_0 values is possible only for very simple molecules, because the vibration–rotation interaction constants must be known for all molecular vibrations for each isotopic species. Other structural parameters denoted by r_s (substitution), introduced by Costain[20] and $\langle r \rangle$ (average), developed by Herschbach and Laurie[21] and by Oka and Morino,[22] have to a large extent helped to overcome these difficulties, and have enabled precise, transferable and yet meaningful parameters to be derived from microwave studies. Average $\langle r \rangle$ values can often conveniently be related to structural parameters obtained from electron diffraction experiments.

Determination of the complete structure of a polyatomic molecule is often a lengthy procedure, owing to the experimental difficulties of obtaining the appropriate isotopically substituted isotopes.

6.2.3 *Low frequency molecular vibrations*

The rotational absorption lines associated with a particular molecular species are nearly always accompanied by weaker satellite lines, which show very similar features to the main lines. These are caused by the presence of vibrationally excited molecules, and the intensities of such lines are related to that of the corresponding ground vibrational-state lines by the Boltzmann factor. Provided that these vibrational satellite lines can be correctly assigned to the appropriate rotational energy transitions (this is often a fairly straightforward procedure if the main species have been assigned correctly), then their intensities, measured relative to those of the corresponding ground-state lines, give the energy differences between the various vibrational states and the ground state of the molecule. This method is most useful for measuring low-frequency modes (e.g. torsions) which are often too weak to be observed in the far infrared spectrum. (The lowest frequency modes also give rise to the strongest microwave satellite lines.) For molecules with planes of symmetry, the sign and magnitude of the inertial defect corresponding to each vibrational satellite often makes it possible to ascertain whether or not a vibration corresponds to an 'in-plane' or an 'out-of-plane' mode.

6.2.4 *Dipole moments*

Microwave spectroscopy offers an elegantly simple but accurate method of measuring molecular dipole moments in the vapour phase where inter-molecular interactions are minimal. The rotational absorption lines of a molecule are displaced by the effect of an accurately known, externally applied electric field; (see §6.3 for more details of the Stark effect). The displacement of the lines is accurately measured, and compared with that observed for rotational lines of a standard substance with a known dipole moment; (usually carbonyl sulphide is used as the standard, with $\mu = 0\cdot71521$ D.)[23] The relative displacement or Stark shift is a function of the rotational constants, of the applied field, and for most molecules, of the square of the dipole moment. Both the magnitude and the orientation of the dipole with respect to the molecular framework can be established. Values as low as $0\cdot1$ D can be measured in this way, and it is also possible to find how the molecular dipole moment changes in different vibrational states of a molecule.

6.2.5 *Quadrupole moments*

The effect of the presence of a nucleus or nuclei with spin quantum number $I > \frac{1}{2}$ on the overall rotational spectrum of a molecule can often be observed as fine structure on some of the lines. The non-spherically symmetrical quadrupole moment (eQ) of a nucleus with $I > \frac{1}{2}$ interacts with the average electric field gradient (q) at the nucleus, due to the extranuclear electrons, and the interaction energy (eQq) is known as the quadrupole coupling constant. During a molecular rotational transition, changes can also occur in the quadrupole coupling energy, and the pure rotation line may be split into several components. This hyperfine splitting pattern is related to the quadrupole coupling constant. (For asymmetric molecules, there are three independent components.) In many instances the measured values of the coupling constants may be usefully related to the distribution of bonding electrons within the molecule.

6.2.6 *Unstable species*

Although beset by considerable experimental difficulties, the detection and identification of small unstable radicals by microwave spectroscopy has been shown to be possible, and much useful structural information has already been amassed.[15] Development of expertise in high-temperature microwave spectroscopy (above 1000°C) has also resulted in many small inorganic species being extensively studied.[14] The equipment required for such studies is necessarily very different from that used in conventional microwave work.[24]

6.2.7 *Chemical reactions*

Some work has been carried out using microwave spectroscopy to follow courses of chemical reactions, and to identify the resulting products. This kind of application utilizes the obvious advantages of the technique as a very precise and selective analytical method.[25]

6.2.8 *Energy transfer and relaxation processes*

The development of microwave double resonance techniques has opened up the field of energy transfer processes, using rotational spectra as a probe. Much of the pioneering work in this area is due to Oka, whose studies on simple molecular systems (e.g. formaldehyde, hydrogen cyanide) have shown, for example, that collision-induced transitions also obey selection rules.[26]

6.2.9 *Magnetic measurements*

Experiments using very high magnetic fields have resulted in observation of molecular Zeeman splittings for molecules in 'non-magnetic' electronic ground states (e.g. $^1\sum$ states). The information potentially available from these hyperfine splittings includes the anisotropy of the magnetic susceptibility, the molecular quadrupole moment and the sign of the electric dipole moment.[27]

These following three topics are discussed in some detail in the following sections of this chapter: (a) barriers to internal rotation, (b) potential energy profiles for internal rotation, and (c) energy differences between rotamers.

6.3 The microwave experiment

Microwave spectroscopy differs from more conventional forms of spectroscopy (e.g. optical, infrared) in several important aspects.

(a) Microwave sources (usually reflex klystrons or backward wave oscillators) emit virtually monochromatic radiation, and thereby eliminate the need for a dispersing element.

(b) Radiation in the frequency range between 10 and 40 GHz ($GHz \equiv 10^9$ Hz) is best transmitted along hollow metal tubes or pipes, and these usually also serve as the sample cell for the experiment.

(c) The high resolving power generally associated with microwave spectroscopy is realized only with gaseous samples at low pressures ($< 10^{-2}$ mm Hg).

A block diagram of a typical Stark-modulated spectrometer is shown in Figure 6.1, while Figure 6.2 shows a recently produced commercial machine which is currently available. Comprehensive accounts of the experimental aspects of microwave spectroscopy are available in specialist texts,[1-4] and so this section will discuss briefly only some of the most important components.

6.3.1 *Sources of microwave power*

The most successful source of radiation for microwave spectroscopy to date is the reflex klystron, whereby a beam of electrons is subjected to an r.f. alternating potential as it traverses a resonant cavity. The beam is also

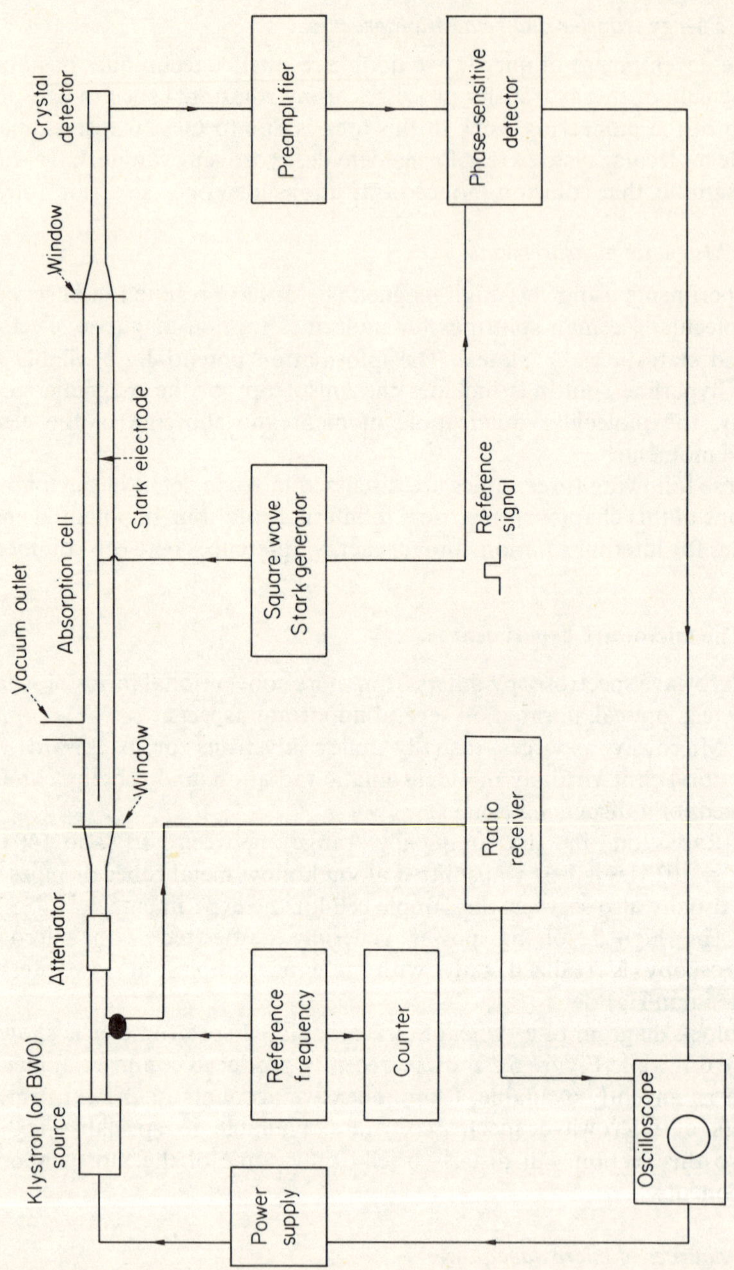

Figure 6.1　A block diagram of a Stark-modulated microwave spectrometer

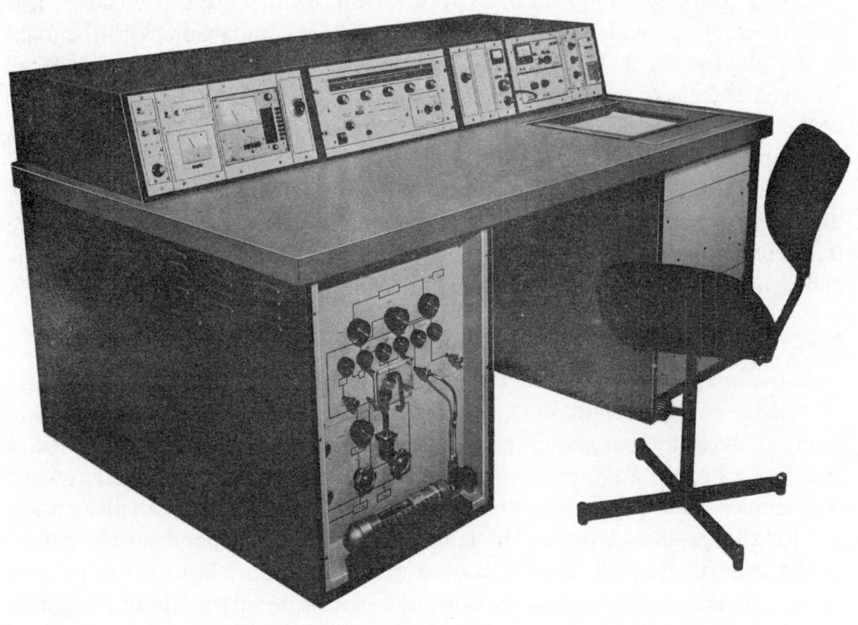

Figure 6.2 'Camspek' commercial microwave spectrometer. (By courtesy of Cambridge Scientific Instruments Ltd.)

made to pass through the cavity in the reverse direction by the action of a negatively charged reflector. The alternating potential breaks up the electron beam into 'bunches' which induce oscillations in the cavity, which in turn becomes a source of microwave radiation. The resonating frequency of a klystron can be varied manually (by mechanical distortion of the cavity) over a continuous range of approximately 15% of the mean frequency. Smaller variations in frequency may also be attained by altering the reflector potential.

Recent improvements in the design of backward-wave oscillators have resulted in BWOs being preferred to klystrons in commercially manufactured microwave spectrometers.

A BWO operates on the principle that when an electron beam travels along the axis of a helically orientated electromagnetic wave, the beam is caused to 'bunch' under the influence of the electric field of the wave, with resulting velocity modulation. The electron beam and the electric field will 'interact', provided that the beam velocity is comparable with the axial velocity of the wave. The circuitry is so designed that the exchange of energy between the electron beam and the wave is such that the 'backward' wave of the system grows (i.e. in the reverse direction to that of the electron beam). When the energy coupling is sufficiently large, the device acts as an oscillator. The oscillating frequency is a direct function of the accelerating voltage, and so can be varied electronically in a continuous manner over a wide frequency range. The power output from backward wave oscillators is generally greater than the few milliwatts obtained from klystrons.

6.3.2 *Microwave absorption cells*

At the low-frequency end of the microwave spectrum, power can be transmitted via coaxial cables, but the loss of power associated with propagation along such cables increases rapidly with increasing frequency, with the result that hollow metal wave-pipes (or wave-guides) are used at normal microwave frequencies. Usually, a section of the wave-guide also serves as an absorption cell for the gaseous sample. The lengths of such cells depend on the nature of the experimental arrangement, and may vary from about 0.3 m to over 10 m. The cross-sectional dimensions of a cell (generally made of copper or aluminium) are largely governed by the wavelength of the radiation required to be transmitted. For conventional spectroscopic work, an X-band copper cell is often used, of about 2–4 m in length, and with approximate cross-sectional dimensions of 2.5×1.3 cm. The radiation is normally propagated using the TE_{10} mode, whereby only the electric field is transverse to the direction of propagation. An absorption cell can be made vacuum tight by means of mica windows (thin pieces of mica are transparent to microwaves) attached to both ends of the wave-guide, using rubber O-rings or other suitable seals.

6.3.3 *Detection of microwaves*

One of the simplest and most popular detectors for use in a microwave spectrometer is a semi-conductor crystal diode. These diodes have a short piece of very fine wire (usually tungsten) in contact with a silicon or germanium crystal, and this acts as a rectifier for the microwave energy giving a d.c. output which is approximately proportional to the square of the amplitude of the a.c. input. The crystal must be correctly mounted in the wave-guide, and its response can be altered for different frequencies by means of a tuning plunger. The rectified output is amplified, and eventually displayed on an oscilloscope or on a pen recorder.

6.3.4 *Stark modulation*

Most conventional microwave spectrometers utilize Stark modulation to increase sensitivity and to facilitate the assignment of absorption lines to specific rotational transitions. A flat metal electrode is carefully arranged inside the absorption cell, running the length of the cell, so that the broad face of the electrode lies parallel to the broad edge of the wave-guide. The Stark electrode is usually rigidly supported within the cell in a highly insulating material (e.g. teflon) so that it lies exactly equidistant from the two sides of the wave-guide. Electrical contact is made at a well-insulated point through the cell wall, and an alternating, zero-based, square wave is applied at a fixed frequency, which is generally chosen to be between 5 and 100 KHz. Voltages ranging from zero to 2,000 V are usually applied. The high-frequency voltage modulation enables use to be made of narrow-banded phase-sensitive detectors. This helps to eliminate much of the low-frequency electrical noise. The other great advantage of using Stark modulation is that the applied electric field shifts the absorption lines, and tends to remove the spatial 'M' degeneracy which is associated with each rotational energy state. (Each level comprises $2J + 1$ degenerate 'M' states, where J is the rotational quantum number, and M is the magnetic quantum number—since M can take any value from $-J$ up to $+J$. For all molecules, except symmetric tops, an external electric field does not resolve $\pm M$ values, and so the total number of components in an applied field becomes $J + 1$.) Since the applied Stark field is arranged parallel to the electric vector of the microwave radiation, the selection rule $\Delta M = 0$ is obeyed (see Figure 6.3). Owing to the high frequency of modulation, spectral patterns both with and without Stark voltage may be displayed together on the oscilloscope. The number of Stark components resolved, their relative intensities and the spacings between them can all help to identify the absorption and assign it to an allowed spectral transition.

6.3.5 *Frequency measurement*

An important feature of spectroscopy in the microwave range is the accuracy with which absorption line frequencies may be measured. The method consists basically of taking the output from a standard quartz crystal, operating say at 5 MHz, and multiplying up this fundamental (using frequency multiplying circuits) to about 500 MHz. This signal is then fed directly on to a silicon crystal diode which is positioned in the wave-guide. The non-linear characteristics of the crystal result in the generation of further frequency harmonics up to the frequency range of the microwave radiation. By simultaneously feeding in many of the harmonics, it is possible to arrange a whole ladder of frequencies (say 50 MHz apart), all of which can 'beat' with the microwave radiation. These beat frequencies can be detected and

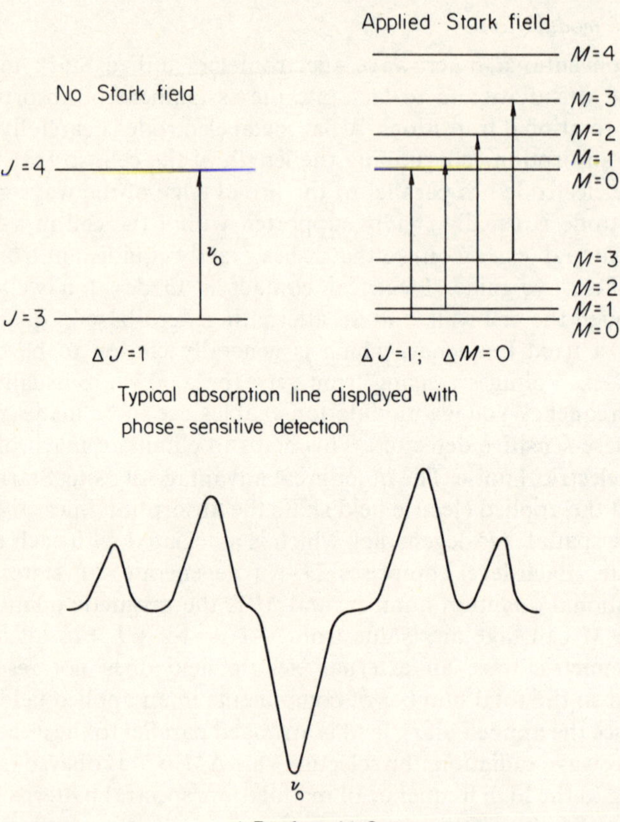

Figure 6.3   The effect of an externally applied electric field on the energy levels of a molecule, and the resulting absorption line pattern

measured by means of a calibrated communications receiver, and the output from the receiver may be displayed simultaneously with the absorption signal on an oscilloscope. The frequency of the fundamental quartz crystal can be accurate to 1 part in 10^8, and can be monitored against a National broadcasted signal standard, and measured with a digital counter. There are very many variations to be found of this general method of frequency measurement, but most systems enable line frequencies to be measured with an accuracy better than 1 part in 10^6.

Approximate frequency measurements (to within a few MHz) are conveniently carried out using resonant cavity wave-meters.

6.3.6 *Commercial spectrometers*

Prior to 1965, microwave spectrometers were built or assembled by individual workers using assorted commercial components. The emergence of commercial interests in the technique (there are at present four companies marketing various models) has coincided with an increased awareness of its analytical potential, as well as with certain developments on the experimental side. The Hewlett Packard (U.S.A.) Company's model 8400 uses a phase-locked BWO as a source, and the operating frequency is read directly from a counter. They have developed a novel method of measuring absorption line intensities, which is based on a 'balanced bridge' principle. Tracerlab (U.S.A.) produce a series of 'modules' for constructing complete spectrometers of varying degree of sophistication; their 4,000 K series models also use phase-locked BWOs as sources. Cambridge Instruments (Campspek) (U.K.) have developed a machine which is primarily designed for accurate analytical applications. The cell design has been so chosen that it can be heated or cooled without difficulty (Figure 6.2). Research Systems Inc. (U.S.A.) produce a microwave spectrometer system suitable for University teaching laboratories. Here the source is a solid state Gunn Microwave oscillator.

One feature common to most of the commercial instruments is that as well as using high resolution conditions, they are able to scan rapidly over wide frequency ranges, to give low resolution spectra. These often show gross features of spectra very clearly (Figure 6.4).

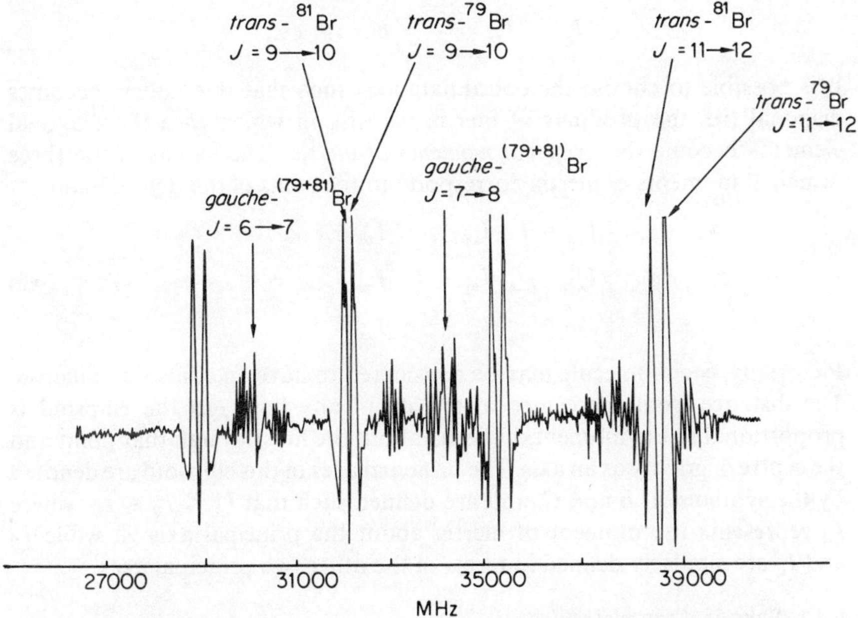

Figure 6.4 The broad band microwave spectrum of n-propyl bromide, showing the near-prolate band heads for *trans* and *gauche* isomers. (From *Molecules and Microwaves*, No. 3, Hewlett Packard, by courtesy of Hewlett Packard.)

6.4 Rotational spectra of molecules

Molecules may conveniently be divided into classes based on their overall shapes, or more specifically, on the relative values of the three principal moments of inertia. For a rigid body rotating about a fixed axis, the moment of inertia about that axis is given by

$$I = \sum_i m_i r_i^2 \tag{6.1}$$

where m_i represents the mass of the ith element, and r is the distance of that element from the axis. For a molecule rotating about its centre of mass, the moment of inertia expressed in cartesian coordinates becomes

$$I_{xx} = \sum_i m_i(y_i^2 + z_i^2)$$

$$I_{yy} = \sum_i m_i(x_i^2 + z_i^2) \tag{6.2}$$

$$I_{zz} = \sum_i m_i(y_i^2 + x_i^2)$$

The complete moment of inertia tensor for a molecule is made up of a symmetric matrix with I_{xx}, I_{yy} and I_{zz} as diagonal elements, and off-diagonal elements I_{xy}, I_{xz} and I_{yz}, where

$$I_{xz} = I_{zx} = - \sum_i m_i x_i z_i \quad \text{etc.}$$

It is possible to choose the coordinate axis such that this matrix becomes diagonal (i.e. the products of inertia vanish), in which case the diagonal elements become the *principal moments of inertia*. The values of the three principal moments of inertia correspond to the roots of the determinant

$$\begin{vmatrix} I_{xx} - I & I_{xy} & I_{xz} \\ I_{xy} & I_{yy} - I & I_{yz} \\ I_{xz} & I_{yz} & I_{zz} - I \end{vmatrix} = 0 \tag{6.3}$$

Pictorially, each molecule may be considered to have an ellipsoid of inertia. The distance from the centre of gravity of any point on the ellipsoid is proportional to the moment of inertia about the line through that point and the centre of gravity as an axis. The principal axes of this ellipsoid are denoted by the symbols A, B and C, and are defined such that $I_A \leqslant I_B \leqslant I_C$, where I_A represents the moment of inertia about the principal axis A, while I_B and I_C are similarly defined in terms of the other two principal axes.

6.4.1 *Spherical top molecules*

When the three principal moments of inertia are equal, a molecule has spherical symmetry, and has no permanent electric dipole. As a result of this, the overall molecular rotational motion does not interact directly with electromagnetic radiation, and the molecules are of little interest to microwave spectroscopists. Examples include carbon tetrachloride CCl_4 and sulphur hexafluoride SF_6.

6.4.2 *Linear molecules*

All atoms in a linear molecule lie on the symmetry axis, and two of the principal moments of inertia are identical.

$$I_B = I_C \quad \text{and} \quad I_A = 0$$

Since the rotational motion is effectively the function of only *one* moment of inertia, linear molecules have very simple spectra. For a rigid molecule, the spectrum consists of equally spaced lines $2B$ apart, where B is given by $h/8\pi^2 I_B$ (h is Planck's constant). The energy levels of a linear molecule may be expressed as

$$E = hBJ(J + 1) \tag{6.4}$$

where J is the quantum number specifying the total angular momentum, and can take all integer values $(0, 1, 2, \ldots, \text{etc.})$. For electric dipole transitions, the selection rules are $\Delta J = \pm 1$, which results in the transition frequencies being expressed simply as

$$\nu = \frac{E(J + 1) - E(J)}{h}$$
$$= 2B(J + 1) \tag{6.5}$$

Examples include carbonyl sulphide OCS and nitrous oxide N_2O. Homonuclear diatomics and symmetrical linear molecules do not show a pure rotational spectrum.

6.4.3 *Symmetric top molecules*

When two of the principal moments of inertia are identical, and the third is non-zero, a molecule is called symmetric top. Two types of symmetric tops are possible; prolate tops where $I_A < I_B = I_C$, and oblate tops where $I_A = I_B < I_C$. These correspond to 'cigar' and 'discoid' shaped inertial ellipses respectively (e.g. CH_3F and $CHCl_3$). For symmetric top molecules, the projection of the total rotational angular momentum on the symmetry axis is a constant of the motion, and this necessitates the introduction of another quantum number, K, in the derived expression for the energy levels

$$E = h[BJ(J + 1) + (A - B)K^2] \tag{6.6a}$$

for a prolate rotor, and

$$E = h[BJ(J + 1) + (C - B)K^2] \tag{6.6b}$$

for an oblate top. To describe each energy level, both J and K must be specified. All levels except those with $K = 0$ are doubly degenerate, since K appears as K^2 in the above expressions. Since K can take any integer value from O up to J, there are $2J + 1$ levels associated with each value of J. The selection rules for symmetric top molecules are $\Delta J = \pm 1$ and $\Delta K = 0$, so that the expression for the absorption line frequencies is identical with that for a linear molecule,

$$v = 2B(J + 1) \qquad (6.7)$$

However, each 'transition' is composed of $2J + 1$ degenerate components. This degeneracy is often removed or partially removed by non-rigidity effects in symmetric rotors.

6.4.4 *Asymmetric top molecules*

When all three moments of inertia are different (which is the case for the vast majority of molecules), a molecule is said to be an asymmetric top. In general, with the exception of some low J levels (see Table 6.1), the energy levels of an asymmetric rotor cannot be described by simple algebraic expressions. The K symmetric top quantum number is no longer a true quantum number, but it is found convenient for asymmetric rotors, for the various energy levels associated with a given value of J, to be labelled using the values of K which the level would have in the corresponding prolate and oblate symmetric top limits; i.e. J_{K_{-1},K_1} where K_{-1} and K_1 represent the K values for the corresponding prolate and oblate top rotors. Thus, the asymmetry of a molecule can be considered as a perturbation which splits the two-fold degeneracy of each K level of the limiting symmetric rotor (Figure 6.5).

The asymmetry of a molecule is most conveniently measured by a parameter 'kappa', which is defined by

$$\kappa = \frac{2B - A - C}{A - C} \qquad (6.8)$$

Kappa has the property of continuously varying from $+1 \cdot 0$ for an oblate rotor to $-1 \cdot 0$ for a prolate molecule, with $\kappa = 0 \cdot 0$ representing the most asymmetric structure.

The quantum-mechanical Hamiltonian form of the rigid rotor energy may be expressed as

$$H = \frac{P_A^2}{2I_A} + \frac{P_B^2}{2I_B} + \frac{P_C^2}{2I_C} \qquad (6.9)$$

where P_A, P_B and P_C represent the angular momentum operators corresponding to the three inertial axes. Alternatively, it may be expressed in terms of

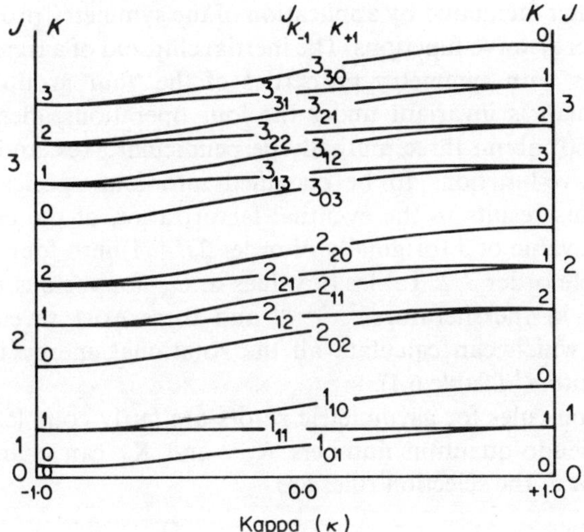

Figure 6.5 The effect of molecular asymmetry on the degeneracy of the energy levels of prolate and oblate symmetric rotors

three rotational constants,

$$H = AP_A^2 + BP_B^2 + CP_C^2 \qquad (6.10)$$

where the rotational constants are redefined, such that $A = h^2/8\pi^2 I_A$ etc., and the angular momenta are expressed in units of $\hbar$ ($h/2\pi$). This equation can be rearranged, making use of equation (6.8) to give

$$H = \frac{(A + C)P^2}{2} + \frac{(A - C)H_{(\kappa)}}{2} \qquad (6.11)$$

where

$$P^2 = P_A^2 + P_B^2 + P_C^2$$

and

$$H_{(\kappa)} = P_A^2 + \kappa P_B^2 - P_C^2$$

With this Hamiltonian the energy of an asymmetric rotor is given by

$$E = \frac{(A + C)}{2}J(J + 1) + \frac{(A - C)}{2}E_{(\kappa)} \qquad (6.12)$$

The term $E_{(\kappa)}$ represents the energy of a hypothetical molecule with inertial constants $+1$, κ and -1, and as such can be calculated from the matrix elements of the full rotational Hamiltonian. Solution of the $E_{(\kappa)}$ energy matrix is further facilitated by application of the symmetry properties of the rotational energy wave-functions. The inertial ellipsoid of a rigid asymmetric rotor behaves with symmetry properties of the 'four group' ($V_{(a,b,c)}$), i.e. the Hamiltonian is invariant under the four operations, identity and the rotation of 180° about three mutually perpendicular axes, and this enables rotational wave-functions to be classified into four species of different symmetry. This results in the eventual factorization of the energy matrix for any given value of J (originally of order $2J + 1$) into four sub-matrices of approximate order $J/2$. Tables of values of $E_{(\kappa)}$ for various intervals of κ are available in the literature,[2,28–30] and there exist several computer programmes which can calculate all the rotational energy levels of any asymmetric rotor[31] (Table 6.1).

The selection rules for asymmetric rotors are fairly complex, since both J and the pseudo-quantum numbers K_{-1} and K_1 can change during a transition. For J, the selection rules are

$$\Delta J = 0, \pm 1$$

Transitions for which $\Delta J = 0$ are called Q-branches, for $\Delta J = +1$, R branches and for $\Delta J = -1$, P branches. Changes in the K_{-1}, K_1 subscripts of J arise because of the symmetry properties of asymmetric rotor wave-functions, with respect to the operations of the 'four' group. Rotation by 180° about the A axis leaves the asymmetric rotor wave-functions unchanged (e) or changed in sign (o), depending on whether the value of K_{-1} is even or odd respectively. Similarly, rotation about the C principal axis is symmetric (e) or antisymmetric (o), depending on whether the value of K_1 is even or odd. The symmetry behaviour of rotation about the B (intermediate) axis depends on whether the sum $(K_{-1} + K_1)$ is even or odd. Table 6.2 summarizes the full selection rules which are derived from the requirement that the transition moment must be non-zero,

$$\int \psi_i \boldsymbol{\mu}_j \psi_k \, d\tau \neq 0 \tag{6.13}$$

$$[\text{or } \langle J, K_{-1}, K_1 | \boldsymbol{\mu}_j | J', K'_{-1}, K'_1 \rangle \neq 0]$$

where ψ represents the asymmetric rotor wave-functions for two levels i and k, and $\boldsymbol{\mu}_j$ is the dipole moment vector ($\boldsymbol{\mu}_j$ may be $\boldsymbol{\mu}_A$, $\boldsymbol{\mu}_B$ or $\boldsymbol{\mu}_C$).

Most molecular species fall into the asymmetric rotor symmetry class, and show rich and complex spectral patterns. Prerequisite, however, for a microwave spectrum to be observed are the following properties: (a) the molecule must have a dipole moment larger than about 0·1 D, and (b) the substance must have a minimum vapour pressure of about 10^{-3} mm Hg at the temperature of the experiment.

Table 6.1 Explicit expressions for some low J rotational levels in terms of the rotational constants and the asymmetry parameter (κ)

$J_{K_{-1}K_1}$	$E(A, B, C)$	$E(\kappa)$
0_{00}	0	0
1_{10}	$A + B$	$\kappa + 1$
1_{11}	$A + C$	0
1_{01}	$B + C$	$\kappa - 1$
2_{20}	$2(A + B + C) + 2[(B - C)^2 + (A - C)(A - B)]^{\frac{1}{2}}$	$2[\kappa + (\kappa^2 + 3)^{\frac{1}{2}}]$
2_{21}	$4A + B + C$	$\kappa + 3$
2_{11}	$A + 4B + C$	4κ
2_{12}	$A + B + 4C$	$\kappa - 3$
2_{02}	$2(A + B + C) - 2[(B - C)^2 + (A - C)(A - B)]^{\frac{1}{2}}$	$2[\kappa - (\kappa^2 + 3)^{\frac{1}{2}}]$
3_{30}	$5(A + B) + 2C + 2[4(A - B)^2 + (A - C)(B - C)]^{\frac{1}{2}}$	$5\kappa + 3 + 2(4\kappa^2 - 6\kappa + 6)^{\frac{1}{2}}$
3_{31}	$5(A + C) + 2B + 2[4(A - C)^2 - (A - B)(B - C)]^{\frac{1}{2}}$	$2[\kappa + (\kappa^2 + 15)^{\frac{1}{2}}]$
3_{21}	$5(B + C) + 2A + 2[4(B - C)^2 + (A - B)(A - C)]^{\frac{1}{2}}$	$5\kappa - 3 + 2(4\kappa^2 + 6\kappa + 6)^{\frac{1}{2}}$
3_{22}	$4(A + B + C)$	4κ
3_{12}	$5(A + B) + 2C - 2[4(A - B)^2 + (A - C)(B - C)]^{\frac{1}{2}}$	$5\kappa + 3 - 2(4\kappa^2 - 6\kappa + 6)^{\frac{1}{2}}$
3_{13}	$5(A + C) + 2B - 2[4(A - C)^2 - (A - B)(B - C)]^{\frac{1}{2}}$	$2[\kappa - (\kappa^2 + 15)^{\frac{1}{2}}]$
3_{03}	$5(B + C) + 2A - 2[4(B - C)^2 + (A - B)(A - C)]^{\frac{1}{2}}$	$5\kappa - 3 - 2(4\kappa^2 + 6\kappa + 6)^{\frac{1}{2}}$
4_{41}	$5(A + B + C) + 5A + 2[4(B - C)^2 + 9(A - C)(A - B)]^{\frac{1}{2}}$	$5\kappa + 5 + 2(4\kappa^2 - 10\kappa + 22)^{\frac{1}{2}}$
4_{31}	$5(A + B + C) + 5B + 2[4(A - C)^2 - 9(A - B)(B - C)]^{\frac{1}{2}}$	$10\kappa + 2(9\kappa^2 + 7)^{\frac{1}{2}}$
4_{32}	$5(A + B + C) + 5C + 2[4(A - B)^2 + 9(A - C)(B - C)]^{\frac{1}{2}}$	$5\kappa - 5 + 2(4\kappa^2 + 10\kappa + 22)^{\frac{1}{2}}$
4_{23}	$5(A + B + C) + 5A - 2[4(B - C)^2 + 9(A - C)(A - B)]^{\frac{1}{2}}$	$5\kappa + 5 - 2(4\kappa^2 - 10\kappa + 22)^{\frac{1}{2}}$
4_{13}	$5(A + B + C) + 5B - 2[4(A - C)^2 - 9(A - B)(B - C)]^{\frac{1}{2}}$	$10\kappa - 2(9\kappa^2 + 7)^{\frac{1}{2}}$
4_{14}	$5(A + B + C) + 5C - 2[4(A - B)^2 + 9(A - C)(B - C)]^{\frac{1}{2}}$	$5\kappa - 5 - 2(4\kappa^2 + 10\kappa + 22)^{\frac{1}{2}}$

Table 6.2   Selection rules for rotational spectra of asymmetric rotors

Dipole component	Permitted changes in K_{-1} and K_1 subscripts during transitions $\Delta J_{K_{-1}K_1} = 0$ and $\Delta J_{K_{-1}K_1} = \pm 1$
μ_A (along axis of smallest moment of inertia)	even, even $\leftrightarrow$ even, odd odd, even $\leftrightarrow$ odd, odd
μ_B (along axis of intermediate moment of inertia)	even, even $\leftrightarrow$ odd, odd odd, even $\leftrightarrow$ even, odd
μ_C (along axis of largest moment of inertia)	even, even $\leftrightarrow$ odd, even even, odd $\leftrightarrow$ odd, odd

The following generalities may in some instances be relevant in assessing the feasibility of successful analysis of a microwave spectrum.

(a) The complexity of a spectrum increases with increasing number of atoms with nuclear spin quantum number $I > \frac{1}{2}$ in a molecule (e.g. Cl, Br, I, N). This may be especially relevant when spectral splittings due to internal rotation are of interest.

(b) Large molecules (e.g. with > 20–30 atoms) tend to have strong high J transitions (often unassignable to specific J values) falling in the most useful microwave region.

(c) Molecules with a large number of methyl groups in different local environments may show very complex patterns, due to internal rotation of the methyl tops.

(d) Substances which react readily on metal surfaces may decompose when using conventional copper wave-guides.

6.5 The effect of molecular vibrations on rotational spectra

No molecule behaves exactly like a rigid rotor, and 'non-rigidity' effects can be observed to some extent in all rotational spectra. Deviations from rigid rotor spectra are observed in the rotational lines of molecules in their ground vibrational states at high J levels, while quite separate spectral patterns can be found for molecules in different vibrational states. The rotational constants of a molecule in any vibrational state (e.g. ground state where $v = 0$) can be related to the equilibrium values as shown.

$$A_v = A_e - \sum_i \alpha_i^a(v_i + \tfrac{1}{2})$$

$$B_v = B_e = \sum_i \alpha_i^b(v_i + \tfrac{1}{2}) \tag{6.14}$$

$$C_v = C_e - \sum_i \alpha_i^c(v_i + \tfrac{1}{2})$$

where v_i represents the vibrational quantum number of the ith vibration, and the coefficients α_i are the vibration–rotation interaction constants. The equilibrium configuration corresponds to the hypothetical 'non-vibrating' position as represented by the minimum of the potential energy curve for a bond. Since the rotational constants vary slightly from one vibrational state to another, so the spectral lines of each vibrational state appear as satellites slightly displaced from the corresponding ground state lines, and often considerably weaker in intensity. When one vibrational satellite spectrum has been correctly identified, it is usually fairly easy to extrapolate and find the lines corresponding to higher states of that vibration, provided that the line intensities are not too low (Figure 6.6), and that the

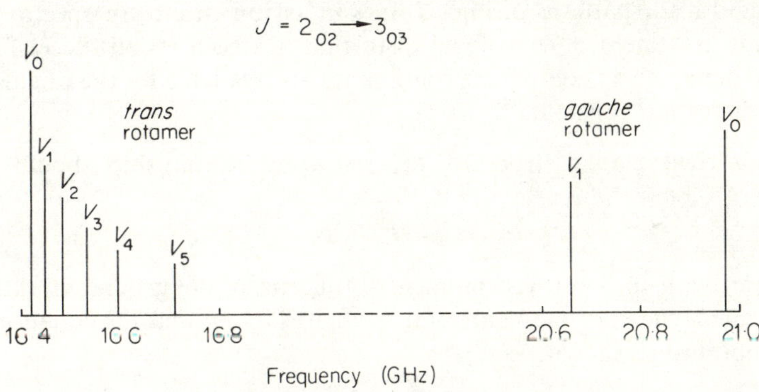

Figure 6.6   The relative intensities of the $2_{02} \rightarrow 3_{03}$ transition for the ground states and for the excited states of the $O\!\!\rightarrow\!\!CH_2CH_3$ torsion for the two rotamers of ethyl formate. (Redrawn from Reference 32 by permission of the authors)

vibration is reasonably harmonic. The intensity of the satellite lines as compared to that of the ground state lines decreases exponentially with increasing vibrational energy as given by the Boltzmann equation

$$\frac{I_i}{I_0} = \frac{N_i}{N_0} = \frac{g_i}{g_0}\exp\left(-\frac{hv_i}{kT}\right) \tag{6.15}$$

where g_i represents the degeneracy of the ith vibrational state, and N_0 and N_i denote the number of molecules in the ground and ith states respectively.

Observation of rotational spectra is often facilitated by cooling the absorption cell (solid CO_2 is a commonly used and convenient refrigerant), and vibrational satellite lines, although somewhat weaker in intensity, are often sharper at low temperatures. Careful intensity studies can lead to deductions regarding the shape of the potential energy well. This is of particular interest when the vibration is a torsional mode, and several investigators have mapped out the shapes of potential energy curves to within a few quantum levels of the barrier maximum.[33, 34]

The 'non-rigidity' effects which may affect the rotational spectrum of the ground vibrational state of a molecule can be considered to be composed of three basic contributions.

(a) Cubic and other higher terms in the expression describing the potential energy curve.

(b) Coriolis forces which are responsible for the coupling of rotational and vibrational energies when degenerate or near degenerate vibrations occur.

(c) Centrifugal distortion which is mainly responsible for the deviation from rigid rotor patterns of high J lines in asymmetric rotor spectra. Perturbation treatment of centrifugal distortion has been found successful for most systems,[35] and computer programmes are available for the calculation of five distortion constants.[36]

For a rigid planar structure, the following relationship should hold exactly:

$$I_C = I_A + I_B \qquad (6.16)$$

However, with the observed moments of inertia of the ground vibrational state (called 'effective' moments of inertia), it is found that the equation has to be modified as shown:

$$I_C - I_A - I_B = \Delta \qquad (6.17)$$

Δ is called the inertial defect, and may have positive or negative values. (Typically, Δ will have values ranging from 0.01–0.1% of the molecular moments of inertia). The non-zero value of the inertial defect arises because of the fact that molecules are not perfectly rigid structures.

The vibration–rotation interaction forces which contribute to the inertial defect can be expressed as $\Delta = \Delta$ (anharmonic terms) + Δ (harmonic terms). For planar molecules the term Δ (anharmonic) is zero, and the harmonic term may be considered to be composed of contributions from 'in-plane' and 'out-of-plane' vibrations

$$\Delta = \Delta_{(i)} + \Delta_{(o)}$$

In general $\Delta_{(i)}$ contributes in a positive sense while $\Delta_{(o)}$ is predominantly negative. For the ground vibrational state of a molecule Δ usually has a small positive value, while for low-lying 'out-of-plane' vibrations it becomes negative, and for in-plane vibrations it becomes more positive. Thus it is often possible to use the value of the inertial defect found for a vibrational satellite to help in the assignment of that vibration. This procedure can be applied to molecules with planes of symmetry if the inertial contributions of the 'out-of-plane' atoms are properly taken into account. For example, for a molecule with a methyl group where two of the hydrogens are out of the molecular plane,

$$I_C - I_A - I_B + 4mx^2 = \Delta'$$

where m represents the mass of the out-of-plane hydrogen atom, x is the atom coordinate in the direction at right angles to the plane and Δ' is a 'pseudo' inertial defect.

6.6 Microwave methods of studying internal rotation

The principal microwave method of studying internal rotation in molecules involves detailed analysis of the resolvable fine structure on the 'rigid-rotor' rotational lines. The internal molecular motion associated with, for example, a methyl group interacts with the overall rotation and the coupling manifests itself as line splittings, which are themselves a function of the potential barrier opposing internal rotation. In this way accurate values of barriers to internal rotation of methyl and other symmetric groups have been found. The method assumes a model comprising a rigid framework attached to a symmetric rotating group having one internal degree of freedom (i.e. the torsional mode is considered separable from the other molecular vibrations). This approach has been found to be successful for three-fold methyl barriers in the range 500–5000 cal/mole, six-fold barriers (with values ranging upwards from a few cal/mole), molecules with more than one methyl group, 'heavy' symmetric groups such as CF_3 and non carbon-containing groups such as SiH_3. Potential barriers may also be calculated from splittings observed in excited vibrational states. For very precise work the moment of inertia of the rotating group and its orientation within the molecular framework must be accurately known. Unfortunately this method is not generally applicable to symmetric top molecules since the interaction terms merely shift all the rotational energy levels by equal amounts.

The most generally applicable method, but one which is much less precise than the former, is that based on relative intensity measurements. The ratio of intensities, at a particular temperature, of the same rotational transition in two different torsional states of a molecule, provides a value for the energy difference between the torsional states. This, in turn, can be related to the potential barrier opposing the torsional motion. In principle, this method may be applied whenever the appropriate rotational lines have been correctly assigned to particular torsional states, but is particularly useful for 'high' barriers ($V_3 > 4000$ cal/mole). It is, of course, related to the more direct infrared practice of calculating barriers from observation of low frequency torsional fundamentals. The same technique enables the energy difference between different rotational isomers to be measured, provided that the rotational spectra of the two separate rotamers can be assigned, and that the dipole moments of the two rotamers are known or can be measured (see § 6.13).

Other methods for calculating barriers in molecules, based on 'non-rigidity' effects, have been suggested and successfully applied in certain instances. The observed changes in rotational constants between the ground and first excited torsional state of a molecule can be related not only to centrifugal distortion and Coriolis forces, but to the potential barrier opposing the torsional motion.[37] For symmetric tops, it has been shown that Coriolis coupling forces in vibrationally excited degenerate states may permit, in some instances, the direct measurement of internal rotation splittings on rotational lines.[38]

By the observation and measurement of intensities of successive satellite lines of a torsional vibration it is possible to map out the potential energy well for the torsion. This has proved particularly interesting when carried out for two rotamers of the same molecule, each at a different minimum (see Figure 6.14).

The phenomenon of inversion in molecules is closely related to that of internal rotation, and inversion can interact with the overall molecular rotation in a manner similar to that found for internal rotation, to give splittings in the pure rotation spectrum. Barriers to inversion have been calculated from microwave line splittings for many substances, and recently the related forces controlling the puckering of small rings have also been studied and analysed through their perturbing effects on rotational spectral lines.

6.7 The potential energy associated with internal rotation

The potential energy opposing internal rotation about single bonds within molecules is a periodic function of α, the angle of rotation. This is illustrated by the fact that in monosubstituted ethanes (CH_3CH_2X) there are found three identical energetically favoured orientations. Such a potential may be expressed in a general form as an expansion of a Fourier series

$$V(\alpha) = \sum_{i=1}^{\infty} a_{iN} \cos iN\alpha + b_{iN} \sin iN\alpha \qquad (6.18)$$

where N represents the number of potential maxima or minima generated in a 360° revolution. For the case of a symmetric barrier where the potential energy is an even function of α:

$$V(\alpha) = \sum_i a_{iN} \cos iN\alpha \qquad (6.19)$$

If this function is expanded we have

$$V(\alpha) = a_N \cos N\alpha + a_{2N} \cos 2N\alpha + a_{3N} \cos 3N\alpha + \cdots \qquad (6.20)$$

Most of the molecules studied by microwave spectroscopy have a three-fold internal rotation barrier ($N = 3$) and thus

$$V(\alpha) = a_3 \cos 3\alpha + a_6 \cos 6\alpha + a_9 \cos 9\alpha + \cdots \qquad (6.21)$$

It has been found (see § 6.10) that to a good approximation the potential is adequately described by the first term only in the above expression.

$$V(\alpha) = a_3 \cos 3\alpha \qquad (6.22)$$

If the frame of reference is chosen to coincide with that shown in Figure 6.7 then the potential may be written as

$$V(\alpha) = \frac{V_3}{2}(1 - \cos 3\alpha) \qquad (6.23)$$

where V_3 represents the maximum value of the potential barrier. For a three-fold barrier, minima occur at $\alpha = 60°$, $180°$ and $300°$. The corresponding

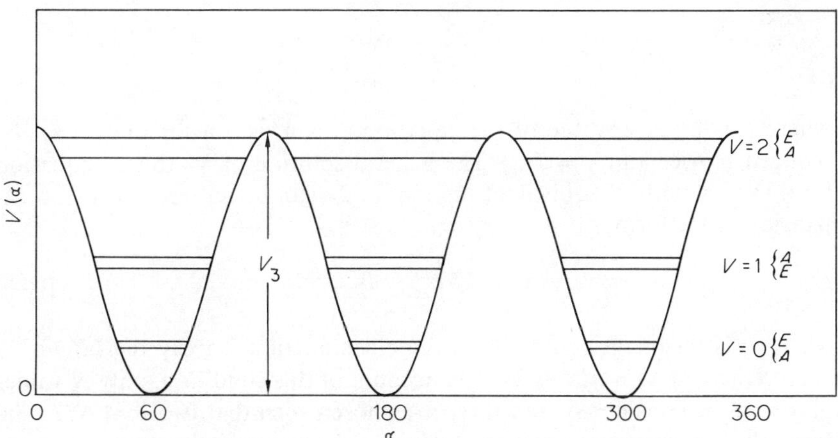

Figure 6.7   A cosine potential energy curve and torsional energy levels for hindered internal rotation with a three-fold barrier

expression for a two-fold potential barrier can be derived in a similar manner:

$$V(\alpha) = \frac{V_2}{2}(1 - \cos 2\alpha) \tag{6.24}$$

If this general expression for the potential energy is substituted into the Schödinger wave-equation for one dimension (α) we have

$$F\frac{d^2\psi(\alpha)}{d\alpha^2} + \left[E - \frac{V_N}{2}(1 - \cos N\alpha)\right]\psi(\alpha) = 0 \tag{6.25}$$

where $F = h^2/8\pi^2 I_r$, and I_r is the reduced moment of inertia for the relative motion of the two groups. $I_r = I_\alpha(1 - \sum_i \lambda_i^2 I_\alpha/I_i)$ for a three-fold symmetric group such as a methyl group, where I_α is the moment of inertia of the methyl top about its symmetry axis and λ_i represent the directional cosines corresponding to the angles which the top makes with the three principal axes I_i. This expression has a very similar form to the Mathieu equation:

$$\frac{d^2y}{dx^2} + (b - s\cos^2 x)y = 0 \tag{6.26}$$

and the two equations can be transformed into one another by means of the substitutions

$$\psi(\alpha) = y$$

$$N\alpha + \pi = 2x$$

$$V_N = \frac{N^2}{4}Fs \tag{6.27}$$

$$E = \frac{N^2}{4}Fb$$

where b is an eigenvalue of the equation, s is a parameter known as the reduced barrier and $y = f(x)$. The general solution of Mathieu's equation for a system with N equivalent minima in 2π can be represented by an expansion of the form

$$y = e^{i\sigma\alpha}\sum_{k=-\infty}^{\infty} B_k e^{iNk\alpha} \tag{6.28}$$

where σ is an integer, and where the eigenfunctions satisfy the boundary requirements $y(x) = y(x + 2\pi)$. On account of this condition, only N values of σ need be considered, and these are chosen such that $0 \leqslant \sigma \leqslant N/2$. The two solutions corresponding to $\pm\sigma$ have the same periodicity in α, and are associated with a degenerate eigenvalue, while $\sigma = 0$ gives rise to a non-degenerate eigenvalue. For $N = 3$, there exist three solutions with $\sigma = 0$,

$\sigma = \pm 1$. The eigenfunctions and eigenvalues of the torsional wave-equation may be evaluated by numerical iterative methods, and may be obtained from published tables of Mathieu functions.[39,40] The eigenvalues $E_{v\sigma}$ (or $b_{v\sigma}$) are subscripted using v, the torsional quantum number and the parameter σ, which defines the symmetry of the torsional wave-functions. Consideration of the symmetry properties of the torsional wave-functions with respect to the symmetry group C_3 shows that when $\sigma = 0$, the eigenvalues correspond to the non-degenerate A levels, and when $\sigma = \pm 1$, they correspond to the doubly degenerate E levels.

By considering both periodic and non-periodic solutions in α, it is possible to express the eigenvalues as a Fourier series of terms:

$$E_{v\sigma} = F \frac{N^2}{4} \sum_{l=0}^{\infty} w_l \cos l\phi \tag{6.29}$$

where ϕ is a function of σ. The coefficients w_l depend only on v and s, and for large values of s, the expression converges rapidly, and only the first few terms are of any importance. These coefficients are found in tabulated form in the literature.[41] Having considered the general case for a V_3 barrier, it may be instructive to consider the two extreme limits.

(a) $V_3 \rightarrow 0$. For a very low barrier, equation (6.25) reduces to:

$$F \frac{d^2\psi(\alpha)}{d\alpha^2} + E\psi(\alpha) = 0 \tag{6.30}$$

A general solution of this type of differential equation is given by

$$\psi(\alpha) = A\,e^{im\alpha} \qquad [\text{or } \psi(\alpha) = A(\cos m\alpha - i \sin m\alpha)]$$

where m is an integer whose values are restricted to $0, \pm 1, \pm 2, \cdots$ by the boundary condition $\psi(\alpha) = \psi(\alpha + 2\pi)$. The eigenvalues of equation (6.30) are given by

$$E = Fm^2 \tag{6.31}$$

The two values $\pm m$, associated with each solution, correspond to clockwise and anticlockwise directions of the free internal rotation (Figure 6.8).

(b) $V_3 \rightarrow \infty$.

For infinitely high barriers, the internal rotation is confined to small torsional oscillations about the potential minima. Equation (6.23) may be rewritten as:

$$V(\alpha) = V_3 \sin^2 \frac{3\alpha}{2} \tag{6.32}$$

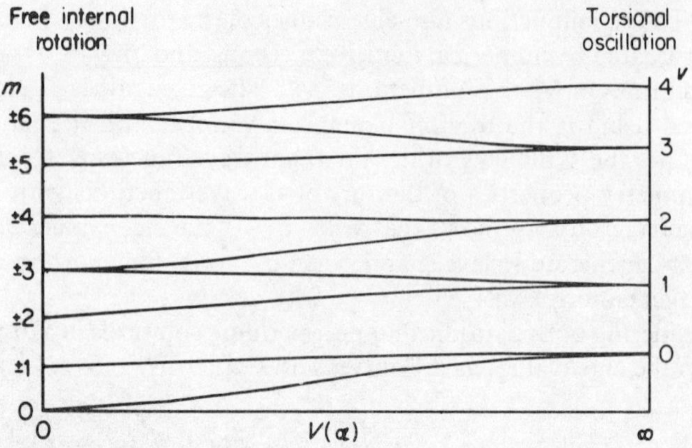

Figure 6.8   The effect of a potential barrier to internal rotation
on the energy levels of a free internal rotor, and the correlation
between m, the quantum number for free rotation, and v, the
harmonic oscillator quantum number

For low amplitude oscillations, $\sin \alpha = \alpha$ (in radians), and so

$$V(\alpha) = V_3\left(\frac{3\alpha}{2}\right)^2 \tag{6.33}$$

Substituting this value of $V(\alpha)$ into equation (6.25) gives

$$F\frac{\mathrm{d}^2\psi(\alpha)}{\mathrm{d}\alpha^2} + \left[E - \frac{9\alpha^2}{4}V_3\right]\psi(\alpha) = 0 \tag{6.34}$$

This is similar to the expression for the harmonic oscillator, and the corresponding solutions are

$$E = 3(V_3 F)^{\frac{1}{2}}(v + \tfrac{1}{2}) \tag{6.35}$$

where $v = 0, 1, 2, \cdots$ etc. The torsional frequency for harmonic oscillations may be written as

$$v = \frac{1}{2\pi}(f/I_\mathrm{r})^{\frac{1}{2}} \tag{6.36}$$

where f (force constant) $= 2V(\alpha)/\alpha^2$. Thus

$$v = \frac{3}{2\pi}\left(\frac{V_3}{2I_\mathrm{r}}\right)^{\frac{1}{2}} \tag{6.37}$$

The potential barrier (in the high barrier harmonic oscillator approximation) may therefore be found from the torsional frequency.

The effect of hindered internal rotation on the overall rotation spectrum of a molecule may be observed as splittings in the microwave spectrum, because the two motions interact with each other. The coupling of energy differs for the two torsional sub-levels $\sigma = 0$ (A), and $\sigma = \pm 1$ (E), and so rotational lines may be split into A and E components, and the separation is a sensitive function of the barrier height.

In some instances, the microwave line splittings in the ground vibrational state may be so large as to make their detection and identification as doublets very difficult. One convenient method which often helps to overcome this problem in the case of methyl groups is to study the deuterated methyl top. The heavier CD_3 group with its larger moment of inertia will result in much smaller splittings. (The torsional energy levels within the potential minima of the internal rotation barrier will be lower, owing to the heavier mass of deuterium, and the barrier to internal rotation will 'appear' to be higher.)

The selection rules governing the appearance of doublets for many of the rotational transitions of an asymmetric molecule with a single internal rotor may be summarized by the expressions:

$$A \leftrightarrow A; \qquad E \leftrightarrow E; \qquad A \not\leftrightarrow E.$$

These rules, which state that only energy levels of the same degeneracy can be connected by quantum transitions, can conveniently be derived using group theoretical arguments.[16] In a similar manner, analysis of the symmetry properties of a methyl group, together with the symmetry properties of nuclear spin functions leads to the conclusion that for protons, the A and E levels should be of equal intensity. On the other hand, a CD_3 group produces A and E doublets with intensity ratio $11:16$ respectively.[42] This fact can help to identify the A and E components for a CD_3 group.

6.8 Solution of the internal rotation Hamiltonian

The molecular model most commonly used for expressing the Hamiltonian energy is that of a rigid molecular framework, which in the general case is asymmetric, with a symmetric rotating top attached to it. (The more complex problem of an asymmetric rotating top is discussed in § 6.15.) The top (for example a methyl group) has one degree of freedom—that of rotation about the bond linking it with the rest of the molecule. The assumption is also made that this hindered rotation may be considered separately from any other internal degree of freedom within the molecule. This is a reasonable assumption, provided the torsion has a much lower frequency than any other mode.

Two different approaches have been utilized to solve the problem of internal rotation, the Principal Axis and the Internal Axis methods. The main points of distinction between the two methods arise from the different axes chosen to describe the molecular system. The Principal Axis method

(PAM) was developed by Wilson,[43,44] Crawford[45] and later Herschbach[46-49] and others, while the Internal Axis method (IAM) was originally formulated by Nielsen[50] and Dennison,[51] and later developed by Woods.[52] For the majority of problems, the two methods lead to the same solution for the internal rotation barrier, although it is probably true to say that the PAM approach is the simplest and quickest for most molecules. The IAM method involves a computational solution, but is superior for examples where the 'framework' is light compared with the 'top'.

6.8.1 *Principal axis method*

The coordinate system is defined in terms of the principal axes of the molecule, while the internal rotation axis coincides with the symmetry axis of the top. (This axis is not in general coincident with any one of the principal axes.) For a symmetric top molecule with a symmetric rotating group attached, there are no observable first-order internal rotation effects on the overall rotational spectrum. For an asymmetric rotor with a symmetric group attached, the kinetic energy may be expressed classically as

$$2T = \sum_i I_i \omega_i^2 + I_\alpha \dot{\alpha}^2 + 2I_\alpha \dot{\alpha} \sum_i \lambda_i \omega_i \tag{6.38}$$

(overall   (internal   (coupling between
rotation)   rotation   internal and overall
            of top)   rotation)

where $i = A$, B and C, and I_A, I_B and I_C correspond to the three principal moments of inertia, I_α is the moment of inertia of the symmetric top about its symmetry axis, λ_i is the directional cosine corresponding to the angle which the top makes with the ith principal axis, ω_i represents the component of the angular velocity of the framework with respect to each principal axis and $\dot{\alpha}$ is the angular velocity of the top relative to the rest of the molecule. If the kinetic energy is expressed in terms of angular momenta, then the Hamiltonian may be written in a very convenient form

$$H = H_r + F(p - \mathscr{P})^2 + V(\alpha) \tag{6.39}$$

where $V(\alpha) = V_N/2(1 - \cos N\alpha)$, H_r is the rigid rotor Hamiltonian ($H_r = AP_A^2 + BP_B^2 + CP_C^2$), $F = h^2/8\pi^2 r I_\alpha$, p is the total angular momentum of the top about its axis, $\mathscr{P} = \sum_i P_i \lambda_i I_\alpha/I_i$ and $r = \sum_i \lambda_i^2 (I_\alpha/I_i)$. (The term $p - \mathscr{P}$ represents the relative angular momentum of the top and framework.)

It is also convenient to consider the total Hamiltonian to be constructed as follows:

$$H = H_R + H_I + H_{IR} \tag{6.40}$$

where H_R represents the rotational contribution, H_I that of the pure internal rotation and H_{IR} the coupling terms between overall and internal rotation.

In terms of equation (6.39), we have

$$H_R = H_r + F\mathscr{P}^2$$

$$H_I = Fp^2 + V(\alpha) \tag{6.41}$$

$$H_{IR} = -2Fp\mathscr{P}$$

The term $F\mathscr{P}^2$ can be incorporated into the rigid rotor Hamiltonian without difficulty by slightly modifying the rotational constants. The expression $Fp^2 + V(\alpha)$ is identical with the form of the Hamiltonian used in equation (6.25), except that the classical symbol for angular momentum has been replaced by its quantum-mechanical equivalent $[p = -i(\partial/\partial\alpha)]$. The internal and overall rotational motions could be considered as completely separate, were it not for the presence of the interaction term $-2Fp\mathscr{P}$.

The interaction terms (H_{IR}) may be treated, for relatively high barriers, as a perturbation on the unperturbed Hamiltonian ($H_R + H_I$). H_{IR} contributes off-diagonal elements in v (the torsional quantum number) to the energy matrix, but these can be reduced to second-order terms by carrying out an operation known as the Van Vleck transformation.[53] Solution of the energy matrix is thus reduced to solving a series of smaller sub-matrices for each torsional state (v). The effective Hamiltonian for each state (v) may be written as

$$H_{v\sigma} = H_r + F \sum_n W_{v\sigma}^{(n)} \mathscr{P}^n \tag{6.42}$$

where $W_{v\sigma}^{(n)}$ represent dimensionless perturbation coefficients, which are functions only of (V_N/F). These may be expanded as a power series in the reduced angular momentum ($\mathscr{P}$) (compare equation 6.29), and are readily available in tabular form.[40,41] The zeroth order term ($n = 0$) represents the pure torsional energy

$$W_{v\sigma}^{(0)} = E_{v\sigma}/F \tag{6.43}$$

where $E_{v\sigma}$ is an eigenvalue of the Mathieu equation (cf. equation 6.27). When $V_N \to \infty$, only the $n = 0$ term need be considered, and the problem reduces to that of the harmonic oscillator. The first term to contribute to the coupling is $W_{v\sigma}^{(1)}$ and for many molecules only terms up to $n = 2$ have to be included. Owing to symmetry requirements, all terms with $n = 1$ are zero for A levels ($\sigma = 0$), and in many cases (e.g. for fairly high barriers, $V_3 > 1000$ cal/mole) they are not significant for the E levels either. (Where the $n = 1$ terms do contribute, the E levels deviate somewhat from a rigid rotor pattern.) The $W_{v\sigma}^{(2)}$ terms differ in magnitude and often in sign for the A and E sub-levels of a given v state, and so the effective Hamiltonians H_{vA} and H_{vE} give rise to slightly different rotational spectral patterns, which in turn can be observed in the microwave spectrum. Herschbach has shown that these perturbation coefficients may be calculated by the 'bootstrap' method in a more efficient manner than using perturbation theory.[41]

The pseudo rigid-rotor Hamiltonians for the A and E levels may be expressed as:

$$H_{vA} = A_{vA}P_x^2 + B_{vA}P_y^2 + C_{vA}P_z^2$$

$$H_{vE} = A_{vE}P_x^2 + B_{vE}P_y^2 + C_{vE}P_z^2 \tag{6.44}$$

where the rotational constants are of the form

$$A_{vA} = \frac{h^2}{8\pi^2 I_x} + F\rho_x^2 W_{vA}^2 \tag{6.45}$$

where $\rho_i = \lambda_i I_\alpha / I_i$.

A value for the potential barrier (V_3) may be obtained directly from the difference between the A and E rotational constants, for example

$$\Delta A = A_{vA} - A_{vE}W_{vA}^{(2)}$$

$$= F\rho_x^2[W_{vA}^{(2)} - W_{vE}^{(2)}] \tag{6.46}$$

Values of $\Delta W_v^{(2)}$ (or $\Delta \eta_{v\sigma}$ where $\eta_{v\sigma}$ represents the coefficients of the quadratic terms in $\mathscr{P}$ in the expansion for $W_{v\sigma}$) are tabulated for various ranges of s (the reduced barrier),[40,41] and thus the corresponding value of V_3 may be found. (Approximate formulae relating the various coefficients to s may be found in Reference 16).

When the differences in rotational constants are very small, a value for the potential barrier may be obtained by measuring the separation between individual transitions. The splitting of the energy levels can be expressed as functions of the rotational constants, as follows:

$$\Delta E = E_A - E_E$$

$$= \left(\frac{\partial E}{\partial A}\right)\Delta A + \left(\frac{\partial E}{\partial B}\right)\Delta B + \left(\frac{\partial E}{\partial C}\right)\Delta C \tag{6.47}$$

For a rigid rotor, however,

$$E = \frac{A + C}{2}J(J + 1) + \frac{A - C}{2}E_{(\kappa)}$$

and thus

$$\left(\frac{\partial E}{\partial A}\right) = \frac{1}{2}\left[J(J + 1) + E_{(\kappa)} - (\kappa + 1)\frac{\partial E_{(\kappa)}}{\partial \kappa}\right]$$

$$\left(\frac{\partial E}{\partial B}\right) = \frac{\partial E_{(\kappa)}}{\partial \kappa} \tag{6.48}$$

$$\left(\frac{\partial E}{\partial C}\right) = \frac{1}{2}\left[J(J + 1) - E_{(\kappa)} + (\kappa - 1)\frac{\partial E_{(\kappa)}}{\partial \kappa}\right]$$

The derivative of the reduced energy $(\partial E_{(\kappa)}/\partial \kappa)$ can be determined from published tables of $E_{(\kappa)}$.[2,28–30] For low J lines with explicit expressions for $E_{(\kappa)}$ (Table 6.1) $\partial E_{(\kappa)}/\partial \kappa$ may be obtained very simply.

6.8.2 *Internal axis method*

The basic differences and similarities in the IAM approach to that of the PAM are outlined in some detail in Lin and Swalen's review,[16] and in other papers.[4,5] The effective Hamiltonian for an asymmetric top molecule is generally more complex than the corresponding expression for the PAM method since one of the coordinate axes is chosen to coincide with the axis of the internal rotor, and thus the three coordinate axes are not in general principal axes. In a recent treatment, however, Woods has shown how the Hamiltonian for any molecule, including those with asymmetric frameworks, can be obtained and expressed in a relatively simple form.[52]

$$H = H_r + \frac{1}{2rI_\alpha}(p - \rho P)^2 + V(\alpha) \tag{6.49}$$

where H_r represents the usual rigid rotor Hamiltonian, $r = 1 - \sum_i \lambda_i^2 I_\alpha/I_i$, $p = I_\alpha \dot{\alpha} + I_\alpha \sum_i \lambda_i \omega_i$, $\dot{\alpha}$ and ω represent the angular velocities of the top relative to the frame, and of overall rotation respectively, P is the total angular momentum and ρ represents $(\sum_i \rho_i^2)^{\frac{1}{2}}$ where $\rho_i = \lambda_i I_\alpha/I_i$.

The parameter ρ defines the direction of an axis, fixed in the framework of the molecule, which is a fundamental one for treating internal rotation by this method. The ρ-axis connects the centre of mass of the internal rotor to the centre of mass of the whole molecule.

Owing to the choice of axes, the coupling terms between the internal rotation and the overall rotation in the IAM treatment are considerably smaller than those found for the PAM method. Solution of the torsional wave-function for the IAM method, however, is complicated by the less simple boundary conditions of the torsional wave-equation. (The parameter σ which has integral values in the PAM treatment is not, in general, integral in the IAM method, but σ may be replaced by $\sigma' = \sigma + KI_\alpha/I_z$). The torsional wave-functions have a dependence on K, the rotational quantum number. By the inclusion of non-periodic Mathieu function solutions, however, the torsional eigenvalues may be expressed as a Fourier series of terms (equation 6.29), which converges very rapidly for fairly high barriers (i.e. when $s > 18$ for the ground vibrational state $v = 0$).[41]

Expressed in the internal axis coordinate system, the total Hamiltonian may be considered to be composed of two separable components, the rigid part and the internal rotation part. (The cross-terms are very small, and do not affect the energy differences between the A and E levels). The calculation of predicted line splittings, using the IAM method, is of necessity a computerized procedure, and consists basically of obtaining the eigenvalues of the torsional wave-function, and incorporating the results with the solution of the rigid part of the Hamiltonian. The main advantage of the IAM procedure over the PAM method is the fact that the former is equally applicable to molecules with 'heavy' rotating groups, as well as to molecules having a heavy framework with a light top.

6.9 Calculation of potential barriers—an example

Methyl fluoroformate ($FCOOCH_3$) represents a typical example taken from a large number of compounds which have been studied in the microwave region, but which have not been subjected to detailed isotopic studies, with the result that the molecular structure is not known with great precision. Provided that the approximate molecular structure is known, however, the potential barrier to internal rotation of the methyl group in such molecules can still be calculated to give meaningful results. The microwave spectrum of methyl fluoroformate was found to be consistent only with a planar heavy atom skeleton, and the most probable molecular conformation was found to have the methyl group *cis* to the carbonyl bond.[54] The approximate molecular structure was derived by fitting the experimentally measured rotational constants to those predicted for models based on accurately known bond parameters found for related molecules (e.g. methyl formate).[55]

The observed microwave spectrum comprises μ_A and μ_B lines (transitions brought about by the components of the overall dipole moment along the A and B principal axes respectively), many of which are split into doublets a few MHz apart by the presence of the methyl group. The component lines of each doublet fit into a rigid rotor spectral pattern, and thus two sets of rotational constants can be derived (Table 6.3).

For the calculation, which uses the PAM method, a three-fold barrier is assumed adequate to describe the internal rotation potential of the methyl group, and the assumption is also made that the 'high barrier' approximations are valid.

From equation (6.46)

$$\Delta A = A_A - A_E$$

$$= F\rho_A^2 [W_{vA}^{(2)} - W_{vE}^{(2)}]$$

For methyl fluoroformate, $\Delta A = 5.05\,\text{MHz}$. All observations were carried out on the ground vibrational state, so that $v = 0$. Thus $[W_{0A}^{(2)} - W_{0E}^{(2)}] = 7.21 \times 10^{-3}$.

From a graphical extrapolation of Herschbach's tables,[41] this corresponds to a value of s (the reduced barrier) of 29.6. Alternatively, the following

Table 6.3 Rotational constants and barrier parameters for methyl fluoroformate

	A(MHz)	B(MHz)	C(MHz)
'A' species	11057·56	4397·31	3208·54
'E' species	11052·51	4397·18	3208·54
Rigid rotor (structural)[a]	11054·19	4397·22	3208·54

[b] I_α (moment of inertia of methyl group) $= 3\cdot15$ amu A^2

[c] θ (angle between the methyl group axis and the A principal axis) $= 22°$

λ_A (directional cosine along A axis) $= 0\cdot9272$

λ_B (directional cosine along B axis) $= 0\cdot3746$

$\lambda_C = 0\cdot0$

$$\rho_A = \frac{\lambda_A I_\alpha}{I_A}$$

$$= 0\cdot06386$$

$$\rho_B = 0\cdot01026$$

[d] $$r = 1 - \frac{\lambda_A^2 I_\alpha}{I_A} - \frac{\lambda_B^2 I_\alpha}{I_B} - \frac{\lambda_C^2 I_\alpha}{I_C}$$

$$= 0\cdot9370$$

[e] $$F = \frac{h}{8\pi^2 r I_\alpha}$$

$$= \frac{505\cdot377}{r I_\alpha}$$

$$= 171\cdot234 \text{ (GHz)}$$

[a] A (rigid rotor 'structural') $= \frac{1}{3}(A_A + 2A_E)$ etc.

[b] The value used for I_α in the literature varies between $3\cdot10$ and $3\cdot20$ amu A^2. A mean value of $3\cdot15$ was used for this calculation.

[c] This value was taken from Table 8 of Reference 54.

[d] Rigid rotor 'structural' constants were used to calculate r.

[e] $505\cdot377$ represents the most recently accepted conversion factor between amu A^2 and GHz.

formulae (from Lin and Swalen) may be used to find s:

$$\log_{10}(W_{OA}^{(2)} - W_{OE}^{(2)}) = 2\cdot07951 + 1\cdot72878 \log_{10} s - 1\cdot75576\sqrt{s}$$

This gives the same value for the reduced barrier.

The three-fold potential is related to the reduced barrier by the following formula:

$$V_3 = \frac{Fs}{4\cdot660} \tag{6.50}$$

where V_3 is expressed in cal/mole and F is in units of GHz (10^9 Hz).

$$V_3 = 1084 \text{ cal/mole}$$

This compares with the quoted value of 1077 ± 30 cal/mole which was obtained using an IAM computer programme.[54] The input data for the IAM calculation included 23 measured A and E splittings, and the uncertainty associated with the final value comes mainly from uncertainties in the values of I_α and of θ, the angle between the methyl group and the A principal axis.

In spite of these uncertainties, the calculated value for V_3 in methyl fluoroformate is found to be significantly different from that found for related substituted methyl formates.[54]

Where the differences in the rotational constants (ΔA etc.) are very small, a value for V_3 may be obtained directly from the splittings of individual lines, using the formulae given in equations (6.47) and (6.48).

6.10 Higher order terms in the potential energy

The majority of barriers for methyl groups have been calculated using the assumption that in the expansion of $V(\alpha)$ (equation 6.20) all terms higher than V_3 can be neglected or assumed zero. The undoubted success of the truncated expression in correlating predicted splitting patterns with experimental findings lends support to this assumption. Values for V_6 terms, however, have been obtained for a limited number of molecules, including some for which the molecular symmetry precludes any three-fold potential (for example p-fluorotoluene and nitromethane). In every instance, the magnitude of the V_6 term is very low compared with typical values for V_3 barriers (see Table 6.4). For the few molecules where the V_6 contribution has been estimated as well as the V_3 term, the ratio V_6/V_3 does not exceed 3%. In general V_6 and V_3 cannot be found from data on a single torsional state; data from excited torsional states must also be obtained. Expressions for the inclusion of a V_6 dependence on the eigenvalues of the torsional wave-equation have been derived by Herschbach.[41]

Since a six-fold symmetric term in $V(\alpha)$ shows a minimum at both the minima and maxima of a three-fold potential barrier, it cannot contribute to the height of the barrier. It does, however, affect the shape of the potential wells (Figure 6.9).

When V_6 is positive, the potential wells become narrower with the energy levels pushed further apart, while a negative V_6 term would make the potential wells broader, and draw the torsional energy levels closer together. Such a distortion on a simple cosine potential barrier can, however, be caused by other factors. Ewig and Harris[91] have suggested that a variation in the reduced moment of inertia of a methyl group with torsional angle (as might be the case if the internal rotation were coupled to another low-lying vibration) would lead to an apparent V_6 contribution to the potential.

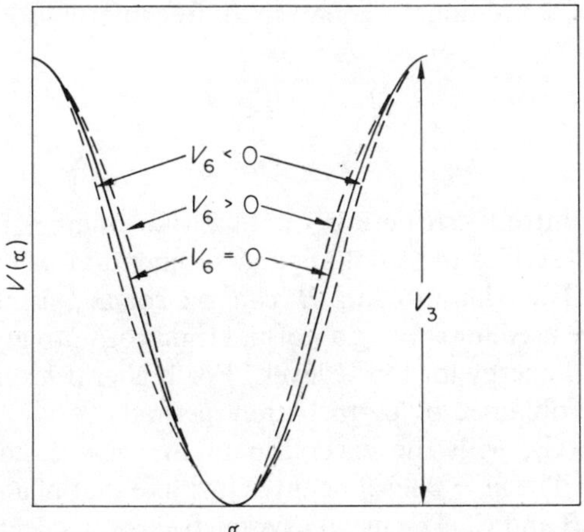

Figure 6.9   The effect of a V_6 term on the shape
of a three-fold potential barrier

6.11 Molecules with low internal rotation barriers

For molecules with internal rotation barriers falling between 500 and
5000 cal/mole, the usual method of solving the internal rotation problem as
outlined in the previous sections consists of treating the rotational and
torsional motions separately, and to consider the interaction terms as a
perturbation on the rotational energy. For molecules with low barriers,
however, the interaction terms become increasingly important, and it is
found more convenient to consider the problem as one of a molecule with
free internal rotation, and the potential barrier is then added as a perturbation
to this system. For an asymmetric molecule with a symmetric, freely rotating
group attached, the total Hamiltonian may be expressed as

$$H = AP_A^2 + BP_B^2 + C'P_C^2 + Fp^2 - 2C'pP_C + V(\alpha) \tag{6.51}$$

Here, A and B represent rotational constants (in units of energy), and C'
is defined as $h^2/8\pi^2(I_C - I_\alpha)$ (i.e. C' represents the rotational constant of the
framework, excluding the internal rotor). $F = h^2 I_C/[8\pi^2 I_\alpha(I_C - I_\alpha)]$, p and P
retain their usual significance. [$V(\alpha) = 0$ for free rotation.] Equation (6.51)
may be rewritten in two parts, illustrating the symmetric and asymmetric
contributions to the energy matrix.

$$H = H^{(0)} + H^{(1)}$$

$$H^{(0)} = \tfrac{1}{2}(A + B)(P_A^2 + P_B^2) + C'P_C^2 + Fp^2 - 2C'pP_C \tag{6.52}$$

$$H^{(1)} = \tfrac{1}{2}(A - B)(P_A^2 - P_B^2)$$

The solution of Schrödinger's equation for free internal rotation was given in § 6.7:

$$\psi(\alpha) = A\,e^{im\alpha}$$

$$E = Fm^2 \tag{6.31}$$

The integer m introduced here is the quantum number for free internal rotation which describes the total angular momentum of the top along its symmetry axis. The Hamiltonian H can be solved, using similar basis functions to those used for solving a normal rigid rotor, to give exact expressions for the total energy for low J levels. For higher J levels, approximate solutions may be obtained by perturbation methods.[44]

The energy levels, with the exception of the $m = 0$ state, are doubly degenerate ($\pm m$). The $m = 0$ level behaves just like that of a rigid rotor with the constants A, B and C'. The usual asymmetric rotor selection rules apply to the rotational spectra, together with the additional rule $\Delta m = 0$. For near symmetric top molecules, K becomes a 'good' quantum number, and transitions for which $\Delta K = 0$ dominate the spectrum. The spectral pattern for a $\Delta J = +1$, $\Delta K = 0$, $\Delta m = 0$ line is a very characteristic one. For example the $J = 1 \rightarrow 2$ (odd K) line of a molecule would appear as in Figure 6.10. Two lines occur for each value of m equally spaced about the band centre

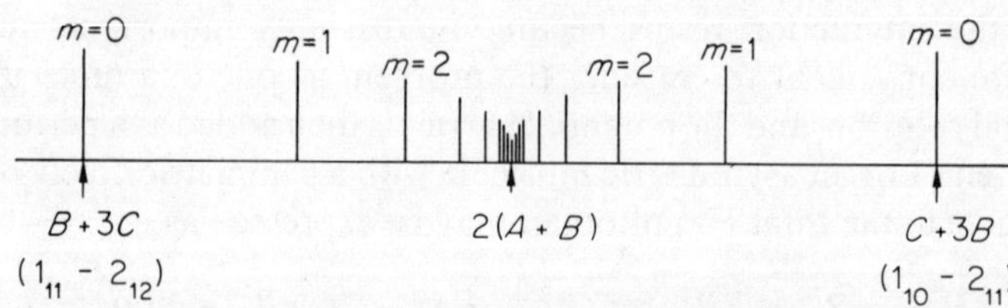

Figure 6.10   Diagrammatic representation of the spectral pattern of the $J = 1 \rightarrow J = 2$, $\Delta K = 0$ (odd K) transition for a molecule with a free internal rotor

[frequency at $2(A + B)$]. The $m = 0$ lines correspond to the asymmetric rotor transitions $1_{10} \rightarrow 2_{11}$ (at $3B + C$) and $1_{11} \rightarrow 2_{12}$ (at $B + 3C$). As m increases, the lines converge towards the centre frequency. For even K transitions, only one half of this pattern, at higher frequency than $2(A + B)$, is seen.

If a finite but small potential barrier is introduced, m is no longer a 'true' quantum number, although it is retained to help label the energy levels. Most of the molecules with low barriers studied are of such symmetry as to preclude the V_3 term in the potential energy expansion leaving the V_6 term as the only important term.

$$V(\alpha) = \tfrac{1}{2}V_6(1 - \cos 6\alpha) \tag{6.53}$$

(Examples include CH_3NO_2, CH_3BF_2, p-fluorotoluene.)

The presence of the V_6 term introduces some extra terms (including off-diagonal terms) to the energy matrix. However, provided that V_6 is small ($V_6 \ll F$), these extra terms can be shown to have a negligible effect on the spectrum, except for the $m = 3$ levels. The correction terms remove the $\pm m$ degeneracy of the $m = 3$ levels. The correction terms remove the $\pm m$ degeneracy of the $m = 3$ state (and also the degeneracy of those states which are multiples of 3).

Thus, the splitting of the $m = 3$ lines in what is otherwise an unperturbed pattern enables an accurate value for V_6 to be found. For example, in CH_3BF_2, the four $m = 3$ lines for the $J = 1 \rightarrow 2$ (K odd) transition were measured and found to be 27904·3, 27625·8, 24278·8 and 23999·7 MHz. These represent very large splittings, and so enable the barrier to be estimated with accuracy ($V_6 = 13\cdot77 \pm 0\cdot03$ cal/mole).[92]

This removal of the degeneracy of the $m = 3$ states does not occur with very low three-fold barriers, and so these barriers cannot be estimated with such accuracy, although approximate values can be derived from small deviations from the spectra of a free internal rotator.[93]

6.12 Molecules with more than one internally rotating group

Molecules with two or three internally rotating methyl groups have been successfully analysed by microwave spectroscopy, despite the added complexity of the spectral features. The approach used for solving the internal rotation problem in such molecules depends largely on the molecular symmetry, and on whether or not the rotating groups can be considered equivalent and/or independent of each other. In the extreme case of a molecule having two methyl groups in different environments and with very different barrier heights and also at a considerable distance apart, so that mechanical coupling is minimized, there is evidence to suggest that it is permissible to consider the two barriers quite independently of each other, and the problem reduces to that of two independent methyl groups, each affecting the rotational spectrum as if the other were not present.[54] In many instances, however, the rotating groups are either in very similar molecular environments, or are equivalent. Theoretical descriptions of the internal rotation in such systems have been developed by several people.[81,94-96] The full rotational Hamiltonian may be expressed as follows:

$$H = H_R + H_T + H_{TT} + H_{RT} \tag{6.54}$$

H_R represents the pure rotational Hamiltonian and H_T the internal rotation energy, H_{TT} describes the interaction between the internal rotors, while H_{RT} is the coupling term between internal and overall rotation.

Most of the examples which have been studied fall into the category of compounds having two equivalent methyl groups, and for such, the respective term of the Hamiltonian may be written as

$$H_R = AP_A^2 + BP_B^2 + CP_C^2 + F(\mathscr{P}_1^2 + \mathscr{P}_2^2) + F'(\mathscr{P}_1\mathscr{P}_2 + \mathscr{P}_2\mathscr{P}_1)$$

$$H_T = F(p_1^2 + p_2^2) + \frac{V_3}{2}(1 - \cos 3\alpha_1) + \frac{V_3}{2}(1 - \cos 3\alpha_2)$$

$$H_{TT} = F'(p_1p_2 + p_2p_1) + V_3' \cos 3\alpha_1 \cos 3\alpha_2 + V_3'' \sin 3\alpha_1 \sin 3\alpha_2$$

$$H_{RT} = -2F(p_1\mathscr{P}_1 + p_2\mathscr{P}_2) - 2F'(p_1\mathscr{P}_2 + p_2\mathscr{P}_1)$$

$$(6.55)$$

Most of the symbols used in equation (6.55) correspond to those used for the 'one-top' expressions; the subscripts 1 and 2 refer to the two separate tops. The extra angular momentum terms are defined as

$$\mathscr{P}_1 = I_\alpha\left(\frac{\lambda_A P_A}{I_A} + \frac{\lambda_B P_B}{I_B}\right) \quad \text{and} \quad \mathscr{P}_2 = I_\alpha\left(\frac{\lambda_B P_B}{I_B} - \frac{\lambda_A P_A}{I_A}\right)$$

whilst the inverse reduced moments of inertia, F and F' are equal to

$$\frac{\hbar^2}{4I_\alpha}\left[\frac{I_A}{(I_A - 2\lambda_A^2 I_\alpha)} + \frac{I_B}{(I_B - 2\lambda_B^2 I_\alpha)}\right] \quad \text{and} \quad \frac{\hbar^2}{4I_\alpha}\left[\frac{I_B}{(I_B - 2\lambda_B^2 I_\alpha)} - \frac{I_A}{(I_A - 2\lambda_A^2 I_\alpha)}\right]$$

respectively. In the expressions for the potential energy in H_T and H_{TT}, the Fourier expansion series in α_1 and α_2 is not taken further than the three-fold terms; V_3' and V_3'' represent the potential energy interaction terms between the two tops. (Only the V_3'' term may be estimated from data on one torsional state alone.)

For the ground vibrational state of molecules, the top–top interaction is often assumed to be zero, or to be so small that it can be neglected (i.e. $F' = V_3' = V_3'' = 0.0$). If, however, the interaction is assumed small but finite, all the coupling terms can be treated as perturbations by the PAM method, and Pierce has shown that the procedure is exactly analogous to the method described by Herschbach for the one-top problem.[95]

$$H_{vv} = H_r + F \sum_n [W_{vv}^{(+n)}\mathscr{P}_+^n + W_{vv}^{(-n)}\mathscr{P}_-^n], \qquad n = 0, 1, 2, \ldots \quad (6.56)$$

where

$$W_{vv}^{(\pm n)} = w_{vv}^{(\pm n)} + xX_{vv}^{(\pm n)} + yY_{vv}^{(\pm n)} + zZ_{vv}^{(\pm n)}$$

and

$$\mathscr{P}_+ = \tfrac{1}{2}(\mathscr{P}_1 + \mathscr{P}_2)$$

$$\mathscr{P}_- = \tfrac{1}{2}(\mathscr{P}_1 - \mathscr{P}_2)$$

The symbols $w_{vv}^{(\pm n)}$ represent linear combinations of the $W_{v\sigma}^{(n)}$ terms given by Herschbach for a single top (equation 6.42)[49], and the parameters x, y and z are functions of F, F', V_3, V'_3 and V''_3. X_{vv}, Y_{vv} and Z_{vv} are also functions of $W_{v\sigma}^{(n)}$.

For acetone $(V_3 = 778\ \mathrm{cal/mole})$,[81] and dimethyl silane $(V_3 = 1647\ \mathrm{cal/mole})$,[95] the top–top coupling terms were found to contribute negligibly in the calculation of the barrier height.

The Internal Axis Method has also been shown to be quite applicable for analysing the spectra of molecules with multiple rotors. A general theory for treating two-top molecules, including those with asymmetric internal rotors, has been developed by Knopp and Quade.[96] The method uses a framework fixed axis system, whereby the central part of the molecule is taken to be the frame with the two tops attached to it. This generalized formulation of internal rotation which includes all types of rotors is likely to become increasingly important as computational methods are standardized, and as more complex molecular systems are studied.

The torsional energy levels in the ground vibrational state of a two-top molecule are nine-fold degenerate in the limits of a very high barrier, but as the barrier becomes lower, this degeneracy is partially removed to give four levels which are labelled AA, EE, AE and EA. The AA level is non-degenerate, the EA and AE states are each doubly degenerate, while the EE state is quadruply degenerate. (The conclusions can be derived by consideration of the symmetry properties of the Hamiltonian.)[97] For intermediate values of the barrier the two levels AE and EA remain degenerate and triplet absorption lines are seen in the microwave spectrum. (For example, triplets occur in the ground state spectrum of dimethyl silane,[95] cis-2,3-epoxy-butane[70] and dimethyl ether,[82] whereas quartets can also be observed in the rotational spectrum of acetone.)[81,94] The separation between the components of the triplets (or quartets) are sensitive functions of the barrier heights.

The first excited torsional state of a two-top molecule is eighteen-fold degenerate in the high barrier limit. There exist two first torsional states corresponding to the separate torsional excitation of the two equivalent tops. In the absence of any top–top coupling these are degenerate, but in practice the interaction is usually sufficient for two distinct torsional state spectra (each consisting mainly of triplets) to be observed. The lower energy state $(v = +1)$ is referred to as the symmetric state and the higher energy state $(v = -1)$ as the antisymmetric state. Because the antisymmetric state has the higher energy and hence a lower population, the rotational lines of the $v = -1$ state will be slightly weaker than the corresponding transitions of the $v = +1$ state. Also, because of its higher energy, the splittings caused by internal rotation in the lines of the $v = -1$ state will be larger than those for the $v = +1$ state.

The microwave spectrum of propane in its first excited state has been analysed fairly thoroughly by Hirota and coworkers.[98] (Propane does not show any internal rotation effects on the rotational spectrum in the ground state.)[99] The values for V_3 and V_3'' were estimated to be 3325 and -170 cal/mole respectively. Hoyland using a slightly different expansion of the potential energy, has also analysed the experimental data of Hirota and has concluded that the most consistent value of V_3 is 3575 cal/mole.[100]

Most of the molecules with three methyl groups that have been studied by microwave spectroscopy have sufficiently large barriers (due probably to large steric effects) to prevent any internal rotation splittings being observed in the spectrum of the ground and first excited torsional states. (For example, $(CH_3)_3CCHO$,[101] $(CH_3)_3CH$[102] and $(CH_3)_3N$.[103]) No detailed analysis of the internal rotation splittings in higher excited torsional states of any three-top molecule has yet been carried out.

Table 6.4 includes a selection of molecules for which internal rotation barriers have been measured from microwave studies.

Table 6.4 Some potential barriers to internal rotation obtained from microwave studies

Molecule	Formula	Potential barrier (cal/mole)	Reference
Ethyl fluoride	CH_3-CH_2F	3306	56
Ethyl chloride	CH_3-CH_2Cl	3685	57
Ethyl bromide	CH_3-CH_2Br	3684	58
Ethyl iodide	CH_3-CH_2I	3220	59
Ethyl cyanide	CH_3-CH_2CN	3050	60
Ethyl isocyanide	CH_3-CH_2NC	3650	61
1,1-Difluoroethane	CH_3-CHF_2	3180	56
Acetaldehyde	CH_3-CHO	1168	49
Acetyl fluoride	CH_3-COF	1041	62
Acetyl cyanide	CH_3-COCN	1270	63
Acetic acid	CH_3-COOH	481	64
1-Fluoropropene (*cis*)	$CH_3-CH{=}CHF$	1057	65
1-Fluoropropene (*trans*)	$CH_3-CH{=}CHF$	2200	66
1-Chloropropene (*cis*)	$CH_3-CH{=}CHCl$	620	67
1-Chloropropene (*trans*)	$CH_3-CH{=}CHCl$	2170	68
Propylene oxide	$CH_3-CH\overset{\diagdown\ \diagup}{\underset{O}{\qquad}}CH_2$	2560	47

Table 6.4   Continued

Molecule	Formula	Potential barrier (cal/mole)	Reference
Propylene sulphide	$CH_3-CH-\!\!-CH_2$ with S bridge	3240	69
2,3-Epoxybutane (*cis*)	$CH_3-CH-\!\!-CH-CH_3$ with O bridge	1607	70
2,3-Epoxybutane (*trans*)	$CH_3-CH-\!\!-CH-CH_3$ with O bridge	2444	71
Methyl alcohol	CH_3-OH	1074	72
Methyl mercaptan	CH_3-SH	1268	73
Methyl hypochlorite	CH_3-OCl	3060	74
Methyl formate	CH_3-OCHO	1190	55
Methyl fluoroformate	CH_3-OCOF	1077	54
Methyl vinyl ether	$CH_3-OCH{=}CH_2$	3830	75
Methyl vinyl sulphide	$CH_3-SCH{=}CH_2$	3230	76
Methylamine	CH_3-NH_2	1976	77
Methyl phosphine	CH_3-PH_2	1960	78
Methyl silane	CH_3-SiH_3	1665	56
Methyl germane	CH_3-GeH_3	1239	79
Methyl stannane	CH_3-SnH_3	650	80
Acetone	$(CH_3)_2CO$	778	81
Dimethyl ether	$(CH_3)_2O$	2720	82
Dimethyl sulphide	$(CH_3)_2S$	2132	83
Dimethyl amine	$(CH_3)_2NH$	3200	84
Fluoral	CF_3-CHO	885	85
Fluorodisilane	SiH_3-SiH_2F	1048	86
Toluene	$CH_3-C_6H_5$	13·9	87
p-Fluorotoluene	$CH_3-C_6H_4F$	13·8	88
Nitromethane	CH_3-NO_2	6·0	89
Trifluoronitromethane	CF_3-NO_2	74·4	90

For a more complete list of potential barriers see Reference 5.

6.13 Potential barriers and energy differences from relative intensity measurements

When the potential barrier opposing internal rotation of a methyl group about a particular bond is greater than say 4–5 kcal/mole, hydrogen tunnelling through the potential barrier becomes more unlikely and no splitting can be resolved on the rotational lines. For such molecules it is often possible to obtain an approximate value for the barrier by comparing the intensities of rotational lines in the vibrational ground state to the intensities of the corresponding lines in the excited torsional states of the molecules (equation 6.15). The experimental difficulties which need to be overcome in order to obtain reliable results with this method have been outlined by Verdier and Wilson[104] and Esbitt and Wilson.[105] It is probably a wise precaution to measure the relative intensities of as many lines as possible so that any anomalous results, due for example to the interfering effects of Stark lobes, be minimized. From the Boltzmann expression,

$$\frac{I_1}{I_0} = \frac{g_1}{g_2} e^{-(h\nu/kT)} \tag{6.15}$$

a value for the torsional frequency (ν) may be obtained. This, in turn, can be related to a potential barrier height by means of one of the following two ways.

(a) The relationship between the torsional wave-equation and Mathieu's equation (§ 6.7) leads to the expression

$$E_v = \frac{9Fb}{4} \tag{6.57}$$

for three-fold symmetric groups (cf. equation 6.27). Here E_v is the energy of a torsional state, and b is an eigenvalue of the Mathieu equation. F depends only on the molecular structure ($F = h^2/8\pi^2 I_r$). For differences between two energy levels

$$\Delta E = \frac{9F}{4} \Delta b \tag{6.58}$$

where $\Delta b = b_{v=1} - b_{v=0}$ ($v = 1$ and $v = 0$ represent the first excited state and the ground state respectively). From the eigenvalues of Mathieu's equation, tabulated as a function of the reduced barrier (s),[49] the potential barrier may be calculated using the equation

$$V_3 = \frac{9Fs}{4} \tag{6.59}$$

(b) For fairly high barriers the torsion may be approximated by the simple harmonic expression (cf. equation 6.37)

$$V_3 = \tfrac{8}{9}\pi^2 \nu^2 I_r \tag{6.60}$$

The relative intensity method is of course merely an indirect means of finding the torsional frequency of a methyl (or similar) group (see Chapters 7 and 8), but these low frequency torsional modes are often very weak in the far infrared and cannot always be observed directly. (This is especially true of gases at low pressure.) Some of the major problems associated with the relative intensity method are listed below.

(a) The satellite lines must be correctly assigned to the appropriate eigenstate of the torsional vibration.

(b) If the coupling between internal and overall motions is considerable, then the satellite lines will be split. This, in turn, will make the measurement of intensities more difficult and the method less reliable.

(c) If the ground state and satellite lines are far removed from each other (this is not often the case for methyl group torsions), the problem of maintaining indentical operating conditions for the measurement of the lines can be very exacting.

(d) The experimental conditions must be such that the lines are fully modulated (i.e. all Stark lobes displaced from the central line), and the operating pressure must be sufficiently high to prevent the saturation of absorption lines by the microwave radiation, but sufficiently low to prevent the pressure-broadening of lines.

For calculations of the difference in energy between two rotamers of the same molecule, allowance must be made for the difference in line strengths of the chosen transitions. The peak intensity of a rotational line depends on several factors including the rotational constants and the square of the dipole moment.[2]

Each line intensity may be calculated using the appropriate formulae or may be obtained from computational procedures.[31] This method has been used for ethyl formate,[32] propionyl fluoride,[34] 1-pentyne[106] and others, whereas for n-propyl fluoride, Hirota obtained the *anti/gauche* energy difference by studying the temperature dependence of lines belonging to the two rotamers.[107] This latter method assumes that all factors which contribute to the line intensities of two rotamers vary with temperature in the same way with the exception of the population of the two forms.

6.14 Inversion and ring puckering

Another form of large amplitude molecular motion which is related to internal rotation is the phenomenon of inversion. In principle, any non-planar molecule possesses the property of inversion, but in practice the internuclear forces which hold together the molecule in its equilibrium conformation ensure that the potential barrier is too high for inversion to take place. Examples of three extreme forms of potential barriers to inversion are shown schematically in Figure 6.11.

When the barrier is very large (Figure 6.11a), the individual vibrational states within each well are exactly degenerate. For intermediate barriers (Figure 6.11b), quantum-mechanical tunnelling results in removal of this

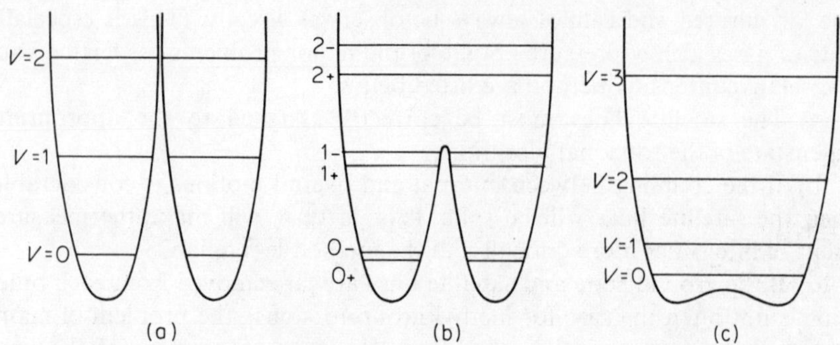

Figure 6.11   The approximate shape of the potential function for inversion, with (a) an infinitely high barrier, (b) a finite barrier and (c) a very low barrier

degeneracy, and each state below the potential maximum splits into a symmetric $(+)$ and an antisymmetric $(-)$ component. In the case of very low barriers (Figure 6.11c) where all the vibrational levels are above the barrier maximum, the spacing between the levels is found to be non-uniform, owing to the slight perturbing effect of the small potential. The phenomenon of 'tunnelling' is largely confined to proton movement only owing to the relatively small mass of the hydrogen atom.

The effects of molecular inversion can often be observed in the microwave region, but the precise nature of the vibration–rotation interaction can vary for different molecules, depending largely on the magnitude of the energy level splittings due to the tunnelling relative to the pure rotational energy spacings. For an absorption to take place, the transition moment $(\int \psi_i^* \mu_g \psi_j \, d\tau)$ must be non-zero (i.e. the total function must be symmetric). For normal rotational spectra, the dipole moment (μ_g) does not change sign with respect to the molecular framework, and so the wave-functions ψ_j and ψ_i must have the same symmetry properties. During an inversion, however, μ_g changes sign, in which case ψ_j and ψ_i must have opposite symmetries for the transition moment to be non-zero. Thus, inversion spectra connect the levels $(+) \leftrightarrow (-)$ (Figure 6.11b), and they can have a very different appearance to the normal spectrum of a rigid rotor.

Ammonia represents a unique example for microwave spectroscopy, as it was the first molecule to be observed in the microwave region,[108] its spectrum is one of the strongest known, and it is composed of pure inversion transitions. The rotational spectrum of ammonia lies in the far infrared region, and the intense microwave lines are caused by the $v = 0^+ \rightarrow v = 0^-$ inversion transition, which is split into a very large number of components corresponding to different rotational (J, K) states of the molecule.

The shape of the potential barrier to inversion in ammonia and related molecules (e.g. amines) may be approximated by various types of expressions, most of which are relatively successful in explaining all the observed spectral features. Some of the most convenient and useful functions are based on a perturbed harmonic oscillator expression. Another function which has been found to be successful is that derived by Manning,[109] and has the form

$$V(x) = A \operatorname{sech}^4 \frac{x}{2\rho} + B \operatorname{sech}^2 \frac{x}{2\rho} \qquad (6.61)$$

where $V(x)$ is the potential height, x represents the height of the nitrogen atom above the plane of the hydrogen atoms, A and B are constants characteristic of the molecule and ρ is a constant which depends on the reduced mass of the inversion motion (μ). [For ammonia, μ is sometimes approximated to $3mM/(3m + M)$.]

For many molecules, the effects of both rotation and inversion are simultaneously observed in the microwave region. If the inversion splittings are small compared with the rotational energy level spacings, the normal rotational spectrum is observed with most of the lines unsplit. In addition, however, transitions where the dipole change lies parallel with the inversion axis will occur, and these will appear as doublets split symmetrically about a hypothetical centre, so that their separation will be twice the inversion splitting of the ground vibrational state (Figure 6.12). When the inversion splittings are very large and fall outside the microwave range (i.e. low barriers), the usual rigid rotor lines are accompanied by strong satellite lines, which may be considered as originating in a low-lying vibrational state associated with the out-of-plane vibration. Figure 6.12 illustrates the differences between the various types of transitions for different inversion barriers.

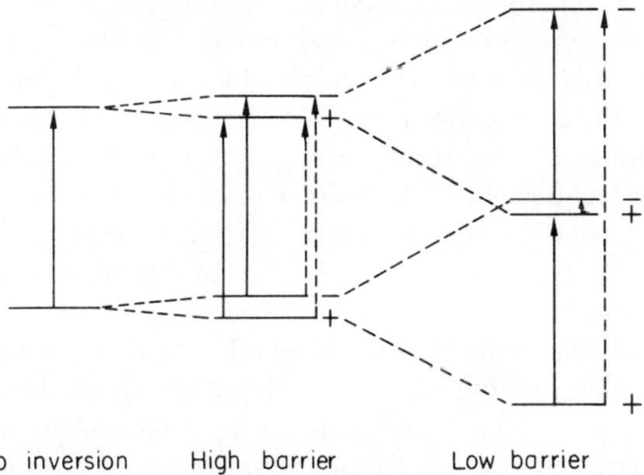

Figure 6.12 Rotation–inversion transitions for (a) an infinitely high barrier, (b) a finite but high barrier and (c) a low inversion barrier

The ease with which inversion can occur is a very sensitive function of the potential barrier, and also of the reduced mass. The latter relationship is illustrated in Table 6.5 where the ground state inversion frequencies are tabulated for ammonia and its deuterated analogues. Also included for comparison are the barrier heights for phosphine and arsine. Similar differences are found for several types of amines; for example, in methylamine[114] the inversion barrier is considerably lower than that found for ammonia (2070 as compared with 5780 cal/mole), while for dimethylamine[115] the value is 4·4 kcal/mole. Monochloramine[116] has an inversion barrier greater than that of ammonia, while difluoramine[117] shows no evidence at all to suggest that the molecule is inverting.

Inversion in methylamine and other amines is complicated by the fact that the motion is directly coupled to the internal rotation of the methyl group. The motion of the NH_2 group causes the methyl hydrogens to change from a stable staggered conformation, with respect to the NH_2 hydrogens, to a relatively unstable eclipsed conformation, and so internal rotation occurs about the C—N bond in order to restore the molecule to its stable minimum energy position. The lack of clear distinction between inversion and internal rotation is also apparent in molecules such as propargyl alcohol,[118] propargyl sulphide[119] and hydroxyacetonitrile.[120] All of these molecular species have a non-planar *gauche* equilibrium configuration. The torsional motion which takes the hydrogen from one *gauche* position to the other results in the μ_C component of the dipole in these molecules changing sign, and hence the similarity of the torsion to inversion. In propargyl alcohol, the normal ground-state rotational lines are accompanied by strong satellites (attributable to the presence of an excited torsional state, $21\cdot5\,cm^{-1}$ above the ground state), and many lines with first-order Stark effects have been assigned to torsional–rotational transitions. In propargyl sulphide, most absorption lines consist of close doublets, strongly indicating that the potential barrier between the two identical *gauche* positions is larger than in the case of the alcohol. The perpendicular torsional–rotational transitions fall in a close series, terminating near 6890 MHz, which corresponds to the frequency of the torsional vibration. For hydroxyacetonitrile, the *gauche–gauche* barrier is higher than for propargyl alcohol (about $360\,cm^{-1}$ as compared with $90\,cm^{-1}$), but probably lower than the barrier in propargyl sulphide.

The equilibrium configurations of simple amides have also been studied fairly thoroughly by microwave spectroscopy. Both formamide[121] and cyanamide[122] have been found to be non-planar, but with a considerably lower barrier than that found for ammonia.

A Hamiltonian describing rotation–inversion phenomena has been derived by Lide,[123] and its general form is similar to the corresponding expressions used to describe internal rotation.

Table 6.5 Inversion frequencies and barriers to inversion of ammonia and related compounds

Molecule	Potential barrier (cal/mole)	Inversion frequency— ground state (MHz)	Reference
NH_3	6056	23800	110, 111
NH_2D	6056	12200	111
NHD_2	6056	5100	111
ND_3	6056	1600	111
PH_3	18240 (37200)	0·14	112 (113)
AsH_3	33640	0·5 cycle/year	112

Closely related to the inversion movement described by ammonia and similar molecules is the puckering motion which four- and five-membered rings execute. Chan and coworkers have extensively studied the microwave spectrum of trimethylene oxide, and have found evidence for the presence of a very small potential barrier associated with the planar ring configuration.[124] However, all the vibrational energy levels, including the ground state, lie above this potential barrier, so that to all intents and purposes the ring is planar. The presence of the small potential is manifested in the vibrational energy spacings, and in the way that the rotational constants vary within the various excited states. The corresponding sulphur analogue, trimethylene sulphide, has a higher potential barrier opposing the attainment of planar ring, and the four lowest vibrational states were all found to lie below the potential maximum.[125] The potential function for ring puckering appears to be described fairly adequately by an expression involving only quadratic and quartic terms:

$$V(q) = k_1 q^4 - k_2 q^2 \qquad (6.62)$$

where q is the ring-puckering coordinate, and k_1 and k_2 are constants. The type of information available from microwave studies of small rings and the subsequent interpretation of the data in terms of molecular forces and geometry has been outlined by Laurie.[126] Table 6.6 lists many of the small-ring compounds which have been studied by microwave spectroscopy to date.

Five-membered rings are capable of executing a motion which has been described as 'pseudo-rotation'. This can occur in molecules where there are two ring-puckering modes sufficiently close in frequency to interact with each other.[144] Values for the barrier to pseudo-rotation in such molecules have been obtained from microwave studies.

Table 6.6   Four- and five-membered ring compounds which have been studied by microwave spectroscopy

Compound	Formula	Conformational deductions	Reference
Trimethylene oxide	H_2C—CH_2 ring with CH_2 and O	Planar[a]	124
Trimethylene sulphide	H_2C—CH_2 ring with CH_2 and S	Non-planar ($V = 784$ cal/mole)	125
Cyclobutyl fluoride	H_2C—CH_2 ring with CH_2 and CHF	Non-planar halogen equatorial	127
Cyclobutyl chloride	H_2C—CH_2 ring with CH_2 and $CHCl$	Non-planar halogen equatorial	127
Cyclobutyl bromide	H_2C—CH_2 ring with CH_2 and $CHBr$	Non-planar halogen equatorial	128
Methylene cyclobutane	H_2C—CH_2 ring with CH_2 and $C{=}CH_2$	Non-planar ($V = 458$ cal/mole)	129
Cyclobutanone	H_2C—CH_2 ring with CH_2 and $C{=}O$	Planar[a]	130
Trimethylene selenide	H_2C—CH_2 ring with CH_2 and Se	Non-planar ($V = 1096$ cal/mole)	131
3-Methyleneoxetane	H_2C—CH_2 ring with O and $C{=}CH_2$	Planar	132
3,3-Difluoroxetane	H_2C—CH_2 ring with O and CF_2	Planar	133
1,1-Difluorocyclobutane	H_2C—CH_2 ring with CH_2 and CF_2	Non-planar ($V = 689$ cal/mole)	134

Table 6.6 Continued

Compound	Formula	Conformational deductions	Reference
Silacyclobutane	CH_2, H_2C CH_2, SiH_2	Non-planar ($V = 1264$ cal/mole)	135
β-Propiolactone	CH_2, O CH_2, C, O	Planar	136
Cyclopentanone	H_2C-CH_2, H_2C CH_2, C, O	Non-planar ($V = 1144$ cal/mole)	137
Tetrahydrofuran	H_2C-CH_2, H_2C O CH_2	Non-planar ($V = 163$ cal/mole)	138
Cyclopentene	$HC=CH$, H_2C CH_2, CH_2	Non-planar ($V = 663$ cal/mole)	139
Cyclopentadiene	$HC=CH$, HC CH_2, CH	Planar	140
Vinylene carbonate	$HC=CH$, O O, C, O	Planar	141
Cyclopent-2-en-1-one	$HC-CH_2$, HC CH_2, C, O	Planar	142
Germacyclopentane	H_2C-CH_2, H_2C CH_2, GeH_2	Non-planar ($V = 1430$ cal/mole)	143

Planar[a] denotes a molecule which has a small potential barrier associated with the planar conformation. The top of the barrier is below the ground vibrational level.

6.15 Asymmetric rotating groups and rotational isomerism

When the three-fold symmetry of an internally rotating group is destroyed, the stable equilibrium configurations are no longer equivalent, and molecules may exist as mixtures of rotational isomers. For monosubstituted ethanes, and for many other molecules, only two distinct rotational isomers exist, since the two *gauche* structures are identical (see Figure 6.13a). In principle, more than two forms can coexist for a molecule (as, for example, in 1,1,2-trisubstituted derivatives of ethane with different substituent groups, Figure 6.13b), but so far, only in one instance (allylamine)[145] have more than two isomers been identified for any one molecular species by microwave spectroscopy.

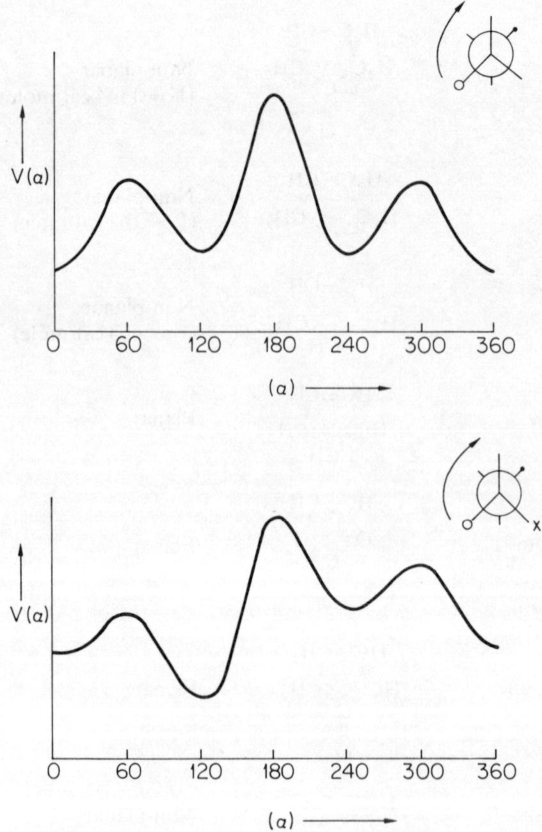

Figure 6.13 The variation of potential energy with dihedral angle (α) for (a) a 1,2-disubstituted ethane and (b) a 1,1,2-trisubstituted ethane, showing the stable conformations of the rotational isomers

Since rotational isomers have lifetimes which are long when compared with rotational transitions, the analysis of the microwave spectrum of a molecule with two rotameric forms proceeds as for a mixture of two quite distinct molecular species. The rotamers may give rise to quite different types of spectra, because their moments of inertia may be very different, and their dipole moments may have different magnitudes and direction. In some instances, one isomer (for example, the *anti* isomer in 1,2 symmetrically substituted ethanes) may have zero dipole moment, and hence no microwave spectrum. In general, changing the orientation of an asymmetric group within a molecule will considerably shift the inertial axes, and the direction of the dipole components, and so the types of transitions observed (i.e. μ_A, μ_B or μ_C) may vary dramatically from one rotamer to another. Often a rotamer with constants near to those of a symmetric top can be identified fairly readily, whereas a more asymmetric form is sometimes more difficult to assign. This may be especially true if the rotational constants change rapidly with the angle of internal rotation. (This was found to be the case for the *gauche* isomer of ethyl formate.)[32]

Much progress has been made recently on the development of a general but quantitative theory of internal rotation in asymmetric molecules with asymmetric rotating groups. Much of this effort is due to the work of Quade and coworkers.[146] In the simplest example of an asymmetric rotating group, CH_2D, the potential function retains its symmetry, but the energy levels within the potential wells are positioned differently, and the kinetic energy of internal rotation varies as a function of the angle of rotation. In the microwave spectra of CD_2HCHO and CHD_2CHO, each transition comprises three components (instead of the usual A and E doublets observed for CH_3CHO).[42] One of the lines arises from the symmetric rotamer, and the other two are caused by tunnelling of the barrier between the two identical asymmetric isomers. The observed splittings have been analysed in terms of a framework fixed axis method (FFAM) which is applicable to molecules with planar frameworks, and were found to be consistent with the barrier height of 1162 cal/mole found for the normal species of acetaldehyde. Similar calculations carried out on 3-fluoropropene have extended this type of analysis to include molecules with completely asymmetric frameworks and internal rotors.[33] This molecule has been the subject of an extensive microwave study, and was found to exist in two rotameric forms, *cis* and *gauche*, with the former being more stable by 166 cal/mole.[147] By combining microwave with far-infrared data on this molecule, the full shape of the potential barrier was predicted. The analysis showed that for very asymmetric molecules, accurate calculations of the shape of the potential barrier require a large amount of experimental information. For molecules where rotational isomerism occurs, microwave spectroscopy provides considerable information relating to the potential minima, but often not sufficient information to predict with certainty the shape near the potential maxima. The potential function is usually expressed as a series of cosine terms (cf. equations 6.21 and 6.23),

$$V(\alpha) = \sum_n \tfrac{1}{2} V_n (1 - \cos n\alpha) \tag{6.63}$$

and the assumption is made that the molecule possesses only one degree of freedom, namely the torsion. The function is also assumed to converge fairly rapidly, since generally only a relatively small number of Fourier coefficients can be determined. From equation (6.63):

$$V(\alpha) = \tfrac{1}{2} V_1 (1 - \cos \alpha) + \tfrac{1}{2} V_2 (1 - \cos 2\alpha) + \tfrac{1}{2} V_3 (1 - \cos 3\alpha)$$
$$+ \tfrac{1}{2} V_4 (1 - \cos 4\alpha) + \cdots \tag{6.64}$$

The term $\cos(n\alpha)$ may be expanded into a series, the first few components of which are given by

$$\cos n\alpha = 1 - \frac{(n\alpha)^2}{2!} + \frac{(n\alpha)^4}{4!} - \frac{(n\alpha)^6}{6!} + \cdots \tag{6.65}$$

If the quadratic and quartic terms only are included in equation (6.64), we have

$$V(\alpha) = \frac{\alpha^2}{4}(V_1 + 4V_2 + 9V_3 + 16V_4 + \cdots)$$
$$- \frac{\alpha^4}{48}(V_1 + 16V_2 + 81V_3 + 256V_4 + \cdots) \tag{6.66}$$

The task of determining the nature and shape of the potential barrier now resolves into the problem of evaluating the Fourier components of $V(\alpha)$. The amount of experimental data available determines how many of the terms in equation (6.66) are retained in the subsequent calculation. Some of the assumptions and the experimental data used to calculate the potential energy terms are listed below.

(a) For harmonic motion about the potential minima, all the quartic components in equation (6.66) are zero.

(b) The variation of the reduced moment of inertia of the torsional motion with angle (α) is assumed negligible.

(c) The curvature of the potential at the minima $d^2V(\alpha)/d\alpha^2$ may be equated to the force constant controlling the harmonic oscillations

$$\left(\frac{\partial^2 V(\alpha)}{\partial \alpha^2} \right)_{\alpha = \min} = \tfrac{1}{2}(V_1 + 4V_2 + 9V_3 + 16V_4 + \cdots)$$
$$= 4\pi^2 v^2 I_r \tag{6.67}$$

Table 6.7 Some compounds exhibiting rotational isomerism which have been studied by microwave spectroscopy

Name	Formula	Most stable isomer (1)	Second isomer (2)	Energy difference $(E_2 - E_1)$ (cal/mole)	Fourier coefficients (cal/mole)				Reference
					V_1	V_2	V_3	V_4	
Fluoroacetyl fluoride		*trans* (as diagram)	*cis*	910	-280	2760	1260	410	150
n-Propyl fluoride		*gauche*	*trans* (as diagram)	470	3220	-3050	6480	-1250	107
3-Fluoropropene		*cis* (as diagram)	*gauche*	166	-707	530	2449	538	33
1-Pentyne		*gauche*	*trans* (as diagram)	77	940	-1045	3518	71	106
Propionyl fluoride		*cis* (as diagram)	*gauche*	1290	930	720	1150	130	34

I_r is the reduced mass of the moving body, and may be calculated (as G^{-1}) using Polo's method,[148] or from Pitzer's formulae,[149] while v is the torsional frequency.

(d) If the potential energy is assumed zero at the minimum of the more stable conformation, then $V(\alpha)$ at the other position of minimum is given by the energy difference between the rotamers (corrected for zero-point energy).

Other data which are often available from microwave studies and which have been used to varying extents are the energy differences between successive torsional states of the rotamers. This information, which is usually obtained from relative intensity studies of the torsional lines, may be related to the anharmonicity terms in the potential function.[32,107] Table 6.7 shows five examples for which two rotational isomers have been identified; also included are the first four Fourier coefficients of the potential barrier. (For more comprehensive lists of rotamers studied by microwave spectroscopy, see References 4 and 5). The first three are seen to be the dominant terms, with V_3 being the largest, except in the case of fluoroacetyl fluoride, where the *cis* conformation, with eclipsed fluorine atoms, is found to exist as the second rotamer, instead of the usual *gauche* structure, due to the large two-fold contribution to the potential.

The consistency in value of the V_3 term in many related molecules has led Stiefvater and Wilson to comment on the possible additivity of the various coefficients.[34] These authors have predicted the approximate shapes

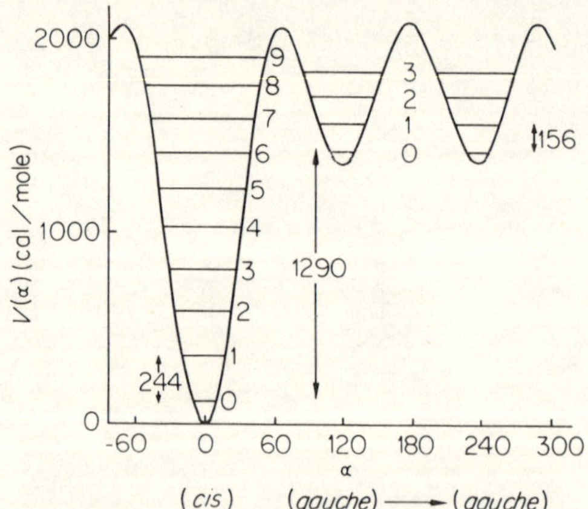

Figure 6.14   The approximate shape of the potential energy barrier to internal rotation about the CH_3CH_2—COF bond in propionyl fluoride, and the relative energies of the torsional states. (Redrawn from Reference 34 by permission of the authors)

of potential barriers to internal rotation for a number of related molecules, basing their suggestions on their conclusion that the V_3 term remains approximately unchanged for a CX_3 group, when $X = H$, CH_3 or F.

The resurgence of interest in *ab initio* calculations of potential barriers to internal rotation over the past few years (see Chapter 11) has been largely paralleled on the experimental side by more extensive microwave spectroscopic analyses of more complex internal rotation potentials. As computational procedures are standardized, and as new experimental devices are developed and improved, so this trend is likely to continue in the future.

6.16 Conclusion

In a comprehensive review, Wilson includes a summary of the data, (obtained by gas-phase microwave spectroscopy up to 1971), available for molecules exhibiting rotational isomerism, and amply illustrates both the utility and the future potential of the microwave technique for studying conformational behaviour in molecules.[151] As more accurate experimental data on internal rotation barriers become available, so efforts are being made to review the basic assumptions in the interpretative theories and to assess the most likely sources of error. The effect of possible coupling between two internally rotating groups or between an internally rotating group and another low-frequency vibration on the resultant value of the hindering potential is discussed by Dreizler.[152] Using data from vibrationally excited states, the V_3'' coupling term between the methyl groups in, for example, dimethyl silane $(CH_3)_2SiH_2$ was found to be -38 ± 2 cal/mole compared to a calculated V_3 of 1646 ± 3 cal/mole (see page 195).

As intimated by Wilson in his review,[151] a powerful reason for producing more accurate experimental data for small molecules is the hope that they will provide stringent tests of current theories of molecular structure. When such theories can successfully account for the internal rotation of small molecules we may then hope that they will provide means of predicting the structure and conformational behaviour of more complex molecules.

6.17 References

1. W. Gordy, W. V. Smith and R. F. Trambarulo, *Microwave Spectroscopy* (New York: Wiley, 1953).
2. C. H. Townes and A. L. Shawlow, *Microwave Spectroscopy* (New York: McGraw-Hill, 1955).
3. T. M. Sugden and C. N. Kenney, *Microwave Spectroscopy of Gases* (London: Van Nostrand, 1965).
4. J. E. Wollrab, *Rotational Spectra and Molecular Structure* (New York: Academic Press, 1967).
5. W. Gordy and R. L. Cook, *Microwave Molecular Spectra* (New York: Wiley, 1970).

6. E. B. Wilson, Jr. and D. R. Lide, Jr., *Determination of Organic Structures by Physical Methods*, Vol. 1, Ed. E. A. Braude and F. C. Nachod (New York: Academic Press, 1955), Chap. 12.

7. D. J. Millen, *Chem. Ind.*, 1472 (1963).

8. T. M. Sugden, *Endeavour*, **22**, 129 (1963).

9. E. B. Wilson, Jr., *Pure Appl. Chem.*, **7**, 23 (1963).

10. J. Sheridan, *Ann. Rept. Progr. Chem. (Chem. Soc. London)*, **60**, 160 (1964).

11. D. R. Lide, Jr., *Advan. Anal. Chem. Instr.*, **5**, 235 (1966).

12. D. J. Millen, *Chem. Brit.*, 202 (1968).

13. E. B. Wilson, Jr., *Science*, **162**, 59 (1968).

14. D. R. Lide, Jr., *Surv. Prog. Chem.*, **5**, 95 (1969).

15. D. R. Lide, Jr., *Ann. Rev. Phys. Chem.*, **15**, 225 (1964); W. H. Flygare, *Ann. Rev. Phys. Chem.*, **18**, 325 (1967); Y. Morino and E. Hirota, *Ann. Rev. Phys. Chem.*, **20**, 139 (1969); H. D. Rudolph, *Ann. Rev. Phys. Chem.*, **21**, 73 (1970).

16. C. C. Lin and J. D. Swalen, *Rev. Mod. Phys.*, **31**, 841 (1959).

17. J. E. Wollrab, *Rotational Spectra and Molecular Structure* (New York: Academic Press, 1967), Chap. 6.

18. H. Dreizler, *Fortschr. Chem. Forsch.*, **10**, 59 (1968).

19. W. Gordy and R. L. Cook, *Microwave Molecular Spectra* (New York: Wiley, 1970), Chap. 12.

20. C. C. Costain, *J. Chem. Phys.*, **29**, 864 (1958).

21. D. R. Herschbach and V. W. Laurie, *J. Chem. Phys.*, **37**, 1668 (1962); V. W. Laurie and D. R. Herschbach, *J. Chem. Phys.*, **37**, 1687 (1962).

22. T. Oka, *J. Phys. Soc. Japan*, **15**, 2274 (1960); T. Oka and Y. Morino, *J. Mol. Spectr.*, **8**, 300 (1962).

23. J. S. Muenter, *J. Chem. Phys.*, **48**, 4544 (1968).

24. F. X. Powell and D. R. Lide, Jr., *J. Chem. Phys.*, **41**, 1413 (1964).

25. K. Hirota and Y. Hironaka, *J. Catalysis*, **4**, 602 (1965); Y. Sakurai, Y. Kaneda, S. Kondo, E. Hirota, T. Onishi and K. Tamaru, *Bull. Chem. Soc. Japan*, **41**, 1496 (1968); S. Saito, *Bull. Chem. Soc. Japan*, **42**, 667 (1969).

26. T. Oka, *J. Chem. Phys.*, **45**, 754 (1966); **47**, 13 (1967).

27. W. Hüttner and W. H. Flygare, *J. Chem. Phys.*, **49**, 1912 (1968); W. H. Flygare, R. L. Shoemaker and W. Huttner, *J. Chem. Phys.*, **50**, 2414 (1969).

28. R. M. Hainer, P. C. Cross and G. W. King, *J. Chem., Phys.*, **17**, 826 (1949).

29. H. C. Allen, Jr. and P. C. Cross, *Molecular Vib-Rotors* (New York: Wiley, 1963).

30. J. W. Blaker, M. Sidran and A. Kaercher, *Grumman Res. Report*, RE-155, Pt. I and II (1962); M. Sidran, F. J. Nolan and J. W. Blaker, *Grumman Res. Reports*, RE-155, Pt. III (1963); RE-172, Pt. IV (1964); RE-178, Pt. V (1964); RE-189, Pt. VI (1964); RE-196, Pt. VII (1964).

31. (a) R. A. Beaudet, *Ph.D. Thesis*, Harvard University (1961).
    (b) K. M. Marstokk and H. Møllendal, *J. Mol. Struct.*, **4**, 470 (1969).

32. J. M. Riveros and E. B. Wilson, Jr., *J. Chem. Phys.*, **46**, 4605 (1967).

33. P. Meakin, D. O. Harris and E. Hirota, *J. Chem. Phys.*, **51**, 3775 (1969).

34. O. L. Stiefvater and E. B. Wilson, *J. Chem. Phys.*, **50**, 5385 (1969).

35. D. Kivelson and E. B. Wilson, Jr., *J. Chem. Phys.*, **21**, 1229 (1953).

36. K. M. Marstokk and H. Møllendal, *J. Mol. Struct.*, **5**, 205 (1970); **8**, 234 (1971).

37. D. Kivelson, *J. Chem. Phys.*, **22**, 1733 (1954); **23**, 2230, 2236 (1955).

38. V. W. Laurie, *J. Mol. Spectr.*, **13**, 283 (1964).

39. *Tables Relating to Mathieu Functions* (New York: Columbia University Press, 1951); R. W. Kilb, *Tables of Degenerate Mathieu Functions* (Cambridge, Mass: Harvard University Press, 1956).

40. M. Hayashi and L. Pierce, *J. Chem. Phys.*, **35**, 1148 (1961).
41. D. R. Herschbach, *Tables for the Internal Rotation Problem* (Dept. of Chemistry, Harvard University, 1957); *J. Chem. Phys.*, **27**, 975 (1957); **31**, 91 (1959).
42. R. W. Kilb, C. C. Lin and E. B. Wilson, Jr., *J. Chem. Phys.*, **26**, 1695 (1957).
43. E. B. Wilson, Jr., *Chem. Rev.*, **27**, 17 (1940).
44. E. B. Wilson, Jr., C. C. Lin and D. R. Lide, Jr., *J. Chem. Phys.*, **23**, 136 (1955).
45. B. L. Crawford, *J. Chem. Phys.*, **8**, 273 (1940).
46. D. R. Herschbach, *J. Chem. Phys.*, **27**, 975 (1957).
47. J. D. Swalen and D. R. Herschbach, *J. Chem. Phys.*, **27**, 100 (1957).
48. D. R. Herschbach and J. D. Swalen, *J. Chem. Phys.*, **29**, 761 (1958).
49. D. R. Herschbach, *J. Chem. Phys.*, **31**, 91 (1959).
50. H. H. Nielsen, *Phys. Rev.*, **40**, 445 (1932).
51. D. G. Burkhard and D. M. Dennison, *Phys. Rev.*, **84**, 408 (1951); J. S. Kohler and D. M. Dennison, *Phys. Rev.*, **57**, 1006 (1940); K. T. Hecht and D. M. Dennison, *J. Chem. Phys.*, **26**, 31 (1957).
52. R. C. Woods, III, *J. Mol. Spectr.*, **21**, 4 (1966); **22**, 49 (1967).
53. J. E. Wollrab, *Rotational Spectra and Molecular Structure* (New York: Academic Press, 1967), Appendix 7; W. Gordy and R. L. Cook, *Microwave Molecular Spectra* (New York: Wiley, 1970), Appendix III.
54. G. Williams, N. L. Owen and J. Sheridan, *Trans. Faraday Soc.*, **67**, 922 (1971).
55. R. F. Curl, Jr., *J. Chem. Phys.*, **30**, 1529 (1959).
56. D. R. Herschbach, *J. Chem. Phys.*, **25**, 358 (1956).
57. R. H. Schwendeman and G. D. Jacobs, *J. Chem. Phys.*, **36**, 1245 (1962).
58. C. Flanagan and L. Pierce, *J. Chem. Phys.*, **38**, 2963 (1963).
59. T. Kasuya, *J. Phys. Soc. Japan*, **17**, 1273 (1960).
60. V. W. Laurie, *J. Chem. Phys.*, **31**, 1500 (1959).
61. R. J. Anderson and W. D. Gwinn, *J. Chem. Phys.*, **49**, 3988 (1968).
62. L. Pierce and L. C. Krisher, *J. Chem. Phys.*, **31**, 875 (1959).
63. L. C. Krisher and E. B. Wilson, Jr., *J. Chem. Phys.*, **31**, 882 (1959).
64. L. C. Krisher and E. Saegebarth, *J. Chem. Phys.*, **54**, 4553 (1971).
65. R. A. Beaudet and E. B. Wilson, Jr., *J. Chem. Phys.*, **37**, 1133 (1962).
66. S. Siegel, *J. Chem. Phys.*, **27**, 989 (1957).
67. R. A. Beaudet, *J. Chem. Phys.*, **40**, 2705 (1964).
68. R. A. Beaudet, *J. Chem. Phys.*, **37**, 2398 (1962).
69. S. S. Butcher, *J. Chem. Phys.*, **38**, 2310 (1963).
70. M. L. Sage, *J. Chem. Phys.*, **35**, 142 (1961).
71. M. R. Emptage, *J. Chem. Phys.*, **47**, 1293 (1967).
72. R. M. Lees and J. G. Baker, *J. Chem. Phys.*, **48**, 5299 (1968).
73. T. Kojima, *J. Phys. Soc. Japan*, **15**, 1284 (1960).
74. J. S. Rigden and S. S. Butcher, *J. Chem. Phys.*, **40**, 2109 (1964).
75. P. Cahill, L. P. Gold and N. L. Owen, *J. Chem. Phys.*, **48**, 1620 (1968).
76. R. E. Penn and R. F. Curl, Jr., *J. Mol. Spectr.*, **24**, 235 (1967).
77. T. Itoh, *J. Phys. Soc. Japan*, **11**, 264 (1956).
78. T. Kojima, E. L. Breig and C. C. Lin, *J. Chem. Phys.*, **35**, 2139 (1961).
79. V. W. Laurie, *J. Chem. Phys.*, **30**, 1210 (1959).
80. P. Cahill and S. Butcher, *J. Chem. Phys.*, **35**, 2255 (1961).
81. R. Nelson and L. Pierce, *J. Mol. Spectr.*, **18**, 344 (1965).
82. P. H. Kasai and R. J. Myers, *J. Chem. Phys.*, **30**, 1096 (1959).
83. L. Pierce and M. Hayashi, *J. Chem. Phys.*, **35**, 479 (1961).
84. J. E. Wollrab and V. W. Laurie, *J. Chem. Phys.*, **54**, 532 (1971).
85. R. C. Woods, III, *J. Chem. Phys.*, **46**, 4789 (1967).

86. A. P. Cox and R. Varma, *J. Chem. Phys.*, **44**, 2619 (1966).
87. H. D. Rudolph, H. Dreizler, A. Jaeschke and P. Wendling, *Z. Naturforsch.*, **22A**, 940 (1967).
88. H. D. Rudolph and H. Seiler, *Z. Naturforsch.*, **20A**, 1682 (1965).
89. E. Tannenbaum, R. J. Myers and W. D. Gwinn, *J. Chem. Phys.*, **25**, 42 (1956).
90. W. M. Tolles, E. Tannenbaum Handelman and W. D. Gwinn, *J. Chem. Phys.*, **43**, 3019 (1965).
91. C. S. Ewig and D. O. Harris, *J. Chem. Phys.*, **52**, 6268 (1970).
92. R. W. Naylor, Jr. and E. B. Wilson, Jr., *J. Chem. Phys.*, **26**, 1057 (1957).
93. V. W. Laurie and D. R. Lide, Jr., *J. Chem. Phys.*, **31**, 939 (1959); W. H. Kirchoff and D. R. Lide, Jr., *J. Chem. Phys.*, **43**, 2203 (1965).
94. J. D. Swalen and C. C. Costain, *J. Chem. Phys.*, **31**, 1562 (1959).
95. L. Pierce, *J. Chem. Phys.*, **34**, 498 (1961).
96. J. V. Knopp and C. R. Quade, *J. Chem. Phys.*, **53**, 1 (1970).
97. R. J. Myers and E. B. Wilson, Jr., *J. Chem. Phys.*, **33**, 186 (1960).
98. E. Hirota, C. Matsumura and Y. Morino, *Bull. Chem. Soc. Japan*, **40**, 1124 (1967).
99. D. R. Lide, Jr., *J. Chem. Phys.*, **33**, 1514 (1960).
100. J. R. Hoyland, *J. Chem. Phys.*, **49**, 1908 (1968).
101. A. M. Ronn and R. C. Woods, III, *J. Chem. Phys.*, **45**, 3831 (1966).
102. D. R. Lide, Jr. and D. E. Mann, *J. Chem. Phys.*, **29**, 914 (1958).
103. D. R. Lide, Jr. and D. E. Mann, *J. Chem. Phys.*, **28**, 572 (1958).
104. P. H. Verdier and E. B. Wilson, Jr., *J. Chem. Phys.*, **29**, 340 (1958).
105. A. S. Esbitt and E. B. Wilson, Jr., *J. Chem. Phys.*, **34**, 901 (1963).
106. F. J. Wodarczyk and E. B. Wilson, *J. Chem. Phys.*, **56**, 166 (1972).
107. E. Hirota, *J. Chem. Phys.*, **37**, 283 (1962).
108. C. E. Cleeton and N. H. Williams, *Phys. Rev.*, **45**, 234 (1934).
109. M. F. Manning, *J. Chem. Phys.*, **3**, 136 (1935).
110. J. D. Swalen and J. A. Ibers, *J. Chem. Phys.*, **36**, 1914 (1962).
111. M. T. Weiss and M. W. P. Strandberg, *Phys. Rev.*, **83**, 567 (1951).
112. C. C. Costain and G. B. B. M. Sutherland, *J. Phys. Chem.*, **56**, 321 (1952).
113. J. M. Lehn and B. Munsch, *J. Chem. Soc.* (*D*), **22**, 1327 (1969).
114. D. R. Lide, Jr., *J. Chem. Phys.*, **27**, 343 (1957).
115. J. E. Wollrab and V. W. Laurie, *J. Chem. Phys.*, **48**, 5058 (1968).
116. G. Cazzoli and D. G. Lister, *J. Mol. Spectr.*, **45**, 467 (1973).
117. D. R. Lide, Jr., *J. Chem. Phys.*, **38**, 456 (1963).
118. E. Hirota, *J. Mol. Spectr.*, **26**, 335 (1968); K. Bolton, N. L. Owen and J. Sheridan, *Nature*, **217**, 164 (1968).
119. K. Bolton and J. Sheridan, *Spectrochim. Acta*, **26A**, 1001 (1970).
120. J. K. Tyler and D. G. Lister, *Chem. Commun.*, 1350 (1971).
121. C. C. Costain and J. M. Dowling, *J. Chem. Phys.*, **32**, 158 (1960).
122. D. J. Millen, G. Topping and D. R. Lide, Jr., *J. Mol. Spectr.*, **8**, 153 (1962).
123. D. R. Lide, Jr., *J. Mol. Spectr.*, **8**, 142 (1962).
124. S. I. Chan, J. Zinn, J. Fernandez and W. D. Gwinn, *J. Chem. Phys.*, **33**, 1643 (1960); S. I. Chan, J. Zinn and W. D. Gwinn, *J. Chem. Phys.*, **34**, 1319 (1961); S. I. Chan, T. R. Borgers, J. W. Russell, H. L. Strauss and W. D. Gwinn, *J. Chem. Phys.*, **44**, 1103 (1966).
125. D. O. Harris, H. W. Harrington, A. C. Luntz and W. D. Gwinn, *J. Chem. Phys.*, **44**, 3467 (1966); H. W. Harrington, *J. Chem. Phys.*, **44**, 3481 (1966).
126. V. W. Laurie, *Acc. Chem. Research*, **3**, 331 (1970).
127. H. Kim and W. D. Gwinn, *J. Chem. Phys.*, **44**, 865 (1966).
128. W. G. Rothschild and B. P. Daily, *J. Chem. Phys.*, **36**, 2931 (1962).

129. L. H. Scharpen and V. W. Laurie, *J. Chem. Phys.*, **49**, 3041 (1968).
130. L. H. Scharpen and V. W. Laurie, *J. Chem. Phys.*, **49**, 221 (1968).
131. M. G. Petit, J. S. Gibson and D. O. Harris, *J. Chem. Phys.*, **53**, 3408 (1970).
132. J. S. Gibson and D. O. Harris, *J. Chem. Phys.*, **52**, 5234 (1970).
133. G. L. McKown and R. A. Beaudet, *J. Chem. Phys.*, **55**, 3105 (1971).
134. A. C. Luntz, *J. Chem. Phys.*, **50**, 1109 (1969).
135. W. C. Pringle, Jr., *J. Chem. Phys.*, **54**, 4979 (1971).
136. D. W. Boone, C. O. Britt and J. E. Boggs, *J. Chem. Phys.*, **43**, 1190 (1965).
137. H. Kim and W. D. Gwinn, *J. Chem. Phys.*, **51**, 1815 (1969).
138. G. G. Engerholm, A. C. Luntz, W. D. Gwinn and D. O. Harris, *J. Chem. Phys.*, **50**, 2446 (1969).
139. L. H. Scharpen, *J. Chem. Phys.*, **48**, 3552 (1968).
140. V. W. Laurie, *J. Chem. Phys.*, **24**, 635 (1956).
141. K. L. Dorris, C. O. Britt and J. E. Boggs, *J. Chem. Phys.*, **44**, 1352 (1966).
142. D. Chadwick, A. C. Legon and D. J. Millen, *Chem. Commun.* (*J. Chem. Soc. D*), 1130 (1969).
143. E. C. Thomas and V. W. Laurie, *J. Chem. Phys.*, **51**, 4327 (1969).
144. D. O. Harris, G. G. Engerholm, C. A. Tolman, A. C. Luntz, R. A. Keller, H. Kim and W. D. Gwinn, *J. Chem. Phys.*, **50**, 2438 (1969).
145. G. Roussy, J. Demaison, I. Botskor and H. D. Rudolph, Contributed Paper at Microwave Conference, Bangor (1970); *J. Mol. Spectr.*, **38**, 535 (1971); I. Botskor, G. Roussy and H. D. Rudolph, Contributed Paper at Microwave Conference, Bangor (1972).
146. C. R. Quade and C. C. Lin, *J. Chem. Phys.*, **38**, 540 (1963); C. R. Quade, *J. Chem. Phys.*, **44**, 2512 (1966); **47**, 1073 (1967); J. V. Knopp and C. R. Quade, *J. Chem. Phys.*, **48**, 3317 (1968).
147. E. Hirota, *J. Chem. Phys.*, **42**, 2071 (1965).
148. S. R. Polo, *J. Chem. Phys.*, **24**, 1133 (1956).
149. K. S. Pitzer, *J. Chem. Phys.*, **14**, 239 (1946).
150. E. Saegebarth and E. B. Wilson, Jr., *J. Chem. Phys.*, **46**, 3088 (1965); E. Saegebarth, *Ph.D. Thesis*, Harvard University (1965).
151. E. B. Wilson, *Chem. Soc. Rev.*, **1**, 293 (1972).
152. H. Dreizler, *Molecular Spectroscopy: Modern Research*, Ed. K. N. Rao and C. W. Mathews (New York: Academic Press, 1972).

7 *The calculation of barriers to internal rotation from torsional frequencies*

A. V. Cunliffe

7.1 Introduction

There are two main methods used in the measurement of barriers to internal rotation in a molecule. In the kinetic method, which includes chemical methods, NMR, dielectric and acoustical relaxation experiments, the essential measurement is of the rate at which a molecule undergoes a transition from one equilibrium conformation to another, and from this, by application of chemical rate theory, activation energies and enthalpies can be calculated. In the alternative spectroscopic method, the energy levels associated with the particular internal rotational mode are investigated and from these are inferred the form of the potential function determining the motion. There are two approaches to this. In the microwave approach transitions associated with the overall rotation of the molecule are measured, so that the essential measurement is of the indirect effect of the torsional motion on the overall rotation. This implies, in particular, that the inter-action between the torsional motion and the overall rotation of the molecule must be carefully treated. In this chapter, however, we are concerned with the second approach in which vibrational frequencies associated with the internal motion are directly measured by infrared, Raman or inelastic neutron-scattering techniques. Historically, much of the relevant theory was initially derived for the indirect microwave method, and as this necessarily concentrates on the rotational aspects of the problem, and particularly on the often small interaction between internal and overall rotation, it is often given in a form which is unnecessarily complicated for the vibrational method. On the other hand, the methods described in this chapter assume that the internal rotational mode can be separated from other vibrations of the molecule and treated in a unique way. This approximation is more likely to be true, the lower the barrier to internal rotation, and the closer the motion is to a rotation rather than a vibration. Thus it may be said that in cases where the rotational aspects of internal rotation are negligible, treatment of the torsional motion within the approximation of a semi-rigid model, in which other vibrational motions are neglected, is simple, but the assumption of no mixing between the torsion and other vibrations is likely to be poor, leading to considerable error in the measured barriers. In contrast, in cases where the rotational aspects of the internal rotation are more important, the more complicated treatment appropriate to microwave spectroscopy may be necessary, but when this is taken into account the barriers obtained are likely to be more reliable.

In order to illustrate the principles involved, we may first consider in detail the case of a molecule with a single top with three-fold symmetry, for instance, a substituted ethane.

7.2 Molecules with a single top with N-fold symmetry

A single-top molecule is one containing one axis for internal rotation, which involves the motion of one part of the molecule, the top, relative to the other, the framework. If one part of the molecule has N-fold symmetry, a common example being a CH_3 group, this is chosen to be the top. The first, often drastic, assumption is that of a semi-rigid model for the molecule. It is assumed that the molecule has only four degrees of freedom, three for overall rotation and one for internal rotation. Any interaction of the torsion with other vibrations of the molecule is neglected, and it is assumed that a transition associated with a purely torsional mode can be identified. Thus, the internal rotation mode is treated in a completely different way from all other internal motions in the molecule. Clearly, this is a better approximation for molecules with a light top, such as ethanes with CH_3 tops, where rotational (tunnelling) effects are more important, than for heavier molecules, such as 1,1,2,2-tetrachloroethane, where the distinction between the torsion and other low frequency modes is less distinct. However, this approximation is necessary in order to calculate barrier heights from the torsional data. In the alternative normal coordinate treatment of molecular vibrations,[1] the torsion is treated simply as a harmonic vibration, and the data is adjusted to give the best fit for all the molecular vibrations.

For a top with N-fold symmetry, the potential function must have N-fold periodicity. Thus, it can be expressed as the general Fourier series:

$$V = \sum_{k=1}^{\infty} \frac{V_{kN}}{2}(1 - \cos kN\alpha) \tag{7.1}$$

It has been shown[2,3] that in most cases only the first term is important, so that, for $N = 3$, a potential function given by equation (7.2) is appropriate, a small V_6 term being introduced as a perturbation, if the available data is sufficiently refined.

$$V = \frac{V_3}{2}(1 - \cos 3\alpha) \qquad (7.2)$$

7.2.1 *Kinetic energy*

The complexity arises in the treatment of the kinetic energy. The total Hamiltonian for the four degrees of freedom model may be formally written

$$H = H_R + H_{RT} + H_T \qquad (7.3)$$

where H_R is the kinetic energy of overall rotation, H_T is the torsional kinetic energy and potential energy, and II_{RT} is a cross-term representing the interaction between internal and overall rotation. It is in the importance of this cross-term that the fundamental difference between the approach for microwave spectroscopy and for vibrational spectroscopy occurs. In microwave spectroscopy, transitions associated with H_R are measured, and the only effect of the torsion on these levels is through the cross-term H_{RT}. Thus, the relevant theory concentrates on treating the cross-term H_{RT} accurately, which is a difficult problem. Microwave investigations are usually restricted to molecules with some simplifying features, for instance a top with N-fold symmetry, the general problem of an unsymmetrical top and framework being extremely complicated.

The exact form of the kinetic energy depends on the particular approach adopted. For the microwave method, two approaches are used, which differ in the choice of axes used to observe the overall rotation of the molecule.[4] In the principal axis method, which is most convenient in the form described by Herschbach,[4–6] the molecular axes are chosen to be the principal axes of the molecule, rigidly attached to the framework with their origin at the centre of mass. This has the advantage that the rotational energy has a simple form. However, the cross-term H_{RT} is not small relative to H_R, so it is necessary to extend the perturbation treatment to high order to treat H_{RT} adequately. It should be noted that, although this method uses a principal axis system, as these are fixed in the framework, it differs from the normal approach to vibrating–rotating molecules, where the axes are chosen so that the net vibrational angular momentum about each axis is zero;[1,4,7] in the latter case, the moving axes allow, to a first approximation, the separation of rotation and vibration, as is usually assumed in vibrational spectroscopy.

In the other method applicable to microwave spectroscopy, known as the internal axis method, the torsional axis is shown to be one of the coordinate axes.[4] It is then possible, by means of the Nielsen transformation,[4] to define a set of internal axes which are not fixed in the framework, but which move relative to the framework and the top, such that the total angular momentum of the top and framework about the torsional axis due to the internal rotation vanishes. Thus, to an observer located in this frame of reference, both the top and the framework appear to be moving, and the overall molecular motion is seen to be a rotation of the molecule, together with a rapid oscillation of the top and framework relative to the internal axis. In this approach, the cross-term H_{RT} is greatly reduced, or, in the case of symmetric top molecules, completely eliminated. The coupling between internal and overall rotation has not been entirely eliminated, however, as it reappears in the boundary conditions for the internal rotation equation, which, because of the transformation to internal axes, depend on the rotational quantum number K. It is the need to be able to work out these boundary conditions which governs the choice of the internal rotation axis as one of the internal axes, so that the axes are not principal axes, and H_R has an inconvenient form, with products of inertia, as compared with the principal axes system.

For the infrared problem, an alternative approach, due to Pitzer and co-workers,[8–10] is possible, which combines the most convenient aspects of both treatments. It affects an approximate separation of the internal rotation and overall rotation, analogous to that used in the normal treatment of molecular vibrations and rotations, and uses an axis system which is best described as a principal internal axis system. Thus, the axes chosen are principal axes, but Nielsen transformations are applied to define a set of principal axes which move relative to both the top and the framework as the molecule undergoes internal rotation, and which eliminate the interaction between internal and overall angular momentum, H_{RT}, completely. As in the internal axis method, the coupling reappears in the boundary conditions for the torsional Hamiltonian. However, because of the choice of principal internal axes, it is not possible to define these boundary conditions, so that the method cannot be used to give a complete solution. Its value lies in the fact that for most problems studied by the vibrational method, the interaction between overall and internal rotation is negligible relative to the torsional energies. Since this interaction is incorporated into the boundary conditions for the torsional problem, it follows that its neglect is equivalent to replacing the non-periodic boundary conditions of the internal axis method with the periodic ones of the principal axis approach.[3–5,8] This is shown by comparing the principal axis equations given by Herschbach[5] and the equations given by Pitzer.[8] Moreover, Pitzer[8] has shown that the energies for non-periodic boundary conditions lie between those for periodic ones. Thus in cases where the energies for periodic boundary conditions of different symmetries are identical to the accuracy of measurement of the vibrational levels, we are justified in neglecting the interaction between internal and overall rotation, and applying the simplified approach of using the equations of Pitzer[8–10] with periodic boundary conditions for the torsional equation. The axis system used is the same as that used in the treatment of normal vibrations,[1] so that the more the torsion approximates to a harmonic vibration, the better the approximation.

7.2.2 Derivation of the kinetic energy

Pitzer and Gwinn[8] give the Hamiltonian in the form applicable to any molecule in which the moments of inertia for overall rotation are independent of torsion angle (or any molecule which may be regarded as a rigid frame with attached symmetric tops). The kinetic energy is then,[8,11] for the case of one top,

$$2T = \sum_{i=1}^{3} I_i(\omega_i'^2) + 2I_\alpha\dot{\alpha}\sum_{i=1}^{3} \lambda_i\omega_i' + I_\alpha\dot{\alpha}^2 \tag{7.4}$$

where ω_i' is the instantaneous angular velocity of the rigid frame about the ith principal axis, with moment of inertia I_i, $\dot{\alpha}$ is the internal angular velocity of the top, with moment of inertia I_α, relative to the framework and λ_i is the direction cosine between the top axis and the ith principal axis. The Nielsen transformation is now introduced:

$$\omega_i = \omega_i' + \lambda_i\frac{I_\alpha}{I_i}\dot{\alpha} \tag{7.5}$$

which gives

$$2T = \sum_i I_i\omega_i^2 + I_\alpha\dot{\alpha}^2 - I_\alpha^2\sum_i\frac{\lambda_i^2}{I_i}\dot{\alpha}^2$$

$$= \sum_i I_i\omega_i^2 + I_\alpha\left(1 - \sum_i\frac{I_\alpha\lambda_i^2}{I_i}\right)\dot{\alpha}^2 \tag{7.6}$$

The quantity rI_α given by equation (7.7) is known as the reduced moment of inertia for internal rotation.

$$rI_\alpha = \left(1 - \sum_i\frac{I_\alpha}{I_i}\lambda_i^2\right)I_\alpha \tag{7.7}$$

Defining the conjugate momenta

$$P_i = \frac{\partial T}{\partial\omega_i}; \qquad p = \frac{\partial T}{\partial\dot{\alpha}}$$

the kinetic energy becomes:

$$2T = \sum_i\frac{P_i^2}{I_i} + \frac{p^2}{rI_\alpha} \tag{7.8}$$

It is instructive to compare this treatment with the principal axis method given by Herschbach.[5] Herschbach uses the same kinetic energy expression (equation 7.4), but does not apply the Nielsen transformation. This leads to different expressions for the conjugate momenta. Thus, the equations of Pitzer and Gwinn give:

$$P_i = \left(\frac{\partial T}{\partial \omega_i}\right) = I_i \omega_i = I_i\left(\omega_i' + \lambda_i \frac{I_\alpha}{I_i}\dot{\alpha}\right)$$

$$p = \left(\frac{\partial T}{\partial \dot{\alpha}}\right) = r I_\alpha \dot{\alpha}$$

whereas those of Herschbach give:

$$P_i' = \left(\frac{\partial T}{\partial \omega_i'}\right) = I_i\left(\omega_i' + \lambda_i \frac{I_\alpha}{I_i}\dot{\alpha}\right)$$

$$p' = \left(\frac{\partial T}{\partial \dot{\alpha}}\right) = I_\alpha \dot{\alpha} + I_\alpha \sum_i \lambda_i \omega_i'$$

Thus we see that although $P_i' = P_i$, $p \neq p'$, so that whereas p' is the total angular momentum of the internal top, including contributions from overall rotation, p is the relative angular momentum of the top and framework. Equation (7.5) transforms to an axis system moving relative to both the top and framework. In this system, the overall rotation is the same as that of the rigid molecule, but the momentum operator for the internal motion is altered. The Nielsen transformation is equivalent to a transformation

$$p = (p' - \pi) = r I_\alpha \dot{\alpha}$$

where

$$\pi = \sum_i \frac{P_i \lambda_i I_\alpha}{I_i}$$

Thus the kinetic energy in the principal axis system, as given by Herschbach[5]

$$2T = \sum_i \frac{P_i^2}{I_i} + \frac{(p' - \pi)^2}{r I_\alpha}$$

is exactly equivalent to equation (7.8).

Because of this relationship, although the treatment of Pitzer and Gwinn achieves a formal separation of the internal and overall rotation, the above relationship for p and p' means that p still depends on the overall rotation, so that the interaction appears in the boundary conditions for the internal rotation equation.[4,8] The relationship shows the equivalence of the kinetic energy given by Pitzer and Gwinn and Herschbach, and also shows that neglect of the interaction terms in the principal axis method is equivalent to replacing the non-periodic boundary conditions in the approach of Pitzer and Gwinn by periodic ones.

Thus the kinetic energy separates, approximately, into a term corresponding to the overall rotation of a rigid molecule, and the term p^2/rI_α. The total torsional Hamiltonian is given by

$$H_{\mathrm{T}} = T_{\mathrm{T}} + V = \frac{p^2}{2rI_\alpha} + V$$

The appropriate quantum-mechanical Hamiltonian is obtained by the substitution

$$p \rightarrow \frac{h}{i}\frac{\partial}{\partial\alpha}$$

Thus, the required Hamiltonian is

$$H = -\frac{h^2}{8\pi^2 rI}\frac{\partial^2}{\partial\alpha^2} + V$$

It is usual to express the energies in wave numbers when equation (7.8) becomes:

$$H = -F\frac{\partial^2}{\partial\alpha^2} + V \tag{7.9}$$

where

$$F = \frac{h}{8\pi^2 rI_\alpha c}$$

where h is Planck's constant and c is the velocity of light.

7.2.2.1 *Solution of the torsional equation*   Assuming that we can neglect terms in equation (7.1) with $k > 1$, the torsional wave-equation for a molecule with a top with N-fold symmetry is given by

$$\left[-F\frac{\partial^2}{\partial\alpha^2} + \frac{V_N}{2}(1 - \cos N\alpha)\right]U(\alpha) = EU(\alpha) \tag{7.10}$$

where $U(\alpha)$ and E are the appropriate eigenfunctions and eigenvalues. This is related to the standard form of the Mathieu equation,[4,5,12,13]

$$\frac{\mathrm{d}^2 M(x)}{\mathrm{d}x^2} + (b - S\cos^2 x)M(x) = 0 \tag{7.11}$$

by the substitutions:

$$2x = N\alpha + \pi \qquad S = \frac{4V_N}{N^2 F}$$

$$b = \frac{4E}{N^2 F} \qquad M(x) = U\left(\frac{N\alpha + \pi}{2}\right)$$

In the fixed axis representation, since the boundary condition is invariant under

$$\alpha \to \alpha + 2\pi$$

the relevant eigenfunctions are periodic, and may be expressed as:

$$U(\alpha) = \exp(i\sigma\alpha) \sum_{k=-\infty}^{+\infty} A_k \exp(iNk\alpha) \tag{7.12}$$

where σ must be an integer. Values of σ differing by N give the same solution, the difference being absorbed by shifting the index k which runs from $-\infty$ to $+\infty$, so there are N values of σ such that $0 \leqslant |\sigma| \leqslant N/2$. The two solutions corresponding to $\pm\sigma$ have the same periodicity and are associated with a degenerate eigenvalue; the exceptions, which give non-degenerate levels, are $\sigma = 0$, which always occurs, and $\sigma = N/2$, which only occurs for even values of N. If $\sigma = 0$, the periodicity of the eigenfunction is $2\pi/N$ in α and π in x; if $\sigma \neq 0$ but a divisor of N, the periodicity is $2\pi/\sigma$ in α and $N\pi/\sigma$ in x; otherwise, the periodicity is 2π in α and $N\pi$ in x.

These considerations are best illustrated by taking the two most common examples. The simplest case is that of a top with a two-fold symmetry, the commonest examples being simple aromatic molecules such as phenol, benzaldehyde etc. In this case two types of solutions are possible, both non-degenerate, corresponding to $\sigma = 0$ and $\sigma = 1$. For $\sigma = 0$ (symmetric), the eigenfunctions are periodic in π, whereas for $\sigma = 1$, they are periodic in 2π. For tables expressed as solutions of equation (7.11), that is with x as variable, solutions periodic in π for $\sigma = 0$, and periodic in 2π for $\sigma = 1$, are required. For $N = 3$, for instance a CX_3 top, which is probably the most important case of all, two types of solution are again possible. For the non-degenerate levels with $\sigma = 0$, the eigenfunctions are periodic in $2\pi/3$ (A levels). For $\sigma = \pm 1$, the levels are doubly degenerate E levels, and the eigenfunctions are periodic in 2π. For solutions in terms of x, solutions periodic in π for $\sigma = 0$, and periodic in 3π for $\sigma = \pm 1$, are required.

The solution of equation (7.10) or (7.11) is described by Lin and Swalen.[4] Substituting equation (7.12) into (7.10), followed by multiplication by $\exp[-i(Nk + \sigma)\alpha]$ and integration, leads to a recursion formula among the A_k's.

$$A_{k-1} + (\lambda - M_k)A_k + A_{k+1} = 0 \tag{7.13}$$

where

$$\lambda = \frac{4}{S}\left(b - \frac{S}{2}\right)$$

and

$$M_k = \frac{16}{N^2 S}(Nk + \sigma)^2 = \frac{16}{S}\left(k + \frac{\sigma}{N}\right)^2$$

It is seen from equation (7.13) that the same solutions are obtained, for the same value of σ/N, for problems of different N. In particular, the totally symmetric solutions $\sigma = 0$ are the same for all problems, irrespective of the periodicity N of the barrier. In practical terms, tables of Mathieu functions will be given in terms of different periodicities in x. The value of σ/N defines which periodicity is required. Thus, if N/σ is integral, the required periodicity is $N\pi/\sigma$ in x; if N/σ is non-integral, the periodicity is $N\pi$ in x, whereas for $\sigma = 0$, the functions periodic in π are required, for all values of N.

Equation (7.13) leads to a continued fraction. Dividing by A_k,

$$\frac{A_{k+1}}{A_k} = (M_k - \lambda) - \frac{A_{k-1}}{A_k} \tag{7.14}$$

Inverting,

$$\frac{A_k}{A_{k+1}} = G_k^- = \frac{1}{M_k - \lambda - (A_{k-1}/A_k)} \frac{1}{M_k - \lambda - (1/M_{k-1} - \lambda - \cdots)}$$

Similarly,

$$G_k^+ = \frac{A_k}{A_{k-1}} = \frac{1}{M_k - \lambda - (1/M_{k+1} - \lambda - \cdots)}$$

Substituting back into (7.14),

$$\lambda = M_k - G_{k+1}^+ - G_{k-1}^- \tag{7.15}$$

This can be solved by successive approximation methods, as described by Lin and Swalen.[4] It is seen that, for a given periodicity, equation (7.13) depends only upon the quantity S, which is proportional to both the barrier height and the reduced moment of inertia for the internal rotation, since

$$S \propto \frac{V_N}{F} \propto V_N r I_\alpha$$

Solution of equation (7.15) gives λ and hence b in terms of S. Thus, for a given periodicity, the eigenvalues may be tabulated as tables of b as a function of S.

The nature of the energy levels depends on the barrier height (or more accurately on S). Thus, for free rotation, the eigenfunctions are simply $A\,e^{im\alpha}$, where m is an integer, which are the basis functions used in equation (7.12). The energies are then simply Fm^2, the levels occurring in degenerate pairs corresponding to $\pm m$, except for $m = 0$. For $N = 3$, the symmetries of the levels are $m = 0(A)$, $m = \pm1(E)$, $m = \pm2(E)$, $m = \pm3(A)$, $m = \pm4(E)$, $m = \pm5(E)$ etc., so that the levels go A, E, E, A, E, E, etc. Thus, for very low barriers, the splitting between the A and E levels is very large. This splitting, in fact, reflects the rotational character of the internal motion. However, for large S, the levels are grouped in a different way. The levels may be labelled $b_{v\sigma}$, where $\sigma = 0$ for A levels and ±1 for E levels; successive symmetric levels are labelled with the index v with $v = 0$ for the lowest eigenvalue (i.e. the symmetric levels are labelled b_{00}, b_{10}, b_{20} etc.). Similarly, successive E levels are labelled $b_{0\pm1}$, $b_{1\pm1}$, $b_{2\pm1}$ etc. For very high barriers, the spacing between the A and E levels for a given value of v is very much smaller than that between levels with different v, for levels well below the barrier height. The index v is called the torsional quantum number, and the A and E levels associated with a given v are torsional sub-levels belonging to the same torsional state. In this situation, we may say that the vibrational nature of the motion predominates, and the rotational aspect, as reflected by the $A–E$ splitting, is negligible. The importance of the latter increases as we go to higher values of v, so that when we approach the top of the barrier it becomes important. Thus, in a number of cases studied experimentally, although the $A–E$ splitting is negligible for the $0 \longrightarrow 1$ transition, it becomes measurable for higher transitions.[2] This is illustrated in Figure 7.1, which shows the variation of energy levels with S. It is seen that for large S the lowest levels are purely vibrational, whereas energy levels above the barrier are predominantly rotational.

7.2.3 *The harmonic oscillator approximation*

For very large values of S (i.e. large barriers), the internal motion eventually approximates to a harmonic oscillation of small amplitude. Thus the potential energy $(V_N/2)(1 - \cos N\alpha)$ may be expanded as a power series, giving as a first approximation $V = N^2 V_N \alpha^2/4$. The corresponding eigenvalues are

$$E_v = N(V_N F)^{\frac{1}{2}}(v + \tfrac{1}{2})$$

With this approach, since we are only considering a harmonic potential near the bottom of a potential well, the wave-functions are localized in one well, so that we have three degenerate functions $H_v^i(\alpha)$, one situated at each of three potential minima. These can be combined into three wave-functions

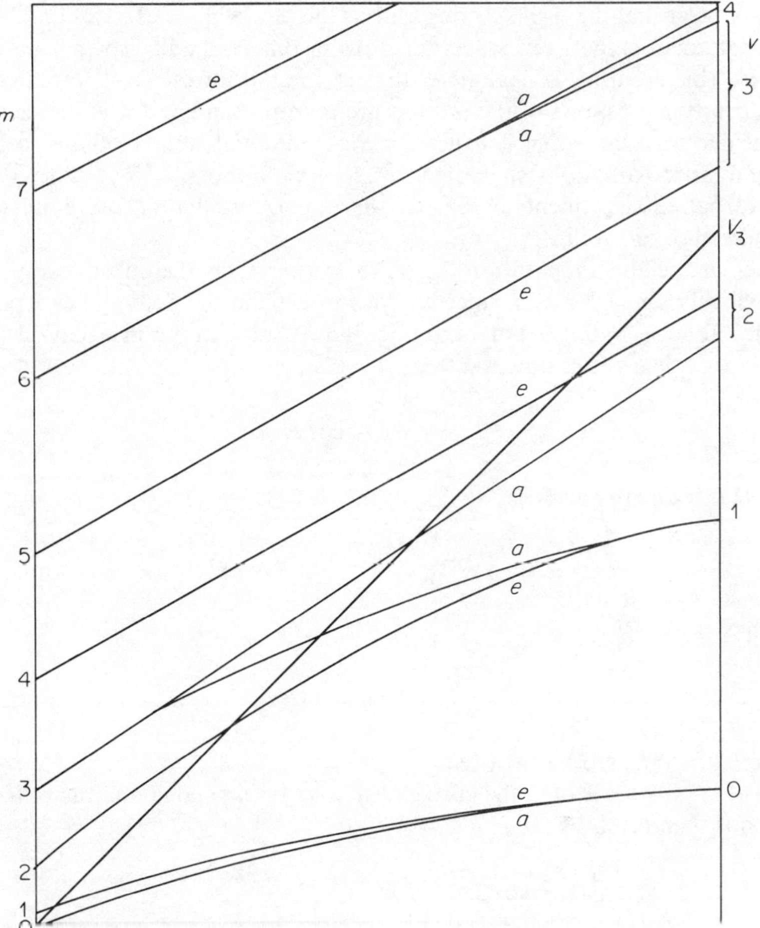

Figure 7.1 Variation of the Mathieu energy levels for a three-fold barrier with barrier height V_3; m is the quantum number associated with the free rotational levels, and v is the vibrational quantum number

of the correct symmetry (one A function U_{v0} and two degenerate E functions U_{v+1} and U_{v-1}) as follows:

$$U_{v0}(\alpha) = 1/\sqrt{3}[H_v^{(1)} + H_v^{(2)} + H_v^{(3)}]$$

$$U_{v+1}(\alpha) = 1/\sqrt{3}[H_v^{(1)} + \omega H_v^{(2)} + \omega^2 H_v^{(3)}]$$

$$U_{v-1}(\alpha) = 1/\sqrt{3}[H_v^{(1)} + \omega^2 H_v^{(2)} + \omega H_v^{(3)}]$$

where $\omega \equiv \exp(2\pi i/3)$.

It is clear that the A–E splitting reflects the tunnelling of the top, and hence the interactions between wave-functions in different wells, so that in cases where this splitting is negligible (that is, for the lowest levels for high S), the rotational aspects of the motion are unimportant, and it is sufficient to consider wave-functions localized in one potential well. In these cases, a useful approximation, suggested by several authors,[14,15,16] is to use a perturbation treatment based on harmonic oscillation wave-functions localized in one well.

We make the substitution $2x = N\alpha$ (rather than the substitution used previously, $2x = N\alpha + \pi$, since we wish to expand the potential as a power series about $\alpha = 0$, the potential minimum). Substituting in equation (7.9), with a single Fourier potential term

$$V = \frac{V_N}{2}(1 - \cos N\alpha)$$

the Hamiltonian becomes

$$\frac{4H}{N^2 F} = \frac{-\partial^2}{\partial x^2} + \frac{S}{2}(1 - \cos 2x)$$

with $S = 4V_N/N^2 F$.

$$\frac{4H}{N^2 F}[U(\alpha)] = bU(\alpha),$$

where $b = (4E/N^2 F)$ as before.

For small values of x, the cosine term may be expanded and the potential term is then given by:

$$\frac{S}{2}(1 - \cos 2x) = Sx^2 - \frac{Sx^4}{3} + \frac{2Sx^6}{45} + \cdots$$

Basis wave-functions are then chosen to be eigenfunctions of the harmonic oscillator Hamiltonian

$$\frac{4H}{9F} = \frac{-\partial^2}{\partial x^2} + Sx^2$$

Applying a perturbation treatment to second order for the terms x^4 and first order for those in x^6, the energies are given by:

$$b_v = (2v + 1)\sqrt{S} - \tfrac{1}{4}\{(v + 1)^2 + v^2\} - \frac{1}{16\sqrt{S}}\{2v^3 + 3v^2 + 3v + 1\} \quad (7.16)$$

Equation (7.16) is sufficiently accurate for most cases where only a single torsional transition, assigned to the $0 \rightarrow 1$ transition, is observed. For $S = 80$, which is about the minimum value for substituted ethanes, the equation gives an accuracy of about 0.1% for the $0 \to 1$ transition, and the accuracy increases fairly rapidly as S increases. However, it is less accurate for higher transitions. Moreover, although it is possible to include V_{2N} and V_{3N} terms in a similar fashion, it is less accurate for these terms, as the higher powers of α are relatively more important because of their higher coefficients. Thus, the full Mathieu approach should be used in cases where hot bands are observed, particularly if splitting between levels of different symmetry (e.g. A–E splittings) are observed.

7.2.4 V_{2N} and higher terms

If only one transition is observed for a molecule with an N-fold symmetric top, it is necessary to assume that the potential function consists of a single Fourier term V_N. However, in certain cases[2,3,17,18] higher transitions (i.e. $1 \to 2$, $2 \to 3$ transitions, also known as hot bands) are observed, and it is then possible to include higher terms in the general Fourier series (equation 7.1). The most common case is the addition of a V_6 term to a barrier with three-fold symmetry. The addition of a V_6 term to a V_3 barrier does not alter the barrier height, but alters the shape of the barrier,[17] slightly affecting the energy levels near the bottom of the well.

The V_{2N} terms are almost invariably much smaller than the V_N terms, and can thus be treated as perturbations.[2,3,17,18] First the values of S and V_N are calculated assuming that V_{2N} is zero. In general, if V_{2N} is not zero, different values of V_N will be obtained for each transition. Using the approximate value of V_N and S, it is possible to evaluate the matrix elements

$$\left\langle U_{v'}(\alpha) \left| \frac{S_{2N}}{2}(1 - \cos 2N\alpha) \right| U_v(\alpha) \right\rangle$$

for trial values of V_{2N}/V_N. Applying perturbation theory, the corrected eigenvalues to second order are given by (using the notation of Herschbach[6])

$$b'_{v\sigma} = b_{v\sigma} + \frac{V_6}{V_3}\frac{S}{2}(1 - 6_{vv}) + \left(\frac{V_6}{V_3}\right)^2 \frac{S^2}{4}\frac{66}{\Delta}$$

The term 6_{vv} is the diagonal matrix element of $\cos 6\alpha$

$$6_{vv} = \langle \psi_{v\sigma} | \cos 6\alpha | \psi_{v\sigma} \rangle$$

and the term $66/\Delta$ represents the second-order corrections of the form

$$\frac{66}{\Delta} = \sum_{v'} \frac{\langle \psi_{v'\sigma} | \cos 6\alpha | \psi_{v\sigma} \rangle \langle \psi_{v\sigma} | \cos 6\alpha | \psi_{v'\sigma} \rangle}{b_{v\sigma} - b'_{v\sigma}}$$

The appropriate matrix elements are tabulated.[3,6] Applying these terms as corrections to the energy levels, new values of V_N are calculated from the values of $b_{v\sigma}$ for each transition. The correct value of V_{2N}/V_N is that which gives the same value of V_N from each transition. Experimental values for V_{2N} terms indicate that they are usually of the order of one hundredth of the V_N terms or less,[2,3,17,18] so that the approximation of a single Fourier term appears to be a very good one in most cases.

Fateley and Miller[2] point out that the effects of the V_{2N} term are also produced by variations in F between different energy levels. However, they obtain a more consistent interpretation of their data for CH_3 tops with a V_{2N} term, and conclude that this is the dominant effect.

In molecules such as ethyl fluoride,[18] ethyl chloride[2,3] and ethyl bromide,[3] where several hot bands are observed, energy levels are seen which are more than three-quarters of the way up the potential well, so that the results should accurately determine the shape and height of the potential barrier.

7.2.5 *Selection rules*

Two factors are involved in a consideration of the selection rules for torsional motions in the high barrier case studied by vibrational spectroscopy, namely: (a) is the torsion allowed in the infrared, Raman etc. and (b) what are the selection rules between the torsional levels of different symmetries (e.g. between the A and E levels for a three-fold barrier)? Question (a) can usually be treated by the usual methods for normal vibrations,[1] treating the torsion as a normal vibration. Usually, the infrared selection rules are obvious from the classical consideration that, in order to be active, the motion must produce a change in direction and/or magnitude of the dipole moment. Thus, for CH_3CCl_3, it is clear that torsion about the C—C bond leaves the dipole moment unchanged, so that the motion is infrared inactive, whereas torsion of the methyl group in ethyl chloride causes a corresponding oscillation in space of the dipole moment, which is fixed in the CH_2Cl group, such that the overall angular momentum is zero, so that the motion is infrared active. Alternatively, the normal group theoretical considerations may be applied, that, for a normal vibration to be infrared active, the normal coordinate must belong to the same irreducible representation as one of the Cartesian coordinates x, y or z, with similar considerations for Raman spectroscopy (the normal coordinate must belong to the same irreducible representation as x^2, y^2, z^2, xy, xz or yz).[1]

In order to treat the second factor, it is necessary to consider the symmetry of the torsional wave-functions in more detail. The discussion is limited to the case of a three-fold barrier with C_s symmetry for the rigid molecule[3] (e.g. CH_3CH_2X). The Hamiltonian is invariant under all products of the operations E, σ_S (reflection in the plane of symmetry) and

$$\delta^1 : \alpha \longrightarrow \alpha + 2\pi/3$$

$$\delta^2 : \alpha \longrightarrow \alpha + 4\pi/3$$

It may be shown that these operators form a group which is isomorphic with the C_{3v} group, so that the Mathieu wave-functions belong to the irreducible representations A_1, A_2, E. The E levels are the wave-functions previously labelled E (i.e. those periodic in 2π, $\sigma = \pm 1$). The A_1 and A_2 levels were previously denoted simply as A ($\sigma = 0$). These levels are, in fact, alternately symmetric and antisymmetric in α, so that the level b_{v0} is A_1 if $v = 0, 2, 4$ etc. and A_2 if $v = 1, 3, 5$ etc. The coordinate α is A_2, so that the selection rules are $A_1 \leftrightarrow A_2$, $E \leftrightarrow E$ ($\Delta\sigma = 0$, $\Delta v = \pm 1$).

7.2.5.1 *Practical analysis of infrared data for a potential function containing a single Fourier term* So far, we have described in detail the relevant theory for the analysis of infrared torsional data, assuming that the potential function solely or predominantly consists of a single Fourier term

$$V = \frac{V_N}{2}(1 - \cos N\alpha)$$

At this stage it is useful to summarize the steps involved in estimating a value of V_N from the observed torsional transitions. It is clear that the observed torsional frequency depends both on the barrier height V_N and the inertial parameter, the torsion, constant F. Thus, in order to measure a value for V_N, we require a value for F. Fateley and Miller[2] have recommended the use of both infrared and microwave data to measure both F and V_N, but in general values for F are assumed, for instance from microwave or electron diffraction data for similar molecules, to measure values of V_N. Considering that the infrared data is often accurate to a fraction of a wave-number in several hundred wave-numbers, this uncertainty in F often represents by far the greatest error in V_N. However, assuming that we have estimates of the bond lengths and angles for a molecule, calculation of F is straightforward.

The first step is to calculate the magnitudes and directions of the principal moments of inertia for the molecule. A convenient Cartesian coordinate system located at the centre of mass of the molecule, is chosen (for instance, for a molecule with a plane of symmetry, one principal axis is perpendicular to the plane, so that it is convenient to choose this as one axis) and the coordinates of the atoms in this axis system are evaluated. The inertial matrix[19] for the whole molecule

$$I = \begin{pmatrix} I_x^2 & I_{xy} & I_{xz} \\ I_{xy} & I_y^2 & I_{yz} \\ I_{xz} & I_{yz} & I_z^2 \end{pmatrix}$$

is then calculated, where

$$I_x^2 = \sum_i m_i(y_i^2 + z_i^2)$$

$$I_{xy} = -\sum_i m_i(x_i y_i) \text{ etc.}$$

An orthogonal transformation is then applied to the coordinates x, y, z[20] so that the new inertial matrix has the form

$$I' = \begin{pmatrix} I_1 & 0 & 0 \\ 0 & I_2 & 0 \\ 0 & 0 & I_3 \end{pmatrix}$$

We thus find the matrix

$$C = \begin{pmatrix} C_{x1} & C_{x2} & C_{x3} \\ C_{y1} & C_{y2} & C_{y3} \\ C_{z1} & C_{z2} & C_{z3} \end{pmatrix}$$

such that

$$CIC = I'$$

and

$$C^+C = E$$

where E is the unit matrix.

C_{jk} is then the direction cosine between the jth axis (x, y, z) of the initial coordinate system and the kth principal axis.

Once the principal moments of inertia of the molecule are found, the reduced moments of inertia may readily be calculated by the methods given by Pitzer and coworkers.[8–10] The equations used depend upon the complexity of the problem. The simplest case is that in which the top has N-fold symmetry with $N \geqslant 3$ (i.e. it is a symmetric top in the rotational sense). In this case, the moment of inertia of the whole molecule is independent of the torsion angle α. The reduced moment of inertia $rI\alpha$ for internal rotation is then given by[4,5,8]

$$rI_\alpha = I_\alpha \left(1 - \sum_i \frac{I_\alpha}{I_i} \lambda_i^2 \right)$$

where I_α is the moment of inertia of the top about its axis, λ_i is the direction cosine between the top axis and the ith principal axis of the whole molecule, and I_i is the moment of inertia of the whole molecule about this axis. Very similar equations apply for the case of several symmetric tops.[5,8]

If the top has two-fold symmetry or less, then the moment of inertia of the whole molecule depends upon the torsion angle α. The equations given by Pitzer[8,9] allow the calculation of rI_α and hence F for any angle. Two approaches are then possible. In principle, the variation of F with α may be allowed for by expressing F as a Fourier series[21,22] or a power series[16] in α,

$$F = F_0 + \sum_n Fn \cos n\alpha$$

and treating the terms with $n > 0$ as perturbations. However, it is usual to calculate a value of F for the equilibrium conformation and to neglect the slight variation of F with α.[17] This is likely to be a good approximation for the lowest levels of high barriers where the effective amplitude of the motion, and hence the variation in F, is likely to be small.

To complete the possible situations, equations for the case of several un-symmetrical tops are given in References 9 and 10.

Given a value for rI_α and hence F from equation (7.9), values of $\Delta b_{v,v+1}$ are calculated from

$$\Delta b_{v,v+1} = \frac{4\Delta E_{v,v+1}}{N^2 F}$$

where

$$\Delta b_{v,v+1} = b_{v+1} - b_v$$

For many molecules with high barriers and large reduced moments of inertia, the motion is essentially harmonic, a single torsional frequency is observed, and equation (7.16) can be used to give a sufficiently accurate value of V_N. If no splitting of the transitions due to levels of different symmetries, e.g. A–E splittings for $N = 3$, are observed, the tables of Mathieu functions periodic in π are applicable for all values of N. Table 7.1 lists several sources of Mathieu functions which are useful in the internal rotation problem.

For the most detailed infrared data, tables of Mathieu functions of the appropriate periodicity are required. The most common examples are molecules with CH_3 tops, where relatively low S values (between 50 and 100) are found,[2,3,17,18,28] so that anharmonicity effects are important. It is often possible to observe several hot bands, and the splitting between A and E transitions may be several wave numbers for the $2 \rightarrow 3$ and $3 \rightarrow 4$ transitions.[2,3,17,18] The observation of several transitions is made possible by the occurrence of sharp Q branches on the otherwise broad gas-phase bands.[3] In these cases, tables of Mathieu functions of the correct symmetry are required. Moreover, for the most detailed calculations, higher Fourier terms can be included, so that, for the three-fold barrier, for instance, tables of the quantities 6_{vv} and $66/\Delta$ are useful. For three-fold barriers with $S \leqslant 100$, the tables given by Herschbach[6] are the most useful, as they list a number of quantities such as b, $6'_{vv}$, $66/\Delta$ etc. connected with the internal rotation problem, together with details of how to calculate these quantities. For $100 < S \leqslant 200$, the tables of Hayashi and Pierce give eigenvalues. Möller and coworkers[3] list the eigenvalues b_v and matrix elements $(p^2)_{vv}$ and 6_{vv} for $100 < S \leqslant 200$ for the A and E states of a three-fold barrier.

For barriers of other periodicities, several tables of eigenvalues are available (Table 7.1), but other useful matrix elements are in general not

Table 7.1   Tables related to the internal rotation problem

Reference	Period in x	Range of v	Range of S	θ_0 (Equation 7.17: $\theta = 0$)	Other information
Mathieu functions[12]	$\pi, 2\pi$	0 (1) 8 9 (1) 14	0 (1) 100 0 (2) 100	$0, \pi$	
Blanch and Rhodes[23]	$\pi, 2\pi$	0 (1) 15	100 to ∞	$0, \pi$	
Kilb[24]	3π	0 (1) 5	2 to 100	$2\pi/3$	Eigenfunctions
Stejskal[25]	$\pi, 3\pi$	0 (1) 4	100 (15) 205	$0, 2\pi/3$	
Pexton[26]	4π	0 (1) 9	1 (1) 40	$\pi/2$	
Herschbach[6]	$\pi, 3\pi$	0 (1) 8	0 (1) 100	$0, 2\pi/3$	$p, p^2, p^3, 6_{vv}, 66/\Delta$
Hayashi and Pierce[27]	$\pi, 3\pi$	0 (1) 15	2 (2) 200	$0, 2\pi/3$	
Möller and coworkers[3]	$\pi, 3\pi$	0 (1) 6	110 (10) 200	$0, 2\pi/3$	$p^2, 6_{vv}$
Herschbach[5]	$\pi, 3\pi$				Fourier coefficients and perturbation coefficients

available. However, these are readily calculated from the equations given by Herschbach for the three-fold case. [5,6]

Finally, Herschbach[5] has shown that the Mathieu eigenvalues can be expressed in a very convenient form which allows the calculation of eigenvalues of any periodicity (periodic or non-periodic) from values of known periodicity. Thus, as first shown by Köhler and Dennison, the energy levels may be expressed as a Fourier series

$$W_{v\sigma} = \frac{N^2}{4} \sum_{e=0}^{\infty} W_e \cos(\theta - \theta_0) \tag{7.17}$$

where

$$\theta = \frac{2\pi}{N} K \frac{I_\alpha}{I_z}, \qquad \theta_0 = -\frac{2\pi\sigma}{N} \quad \text{and} \quad E = FW_{v\sigma}$$

The term θ reflects the non-periodic boundary conditions found in the internal axis method (the boundary conditions are functions of the overall rotational quantum number K), and may be equated to zero in the vibrational problem when interactions between internal and overall rotation are neglected. The coefficients W_e do not depend on σ or N, but only on v and the barrier parameter S. Thus, once a sufficient number of the W_e are determined, for given v and S, the eigenvalues for any periodicity, as specified by θ_0, can be calculated.[5] Moreover, the tables given by Herschbach[5] show that for $S > 40$ only the first two terms in the Fourier series need be considered, as the series in equation (7.17) converges rapidly. Thus, values for only two periodicities are necessary for most high barrier cases of interest in vibrational spectroscopy. The Fourier coefficient for values of S up to 100 are given in Reference 5. This approach can also be used to treat the V_6 potential term for small values of V_6. It may be shown that the effect of a V_{2N} term is simply to alter the coefficients W_e in equation (7.17). Thus, Herschbach[5] lists coefficients W_e such that, for a value of V_{2N} which may be treated as a perturbation, the Fourier coefficients W_e are given by:

$$W_e = W_e^0 + \frac{V_{2N}}{V_N} \cdot W_e'$$

where W_e^0 are the coefficients for a pure V_N barrier.

7.3 Single-top molecules for which neither the top nor framework has a rotational symmetry axis

So far, we have considered molecules for which the top has an N-fold rotation axis, so that the barrier is periodic in $2\pi/N$, and is dominated by a single Fourier term. However, for many molecules of chemical interest both the top and framework are asymmetric. As before, we will be mainly concerned with predominantly three-fold barriers (substituted ethanes) although extension to barriers of different symmetry is straightforward. The classic example of this type of molecule is 1,2-dichloroethene. The barrier is expected to be predominantly three-fold. Symmetry considerations show that the *trans* isomer (I) is one equilibrium conformation. However, rotation of the molecule by $2\pi/3$ from this conformation does not produce an equivalent isomer. Thus the molecule can exist as a *trans* isomer, which has a plane of symmetry and two *gauche* isomers (II), which are mirror images and are obtained from the *trans* isomer by rotations of approximately, though in general not exactly, $2\pi/3$ and $-2\pi/3$. In general, there is an energy difference ΔE between the *trans* and *gauche* isomers, so that the potential well is as shown in Figure 7.2. The *trans* well is symmetrical about the

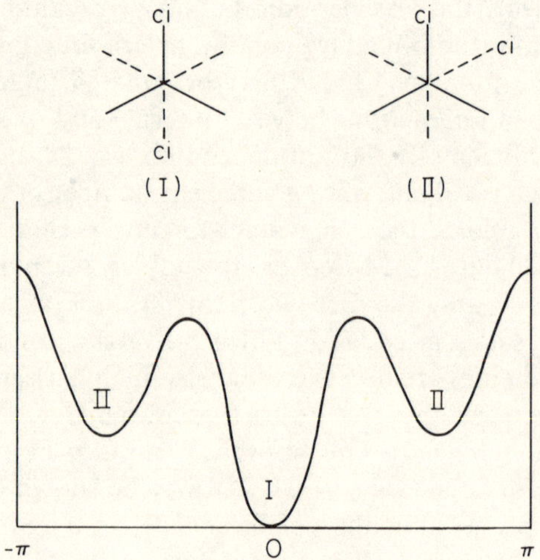

Figure 7.2   Potential   function   for   1,2-
dichloroethane

equilibrium position, but the *gauche* well is in general unsymmetrical. The potential function has the general form of equation (7.1) with $N = 1$, and as there are no symmetry restrictions, all Fourier terms are allowed. (In the most general case, in which neither the top nor the framework has a plane of symmetry, sine terms are also necessary, but this does not introduce any new features into the analysis.)

In addition to the more complicated potential function, a second complication arises because the inertial parameters are now functions of the torsion angle α. The equations of Pitzer[9] still apply for the instantaneous motion of the system, and the kinetic energy associated with the torsion can still be written in the form

$$2T = \frac{p^2}{I_r(\alpha)}$$

where $I_r(\alpha)$, the reduced moment of inertia for the torsional motion, is a function of α. This angular dependence introduces extra complexity into the quantum-mechanical expression for the kinetic energy T. It may be shown that the correct quantum-mechanical expression for the kinetic energy, expressed in wave numbers, is given by[16,29,30]

$$T = g^{-\frac{1}{4}}pg^{\frac{1}{2}}Fpg^{-\frac{1}{4}} \qquad (7.18)$$

where

$$F(\alpha) = \frac{h}{\delta\pi^2 cI_r(\alpha)}; \qquad p = \frac{1}{i}\frac{\partial}{\partial\alpha}; \qquad g(\alpha) = (I_x I_y I_z I_r)$$

The most convenient way of treating this variation of T with α is to calculate F and g for discrete values of α from the equations of Reference 9, and then to fit them either to power series or Fourier series, depending on the method adopted to solve the overall Hamiltonian, as described below. The total Hamiltonian is given by

$$H = T + \sum_{k=1}^{n} \frac{V_k}{2}(1 - \cos k\alpha) \qquad (7.19)$$

Very frequently, there are only three independent pieces of experimental information, so that, assuming by comparison with the symmetrical case that the lowest Fourier terms are the most important ones, we restrict ourselves to the first three Fourier terms so that the Hamiltonian becomes

$$H = T + \sum_{k=1}^{3} \frac{V_k}{2}(1 - \cos k\alpha) \qquad (7.20)$$

Hunt and coworkers[31] found that this potential gave a satisfactory interpretation of the very complex far infrared and microwave spectra of H_2O_2 and D_2O_2, where a great deal of experimental information was observed.

Several approaches are possible to the problem of the asymmetric molecules, depending on the amount of experimental data available and the degree of sophistication attempted. The simplest approach, which is the only one possible when a single isomer is observed, so that only a single torsional frequency is measured, is to assume that the barrier is dominated by a single Fourier term (the V_3 term for a predominantly three-fold barrier in, for instance, a substituted ethane). This gives an approximate barrier height, which may, however, be useful in comparing a series of molecules.

This approach actually measures, in the case of high barriers, the restriction to motion near the potential minimum, rather than the actual barrier height.

7.3.1 *High barriers—harmonic oscillation approximation*

Many molecules of this type contain heavy atoms off the torsion axis in both top and framework, so that the reduced moment of inertia I_r is relatively large, and F is small. Moreover, these bulky atoms result in relatively high barriers (compared, for instance, with molecules with CH_3 tops), so that the net effect is that the barrier parameter S is large. Thus, rotational effects are unimportant and it is a reasonable approximation to treat the potential as harmonic. The variation of T with α is usually neglected, a value of F for the equilibrium conformation (or usually, for $\alpha = 120$ for the *gauche* isomer) being used. Again, this is a better approximation the higher the value of S and hence the lower the amplitude of vibration. The three pieces of information which we expect to measure for this type of molecule are the torsional frequencies for the two isomers and the energy difference ΔE between the two isomers. Usually, only one transition is observed for each isomer, and this is assumed to be the $0 \rightarrow 1$ transition (in fact, it will in general be some weighted average of this and higher transitions). Assuming a V_1, V_2 and V_3 term potential (equation 7.20), and a predominantly three-fold barrier, the three experimental quantities are approximately given by:

$$\frac{v_1^2}{F} = V_1 + 4V_2 + 9V_3 \tag{7.21}$$

$$\frac{v_2^2}{F} = -\tfrac{1}{2}(V_1 + 4V_2) + 9V_3 \tag{7.22}$$

$$\Delta E = \tfrac{3}{4}(V_1 + V_2) \tag{7.23}$$

v_1, F_1 and v_2, F_2 are the observed frequencies and the equilibrium F values in the *trans* and *gauche* wells respectively, and ΔE is equated to the difference in potential between $V = 0$ and $V = 120°$. This type of approach has been applied by Fateley, Miller and coworkers to a number of cases with predominantly two-fold symmetry.[32,33] In this case, assuming potential minima at $\alpha = 0$ and $\alpha = 180°$, the appropriate equations are:

$$\frac{v_1^2}{F} = V_1 + 4V_2 + 9V_3$$

$$\frac{v_2^2}{F} = -V_1 + 4V_2 - 9V_3$$

$$\Delta E = (V_1 + V_3)$$

The next approximation is to allow for the higher powers of α in the Fourier potential. Thus, expanding the potential term in equation (7.20) for small values of α about the symmetric well ($\alpha = 0$), we have:

$$V = \left(\frac{V_1}{4} + V_2 + \frac{9V_3}{4}\right)\alpha^2 - \left(\frac{V_1}{48} + \frac{V_2}{3} + \frac{27V_3}{16}\right)\alpha^4$$
$$+ \left(\frac{V_1}{1440} + \frac{2V_2}{45} + \frac{81V_3}{160}\right)\alpha^6 + \cdots \tag{7.24}$$

Wave-functions are chosen which are eigenfunctions of the harmonic oscillator Hamiltonian, neglecting the variation of F with α,

$$H_0 = F_0 p^2 + \left(\frac{V_1}{4} + V_2 + \frac{9V_3}{4}\right)\alpha^2$$

The wave-functions have the well-known form (see, for instance, Schiff[34])

$$\psi_v = N_v\, e^{-\gamma^2/2} H_v(\gamma) \tag{7.25}$$

where $H_v(\gamma)$ is the vth Hermite polynomial, N_v is a normalization factor given by $N_v = 1/(2^v v!\,\sqrt{\pi})^{\frac{1}{2}}$ and γ is defined by

$$\gamma = \left(\sum_k \frac{V_k k^2}{4F_0}\right)^{\frac{1}{4}} \alpha = \delta\alpha \tag{7.26}$$

The eigenvalue is given by $E/F_0 = (2n + 1)\delta^2$. The higher terms in equation (7.24)

$$H_1 = A\alpha^4 + B\alpha^6$$

are treated as perturbations.

The matrix elements $\langle\psi_j|\alpha^n|\psi_k\rangle$ are evaluated by repeated application of the recursion formula for Hermite polynomials[34]

$$\gamma H_n(\gamma) = n H_{n-1}(\gamma) + \tfrac{1}{2} H_{n+1}(\gamma)$$

$$\alpha H_n(\gamma) = \frac{\gamma}{\delta} \cdot H_n(\gamma)$$

Thus, for the term in α^4, we obtain

$$\alpha^4 H_n(\gamma) = \frac{1}{\delta^4}\left(\tfrac{1}{16}H_{n+4}(\gamma) + \frac{(2n+3)}{4}H_{n+2}(\gamma) + \tfrac{3}{4}(2n^2 + 2n + 1)H_n(\gamma)\right.$$

$$\left. + (2n^3 - 3n^2 - n)H_{n-2}(\gamma) + n(n-1)(n-2)(n-3)H_{n-4}(\gamma)\right)$$

The matrix elements $\langle \psi_j | \alpha^4 | \psi_k \rangle$ may thus be evaluated. For instance

$$\langle \psi_n | \alpha^4 | \psi_n \rangle = (\alpha^4)_{nn} = \frac{N_n^2}{\delta^4} \int_{-\infty}^{+\infty} H_n(\gamma)\gamma^4 H_n(\gamma)\, e^{-\gamma^2}\, d\gamma$$

$$= \frac{N_n^2}{\delta^4} \int_{-\infty}^{+\infty} \tfrac{3}{4}(2n^2 + 2n + 1)[H_n(\gamma)]^2\, e^{-\gamma^2}\, d\gamma$$

$$= \frac{3}{4\delta^4}(2n^2 + 2n + 1)$$

since the ψ_n form an orthonormal set.

The first-order perturbation correction for the α^4 term is given by:

$$\left\langle \psi_v \left| -\sum_k \frac{V_k k^4}{2 \cdot 4!} \alpha^4 \right| \psi_n \right\rangle = E^{(1)}(\alpha^4) = -\sum_k \frac{V_k k^4}{2 \cdot 4!} \cdot \frac{3}{4\delta^4}(2n^2 + 2n + 1)$$

Expressions can be similarly obtained for the next perturbation terms, which are the second-order terms in α^4 and the first-order terms in α^6. Thus, the second-order correction in α^4 is given by:

$$E^{(2)}(\alpha^4) = -\frac{1}{F_0\delta^{10}}\left(\sum_k \frac{V_k k^4}{2 \cdot 4!}\right)^2 \tfrac{1}{16}(34n^3 + 51n^2 + 59n + 21)$$

The first-order correction in α^6 is given by:

$$E^{(1)}(\alpha^6) = \left(\sum_k \frac{V_k k^6}{2 \cdot 6!}\right)\frac{5}{8\delta^6}(4n^3 + 6n^2 + 8n + 3)$$

A similar treatment is possible for the unsymmetrical *gauche* well. In this case, as the barrier is unsymmetrical, sine terms also occur when the potential is expanded about the equilibrium position. The term in α is zero, because of the equilibrium condition, but terms in α^3 and α^5 occur, giving off-diagonal corrections to the energy. These may readily be treated in a similar manner to the even power terms above. However, in a number of halogenated ethanes, we have found[35] that the second-order perturbation theory does not give very good results for the unsymmetrical well, as the deviation from a harmonic potential is considerable; the more complete treatment described below is therefore preferable. A similar perturbation treatment may also be applied to the predominantly two-fold barrier case, and here, as both wells are symmetrical, the method should give reasonable results for both isomers provided the barrier is reasonably high.

7.3.2 *Complete treatment*

A number of authors[21,22] have treated the general problem of an asymmetric top and framework from the point of view of microwave spectroscopy. As in the case of the N-fold symmetrical top, these treatments are unnecessarily complicated for most infrared data, as they are concerned with the effects of the internal rotation on the overall rotation of the molecule, which is unimportant in vibrational spectroscopy. In fact, in general, the rotational aspects of the internal motion are usually less important in the asymmetric case than in the symmetrical one.

Wood[36] has described the exact solution of the Hamiltonian of equation (7.20) with the simplifying assumption that F is independent of angle, so that the kinetic energy is given by

$$T = F_0 p^2$$

It is clear that the exact eigenfunctions of equation (7.20) can be expressed as a Fourier series, and the method therefore constructs the Hamiltonian matrix for a basis set of cosine (even solutions) and sine (odd solutions) wave-functions. However, the method requires relatively large matrices for the high barriers with which we are usually concerned, the size of the matrix required increasing with the barrier height. This is readily understandable when it is considered that the cosine and sine basis functions, and the basis functions used in the complete solution of the Mathieu equation, are closely related to those for free rotation, which is expected to be a poor approximation for the lowest eigenvalues of a high potential barrier. Wood[36] shows that, even for the relatively low barriers which he considers, the variation of energies with boundary conditions, and hence tunnelling effects, are negligible, and the wave-functions are essentially localized in either the *trans* or *gauche* wells. The treatment can easily be extended to include the variation of F and g with α if these quantities are expressed as Fourier series, when the matrix elements of the kinetic energy T of equation (7.18) can readily be evaluated. It may be necessary to apply this treatment in cases where the potential barrier and the reduced moment of inertia for the internal rotation are low.

For a high barrier a much better approximation to the lowest wave-functions is expected to be given by the harmonic oscillator wave-functions.[16] It is assumed that each potential well can be treated separately, so that the wave-functions are localized in one or other of the potential wells, any interaction between wave-functions in separate wells is neglected, and tunnelling effects are therefore ignored.[16]

7.3.3 *Symmetric well*

The symmetric well is symmetrical about the potential minimum, $\alpha = 0$. We have:

$$H = T + \sum_{k=1}^{3} \frac{V_k}{2}(1 - \cos k\alpha) \equiv H_0 + H_1 \tag{7.27}$$

$$H_0 = F_0 p^2 + \sum_{k=1}^{3} k^2 \frac{V_k \alpha^2}{4} \tag{7.28}$$

The basis functions are chosen to be eigenfunctions of the harmonic oscillator Hamiltonian H_0. Thus, the basis functions are given by equation (7.25). If one wishes to include the variation of F and g with α, F and g may be expressed as Fourier series, or, more conveniently for the harmonic oscillator approach, as power series in α for small values of α. Thus, we have:

$$F = F_0 + \sum_{n=1}^{\infty} F_n \alpha_n$$

$$g = g_0 + \sum_{n=1}^{\infty} g_n \alpha_n \qquad (7.29)$$

The coefficients F_n and g_n may be evaluated by least-squares fitting procedures from values of F and g calculated for a series of angles α. If these equations are substituted into equation (7.18), we obtain:

$$T = F_0 p^2 + \sum_{n=1}^{\infty} p\alpha^n p F_n + \tfrac{3}{16} F g^{-2}(pg)^2 - \tfrac{1}{4} g^{-1}(pF)(pg) - \tfrac{1}{4} F g^{-1}(p^2 g)$$

where if p is in a term in brackets, it operates only on the term in brackets. In fact, in all cases for a series of halogenated ethanes which we have studied so far the last three terms (those involving g) have been negligible.[35]

The Hamiltonian matrix H given by equation (7.27) is constructed with the harmonic oscillator wave-functions. The integrals $\langle \psi_j | \alpha^n | \psi_k \rangle$, $\langle \psi_j | p\alpha^n p | \psi_k \rangle$ and $\langle \psi_j | \cos n\alpha | \psi_k \rangle$ are required. The first two integrals are generated from the well-known recurrence relationships[34] for the Hermite polynomials, as described previously:

$$\gamma H_n(\gamma) = n H_{n-1}(\gamma) + \tfrac{1}{2} H_{n+1}(\gamma)$$

$$\frac{\partial}{\partial \gamma}[H_n(\gamma)] = 2n H_{n-1}(\gamma)$$

These lead to the recurrence equations:

$$\alpha | \psi_k \rangle = \frac{1}{\delta}\sqrt{\frac{k+1}{2}} | \psi_{k+1} \rangle + \frac{1}{\delta}\sqrt{\frac{k}{2}} | \psi_{k-1} \rangle$$

$$p | \psi_k \rangle = i\delta\sqrt{\frac{k+1}{2}} | \psi_{k+1} \rangle - i\delta\sqrt{\frac{k}{2}} | \psi_{k-1} \rangle$$

The cosine (and sine) matrix elements are obtained from equations of the type:

$$\int_{-\infty}^{+\infty} H_0(\gamma) H_n(\gamma)\, e^{ib\gamma} e^{-\gamma^2}\, \mathrm{d}\gamma = \sqrt{\pi}\, e^{-b^2/4}(ib)^n \qquad (7.30)$$

Given the equation for $H_j(\gamma)$ and $H_n(\gamma)$, the equation for $H_{j+1}(\gamma)$ and $H_n(\gamma)$ is obtained by differentiating both sides with respect to b and applying the recurrence relationship for $xH_j(x)$. Thus, for the integral $\cos n\alpha$, the real part of the integrals for $e^{ib\gamma}$ is taken with $b = n/\delta$. The Hamiltonian matrix H is then set up and diagonalized numerically to solve for the required energy levels.

7.3.4 Unsymmetrical well

The treatment for the unsymmetrical well is only slightly more complicated. It is convenient to measure the torsion angle relative to the *gauche* potential minimum α_0 which is given by:

$$\cos \alpha_0 = \frac{-V_2 - \sqrt{[V_2^2 - 3V_3(V_1 - V_3)]}}{6V_3}$$

α' is defined as $\alpha - \alpha_0$ and the potential function becomes:

$$H = T + \sum_k \frac{V_k}{2}\left[1 - \cos k(\alpha' + \alpha_0)\right]$$

$$= T + \sum_k \frac{V_k}{2}\left[1 - \cos k\alpha' \cos k\alpha_0 + \sin k\alpha' \sin k\alpha_0\right]$$

The harmonic part of the Hamiltonian is given by:

$$H_0' = F_0' p^2 + \sum_k \left(\frac{V_k k^2}{4} \cos k\alpha_0\right)\alpha'^2$$

In this case, the harmonic oscillator basis functions are given by:

$$\psi_j = N_j^2 \, e^{-\varepsilon^2/2} H_j(\varepsilon)$$

where

$$\varepsilon = \left(\sum_k \frac{V_k k^2 \cos k\alpha_0}{4F_0'}\right)^{\frac{1}{4}} \alpha'$$

$$= \rho\alpha'$$

The sine terms occur because the well is no longer symmetrical about the potential minimum. The appropriate matrix elements are calculated as before.

The accuracy of the method has been investigated[16] by comparison with values calculated using the exact method of Wood[36] neglecting the variation of F with angle, which generally only makes a small correction to the energy levels. It was found that the method gave accurate values for the first four energy levels in each well for values appropriate to 1,2-difluoroethane, which is expected to be the least favourable unsymmetrical halogenated ethane, having the lowest barrier and reduced moment of inertia.

The final problem is to perform the reverse calculation: that is, to calculate the V_k from the experimental measurements. Normally, one expects to measure two $0 \rightarrow 1$ transitions, one for each isomer, and the energy difference ΔE between the two isomers, which is assumed to be equal to the energy difference between the lowest energy levels of the two isomers. Calculation of V_1, V_2 and V_3 from these quantities is most conveniently performed by an iterative method. The above theory is particularly suitable for this as the basis functions are approximately diagonal for the lowest levels, and the calculation of energy levels, which is performed each iteration, is extremely rapid. The standard least-squares fitting method[37] requires a knowledge of the derivatives $(\partial v_i / \partial V_k)$, where v_i is the ith observed frequency and V_k is the kth potential term. It is usually found[16] that it is sufficient to use the harmonic oscillator approximation for the energy levels to calculate these derivatives. The method converges rapidly, to fit the experimental data to better than 0.1 cm^{-1} (which is usually appreciably better than the experimental data) in three or four iterations. Initial estimates of V_1, V_2 and V_3 can be obtained from the approximate equations (7.21), (7.22) and (7.23).

7.3.5 *Mixing of the torsion with other vibrations*

We have seen that it is possible to calculate the torsional energy levels for any molecule consisting of a top and framework to an accuracy sufficient for infrared measurements. Given that the model chosen is satisfactory, the greatest source of errors is likely to be in the calculation of the inertial parameter F. However, in a number of cases, a greater source of error is likely to be in the simple model chosen, namely that other vibrations may be neglected. If appreciable mixing can occur, then the potential function is no longer the simple cosine function assumed in the calculation. In some cases, we can be confident that the mode is a pure torsion, namely if: (a) the torsion is the only fundamental in its symmetry species, or (b) the next nearest fundamental in the same species has a very different frequency. This is the case of ethyl chloride, where the torsion occurs at 250 cm^{-1}, and the next nearest frequency of the same symmetry is at 969 cm^{-1}.

If data are available for deuterated derivatives, it is possible to test the mixing experimentally. Deuteration will appreciably alter the torsional frequency, while the nearest modes, which are likely to be skeletal vibrations, will be little affected. Thus, the mixing between the two should be greatly reduced, so that, if considerable mixing occurs, a different value of V_N should be obtained for the deuterated and undeuterated molecules. If V_N is the same for both deuterated and undeuterated molecules, then mixing of the torsion with other vibrations is negligible. This has been verified in a number of cases.[3,5,17,31] In particular, Fateley and Miller[17] have obtained practically identical V_3 terms from ethyl chloride and three deuterated ethyl chlorides. It appears that, for molecules with CH_3 tops, mixing of the torsion with other vibrations is often negligible. However, in the case of 2-chloropropene, Hunziker and Günthard[38] found that the torsion is appreciably mixed with a low bending mode, obtaining $V_3 \sim 935\,cm^{-1}$ for the CH_3 compound and $\sim 850\,cm^{-1}$ for the CD_3 one.

In the case of heavy molecules of low symmetry, which have low skeletal vibrations of the same symmetry as the torsion, it seems likely that mixing of the torsion with other vibrations is appreciable. In normal coordinate calculations on the *gauche* isomer of 1,2-dichloroethane, Miyazawa and Fukushima[39] found appreciable mixing of the torsion with other vibrations. In general, in molecules of this type, where there is little difference between the torsional motion and other low frequency bending modes, it seems unreasonable to treat the torsion completely separately from the other motions in a one internal degree of freedom model. Thus, it is doubtful if a sophisticated treatment of this problem with a semi-rigid model is appropriate. It may, in fact, be more reasonable to use the harmonic oscillator approximations for the torsions and to allow for mixing of the torsion with other vibrations, rather than to treat the anharmonicity of the torsion but neglect the mixing. In general, it appears that, the lighter the top and the lower the barrier, that is the greater the rotational nature of the motion, the more accurate are the barrier parameters, given that a sufficiently accurate treatment is applied.

7.4 Molecules with more than one top

7.4.1 *Molecules with two CX_3 tops*

By far the most important case appears to be that of molecules with two CX_3 tops (usually two CH_3 tops). Treatments applicable to microwave spectroscopy for this type of molecule have been given by Swalen and Costain[40] and Pierce[41] and infrared investigations have been made by Fateley and Miller[42] and by Möller and coworkers.[3,28] The papers by Möller and coworkers[3] give the most comprehensive account for infrared measurements and show in particular that the treatment in Reference 42 is open to criticism.

7.4.1.1 *Kinetic energy* The kinetic energy for the general case of several tops is given by Pitzer and coworkers[8,10] and Herschbach.[5] Swalen and Costain[40] and Möller[3,28] give explicit equations for two-top molecules with C_{2v} and C_s symmetries. (Möller and Andressen[43] and Fateley and Miller[42] give analogous equations for molecules with three CX_3 tops and C_{3v} symmetry.)

Following the treatment of Pitzer and Gwinn,[8] the kinetic energy for N symmetrical tops is given by:

$$2T = \sum_{i=1}^{3} I_i(\omega_i')^2 + 2\sum_{i=1}^{3}\sum_{m=1}^{N} I_m\lambda_{mi}\dot{\alpha}_m\omega_i' + \sum_{m=1}^{N} I_{\alpha m}\dot{\alpha}_m^2$$

The notation is analogous to that in equation (7.4).

Applying the transformation

$$\omega_i = \omega_i' + \sum_{m=1}^{N} \frac{I_m\lambda_{mi}\dot{\alpha}_m}{I_i}$$

the kinetic energy becomes

$$2T = \sum_i I_i\omega_i^2 + \sum_m I_m^2 - \sum_m\sum_{m'} \Lambda_{mm'}\dot{\alpha}_m\dot{\alpha}_{m'}$$

This has the form:

$$2T = \omega^+ I\omega$$

where

$$\omega^+ = (\omega_x, \omega_y, \omega_z, \dot{\alpha}_1, \dot{\alpha}_2 \ldots)$$

and

$$I = \begin{pmatrix} I_x & 0 & 0 & \\ 0 & I_y & 0 & O \\ 0 & 0 & I_z & \\ & O & & I_m\delta_{mm'} - \Lambda_{mm'} \end{pmatrix}$$

where

$$\Lambda_{mm'} = I_m I_{m'} \sum_{i=1}^{3} \frac{\lambda_{mi}\lambda_{m'i}}{I_i}$$

Introducing the momenta:

$$P_i = \frac{\partial T}{\partial \omega_i}; \qquad p_m = \frac{\partial T}{\partial \dot{\alpha}_m}$$

the kinetic energy becomes:

$$2T = \sum_i \frac{P_i^2}{I_i} + p^+\beta p$$

where $p^+ = (p_1, p_2, \ldots)$ and β is the inverse of the matrix $(I_m\delta_{mm} - \Lambda_{mm'})$. Given that, as in the single-top case, p_m of Pitzer and Gwinn equals $(p_m' - \pi_m)$

of Herschbach, this is exactly equivalent to the equation given by Herschbach[5] from the principal axis method.

The quantum-mechanical equation for the internal kinetic energy is given by

$$T = p^+ F p$$

where

$$p^+ = \left(\frac{1}{i}\frac{\partial}{\partial \alpha_1}, \frac{1}{i}\frac{\partial}{\partial \alpha_2} \cdots \frac{1}{i}\frac{\partial}{\partial \alpha_m} \cdots \right)$$

and

$$F_{ij} = \frac{h}{8\pi^2 c}\beta_{ij}(\text{cm}^{-1})$$

The most common cases of molecules with two CX_3 tops are those with either C_{2v} symmetry (acetone, dimethyl ether, 2,2-dihalopropanes etc.) or C_s symmetry (dimethyl amine, 2-halopropanes etc.). The Hamiltonian then has the general form[3,28] neglecting the terms due to overall rotation

$$H = F(p_1^2 + p_2^2) + F'(p_1 p_2 + p_2 p_1) + \tfrac{1}{2}V(2 - \cos 3\alpha_1 - \cos 3\alpha_2)$$

$$+ \tfrac{1}{2}V^*(\cos 6\alpha_1 + \cos 6\alpha_2) + V_{12}\cos 3\alpha_1 \cos \alpha_2 \qquad (7.31)$$

$$+ V'_{12}\sin 3\alpha_1 \sin 3\alpha_2$$

the explicit equations for F and F' depending on the symmetry.[3,28,40]

V and V^* are analogous to the potential for a single top, and V_{12} and V'_{12} represent the top–top interactions. Following Möller and coworkers[3] it is useful to consider the solution in the harmonic oscillator limit. The Hamiltonian may then be separated into the two equations:

$$H_+ = (F + F')p_+^2 + M\alpha_+^2$$

$$H_- = (F - F')p_-^2 + N\alpha_-^2$$

where α_+ and α_- are the normal coordinates

$$\alpha_\pm = \frac{1}{\sqrt{2}}(\alpha_1 \pm \alpha_2)$$

and p_+ and p_- are the momenta corresponding to α_+ and α_- respectively

$$p_\pm = \frac{1}{\sqrt{2}}(p_1 \pm p_2)$$

The coefficients M and N are given by:

$$M = \tfrac{9}{4}[V - 2(V_{12} - V'_{12}) - 4V^*]$$

$$N = \tfrac{9}{4}[V - 2(V_{12} + V'_{12}) - 4V^*] \qquad (7.32)$$

The torsional frequencies are then given by:

$$V_+ = [4M(F + F')]^{\frac{1}{2}} \tag{7.33}$$

$$V_- = [4N(F - F')]^{\frac{1}{2}} \tag{7.34}$$

It is clear that, in this limit, it is only possible to calculate $V - 2V_{12} - 4V^*$ and V'_{12} even if both transitions are observed (see below). This is to be expected as the terms V_{12} and V^* make only harmonic contributions to the potential whereas the V'_{12} term appears as $V'_{12}\alpha_1\alpha_2$ in the harmonic limit, and thus has the effect of mixing the α_1 and α_2 motion, and hence of splitting the α_+ and α_- energy levels. Its effect is exactly analogous to the terms $F'p_1p_2$ in the kinetic energy, so that, as pointed out by Möller, it is unreasonable to include the F' term while neglecting the V'_{12} term, as in the treatment of Fateley and Miller.[42] In order to obtain values for V_{12} and V^* it is necessary to observe anharmonicity effects in several transitions (i.e. $0 \rightarrow 1$, $1 \rightarrow 2$, $2 \rightarrow 3$ etc.) as for the V_6 term in the single-top case.

7.4.2 *Selection rules*

For molecules with C_{2v} symmetry, the normal coordinates α_+ and α_- belong to the symmetry species A_2 and B_1 respectively. Classically, in the α_+ motion the two tops rotate in the same sense when viewed along the two-fold axis, and the resultant net angular momentum is along the two-fold axis. In order that the overall angular momentum for the internal motion is zero, the framework also rotates about the two-fold axis to compensate for the top angular momentum. However, as the dipole moment also lies along the two-fold axis, the motion does not alter the magnitude or direction of the dipole moment, so that the transition is infrared inactive. The α_- motion produces a net angular momentum along an axis in the symmetry plane containing the heavy atoms of the two tops, and perpendicular to the two-fold axis. Thus, the motion produces a resultant oscillation of the dipole moment about this axis, and the transition is infrared active. (Both transitions are Raman active.)

For molecules with C_s symmetry the α_+ and α_- transitions are A' and A'', so that both are allowed in the infrared and Raman. Classically, the α_+ and α_- motions produce resultant angular momenta along the perpendicular to the axis bisecting the top axes, and rotation of the framework about either of these axes causes a change in direction of the dipole moment. Although both transitions are allowed, they are not necessarily of the same intensity, and simple calculations based on the classical amplitude of motion of the dipole moment for given rotations of the top[35] suggest that the two intensities may be very different. For instance, results for 2-chloropropane suggest that the α_- transition is approximately six times as intense as the α_+ transition. (However, the ease of observation of the two transitions will also depend on the rotational band shape, transitions with sharp Q-branches being readily observed.) This may explain why there are very few cases of the observation of both torsions in C_s molecules.[3,42]

7.4.3 *Complete treatment*

If we do not make the harmonic oscillator approximation, but retain the sine and cosine terms in equation (7.31), then although the kinetic energy terms separate into two terms corresponding to α_+ and α_- motion, the potential terms do not.[3,28] A complete treatment has been given by Möller and Andressen.[28] The correct wave-functions are expressed as symmetric and antisymmetric combinations of the wave-functions for the two tops. Using perturbation theory, they show that the two $0 \longrightarrow 1$ transitions are given by:

$$W(0 \longrightarrow 1)^{\pm} = E_1 - E_0 + f_{01} \pm h$$

E_1 and E_0 are the single-top energy levels, f_{01} contains the terms involving V_{12} and V^*, which shift all the energy levels with a given principal quantum number V (0, 1, 2 etc.) in the same way, and the term h, which depends on V'_{12} and F', represents the splitting of the $0 \longrightarrow 1$ transitions. Thus, as in the harmonic oscillator limit, in order to obtain independent values of V, V^* and V_{12}, it is necessary to observe several hot bands in addition to the $0 \longrightarrow 1$ transition. However, in the C_s case, if both transitions are observed, a value of V'_{12} can easily be calculated, since F' can be calculated from standard data. (In cases where insufficient data to determine V_{12} and V^* is available, Hoyland[44] has suggested that it is assumed that $V_{12} = -V'_{12}$.)

The next highest transitions are given by:

$$W(1 \longrightarrow 2) \pm \pm = \tfrac{1}{2}(E_2 - E_0) + f_{12} \pm h_2 \pm h$$

$$W(1 \longrightarrow 2) \pm = (F_2 - F_1) + f \pm h$$

In these equations (and those for $W(0 \longrightarrow 1)\pm$), the quantities E_V are the single-top energy levels with a simple V_3 potential, f_{01}, f, f_{12} etc. are functions of V_{12}, V^* and matrix elements of $\cos 6\alpha$ and $\cos 3\alpha$, $\cos 3\alpha_2$, and the quantities h etc. are given by:

$$h = p_{10}^2 \left[2F' + V_{12}\left(\frac{4}{9s^2}\right)(b_1 - b_0)^2 \right]$$

$$h_1 = p_{21}p_{01}\left[2F' + V'_{12}\left(\frac{4}{9s^2}\right)(b_1 - b_0)(b_2 - b_1) \right]$$

$$h_2 = [\tfrac{1}{4}\{2E_1 - E_2 - E_0 + V_{12}[(3)_{11}^2 - (3)_{12}(3)_{00}]$$
$$+ \tfrac{1}{2}V^*[2(6)_{11} - (6)_{22} - (6)_{00}]\}^2 + 2h_1^2]^{\frac{1}{2}}$$

where

$$(6)_{vv'} = \langle \psi_v | \cos 6\alpha | \psi_{v'} \rangle \text{ etc.}$$

Möller and Andressen[28] have also considered the selection rules in more detail. They have used symmetry groups which include the internal rotation operations as well as the overall symmetry of the rigid molecule. They show that even for the molecule with C_{2v} symmetry, both $0 \rightarrow 1$ transitions are in principle observable. However, consideration of their equations shows that the α_+ transition is only allowed because there is an overall rotational (tunnelling) component to the internal rotation of the tops. The transition moment is non-zero only to the extent that there is a slight difference between integrals involving A and E solutions of the Mathieu equation. Thus, in cases where the difference between A and E solutions is negligible, that is for most cases of interest in vibrational spectroscopy, we only expect one $0 \rightarrow 1$ transition to be allowed, and this will usually be the transition from the ground state to the upper level of the first excited state.[28] However, even in this case, we expect to observe $1 \rightarrow 2$, $2 \rightarrow 3$ transitions for molecules with CH_3 tops, so that the spectra are still likely to be complicated, and in fact, detailed analyses of these types of molecules have not so far been achieved.[3,28,42]

Möller and Andressen[43] have also treated the case of three CX_3 tops in an analogous manner. For molecules with C_{3v} symmetry, the harmonic oscillator limit gives two normal coordinates, a non-degenerate A_2 species in which all three tops rotate in the same sense and in plane, which is forbidden in both the infrared and Raman, and a doubly degenerate E species, which is infrared and Raman active. Again, both transitions become allowed in the low barrier limit, but only one transition is likely to be observed in the high barrier case. The complete treatment[43] gives, for the $0 \rightarrow 1$ transition for the A and E species, $W_A(0 \rightarrow 1)$ and $W_E(0 \rightarrow 1)$ respectively,

$$W_E(0 \rightarrow 1) = (f_1 - f_0) + |h_1|$$
$$W_A(0 \rightarrow 1) = (f_1 - f_0) - 2|h_1|$$

where $(f_1 - f_0)$ represents the terms due to the potential of the individual tops V and V^*, and the cosine interaction term V_{12}, which again cannot be separated from the $0 \rightarrow 1$ transitions, and h_1 contains the sine interaction term V'_{12} and the kinetic energy interaction term F'. The harmonic approximations, analogous to (7.33) and (7.34) are:

$$V_E = 3[(F - F')(V - 4V_1 - 2V'_1 - 4V^*)]^{\frac{1}{2}}$$
$$V_{A_2} = 3[(F + 2F')(V - 4V_1 + 4V'_1 - 4V^*)]^{\frac{1}{2}}$$

for a Hamiltonian of the form

$$H = \sum_{k=1}^{3} F p_k^2 + \tfrac{1}{2}V\left(1 - \sum_{k=1}^{3} \cos 3\alpha_k\right) + F' \sum_{i \neq k}\sum p_i p_k + V^* \sum_{k=1}^{3} \cos 6\alpha_k$$

$$+ V_1(\cos 3\alpha_1 \cos 3\alpha_2 + \cos 3\alpha_1 \cos 3\alpha_3 + \cos 3\alpha_2 \cos 3\alpha_3)$$

$$+ V_1'(\sin 3\alpha_1 \sin 3\alpha_2 + \sin 3\alpha_1 \sin 3\alpha_3 + \sin 3\alpha_2 \sin 3\alpha_3)$$

The final problem to be considered is that of two dissimilar tops. Pitzer[9,10] has given the kinetic energy for the general case where both tops are asymmetric (for symmetric tops, the equations of Reference 8 may be used). The kinetic energy is given by

$$2T = \dot{\alpha}^{+} I \dot{\alpha}$$

where $(\dot{\alpha}) = (\dot{\alpha}_1 \dot{\alpha}_2)$ and I is the inertial matrix.

$$2T = p^{+} I^{-1} p$$

$$T = F_1 p_1^2 + F_2 p_2^2 + F_{12}(p_1 p_2 + p_2 p_1)$$

$$\tag{7.35}$$

where

$$F_i = \frac{h}{8\pi^2 c}(I^{-1})_{ii}(\text{cm}^{-1})$$

$$F_{ij} = \frac{h}{8\pi^2 c}(I^{-1})_{ij}(\text{cm}^{-1})$$

and

$$p_i = \frac{1}{i}\frac{\partial}{\partial \partial_i}$$

By analogy with the case of two CX_3 tops, one does not expect to be able to distinguish the V^* and V_{12} terms, so one chooses a potential of the form:

$$V = \frac{V_1}{2}(1 - \cos n_1 \alpha_1) + \frac{V_2}{2}(1 - \cos n_2 \alpha_2) + V_{12}' \sin n_1 \alpha_1 \sin n_2 \alpha_2$$

In the harmonic oscillator approximation:

$$V = \frac{V_1 n_1^2 \alpha_1^2}{4} + \frac{V_2 n_2^2 \alpha_2^2}{4} + V_{12}' n_1 n_2 \alpha_1 \alpha_2$$

$$2V = (\alpha_1 \alpha_2)\begin{pmatrix} \dfrac{V_1 n_1^2}{2} & V_{12}' n_1 n_2 \\[2ex] V_{12}' n_1 n_2 & \dfrac{V_2^2 n_2^2}{2} \end{pmatrix}\begin{pmatrix} \alpha_1 \\[1ex] \alpha_2 \end{pmatrix} \tag{7.36}$$

and equation (7.35) can be written:

$$2T = (p_1 p_2)\begin{pmatrix} 2F_1 & 2F_{12} \\ 2F_{12} & 2F_2 \end{pmatrix}\begin{pmatrix} p_1 \\ p_2 \end{pmatrix}$$

These equations are in the form of the FG matrix method for the calculation of normal coordinates[1] and it may be shown that the harmonic oscillator energy levels are given by:

$$E_v = (v + \tfrac{1}{2})\lambda^{\frac{1}{2}}$$

where λ is a solution of the secular equation

$$\begin{vmatrix} F_1 V_1 n_1^2 + 2F_{12}n_1 n_2 V_{12}' - \lambda & 2n_1 n_2 V_{12}' F_1 + F_{12}n_2^2 V_2 \\ F_{12}n_1^2 V_1 + 2F_2 n_1 n_2 V_{12}' & 2F_{12}n_1 n_2 V_{12}' + F_2 n_2^2 V_2 - \lambda \end{vmatrix} = 0$$

E and λ are in wave numbers for F and V in wave numbers.

7.5 References

1. E. B. Wilson, J. C. Decius and P. C. Cross, *Molecular Vibrations* (New York: McGraw-Hill, 1955).
2. W. G. Fateley and F. A. Miller, *Spectrochim. Acta*, **19**, 611 (1963).
3. K. D. Möller, A. R. de Meo, D. R. Smith and L. H. London, *J. Chem. Phys.*, **47**, 2609 (1967).
4. C. C. Lin and J. D. Swalen, *Rev. Mod. Phys.*, **31**, 841 (1959).
5. D. R. Herschbach, *J. Chem. Phys.*, **31**, 91 (1959).
6. D. R. Herschbach, *Tables for the Internal Rotation Problem*, Dept. of Chemistry, Harvard University; described in *J. Chem. Phys.*, **27**, 975 (1957).
7. E. B. Wilson and J. B. Howard, *J. Chem. Phys.*, **4**, 260 (1936).
8. K. S. Pitzer and W. D. Gwinn, *J. Chem. Phys.*, **10**, 428 (1942).
9. K. S. Pitzer, *J. Chem. Phys.*, **14**, 239 (1946).
10. J. E. Kilpatrick and K. S. Pitzer, *J. Chem. Phys.*, **17**, 1064 (1949).
11. K. L. Kassell, *J. Chem. Phys.*, **4**, 276 (1936); B. L. Crawford, *J. Chem. Phys.*, **8**, 273 (1940).
12. *Tables Relating to Mathieu Functions* (New York: Columbia University Press, 1951).
13. R. W. Kilb, *Tables of Degenerate Mathieu Functions* (Cambridge: Harvard University Press, 1956).
14. T. Miyazawa and K. S. Pitzer, *J. Chem. Phys.*, **30**, 1076 (1959).
15. G. Lane, *Ph.D. Thesis*, University of Manchester (1966).
16. A. V. Cunliffe, *J. Mol. Struct.*, **6**, 9 (1970).
17. F. A. Miller, *Molecular Spectroscopy*, Ed. P. Hepple (London: Institute of Petroleum, 1968).
18. G. Sage and W. Klemperer, *J. Chem. Phys.*, **39**, 371 (1963).
19. H. Margenau and G. M. Murphy, *The Mathematics of Physics and Chemistry* (New Jersey: Van Nostrand, 1956), p. 286.
20. H. Margenau and G. M. Murphy, *The Mathematics of Physics and Chemistry* (New Jersey: Van Nostrand, 1956), p. 324.
21. C. R. Quade and C. C. Lin, *J. Chem. Phys.*, **38**, 540 (1963).

22. C. R. Quade, *J. Chem. Phys.*, **47**, 1073 (1967).

23. G. Blanch and I. Rhodes, *Wash. Acad. Sci.*, **45**, 166 (1955).

24. R. W. Kilb, *Tables of Mathieu Eigenvalues and Eigenfunctions for Special Boundary Conditions*, Dept. of Chemistry, Harvard University (1956).

25. E. O. Stejskal and H. S. Gutowsky, *J. Chem. Phys.*, **28**, 388 (1958).

26. J. C. M. Li and K. S. Pitzer, *J. Phys. Chem.*, **60**, 466 (1956).

27. M. Hayashi and L. Pierce, *Tables for the Internal Rotation Problem* (Dept. of Chemistry, University of Notre Dame, Indiana, 1961).

28. K. D. Möller and H. G. Andressen, *J. Chem. Phys.*, **37**, 1800 (1962).

29. E. B. Wilson and J. B. Howard, *J. Chem. Phys.*, **4**, 260 (1936).

30. P. M. Dennison and B. T. Darling, *Phys. Rev.*, **57**, 128 (1940).

31. R. Hunt, R. A. Leacock, C. W. Peters and K. T. Hecht, *J. Chem. Phys.*, **43**, 1931 (1965); R. Hunt and R. A. Leacock, *J. Chem. Phys.*, **45**, 3141 (1966).

32. W. G. Fateley, R. K. Harris, F. A. Miller and R. E. Witkowski, *Spectrochim. Acta*, **21**, 231 (1965).

33. F. A. Miller, W. G. Fateley and R. E. Witkowski, *Spectrochim. Acta*, **23A**, 891 (1967).

34. L. I. Schiff, *Quantum Mechanics* (New York: McGraw-Hill, 1955).

35. G. Allen, A. V. Cunliffe and G. Dew (Unpublished results).

36. J. L. Wood, *Trans. Faraday Soc.*, **62**, 1411 (1966).

37. H. Cramer, *Mathematical Methods of Statistics* (Princeton: University Press, 1956).

38. H. Hunziker and H. H. Günthard, *Spectrochim. Acta*, **21**, 51 (1965).

39. T. Miyazawa and K. Fukushima, *J. Mol. Spectr.*, **15**, 308 (1965).

40. J. D. Swalen and C. C. Costain, *J. Chem. Phys.*, **31**, 1562 (1959).

41. L. Pierce, *J. Chem. Phys.*, **34**, 498 (1961).

42. W. G. Fateley and F. A. Miller, *Spectrochim. Acta*, **18**, 977 (1962).

43. K. D. Möller and H. G. Andressen, *J. Chem. Phys.*, **39**, 17 (1963).

44. J. R. Hoyland, *J. Chem. Phys.*, **49**, 1908 (1968).

8 Torsional vibrations and rotational isomerism

G. Allen and S. Fewster

8.1 Introduction

This chapter is concerned with the study of torsional vibrations in simple molecules, particularly with the observation of torsional frequencies and the information which can be obtained from vibrational spectra about internal rotation and rotational isomerism. There is already a vast literature in this field and in order to highlight the successes and limitations of the method the chapter will be confined to results obtained for a restricted range of simple compounds. The origin of the barrier to internal rotation will be dealt with in Chapter 11.

8.1.1 *Molecules possessing symmetrical tops*

In Chapter 7 the reader will see that the most complete theoretical analysis of torsional vibrations can be carried out on simple molecules containing a symmetric top attached to a symmetrical or asymmetrical frame, which have a single axis for internal rotation.

Typical of the molecules containing two symmetrical tops, and ones which will be chosen for special consideration in this chapter are:

$$CH_3.CH_3 \qquad CH_3.CCl_3 \qquad CCl_3.CCl_3$$

$$CH_3.CF_3 \qquad CF_3.CF_3$$

A normal coordinate analysis of these molecules immediately reveals two very important points:

(a) In each molecule the torsional vibrational mode is in a class by itself and thus mixing of the torsional mode with other normal modes of the molecule is unlikely to be serious.

(b) The torsional mode is inactive in both infrared and Raman spectroscopy since $|\partial\mu/\partial q_i|^2$ and $|\partial\alpha/\partial q_i|^2$ are both zero, and so a technique other than optical spectroscopy is required for the direct observation of the torsional frequency or the spectra must be obtained under extreme conditions where the selection rules are likely to break down.

The chloro- and fluoroethanes also provide suitable examples of molecules which contain a single symmetrical top attached to an asymmetrical framework:

$$CH_3.CH_2Cl \qquad CH_3.CHCl_2 \qquad CH_2Cl.CCl_3 \qquad CHCl_2.CCl_3$$

$$CH_3.CH_2F \qquad CH_3.CHF_2 \qquad CH_2F.CF_3 \qquad CHF_2.CF_3$$

Torsional vibrational frequencies in this group are very much easier to observe because the vibrational modes are active in both infrared and Raman spectra.

For all these molecules the potential function governing internal rotation can be represented by a simple Fourier series of the form

$$V(\alpha) = \sum_k \frac{V_{kN}}{2}(1 - \cos kN\alpha)$$

where α is the angle defining the rotation of the top with respect to its *frame*, N is the symmetry number of the top, e.g. $N = 3$ for CH_3, CCl_3, CF_3, and $k = 1, 2, 3 \ldots$ etc.

Usually the higher terms in the expansion are relatively small and $V_N \gg V_{2N} > V_{3N}$ etc. Thus for approximate treatments the potential function is written as

$$V(\alpha) = \frac{V_N}{2}(1 - \cos N\alpha)$$

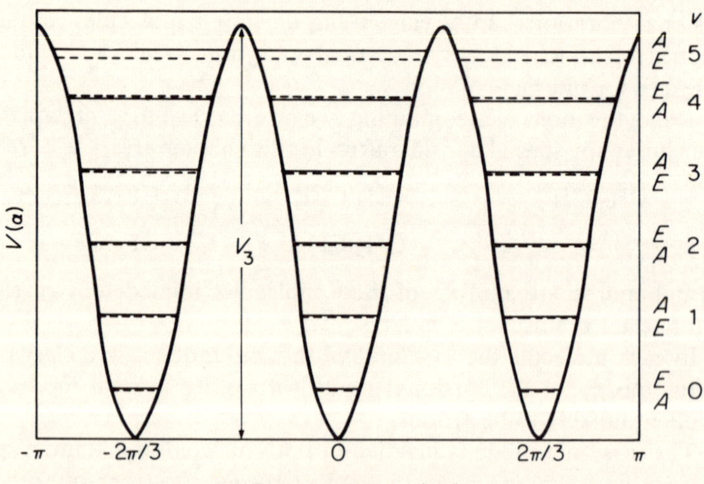

Figure 8.1   Potential energy curve for a three-fold symmetric rotor; v is the torsional energy level quantum number

The form of this function is shown in Figure 8.1 for the case $N = 3$. Also drawn in the figure are the torsional energy levels obtained from a solution of the Schrödinger wave-equation (see Chapter 7):

$$Fp^2 + \frac{V_N}{2}(1 - \cos N\alpha)\psi_{v\sigma}(\alpha) = E_{v\sigma}\psi_{v\sigma}(\alpha)$$

where $p = i(d/d\alpha)$, v is the vibrational quantum number and F is a structure factor, the reduced constant for internal rotation $\hbar^2/2rI_\alpha$, where rI_α is the reduced moment of inertia for internal rotation of the top and frame.

For $N = 3$ the allowed values of σ are $0, \pm 1$. The energy levels with $\sigma = 0$ are totally symmetric non-degenerate A species of the C_3 group. Those with $\sigma = \pm 1$ are doubly degenerate of the E species. Levels of species A correspond approximately to undisturbed torsional levels localized in one potential well while the E levels can be thought of as representing tunnelling through the barrier between wells in the two possible directions.

The fundamental torsional frequency observed in vibrational spectroscopy corresponds to the energy of the $v = 0 \to 1$ transition. For most of the molecules we will consider, the splitting of the lower torsional levels is very small and thus the transitions $v = 0 \to 1(A')$ and $v = 0 \to 1(A'')$ will have undetectably different frequencies. Thus

$$E_1 - E_0 = hc\omega_{0-1}$$

where ω_{0-1} is the wave number (cm^{-1}) of the fundamental torsional vibration. In order to relate ω_{0-1} to the potential constant V_N, i.e. V_3 in this particular case, the wave-equation has to be recast in the form of a Mathieu equation (see Chapter 7). Thus by the following substitutions

$$2x = n\alpha + \pi; \qquad y_{v\sigma}(x) = \psi_{v\sigma}(\sigma);$$

$$V_N = N^2 Fs/4; \qquad E_{v\sigma} = N^2 Fb_{v\sigma}/4$$

the equation becomes

$$d^2y/dx^2 + (b - s\cos^2 x)y = 0$$

in which s and b are dimensionless parameters corresponding to the potential barrier height V_N, and the torsion energy levels $E_{v\sigma}$ respectively. Thus

$$b_1 - b_0 = \frac{4(E_1 - E_0)}{N^2 F} = \frac{4hc\omega_{0-1}}{N^2 F}$$

Tables are available giving $b_{v\sigma}$ as a function of s, and so the value of s corresponding to $b_1 - b_0$ can be estimated by interpolation between the tabulated values. The barrier height V_N in the potential function can then be calculated from the relation given above, which for a three-fold rotor becomes

$$V_3 = 9Fs/4$$

Thus a knowledge of molecular geometry and the fundamental torsional vibration frequency allows V_3 to be calculated on the assumption that the potential function is of simple three-fold cosine form. In order to probe the validity of this approximation the frequencies of other transitions, e.g. $v = 1 \rightarrow 2$, $v = 2 \rightarrow 3$ etc. would have to be observed.

To demonstrate this type of analysis in more complicated molecules, we shall refer to molecules containing two methyl tops and to the substituted anisoles:

Substituted anisoles are included because they will illustrate two important points. First of all, the observed torsional frequencies in halogenated ethanes and molecules containing two methyl tops are not very strongly dependent on whether a liquid or a gaseous sample is used. However, the torsional frequencies of the anisoles observed for torsional vibrations about the C_6H_5—O axis are strongly dependent on whether the molecule is in the liquid or vapour phase and this is due presumably to strong intermolecular correlations and is a feature commonly encountered with relatively simple aromatic molecules. In principle, it is ideal to use the vapour phase value of the torsional frequency in the calculation of V_N so that intermolecular perturbations are minimized.

The second feature illustrated by the anisoles is that the simple model possessing one degree of torsional freedom which is satisfactory for simple ethanes tends to break down when used on these more complicated molecules. Thus, as we shall see, a comprehensive torsional analysis of even a relatively simple two-top system such as

becomes extremely difficult.

8.1.2 *Molecules possessing asymmetric tops*

Formidable problems are presented by the analysis of even the simplest molecule having a single top for which neither the top nor the framework has a rotational symmetry axis. These problems are discussed by Cunliffe in Chapter 7 and they are often compounded by experimental difficulties too. The molecules chosen for consideration in this chapter include:

$CH_2Cl.CH_2Cl$	$CH_2Cl.CHCl_2$	$CHCl_2.CHCl_2$
$CH_2F.CH_2F$		$CHF_2.CHF_2$

and also $CH_2F.CH_2Cl$ and $CHCl_2.CCl_2.CH_3$

For this group of molecules the potential function governing internal rotation is also a Fourier series:

$$V(\alpha) = \sum_k \frac{V_k}{2}(1 - \cos k\alpha)$$

and even for the simplest molecules the potential constants V_1, V_2 and V_3 must be determined. The whole analysis is complicated by the fact that the reduced moment of the molecule and hence the structure factor $F(\alpha)$ is now dependent on the torsion angle α. Fortunately Cunliffe has shown that the torsional constant $F(\alpha)$ can be written as a power series in α, rather than a Fourier series and this results in a useful computational simplification. Of more immediate importance is the existence of at least two rotational isomers in these systems. We have deliberately chosen examples where there arc only two possible isomers, but some of the mixed halogenated ethanes have three isomers present in equilibrium concentrations determined by the temperature of the sample. A schematic potential function appropriate to the molecules which we have selected for discussion is shown in Figure 8.2. There are two

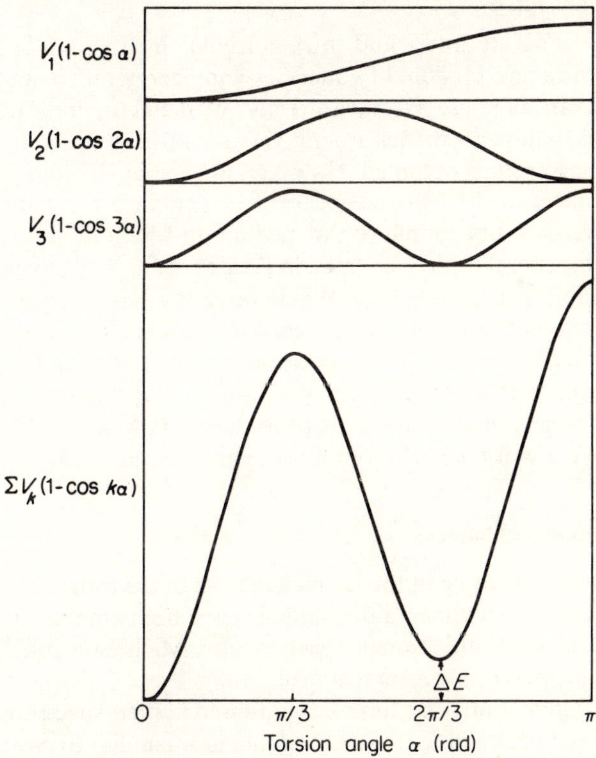

Figure 8.2 Potential energy curve for coupled asymmetric rotors. The symmetrical (or *trans*) well is at $\alpha = 0$ and the asymmetric wells are at $\alpha = \pm 2\pi/3$

rotational isomers, the symmetric or *trans* and asymmetric or *gauche*, with an energy difference ΔE between the isomers and non-equivalent potential wells. Three potential constants V_1, V_2 and V_3 are needed to describe $V(\alpha)$. In order to calculate these parameters it is necessary to determine the spacings of the energy levels in each well, i.e. to observe the torsional frequencies for *each* isomer and also to measure the energy difference between the ground states of the two isomers.

In practice ΔE is obtained from the temperature dependence of the equilibrium constant for the interconversion between the two rotamers, which is in turn being measured from the relative intensities of vibrational bands characteristic of each isomer. The torsional frequencies are observed in either the infrared or Raman spectrum, since all torsional modes are optically active (although the torsional modes for the 1, 2- and 1, 1, 2, 2-substituted *trans* isomers listed above are only active in the infrared spectrum). Unfortunately the overlapping of the torsional bands for the two isomers can prove to be a troublesome source of experimental uncertainty.

8.1.3 *Effect of medium*

In order to separate inter- and intramolecular interactions it is clearly preferable to measure torsional frequencies and energy differences between isomers in the vapour phase. We have already noted that torsional frequencies in many simple halogenoethanes are not very sensitive to the state of matter in which the spectrum is obtained. However, the energy differences between rotational isomers of the halogenated ethanes containing asymmetric tops and frames are *very* susceptible to the medium in which the measurements are made. This general topic is covered in greater detail in Chapter 13, but it is of considerable importance here. Where there is a large energy difference between isomers in the vapour phase, accurate values of ΔE cannot be obtained from intensity measurements because the concentration of the isomer of higher energy will be small. Thus the value of ΔE has to be estimated from solution measurements using the procedures in Chapter 13. The results presented in § 8.3 will refer mainly to gas-phase measurements.

8.2 Spectroscopic techniques

Torsional frequencies lie in the far infrared, i.e. in the range 50–400 cm^{-1} and so a variety of experimental techniques have been employed including conventional infrared and Raman spectroscopy, Michelson interferometry and neutron incoherent inelastic scattering.

The use of conventional far infrared and laser Raman spectrometers will not be described in detail since the methods are now familiar to most chemists and physicists. It is worth pointing out, however, that the introduction of double and triple monochromators in Raman spectroscopy has made much easier the observation of these vibrational modes of low frequency, and the use of lasers of high power has facilitated the study of torsional modes in the vapour phase.

Recently Weiss and Leroi[1] were able to observe the inactive torsion in C_2H_6, C_2D_6 and CH_3CD_3 in the far infrared spectrum by using a high-pressure (30–70 atm) cell with a path-length of the order of 1 m. Thus there is now the possibility of applying this technique to other molecules where the torsional frequency is normally inactive.

8.2.1 Far infrared interferometers

One of the more recent advances in instrumentation has been the application of interferometers to obtain very weak emission and absorption spectra. They have proved very useful in the far infrared region where conventional spectrometers are handicapped by the low energy output of available sources and the lack of sensitivity of detectors. The most important advantages of using an interferometer as a spectrometer are: (a) that all wavelengths present in the radiation fall on the detector for the total observation time (this improves the signal-to-noise ratio of the output from the detector) and (b) that the axial symmetry of an interferometer allows circular entrance and exit holes to be used, instead of narrow slits, for the light passing through the instrument. This second factor greatly enhances the amount of light collected by the detector.

A typical Michelson interferometer for spectroscopic measurements in absorption is shown in Figure 8.3. Instead of the monochromatic source used

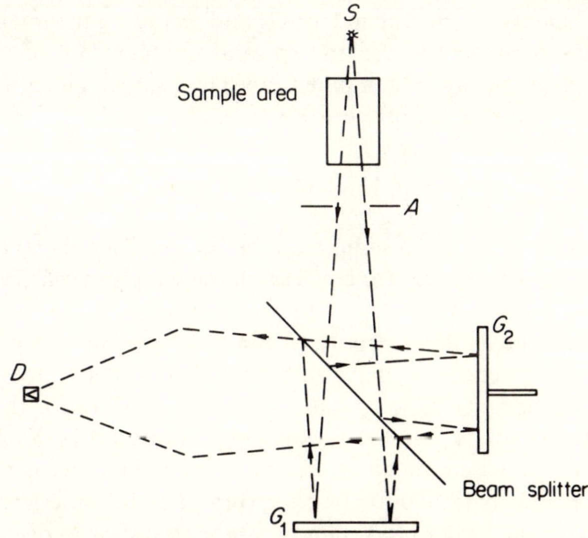

Figure 8.3 Schematic diagram of a Michelson interferometer:
S = mercury arc, A = aperture, D = detector,
G_1 = fixed optical flat, G_2 = movable optical flat

in the determination of the velocity light, a high pressure mercury arc S is used as a source of continuous wavelengths for the far infrared region. The radiation is collimated, passed through the sample cell and the circular entrance aperture A and divided into two beams which fall on optically flat mirrors G_1 and G_2. After reflection at G_1 and G_2 the beams come together again and are directed on the detector D. They interfere constructively or destructively depending on the frequency of the spectral component and the path difference X traversed by the two beams. The mirror G_2 is mounted on a precision screw so that X can be varied. Thus the total intensity of radiation falling on the detector will be a function of X of the form

$$F(X) = 2 \int_0^\infty B(v)\, dv + 2 \int_0^\infty B(v) \cos(2\pi X v)\, dv$$

where $B(v)$ is the component of radiant energy lying between v and $(v + dv)$ which reaches the detector. The first integral is a function of the source characteristics and the reflection properties of the beam splitter and mirrors: it is therefore a constant for a given experiment. If monochromatic radiation were used the second term would be simply the normal cosine interferogram. If an interferogram function $I(X)$ for the polychromatic radiation is defined

$$I(X) = F(X) - 2 \int_0^\infty B(v)\, dv = 2 \int_0^\infty B(v) \cos(2\pi X v)\, dv$$

then $I(X)$ can be obtained experimentally from the detector signal as a function of X simply by moving mirror G_2 and recording the interferogram. Once $I(X)$ has been determined the spectral contour $B(v)$ of the energy falling on the detector can be computed, since by Fourier's integral theorem there is a reciprocal relation between $I(X)$ and $B(v)$:

$$B(v) = 2 \int_0^\infty I(X) \cos(2\pi k v)\, dX$$

The absorption spectrum of a sample can be obtained simply by comparing $B(v)$ as a function of frequency (or cm^{-1}) with the sample in cell C and with C evacuated.

The equations given above apply to the ideal case in which there is a point source; in practice they have to be modified to take into account its finite size. A further modification has to be made to allow for the fact that only finite values of X can be investigated. Fortunately a high speed computer resolves these difficulties. In any event a computer is required to transform the information into the spectrum after the interferogram has been obtained in the form of, say, punched-tape recording at discrete values of path difference.

The resolution of the instrument is inversely related to the distance through which the mirror G_2 is moved, and for the observation of torsional frequencies a movement of 1–5 mm would be adequate to produce a resolving power of the order of 1 cm^{-1}. Liquid samples are used in cells of 0·1–2 mm thickness but for gaseous samples a multipass cell with a path-length of the order of 10 m is required.[2]

8.2.2 *Neutron incoherent inelastic scattering*

In the past six years neutron incoherent inelastic scattering (NIIS) spectroscopy has been applied to a variety of problems in chemical physics.[3] In the particular application to the observation of torsional frequencies the method can be regarded as Raman spectroscopy performed with neutrons rather than photons, but with the important difference that optical selection rules do not operate. In chemical applications advantage is taken of the following features of NIIS:

(a) For low energy neutrons the scattering cross-section of a proton is an order of magnitude greater than any other nucleus including the deuteron. Thus the scattering of neutrons from a target molecule containing several hydrogen atoms comes overwhelmingly from neutron–proton collisions.

(b) The scattering is also isotropic and is described in terms of a scattering length b, analogous to the atomic scattering factor $f(\theta)$ used in X-ray studies, except that the latter is angle-dependent. In practice the scattering length depends, both in magnitude and sign, on whether the spins of the proton and incoming neutron are parallel (total spin = 1), or antiparallel (total spin = 0). For an unpolarized neutron beam incident on a sample containing protons with random spin orientations, the average scattering length at each scattering centre determines the coherent scattering length $b_{coh.}$. The incoherent scattering length $b_{inc.}$ is determined by the r.m.s. deviation of scattering length among the scattering centres. For bound protons and low energy neutrons (the cross-section varies with neutron energy) $\lambda \sim 5$ Å:

$$b_{coh.} = -0.38 \times 10^{-12} \text{ cm}$$

$$b_{inc.} = 2.5 \times 10^{-12} \text{ cm}$$

Thus the scattering is predominantly incoherent since the cross-section is proportional to b^2.

Time-of-flight spectrometers are usually employed for this kind of work. The source of neutrons is a nuclear reactor and the neutrons are moderated with liquid hydrogen in order to enhance the flux at the required wavelength usually $\lambda = 5.3$ Å corresponding to an energy of 23 cm^{-1}. Bursts of approximately monochromatic neutrons are produced by mechanical selection by passage through two phased rotors. The neutrons are scattered from liquid or gaseous samples in aluminium containers. The energy spectrum of the scattered neutrons is analysed by time-of-flight methods at several angles of scatter between 18° and 90°. The resultant spectrum consists of an intense peak corresponding to elastic scattering of the incident neutrons together with other peaks which correspond to neutrons gaining a quantum of energy in the scattering event from a molecular vibrational transition. This inelastic scattering is an anti-Stokes process since the incident neutron energy is too low to excite a transition which would correspond to a Stokes process. Figure 8.4 shows the time-of-flight neutron spectrum for $CH_3.CCl_3$; the torsional frequency can be seen clearly at $300\ cm^{-1}$ despite the fact that the torsional mode is forbidden in optical spectra.

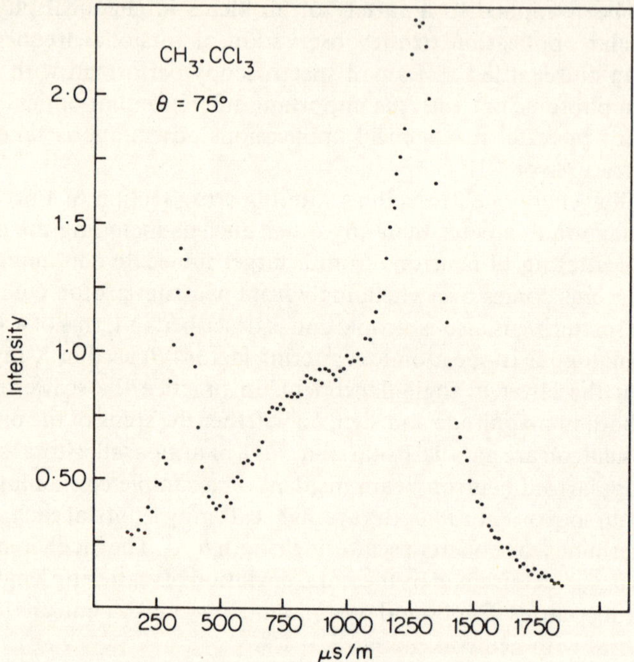

Figure 8.4   Neutron incoherent inelastic scattering spectrum of $CH_3.CCl_3$(liquid) obtained at a scattering angle of 75°. The intense peak at 1250 $\mu s/m$ is the quasi-elastic peak and the peak at 350 $\mu s/m$ is the torsional peak corresponding to a torsional frequency of $310 \pm 15\ cm^{-1}$

There are still severe practical limitations on the NIIS method. Present neutron spectrometers are limited in resolution due to the available neutron flux. Typical values are 5–10 % energy resolution (i.e. 15–$30\ cm^{-1}$ at $300\ cm^{-1}$ frequency shifts) with neutron counting times of about 24 hours to give adequate noise statistics. In addition, the use of low energy neutrons is best applied at frequency shifts below $500\ cm^{-1}$ not only because the constant energy resolution becomes progressively worse, but also because the thermal population factor adversely affects the intensity of the vibrational band and because time-of-flight spectra are non-linear in energy, which leads to a very small number of time channels representing a considerable wavenumber interval.

The intensity of the torsional mode in NIIS is determined by the probability that a neutron will gain a quantum of molecular vibrational energy in the scattering event. The corresponding expression for the intensity of infrared absorption involves the square of the matrix element of the electric dipole operator between the initial and final states of the molecule, the absorption being forbidden when this quantity is zero. A similar expression is used for the intensities in NIIS except that the operator involved is different. The neutron–nucleus interaction potential causing scattering is of such short range that mathematically it can be represented as a δ-function (Fermi pseudo-potential) with a strength proportional to the scattering length of the nucleus. (The δ-function implies scattering is via direct collisions). The interaction potential between a neutron and a target molecule is therefore a sum of such terms over the nuclei in the molecule, i.e.

$$V(\mathbf{r}) = \sum_k \frac{-2\pi\hbar^2}{m} b_k \delta(\mathbf{r} - \mathbf{R}_k)$$

where $\mathbf{r}$ is the position vector of the neutron of mass m, and $\mathbf{R}_k$ is the position vector of the kth nucleus with scattering length b_k. The probability that a neutron of incident energy $E_0 = \hbar^2 k_0^2/2m$ (where $k = 2\pi/\lambda$ and $\hbar\mathbf{k}$ is the neutron momentum) will be scattered with energy $E = \hbar^2 k^2/2m$ in a direction $\mathbf{k}$ depends on the factor

$$|\langle \Psi_i\, e^{i\mathbf{k}_0 \cdot \mathbf{r}} | V(\mathbf{r}) | \Psi_f\, e^{i\mathbf{k} \cdot \mathbf{r}} \rangle|^2 \tag{8.1}$$

where the initial and final wave-functions of the system (neutron plus molecule) are products of molecular wave-functions Ψ and plane waves of the appropriate momenta for the neutron.

Inserting the expression for $V(\mathbf{r})$ into equation (8.1) and integrating over the neutron coordinates, we get, apart from constant factors

$$\left| \left\langle \Psi_i \left| \sum_k b_k \exp(i\mathbf{K} . \mathbf{R}_k) \right| \Psi_f \right\rangle \right|^2$$

where $\mathbf{K} = \mathbf{k}_0 - \mathbf{k}$ and therefore $\hbar\mathbf{K}$ represents the momentum lost by the neutron in the scattering process. We are interested in the case when the change in neutron energy corresponds to a quantum of molecular vibrational energy of frequency ω_0 say, i.e. when

$$\frac{\hbar^2 k^2}{2m} - \frac{\hbar^2 k_0^2}{2m} = \hbar\omega_0$$

For a vibrating but fixed molecule, writing the position vectors of the nuclei as their equilibrium values $\mathbf{R}_k^0$ plus their displacements $\mathbf{U}_k$ due to vibrations, and for purely incoherent scattering the above expression becomes

$$\sum_k b_k^2 |\langle \Psi_i | \exp(i\mathbf{K} . \mathbf{U}_k) | \Psi_f \rangle|^2 \tag{8.2}$$

Thus instead of a dipole operator as in the infrared, there is an exponential factor involved, which contains the displacements of the nuclei from their equilibrium positions. This sum will be dominated by the contribution from the hydrogen atoms in the molecule for which b_k is large.

For harmonic vibrations, the molecular states are specified by the quantum numbers of all the normal modes, i.e. $\Psi = \Psi(n_1, n_2, \ldots n_j, \ldots)$ and the nuclear displacements are expanded in terms of the normal coordinates Q_j

$$\sqrt{M_k} \cdot \mathbf{U}_k = \sum_j \mathbf{c}(k, j) Q_j \tag{8.3}$$

the vectors $\mathbf{c}(k, j)$ being the usual normalized displacement vectors of molecular vibration theory. The expansion (8.3) and products of harmonic oscillator wave-functions are inserted in (8.2) and summing over all possible final states and averaging over the thermal population of the initial states, one obtains the following.

(a) Elastic scattering:

$$\text{intensity} \sim \sum_k b_k^2 \exp \left| \frac{-\hbar}{2M_k} \left\{ \sum_j [\mathbf{K} \cdot \mathbf{c}(k, j)]^2 \frac{1}{\omega_j} \coth \frac{\hbar \omega_j}{2k_B T} \right\} \right|$$

i.e. the elastic peak intensity decreases with increasing angle of scatter $[\mathbf{K} = 2k_0 \sin (\theta/2)]$ due to a Debye–Waller factor involving a sum over all the normal modes (cf. X-ray scattering theory). We shall write the exponential factor as e^{-2W}.

(b) Inelastic scattering: if the neutron gains a single quantum of energy from the vibrational mode of frequency ω_0,

$$\text{intensity} \sim \sum_k b_k^2 \exp(-2W) \exp\left(\frac{-\hbar\omega_0}{k_B T}\right)$$

$$\times \frac{1}{1 - \exp\left(\frac{-\hbar\omega_0}{k_B T}\right)} \cdot \left\{ [\mathbf{K} \cdot \mathbf{c}(k, 0)]^2 \frac{\hbar}{2M_k \omega_0} \right\}$$

Apart from the factor e^{-2W} and a thermal population factor, the intensity depends on the quantity $[\mathbf{K} \cdot \mathbf{c}(k, 0)]^2$. The whole expression should be averaged over all molecular orientations for comparison with experimental results on liquids. Then, approximately, the intensity depends on K^2 and c^2. For a mode which involves large displacements of the hydrogen atoms the intensity will therefore be large; conversely it will be small when no hydrogen motions are involved. This fact is an important aid in making assignments. Two quanta transitions are not forbidden for the harmonic oscillator, but their probability depends on $\sim [\mathbf{K} \cdot \mathbf{c}(k, 0)]^4$ and is much less than single quantum events.

The problems for which NIIS spectroscopy is ideally suited therefore include the observation of vibrational modes of low frequency which involve large displacements of hydrogen atoms. Since the resolution of the apparatus is poor, such frequencies should be well separated and thus few in number. The observation of torsional frequencies in the chloro- and fluoroethanes is such a problem and the frequencies are observed irrespective of the optical selection rules.

8.3 Spectroscopic measurements

8.3.1 *Molecules possessing symmetrical tops*

For this group of molecules the principal measurements involve the use of infrared, Raman and NIIS spectroscopy to determine directly the torsional frequency of the symmetrical top with respect to the rigid frame which itself may, or may not, be a symmetric top. Until recently torsional frequencies in molecules like $CH_3.CH_3$, $CH_3.CCl_3$ and $CH_3.CF_3$ were determined indirectly[4,5] from a statistical-mechanical analysis of heat capacity data combined with the fundamental frequencies of all other normal modes of vibration. The advent of NIIS spectroscopy[3] and, in the case of ethane, the breakdown at high pressure of the infrared selection rule governing the torsional mode[1] have eliminated the need to rely on indirect estimates.

The assignment of the torsional mode to an inelastic neutron scattering peak is unambiguous in molecules for which displacements of protons are involved. Since the torsional mode is the only low frequency vibration in which a significant displacement of the protons occur, it is by far the most intense inelastic peak. The disadvantage of NIIS spectroscopy is that the position of the peak is determined with relatively low precision compared with, for example, infrared or Raman spectroscopic measurements. An additional advantage of infrared absorption spectroscopy is that spectra can be obtained readily in the vapour phase; this is an important factor in view of the desirability of eliminating intermolecular interactions which may perturb the torsional frequency. Unfortunately there is no clear cut intensity rule which applies to torsional bands in infrared and Raman vibrational spectroscopy and the only guide to assignment comes from normal calculations of the torsional frequency.

8.3.1.1 *Chloroethanes* Torsional frequencies have now been observed directly for all chloroethanes with the notable exception of C_2Cl_6. For this molecule the torsional mode is optically inactive and NIIS spectroscopy is not as useful as for molecules containing protons. Unfortunately not all molecules have been studied in the vapour phase, though where results are available it is clear that only a small shift in torsional frequency ($\sim 5\ cm^{-1}$) occurs on transfer from the liquid to vapour state. Even so, since the potential constant V_3 which defines the height of the potential barrier is approximately proportional to ω_{0-1}^2, and ω_{0-1} lies in the range 70–300 cm^{-1}, these shifts can be significant.

Table 8.1 summarizes the fundamental frequencies which have been observed for the chloroethanes possessing symmetrical tops. Gas phase results are quoted where possible. In cases when only liquid phase results are available an empirical correction for the expected shift in going to the vapour has been made. The structure factor F calculated from known geometries, and the potential constant determined with the aid of tables of Mathieu constants s and b are also included in Table 8.1.

Table 8.1   Chloroethanes

Molecule	Reference	Method	ω_{0-1} (cm^{-1})	F (cm^{-1})	V_3 (cal/mole)
$CH_3.CH_3$	1	IR (gas)	289	10·33	2930 ± 25
CH_3CH_2Cl	6	IR (gas)	251	6·09	3700 ± 40
	7	Microwave splitting			3690 ± 100
$CH_3.CHCl_2$[a]	8	NIIS (liquid	293 ± 15		
		(gas)[b]	288 ± 15	5·39	5300 ± 500
$CH_3.CCl_3$[c]	8	NIIS (liquid)	310 ± 15		
		(gas)[b]	305 ± 15	5·32	6000 ± 600
$CH_2Cl.CCl_3$	9	IR (liquid)	116		
		(gas)	108	0·429	8720 ± 100
$CHCl_2.CCl_3$	9	IR (liquid)	85	0·163	14200 ± 300
		Raman (gas)[b]	80		12500 ± 600
C_2Cl_6	10	IR (liquid)	79	0·113	17500 ± 500
		(gas)[b]	74		15000 ± 1000

[a] Recent infrared gas-phase measurement[11] gives $\omega_{0-1} = 231$ cm^{-1}, $V_3 = 3500$ cal/mole.
[b] Estimated value: gas frequency = liquid frequency $- 5$ cm^{-1}.
[c] Recent infrared solid phase measurement[11] gives $\omega_{0-1} = 290$ cm^{-1}.

The NIIS value for the torsional frequency in $CH_3.CHCl_2$ is preferred to the more recent value obtained from the infrared spectrum,[11] not only because the peak at 293 cm^{-1} is the most intense in the NIIS spectrum but also because the intensity relative to the intensities of other torsional modes in the series is in accord[8] with the results of normal coordinate analyses. The frequency listed for $CHCl_2.CCl_3$ must be considered uncertain since the infrared and Raman estimates in the liquid phase[9] differ by 6 cm^{-1}. It is difficult to decide whether the quoted values should be adjusted to a lower frequency to represent the gas phase value, since mixing with other normal modes in the same a'' symmetry class will have already depressed the observed wave number for the torsional frequency. The wave number quoted for the torsion in C_2Cl_6 is perhaps the most uncertain in Table 8.1, since it has been estimated from analysis of combination bands—a method which has proved unreliable for other halogenated ethanes. Nevertheless this liquid phase value is the best estimate available. The value of V_3 (~ 15 kcal/mole) is to be compared with 11·0 kcal/mole obtained by Karle and Karle[12] from an electron diffraction study.

Accurate results for CH_3CH_3, CH_3CCl_3 and C_2Cl_6 are especially desirable because for these molecules the torsional mode is alone in a symmetry class and therefore will not be mixed with other low frequency modes.

8.3.1.2 *Fluoroethanes* Results for this series must be considered somewhat more reliable than for the corresponding chloroethanes because corroborative evidence on the magnitude of the potential constant V_3 can be obtained, as described in Chapter 4, from microwave splitting measurements for the lower members of the fluoroethane series. In addition, gas phase infrared measurements are available for every molecule except $CH_3.CH_3$ and $CF_3.CF_3$. Table 8.2 gives selected values of ω_{0-1}, F and V_3

Table 8.2 Fluoroethanes

Molecule	Reference	Method	ω_{0-1} (cm^{-1})	F (cm^{-1})	V_3 (cal/mole)
$CH_3.CH_3$	1	IR	289	10·33	2930 ± 25
$CH_3.CH_2F$	13	IR	243	6·43	3330 ± 50
	14	Microwave splitting	—	—	(3310 ± 100)
$CH_3.CHF_2$	15, 16	IR	222	5·47	3210 ± 50
	14	Microwave	—	—	(3180 ± 100)
$CH_3.CF_3$	16, 17	NIIS	220 ± 10	5·47	3150 ± 300
$CH_2F.CF_3$	16, 18	IR	111	1·13	3650 ± 100
$CHF_2.CF_3$	16	IR	75·5	0·504	3700 ± 100
$CF_3.CF_3$	19	Thermodynamic	64·5	—	3920

which the authors consider to be the most reliable results available. The torsional frequencies for $CH_3.CH_2F$ and $CH_3.CHF_2$ are well established and the derived values of V_3 are in very satisfactory agreement with microwave splitting measurements. $CH_3.CF_3$ has previously been investigated by microwave intensity measurements[20] and also using thermodynamic measurements;[5] both methods gave $V_3 = 3500$ cal/mole with large experimental errors in each case. Two estimates of ω_{0-1} from analyses of infrared combination bands[21,22] gave respectively 238 cm^{-1} and 218 cm^{-1}. Fortunately $CH_3.CF_3$ is an ideal case for study by NIIS scattering spectroscopy, although scattering from the gaseous sample does lead to inelastic peaks broadened by Doppler effects. The result quoted in Table 8.2 was obtained at 4 atm pressure. Dew[16] used the technique of Weiss and Leroi[1] in an attempt to observe ω_{0-1} in the far infrared. Unlike $CH_3.CCl_3$ which has other much more intense low frequency modes near the torsional frequency, only the torsional mode occurs in the 250 cm^{-1} region in $CH_3.CF_3$. A band was observed in the infrared spectrum of $CH_3.CF_3$ at 237 cm^{-1} with a path-length of 12 m and a vapour pressure of 40 cm, but this has been assigned as a difference band. This difference band is sufficiently intense to obscure the torsional frequency at the higher pressures which would be required to detect the weakly allowed torsion.

The results quoted for $CF_3.CH_2F$ and $CF_3.CHF_2$ are taken from the work of Dew.[16] The spectra are of very good quality and the frequencies quoted in Table 8.2 are the Q-branch centres. Comparison with other work provides good examples of the influence of minor differences in ω_{0-1} and geometrical data (i.e. differences in F) on the estimates of V_3. Danti and Wood [18] have studied the infrared spectrum of $CF_3.CH_2F$ in the vapour phase using different geometrical data with the result: $\omega_{0-1} = 120 \text{ cm}^{-1}, F = 1.11 \text{ cm}^{-1}$ and $V_3 = 4200 \text{ cal/mole}$. Brown and coworkers[23] studied the infrared spectrum of pentafluoroethane in the vapour phase. Despite corrections made for underlying hot bands, their estimate of the torsional frequency is lower than the value quoted in Table 8.2. Their value of $F = 0.483$ is not in accord with the value used in Table 8.2 or with the value used in microwave measurements, but the resulting value of V_3 is fortuitously only slightly changed, i.e. $\omega_{0-1} = 74.2 \text{ cm}^{-1}$, $F = 0.483 \text{ cm}^{-1}$ and $V_3 = 3750 \text{ cal/mole}$. The most unsatisfactory entry in Table 8.2 is for $CF_3.CF_3$. The original thermodynamic estimate was $V_3 = 4350 \text{ cal/mole}$ but more recent geometrical data has revised this value to $V_3 = 3920 \text{ cal/mole}$ quoted in Table 8.2. The result quoted from incoherent inelastic neutron scattering is too imprecise to afford a satisfactory direct estimate of ω_{0-1}.

8.3.2 *Molecules possessing two symmetric tops*

The theory relating the torsional energy levels to the barrier heights for molecules with two tops is much more complex than for the one-top case as will be seen in Chapter 7. Coupling between the internal rotational potential and kinetic energy of the tops changes the Hamiltonian in such a way that the energy levels are not easily obtained as solutions of the Mathieu equation. The theory for molecules of the $(CH_3)_2-X$ class has been developed by Möller and Andresen[24] using first-order perturbation techniques and predicts two fundamental transitions and six active transitions between the 1 and 2 levels. This is in good agreement with the complex nature of the torsional band of small molecules such as dimethyl ether[25,26] and dimethyl sulphide.[25] The model chosen by Möller and Andresen is based on the assumption that interactions between the torsion and other vibrations can be neglected. This assumption has recently been tested and found to be valid for a molecule with only one methyl top, ethyl chloride.[27] In this case the nearest normal mode of the same symmetry species as the torsion has a frequency three times greater than that of the torsion. The model is not expected to be so good in two-top molecules where there will be at least three heavy atoms and the lowest fundamental vibrational frequencies will be much closer to the torsion, indeed in some cases the torsion may not be the lowest frequency fundamental. However, in view of the complexity of the model with only two torsional degrees of freedom, it is necessary to make this assumption.

The Hamiltonian for this model can be split into two parts:

$$H = H_T + H_I$$

where H_T describes the torsion vibration of one top alone and H_I describes the potential and kinetic energy interactions between the tops. For the case of a two-top molecule with two methyl rotors of three-fold symmetry the terms of the Hamiltonian can be taken to be:

$$H_T = F(p_1^2 + p_2^2) + \tfrac{1}{2}V_3(2 - \cos 3\alpha_1 - \cos 3\alpha_2) \tag{8.4}$$

$$H_I = F'(p_1 p_2 + p_2 p_1) + \tfrac{1}{2}V^*(\cos 6\alpha_1 + \cos 6\alpha_2)$$

$$+ V_{12}\cos 3\alpha_1 \cos 3\alpha_2 + V'_{12}\sin 3\alpha_1 \sin 3\alpha_2 \tag{8.5}$$

In equations (8.4) and (8.5) the notation used is that of Möller and Andresen.[24] The term in F' is the kinetic energy coupling between the tops, the V_3 term is the barrier height for each of the tops considered separately and the V_{12} and V'_{12} terms represent the interaction between the tops. For the purposes of this discussion it is convenient to divide the two-top molecules into two groups. The first group comprises molecules with two methyl tops and the second group includes molecules in which one of the tops is a phenyl group.

8.3.2.1 *Molecules with two methyl tops*   For molecules which possess two identical tops the two torsional frequencies would be degenerate in the absence of the top interactions V'_{12} and F'. The effect of these terms is to remove the degeneracy and produce two $0 \rightarrow 1$ transitions (usually labelled α^+ and α^- transitions) which are split symmetrically about the unperturbed frequency. The molecules considered in this section are of two types: (1) $CH_3.CX_2.CH_3$ or $CH_3{-}X{-}CH_3$ or (2) $CH_3.CXY.CH_3$. For molecules of type (1) the α^+ fundamental is forbidden on symmetry grounds; only the α^- torsional frequency is observed for the high barrier limit with which we have been concerned. Both the α^+ and the α^- torsional frequencies are observed for molecules of type (2).

Dew[16] has investigated in detail the far infrared spectra of three substituted propanes and deuterated derivatives:

  (a) $CH_3.CHF.CH_3$ $CD_3.CHF.CD_3$
  (b) $CH_3.CF_2.CH_3$ $CD_3.CF_2.CD_3$
  (c) $CH_3.CHCl.CH_3$

Several hot bands as well as the allowed fundamental transitions were observed and for all three molecules all four potential constants V_3, V^*, V_{12}, V'_{12} have been determined.

Dimethyl sulphide is another molecule for which all four potential constants V_3, V^*, V_{12} and V'_{12} have been evaluated. The far infrared spectrum of this molecule shows at least 5 peaks in the 160–190 cm^{-1} region and the torsional frequency is assigned at 180 cm^{-1}. The values of the potential constants calculated by Möller and Andresen[24] are given in Table 8.3, together with the data for some other two-top molecules. The values of V^*, V_{12} and V'_{12} are small but they have a considerable effect on the value of V_3, since if the interaction terms are neglected, the calculated value of V_3 is 2120 cal/mole. The difficulty of this type of analysis is emphasized by the fact that it cannot be applied to two molecules which are very similar to dimethyl sulphide, i.e. dimethyl ether and deuterated dimethyl ether $(CD_3)_2O$. In both cases the far infrared torsion band is too complicated to permit definite assignments of the torsional fundamentals and this is due, probably, to interaction of the torsions with overall rotation. The best that can be done is to assume that there is only one fundamental transition and assign this to the strongest peak in the torsion band.

An estimate of V'_{12} is available for hexafluoroacetone.[26] Two bands are observed in the far infrared spectrum, at 72·5 cm^{-1} and 59·0 cm^{-1}, and with

Table 8.3 Barriers to internal rotation of $(CX_3)_3$ Y type molecules

Molecule	Reference	Torsional fundamental[a] (cm^{-1})	V_3	V^*	V_{12}	V'_{12}
				(cal/mole)		
$(CH_3)_2CH_2$	16, 18	—	>2700	—	—	—
$(CH_3)_2CHF$	16	247·5⎱				
$(CD_3)_2CHF$	16	182·5⎰	4160	34	370	270
$(CH_3)_2CF_2$	16	242⎱				
$(CD_3)_2CF_2$	16	175⎰	4160	57	370	−286
$(CH_3)_2CHCl$	16, 26	262	3910	−57	400	486
$(CH_3)_2CHBr$	26	256	4260	—	—	—
$(CH_3)_2CHI$	26	251	4140	—	—	—
$(CH_3)_2O$	26	241	<3120	—	—	—
$(CD_3)_2O$	26	195·5	<3480	—	—	—
$(CH_3)_2S$	24	180⎱				
		183⎰	2540	83	143	46
$(CH_3)_2NH$	24	257	3620	—	—	—
$(CD_3)_2NH$	24	198	3620	—	—	—
$(CH_3)_2CO$	25	109	830	—	—	—
$(CF_3)_2CO$	26	59·0⎱				
		72·5⎰	2950	—	—	−322
$(CH_3)_2C{=}CH_2$	25	181	2120	—	—	—
$(CH_3)_2C{=}C{=}O$	28	—	2120	—	—	—

[a] $\omega_{0-1}(\alpha^-)$ is quoted for molecules of type (1), $\omega_{0-1}(\alpha^+)$ for type (2).

the assumption that V_{12} is negligibly small, values of $V_3 = 2950$ cal/mole and $V'_{12} = -322$ cal/mole result. For the remaining molecules listed in Table 8.3, the assignments of the far infrared spectra were only sufficient to allow an estimate of V_3 to be made and the interest in these results is in the changes in the barriers with changes in chemical structure. The data can be separated into groups with the general structure

$$\underset{CX_3 \qquad CX_3}{\overset{Y}{\diagup \diagdown}}$$

where X is either H, D or F and the framework atom or group Y can be one of the following series:

=O;	=S;	=NH			
=CH$_2$;	=CHF;	=CF$_2$;	=CHCl;	=CHBr;	=CHI
=CO;	=C=CH$_2$;	=C=C=O			

The first series consists of dimethyl ether and dimethyl sulphide. The decrease in the barrier when the oxygen atom in dimethyl ether is replaced by sulphur can be interpreted as being due to the increase in length of the torsion bonds and the decreased electronegativity of the framework atom. The barrier in dimethyl amine is a little larger than that for dimethyl ether but the difference is not sufficient to warrant speculation.

The second series, based on propane, also shows behaviour which cannot be simply explained in terms of either the size of the substituent or effects such as electronegativity. Substitution at the 2-carbon atom increases the barrier considerably but it is possible also that neglect of the higher potential energy constants has produced results which mask the effects of changes in the chemical structure.

The third series, based on acetone, shows considerable variations. The barrier in acetone is very low and this is attributed to the strong polarity of the C=O bond which reduces the electron density in the C—C bonds. Replacing the C=O group by less polar groups such as C=CH$_2$ or C=C=O increases the barrier towards the value for propane. Replacement of CH$_3$ by CF$_3$ in acetone increases the barrier considerably. This is to be expected since the increased electronegativity of the fluorine atoms will offset the effect of the polarity of the C=O bond. Steric effects are also likely to play an important part as is evidenced by the relatively large value of the interaction constant V'_{12}.

Thus, although it has proved possible to calculate the value of V_3 for molecules possessing two methyl tops, it has only proved possible to evaluate the interaction potential constants for a few molecules. Higher resolution far infrared spectra may allow, for a few more molecules, analysis of the complex torsional bands which are observed.

8.3.2.2 *Molecules with two non-identical tops*   Only a few molecules of this class have been investigated using far infrared spectroscopy. There are, of course, considerable theoretical and experimental difficulties associated with the determination of barriers to internal rotation in a molecule where the two tops are not equivalent and may not even have the same symmetry. The general theory of the torsional behaviour of a molecule with two non-equivalent tops of different symmetry has not been investigated and approximate methods must be used. Initially the two torsional vibrations can be treated as independent and the kinetic and potential energy coupling between them can be introduced as perturbations. This approach is suitable as long as the coupling terms are small which is usually the case in molecules, such as anisole $C_6H_5OCH_3$, with one very heavy top and one very light top. In cases where the coupling is not small, the situation becomes more like that in the case of two identical tops and is best treated by the choice of suitable normal co-ordinates for the two torsions. Accurate barrier calculations require that the equilibrium bond lengths and angles of the molecule be known and in many cases these data are not available. A suitable structure must then be inferred from data for other molecules and this leads to further uncertainties in the barrier calculations. Experimentally, the main difficulty is the observation of the far infrared gas-phase spectrum usually because two-top molecules in this class have low vapour pressures at ambient temperature.

For the most part, the molecules to be discussed in this section are of the general structure

$$\text{Ph—X} \diagdown_{\text{Y}}$$

where Ph represents a phenyl or substituted group, X is the framework and Y is the other top. Anisole, for which $X = O$ and $Y = CH_3$, has been most intensively investigated, mainly because it is possible to study a series of related molecules by making various substitutions on the phenyl group. In theory, it should then be possible to correlate changes in the barrier to internal rotation about the phenyl–oxygen bond with the position and type of phenyl substitution. Some single-top aromatic molecules have also been studied with this end in view and the results are summarized in Table 8.4, together with the data for the two-top molecules.

In anisole, the phenyl torsion has been observed at $81.5\ \text{cm}^{-1}$ in the gas phase.[2] Studies of the far infrared spectra of anisole and two deuterated species, $C_6H_5OCD_3$ and $C_6D_5OCH_3$, in the 180–260 cm^{-1} region, where the methyl torsion is expected, failed to indicate any absorption which could be definitely assigned to this torsional mode. The far infrared and low frequency Raman spectra are shown in Table 8.5.

The spectra of anisole and anisole-d_5 are similar, while that of anisole-d_3 shows a difference as the band near $200\ \text{cm}^{-1}$ disappears altogether. The barriers calculated from the gas-phase torsion frequencies are similar for anisole and anisole-d_5 ($V_2 = 3610$ cal/mole) but the value for anisole-d_3 is 4080 cal/mole. The discrepancy could not be improved by including in the calculation potential and kinetic energy coupling between the phenyl and

Table 8.4 Barriers to internal rotation in some one- and two-top molecules with phenyl tops

Molecule	Reference	ω_{0-1}(cm^{-1})	V_2 (cal/mole)
Acetophenone (gas)	29	48 ± 2	3100
p-Fluoroacetophenone (gas)	29	51 ± 2	3500
Anisole (liquid)	30	108	6050
p-Fluoroanisole (liquid)	30	102	6100
p-Chloroanisole (liquid)	30	97	5550
p-Bromoanisole (liquid)	30	~97	6200
Anisole (gas)	2	81·5	3610
Anisole-d_3 (gas)	2	77·5	4080
Anisole-d_5 (gas)	2	80·0	3600
p-Fluoroanisole (gas)	2	71	2950
p-Chloroanisole (gas)	2	70	2990
p-Methylanisole (gas)	2	72	3060
Pentafluoroanisole (gas)	2	34	860
2,6-difluoroanisole (gas)	2	36	870
Benzaldehyde (gas)	29	111	4660
p-Fluorobenzaldehyde (gas)	29	93·5	3580
p-Chlorobenzaldehyde (gas)	29	81·5	2800
p-Bromobenzaldehyde (gas)	29	73·5	2370
p-Methylbenzaldehyde (gas)	29	89·5	3480
Phenol (gas)	31	310	3260
p-Fluorophenol (gas)	31	280	2700
p-Chlorophenol (gas)	31	302	3080
Pentafluorobenzaldehyde (gas)	2	58	1790
Pentafluorophenol (gas)	2	297	3320

Table 8.5 Far infrared and Raman spectra of anisole[a]

Molecule	Infrared (cm^{-1}) Gas	Liquid	Raman (cm^{-1}) Liquid
$C_6H_5OCH_3$	253(w)	261(m)	259(m)
			209(m)
	81·5(w)	110(w)	Phenyl torsion
$C_6H_5OCD_3$	239(w)	244(w)	244(w)
	77·5(w)	105(w)	Phenyl torsion
$C_6D_5OCH_3$	254(w)	256(w)	254(m)
			203(m)
	80(w)	105(w)	Phenyl torsion

[a] m = medium intensity, w = weak intensity.

methyl tops. It must therefore be concluded that mixing between the low frequency out-of-plane ring vibrations (which will be sensitive to the deuteration of the methyl group) and the torsions is considerable and the model with two degrees of freedom is inadequate in this case. Anisole also provides a good example of the errors which accrue if torsion frequencies taken from liquid-phase data are used to calculate barriers. The barrier calculated for anisole, using the liquid-phase torsional frequency of 108 cm^{-1} and a one-degree-of-freedom model, is 6050 cal/mole, nearly double the value for the gas phase. The intermolecular contribution to the barrier in the liquid phase is thus of the same order as the intramolecular barrier. This enormous increase associated with a change of state from gas to liquid has not been adequately explained, but it could possibly be due to some short range correlational ordering of the phenyl groups of neighbouring molecules in the liquid state.

Substitution of phenyl groups in the *para* position lowers the calculated value of the barrier, but no definite trend is observed which can be related to a property of the substituent group. The values of V_2 for *p*-fluoroanisole, *p*-chloroanisole and *p*-methylanisole are all similar in magnitude, 3000 $\pm$ 60 cal/mole and are about 20% lower than in anisole itself. Large changes with substitution have been observed in a series of *para*-substituted benzaldehydes[29] and to a lesser extent in phenols.[31] In the case of the benzaldehydes the variation in the barrier height with *para* substitution is in the order H > F $\simeq$ CH$_3$ > Cl > Br. However, this sequence cannot be correlated with any series describing *ortho–para* directing ability such as electronegativity or Hammett's σ-constants. The variations with *para* substitution in the phenols are rather smaller and the decrease is of the same order as in the case of anisole, about 20%.

Substitution in the *ortho* position affects the barrier considerably as might be expected because of the possibility of steric effects, which are absent in the case of *para* substitution. The molecules discussed here are all symmetrically disubstituted in the 2- and 6-positions to avoid the complications introduced by rotational isomerism. A striking decrease occurs in the value of V_2 when anisole is substituted with fluorine in the 2- and 6-positions. In the two molecules studied, 2,6-difluoroanisole and pentafluoroanisole, the torsions are observed at 36 cm^{-1} and 34 cm^{-1}. The values of V_2 resulting from these assignments are 860 and 870 cal/mole respectively. Although these are low barriers in absolute terms, because of the low fundamental frequency there will be at least 14 levels below the top of the barrier where the energy levels go over to those of a free rotor modified by the underlying torsional potential. The low barrier relative to the value of kT at ambient temperature means that transitions between levels near the top of the barrier will have considerable intensity. This is a possible explanation of the high frequency shoulder on the torsion bands of these molecules[2] which could be due to free rotor transitions at frequencies higher than the fundamental.

A similar decrease in the barrier with di-*ortho* substitution is seen in pentafluorobenzaldehyde. The torsional frequency decreases from 111 cm^{-1} in benzaldehyde to 58 cm^{-1} in pentafluorobenzaldehyde, the corresponding two-fold barriers being 4460 and 1790 cal/mole. The decreases observed with *ortho* substitution can be attributed in part to the electronegativity of the fluorine substituent which reduces the electron density in the bond which is the internal rotation axis.

It is also possible that steric effects can lower the barrier, as has been observed in *cis*- and *trans*-substituted propenes.[28]

Propene: $V_3 = 1990$ cal/mole

trans-1-Fluoropropene: $V_3 = 2200$ cal/mole

cis-1-Fluoropropene: $V_3 = 1060$ cal/mole

cis-1-Chloropropene: $V_3 = 620$ cal/mole

Substitution in the *trans* position on the 1-carbon atom raises the barrier slightly but substitution in the *cis* position decreases the barrier to less than half the value in propane. In the equilibrium position, the methyl group of propene is oriented so that two hydrogens are staggered with respect to the hydrogen on the 2-carbon atom and therefore the third hydrogen on the methyl group must be eclipsed by the *cis* hydrogen on the 1-carbon atom. When the hydrogen in the *cis* position is substituted by either fluorine or chlorine, the increased steric hindrance has the effect of raising the level of the bottom of the barrier but leaves the top of the barrier unchanged. This results in a decrease in the barrier height. Similarly, in anisole, evidence from X-ray studies of *p*-dimethoxybenzene[32] indicates that the equilibrium position of the methyl group is in the plane of the benzene ring. The barrier to rotation about the phenyl–oxygen axis is thought to be due to overlap between the lone pair of electrons on the oxygen atom and the π-electrons of the benzene ring and this is maximized when the methyl group is in the plane of the ring. Substitution of the ring in the *ortho* position will have the same effect as in the propenes, raising the bottom of the barrier by steric hindrance without having much effect on the top of the barrier.

8.3.3 *Molecules possessing two coupled asymmetric rotors*

In § 8.1.2 we noted that in addition to the fundamental torsional frequencies for the two isomers of the asymmetric halogenated ethanes to be considered now, it is also necessary to know the energy difference $\Delta E_{(sym-asym)}$ in order to calculate the potential constants V_k. Fortunately gas-phase data are available for seven molecules. The molecules fall into two groups:

(a) $CH_2Cl.CH_2Cl$, $CHCl_2.CHCl_2$, $CH_2F.CH_2F$ and $CHF_2.CHF_2$ for which the *trans* isomer has C_{2h} symmetry; the torsional mode of this isomer is A_u and is infrared active but inactive in the Raman spectrum. The *gauche* isomer is of lower symmetry C_2, its torsional mode is A and is both Raman and infrared active.

(b) $CH_2Cl.CHCl_2$, $CH_2F.CH_2Cl$ and $CHCl_2.CCl_2.CH_3$ for which the symmetrical isomer has C_3 symmetry and the asymmetrical isomer has symmetry C_1. The torsional mode in each isomer is both Raman and infra-red active.

The ratio of the intensities of the two torsional fundamentals for a given molecule is a function of the Boltzmann distribution arising from the energy difference between the isomers and thus ΔE can be obtained from the tem-perature dependence of this ratio.

$$\frac{I_{sym}}{I_{asym}} \text{ or } \frac{I_{trans}}{I_{gauche}} = \text{constant } e^{-(\Delta E/RT)}$$

The experimental data are summarized in Table 8.6. Most of the results are drawn from the work of Klaboe and Nielsen[33] or more recent work in Manchester.[16,35,36] The exact values of the potential constants are calculated using Cunliffe's method which was developed in Manchester. Expansion of the potential function in § 8.1.2 as far as the term in α^2 gives the following approximate relations which give potential functions sufficiently accurate for input into the final accurate calculation.

$$\omega^2_{0-1(sym,trans)} = F(V_1 + 4V_2 + 9V_3)$$

$$\omega^2_{0-1(asym,gauche)} = F(-0{\cdot}5V_1 - 2V_1 + 9V_3)$$

and if we assume that the potential minima occur at $0°$ (sym.) and $120°$ (asym.)

$$E_{(s-a,t-g)} = 0{\cdot}75(V_1 + V_2)$$

The values of V_1, V_2 and V_3 listed in Table 8.6 are calculated by Cunliffe's method.

From the data in Table 8.6 the function $V(\alpha)$ can be constructed for each molecule. Of especial interest are the locations of the maxima and minima in each case. These are tabulated in Table 8.7; $\alpha = 0°$ defines the symmetric isomer, the symmetric well-height is close to $\alpha = 60°$, the asymmetric isomer is close to $\alpha = 120°$ and the asymmetric well-height is defined by $\alpha = 180°$.

Table 8.6 Torsional frequencies, energy differences and V_k for halogenoethanes in the gas phase

Molecule	Reference	$\omega_{0-1(sym., trans)}$ (cm^{-1})	$\omega_{0-1(asym., gauche)}$ (cm^{-1})	$\Delta E_{(s-a, t-g)}$	V_1	V_2 (cal/mole)	V_3
$CH_2F.CH_2F$	16, 33	138	152	0	1160	-1100	3530
$CHF_2.CHF_2$	16, 33	83	83	1160	2950	-1180	3560
$CHF.CHCl$	16, 34	128	133	450	1660	-970	3770
$CH_2Cl.CH_2Cl$	16, 35	129	126	1100	3630	-1900	5680
$CH_2Cl.CHCl_2$	16, 35	100	113	-3100	-5300	1320	7250
$CHCl_2.CHCl_2$	16, 35	88	88	0	2040	-1960	11800
$CHCl_2.CCl_2.CH_3$	36	75	—	-1000	—	—	—

Table 8.7 α and $V(\alpha)$ for the maximum and minimum values of the potential energy function

Molecule	Sym., *trans* isomer		Symmetric well-height		Asym., *gauche* isomer		Asymmetric well-height	
	α	$V(\alpha)$ (cal/mole)	α	$V(\alpha)$ (cal/mole)	α	$V(\alpha)$ (cal/mole)	α	$V(\alpha)$ (cal/mole)
$CH_2F.CH_2F$	0	0	60	3000	116·5	0	180	4700
$CHF_2.CHF_2$	0	0	63	3400	113·5	1150	180	6500
$CH_2F.CH_2Cl$	0	0	61	3450	116	450	180	5450
$CH_2Cl.CH_2Cl$	0	0	62	5150	114·5	1150	180	9300
$CH_2Cl.CHCl_2$	0	0	57	6950	118	-3100	180	2000
$CHCl_2.CHCl_2$	0	0	60	10850	125	0	180	13850
$CHCl_2.CHCl_2.CH_3$	0	0	60	13000	120	-1000	180	>14000

The symmetric well-height is most sensitive to $\omega_{0-1(\text{sym})}$ and the asymmetric well-height is most affected by $\omega_{0-1(\text{asym})}$ but it is less well defined and is more sensitive to other variations than is the symmetric well-height. In the case of the most heavily substituted molecules it must be borne in mind that there is an increasing likelihood of mixing with other vibrations.

There is not space to discuss in detail the assignments preferred in Table 8.6, the accuracy of which are so important in fixing the values of $V(\alpha)$ listed in Table 8.7. The uncertainties in $V(\sim 60)$ and $V(\sim 120)$ are of the order of ± 100 to ± 300 cal/mole. $V(180)$ is less accurate with an error ± 500 to ± 900 cal/mole except in the case of $CHCl_2.CHCl_2$ (± 1500 cal/mole) and $CHCl_2.CCl_2.CH_3$ (unknown).

Finally it is worthwhile comparing the results reported in § 8.3.1 and 8.3.3 for all the chloro- and fluoroethanes. Figure 8.5 shows that the values of V_3

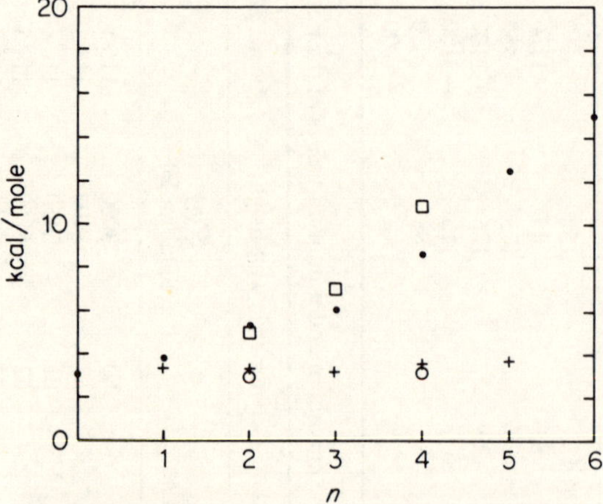

Figure 8.5   Barrier heights in chloro- and fluoro-ethanes plotted as a function of the degree of substitution, where n is the number of halogen atoms in the molecule ;
  + : V_3 for fluoroethanes with symmetric tops,
  ● : V_3 for chloroethanes with symmetric tops,
  ○ : height of *trans–gauche* barrier for asymmetric fluoroethanes,
  □ : height of *trans–gauche* barrier for asymmetric chloroethanes

for the molecules containing symmetric tops fall on a smooth curve which rises as the degree of substitution increases. In the case of the fluoroethanes there is only a very slight increase in barrier height, but the chloroethane values rise very rapidly. A similar pattern occurs for the asymmetric molecules listed in Table 8.7, provided that the depth of the symmetric (*trans*) well is taken as a guide to the effective barrier. Again fluorine substitution produces little change but chlorine substituents raise the barrier. Comparison of the V_3 terms again shows similar trends in both sets of molecules, irrespective of the symmetry.

8.4 Limitations in the use of torsional frequencies

Apart from the certainty with which a torsional mode can be assigned to the particular frequency of a band in the molecular spectrum of a molecule and the accuracy with which this frequency can be determined, several other factors combine to limit the usefulness of torsional frequencies in the calculation of the potential constants which define the contour of the potential curve generated by internal rotation. These limitations are common to all torsional frequencies listed in § 8.3.

8.4.1 *Theoretical assumptions*

The limitations imposed by theoretical assumptions are fully discussed in Chapter 7. In particular they begin with the assumption that the Hamiltonian for internal rotation can be separated from overall rotation and the remainder of the vibrational Hamiltonian. For light molecules containing a CH_3 top attached to a rigid frame this assumption will hold more rigorously than in the cases of heavier tops or more than one top. Ultimately the problem of the torsional frequency mixing with other vibrations arises and this gives uncertainties which are difficult to assess quantitatively for molecules containing so many atoms.

8.4.2 *Structural parameters*

Even when the assignment of the torsional frequency is certain and the frequency is measured with precision it is essential to have accurate bond lengths and bond angles in order to calculate the structure factor F. Regrettably, even for some of the simpler fluoro- and chloroethanes where the torsional frequencies are well established, the accuracy of the structural parameters is sufficiently low to be a major source of error in the calculation of V_3. For some coupled asymmetric tops and for most of the aromatic molecules the parameters have not been measured. Clearly there is now a strong case for a more detailed electron diffraction study of the gases for which good torsional frequencies have been established.

8.4.3 *Potential function*

Although the potential function governing internal rotation is expressed as a Fourier series, only for a few selected molecules have terms higher than V_3 been evaluated. In most instances the fundamental torsional frequency ω_{0-1} is observed and then the spacing of the levels $V = 0$ and $V = 1$ is used to calculate the height of relatively large barriers. Hot bands are often observed but only in a few instances (eg. $CH_3.CH_2Cl$) have they been used to investigate the more detailed shape of the barrier. It is likely that for a three-fold rotor with a large value of V_3 the $\cos 3kN\alpha$ function is a poor representation of the top of the barrier which in fact might be flatter than this function indicates or, indeed, possess minor undulations. We simply do not know very much about the detailed shapes of potential barriers and our information based on fundamental torsional frequencies is likely to be less precise as the height of the barrier increases.

8.5 Acknowledgment

The authors wish to acknowledge their debt of gratitude to their colleagues, particularly P. N. Brier, A. R. Gee, G. Lane, G. Dew and A. V. Cunliffe, on whose research work over the past ten years much of this chapter is based.

8.6 References

1. S. Weiss and G. Leroi, *J. Chem. Phys.*, **48**, 962 (1968).
2. S. Fewster, *Ph.D. Thesis*, University of Manchester (1970); G. Allen and S. Fewster, to be published.
3. H. Boutin and S. Yip, *Molecular Spectroscopy with Neutrons* (Cambridge, Mass: MIT Press, 1968).
4. K. S. Pitzer, *Discussions Faraday Soc.*, No. **10**, 66 (1951).
5. H. Russell, D. R. Golding and D. M. Yost, *J. Am. Chem. Soc.*, **66**, 16 (1964).
6. W. G. Fateley and F. A. Miller, *Spectrochim. Acta*, **19**, 611 (1963).
7. R. H. Schwendeman and J. D. Jacobs, *J. Chem. Phys.*, **36**, 1245 (1962).
8. P. N. Brier, J. S. Higgins and R. H. Bradley, *Mol. Phys.*, **21**, 72 (1971).
9. G. Allen, P. N. Brier and G. Lane, *Trans. Faraday Soc.*, **63**, 824 (1967).
10. T. Shimanouchi and T. Fujiama (Private communication).
11. J. R. Durig, S. M. Craven, K. K. Lane and J. Bragin, *J. Chem. Phys.*, **54**, 481 (1971).
12. I. L. Karle and J. Karle, *J. Chem. Phys.*, **17**, 1052 (1949).
13. G. Sage and J. Klemperer, *J. Chem. Phys.*, **39**, 371 (1963).
14. D. R. Herschbach, *J. Chem. Phys.*, **25**, 358 (1956).
15. W. G. Fateley and F. A. Miller, *Spectrochim. Acta*, **17**, 857 (1961).
16. G. Dew, *Ph.D. Thesis*, University of Manchester (1969); G. Allen, A. V. Cunliffe, and G. Dew, to be published.
17. P. N. Brier and J. S. Higgins, *Mol. Phys.*, **19**, 645 (1970).
18. A. Danti and J. L. Wood, *J. Chem. Phys.*, **30**, 582 (1959).
19. D. E. Mann and E. K. Plyler, *J. Chem. Phys.*, **21**, 1116 (1953).
20. H. T. Minden and B. P. Dailey, *Phys. Rev.*, **A338**, 82 (1951).
21. J. R. Nielsen, H. H. Claasen and D. C. Smith, *J. Chem. Phys.*, **18**, 1471 (1950).
22. H. S. Gutowsky and H. B. Levine, *J. Chem. Phys.*, **18**, 1297 (1950).
23. F. B. Brown, A. D. H. Clague, N. D. Heitkemp, D. F. Koster and A. Danti, *J. Mol. Spectr.*, **24**, 163 (1967).
24. K. D. Möller and H. G. Andresen, *J. Chem. Phys.*, **37**, 1800 (1962).
25. W. G. Fateley and F. A. Miller, *Spectrochim. Acta*, **18**, 977 (1962).

26. K. D. Möller, A. R. De Meo, D. R. Smith and L. H. London, *J. Chem. Phys.*, **47**, 2609 (1967).
27. W. G. Fateley, F. E. Kiviat and F. A. Miller, *Spectrochim. Acta*, **26A**, 315 (1970).
28. J. Dale, *Spectrochim. Acta*, **22**, 3373 (1966).
29. F. A. Miller, W. G. Fateley and R. E. Witowski, *Spectrochim. Acta*, **23A**, 891 (1967).
30. N. L. Owen and R. E. Hester, *Spectrochim. Acta*, **25A**, 343 (1969).
31. W. G. Fateley, F. A. Miller and R. E. Witkowski, *Technical Report AFML-TR-66-408*, Wright Patterson Air Force Base, Ohio.
32. T. H. Goodwin, M. Przybylska and J. Monteath Robertson, *Acta Cryst.*, **3**, 279 (1950).
33. P. Klaboe and J. R. Nielsen, *J. Chem. Phys.*, **32**, 899 (1960); **33**, 1764 (1960).
34. A. D. Giacomo and C. P. Smyth, *J. Am. Chem. Soc.*, **77**, 1361 (1955).
35. G. Lane, *Ph.D. Thesis*, University of Manchester (1966).
36. F. Heatley, G. Allen, S. Hameed and P. W. Jones, *J. Chem. Soc., Faraday Trans. II*, **68**, 1547 (1972).

9 Molecular acoustics and conformational behaviour

S. M. Walker

9.1 Introduction

The propagation of sound waves through a body produces alternating compression and rarefaction and, since the period of fluctuation is short compared with the time required for thermal equilibration with the surroundings, the process may be assumed to be reversible and adiabatic. However, at very high frequencies ($\sim 5\,\text{GHz}$) the sound wavelength approaches the same order of magnitude as the molecular mean free path and the propagation becomes isothermal. Behaviour under these conditions is beyond the scope of this review. In all fluids with a specific heat ratio γ greater than unity, these perturbations may also be visualized as a temperature wave and the system can respond either to the pressure fluctuations (volume response) or to the alternating temperature (specific heat response). Water is an important exception at about 4°C when the temperature coefficient of volume expansion and hence $C_p - C_v$ becomes zero and sound waves propagate without producing temperature changes.

As an idealization, a plane harmonic wave will travel through a medium unattenuated, with a velocity determined, through compressibility, from the thermodynamic equation of state of the fluid. In practice energy is extracted from the sound wave resulting in exponential attenuation of the wave intensity and change in its phase velocity c. The acoustic pressure p at a distance x from the source (amplitude p_0) is given by

$$p = p_0 \exp\left(-\alpha x\right) \exp\left[i(\omega t - kx)\right] \tag{9.1}$$

where α is the absorption coefficient of the propagating medium and $k = \omega/c = 2\pi/\lambda$ is the propagation constant. Experimentally it is found that attenuation changes are several orders of magnitude greater than the accompanying velocity changes. For the *cis/trans* isomerization equilibrium in ethyl acetate, for example, the absorption changes by a factor of 500 while the velocity dispersion is less than 1 %. Consequently velocity changes will be largely ignored in this review.

There are several mechanisms by which energy can be extracted from the acoustic wave. Firstly, there is a contribution to the total absorption due to the shearing motion inherent in the progression of a plane wave. Calculations based on the Stokes–Navier equations give this contribution (from the shear viscosity η_s) as[1]

$$\alpha_s = \frac{2\pi^2}{\rho c^3} \cdot \frac{4\eta_s}{3} \cdot f^2 \tag{9.2}$$

where f is the sound frequency. This particular aspect of fluid behaviour, namely the capacity to support viscous waves, can be studied more thoroughly by introducing shear waves into the liquid. In this chapter we are concerned with conformational studies which respond mainly to longitudinal waves and no further discussion of shear behaviour will be given. In addition, since all fluids possess thermal conductivity κ, there will be some heat transfer notwithstanding the adiabatic character of the propagation. Absorption due to this mechanism is small except for liquid helium and liquid metals and is given by

$$\alpha_\kappa = \frac{2\pi^2}{\rho c^3}(\gamma - 1) \cdot \frac{\kappa}{C_p} \cdot f^2 \tag{9.3}$$

It is convenient to combine the two classical causes of sound attenuation (collectively denoted viscothermal absorption) and express the result in frequency-independent form:

$$\left(\frac{\alpha}{f^2}\right)_{\text{class.}} = \frac{2\pi^2}{\rho c^3}\left\{\frac{4\eta_s}{3} + \frac{(\gamma - 1)\kappa}{C_p}\right\} = \text{constant} \tag{9.4}$$

This treatment assumes that viscous losses are absent in compression, i.e. the volume viscosity η_v is zero. This is not generally true and measured absorption coefficients are often much greater than predicted by equation (9.4). The 'excess' absorption is considered to originate entirely from the volume viscosity associated with changes in molecular equilibria produced by the temperature and volume alternations of the wave.

$$\alpha_{\text{excess}} = \alpha_{\text{measured}} - \alpha_{\text{classical}}$$

$$= \frac{2\pi^2}{\rho c^3}\eta_v f^2 \tag{9.5}$$

Several distinct effects are responsible for excess absorption in liquids and all are believed to be relaxational in character. Large relaxation effects can be observed in structural relaxation where the different equilibrium states under study possess different volumes. When the volume viscosity arises in response to temperature fluctuations in the wave, then thermal relaxation is observed. Examples of this latter behaviour are vibrational specific heat relaxation and relaxation due to perturbation of the equilibrium between rotational isomers.

9.2 Relaxation theory

Consider a simple fluid containing a single equilibrium described in terms of irreversible thermodynamics by three independent variables. These are usually chosen for convenience to be P, T and an internal ordering parameter ζ with entropy, volume and affinity as the corresponding dependent variables.

$$\delta G = -S\delta T + V\delta P + Z\delta\zeta \tag{9.6}$$

The second derivatives of the Gibbs free energy are relevant.

$$\text{Isothermal compressibility} \quad K_T^\infty = -\frac{1}{V}(\delta V/\delta P)_{T,\zeta}$$

$$\text{Adiabatic compressibility} \quad K_S^\infty = -\frac{1}{V}(\delta V/\delta P)_{S,\zeta}$$

$$\text{Heat capacity} \quad C_P^\infty = T(\delta S/\delta T)_{P,\zeta} \tag{9.7}$$

$$\text{Expansivity} \quad \beta^\infty = -\frac{1}{V}(\delta S/\delta P)_{T,\zeta}$$

The superscript ∞ refers to the 'frozen' values of these coefficients implying correspondence to processes occurring at too fast a rate (infinite frequency) for the ordering parameter to attain its equilibrium value. Constants measured at equilibrium refer to slow processes measured, by definition, at minimum free energy.

$$(\delta G/\delta\zeta)_{P,T} = Z(T, P, \zeta) = 0 \tag{9.8}$$

These are written without superscription. Using equations (9.7) we have, for infinitesimal changes in S, V and Z, treating them as functions of P, T and ζ,

$$\delta S = (C_P^\infty/T)\delta T - V\beta^\infty\delta P + (\delta S/\delta\zeta)_{P,T}\delta\zeta \tag{9.9a}$$

$$\delta V = V\beta^\infty\delta T + (\delta V/\delta P)_{T,\zeta}\delta P + (\delta V/\delta\zeta)_{P,T}\delta\zeta \tag{9.9b}$$

$$\delta Z = (\delta Z/\delta T)_{P,\zeta}\delta T + (\delta Z/\delta P)_{T,\zeta}\delta P + (\delta Z/\delta\zeta)_{P,T}\delta\zeta \tag{9.9c}$$

It is convenient to replace entropy and volume by their changes per unit change of ζ, and to introduce enthalpy changes.

$$\Delta S = (\delta S/\delta\zeta)_{P,T} \qquad \Delta V = (\delta V/\delta\zeta)_{P,T} \qquad \Delta H = T(\delta S/\delta\zeta)_{P,T}$$

Introducing the appropriate Maxwell relations,

$$(\delta Z/\delta T)_{P,\zeta} = (\delta S/\delta\zeta)_{P,T} = \Delta H/T$$

$$(\delta Z/\delta P)_{T,\zeta} = -(\delta V/\delta\zeta)_{P,T} = -\Delta V \tag{9.10}$$

combination of equations (9.9) and (9.10) yields

$$dS^\infty = (C_p^\infty/T)\,dT - V\beta^\infty\,dP$$
$$dV^\infty = V\beta^\infty\,dT - VK_T^\infty\,dP \tag{9.11}$$

Equations (9.11) assume a constant ζ, i.e. they describe the behaviour at infinitely high frequencies. The equilibrium conditions can be derived from equations (9.9) by putting $Z = dZ = 0$. Thus

$$\delta\zeta = -\frac{(\delta Z/\delta P)_{T,\zeta}\,\delta P - (\delta Z/\delta T)_{P,\zeta}\,\delta T}{(\delta Z/\delta\zeta)_{P,T}}$$

which, using equations (9.10) becomes

$$d\zeta = \frac{1}{\phi}\left(\Delta V\,dP - \frac{\Delta H}{T}\,dT\right) \tag{9.12a}$$

where

$$\phi = (\delta Z/\delta\zeta)_{P,T} = (\delta^2 G/\delta\zeta^2)_{P,T}$$

and

$$dS = (C_p/T)\,dT - V\beta\,dP \tag{9.12b}$$
$$dV = V\beta\,dT - VK_T\,dP \tag{9.12c}$$

The 'excess' values of these properties as a result of the presence of a relaxation may be written

$$\delta C_p = C_p - C_p^\infty = \Delta H^2/T\phi \tag{9.13a}$$

$$\delta\beta = \beta - \beta^\infty = \Delta H\Delta V/T\phi V \tag{9.13b}$$

$$\delta K_T = K_T - K_T^\infty = \Delta V^2/\phi V \tag{9.13c}$$

It has been mathematically convenient to employ the isothermal compressibility but it is much more sensible to use the adiabatic parameter and thus reflect the true nature of the propagation. Now,

$$K_S = K_T - TV\beta^2/C_p$$

or, for infinitesimally small changes,

$$\delta K_S = \delta K_T - TV\delta(\beta^2/C_p) \tag{9.14}$$

Using equations (9.13) and (9.14) and the thermodynamic identity

$$(\gamma - 1) = \frac{TV\beta^2}{C_p K_S}$$

Then,

$$\frac{\delta K_S}{K_S} = r_s = \frac{(\gamma - 1)\delta C_p}{C_p^\infty}\left\{1 - \frac{\Delta V}{V}\cdot\frac{C_p}{\Delta H\beta}\right\}^2 \tag{9.15}$$

The quantity r_s is known as the relaxation strength. For thermal relaxations in which the volume change can be regarded as negligible ($\Delta V/V < 1\%$) then equation (9.15) reduces to the form

$$r_s = \frac{(\gamma - 1)\delta C_p}{C_p^\infty} \tag{9.16}$$

It will be seen later that this is a critical trnasformation since it renders the calculation of enthalpy differences in equilibria very much simpler.

So far only the 'unrelaxed' (equilibrium) and 'relaxed' (frozen) states have been defined. In order to account for relaxational behaviour it is necessary to specify the nature of the transition from one situation to the other. In the equilibrium state $Z(P, T, \zeta) = 0$, the ordering parameter takes the value $\bar{\zeta}$. If P and T are now changed by small increments δP and δT then ζ will not, in general, instantaneously attain its new equilibrium value $\bar{\zeta} + \delta\bar{\zeta}$. For small perturbations the rate of change of ζ is linearly proportional to the ordering force Z.

$$\frac{\mathrm{d}\delta\zeta}{\mathrm{d}t} = L\delta Z \tag{9.17}$$

If the pressure and the temperature remain constant at their new values $P + \delta P$ and $T + \delta T$ respectively then, using equations (9.9c) and (9.12a), equation (9.17) may be rewritten as

$$\frac{\mathrm{d}\delta\zeta}{\mathrm{d}t} = \frac{\delta\zeta - \delta\bar{\zeta}}{\tau_{TP}} \tag{9.18}$$

where $\tau_{TP}^{-1} = -L(\delta Z/\delta\zeta)_{T,P}$. Equation (9.18) is a typical relaxation equation. In practice pressure and temperature are inconvenient variables and it becomes more appropriate to use pressure and entropy as the independent parameters. The enthalpy is now the primary thermodynamic quantity (cf. equation 9.6),

$$\delta H = T\delta S + V\delta P + Z\delta\zeta \tag{9.19}$$

and equilibrium is defined by the condition

$$(\delta H/\delta\zeta)_{S,P} = Z(P, S, \zeta) = 0 \tag{9.20}$$

A similar treatment leads to the analogues of equations (9.9b) and (9.18).

$$\delta V = (\delta V/\delta S)_{P,\zeta}\delta S - VK_S^\infty\delta P + (\delta V/\delta\zeta)_{P,S}\delta\zeta \tag{9.21}$$

$$\frac{\mathrm{d}\delta\zeta}{\mathrm{d}t} = \frac{\delta\zeta - \delta\bar{\zeta}}{\tau_{SP}} \tag{9.22}$$

Combination of equations (9.21) and (9.22) with (9.12) yields

$$\left(1 + \tau_{SP}\frac{d}{dt}\right)(\delta V/V) = K_S\left(1 + \frac{K_S^\infty}{K_S}\tau_{SP}\frac{d}{dt}\right)\delta P \qquad (9.23)$$

The steady state solution of this equation must be proportional to the real part of the exponent in equation (9.1) if it is to have the correct sinusoidal form of the compressional wave. Such a solution is of the form

$$\frac{k^2}{\omega^2} = \rho K_S\left\{\frac{1 + i\omega\tau_{SP}K_S^\infty/K_S}{1 + i\omega\tau_{SP}}\right\} \qquad (9.24)$$

where the density $\rho = K_S/c_0^2$. Here, c_0 refers to the velocity at infinitely low frequencies and since velocity dispersion can be neglected it is common to put $c_0 = c$. If several independent relaxations are being dealt with, each with strength r_j and relaxation time τ_j (at constant P and S) then equation (9.24) may be generalized to

$$\frac{k^2}{\omega^2} = \rho K_S\left\{1 - \sum_j \frac{ir_j\omega\tau_j}{1 + i\omega\tau_j}\right\} \qquad (9.25)$$

In principle equations (9.15) and (9.25) contain all the necessary information relating the sound wave propagation to chemical equilibria. For reasons indicated in the derivation of equation (9.4) experimental measurements of the sound absorption coefficient α as a function of the sound frequency f are presented in the form of graphs of α/f^2 versus f. Rearrangement of equation (9.25) in terms of the first ($j = 1$) and subsequent ($j > 1$) relaxations, and equating real and imaginary parts gives

$$k = k_r - i\alpha$$

$$\frac{2\alpha k_r}{\omega^2} = \frac{2\alpha c_0^2}{\omega c} = \frac{r_1\omega\tau_1}{1 + \omega^2\tau_1^2} + \sum_{j>1} r_j\omega\tau_j \qquad (9.26)$$

This may be written in a more general way as

$$\frac{\alpha}{f^2} = \frac{A}{1 + (f/f_r)^2} + B \qquad (9.27)$$

in which the constants A and B are given by

$$A = \frac{2\pi r_1\tau_1}{c} \qquad B = \frac{2\pi^2}{c}\sum_{j>1} r_j\tau_j \quad \text{and} \quad f_r = 1/2\pi\tau_1$$

using $c_0 = c$ as before. The constant A is a parameter of the particular single relaxation under investigation, with relaxation frequency f_r, while visco-thermal absorption in addition to contributions from all higher frequency relaxations is included in B. Experimentally it is found that all single thermal

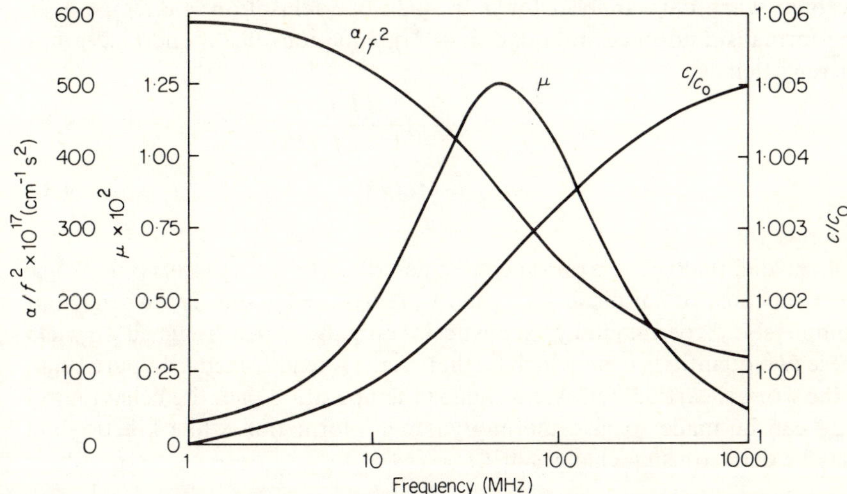

Figure 9.1   Calculated curves of α/f^2, μ and c/c_0 against frequency for a single relaxation process with the parameters $A = 500 \times 10^{-17}\,\text{s}^2\text{cm}^{-1}$, $B = 100 \times 10^{-17}\,\text{s}^2\text{cm}^{-1}$, $f_r = 50\,\text{MHz}$, $c_0 = 1 \times 10^5\,\text{cm s}^{-1}$ and $r = 0.01$

relaxations follow closely the behaviour predicted by equation (9.27). At very low frequencies ($f \ll f_r$), α/f^2 tends to $A + B$ while at higher frequencies ($f \gg f_r$) it tends to B. In intermediate regions ($f \sim f_r$) the value of α/f^2 falls from $A + B$ to B more or less sigmoidally as shown in Figure 9.1. The absorption per wavelength μ is a useful quantity given by

$$\mu = \alpha\lambda = \frac{r\pi\omega\tau}{1 + (1 - r)\omega^2\tau^2} \tag{9.28}$$

Figure 9.1 shows that this quantity displays a characteristic bell shape as a function of frequency, passing through its maximum value at the relaxation frequency

$$\mu_{\max} = \frac{r\pi}{2(1 - r)^{\frac{1}{2}}} \tag{9.29}$$

The μ curve illustrates an extremely important feature, one that is common to all relaxation techniques, namely the very wide frequency range over which the relaxation extends. This may be demonstrated by calculating the width of the μ curve between the points at which the absorption is one half of the maximum value. Putting $\mu = r\pi/4(1 - r)^{\frac{1}{2}}$ into equation (9.28) yields a quadratic in $\omega\tau$ whose roots are $(2 \pm \sqrt{3})/(1 - r)^{\frac{1}{2}}$. Thus the ratio of the two frequencies at which $\mu = \mu_{\max}/2$ is $(2 + \sqrt{3})/(2 - \sqrt{3})$ or 13·93. Consequently the minimum width at half-height for a single relaxation is 1·144 decades in frequency. This extremely important result must be taken into account when data analysis for more than one relaxation is attempted. In the formalism adopted in equation (9.27), equations (9.28) and (9.29) may be rewritten as

$$\mu = \frac{\alpha_r C}{f} = (Acf_r)\frac{(f/f_r)}{1 + (f/f_r)^2}$$

(9.30)

$$\mu_{max} = Acf_r/2$$

(9.31)

where $\alpha_r/f^2 = \alpha/f^2 - B$.

Provided that the relaxation can be described by a single relaxation time then only the two parameters μ_{max} and f_r are required to specify the behaviour completely. Experimentally, acoustic investigations are designed to yield these two quantities from which further analysis readily stems. In particular, if the work is carried out over a range of temperature then the behaviour of μ_{max} can be made to give thermodynamic information while kinetic data may be derived using changes in f_r.

9.3 Kinetics

The kinetic analysis of acoustic data from conformational processes is a relatively simple procedure since only unimolecular reactions need be considered. For the reaction illustrated in Figure 9.2 the kinetic equation is

$$\text{Conformer 1} \underset{k_{21}}{\overset{k_{12}}{\rightleftharpoons}} \text{Conformer 2}$$

where k_{12} and k_{21} represent the forward and reverse rate constants respectively. The rate equation for this equilibrium is

$$dn_2/dt = k_{12}n_1 - k_{21}n_2$$

(9.32)

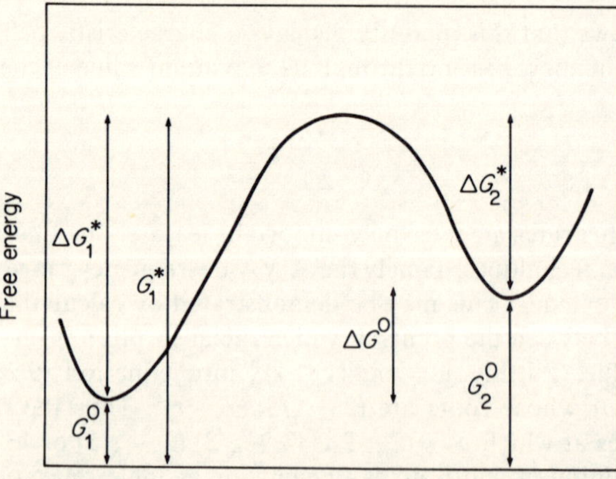

Figure 9.2　　Free energy profile for a unimolecular reaction

where n_1 and n_2 are the populations $(n_1 + n_2 = N)$. Denoting the equilibrium situation by bar symbols then the equilibrium constant is written as

$$K = \bar{k}_{12}/\bar{k}_{21} \qquad (9.33)$$

Since the ordering parameter is given by

$$\zeta = n_2/N \qquad (9.34)$$

then

$$\frac{d\zeta}{dt} = \frac{-\bar{k}_{21}N}{\bar{n}_1} [\zeta - K/(1 + K)] \qquad (9.35)$$

and

$$\bar{\zeta} = K/(1 + K)$$

Thus

$$\frac{d\zeta}{dt} = \frac{-\bar{k}_{21}N(\zeta - \bar{\zeta})}{\bar{n}_1} \qquad (9.36)$$

Comparison with equation (9.18) leads directly to an expression for the relaxation time

$$\tau = \frac{\bar{n}_1}{\bar{k}_{21}N} = \frac{1}{\bar{k}_{12} + \bar{k}_{21}} \qquad (9.37)$$

Since $f_r = (2\pi\tau)^{-1}$ then

$$f_r = (1/2\pi)(\bar{k}_{12} + \bar{k}_{21}) = \frac{\bar{k}_{21}(1 + K)}{\pi} \qquad (9.38)$$

If the standard postulate is applied that the backward reaction rate is an exponential function of the barrier height

$$\bar{k}_{21} = \left(\frac{kT}{h}\right) \exp\left(\frac{-\Delta G_2^*}{RT}\right) \qquad (9.39)$$

then,

$$f_r = \frac{1}{2\pi}\left(\frac{kT}{h}\right) \exp\left(\frac{\Delta S_2^*}{R}\right) \exp\left(\frac{-\Delta H_2^*}{RT}\right)\left\{1 + \exp\left(\frac{-\Delta G^0}{RT}\right)\right\} \qquad (9.40)$$

In the majority of cases the term $\exp(-\Delta G^0/RT)$ is much less than unity and therefore equation (9.40) may be approximated as

$$f_r = \frac{1}{2\pi}\left(\frac{kT}{h}\right) \exp\left(\frac{\Delta S_2^*}{R}\right) \exp\left(\frac{-\Delta H_2^*}{RT}\right) \qquad (9.41)$$

It follows that ΔH_2^* can be determined from the slope of a plot of $\log(f_r/T)$ against T^{-1}. Experimentally it is necessary simply to measure the change in relaxation frequency as a function of temperature. The kinetic information thus results directly from the acoustic measurements in contrast to the often laborious methods employed using other techniques for obtaining the activation energy for conformational interchange.

Relaxation frequencies measured as a function of concentration (in an inert diluent) provide valuable diagnostic information since unimolecular equilibria are the only examples in which f_r is independent of concentration. Furthermore all the unimolecular reactions investigated by molecular acoustics have been satisfactorily assigned to conformational processes. In studies of this kind it is essential therefore that the concentration behaviour be tested. A constant relaxation frequency indicates, to the limit of present knowledge, the existence of a conformational equilibrium. Concentration-dependent relaxation frequencies completely rule out this type of mechanism —a fact that has occasionally been ignored in the literature.

9.4 Thermodynamic properties

The thermal characteristics of relaxation can also be used to obtain ΔG^0. Remembering that $\phi = (\delta Z/\delta \zeta)_{P,T}$ and $Z = -\sum_j \mu_j n_j$ where μ_j is the chemical potential of species j, then substitution of these identities into equation (9.34) gives

$$\phi = -\sum_j n_j(\delta \mu_j/\delta \zeta)_{P,T} \tag{9.42}$$

and

$$\frac{\delta \mu_j}{\delta \zeta} = \sum_l \left(\frac{\delta \mu_j}{\delta N_l}\right)_{P,T}\left(\frac{\delta N_l}{\delta \zeta}\right)_{P,T} = \sum_l \left(\frac{\delta \mu_j}{\delta N_l}\right)_{P,T} n_l \tag{9.43}$$

Thus

$$\phi = -\sum_j \sum_l n_j n_l(\delta \mu_j/\delta N_l)_{P,T} \tag{9.44}$$

But since

$$u_j = \mu_j^0 + RT \ln x_j \tag{9.45}$$

where x is the mole fraction, then

$$(\delta \mu_j/\delta N_l) = (RT/x_j)(\delta x_j/\delta N_l)_{P,T} \tag{9.46}$$

Using $x_j = N_j/N$, $n = \sum_l n_l$ and $N = \sum_l N_l$ together with equation (9.46) yields, after substitution into equation (9.44)

$$\phi = -RT\left\{\sum_j n_j^2/N_j - n^2/N\right\} \tag{9.47}$$

The mole numbers N_j have changed from their initial values N_j^0 by an amount depending on the degree of reaction ζ. Thus,

$$N_j = N_j^0 + n_j \zeta \tag{9.48}$$

Now at equilibrium, $Z = 0$ by definition, and therefore equation (9.45) becomes

$$\sum_j n_j \mu_j^0 + RT \sum_j n_j \ln x_j = 0$$

or

$$\Delta G^0 + RT \sum_j n_j \ln x_j = 0 \tag{9.49}$$

Furthermore, since $x_j = (N_j/N) = (N_j^0 + n_j\zeta)/(N^0 + n\zeta)$ from equation (9.48) equation (9.49) becomes,

$$\sum_j n_j \ln (N_j^0 + n_j \zeta) = -\Delta G^0/RT + n \ln (N^0 + n\zeta)$$

or

$$\prod_j (N_j^0 + n_j\zeta)^{n_j} = \exp(-\Delta G^0/RT)(N^0 + n\zeta)^n \tag{9.50}$$

For a unimolecular reaction

$$-A_1 + A_2 = 0$$

then

$$n_1 = -1 \qquad n_2 = 1 \qquad N_1^0 = 1 \qquad N_2^0 = 0$$

$$N_1 = 1 - \zeta \qquad N_2 = \zeta \qquad N^0 = 1 \qquad n = 0$$

Substitution of these values into equation (9.50) produces the results

$$(1 - \zeta)/\zeta = \exp(\Delta G^0/RT) \tag{9.51}$$

and

$$\phi = \frac{-RT}{\zeta(1 - \zeta)} = \frac{-RT[1 + \exp(-\Delta G^0/RT)]^2}{\exp(-\Delta G^0/RT)} \tag{9.52}$$

Use of equation (9.13a) at this stage gives direct meaning to the relaxing specific heat

$$\delta C_p = R\left(\frac{\Delta H^0}{RT}\right)^2 \frac{\exp(-\Delta G^0/RT)}{[1 + \exp(-\Delta G^0/RT)]^2} \tag{9.53}$$

The simplest form of this equation occurs when the entropy difference between the two states ΔS^0 is negligible. Then

$$\delta C_p = R\left(\frac{\Delta H^0}{RT}\right)^2 \frac{\exp\left(-\Delta H^0/RT\right)}{[1 + \exp\left(-\Delta H^0/RT\right)]^2} \tag{9.54}$$

Equation (9.54) is known as Shottky's function and essentially it gives the contribution to the specific heat arising from a single excited state above a non-degenerate ground state. Figure 9.3 shows the form of this function together with three-state solutions for degenerate excited states or degenerate ground states. The maximum in δC_p occurs at $\Delta H^0/RT = 2\cdot4$, $2\cdot65$ and $2\cdot23$ respectively. Further discussion of the significance of Figure 9.3 will be deferred until the end of this section.

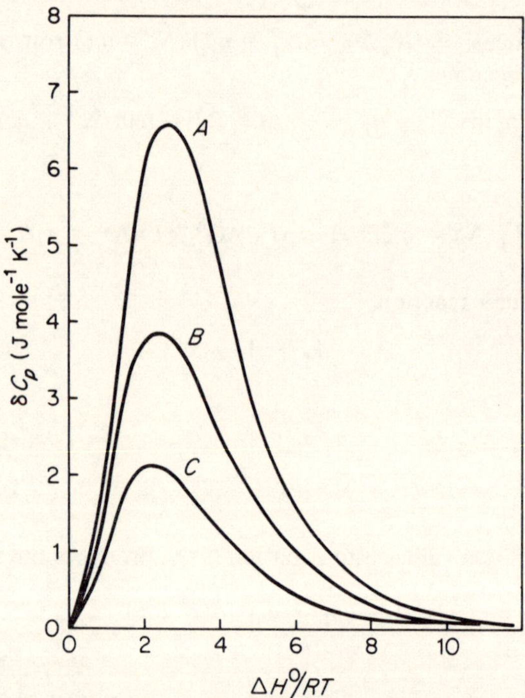

Figure 9.3   The relaxing specific heat due to rotational isomerism as a function of $\Delta H^0/RT$. Curve A is a three-state solution with a degenerate excited state. Curve B is the two-state solution and curve C is the three-state solution for a degenerate ground state

For a two-state equilibrium, assuming that the thermal relaxation involves no volume change, combination of equations (9.29) and (9.16) gives

$$r = \frac{2\mu_{\max}}{\pi} \tag{9.55}$$

and equation (9.54) becomes

$$\frac{2\mu_{\max}}{\pi}\frac{C_p}{(\gamma-1)} = R\left(\frac{\Delta H^0}{RT}\right)^2 \frac{\exp\left(-\Delta H^0/RT\right)}{[1+\exp\left(-\Delta H^0/RT\right)]^2} \tag{9.56}$$

Figure 9.3 shows quite clearly that, provided $\Delta H^0 > 2 \cdot 4\,RT$, then $C_p\mu_{\max}/(\gamma-1)$ increases with increasing temperature and equation (9.56) may be approximated as

$$\frac{2\mu_{\max}}{\pi}\frac{C_p}{(\gamma-1)} = R\left(\frac{\Delta H^0}{RT}\right)\exp\left(-\Delta H^0/RT\right) \tag{9.57}$$

and ΔH^0 is derived from the slope of a plot of $\log\left[T^2 C_p\mu_{\max}/(\gamma-1)\right]$ versus T^{-1}. Unfortunately C_p and γ are rarely well documented or easy to measure and use is often made of the relation

$$c^2 = \frac{(\gamma-1)C_p}{\beta^2 T} \tag{9.58}$$

Plots of $\log\left(T\mu_{\max}/c^2\right)$ against T^{-1} are used to calculate ΔH^0 since $(C_p/\beta)^2$ is not strongly temperature dependent. Substitution of the result back into equation (9.57) indicates the validity of assuming $\Delta S^0 = 0$, and in most cases $(\Delta G^0 > 3RT)$ the entropy difference can be determined with only a small error by the approximate form of equation (9.53).

$$\frac{2\mu_{\max}}{\pi}\frac{C_p}{(\gamma-1)} = R\left(\frac{\Delta H^0}{RT}\right)^2\exp\left(\frac{-\Delta H^0}{RT}\right)\exp\left(\frac{\Delta S^0}{R}\right) \tag{9.59}$$

If $\Delta H^0 < 2 \cdot 4\,RT$ then the solution for ΔH^0 can only be obtained by direct iterative analysis of equation (9.56).

Similar expressions are readily derived for three-state equilibria. For example, for a doubly degenerate ground state,

$$\frac{2\mu_{\max}}{\pi}\frac{C_p}{(\gamma-1)} = R\left(\frac{\Delta H^0}{RT}\right)^2 \frac{2\exp\left(-\Delta H^0/RT\right)}{[2+\exp\left(-\Delta H^0/RT\right)]^2} \tag{9.60}$$

The foregoing derivation deserves discussion in respect of two important features. Firstly, it will be noted from Figure 9.3 that substantial contributions to the relaxing specific heat are made by the excited state even at relatively large values of $\Delta H^0/RT$. More specifically, for a two-state equilibrium, values for δC_p of less than $0 \cdot 2\,\text{J mole}^{-1}\,\text{K}^{-1}$ can be detected corresponding to ΔH^0 values of about $20\,\text{kJ mole}^{-1}$. This implies a population of about $0 \cdot 03\%$ in the excited state and even smaller values have been measured. Thus the molecular acoustic technique is probably the most sensitive method available for the study of conformational equilibria. Compare the conventional resonance technique typically requiring a 10% excited state occupancy.

The most important of the many approximations used in ΔH^0 calculations is undoubtedly the neglect of any volume changes accompanying the relaxation process. While the principal response of conformational equilibria is unquestionably to the temperature fluctuations of the compressional wave, incidental volume differences between conformers are almost inevitable. Consequently the magnitude of $\Delta V/V$ in equation (9.15), or more strictly, obedience to the condition $(\Delta V/\Delta H)(C_p/V\beta) \ll 1$ is of paramount importance. Until fairly recently it has always been assumed that $\Delta V/V < 1\%$ for the majority of thermal relaxations involving rotational isomers and this view was supported by excellent experimental evidence which gave values of about 0.4% for a series of alkanes[5] and cycloalkanes,[6] and 0 for ethyl acetate;[7] it was shown that under these conditions, the neglect of volume changes was a valid assumption and good straight lines were obtained for plots of $\log(T\mu_{\max}/c^2)$ versus T^{-1}. There are two basic methods used in the measurement of $\Delta V/V$. The more obvious, and perhaps more satisfactory, is to study the effect of changes in pressure on the relaxation frequency. Of course, if there is no volume change accompanying the equilibrium then f_r is independent of pressure. The experiment is essentially a test of the alternative condition $(\Delta V/R)(\delta P/\delta T)_S \leqslant \Delta H/RT$. In the work of Slie and Litovitz[7] the volume change in ethyl acetate was found to be negligible using this technique. More recently, estimates of $\Delta V/V$ have been made using complementary ultrasonic and NMR data on 1,1,2-trichloroethane. Now this molecule has been studied extensively in the neat state by Lamb[8] and Padmanaban[9] using the acoustic technique and excellent linearity was obtained for the ΔH^0 graphs (see Figure 9.19). There seemed no reason to doubt the volume assumption. However, extensive and thorough measurements of the relaxation over a wide concentration range in a large number of solvents, together with well-established ΔH^0 figures from NMR data, showed that $\Delta V/V$ varied in the range $1-3\%$ depending on the solvent[10] (see Table 9.2). This work has thrown much doubt on the value of the acoustic method as a means of deriving ΔH^0 data. Often it is found that $\log(T\mu_{\max}/c^2)$ versus T^{-1} graphs are distinctly non-linear, probably as a result of volume changes, and the technique is clearly inapplicable. Unfortunately, even when good straight lines are obtained, the analysis may still be in error.

9.5 Apparatus

A considerable variety of methods exists for the study of acoustic propagation through fluids. This chapter will be confined to those methods that have found general use in many laboratories and which cover the frequency range of interest (10 kHz–300 MHz) for conformational studies.

One important factor that all the techniques possess in this frequency region is the use of piezoelectric transducers to generate the ultrasonic wave.

The theory of these devices is beyond the scope of this review. It is sufficient to note that quartz crystals, when cut along a direction perpendicular to the principal axis and subjected to alternating current, will transform the electrical input into mechanical compression waves at the same alternating frequency, providing it is of the correct thickness. A quartz crystal of thickness x cm vibrates in its fundamental mode at a frequency of approximately $290/x$ kHz. Working at more than one frequency implies either the use of several crystals of differing thickness, or tuning the apparatus to detect high frequency harmonics of the fundamental. Other piezoelectric devices may be employed but quartz is the preferred material, except possibly at very low frequencies where its thickness (3 cm at 100 kHz) is awkward and barium titanate may be substituted.   Table 9.1 summarizes the principal features of the experimental methods to be described.

Table 9.1   Experimental methods for acoustic studies on conformational processes

Technique	Frequency range	Comments
Reverberation	10 kHz–1 MHz	Fairly accurate (5 %). Important for very low frequencies.
Streaming	100 kHz–1 MHz	Poor accuracy (10 %).
Optical methods	1–100 MHz	Classical visualization of sound field. Now largely obsolete in this form, although fairly accurate.
Resonance	100 kHz–10 MHz	Very new technique. Capable of high accuracy (1 %) and low frequency limit.
Interferometry	1–100 MHz	Poor accuracy for highly absorbing media. Very accurate velocity measurements.
Pulse methods	10–500 MHz	Principal present-day technique. Accuracy 1–3 %.

9.5.1 *Reverberation*

For work in the low-frequency region of the ultrasonic spectrum the reverberation technique shown in Figure 9.4 is the most common method. The quoted high-frequency limit in Table 9.1 is less than that obtainable with the apparatus, but above this value the accuracy falls below that obtainable by other methods.

A large spherical vessel (0·5–10 l.) is completely filled with the specimen liquid and suspended. Originally, fine wire suspensions were used but more recently air bearings inside a continuously evacuated chamber have been found necessary to excite pure radial modes.[12] These modes are generated by oscillations from a transmitting transducer supplied by a pulsed sinusoidal voltage. At the termination of each pulse the oscilloscope is gated to receive

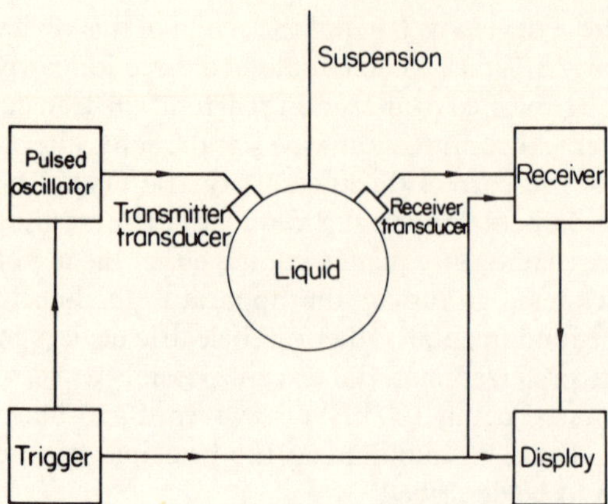

Figure 9.4 Reverberation technique for the measurement of absorption in liquids[11]

information from the second crystal. The rate of decay of the oscillations yields the absorption coefficient very simply if the time interval T is measured between two points on the decay envelope whose signal intensities are in the ratio 1/e.

$$T = 1/\alpha c \tag{9.61}$$

Measured losses include contributions arising from absorption by the vessel and suspension system. Consequently the apparatus requires prior calibration with liquids of known absorption. One serious drawback of the device is the uncertainty in system losses due to the propagation of non-radial modes, since different modes decay at different rates. Difficulties of this kind have been partially overcome by deliberately introducing constant multiple vibration modes rather than allowing random-mode generation. Vessels with very low symmetry excited by pulsed high-frequency noise bands are employed in this variation.

9.5.2 *Streaming*

When a fluid absorbs acoustic radiation, pressure is generated that can cause the liquid to flow. Figure 9.5 illustrates the type of apparatus used to obtain the absorption coefficient by measuring the volume flow, or stream.

A transducer radiates directly into a column of liquid after which it is totally absorbed to prevent reflection. The apparatus contains a side-tube in which the movement of the flowing liquid can be observed by using finely suspended aluminium particles. Under conditions of low absorption

$$Q = \frac{\pi r^4 E \alpha}{\eta c} \tag{9.62}$$

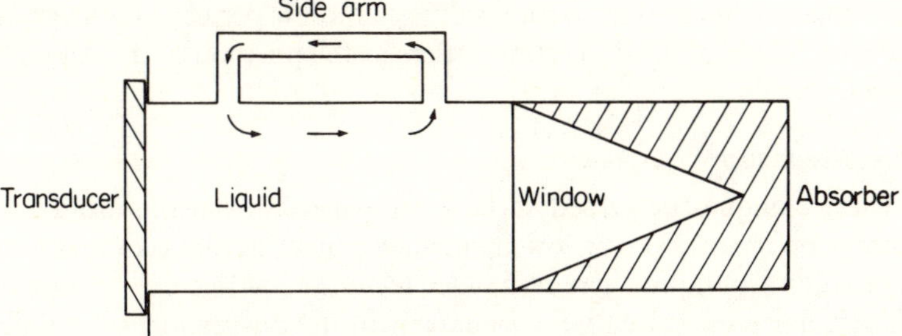

Figure 9.5 Acoustic streaming apparatus

where Q is the quantity of liquid flowing per unit time, r the radius of the side-tube, η the liquid viscosity and E the radiation energy density. Corrections are usually made to account for end-effects where the side-tube joins the main vessel. These are most simply effected by calibration with standard liquids. One method of measuring the radiation energy is to compare the admittance of the transducer when loaded by the liquid and unloaded.[13]

9.5.3 Optical methods

There have been many optical methods of studying sound-wave propagation originating with the Schlieren technique of Töpler. Basically the system relies on the fact that the passage of a sound beam alters the local density in the medium and hence the refractive index. Present-day workers utilize direct visual observation of the travelling wave rather than the Schlieren system.

Figure 9.6 illustrates one form of the technique whereby the apparatus measures a resultant diffraction pattern,[14,15] since the refractive index of the fluid is sinusoidal, alternating with wavelength λ. Consequently the optical diffraction formula may be applied using λ as the grating spacing, the incident light wavelength being λ_1:

$$\sin \theta = n\lambda_1/\lambda \tag{9.63}$$

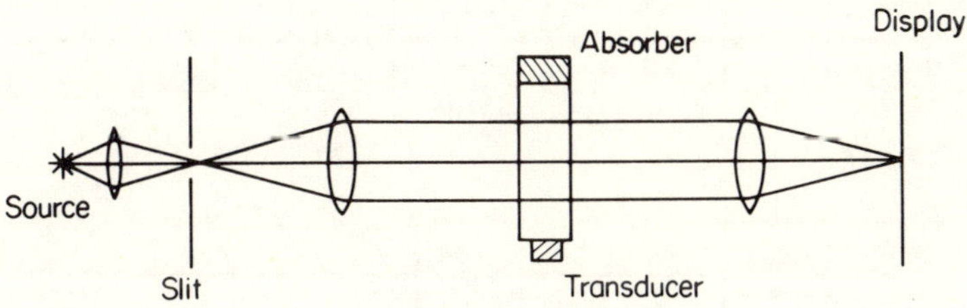

Figure 9.6 Apparatus for the study of light diffracted by an acoustic beam

Losses can be measured to an accuracy of a few per cent by measuring the intensity of one of the diffraction orders as the position of the sound beam is varied.

9.5.4 *Resonance*

The techniques described so far have suffered from the disadvantage of either a poor accuracy at low frequencies, or of needing an inconveniently large volume of liquid. The resonance technique of Eggers,[16] although in its infancy, has great potential, particularly in the low frequency regime where few methods are available. Accuracies for absorption measurements of less than 1 % are feasible and only 1–2 cm^3 of test liquid are required. In addition resonances may be observed every 300–500 kHz instead of being limited to multiples of the fundamental quartz resonance (typically 5 MHz).

A diagram of the apparatus is given in Figure 9.7. It consists essentially of a resonance cavity formed by two accurately aligned quartz crystals containing the test liquid. Standing waves are generated when the sound wavelength is an integral multiple of the distance separating the crystals, with a node or antinode at the centre. The position of the resonance so formed is proportional to the sound velocity and its width to the absorption. In practice the equations governing the velocity are complex and calibration with a standard liquid is used:

$$\delta c/c = \delta f/f \qquad (9.64)$$

where δc and δf are the respective differences in velocity and frequency between the specimen and comparison liquids. The loss is measured from the width (Δf) of the resonance at half its height.

$$\text{quality } Q = f/\Delta f = \pi/\alpha\lambda \qquad (9.65)$$

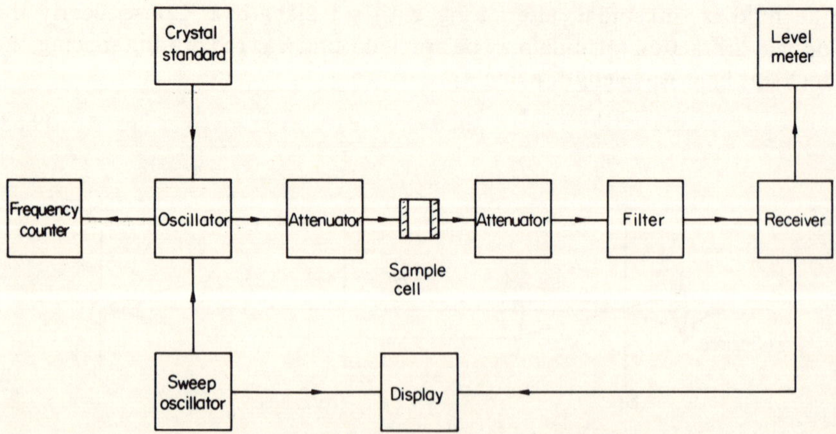

Figure 9.7   Block diagram of the resonance apparatus

Again, parasitic losses in the apparatus can be estimated using comparison liquids since the reciprocal qualities are additive

$$Q_{\text{measured}}^{-1} = Q_{\text{liquid}}^{-1} + Q_{\text{losses}}^{-1} \tag{9.66}$$

9.5.5 *Interferometry*

Interferometry has been used extensively in the study of sound-wave propagation[17] and is still the most important source of accurate velocity data.

The quartz transducer is operated under resonant mode conditions, emitting a continuous wave into the liquid specimen which is reflected back by a movable plate (Figure 9.8). Velocity and attenuation are calculated from changes in the electrical behaviour of the transmitting crystal

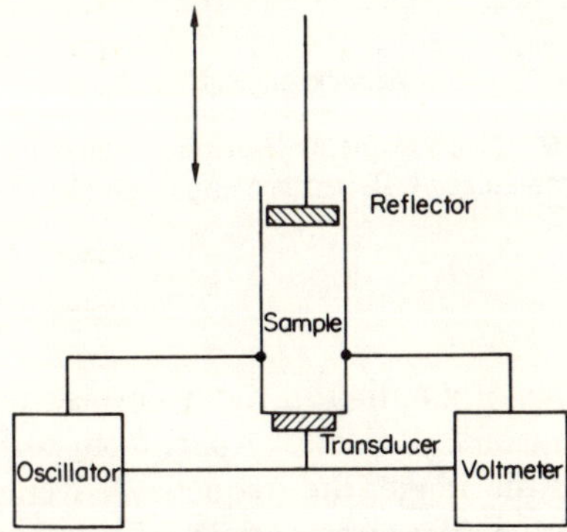

Figure 9.8   The acoustic interferometer

as a function of the transducer/reflector separation. For example, the current delivered from the generator shows the periodicity as a function of reflector position in a liquid depicted in Figure 9.9. A maximum occurs whenever the condition $l/\lambda = (2n + 1)/4$ is satisfied, where l is the spacing and n an integer. Minima occur whenever $l/\lambda = n/2$. Attenuation is measured from the widths of the peaks in a similar manner to the resonance apparatus

$$\tanh(\alpha l) = \left\{ \frac{i_m}{i_{\max}} \cdot \frac{(i_m - i_{\max})}{(i_m - i_0)} \cdot \frac{(i_{\min} - i_0)}{(i_m - i_{\min})} \right\}^{\frac{1}{2}} \tag{9.67}$$

where i_m and i_0 are indicated in Figure 9.9 and $i_{\max}$ and $i_{\min}$ are the maximum and minimum currents at l. In principle, if measurements can be made over a distance of about 1 cm using a 4 in. drum micrometer accurate to 4×10^{-5} cm then velocity measurements ought to be made to 1 part in 20,000. In practice, although broadening of the peaks due to absorption reduces this figure, interferometry is the principal technique for velocity measurements.

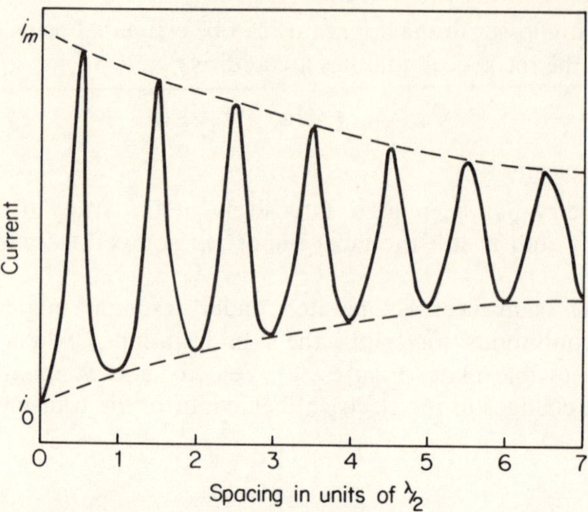

Figure 9.9   Interferometer transducer current as a function of transducer/reflector spacing in an absorbing liquid

9.5.6 *Pulse methods*

The pulse techniques originating in radar development are probably the most widely used ultrasonic devices.[18] Apart from some deficiencies at the low-frequency end of the range, the frequency coverage coupled with the accuracy for attenuation measurements (1–5%) is ideal for the study of conformational equilibria. The relatively poor accuracy for velocity measurements (1%) is no drawback since velocity dispension is negligible for thermal relaxations. Figure 9.10 shows the block diagram of a typical arrangement and Figure 9.11 one of the sample cells used in the author's laboratory.

A pulse ($\sim 5\,\mu$s) of sinusoidal voltage (~ 100 V) is applied to a transducer in contact with the sample at a repetition rate of about $500\,\text{s}^{-1}$. (Simple arithmetic shows that the signal is present for less than 0·1% of the total time, thus avoiding some of the troubles such as streaming and heating attending the continuous wave interferometer). After passage through the sample, the ultrasonic pulse is reflected back to the transducer which also serves as receiver. The reflector can be either a solid metal plate or a quartz rod. This latter device imposes a delay on the pulse due to the time taken for the sound to traverse the rod (Figure 9.12b). This is used to separate the received pulse from the original transmission for the very small (< 1 cm) liquid path-lengths necessary with highly absorbing material. In either case the reflecting face must be accurately flat and parallel to the transmitter, conditions that become increasingly important with increasing frequency. When the reflected pulse reaches the transducer, a small part is reconverted

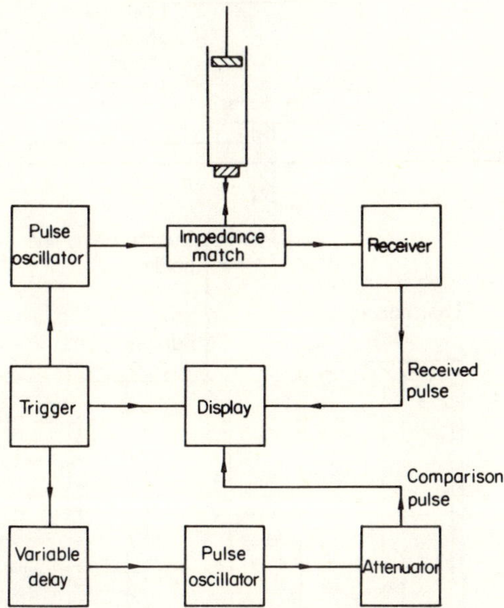

Figure 9.10 Block diagram of typical pulse apparatus for ultrasonic absorption

back to electrical energy and appears on the oscilloscope as an 'echo'. The remainder is reflected back into the liquid and further reflections produce a series of echoes, each of which has made one round trip through the liquid more than its predecessor. The velocity and attenuation are determined by measuring the position (time) and height respectively of any one echo as a function of reflector distance. This can be facilitated by the use of a second pulse generator (driven by the same oscillator to maintain exactly identical frequencies) to produce a comparison pulse whose position and size can be accurately controlled by precision time delay and attenuator mechanisms. Alternatively an exponential generator may be used to fit the whole envelope of echo peaks rather than study any one single echo. The experimental similarity between this technique and interferometry is utilized by a number of workers who include a facility for conversion from pulse to interferometry.[4] Such versatility leads to much improved accuracy in velocity measurements.

Alternative transducer/reflector arrangements are common (Figure 9.12). The double transducer configuration of system (d) is employed in the 'sing-around' technique whereby the received signal is used to retrigger the transmitter. Velocity changes of 1 part in 10^7 can be detected by accurate measurement of the repetition rate.

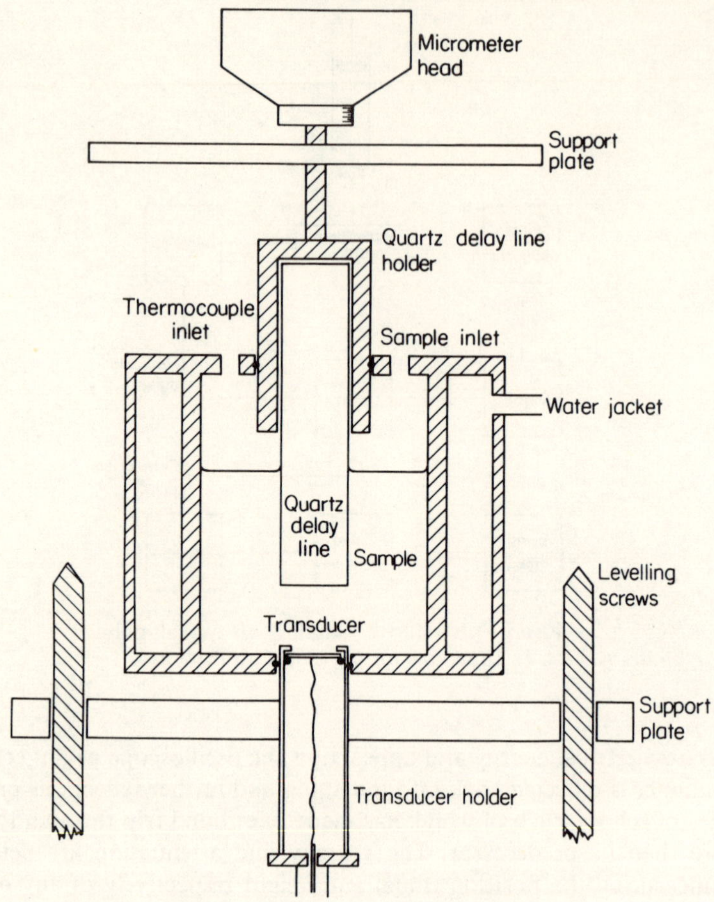

Figure 9.11   Absorption sample cell

Careful thermostatting is necessary so that temperature changes and hence density gradients do not cause excessive scatter of the radiation. A more important source of error originates in the non-planarity of the wave-front. In the low-frequency regime a great deal of beam spreading occurs, leading to diffraction and apparent attenuation values greatly in excess of those expected. Figure 9.13 shows the decibel loss as a function of a^2/λ, where a is the transducer radius. The effect is minimized by using large area transducers (with a corresponding increase in test-liquid volume) or by working at high frequencies. Probably all measurements below 5 MHz using the pulse technique are suspect and it is usual practice to restrict the frequency range to 15 MHz and above using 5 MHz fundamental transducers.

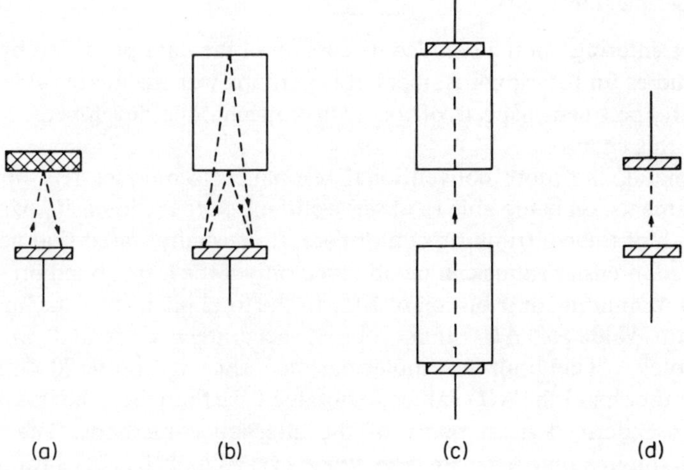

Figure 9.12   Common transducer/reflector configurations used in ultrasonic pulse techniques

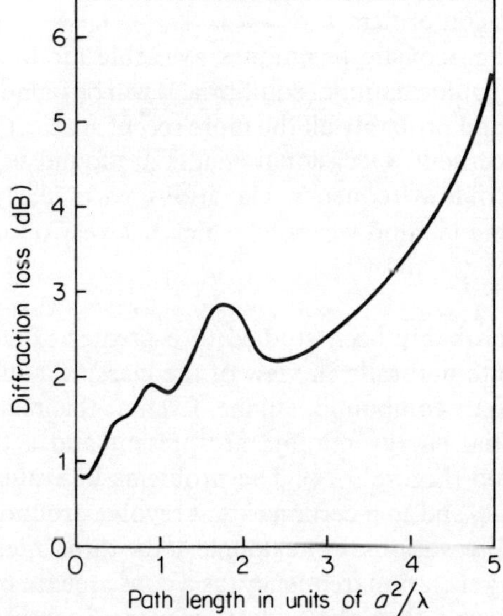

Figure 9.13   Additional absorption (in decibels) due to diffraction as a function of frequency and transducer size

9.6 Experimental results

Before entering upon a detailed discussion of the data produced by ultrasonic studies on rotational isomers, it is perhaps worthwhile to review very briefly the pertinent aspects of the acoustic technique developed in earlier parts of this chapter.

In general, the more conventional resonance techniques rely, for ΔH^0 measurements, on being able to observe and measure the intensity of spectra from each of the contributing conformers. It is usually stated that accurate integrated intensity ratios can be obtained only if the least abundant species makes a minimum contribution of 10 % to the total intensity. This limits the maximum value of ΔH^0 that can be accurately estimated to about $5 \cdot 5 \, \text{kJmole}^{-1}$. The limit for molecular acoustics is about $20 \, \text{kJmole}^{-1}$. On the other hand the ΔH^0 values so obtained are thought to be less reliable than those derived from many of the alternative methods. The energy barrier to interchange ΔH^* can be estimated very simply by acoustic methods provided the interchange frequency lies within or close to the experimentally observable region. Many alternative techniques require detailed curve fitting before this type of information can be derived.

It must be remembered, that in common with other relaxation studies, the data *per se* yield no information on the nature of the species involved in the equilibrium. Contrastingly, resonance spectra can usually be assigned directly to specific conformers.

Summarizing the acoustic techniques available for the study of relaxations arising from conformational equilibria, it will be found that the majority of investigations (and probably all the more recent studies) have been carried out using pulse methods. Occasional results are found using the interferometer. In the case of low-frequency relaxations, particularly the cyclohexane series, the streaming method was used which has very poor accuracy.

9.6.1 *Alkanes*

Alkanes have probably been studied to a greater extent than any comparable series, quite naturally in view of the classical simplicity of the isomerism in the parent compound, ethane. Even at this relatively elementary level three potential energy minima are present and a three-state system must be considered (Figure 9.14). The problems in evaluating this type of behaviour are large, and to a certain extent revolve around the experimental adequacy of the observations. For example, if the three interchange processes differ markedly in relaxation frequency (as a consequence of large differences in activation energies) then the widest frequency coverage, together with extensive temperature variation, will be necessary to characterize the mechanisms. Alternatively if the relaxation frequencies are close then, in view of the statements in § 9.2 on the widths of relaxation curves (Figure 9.1),

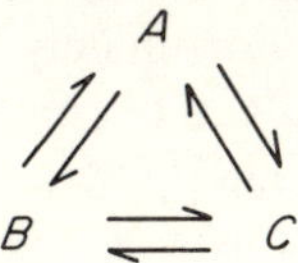

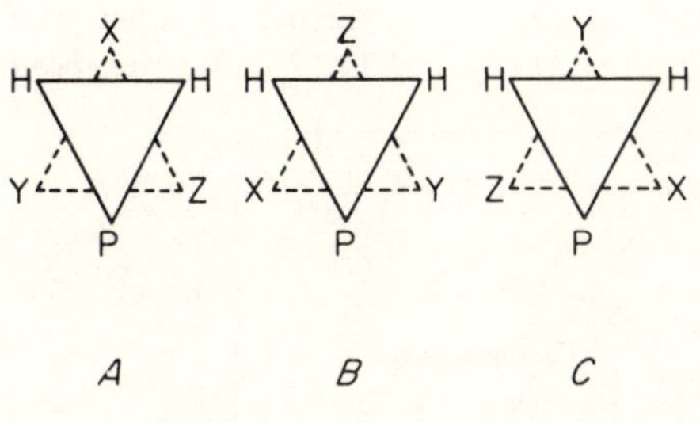

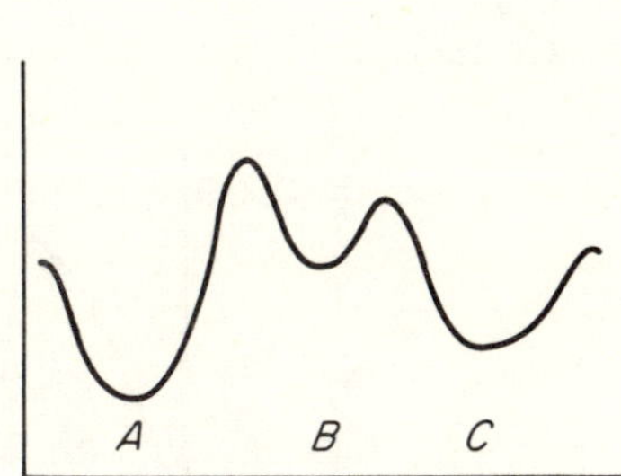

Figure 9.14　Potential energy profile for conformational isomerism in a substituted ethane

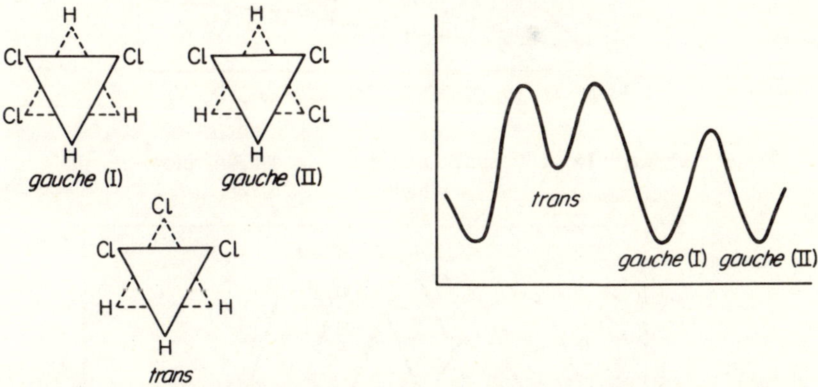

Figure 9.15　Potential energy profile for conformational isomerism in 1,1,2-trichloroethane

it will be impossible to resolve the individual parameters. Under these circumstances the thermal relaxation equation (9.27) becomes

$$\frac{\alpha}{f^2} = \sum_i \frac{A_i}{1 + (f/f_{ri})^2} + B \tag{9.68}$$

Curve fitting for this type of equation may be acceptable for two processes (a five-parameter equation) but a seven-parameter fit over a limited frequency range is probably not significant. Fortunately, the majority of conformational equilibria that have been studied can be treated as two-state systems, either because the population of the highest energy state is negligible, or, because degeneracy of some of the conformers occurs.

Acoustically, the most intensively studied alkane is 1,1,2-trichloroethane which possesses a doubly degenerate ground state (equation 9.60) as shown in Figure 9.15. The ultrasonic absorption data, showing the presence of a single relaxation, are depicted in Figures (9.16) and (9.17). The observed

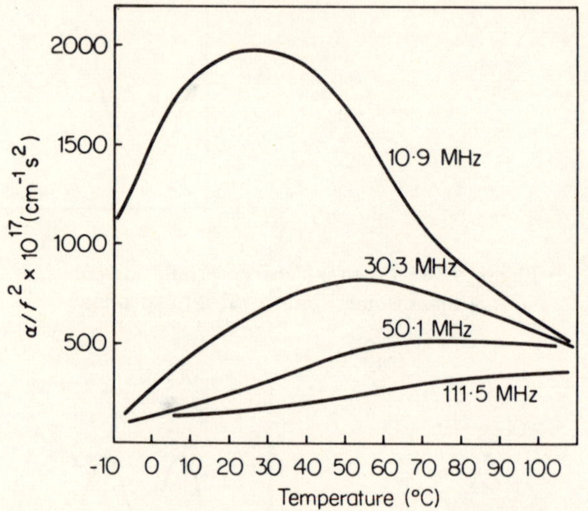

Figure 9.16   Ultrasonic absorption in 1,1,2-trichloro-ethane[8]

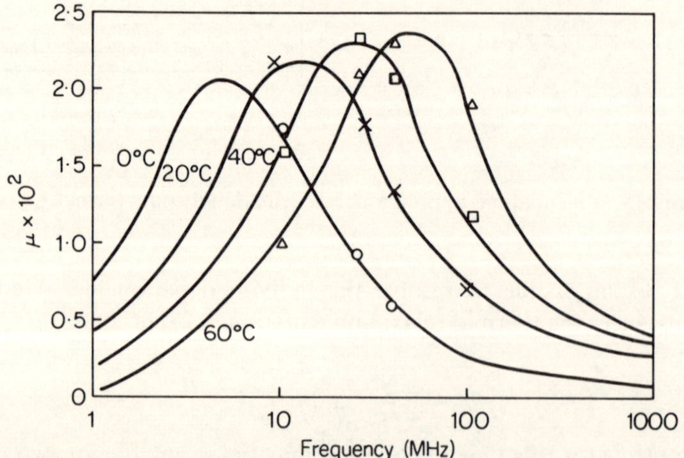

Figure 9.17   Absorption per unit wavelength for the relaxation in 1,1,2-trichloroethane.[8] The full line represents the fit to equation (9.27)

velocity dispersion, which was less than 0·75 %, was neglected. The relaxation is presumed to arise from perturbation of the equilibrium between the *gauche* (degenerate ground state) and *trans* (excited state) conformers. The kinetic and thermodynamic data plots for this molecule are shown in Figures (9.18) and (9.19) respectively. Excellent straight lines are obtained

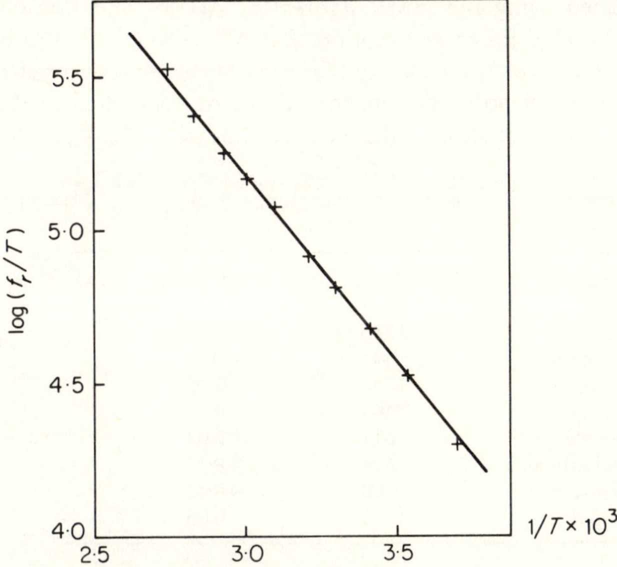

Figure 9.18   Plot of log (f_r/T) against T^{-1} for
1,1,2-trichloroethane[8]

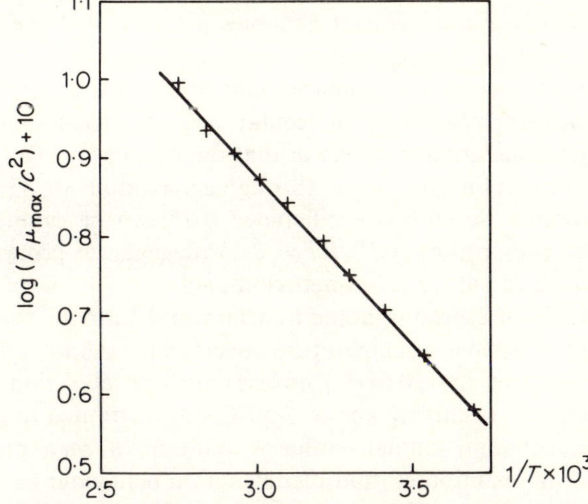

Figure 9.19   Plot of log $(T\mu_{max}/c^2)$ against T^{-1}
for 1,1,2-trichloroethane[8]

yielding a value for $\Delta H_{\frac{1}{2}}^{*}$ of $24\cdot2\,\text{kJmole}^{-1}$ and for ΔH^0 the answer $8\cdot8$ kJmole^{-1}. The energy barrier in the forward direction is thus $33\cdot0\,\text{kJmole}^{-1}$.

This same molecule has been chosen by Crook and Wyn-Jones[10] for a detailed study of the validity of the volume assumption in a variety of solvents. Table 9.2 lists the relevant information. Values for $\Delta V/V$ were calculated using the NMR figures for ΔH^0,[19] and using equation (9.60) with the reasonable assumption that $\Delta S^0 = -R\ln 2$. The final column in Table 9.2 shows quite clearly that the critical condition $(\Delta V/V) \times (C_p/\beta) \ll \Delta H^0$ does not apply despite the 50% error estimated in the derivation of $\Delta V/V$. This is so despite the excellent linearity of the graph in Figure 9.19.

Table 9.2   Data used to test the volume changes in 1,1,2-trichloroethane in a number of solvents

Solvent	$\Delta H^0\,(\text{Jmole}^{-1})^{19}$	$\Delta V/V \pm 50\%$	$(\Delta V/V) \times (C_p/\beta)(\text{Jmole}^{-1})$
p-Xylene	1280	$-0\cdot03$	-1715
Mesityl oxide	941	$-0\cdot01$	-1464
n-Heptane	1752	$-0\cdot01$	-1338
Ethyl acetate	448	$-0\cdot01$	-502
Acetonitrile	611	$-0\cdot03$	-1548
Trichloroethylene	2614	$-0\cdot01$	-247
Nitrobenzene	100	$-0\cdot02$	-3180
Pure liquid	1761	$-0\cdot06$	-5438

Clearly the whole basis for the calculation of enthalpy differences by acoustic methods must be in doubt. Interestingly, the negative volume change implies that the *trans* isomer occupies a greater volume than the *gauche*, as predicted from simple models.

Table 9.3 lists all the available kinetic and thermodynamic parameters in alkanes derived using molecular acoustic techniques. The data for 2-methylbutane are interesting in that the plot of $\log{(T\mu_{\text{max}}/c^2)}$ versus T^{-1} is not a straight line, but passes through a maximum at $185\,\text{K}$ (cf. Figure 9.3). Consequently the enthalpy difference ΔH^0 can be estimated very simply using the prescription $\Delta H^0/RT = 2\cdot23$ (degenerate ground state). Similar behaviour is found for 2,3-dimethylbutane.

The study of 2-bromobutane by Clark and Litovitz[20] is unusual in that, by virtue of extensive temperature coverage $(-166$ to $+50°\text{C})$ and a large frequency range $(25–150\,\text{MHz})$, no less than four relaxations were postulated. The relaxation occurring above $-50°\text{C}$ was attributed to a rotameric equilibrium. Although similar enthalpy data have been reported by other workers,[1] no mention of multiple relaxation behaviour has been made.

Work on a series of normal alkanes by Piercy and Seshagiri Rao[5] has revealed the existence of thermal relaxations, whereas previous studies

Table 9.3 Molecular acoustic data for some alkanes

Compound	f_r (MHz)	ΔH_2^* (kJmole^{-1})	ΔH (kJmole^{-1})	Reference
1,1,2-Trichloroethane	14 at 20°C	24·3	8·8	23, 9
1,1,2-Tribromoethane	—	26·8	6·7	23, 9
sym-Tetrabromoethane	—	17·9	3·8	23
1,1,2,2-Tetrafluoro-1,2-dibromoethane	23·7 at 25°C	31·4	6·8	24
1-Bromopropane	—	15·1	5·4(1·7)	23, (9)
1,2-Dichloropropane	—	19·6	4·6	23, 9
1,2-Dibromopropane	—	20·5	3·8	23, 9
1-Chloro-2-bromopropane	50·4 at 20°C	23·0	—	25
1,2-Dichloro-2-methylpropane	36·6 at 25°C	—	8·4	26
1,2-Dibromo-2-methylpropane	18·3 at 25°C	23·0	3·8	26
Butane	20·5 at −120°C	14·2	—	5
2-Chlorobutane	—	18·8	3·8(5·4)	1, (20)
2-Bromobutane	—	21·3 (19·6)	3·8(2·6)	1, (20)
1,2-Dibromobutane	29·3 at 20°C	—	—	25
2-Methylbutane	∼24 at −50°C	19·6	3·8	22
2,3-Dimethylbutane	∼35 at −40°C	11·7	4·2	21
n-Pentane	15·0 at −120°C	17·6	—	5
2-Chloropentane	125·5 at 20°C	18·4	—	25
2-Bromopentane	90·5 at 20°C	20·1	—	25
3-Chloropentane	103·8 at 20°C	14·2	—	25
3-Bromopentane	72·7 at 20°C	15·1	—	25
2-Methylpentane	∼35 at −50°C	16·3	4·0	21
3-Methylpentane	∼35 at −40°C	17·1	4·0	21
1,2-Dibromopentane	15·4 at 20°C	14·6	—	25
1,4-Dibromopentane	89·2 at 20°C	19·7	—	25
n-Hexane	—	13·0	—	5
2-Bromohexane	84·6 at 20°C	17·6	—	25
3-Bromohexane	55·8 at 20°C	15·1	—	25
n-Heptane	—	15·9	—	5
4-Chloromethyl-1,3-dioxolan-2-one (see text)	48 at 25°C	21·3	<4·6	27

(above −70°C) have failed to reveal any unusual absorption,[21,22] The evidence was obtained over a limited frequency range (5–45 MHz) at temperatures down to the freezing points (∼ −130°C) of the materials. Anomalous absorption was detected below −90°C. Unfortunately, the temperatures necessary were too close to the freezing points to give an accurate estimation of the energy parameters and the authors developed a novel method of treating the data to allow use of the results at higher temperatures. This was based essentially on the assumption that the parameter α/f^2 is exponentially related to the temperature for thermal relaxations. The authors place less confidence in the ΔH_2^* values for the higher alkanes since multiple relaxations were thought to be present.

The relaxation in 3,3-diethylpentane (Et_4C) reported by Blandamer and coworkers[28a] is not included in Table 9.3 since the data show quite clearly that f_r is not independent of concentration and the assignment to an internal rotation mechanism must be suspect. Included in Table 9.3 are data from the molecule 4-chloromethyl-1,3-dioxolan-2-one in which alkane-like rotamers are possible:

Only one relaxation was detected implying either very different relaxation times or degeneracy. In view of the low ΔH^0 value, which indicates an excited state population of 10% and which is certainly detectable by other techniques, then widely separated relaxations seem likely.

9.6.2 *Esters*

Historically, the study of esters is one of the earliest applications of acoustic techniques to relaxations associated with conformational equilibria.[28] Unfortunately, although the literature on these molecules is varied and extensive, there are few systematic studies and much of the older work is subject to relatively large experimental errors by today's standards. Indeed, very few workers have reported results over a range of temperatures, largely because many of the theoretical relationships had not been derived and the importance of this type of study was unrecognized.

The conformational problem in the case of esters arises from restricted rotation about the C—O bond due to its partial double-bond character. Planar *cis* and *trans* isomers are postulated.

A compilation of the available data from acoustic measurements on a number of monoesters is presented in Table 9.4, together with some vinyl ether results for comparison where similar conformational equilibria are possible. The cause of the energy difference between the isomers was fully

Table 9.4 Molecular acoustic data for monoesters and vinyl ethers

Compound	f_r (MHz)	ΔH_2^* (kJmole^{-1})	ΔH^0 (kJmole^{-1})	ΔH_1^* (kJmole^{-1})	Reference
Ethyl acetate	11·8 at 20°C	23·8	~12	—	29
Ethyl formate	0·6 at 60°C	33·5	9·6	—	30
	(0·14 at 25°C)	(34·3)	(13·4)	—	32
Methyl formate	0·7 at 60°C	32·6	9·6	—	30
Methyl formate	1·7 at 23°C	32·6	1·7	34·3	31
Ethyl formate	0·6 at 23°C	25·5	2·1	27·6	31
Methyl acctate	8·8 at 23°C	24·7	17·6	42·3	31
n-Propyl formate	0·5 at 23°C	28·0	15·5	43·5	31
Ethyl acetate	10·5 at 23°C	17·6	18·8	36·4	31
Methyl propionate	1·4 at 23°C	20·5	21·3	41·8	31
n-Propyl acetate	13 at 23°C	4·6	32·6	37·2	31
Isopropyl acetate	15 at 23°C	3·8	29·7	33·5	31
Ethyl propionate	12·2 at 23°C	5·1	24·3	29·4	31
Ethyl n-butyrate	—	2·1	—	—	31
Methyl vinyl ether	217 at −25°C	—	—	—	33
Ethyl vinyl ether	200 at −25°C	—	—	—	33
	(344 at −6°C)				
2-Chloroethylvinyl ether	305 at −25°C	—	—	—	33

investigated by Bailey and coworkers.[31,34] Three possible reasons had been proposed:

(a) Stabilization of the *cis* state by hydrogen bonding between $C=O$ and R_2. Thus ΔH^0 will be influenced more by the nature of R_2 than of R_1.

(b) The *trans* state, with higher dipole moment, is less stable in media of low dielectric constant. The energy will be only slightly affected by changing alkyl groups R_1 or R_2.

(c) The *trans*-state energy is raised by steric interaction between R_1 and R_2. The combined size of $R_1 + R_2$ will be the dominant feature under these circumstances.

All of these explanations place the *cis* isomer as the most stable conformation. Hall and Lamb[35] working on ethyl formate in acetone and n-hexane solvents, and in contrast with the work of Tabuchi,[36] found that the *cis/trans* stabilities in the compound were consistent with explanation (b). However, a glance at Table 9.4 for the series from methyl formate to ethyl n-butyrate, clearly reveals a steady increase in ΔH^0 with the size of $R_1 + R_2$, from under 4 kJmole^{-1} for the formates to about 33 kJmole^{-1} for the C_4 groups. In addition ΔH_2^* falls from nearly 33 kJmole^{-1} to less than 4 kJmole^{-1} for the same series. Thus, ΔH_1^* ($= \Delta H^0 + \Delta H_2^*$) is almost constant at a mean value of about 35·9 kJmole^{-1}. The increasing size of the groups raises the energy of the upper state (*trans*), but has little influence on the energies of either the lower or transition states. Further, the effect is roughly dependent on the size of $R_1 + R_2$ rather than these groups alone. Clear support is given to postulate (c) as the dominant factor in ΔH^0.

Bailey and coworkers have also studied the energetics of a number of diesters (Table 9.5) in order to investigate the importance of correlations

Table 9.5 Molecular acoustic data for some diesters[34]

Compound	n	f_r (MHz) at 23°C	ΔH_2^* (kJmole^{-1})	ΔH^0 (kJmole^{-1})	ΔH_1^* (kJmole^{-1})
Dimethyl oxalate	0	13	16·7	15·9	32·6
Dymethyl malonate	1	5	9·6	18·0	27·6
Dimethyl succinate	2	12	6·3	21·7	28·0
Dimethyl glutarate	3	27	4·6	22·2	26·8
Dimethyl adipate	4	26	3·8	7·1	10·9
Diethyl oxalate	0	5	25·1	2·1	27·2
Diethyl malonate	1	14·5	9·6	20·5	30·1
Diethyl succinate	2	15	6·3	21·3	27·6
Diethyl adipate	4	10·5	4·6	8·0	12·6

in the conformation of neighbouring groups.[34] Acoustic studies were made on dimethyl or diethyl esters of the type $ROOC(CH_2)_nCOOR$ where n varied between zero and eight. No relaxations were observed at normal temperatures for $n > 4$ (pimelate, suberate and sebacate) but anomalous absorption was detected for the oxalates, malonates, succinates, glutarates and adipates. The data were analysed assuming a two-state process.

For the esters with $n = 1$ and 2, comparison of Tables 9.4 and 9.5 shows that ΔH^0 values are comparable between the corresponding mono- and diesters, while the ΔH_2^* values are slightly smaller in the latter case. It must be concluded that the ester group conformations are reached with some rapidly equilibrated stable arrangement of the intervening $(CH_2)_n$ chain so that energy differences are virtually those for the unperturbed groups. In the adipates it is possible to bring the $> O$ and $C{=}O$ groups into a 'cyclic' configuration, stabilized by dipole–dipole interactions

and *cis/trans* isomerism becomes very hindered giving anomalous energy values.

The work of Slie and Litovitz[7] and Kal'yanov and Nozdrev[37] on ethyl acetate provides rare information on the effect of pressure or density changes on acoustic relaxation processes. The former work, mentioned earlier, showed that $\Delta V/V = 0$ for the conformational equilibrium in this compound.

The data on isoamyl salicylate[38] assigned to internal rotation must be viewed with suspicion since f_r is concentration dependent.

9.6.3 Aldehydes and ketones

The majority of the investigations on these types of compounds have been carried out by Lamb and coworkers[33] and by Pethrick and Wyn-Jones.[39]

Table 9.6 records the energy parameters as determined by acoustic methods for a number of α,β-unsaturated molecules. In this type of system

Table 9.6 Molecular acoustic data for some aldehydes and ketones

Compound	f_r (MHz)	$\Delta H_{\frac{1}{2}}^{*}$ (kJmole^{-1})	ΔH^{0} (kJmole^{-1})	Reference
Acrolein	176 at 25°C	20·9	8·8	33
Methacrolein	174 at 25°C	22·2	13·0	33
α,β-Dimethylacrolein	10·6 at 20°C	23·2	—	39
α-Ethyl-β-propylacrolein	24·6 at 25°C	21·6	—	33, 39
Cinnamaldehyde	15·7 at 25°C	23·4	6·3	33
α-Methylcinnamaldehyde	12·0 at 20°C	21·5	7·5	39
α-n-Hexylcinnamaldehyde	18·2 at 25°C	19·7	—	33, 39
Crotonaldehyde	30·3 at 25°C	23·0	8·0	33
β-2-Furylacrolein	17·3 at 65°C	30·7	8·5	39
β-2-Furylidene acrolein	7·5 at 40°C	29·7	—	39
β-2-Furylacrylophenone	14·8 at 55°C	25·5	—	39
Citral	59·5 at 20°C	18·7	—	39

a planar structure is favoured where π-electron-repulsive interactions are at a minimum. In aldehydes and ketones, two isomers will exist as a result of restricted rotation about the C—C single bond:

$$\underset{\text{s-}trans}{\begin{array}{c} X \quad\quad Y \\ \diagdown\quad\diagup \\ C{=}C \\ \diagup\quad\diagdown \\ H \quad\quad C \\ | \\ Z \end{array}} O \rightleftharpoons \underset{\text{s-}cis}{\begin{array}{c} X \quad\quad Y \\ \diagdown\quad\diagup \\ C{=}C \quad Z \\ \diagup\quad\diagdown \\ H \quad\quad C \\ \| \\ O \end{array}}$$

De Groot and Lamb showed clearly[33] (Table 9.6) that the activation energy for the *cis/trans* interconversion increased from acrolein (X = Y = Z = H) through crotonaldehyde ($X = CH_3$) to cinnamaldehyde ($X = C_6H_5$). The successive increments in ΔH_2^* are attributed to the increasing double-bond character of the $C-C$ bond due to the respective electron donating properties of the X substituent. The effect is more noticeably demonstrated in the value for f_r at 25°C which drops from 176 MHz in acrolein through 30·3 MHz for crotonaldehyde to 15·7 MHz for cinnamaldehyde. On the other hand Wyn-Jones has investigated the series α-methylacrolein ($Y = CH_3$), α,β-dimethylacrolein and α-methylcinnamaldehyde where the periodicity is not so straightforward and the steric influence of the α-methyl group is relevant. However, the dominant feature in these molecules is undoubtedly the bond order of the $C-C$ bond. In the same investigation[39] Wyn-Jones has correlated ΔH_1^* for the relaxation with the $^{13}C-H$ coupling constant for the aldehyde group. This latter parameter is known to reflect the degree of s character in the carbon atom. A good correspondence was achieved despite the need to convert the ultrasonic activation energy which is measured in the reverse direction (*trans* to *cis*) to the forward equivalent ($J^{13}C-H$ is measured in the more abundant *trans* isomer) using inaccurate acoustic ΔH^0 data.

In the furyl-substituted molecules a second possible relaxation process is possible as a result of rotation of the furan ring with respect to the α,β-unsaturated skeleton. The observed results, however, are entirely consistent with the presence of only one relaxation which is assigned to furyl group rotation and not to aldehyde group motion. Table 9.6 shows that the relaxation frequencies are essentially identical when normalized to the same temperature and since the structural differences are due entirely to substitution at the Z position, the equilibrium is most unlikely to involve $Z-C=O$ motion.

No relaxation is observed in mesityl oxide over the frequency range 200 kHz to 200 MHz[33] since the *trans* configuration is blocked.

In methyl vinyl ketone ($X = H, Y = H, Z = CH_3$), acoustic measurements indicate a lower ΔH^0 than acrolein, with f_r values of about 80 MHz at $-25°C$ and 190 MHz at 0°C.[33]

In saturated aldehydes where no contribution to the π-bond character of the $C-C$ bond can come from conjugation, it would be expected that relaxation should occur at higher frequencies. In propionaldehyde f_r is about 200 MHz at $-25°C$ and of similar magnitude in n-butyraldehyde. Contrastingly in thiophene-2-aldehyde, which possesses a mobile π-electron system entering into strong conjugation with the aldehyde group and where steric effects are of little consequence, the relaxation frequency is below 10 MHz even at 75°C. In furfural, with even stronger conjugation, f_r is below 1 MHz and in the ketone 2-methyl-5-acetyl furane anomalous absorption at low frequencies can be detected only above 100°C.

9.6.4 Amines

Triethylamine and tri-n-butylamine have been studied by Krebs and Lamb.[40] Ultrasonic relaxation was detected and the data are shown in Table 9.7. The relaxation is presumed to arise from perturbation of the equilibrium involving the three isomers where $X = CH_3$ or $CH_2CH_2CH_3$.

$$
\begin{array}{ccc}
\text{(I)} & \text{(II)} & \text{(III)}
\end{array}
$$

Isomer (III) is considered improbable in the butyl derivative on steric grounds. In any case (II) and (III) become degenerate if nitrogen inversion takes place. In the derivative the freedom of movement of the ethyl group is greater in (II) than in (I) but the entropy value indicates that state (I) has a lower energy. Consequently the siting of the nitrogen lone pair takes precedence over repulsive forces between ethyl groups. In the butyl derivative larger repulsive forces are encountered leading to an increase in the population of (II). The sign of ΔS^0 indicates that isomer (II) is more stable.

Table 9.7   Molecular acoustic data for triethylamine and tri-n-butylamine[40]

Compound	$\Delta H_2^{\ddagger}$ (kJmole^{-1})	ΔH^0 (kJmole^{-1})	ΔS^0 (kJmole^{-1})
NEt_3	28·4	14·2	$+19·6$
$Nn\text{-}Bu_3$	18·0	3·6	$-3·8$

The work of Aliev and coworkers[41] on NEt_3, obtained over a wide frequency range (8–1000 MHz), contains an unusual method of presentation. Very accurate velocity values were obtained enabling precise characterization of the velocity dispersion. A plot of $\mu(c/c_0)^2$ versus $\pi(c/c_0)^2$ has a semi-circular form for a single relaxation. From equation (9.24)

$$
\frac{k^2}{\omega^2} = \frac{c_0^2}{c^2} = \frac{1 + (1 - r)i\omega\tau}{1 + i\omega\tau} \tag{9.69}
$$

After some manipulation equation (9.69) may be rearranged to

$$
\left(\frac{c_0}{c}\right)^2 + i\left(\frac{c_0^2}{c^2} \times \frac{\mu}{\pi}\right) = \frac{1 + (1 - r)\omega\tau}{1 + \omega^2\tau^2} - \frac{ir\omega\tau}{1 + \omega^2\tau^2} \tag{9.70}
$$

Thus a graph of $(c_0/c)^2$ versus $(c_0^2/c^2 \times \mu/\pi)$ is a semi-circle with radius $r/2$ centred on $(c_0/c)^2 = 1 - (r/2)$ and $(c_0^2/c^2 \times \mu/\pi) = 0$. The symmetrical regular semi-circle obtained by Aliev is excellent evidence for a single relaxation in NEt_3.

9.6.5 *Cyclic molecules*

Substituted cyclohexanes have received very little attention from acoustic investigations. Unfortunately the chair/chair relaxation occurs at very low frequencies and the information is rather inaccurate. In consequence of this, very few workers have thought it worthwhile to carry out experiments at more than one temperature and therefore the energy data is very sparse. Table 9.8 records the data available. Hall[45] has reported the possible

Table 9.8   Molecular acoustic data for some substituted cyclohexanes

Compound	f_r (kHz)	ΔH_2^* (kJmole^{-1})	ΔH^0 (kJmole^{-1})	Reference
Methylcyclohexane	140 at 16°C	45·5, 43·1	14·6, 12·1	6, 42
	(105 at 25°C)	(26·8)	(8·0)	(43)
Ethylcyclohexane	60 at 16°C	—	—	44
Bromocyclohexane	—	~50	—	6
Chlorocyclohexane	—	~50	—	6
1,2-*trans*-Dimethylcyclo-hexane	120 at 16°C	—	—	44
1,4-*trans*-Dimethylcyclo-hexane	150 at 16°C	—	—	44
1,1,3-Trimethylcyclohexane	>1 MHz at 16°C	—	—	44

existence of two relaxations for solutions of methylcyclohexane in n-hexane. Chair/chair interconversion was proposed for the relaxation at 100 kHz while a chair/boat equilibrium was postulated at much lower frequencies. Subrahmanyam and Piercy[42] have reworked this data and showed that a single relaxation at 65 kHz could explain the results equally well.

Cyclohexene derivatives have received some attention and relaxations have been detected at higher frequencies than those found in the corresponding saturated molecules. A transition between half-chair states is proposed for the mechanism. Acoustic data are shown in Table 9.9.

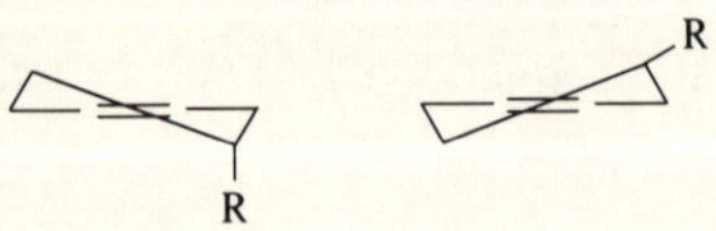

Karpovitch[47] reported the existence of a relaxation in cyclohexene at 100 kHz to 1 MHz which he attributed to chair/boat interconversion. No

Table 9.9   Molecular acoustic data for some substituted cyclohexenes[46]

Compound	f_r (MHz)	ΔH_2^* (kJmole^{-1})	ΔS_2^* (kJmole^{-1})
4-Bromocyclohexene	18·5 at 29°C	21·7	− 20·3
4-Methylcyclohexene	23·7 at 4°C	14·2	− 17·2
3-Methylcyclohexene	28·4 at 27°C	13·8	—
4-Vinylcyclohexene	31·1 at 26°C	19·2	—

similar behaviour has been reported by other workers[46,48] and this anomalous behaviour has been attributed to a cyclohexane derivative impurity (20%).[48] Indeed the entropy data in Table 9.9 are even inconsistent with the existence of a boat-type transition state.

A number of heterocyclic molecules have been investigated recently as shown in Table 9.10. The behaviour of caprolactam is attributed to interconversion between 'boat' and 'chair' states:

In the sulphites more conventional chair/chair isomers are postulated with the S=O bond occupying an axial position in the stable conformer.

Table 9.10   Molecular acoustic data for some heterocyclic molecules

Compound	f_r (MHz)	ΔH_2^* (kJmole^{-1})	ΔH^0 (kJmole^{-1})	Reference
N-Methylcaprolactam	8·6 at 29°C	11·8	16·9	49
4-Methyl-1,3-dioxan	7·8 at 25°C	38·5	10·5	50
4-Phenyl-1,3-dioxan	10·7 at 25°C	18·8	22·2	50
Trimethylenesulphite (TMS)	25·6 at 20°C	23·0	5·4	51
4-Methyl TMS	90 at 15°C	20·1	⩽5·0	51
5,5-Dimethyl TMS	21·6 at 25°C	18·0	6·3	51
5,5-Diethyl TMS	37·4 at 30°C	17·6	⩽5·4	51
5-Methyl-5-ethyl TMS	30·2 at 30°C	18·0	⩽5·9	51
4,6-Dimethyl TMS (*meso*)	61·5 at 20°C	24·3	⩽5·0	51
4,5,6-Trimethyl TMS	77·5 at 20°C	15·1	⩽5·0	51

9.7 Bibliography and references

1. J. Lamb, 'Thermal relaxation in liquids', *Physical Acoustics*, Vol. IIA, Ed. W. P. Mason (New York: Academic Press, 1965) Chap. 4.
2. R. T. Beyer and S. V. Letcher, *Physical Ultrasonics* (New York: Academic Press, 1969).
3. A. B. Bhatia, *Ultrasonic Absorption* (Oxford: University Press, 1967).
4. G. L. Gooberman, *Ultrasonics* (London: English Universities Press, 1968).
5. J. E. Piercy and M. G. Seshagiri Rao, *J. Chem. Phys.*, **46**, 3951 (1967).
6. J. E. Piercy, *J. Acoust. Soc. Am.*, **33**, 198 (1961).
7. W. M. Slie and T. A. Litovitz, *J. Chem. Phys.*, **39**, 1538 (1963).
8. J. Lamb, *Z. Elektrochem.*, **64**, 136 (1960).
9. R. A. Padmanaban, *J. Sci. Ind. Res.*, **19**, 336, 457 (1959).
10. K. R. Crook and W. Wyn-Jones, *J. Chem. Phys.*, **50**, 3445 (1969).
11. J. Karpovitch, *J. Acoust. Soc. Am.*, **26**, 819 (1955).
12. J. H. Andrae and P. D. Edmonds, *Proc. 3rd Int. Cong. Acoust.*, 556 (1959).
13. D. N. Hall and J. Lamb, *Proc. Phys. Soc.*, **B73**, 354 (1959).
14. P. Debye and F. W. Sears, *Proc. Nat. Acad. Sci., U.S.*, **18**, 410 (1932).
15. R. Lucas and P. Biquard, *J. Phys. Radium*, **3**, 464 (1932).
16. F. Eggers, *Acoustics*, **19**, 323 (1968).
17. J. C. Hubbard, *Phys. Rev.*, **46**, 525 (1934).
18. J. H. Andrae, R. Bass, E. L. Heasell and J. Lamb, *Acustica*, **8**, 3 (1958).
19. R. J. Abraham and M. A. Cooper, *J. Chem. Soc.* (*B*), 202 (1967).
20. A. E. Clark and T. A. Litovitz, *J. Acoust. Soc. Am.*, **32**, 1221 (1960).
21. J. H. Chen and A. A. Petrauskas, *J. Chem. Phys.*, **30**, 304 (1959).
22. J. M. Young and A. A. Petrauskas, *J. Chem. Phys.*, **25**, 934 (1956).
23. J. Lamb, *Z. Elektrochem.*, **64**, 135 (1960).
24. K. R. Crook, P. J. D. Park and E. Wyn-Jones, *J. Chem. Soc.* (*A*), 2910 (1969).
25. T. H. Thomas, E. Wyn-Jones and W. J. Orville-Thomas, *Trans. Faraday Soc.*, **65**, 974 (1969).
26. E. Wyn-Jones and W. J. Orville-Thomas, 'Molecular Relaxation Processes', *Chem. Soc.* (*London*) *Spec. Publ.*, No. **20**, 209 (1966).
27. R. A. Pethrick, E. Wyn-Jones, P. C. Hamblin and R. F. M. White, *J. Chem. Soc.* (*A*), 1852 (1969).
28. P. Biquard, *Ann. Phys.* **6**, 195 (1936).
28a. M. J. Blandamer, M. J. Foster, N. J. Hidden and M. C. R. Symons, *J. Phys. Chem.*, **72**, 2268 (1968).
29. J. Karpovitch, *J. Chem. Phys.*, **22**, 1767 (1954).
30. S. V. Subrahmanyam and J. E. Piercy, *J. Acoust. Soc. Am.*, **37**, 340 (1965).
31. J. Bailey and A. M. North, *Trans. Faraday Soc.*, **64**, 1499 (1968).
32. M. Tannaka, *Acustica*, **23**, 328 (1971).
33. M. S. de Groot and J. Lamb, *Proc. Roy. Soc.*, **A242**, 36 (1957).
34. J. Bailey, A. M. North and S. M. Walker, *J. Mol. Struct.*, **6**, 53 (1970).
35. D. N. Hall and J. Lamb, *Trans. Faraday Soc.*, **55**, 784 (1959).
36. D. Tabuchi, *J. Chem. Phys.*, **28**, 1014 (1958).
37. B. I. Kal'yanov and V. F. Nozdrev, *Soviet Phys. Acoust.* (*English Transl.*), **5**, 377 (1959).
38. S. Rao and V. Raju, *J. Phys. Soc. Japan*, **29**, 1652 (1970).
39. R. A. Pethrick and E. Wyn-Jones, *Trans. Faraday Soc.*, **66**, 2483 (1970).
40. K. Krebs and J. Lamb, *Proc. Roy. Soc.*, **A244**, 558 (1958).
41. S. S. Aliev, K. Parpiev and P. K. Khabibullaev, *Sov. Phys. Acoust.* (*English Transl.*) **15**, 444 (1969).

42. S. V. Subrahmanyam and J. E. Piercy, *J. Chem. Phys.*, **42**, 4011 (1965).
43. M. E. Pedinoff, *J. Chem. Phys.*, **36**, 777 (1962).
44. J. Lamb and J. Sherwood, *Trans. Faraday Soc.*, **51**, 1674 (1955).
45. D. N. Hall, *Trans. Faraday Soc.*, **55**, 1319 (1959).
46. K. R. Crook and E. Wyn-Jones, *Trans. Faraday Soc.*, **67**, 660 (1971).
47. J. Karpovitch, *J. Chem. Phys.*, **22**, 1767 (1954).
48. S. V. Subrahmanyam and J. E. Piercy, *J. Chem. Phys.*, **42**, 1845 (1965).
49. A. N. Moran and S. M. Walker, *J. Mol. Struct.*, **9**, 299 (1971).
50. P. C. Hamblin, R. F. M. White and E. Wyn-Jones, *J. Mol. Struct.*, **4**, 275 (1969).
51. G. Eccleston, R. A. Pethrick, E. Wyn-Jones, P. C. Hamblin and R. F. M. White, *Trans. Faraday Soc.*, **66**, 310 (1970).

10 Electron diffraction studies and rotational isomerism

A. H. Clark

10.1 Introduction

The vapour-phase electron diffraction technique was first applied to the study of internal rotation in the early 1930s. Initially, only rather crude estimates could be made of the geometries and relative amounts of the rotamers in a gas sample, but, as the method improved, its effectiveness in providing data of this kind increased. Today, with the aid of electronic computers, it is not only possible to measure rotamer geometries and proportions fairly accurately, but also to obtain rough values for barrier heights, and even, in favourable cases, to compare analytical forms for the hindering potential energy function.

Before considering the various refinements which must be made to the standard electron diffraction technique to obtain such information, three sections of this chapter will be devoted to accounts of the history, the experimental aspects and the basic theory of the method. Its application to internal rotation problems will then be discussed and illustrated by a number of examples taken from the literature of the last twenty years.

10.2 The history of gas-phase electron diffraction

The first electron diffraction patterns from gases were recorded photo-graphically by Wierl[1] some forty years ago, and, to the eye, each of these seemed to consist of a series of diffuse concentric rings. Photometer measurement indicated, however, that in most cases these rings were not true maxima but were slight inflexions appearing on a steeply falling background. This was confirmed by theoretical treatment of the scattering process, for it was shown that the scattered intensity should be the sum of a smooth atomic and an oscillating molecular component, both terms being multiplied by s^{-4}, where $s = 4\pi \sin(\theta/2)/\lambda$, θ is the angle of scattering and λ the electron wavelength.

It appeared that in viewing the pattern the eye effectively subtracted out the atomic background, and then divided by it, giving the impression that the molecular component was present alone. Since this latter function is strongly dependent on the geometry of the molecule studied, and since it could not be accurately determined from featureless photometer traces, the positions and relative intensities of the diffraction rings were estimated visually. This visual approach, which is associated with Wierl himself, and with Pauling and Brockway,[2] persisted throughout the 1930s and yielded a vast amount of structural information, but eventually the photometer method was reintroduced as a result of an important experimental advance.

Photometer measurement of the original diffraction photographs had been unsatisfactory on account of the large gradients of blackness involved, but by allowing a specially-cut metal shutter or sector (see Figure 10.1) to spin just above the photographic plate, it was possible to multiply the total scattered intensity by between s^2 and s^3 before it was recorded, and so obtain low gradients of optical density across the pattern.

This experimental innovation, which was suggested by Finbak[3] in 1937, and later, independently, by Debye[4] in 1939, was successfully applied in Norway from 1940 onwards, and subsequently in other countries. With the greater accuracy achieved it became possible to extract more information from electron diffraction data, and in the 1940s and early 1950s a number of papers were published by Debye,[5] and J. Karle and coworkers,[6–12] building on the work of James,[13] and showing the effects of molecular vibration, and particularly internal rotation, on the molecular intensity function. These theoretical advances, together with the demonstration[14,15] in 1952 that scattering factors based on the First Born Approximation were inadequate for molecules containing heavy atoms, enabled the electron diffraction technique to be applied not only to the accurate determination of molecular structure, but also to the study of intramolecular motion.

In 1957 the power of the method was further extended by the application[16] of least-squares variation to the task of fitting theory to experiment, and in recent years this approach has become of fundamental importance as a result of the increasing availability of electronic computers. Developments in computation have particularly assisted the study of internal rotation, as in this application of the electron diffraction technique extra long calculations are often involved.

10.3 The electron diffraction experiment

Modern electron diffraction instruments[17,18] normally include the following components: (a) a high-tension power supply (30–60 kV) stabilized to within at least one part in ten thousand, (b) an electron gun consisting of a cathode, anode, condenser lens and apertures, and designed to produce a narrow monochromatic electron beam, (c) a diffraction chamber, (d) a gas nozzle operating at sample pressures of around 20 mm Hg, and designed to ensure correct gas flow into the diffraction chamber, (e) a cooled metal surface or cold trap situated close to the scattering centre, and designed to extract the sample after diffraction, (f) a powerful pumping system capable of maintaining a continuous vacuum of 10^{-5} mm Hg or less throughout the equipment, (g) a sector-bearing designed to allow rapid rotation of the sector above the photographic plate and (h) a photographic-plate carriage and storage box enabling repeated recording of the diffraction pattern.

These various items are usually arranged as indicated in Figure 10.1, and with care, such an apparatus can be used to measure internuclear distances with an accuracy of several thousandths of an Ångstrom unit. For

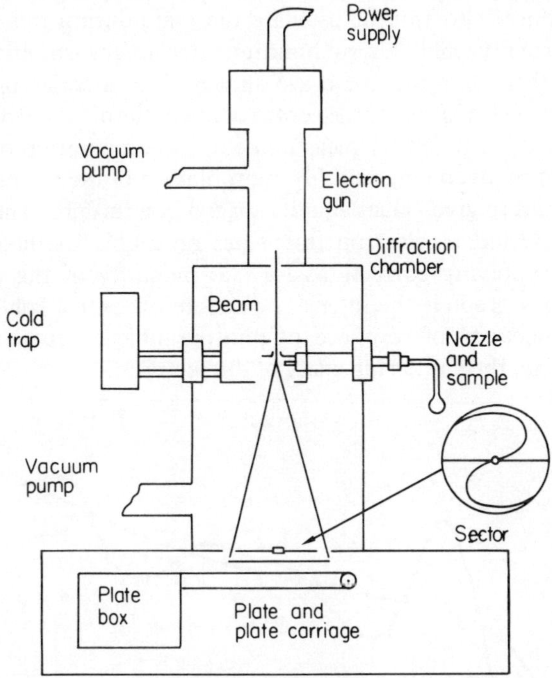

Figure 10.1   A schematic drawing of a typical electron
diffraction instrument

this accuracy to be achieved, however, the wavelength of the electron beam must be determined even more precisely, and this quantity, which usually lies in the range 0·07–0·04 Å, is measured by recording diffraction patterns from a polycrystalline standard such as thallous chloride (TlCl), or from a gas standard such as carbon dioxide.

Precise measurements of the sector geometry and the jet-to-photographic plate distances are also required, and in a thorough electron diffraction study recordings of the diffraction pattern are made at a minimum of three camera lengths. For example, distances of 100, 50 and 18 cm might be used, and three plates obtained at each. This procedure enables accurate measurement of electron intensity over a wide range of scattering angle, as data at low angles are largely determined from 100 cm and 50 cm photographs, whilst high-angle scattering is measured at the shortest distance.

When a full set of plates has been obtained in this way, all recordings are scanned by a microphotometer, and today it is common to have an instrument capable of outputting readings on paper tape ready for computer treatment.[19] In some instruments the plate remains fixed during measurement and the apparatus scans across one diameter of the pattern, but a better procedure is to spin or oscillate the plate during this process. One such scan normally yields several hundred intensity measurements, and, as a first step in their analysis, these are subjected to a series of calculations involving:[20,21] (a) location of the centre of the pattern, (b) correction of each reading to an optical density measurement, (c) combination of both halves of the pattern by averaging and (d) interpolation of the resulting radius of optical densities to give values equally spaced in terms of s. These quantities are then converted to electron intensities amenable to theoretical interpretation by applying corrections for: (a) planarity of the photographic plate, as theory predicts the intensity distribution over a spherical surface, (b) the non-linearity of response of photographic emulsions to incident electrons and (c) the shuttering effect of the sector.

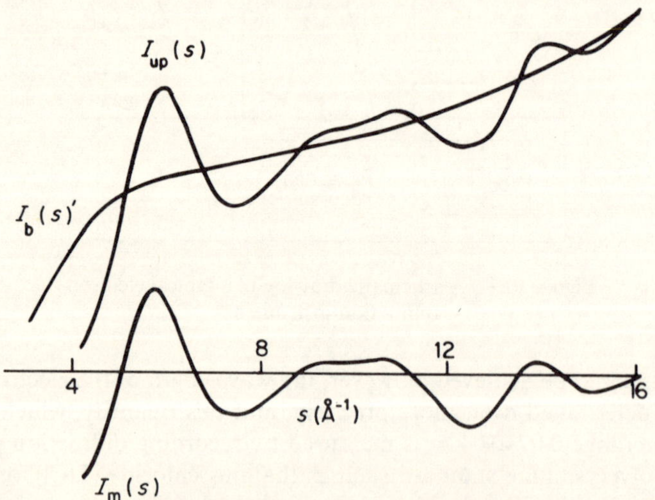

Figure 10.2   Examples of uphill, $I_{up}(s)$, background, $I_b(s)'$, and unmodified molecular intensity, $I_m(s)$, curves

As expected from theoretical considerations the experimental function of 's' derived in this way declines rapidly as s increases, and a final step is to multiply by s^4 to produce an increasing function, sometimes called an uphill curve,[20] which shows clear indications of the oscillating molecular component (Figure 10.2). In an accurate investigation at least two, and preferably three, overlapping curves of this type are necessary, each one being an average derived from several plate scans, and each one corresponding to a particular camera distance.

10.4 The theoretical treatment of electron scattering by semi-rigid molecules

10.4.1 *Preliminary definitions and relationships*

To determine structural information from uphill data it is necessary to employ theoretical expressions based on the geometrical and vibrational properties of molecules, and it is the purpose of this introductory section to define the generalized mean-square amplitudes of vibration of an internuclear distance, as these feature prominently in any discussion of the effects of intramolecular motion on the diffraction pattern.

The generalized amplitudes of vibration may be defined for any atom pair ij in a molecule by setting up the localized Cartesian reference frame shown in Figure 10.3. In this system the z axis runs along the line connecting the equilibrium positions of the two atoms, and the x and y axes are at right

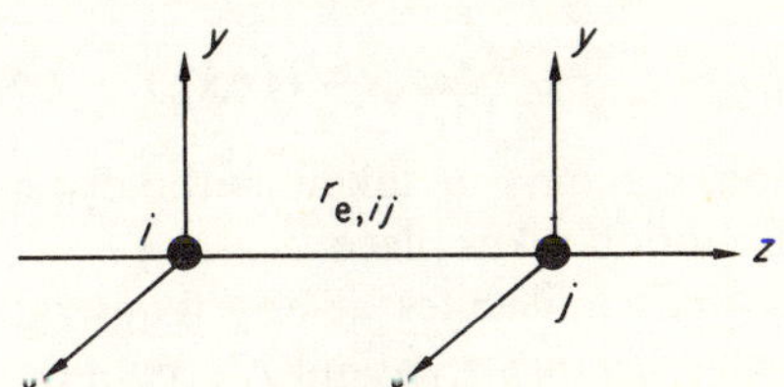

Figure 10.3 A localized Cartesian reference system suitable for describing the intramolecular motion of an atom pair ij

angles to this direction. If Δx_i, Δy_i, Δz_i and Δx_j, Δy_j, Δz_j specify the instantaneous positions of atoms i and j at any time during vibration, then generalized amplitudes of displacement for the distance r_{ij} may be defined by the equations

$$\Delta x_{ij} = \Delta x_j - \Delta x_i \tag{10.1}$$

$$\Delta y_{ij} = \Delta y_j - \Delta y_i \tag{10.2}$$

$$\Delta z_{ij} = \Delta z_j - \Delta z_i \tag{10.3}$$

The generalized mean-square amplitudes of vibration are then the average quantities: $\langle\Delta x_{ij}^2\rangle$, $\langle\Delta y_{ij}^2\rangle$, $\langle\Delta z_{ij}^2\rangle$, $\langle\Delta x_{ij}\Delta y_{ij}\rangle$, $\langle\Delta x_{ij}\Delta z_{ij}\rangle$ and $\langle\Delta y_{ij}\Delta z_{ij}\rangle$, the first two being called the mean-square perpendicular amplitudes, the third the mean-square parallel amplitude, and the remaining three the mean cross-products.

If the instantaneous distance between atoms i and j is r_{ij}, and the corresponding equilibrium value is $r_{e,ij}$, it is clear that

$$r_{ij} = \{(r_{e,ij} + \Delta z_{ij})^2 + \Delta x_{ij}^2 + \Delta y_{ij}^2\}^{\frac{1}{2}} \tag{10.4}$$

and, for very small displacements,

$$r_{ij} \sim r_{e,ij} + \Delta z_{ij} + (\Delta x_{ij}^2 + \Delta y_{ij}^2)/2r_{e,ij} \tag{10.5}$$

If the internal coordinate $r_{ij} - r_{e,ij}$ is written as Δr_{ij} it follows on taking averages that

$$\langle\Delta r_{ij}^2\rangle \sim \langle\Delta z_{ij}^2\rangle \tag{10.6}$$

This is valid if all displacements are sufficiently small that terms of power greater than two may be neglected. The quantity $\langle\Delta r_{ij}^2\rangle$ is often symbolized u_{ij}^2 in electron diffraction literature and its square root $\langle\Delta r_{ij}^2\rangle^{\frac{1}{2}}$, is usually referred to as the root-mean-square amplitude of vibration, u_{ij}, of the distance concerned.

It also follows from equation (10.5), by averaging both sides, that

$$r_{g,ij} = \langle r_{ij}\rangle \sim r_{e,ij} + \langle\Delta z_{ij}\rangle + (\langle\Delta x_{ij}^2\rangle + \langle\Delta y_{ij}^2\rangle)/2r_{e,ij} \tag{10.7}$$

and, in this expression, $r_{g,ij}$ may be taken as the distance r_{ij} averaged over whatever vibrational motion takes place.

For a collection of gas molecules at any temperature it is possible to define the spread of the distance r_{ij} about $r_{e,ij}$ by a normalized probability density function $P_{ij}(r)$, and in terms of this

$$r_{g,ij} = \int_0^\infty rP_{ij}(r)\,dr \tag{10.8}$$

and

$$u_{ij}^2 = \int_0^\infty (r - r_{e,ij})^2 P_{ij}(r)\,dr \tag{10.9}$$

It will later be shown that the molecular component of the total scattered intensity is strongly dependent on the $P_{ij}(r)$ distributions, and indeed the whole problem of predicting the effects of intramolecular motion on the diffraction pattern centres on finding expressions for these in terms of the constants which define the molecular potential energy function.

10.4.2 *Theoretical expressions for the scattered intensity*

The basic data obtained from microphotometer measurements are, after correction, a set of numbers proportional to s^4 times the total intensity of scattered electrons measured at various scattering angles, and at a constant distance from the scattering centre.

When attempting to obtain a mathematical expression for this the uphill data, it is necessary to consider the scattering which would be produced by a completely rigid molecule, and then to introduce such modifications as may be necessary to take into account the effects of intramolecular motion.

If bonding is neglected, this rigid configuration may be assumed to consist of spherical atoms, and each atom may be treated as a separate scattering centre. For an atom the process of electron scattering is described by a complex-number scattering factor,[22]

$$F(s) = |F(s)| \, e^{i\eta(s)} \tag{10.10}$$

and in most cases it is possible to set $|F(s)|$ equal to $Z - f(s)$, where Z is the atomic number and $f(s)$ is the X-ray scattering factor.

If each atom is assumed to scatter in this way, it is easy to show[22] that a collection of rigid gas molecules, arranged in all orientations relative to the incident beam, should give rise to an uphill function of the form

$$
\begin{aligned}
I_{\text{up}}(s)^{\text{rigid}} = \; & k\left\{ \sum_i \left([Z - f(s)]_i^2 + S_i(s) \right) \right. \\
& \left. + \sum_{\substack{i,j \\ (i \neq j)}} \left([Z - f(s)]_i [Z - f(s)]_j \cos \Delta\eta_{ij}(s) \frac{\sin [sr_{e,ij}]}{sr_{e,ij}} \right) \right\} \\
= \; & k\{I_b(s) + I_m(s)^{\text{rigid}}\}
\end{aligned}
\tag{10.11}
$$

In this expression k is a proportionality constant, $\Delta\eta_{ij}(s)$ equals $\eta_j(s) - \eta_i(s)$, and is the phase difference between waves scattered coherently by atoms i and j, and $S_i(s)$ is the incoherent scattering factor for the ith atom.

The uphill curve therefore consists of two components, $I_b(s)$ and $I_m(s)$, examples of which are included in Figure 10.2. The term $I_b(s)$ is the atomic background, which theory predicts to be a smooth function steadily increasing with s, but levelling off at high s, whilst $I_m(s)$ is the molecular intensity which oscillates in an irregular way with increasing s. This latter component is responsible for the wave-like qualities of electron diffraction patterns, and whilst the background curve gives no information about the molecular geometry or motion, the molecular intensity depends very much on the spectrum of internuclear distances involved in the structure.

In practice it is found that the background function, though non-oscillating, does not conform exactly to the theoretical form assigned to it in equation (10.11), and it is usual to add a more or less empirical function, $I_e(s)$, to this expression, to compensate for any discrepancy. The need for this addition seems to arise from neglect in equation (10.11) of certain effects such as multiple scattering and scattering from parts of the apparatus. A modified expression for the uphill curve is therefore

$$I_{up}(s)^{rigid} = k\{I_b(s) + I_e(s) + I_m(s)^{rigid}\}$$
$$= k\{I_b(s)' + I_m(s)^{rigid}\} \tag{10.12}$$

As might be expected, the rigid approximation on which this equation is based is a poor one, and in practice equation (10.11) must be modified to take into account the normal flexibility of molecules. To achieve this it is necessary to replace $I_m(s)^{rigid}$ by

$$I_m(s) = k \sum_{\substack{i,j \\ (i \neq j)}} g_{ij}(s) \int_0^\infty P_{ij}(r) \frac{\sin (sr)}{sr} dr \tag{10.13}$$

in which $g_{ij}(s)$ is equal to $[Z - f(s)]_i[Z - f(s)]_j \cos \Delta\eta_{ij}(s)$, and $P_{ij}(r) \, dr$ is the probability that distance r_{ij} lies between r_{ij} and $r_{ij} + dr_{ij}$, that is, $P_{ij}(r)$ is the probability density function for distance r_{ij} consistent with the gas temperature and the type of intramolecular motion taking place.

If this latter is assumed to be confined to vibrations of very small amplitude and only quadratic terms in the potential energy expression are retained then harmonic vibration results, and each $P_{ij}(r)$ function may be calculated fairly rigorously. The analysis involved has been discussed at length in papers by James,[13] Karle and Karle,[8,10] Morino and Hirota,[23] and in a review by Rambidi and coworkers;[24] hence what follows is only a summary of the relevant mathematical treatment.

The first step is to apply quantum mechanics to the problem of harmonic vibration and so obtain wave-functions and energy levels for the motion involved. If the molecule contains N atoms, and is non-linear, these wave-functions may be expressed in terms of $3N - 6$ simple harmonic oscillator equations, one for each normal mode. Each total wave-function is then obtained as a product of solutions to these individual equations, and can be used to formulate a probability density for the $3N - 6$ normal co-ordinates. If the gas temperature is specified, it is also possible to calculate an average probability density function of this type, by summing up individual contributions from each energy state, each one being given the weight predicted for it by the Boltzmann equation. This average distribution is then transformed from a function of the normal coordinates, to one expressed in terms of the interatomic distance displacements Δr_{mn}, by applying the theory of normal coordinate analysis which enables the displacements Δr_{mn} to be expressed linearly in terms of the normal coordinates. Finally this function is reduced to one applicable to one distance displacement

Δr_{ij} only, that is, to $P_{ij}(r)$ the required distribution. The final result is

$$P_{ij}(r) = [1/\sqrt{(2\pi\langle\Delta z_{ij}^2\rangle)}]\exp{(-\Delta r_{ij}^2/2\langle\Delta z_{ij}^2\rangle)}[1 + A_{ij}(r)] \qquad (10.14)$$

in which

$$\begin{aligned}
A_{ij}(r) = (\Delta r_{ij}/r_{e,ij})&\{((\langle\Delta x_{ij}^2\rangle + \langle\Delta y_{ij}^2\rangle)/2\langle\Delta z_{ij}^2\rangle \\
&-(1/2\langle\Delta z_{ij}^2\rangle)(\langle\Delta x_{ij}\Delta z_{ij}\rangle^2 + \langle\Delta y_{ij}\Delta z_{ij}\rangle^2)(3 - \Delta r_{ij}^2/\langle\Delta z_{ij}^2\rangle)\}
\end{aligned} \qquad (10.15)$$

and equation (10.14) reduces to the more familiar Gaussian form

$$P_{ij}(r) = [1/\sqrt{(2\pi\langle\Delta z_{ij}^2\rangle)}]\exp{(-\Delta r_{ij}^2/2\langle\Delta z_{ij}^2\rangle)} \qquad (10.16)$$

if $\langle\Delta x_{ij}^2\rangle$, $\langle\Delta y_{ij}^2\rangle$, $\langle\Delta x_{ij}\Delta z_{ij}\rangle$ and $\langle\Delta y_{ij}\Delta z_{ij}\rangle$ are small enough to be neglected.

From equations (10.14) and (10.15) it is clear that if the perpendicular amplitudes of vibration are sizeable, $P_{ij}(r)$ will deviate considerably from a Gaussian form, and indeed in polyatomic molecules an atom pair do not in general confine their displacements to the line joining their equilibrium positions.

This is particularly important where linear and planar molecules are concerned, for in a linear XOX system it is often found that the r_g distance measured by electron diffraction for r_{XX} is smaller than the sum of the r_g values for r_{OX}. This so-called shrinkage effect depends very much on the size of the perpendicular amplitudes of vibration, as is evident from equation (10.7), for if $\langle\Delta x_{ij}^2\rangle$ and $\langle\Delta y_{ij}^2\rangle$ are large, $r_{g,ij}$ will deviate considerably from $r_{e,ij}$ even when $\langle\Delta z_{ij}\rangle$ is zero.

If equations (10.14) and (10.15) are assumed, with the simplifying assumption that the terms $\langle\Delta x_{ij}\Delta z_{ij}\rangle$ and $\langle\Delta y_{ij}\Delta z_{ij}\rangle$ may be ignored, then the integral in equation (10.13) may be evaluated, and a more realistic expression for the molecular intensity component obtained. This is

$$I_m(s)^{\text{harm}} = k \sum_{\substack{i,j \\ (i \neq j)}} g_{ij}(s)[\sin{(sr_{a,ij})}/sr_{a,ij}]\exp{(-\tfrac{1}{2}s^2\langle\Delta z_{ij}^2\rangle)} \qquad (10.17)$$

in which

$$r_{a,ij} = r_{e,ij} - \langle\Delta z_{ij}^2\rangle/r_{e,ij} + (\langle\Delta x_{ij}^2\rangle + \langle\Delta y_{ij}^2\rangle)/2r_{e,ij} \qquad (10.18)$$

In practice it is common to replace $\langle\Delta z_{ij}^2\rangle$ in (10.17) by u_{ij}^2, (see equation 10.6), and if this modified version is applied to the experimental molecular intensity data, the distances $r_{a,ij}$, and the vibrational amplitudes, u_{ij}, can be calculated by least-squares refinement.

Comparison of equation (10.18) with equation (10.7) shows that if anharmonicity can be neglected, and $\langle\Delta z_{ij}\rangle$ assumed equal to zero, it is possible to relate $r_{g,ij}$ to $r_{a,ij}$ by the relationship

$$r_{g,ij} \sim r_{a,ij} + \langle\Delta z_{ij}^2\rangle/r_{e,ij} \sim r_{a,ij} + u_{ij}^2/r_{a,ij} \qquad (10.19)$$

and in electron diffraction work it is common to use this expression to correct r_a distances to corresponding r_g values.

Equation (10.17) can in fact be extended to include the effects of a slight amount of anharmonicity.[25] The new form required is

$$I_m(s)^{\text{anharm}} = k \sum_{\substack{i,j \\ (i \neq j)}} g_{ij}(s)\{\sin\,[s(r_{a,ij} - x_{ij}s^2)]/sr_{a,ij}\}\exp\,(-\tfrac{1}{2}u_{ij}^2 s^2) \quad (10.20)$$

in which x_{ij} is a small constant for each distance which may be included in least-squares calculations, though it is in fact rarely determined in this way on account of the tendency for it to be strongly correlated with $r_{a,ij}$. It can be estimated from spectroscopic data, particularly for bonded distances, and, in practice, such estimates are sometimes included as unrefined parameters in least-squares calculations.

A general expression for the uphill data corresponding to a molecule of reasonable rigidity is therefore

$$I_{\text{up}}(s) = k\{I_b(s)' + I_m(s)^{\text{anharm}}\} \quad (10.21)$$

10.4.3 *Subtraction of the atomic background*

Since $I_b(s)'$ is partly empirical it is usual to attempt to subtract it from the uphill data, and to confine least-squares calculations to the molecular component only. One rather direct method of doing this is to subtract a smooth curve drawn by hand through the uphill data (see Figure 10.2), but more sophisticated approaches involve calculation of the theoretical atomic background $I_b(s)$.

Thus, in one procedure, the uphill curve is divided by $I_b(s)$ to give a roughly horizontal function of the form

$$I(s)_{\text{level}} = k\{1 + I_e(s)/I_b(s) + I_m(s)/I_b(s)\} \quad (10.22)$$

after which it is necessary to draw in an empirical curve to approximate the component $k\{1 + I_e(s)/I_b(s)\}$. In this case the operation of curve drawing is often easier than in the direct method, but after subtraction, multiplication by $I_b(s)$ is required to obtain the molecular intensity function.

In both approaches $I_e(s)$ is largely guessed in the first instance, and so early estimates of $I_m(s)$ will contain background errors which must be removed in stages as the structure determination proceeds. Such corrections are usually based on examination of the results of intermediate least-squares refinements, as by comparing experimental and theoretical versions of the molecular intensity it is possible to detect the type of slowly-varying difference function characteristic of a background error. Hand-drawn curves are then altered accordingly, two or three corrections being usually sufficient to produce a satisfactory result.

10.4.4 *Modification of the molecular intensity*

Although least-squares calculations can be based on the $I_m(s)$ component obtained after background subtraction, it is normal to multiply this by a so-called modification function, particularly if a radial distribution curve is required. The object of this is to produce a function $I_m(s)'$ consisting of a sum of damped sine waves each with an amplitude as nearly as possible independent of s.

If all atoms present are light, all $\Delta\eta_{ij}(s)$ functions will be close to zero, and in such a case modification is achieved by multiplying by $s/\{[Z - f(s)]_A \times [Z - f(s)]_B\}$, where A and B refer to two atoms commonly occurring in the molecule. A general expression for $I_m(s)'$ based on equation (10.13), is then

$$I_m(s)' = k \sum_{\substack{i,j \\ (i \neq j)}} g'_{ij} \int_0^\infty P_{ij}(r)[\sin (sr)/r]\, dr \tag{10.23}$$

in which g'_{ij} is equal to $[Z - f(s)]_i[Z - f(s)]_j \cos \Delta\eta_{ij}(s)$ divided by $[Z - f(s)]_A \times [Z - f(s)]_B$, and if for convenience all n_i symmetrically equivalent distances r_{mn} are grouped into a set labelled i, this equation becomes

$$I_m(s)' = k \sum_i n_i g'_i \int_0^\infty P_i(r)[\sin (sr)/r]\, dr \tag{10.24}$$

The summation is now over all non-equivalent distances in the molecule, and if near-harmonic vibration can be assumed, an explicit version of (10.24) is

$$I_m(s)' = k \sum_i n_i g'_i \{\sin [s(r_{a,i} - x_i s^2)]/r_{a,i}\} \exp\left(-\tfrac{1}{2} u_i^2 s^2\right) \tag{10.25}$$

in agreement with equation (10.20). This expression is generally the one used to carry out least-squares calculations if a modified intensity curve is available.

10.4.5 *The radial distribution curve*

Equation (10.24) may be rewritten in the form

$$I_m(s)' = k \int_0^\infty \left[\sum_i n_i g'_i P_i(r)/r \right] \sin (sr)\, dr \tag{10.26}$$

and, if all the coefficients g'_i are assumed independent of s, sine-Fourier transformation of $I_m(s)'$ gives the radial distribution function $\sigma(r)/r$. That is

$$\sigma(r)/r = k' \int_0^\infty I_m(s)' \sin (sr)\, ds$$

$$= k'' \sum_i n_i g'_i P_i(r)/r \tag{10.27}$$

$$\approx k'' \sum_i \left(\frac{n_i g'_i}{r_i}\right) P_i(r)$$

and is a weighted sum of probability density functions, one for each of the distance types involved. If the molecule vibrates harmonically, each of these distributions will be almost Gaussian, and hence, ideally, the radial distribution curve should constitute a kind of distance spectrum and so provide direct information about the structure of the molecule studied. In practice, however, certain difficulties prevent rigorous application of equation (10.27), and troublesome deviations of $\sigma(r)/r$ from ideality are always found.

A main difficulty results from the restricted s-range of the experimental data, for an electron diffraction experiment can only provide intensities within a limited interval of s, such as $1\,\text{Å}^{-1}$ to $45\,\text{Å}^{-1}$, and this range falls far short of that required if equation (10.27) is to be applied correctly.

Fourier transformation of this finite amount of data can of course be attempted, but the missing low-s information produces an envelope effect, or deviation of the base-line of the resulting radial distribution curve from the $\sigma(r)/r = 0$ axis, whilst the omission of high-s data causes spurious ripples to extend from the base of each $P_i(r)$ peak.

Since, at the present time, these distortions cannot be avoided completely, it is usual to calculate a damped radial distribution curve according to the equation

$$\sigma(r)/r = k' \int_{s_{\min}}^{s_{\max}} I_{\mathrm{m}}(s)'\, e^{-as^2} \sin(sr)\, ds \tag{10.28}$$

and to choose the damping constant 'a' so that $I_{\mathrm{m}}(s)'$ is reduced to zero at its upper s-limit. In this way the ripple effect can be removed, but at the same time some resolution of the peaks in the radial distribution curve is lost as each $P_i(r)$ function is artificially broadened by the exponential factor.

Since the integral in equation (10.28) has a non-zero lower limit, the envelope effect is also present in the damped radial distribution, and indeed in the early stages of an investigation, there is little that can be done to eliminate this. As information about the molecular structure becomes available, however, it is possible to introduce a set of approximate theoretical intensities for values of s below the lower s-limit, and, by an iterative process, to proceed to a more correct molecular structure, and to more accurate values for these added intensities.

Even at this stage the experimental radial distribution curve can deviate from the simple form predicted by equation (10.27), for in practice, despite modification, many of the coefficients g_i' will vary with s. If light atoms alone are involved, only a slight distortion of the peaks in the radial distribution curve is produced, but if heavy atoms are also present many of the $\cos \Delta\eta_i$ functions will show a strong dependence on s, and certain peaks will be broadened or even split into two components. This effect is difficult to remove by any modification procedure applied to the $I_{\mathrm{m}}(s)$ data, but if some knowledge is available about the molecular structure under investigation, the experimental intensities can be corrected[26] for non-nuclear scattering before Fourier transformation, and improvement of the molecular structure, and of this correction function, achieved in stages.

When all of these steps have been carried out, the resulting radial distribution curve approaches the ideal form specified by equation (10.27), and constitutes a distance spectrum for the molecule studied. Nonetheless it may still be difficult to gain a great deal of structural information from this on account of the overlapping of peaks corresponding to very similar internuclear distances. Instead of being able to pick out distinct $P_i(r)$ functions it is more common to be faced with the problem of trying to resolve a compound peak into Gaussian components, and though calculation[27] of spectroscopic values for the amplitudes of vibration (u_i) can reduce these difficulties in simple cases, resolution problems increase as the size of the molecule investigated increases.

Despite this basic limitation, the radial distribution curve is widely used during structure determination, and in some laboratories structural parameters are calculated directly from it by least squares refinement, whilst in others it is used in conjunction with an approach based on least-squares fitting of the intensity data. In this latter procedure a visual impression of the progress made at each stage of refinement is achieved, and this enables errors in the molecular model to be quickly identified. In cases where rotational isomerism is possible, this is particularly useful, as even a crude radial distribution curve can indicate the presence or absence of a multiplicity of conformers.

10.4.6 *Least-squares refinement*

Least-squares refinement of internuclear distances and vibrational amplitudes is normally achieved by applying equation (10.20) or (10.25) to unmodified or modified molecular intensity data. In modern work it is necessary to have at least two overlapping intensity curves available, and the first step in the calculation is to choose a reasonable starting model for the molecule under investigation. This choice may be based on a preliminary radial distribution curve, on chemical experience, or on results for related systems, and once it has been made, a number of distances and angles can be selected as refinable parameters, and the remaining distances formulated in terms of these according to the symmetry requirements of the model assumed. Though commonly applied, this procedure is not entirely valid, as there is no reason why r_a or r_g distances should conform exactly to the geometrical constraints of the equilibrium structure, and indeed, if certain perpendicular amplitudes of vibration are large, the centres of gravity of outer peaks in the radial distribution curve may not conform geometrically to those of the inner ones. In such a situation shrinkage effects are involved, and must be introduced as refinable quantities, or as constraints calculated from spectroscopic data.

It is also necessary, when starting refinement, to deal satisfactorily with the scale differences existing between the different data sets, and to weight each intensity measurement correctly. Scaling is usually achieved by introducing a number of scale factors into the refinement, or by putting all data curves on a common scale before refinement, but realistic weighting of the data is more difficult. Until recently, it was common to assume a diagonal weight matrix defined by a trapezoidal function of s,[28] but more reliable values for parameters and standard deviations have been obtained[29] by employing non-diagonal matrices determined by detailed analysis of the data spreads obtainable from a particular apparatus and microphotometer. It has in fact been demonstrated that standard deviations obtained by this latter procedure, which takes into account correlations of adjacent intensity measurements, are often two or three times those calculated by the older approach.

For molecules of low rigidity, such as those capable of loose torsional motions or internal rotation, least-squares analysis of the intensity data is still possible, but equations (10.20) and (10.25) must be abandoned, and a somewhat more sophisticated approach adopted. This modified procedure is discussed at length in the section which follows.

10.5 Electron diffraction and internal rotation

10.5.1 *Introduction*

When a molecule is capable of hindered internal rotation, its study by electron diffraction often involves the application of theoretical equations of a more complex type than those appropriate to a semi-rigid species. The exact treatment required usually depends on the nature of $V(\phi)$, the hindering potential energy function, that is, on the number of non-equivalent minima in this function, and on the heights of the intervening potential energy barriers.

The simplest situation arises when $V(\phi)$ contains several minima, separated by high barriers, for if these are all equivalent the molecule behaves as a single semi-rigid structure, whilst if they are not, the gas sample can be assumed to contain a number of different semi-rigid molecules. In this latter case the only new difficulty is to calculate the proportions of the isomers present, that is the weights which must be given to the various intensity curves added together to fit the experimental data. Often this problem is solved by measuring the relative areas of peaks in the outer part of the radial distribution curve, but more commonly trial-and-error, or least-squares, variation of the isomer proportions is attempted.

When known, these proportions provide useful thermodynamic information, for if, for example, two species A and B are involved, and n_A and n_B

are the numbers present at equilibrium, then the free-energy difference between A and B at absolute zero, can be calculated from the formula[30]

$$n_A/n_B = m_A/m_B(f_A^{vib}f_A^{rot}/f_B^{vib}f_B^{rot})\exp(-\Delta G^0/RT) \qquad (10.29)$$

in which m_A and m_B are the numbers of potential energy minima corresponding to each isomer, that is their multiplicities, and the quantities in parentheses are their partition functions for rotation and vibration.

As barrier heights fall, however, the study of rotational isomerism by electron diffraction becomes more difficult, as each of the isomers involved may develop a considerable flexibility, and the molecular intensity expression applied to semi-rigid structures ceases to be valid. In many cases it is still possible to think in terms of a mixture of individual molecules, but each of these may exhibit a torsional mode of motion of large amplitude, the effect of which will be to broaden or to introduce shrinkage effects into peaks in the radial distribution curve.

When barriers are very low, a situation approaching free rotation occurs, and it is no longer realistic to describe the vapour in terms of distinct rotamers. In this case the theoretical intensity and radial distribution functions are best calculated by integration round the circle of rotation, that is by forming a weighted sum of contributions from a large number of semi-rigid conformers corresponding to all possible values for the angle of rotation about the central axis. In such calculations the weight function which is required, $w(\phi)$, is a constant for free rotation only, and is otherwise determined by the hindering potential energy $V(\phi)$.

10.5.2 *Molecular intensity and radial distribution functions for a molecule executing hindered or free internal rotation*

In Figure 10.4 the two atoms A and B may change their separation as a result of an angular displacement ϕ from the position of equilibrium defined

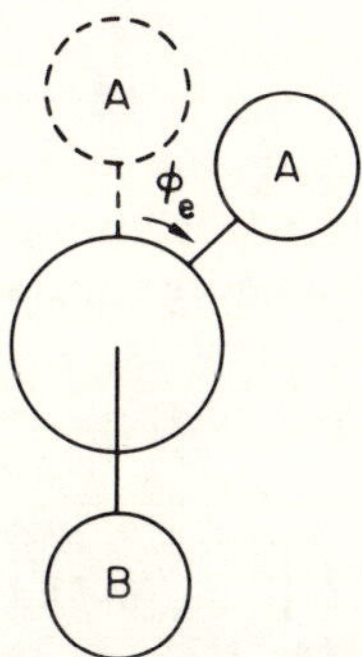

Figure 10.4   The dihedral angle of an AOOB system as measured
from the *trans* configuration

by ϕ_e. If this latter angle is measured from the *trans* configuration it is easy to prove that

$$r_e = [c_1 + c_2 \cos(\phi_e)]^{\frac{1}{2}} \tag{10.30}$$

where $c_1 = \frac{1}{2}(l_2^2 + l_1^2)$, $c_2 = \frac{1}{2}(l_2^2 - l_1^2)$, and l_1 and l_2 are the minimum and maximum separations of A and B during one revolution.

For a displacement ϕ from equilibrium, it is clear that r_e changes to a new value r_o defined by the equation

$$r_o(\phi) = [c_1 + c_2 \cos(\phi_e + \phi)]^{\frac{1}{2}} \tag{10.31}$$

and if the torsional motion involving ϕ is much slower than the other molecular vibrations, it may be treated as separate from these, and each $r_o(\phi)$ assigned a probability distribution function of the form

$$\pi(r)\,dr = \sqrt{(h(\phi)/\pi)} \cdot \exp\left\{-h(\phi)[r_o(\phi) - r]^2\right\} dr \tag{10.32}$$

arising from the rapid harmonic oscillations of the molecular frame. In this equation $h(\phi)$ is related to the frame vibrational amplitude of distance r_{AB} by the expression

$$h(\phi) = 1/2u_{\text{fr}}^2(\phi) \tag{10.33}$$

and is normally a slowly-varying function which can be calculated from force-constant data.

The displacement angle ϕ can also be assigned a probability distribution $w(\phi)$ dependent on the hindering potential energy $V(\phi)$, and in view of the assumed separability of the frame and torsional vibrations, an overall probability distribution can be calculated for the distance r_{AB} by the integral

$$P(r)\,dr = \int_0^{2\pi} \pi(r)w(\phi)\,dr\,d\phi$$

$$\tag{10.34}$$

$$= \int_0^{2\pi} \sqrt{[h(\phi)/\pi]} \cdot \exp\left\{-h(\phi)[r_o(\phi) - r]^2\right\}w(\phi)\,dr\,d\phi$$

This last result is of basic significance as it can be used to formulate the contribution may by an atom pair to the total molecular intensity, for if, as in the previous section, $g(s)$ depends on the scattering factors of the atoms A and B, and k is a proportionality constant:

$$i_m(s)_{AB} = k \cdot g(s) \int_{-\infty}^{\infty} P(r)[\sin(sr)/sr]\,dr$$

$$= k \cdot g(s) \int_0^{2\pi} w(\phi)\,d\phi \int_{-\infty}^{\infty} \sqrt{[h(\phi)/\pi]}\,[\sin(sr)/sr]$$

$$\times \exp\left\{-h(\phi)[r_o(\phi) - r]^2\right\} dr$$

$$= k \cdot g(s) \int_0^{2\pi} \{\sin\,[sr_0(\phi)]/sr_0(\phi)\} \exp\,[-s^2/4h(\phi)]w(\phi)\,\mathrm{d}\phi$$

$$= k \cdot g(s) \int_0^{2\pi} \{\sin\,[sr_0(\phi)]/sr_0(\phi)\} \exp\,[-\tfrac{1}{2}u_{\mathrm{fr}}^2(\phi)s^2]w(\phi)\,\mathrm{d}\phi \qquad (10.35)$$

and is clearly a summation of intensity contributions from a continuous series of frame structures each receiving a weight $w(\phi)$.

In a similar way the probability distribution function for r_{AB} can be used to calculate the peak $t(r)$ which this distance produces in the experimental radial distribution curve, for if $P(\rho)\,\mathrm{d}\rho$ is substituted into

$$t(r) = kg'\sqrt{(\pi/a)}/4 \int_{-\infty}^{\infty} [P(\rho)/\rho] \exp\,[-(\rho - r)^2/4a]\,\mathrm{d}\rho \qquad (10.36)$$

the effect of damping the intensity data before Fourier transformation is taken into account, and integration gives the result

$$t(r) = k\pi g'/2 \int_0^{2\pi} \sqrt{[H(\phi)/\pi]}[r_0(\phi)]^{-1}\exp\,\{-H(\phi)[r_0(\phi) - r]^2\}w(\phi)\,\mathrm{d}\phi \qquad (10.37)$$

which is also a weighted sum of frame contributions. In this latter expression

$$H(\phi) = h(\phi)/[4ah(\phi) + 1] \qquad (10.38)$$

and 'a' is the damping constant.

When attempting to apply equations (10.35) and (10.37) integration is normally carried out numerically, by means of an electronic computer, as analytical expressions are usually approximate, and cumbersome, even for simple forms of $w(\phi)$. In all cases the latter is determined by the potential energy $V(\phi)$, and if barrier heights are large, a full quantum-mechanical calculation may be necessary. For intermediate and low barriers, however, and moderate gas temperatures, the classical equation

$$w(\phi) = N \exp\,[-V(\phi)/kT] \qquad (10.39)$$

is normally assumed, N being a normalization constant, and T the temperature in question. Thus it is common when treating a molecule with a barrier height of 2 kcal/mole or less to substitute a cosine potential of the general form

$$V(\phi) = \sum_i (V_i/2)(1 - \cos i\phi) \qquad (10.40)$$

into equation (10.39), and to vary certain of the constants V_i until a best fit to the experimental data is achieved.

Some practical illustrations of the foregoing theory will now be considered.

10.6 Electron diffraction studies of ethane derivatives and related molecules

10.6.1 *Compounds with only one staggered conformation*

For a derivative of ethane, or a similar molecule, to be included in this category at least one of its two constituent rotors must retain three-fold symmetry, as this is necessary if the potential energy function restricting internal rotation is to have three identical minima corresponding to equivalent staggered conformations.

Because of this equivalence such a molecule behaves as a single species when studied by electron diffraction, and if the potential energy barrier V_0 is high, the usual treatment appropriate to a semi-rigid structure may be applied. When this barrier falls, however, an increasing amplitude of torsional motion develops, and this growing flexibility may have a sufficiently large effect on the scattered intensity data to allow V_0 to be estimated quantitatively.

Information of this kind has in fact been obtained for a number of simple examples including hexafluoroethane, C_2F_6,[31] hexachloroethane, C_2Cl_6,[32,33] and hexachlorodisilane, Si_2Cl_6,[31,33] and in this section results published for the last of these will be considered in detail, as hexachlorodisilane has been studied by two different approaches, which, when considered together, illustrate most of the principles involved in determinations of this sort.

The first electron diffraction investigation of this molecule was carried out by Swick and Karle[31] in 1955, and the experimental radial distribution curve determined by these authors is reproduced in Figure 10.5. It contains

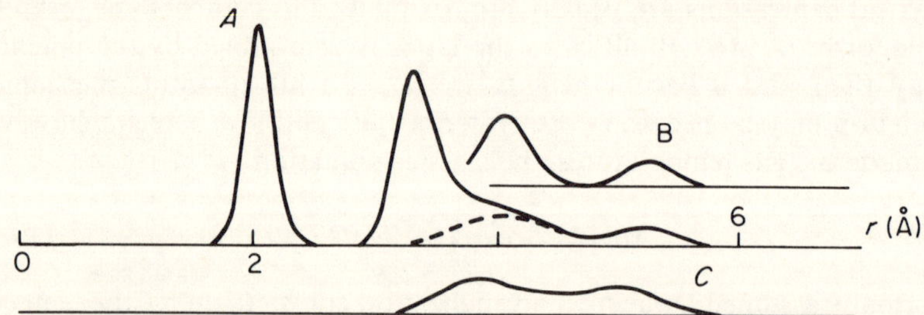

Figure 10.5　　The experimental radial distribution curve, A, for Si_2Cl_6, and two theoretical curves, B and C, as published by Swick and Karle[31]

four peaks, one at 2·0 Å, corresponding to the two types of bonded distance, one at 3·3 Å, corresponding to the edges of the Si_2Cl_3 tetrahedra, and two at 4·1 and 5·2 Å, corresponding to the *gauche* and *trans* Cl ... Cl distances characteristic of a staggered conformation.

These latter two peaks are of basic importance when calculating the barrier height, for increased torsional oscillation about the central axis has a marked effect upon them, the *gauche* peak being substantially broadened, and the *trans* peak distorted from a Gaussian to a skew-Gaussian form. The first of these effects is well demonstrated by the two theoretical curves included in the figure, for in the upper of these, which was calculated on the assumption of a high barrier to internal rotation, the *gauche* and *trans* peaks are well defined, and have virtually the same width, whereas in the lower, corresponding to free rotation, resolution of the Cl . . . Cl distances is almost completely lost. It is clear that the experimental curve lies between these extremes, and that the barrier to internal rotation in hexachlorodisilane is low, but significantly different from zero.

One way of determining this barrier quantitatively is to carry out an analytical treatment of the effect of twisting vibration on the radial distribution curve of a molecule of this type. Such a treatment has been published by Karle,[9,12] and rests on the assumption that the torsional motion is of a sufficiently low frequency to be considered separate from the other molecular vibrations. If this assumption is made, equation (10.37) can be applied to the problem of calculating peaks in the radial distribution curve corresponding to the *gauche* and *trans* distances.

The first step is to define $V(\phi)$, the potential energy function restricting rotation about the central axis, and from it to estimate $w(\phi)$ by classical or quantum-mechanical means. This is particularly simple for the case of a twisting vibration, for the angle of displacement ϕ remains small throughout, and the minimum of the usual cosine potential can be approximated by the quadratic expression

$$V(\phi) = K\phi^2 \tag{10.41}$$

in which

$$K \sim 9V_0/4 \tag{10.42}$$

if $V(\phi)$ is regarded as having been derived from the complete cosine form

$$V(\phi) = V_0(1 - \cos 3\phi)/2 \tag{10.43}$$

by expansion of the latter in powers of ϕ.

For the particular case of a quadratic potential energy function, $w(\phi)$ can be calculated quantum mechanically by solving the Schrödinger equation

$$(-h^2/8\pi^2 I)\frac{d^2\psi}{d\phi^2} + [E - V(\phi)]\psi = 0 \tag{10.44}$$

in which I is the reduced moment of inertia equal to $I_1 I_2/(I_1 + I_2)$, and I_1 and I_2 are the individual moments of inertia of the two rotors about the central axis. By squaring the wave-functions obtained in this way, and summing them according to the Boltzmann equation, the angular probability distribution function

$$w(\phi) = \sqrt{(b^2/\pi)}\, e^{-b^2\phi^2} \tag{10.45}$$

may be derived, in which

$$b^2 = \alpha \tan h(h\nu/2kT) = 1/2\langle\phi^2\rangle \tag{10.46}$$

$$\alpha^2 = 8K\pi^2 I/h^2 \tag{10.47}$$

T is the gas temperature, and

$$\nu^2 = K/2\pi^2 I \tag{10.48}$$

Of these quantities the last is the square of the frequency of the twisting vibration, and if this is sufficiently low that $h\nu \ll kT$

$$b^2 \sim K/kT \tag{10.49}$$

and

$$w(\phi) = \sqrt{(K/\pi kT)} \exp\left[-K\phi^2/kT\right] = N \exp\left[-V(\phi)/kT\right] \tag{10.50}$$

which is the usual classical relationship between $w(\phi)$ and the potential energy $V(\phi)$. From equations (10.42), (10.46) and (10.49) it also follows that

$$b^2 \sim 9V_0/4kT \tag{10.51}$$

and

$$\langle\phi^2\rangle^{\frac{1}{2}} \sim \sqrt{(2kT/9V_0)} \tag{10.52}$$

the latter being the root-mean-square amplitude of angular displacement from the staggered configuration, and a quantity potentially available from electron diffraction data.

Substitution of equation (10.45) into the general formula for a peak $t(r)$ in the radial distribution curve, that is into equation (10.37), gives the result

$$t(r) = k(bg'H^{\frac{1}{2}}/2r_e)\int_{-\infty}^{\infty} \exp\left\{-H[r_0(\phi) - r]^2 - b^2\phi^2\right\} d\phi \tag{10.53}$$

if the limits of integration in (10.37) are changed to correspond to the Gaussian nature of $w(\phi)$, $r_0(\phi)$ in the denominator is replaced by r_e, and the dependence of H on ϕ is neglected. In this expression k is a proportionality constant, and g' depends on atomic scattering factors.

It has been shown[12,31] that if V_0 is large enough to ensure that b is greater than three, equation (10.53) may be manipulated to give approximate forms for peaks in the radial distribution curve corresponding to the *gauche* ($\phi_e = 120°$) and the *trans* ($\phi_e = 0°$) distances. These results are

$$t(r)_{gauche} = k' \exp\left\{-(H/[1 + 3c_2^2H/16r_e^2b^2])(r_e - r)^2\right\} \tag{10.54}$$

and

$$t(r)_{trans} = k' \exp\left[-H(r_e - c_2/8r_e b^2 - r)^2\right] \qquad (10.55)$$

where k' is a proportionality constant, c_2 is defined by equation (10.30), and H is related to the frame vibrational amplitude and the damping constant 'a' of the radial distribution curve by the expression

$$H = 1/2(2a + u_{fr}^2) \qquad (10.56)$$

which follows from equations (10.33) and (10.38).

Examination of equation (10.54) shows that the *gauche* distribution is Gaussian about r_e but has a larger width than would be expected if frame vibrations alone were involved; that is the total amplitude of vibration, u_{tot}, is related to H by the equation

$$1/2u_{tot}^2 = H/(1 + 3c_2^2 H/16r_e^2 b^2)$$

or more explicitly

$$u_{tot}^2 = 1/2H + 3c_2^2/32r_e^2 b^2$$

and by assuming equations (10.51) and (10.56) it follows that

$$u_{tot}^2 = (2a + u_{fr}^2) + c_2^2 kT/24r_e^2 V_0 = A + B/V_0 \qquad (10.57)$$

Thus the square of the total vibrational amplitude of the *gauche* distance can be written linearly in terms of a frame contribution and a torsional contribution dependent on V_0 the barrier height.

The effect of torsional vibration on the *trans* distribution is quite different, however, as is shown by equation (10.55), for there is no broadening effect, only a tendency for the centre of gravity of this peak to be shifted by an amount δ to a lower r value, this shrinkage effect being related to the barrier height by the equation

$$\delta = c_2/8r_e b^2 = c_2 kT/18r_e V_0$$

It is clear from the above analysis that measurement of u_{tot} for the *gauche* distance, and δ for the *trans*, both give information about the barrier height, but in applying equation (10.57) it is necessary to have available a value for the frame contribution to the overall amplitude of vibration, otherwise the maximum information which can be obtained is a relationship between u_{fr} and V_0.

This limitation was in fact experienced by Swick and Karle during their investigation of hexachlorodisilane, for they found it difficult to measure δ accurately, and so based their entire determination of V_0 on the experimental amplitude of vibration of the *gauche* distance. This was estimated to be 0.35 ± 0.05 Å, and, by substituting this result into equation (10.57), it was possible to plot the relationship between the barrier height V_0 and the contribution u_{fr}, of frame vibrations, to the total broadening of the *gauche* peak. The graph obtained in this way is reproduced in Figure 10.6 and shows that if the frame contribution is close to zero, V_0 has a minimum value of around 700 cal/mole, whilst if frame vibrations account for the entire amplitude of the *gauche* distance V_0 is very large. The true situation naturally lies between these extremes, and it was suggested by Swick and Karle[31] that if u_{fr} can be approximated by the experimental vibrational amplitude of the *trans* distance, the barrier height in hexachlorodisilane must be close to 1 kcal/mole.

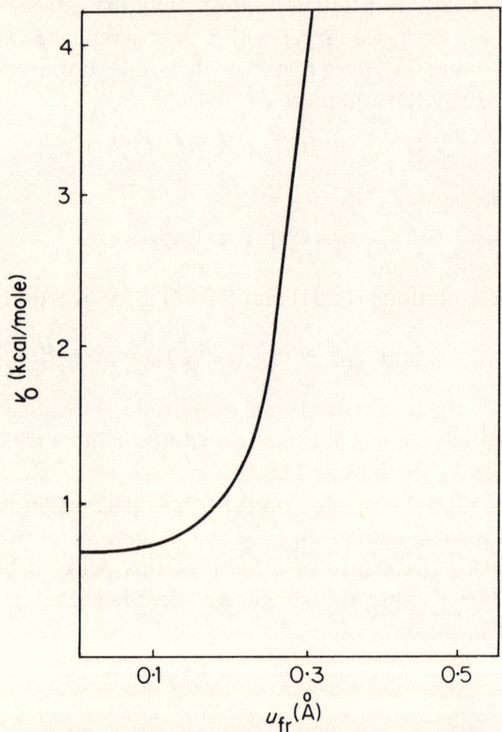

Figure 10.6 The functional relationship between the barrier height V_0, and the framework amplitude of vibration of the *gauche* Cl$\cdots$Cl distance in Si_2Cl_6, as determined by Swick and Karle[31]

In 1958 Morino and Hirota[33] reinvestigated this molecule and determined the barrier height by a somewhat different procedure from that just described. In this second approach hexachlorodisilane was treated as a low-barrier problem and theoretical intensity and radial distribution curves were calculated by applying equations (10.35) and (10.37) of the previous section. When computing these integrals, $w(\phi)$, the angular probability distribution was related classically by equation (10.39) to the simple cosine potential $(V_0/2)(1 - \cos 3\phi)$, and the framework amplitude of vibration was calculated spectroscopically, and given its proper dependence on ϕ. This latter function is reproduced in Figure 10.7 and was determined by normal coordinate analysis assuming separability of the torsional and frame motions of the system. By choosing different values for V_0, until a good fit to the experimental intensity and radial distribution curves was achieved, it was possible to show that the barrier to internal rotation in hexachlorodisilane is roughly 1 kcal/mole, a result in complete agreement with the earlier estimate made by Swick and Karle.[31]

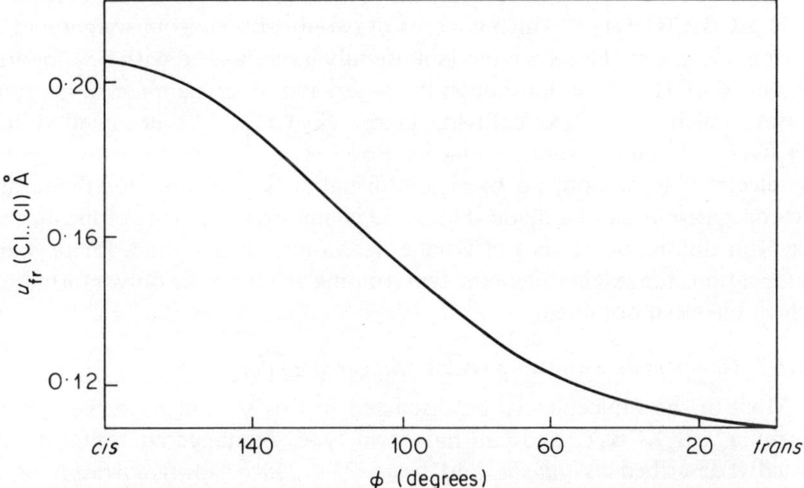

Figure 10.7 The dependence of the framework amplitude of vibration of the rotation-dependent $Cl\cdots Cl$ distances in Si_2Cl_6 on the dihedral angle ϕ, as calculated from spectroscopic data by Morino and Hirota[33]

This agreement seems to suggest that the vibrational amplitude approach may be applied with success even when V_0 is low, and hence its application to molecules with higher barriers may be attempted with some confidence. One such example is hexafluoroethane for which Swick and Karle[31] have determined a barrier height of 4 kcal/mole by applying the same procedure as was adopted for hexachlorodisilane; other examples are hexachloroethane[33] and trichloromethyltrichlorosilane, CCl_3SiCl_3,[33] for which barrier heights of 10·8 and 4·3 kcal/mole have been determined by Morino and Hirota by an improved approach involving calculation of frame vibrational amplitudes from force-constant data.

These last two results may be usefully compared with the barrier of 1 kcal/mole appropriate to hexachlorodisilane, for in the series so formed, the steady fall of V_0 demonstrates how important a factor the spatial separation of a pair of rotating groups is in determining the hindering potential energy function. The truth of this may also be seen by comparing electron diffraction results for hexafluoroethane and hexafluorobut-2-yne, CF_3CCCF_3,[34] for although the former has a barrier of around 4 kcal/mole, the two trifluoromethyl groups in the latter appear to undergo almost free rotation presumably on account of their greatly increased separation. It is also of interest that spectroscopic work has revealed a very similar relationship between the parent molecules ethane, C_2H_6 (V_0 equal to 3 kcal/mole),[35] and but-2-yne, CH_3CCCH_3 (V_0 less than 0·1 kcal/mole).[36]

Another molecule which has been shown to undergo virtually unhindered rotation by an electron diffraction experiment is di-*t*-butylberyllium, $(CH_3)_3CBeC(CH_3)_3$,[37] which consists of two di-*t*-butyl groups separated by a long linear axis. This system could usefully be compared with hexamethylethane, $C_2(CH_3)_6$,[38] but although the latter, and other members of the same series, such as hexamethyldisilane, $Si_2(CH_3)_6$,[39] hexamethylditin, $Sn_2(CH_3)_6$,[40] and hexamethyldilead, $Pb_2(CH_3)_6$,[41] have been investigated by electron diffraction, no barrier information is available for them, and no comparisons can be made. The same is unfortunately true of most other electron diffraction studies of ethane derivatives of this kind, for in many publications the staggering of the two rotating groups is the only information which has been obtained.

10.6.2 *Compounds with two possible staggered conformations*

Most of the molecules to be discussed in this section have the general formula, CB_2XCB_2Y, and can have two types of staggered conformation usually described as *gauche* and *trans*. The *gauche* form corresponds to an XCCY dihedral angle of roughly 60°, and has a multiplicity of two, whilst the *trans* species allows maximum separation of the atoms or groups X and Y and has a multiplicity of one. A number of examples have been studied including the following.

10.6.2.1 *n-Butane, n-propyl chloride, and 1,2-dichloroethane*   These molecules form an interesting series, since a chlorine atom and a methyl group are sterically rather similar, and gas samples of these compounds, at the same temperature, might be expected to contain similar proportions of the *gauche* and *trans* isomers, the *trans* form being likely to be the more stable, on account of the reduced non-bonded interaction it allows.

All three molecules have been studied by electron diffraction,[28,42–45] and a summary of the information available for them is given in Table 10.1. From this table it is clear that the expectations mentioned above are not wholly fulfilled, for whereas the *trans* forms of n-butane, $(CH_3)CH_2CH_2(CH_3)$, and 1,2-dichloroethane, $ClCH_2CH_2Cl$, are indeed several hundred calories per mole more stable than the *gauche*, the *gauche* species is the more abundant, and is the more stable isomer of n-propyl chloride $(CH_3)CH_2CH_2Cl$,

Table 10.1   Electron diffraction results for n-butane, n-propyl chloride and 1,2-dichloroethane. Error limits where published are in parentheses

	$CH_2(CH_3)CH_2(CH_3)$		$CH_2(CH_3)CH_2Cl$	CH_2ClCH_2Cl	
Reference	42	43	44	45	28
T(°C)	14	—	20	22	20
% *gauche*	40(15)	35–40	81(5)	27(5)	27(7)
% *trans*	60(15)	65–60	19(5)	73(5)	73(7)
More stable isomer	*trans*	*trans*	*gauche*	*trans*	*trans*
ΔG^0 (kcal/mole)	0·63(0·35)	0·65	0·3(0·2)	0·89	0·89
ΔE^0 (other sources[46–48]) (kcal/mole)	0·8	0·8	0·05(0·15)	1·14	1·14

there being apparently some kind of attraction between the methyl group and the chlorine atom.

In addition to these determinations of isomer proportions, barrier information has also been obtained for n-butane, for by calculating the root-mean-square amplitude of angular displacement of the *trans* rotamer from a true 180° form, Bonham and Bartell[42] were able to make a very rough estimate of 3 kcal/mole for the height of the barrier restricting internal rotation from the *trans* to *gauche* configurations.

The existence of two such conformations in samples of n-butane is particularly well demonstrated by a radial distribution curve published for this molecule by Kuchitsu.[43] This is reproduced in Figure 10.8, and was calculated from the experimental data by subtracting out contributions

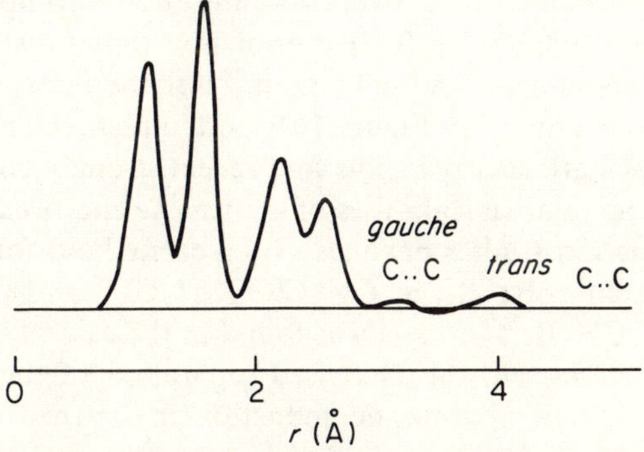

Figure 10.8   A modified experimental radial distribution curve for n-butane as published by Kuchitsu.[43] This shows the presence of both *gauche* and *trans* isomers

from non-bonded distances involving hydrogen atoms, for in this way the peaks corresponding to the *gauche* and *trans* carbon–carbon distances were exposed and the extra stability of the *trans* isomer demonstrated.

In Table 10.1 energy differences obtained for n-butane, n-propyl chloride and 1,2-dichloroethane, by other physical methods,[46–48] are also included, and these show a similar trend to that suggested by the electron diffraction results.

10.6.2.2 *1,2-Difluoroethane, 1,2-dichloroethane and 1,2-dibromoethane*   Results for this series have been published in an electron diffraction review article by Almenningen and coworkers[28] and are summarized in Table 10.2.

Table 10.2   Electron diffraction results for 1,2-difluoroethane,
1,2-dichloroethane and 1,2-dibromoethane. Error limits where
published are in parentheses

	CH_2FCH_2F	CH_2ClCH_2Cl	CH_2BrCH_2Br
Reference	28	28	28
T(°C)	20	20	20
% gauche	>67	27(7)	11(5)
% trans	<33	73(7)	89(5)
More stable isomer	gauche	trans	trans
ΔG^0 (kcal/mole)	0·59–1·42	0·89	1·63

This table reveals a trend in the relative stabilities of the *gauche* and *trans* isomers which is easily understood in terms of the sizes of the three halogen substituents. Thus, in samples of 1,2-difluoroethane, FCH_2CH_2F, the *gauche* rotamer predominates, and appears to be slightly more stable than the *trans*, whilst the bulk of the chlorine and bromine atoms ensures an excess of the *trans* forms in samples of 1,2-dichloro- and 1,2-dibromoethane, that is $ClCH_2CH_2Cl$ and $BrCH_2CH_2Br$. Experimental radial distribution curves published by Almenningen and coworkers[28] for the fluorine and bromine compounds are reproduced in Figure 10.9, and in these the *gauche* and *trans* interhalogen peaks are clearly visible and reveal strongly contrasting situations regarding the relative stabilities of the *gauche* and *trans* isomers.

Electron diffraction studies have also been carried out for the analogous acetylenes, 1,4-dichlorobut-2-yne, $CH_2ClCCCH_2Cl$,[49] and 1,4-dibromobut-2-yne, $CH_2BrCCCH_2Br$,[50] and as was found in the case of the symmetrical ethanes already discussed, the increased separation of the end-groups in these molecules allows a dramatic reduction in the restricting potential energy barrier. The investigation of the chlorine compound has established this point particularly clearly, for intensity data collected at low scattering angles was subjected to the same kind of low-barrier analysis described for hexachlorodisilane, and shown to be consistent with almost completely unhindered internal rotation of the CH_2Cl end-groups.

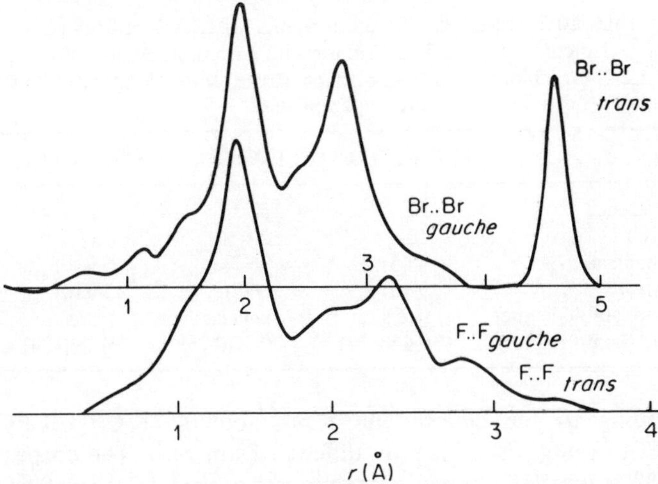

Figure 10.9 Experimental radial distribution curves for 1,2-difluoro- and 1,2-dibromoethane, as published by Almenningen and coworkers.[28] These show the presence and relative amounts of *gauche* and *trans* isomers in samples of these compounds

10.6.2.3 *Ethylene fluorohydrin, chlorohydrin, cyanohydrin and ethyleneglycol* The series ethylene fluorohydrin, CH_2FCH_2OH, ethylene chlorohydrin, CH_2ClCH_2OH, and ethylene cyanohydrin, CH_2CNCH_2OH, was examined by electron diffraction in 1956.[51-53] In each case the *gauche* isomer was found to predominate, a fact which was rationalized in terms of intramolecular hydrogen bonding. More recently, one of these molecules, ethylene chlorohydrin, has been reinvestigated,[54] and this modern study has confirmed that at 37°C around 90 % of the molecules present have the *gauche* conformation. This figure implies that the *gauche* species is some 2·6 kcal/mole more stable than the *trans*.

A similar property has been found for ethylene glycol, CH_2OHCH_2OH,[55] for recent electron diffraction work has suggested that at 100°C the proportion of the *gauche* species is around 80%. Once again intramolecular hydrogen bonding provides a likely explanation for this preference.

10.6.2.4 *1,2-Dichloro-1,1,2,2-tetrafluoroethane, 1,1,2-trifluoro-1,2,2-trichloroethane and 1,2-difluoro-1,1,2,2-tetrachloroethane* This series, CF_2ClCF_2Cl, $CF_2ClCFCl_2$ and $CFCl_2CFCl_2$, was thoroughly examined by electron diffraction[56-59] between the years 1957 and 1959, the main object being to determine isomer proportions for each compound, and so compare free-energy differences with corresponding results for the series CH_2ClCH_2Cl, $CH_2ClCHCl_2$ and $CHCl_2CHCl_2$, in which hydrogen replaces fluorine.

Results for the fluorine derivatives are listed in Table 10.3, and from this summary it is clear that all free-energy differences are small, the values

Table 10.3   Electron diffraction results for 1,1,2,2-tetrafluoro-
1,2-dichloroethane, 1,1,2-trifluoro-1,2,2-trichloroethane and
1,1,2,2-tetrachloro-1,2-difluoroethane. Error limits where published
are in parentheses

	CF_2ClCF_2Cl	$CF_2ClCFCl_2$	$CFCl_2CFCl_2$
Reference	57	58	56
T(°C)	10	20	15
% *gauche*	48(5)	76(7)	55(10)
% *trans*	52(5)	24(7)	45(10)
More stable isomer	*trans*	*gauche*	*trans*
ΔG^0 (kcal/mole)	0·44(0·11)	0·27(0·25)	0·28(0·24)

corresponding to the *gauche* and *trans* isomers of $CF_2ClCFCl_2$ and
$CFCl_2CFCl_2$ being insignificantly different from zero. The corresponding
results[46,60–62] for the hydrogen-containing series are 1·2, 2·9, and 0 kcal/mole,
respectively, and are for the most part considerably larger, a fact which has
been explained[59] in terms of electrostatic, rather than steric interaction, on
account of the rather similar steric behaviour of hydrogen and fluorine atoms.

10.6.2.5 *Derivatives of hydrazine and biphosphine*   These compounds are
analogous to the ethane derivatives discussed above if lone pairs of electrons
are treated as substituents.

Both parent molecules have been investigated by electron diffraction,[63,64]
but in view of the low scattering power of the hydrogen atom, no accurate
information was obtained about their conformational properties. From
high-resolution infrared data[65,66] it has been concluded, however, that
gauche isomers predominate, one half of the molecule being rotated through
roughly 90% from the *cis* configuration.

A number of hydrazine derivatives have also been studied by electron
diffraction. Thus, tetrafluorohydrazine, N_2F_4,[67] was investigated at a
temperature somewhere in the range $-120°C$ to $-20°C$ and shown to
contain almost equal amounts of *gauche* and *trans* isomers, whilst tetra-
silylhydrazine, $N_2(SiH_3)_4$,[68] and tetrakis(trifluoromethyl)hydrazine,
$N_2(CF_3)_4$,[69] both have planar configurations at nitrogen, and exist mainly
as *gauche* species with the two planar groups at right angles. A preferred
gauche geometry has also been found[70] for 1,1-dimethylhydrazine, though
the exact conformation of this molecule was difficult to determine on account
of the low scattering power of the hydrogen atoms in the NH_2 group.

Much less information is available for substituted biphosphines, tetra-
methylbiphosphine being the only example so far studied by the diffraction
method.[71] It appears that in gas samples between 150 and 200°C this molecule takes up a *trans* conformation, a finding which agrees well with spectroscopic results[72–74] for the series of halogen analogues, P_2F_4, P_2Cl_4 and
P_2Br_4.

Electron diffraction data are also available[75,76] for the mixed compounds, dimethylaminodichlorophosphine, $(CH_3)_2NPCl_2$, and dimethylaminodifluorophosphine, $(CH_3)_2NPF_2$. The first of these seems to have a planar configuration at nitrogen and an overall *trans* conformation, whilst the second is non-planar in this respect, and adopts a *gauche* geometry.

10.6.2.6 *Derivatives of hydrogen peroxide and disulphane* Both parent molecules, H_2O_2 and H_2S_2, have been investigated by the diffraction technique,[77,78] but in neither case was conformational information reliably obtained. From high-resolution infrared work[79,80] it seems likely, however, that they take up configurations in which one bond to hydrogen is situated roughly at right angles to the other.

This preference also appears to apply to derivatives of these molecules, for electron diffraction investigations have indicated similar conformations for sulphur monochloride, S_2Cl_2,[81,82] and monobromide, S_2Br_2,[81] dimethyl disulphide, $(CH_3)_2S_2$,[83] and dimethyl diselenide, $(CH_3)_2Se_2$,[84] as have microwave studies of O_2F_2, dioxygen difluoride,[85] and sulphur monofluoride, S_2F_2.[86]

No barrier information was obtained by these diffraction studies but it is to be expected that the hindering potential energy function will have two equivalent minima situated at 90° and 270° from the *cis* position, and barriers at 0° and 180°.

10.6.3 *Compounds with three possible staggered conformations*

Many compounds with three possible staggered conformations exist. A good example of this type of molecule is *s*-butyl chloride, $CH_2(CH_3)CH(CH_3)Cl$, a compound which has been thoroughly investigated[87] by electron diffraction, and shown to exist in three conformations in the gas phase. These are shown in projection in Figure 10.10, and are labelled *TG*, *GT*, and *GG*, the first symbol referring to the chlorine atom, *G* standing for the word *gauche*, and *T* meaning *trans*.

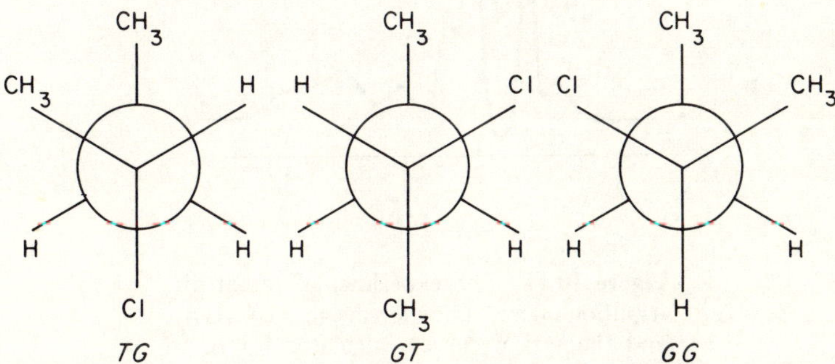

Figure 10.10 The three staggered conformations possible for *s*-butyl chloride

It was found experimentally, that at 25°C the *GT* conformer is present in excess (48 %), whilst the other two are present in roughly equal proportions. These figures imply that the *GT* form is some 400 cal/mole more stable than the other two, a fact which may possibly be a result of the maximum separation of the methyl groups which this conformation achieves and the *gauche* situation of the chlorine atom relative to one of the methyl groups.

10.7 Electron diffraction studies of sandwich compounds

10.7.1 *Ferrocene, ruthenocene and nickelocene*

An early electron diffraction study of ferrocene vapour, $(C_5H_5)_2Fe$,[88] at 400°C, confirmed the sandwich structure expected for this molecule and indicated almost free internal rotation of the two five-membered rings. In 1965 Bohn and Haaland[89] carried out a reinvestigation, this time at 140°C, and showed that both an eclipsed D_{5h} model, and one based on free rotation, could be made to fit the experimental data, whilst a staggered D_{5d} model could not (see Figure 10.11). These authors concluded that a low barrier, probably less than 1 kcal/mole, opposes internal rotation.

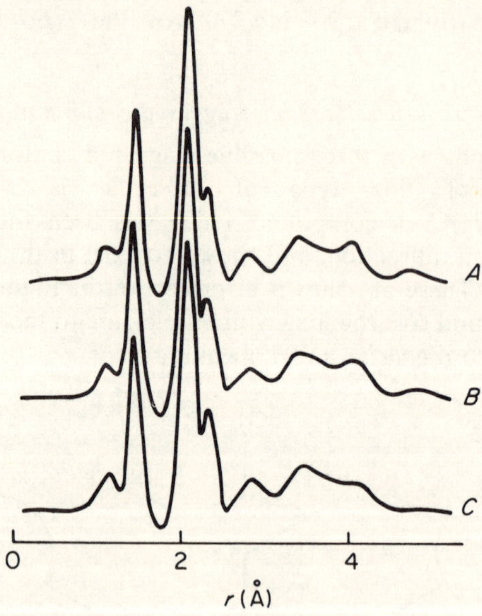

Figure 10.11   The experimental radial distribution curve, *A*, for ferrocene, $Fe(C_5H_5)_2$, and theoretical curves calculated for *B*, an eclipsed, and *C*, a staggered model, by Haaland and Nilsson[90]

Later, in 1968, Haaland and Nilsson[90] re-examined the 1965 data with a view to making a more accurate assessment of the barrier height, and, as a first step, tested the validity of assuming a classical relationship between the angular probability distribution $w(\phi)$ and the cosine potential

$$V(\phi) = V_0(1 - \cos 5\phi)/2 \tag{10.58}$$

applicable to the ferrocene problem. To examine the reliability of this approximation, equation (10.58) was substituted into the same Schrodinger equation (equation 10.44) assumed earlier for hexachlorodisilane, and the wave-functions, $\psi_i(\phi)$, and energy levels, E_i, calculated by a suitable mathematical procedure. From these results $w(\phi)$ was determined quantum-mechanically for a series of choices of V_0, increasing up to 7 kcal/mole, by carrying out the summation

$$w(\phi) = N \sum_{i=1}^{50} \psi_i(\phi)^2 \exp\left(-E_i/kT\right) \tag{10.59}$$

which extends over the first fifty levels. Each result was then carefully compared with a corresponding estimate based on the classical relationship formulated in equation (10.39), and in this way it was established that for a potential of cosine form, with barriers less than 7 kcal/mole, there is no significant discrepancy between the quantum-mechanical and classical approaches.

In view of this agreement the classical method was used to evaluate $w(\phi)$ for the ferrocene problem, but instead of restricting $V(\phi)$ to the simple form tested, the more general expression

$$V(\phi) = V_0(1 - \cos 5\phi - \beta_1 \cos 10\phi)/2 \tag{10.60}$$

was assumed, and the contributions made to the total molecular intensity by the rotation-dependent carbon–carbon distances were calculated according to equation (10.35).

In this integration, which was carried out numerically, the framework amplitudes of vibration were treated as variables, and given the dependence

$$u_{\text{fr}}(\phi) = C_1 + D_1 \cos \phi \tag{10.61}$$

suggested by Morino and Hirota's results for hexachlorodisilane.[33]

When fitting the intensity data by least-squares refinement C_1 and D_1 were varied together with the barrier height V_0, but β_1 was kept fixed at a suitable trial value. Refinements involving these parameters, and various rotation-independent quantities, were carried out for several choices of β_1, and the results obtained demonstrated that these variables were rather insensitive to the slight changes in barrier shape caused by allowing the $\cos 10\phi$ term to contribute. It was evident that although the height of the barrier could be determined from the experimental data, the exact shape could not.

Other least-squares calculations assuming different analytical forms for $V(\phi)$ and $u_{fr}(\phi)$ did nothing to alter this conclusion and a survey of the numerous estimates obtained for V_0 led to a final value of 0.9 ± 0.3 kcal/mole, the equilibrium conformation of ferrocene being, as previously concluded, an eclipsed one.

Haaland and Nilsson also carried out similar calculations for the related molecule, ruthenocene $(C_5H_5)_2Ru$, and although they concluded that this system also prefers an eclipsed conformation it was impossible to determine a well-defined value for V_0 from the intensity data available. This failure seemed partly due to the slightly poorer quality of the data collected, and was partly a consequence of the smaller contributions made by the rotation-dependent distances to the total scattered intensity.

An eclipsed conformation has also been suggested for nickelocene, $(C_5H_5)_2Ni$, as a result of an electron diffraction study,[91] but, as in the case of ruthenocene, no details of the hindering potential energy function were obtained.

10.7.2 *Biscyclopentadienylmanganese and -beryllium*

These sandwich compounds differ from the iron, ruthenium and nickel analogues in that they are more chemically reactive in their behaviour, and apparently more ionically bound. This latter property has in fact been detected by electron diffraction experiments[92-94] for it has been shown that in these systems the metal-to-ring distances have vibrational amplitudes almost twice those found for the corresponding distances in members of the covalent iron series.

Biscyclopentadienylmanganese has a regular sandwich structure but the beryllium compound is peculiar in that the beryllium atom, while lying on the central axis, is nearer to one ring than to the other (see Figure 10.12). The two distances involved are 1.485 ± 0.005 Å and 1.980 ± 0.010 Å, respectively, and in such a situation it seems likely that there are two equivalent equilibrium positions for the beryllium atom situated on either side of the centre of the molecule and separated by an energy barrier. This asymmetry has in fact been rationalized in terms of an ionic model.[93]

For both the manganese and beryllium compounds there is an absence of information about the nature of the hindering potential energy function, but it does seem probable that both molecules have staggered arrangements of the five-membered rings in contrast to what was found for the covalent structures already mentioned. It is of interest that an electron diffraction study of dibenzene chromium[95] has also indicated this conformation for the benzene rings.

10.7.3 *Biscyclopentadienyltin and -lead*

These molecules also seem to be ionic in character and differ from the other systems discussed in that electron diffraction experiments[96] have shown

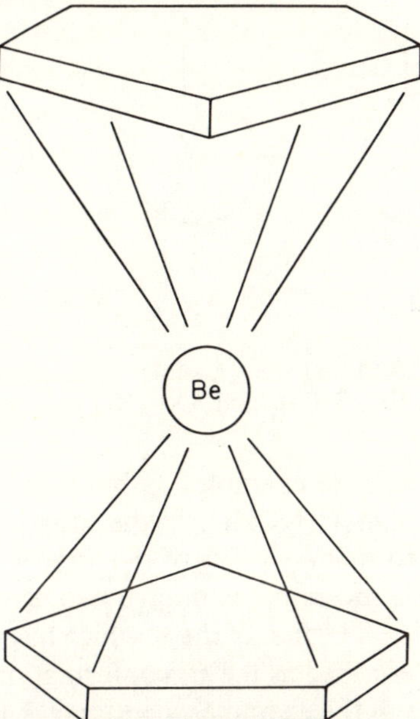

Figure 10.12 The staggered, un-symmetrical, sandwich structure proposed for beryllocene, $Be(C_5H_5)_2$, by Almenningen and coworkers.[93] One ring is closer to the central atom than is the other

them to be wedge-shaped with angles between the ring planes of 45° and 55° respectively. No reliable information is available, however, about the relative orientations or motions of these rings, nor about hindering potential energy barriers.

10.8 Electron diffraction studies of compounds containing vinyl, carbonyl and phenyl groups

In this section molecules having a cylindrical rotor attached to a planar frame will be considered, and classification made according as the latter is a vinyl group, carbonyl group or a benzene ring.

Three conformational possibilities for the first two types of molecule are shown in Figure 10.13, it being assumed that the attached rotor has three-fold symmetry. Of these, the second configuration might seem the most stable on account of the lack of eclipsing it allows, but in fact a microwave investigation of a very simple example, prop-1-ene, CH_3CHCH_2,[97] has shown that this molecule exists almost entirely in conformation (A), and similar studies of acetaldehyde, CH_3CHO,[98] and acetyl fluoride, CH_3CFO,[99] confirm that in these molecules also, one hydrogen atom of the methyl group eclipses the double bond.

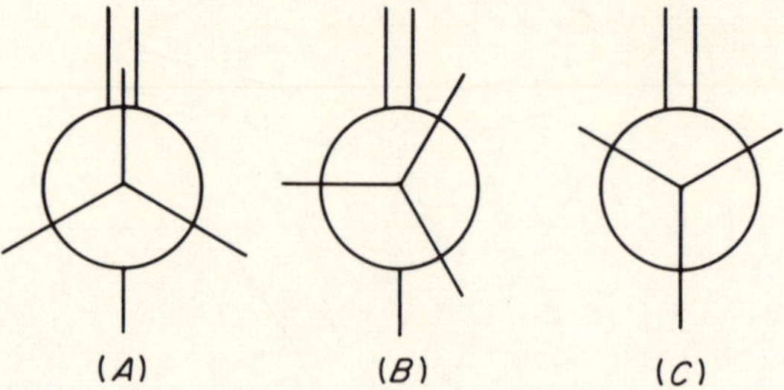

Figure 10.13 Three possible conformations for
AX_3CHCH_2 and AX_3CHO systems

For such less symmetrical examples as but-1-ene, $CH_3CH_2CHCH_2$,[100] and propionaldehyde, CH_3CH_2CHO,[101] the situation is more complicated, for even if it is accepted that eclipsing of the attached rotor and the double bond is preferred, these molecules can have two conformations of this sort as shown in Figure 10.14. The first of these, which has the methyl group in the eclipsing position, is described as the *cis* conformation, the other the *gauche*, and microwave studies of the examples mentioned have suggested that both isomers are present in gas samples. It seems that for propionaldehyde at least, the *cis* rotamer is significantly the more stable.

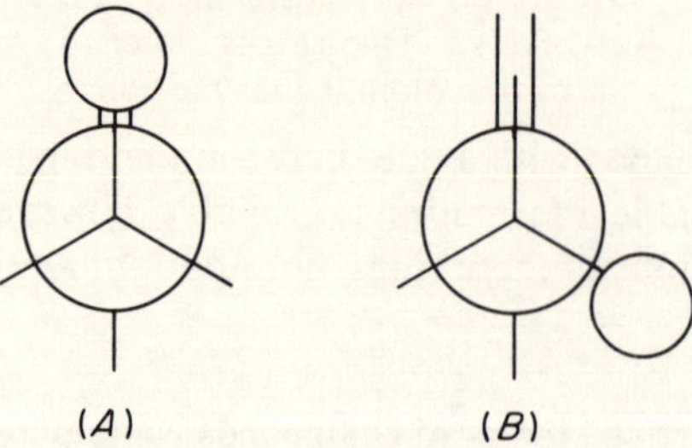

Figure 10.14   The two eclipsed conformations possible for AX_2YCHCH_2 and
AX_2YCHO systems

There are a number of electron diffraction results available for vinyl and carbonyl compounds of this kind, and in general these confirm the spectroscopic findings just described.

10.8.1 *Results for vinyl derivatives*

Only one molecule of this type, with a three-fold symmetric rotor, has been studied. This is vinyltrichlorosilane, $SiCl_3CHCH_2$,[102] which has been shown to take up the same eclipsed conformation adopted by prop-1-ene.

The latter is of course a bad subject for electron diffraction investigation, on account of the low scattering power of hydrogen atoms, but in the silicon compound the presence of chlorine substituents makes conformational study particularly easy.

A good example of a vinyl derivative with a less symmetrical rotor, and two possible eclipsed conformations, is 2-methylbut-1-ene, $CH_3CH_2C(CH_3)CH_2$, which has been investigated[103] by electron diffraction at 20°C and shown to contain roughly equal amounts of the *gauche* and *cis* rotamers. These proportions imply that the *cis* form, with an eclipsing methyl group, is the more stable by some 300 ± 120 cal/mole.

Another vinyl compound for which data are available is the cyclic molecule vinylcyclopropane, $C_3H_5CHCH_2$, which has been shown[104] to exist in two conformations in the vapour at room temperature. These are shown in Figure 10.15, and may be described as *s-trans* and *s-gauche*, if these terms

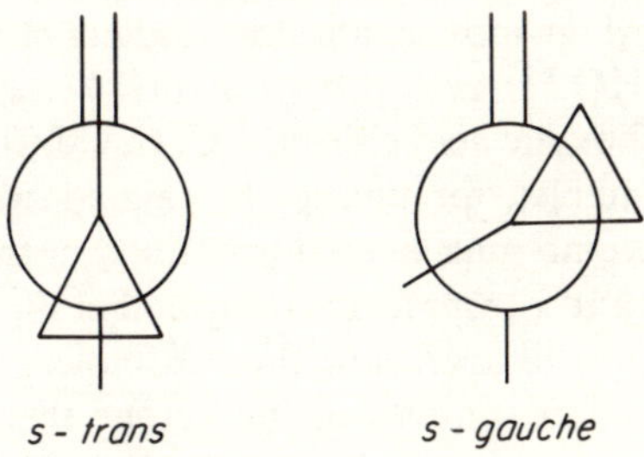

Figure 10.15   The preferred conformations of vinylcyclopropane[104]

refer to the orientation of the cyclopropane ring relative to the double bond. It seems that the first of these rotamers is the more stable, being present to the extent of 75% in the sample investigated.

If lone pairs of electrons are considered equivalent to attached atoms, the molecules methyl vinyl ether, CH_3OCHCH_2, and methyl vinyl sulphide, CH_3SCHCH_2, are analogous to but-1-ene. Both have recently been studied[105,106] in the vapour phase by electron diffraction, and shown to exist as mixtures of *cis* and *gauche* conformations. In the ether, at 200°C, the *cis* form, with the carbon atom of the methyl group in the plane of the other two carbon atoms, accounts for roughly 64% of the molecules present, whilst the rest have a *gauche* conformation with an out-of-plane angle somewhere in the range 80–111°. Corresponding results for the sulphide, studied at a similar temperature, were 33% for the proportion of the *cis* isomer, and a dihedral angle of roughly 107° for the *gauche*. These results indicate that for the ether the *cis* form is the more stable by about 1·2 kcal/mole, whilst the energy difference between the *cis* and *gauche* conformations of the sulphide is much smaller.

10.8.2 *Results for carbonyl derivatives*

Propionic acid, CH_3CH_2COOH,[107] and methyl ethyl ketone, $CH_3CH_2COCH_3$,[108,109] have both been studied by electron diffraction. They are similar to propionaldehyde in that two eclipsed conformations of the ethyl group are possible, but the ketone has been demonstrated to exist almost entirely in the *cis* form at 20°C, and a similar preference for eclipsing by the methyl group has been found for isopropyl carboxaldehyde, $(CH_3)_2CHCHO$:[110] indeed in samples of the aldehyde at -7°C it appears that some 90% of the molecules present are *gauche* species. It seems that in the case of carbonyl compounds a methyl group is strongly preferred to a hydrogen atom as an eclipsing substituent.

Electron diffraction results are also available for carbonyl compounds containing cyclopropyl groups, examples studied being cyclopropyl carboxaldehyde, C_3H_5CHO,[111] cyclopropyl methyl ketone, $C_3H_5COCH_3$,[112] and cyclopropane carboxylic acid chloride, C_3H_5COCl.[112] These molecules are of considerable interest, for unlike the related system, isopropyl carboxaldehyde, they have no *gauche* conformations in the vapour phase, but are present in the *s-cis* and *s-trans* forms indicated in Figure 10.16. In samples of the aldehyde at room temperature these rotamers are present in equal proportions, but in similar samples of the ketone there is roughly 80% of the *cis* form, and possibly an even larger amount of this conformer in the case of the acid chloride. It seems that as the size of the substituent X in Figure 10.16 increases the proportion of the *s-trans* isomer declines.

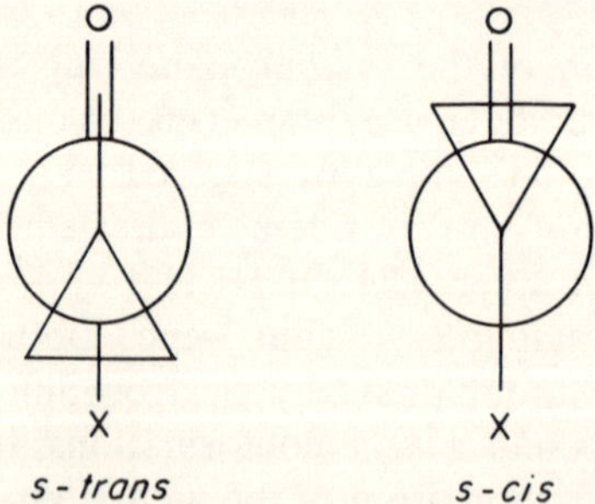

Figure 10.16 The *s-cis* and *s-trans* conformations found for cyclopropyl carboxaldehyde, cyclopropyl methyl ketone and cyclopropyl carboxylic acid chloride by electron diffraction experiments[110–112]

The existence of an *s-cis* conformation for these molecules requires explanation, and raises the entire question of what factors determine the conformational stabilities of the unsaturated systems discussed in this section. For most examples, such as prop-1-ene, propionaldehyde, etc., it has been suggested[113] that if the double bond is considered equivalent to two bent single bonds, conformations of type (*A*) in Figure 10.13 are in reality staggered and are stable for this reason. Such an explanation is clearly invalid for the cyclopropyl carbonyl compounds mentioned above, and a special theory has been proposed to rationalize their behaviour. As pointed out by Walsh,[114] a cyclopropane ring can in certain cases behave like a vinyl group, that is, as an unsaturated rather than saturated substituent, and if this is assumed, cyclopropyl carboxyaldehyde and the others should behave like ene-ones whose preference for planar conjugated conformations is well known. This latter explanation does not unfortunately account for the *s-gauche* conformation of vinylcyclopropane which was discussed earlier and which should, according to the above ideas, be similar to butadiene.

It is of interest that cyclobutanecarboxylic acid chloride, C_4H_7COCl,[115] does not show the same anomalous behaviour as the cyclopropyl derivatives, for a recent electron diffraction investigation of this compound has revealed a preference for the *s-gauche* geometry consistent with the simple bent-bond theory.

10.8.3 *Results for benzene derivatives*

There are two likely conformations for a molecule of the general type $C_6H_5AX_3$. These are shown in Figure 10.17, and in view of what has been found for vinyl and carbonyl derivatives, it would be expected that the eclipsed form (*A*) would be the more stable.

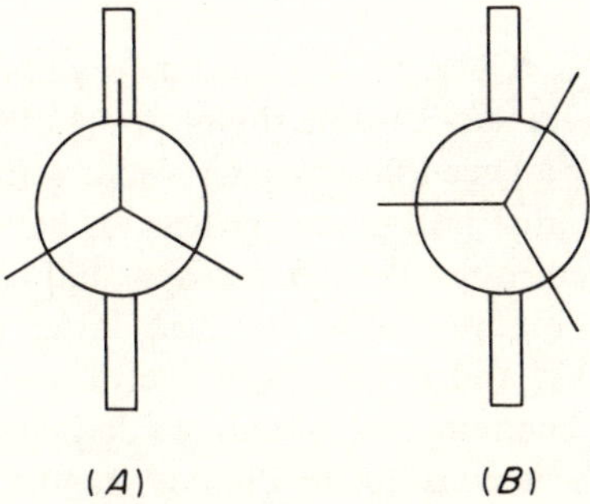

Figure 10.17   Two possible conformations for $C_6H_5AX_3$ systems

Electron diffraction studies of such examples as phenyltrichlorosilane, $C_6H_5SiCl_3$,[116] and phenylchlorosilane, $C_6H_5SiH_2Cl$,[117] have not proved helpful in deciding this point, however, as in both cases internal rotation seems to be fairly free, but a study of cumene, $C_6H_5CH(CH_3)_2$,[118] has shown that an eclipsed *cis* geometry is preferred, and investigations of the related cyclic molecules phenyl cyclopropane, $C_6H_5C_3H_5$,[118,119] and phenyl cyclobutane, $C_6H_5C_4H_7$,[118] have also indicated conformations with a hydrogen atom eclipsing the benzene ring (see Figure 10.18). Another

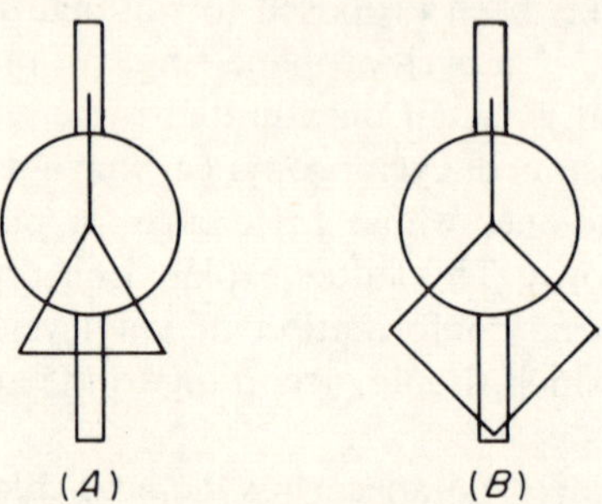

(A)   (B)

Figure 10.18 The preferred conformations of phenyl cyclopropane and phenyl cyclobutane[118,119]

example studied, *NN*-dimethylaniline, $C_6H_5N(CH_3)_2$,[120] appears to be exceptional, however, in that this compound has been shown to have an almost planar configuration at the nitrogen atom, and to exist in a conformation with the plane of the benzene ring at right angles to the lone pair of electrons.

10.9 Electron diffraction studies of dienes, polyenes and related compounds

10.9.1 *Dienes*

Butadiene, $CH_2CHCHCH_2$,[121,122] 2,3-dimethylbutadiene, $CH_2C(CH_3)$ $.C(CH_3)CH_2$,[123] and *cis,cis*-3,4-dimethylhexa-2,4-diene, $CH_3CH(CH_3)C$ $.(CH_3)CHCH_3$,[124] have all been shown to be planar with a *trans* arrangement of the double bonds, whilst non-planar structures have been found for the more sterically hindered molecules, *trans,trans*- and *cis,trans*-3,4-dimethyl-hexa-2,4-diene[124] (see Figure 10.19). In these latter two compounds the dihedral angles are 113·3° and 114·3° respectively.

In rationalizing the behaviour of butadiene derivatives, conjugative and steric effects must both be considered. Planar structures obviously give the best overlap between the π-clouds of both double bonds, but the stability allowed by this arrangement must occasionally be sacrificed if the planar structure involves considerable steric repulsion. Fateley and coworkers have

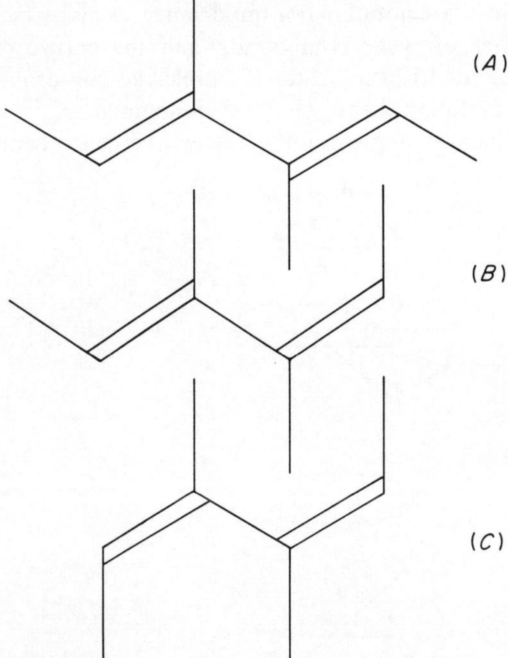

Figure 10.19   Three isomeric butadienes: (A) *cis,cis-*, (B) *cis, trans-* and (C) *trans, trans-*3,4-dimethyl hexa-2,4-diene, investigated by electron diffraction.[124] Only the first of these molecules is planar

in fact suggested[125,126] a potential energy function for dienes of the form

$$V(\phi) = V_1(1 - \cos \phi)/2 + V_2(1 - \cos 2\phi)/2 \qquad (10.62)$$

and it would be expected that whereas steric effects could contribute to both terms, conjugative effects should give rise to a two-fold barrier with minima at the *cis* and *trans* positions. Unfortunately the electron diffraction data so far available for butadiene derivatives contain little information about the nature of $V(\phi)$ in these compounds, and even in the simple case of butadiene itself, neither the number of minima, nor the heights of the intervening barriers, have been established. It is of interest, however, that when steric interactions do force a breakdown of conjugation there is no significant change in the length of the central carbon–carbon bond, this distance being 1·473 Å in the planar *cis,cis-*3,4-dimethylhexa-2,4-diene, and 1·479 Å in the non-planar *trans,trans* isomer.

Electron diffraction investigations have also been made of a number of cyclic dienes but, from the point of view of the study of internal rotation, the results obtained are not of major importance, as many factors determine the conformations of cyclic compounds, and the nature of $V(\phi)$ is only one of these. Figure 10.20 indicates the preferred conformations of cyclopentadiene,[127] cyclohexadiene,[128,129] cycloheptadiene,[130] and cyclooctadiene,[131] the dihedral angles of the diene fragments being respectively,

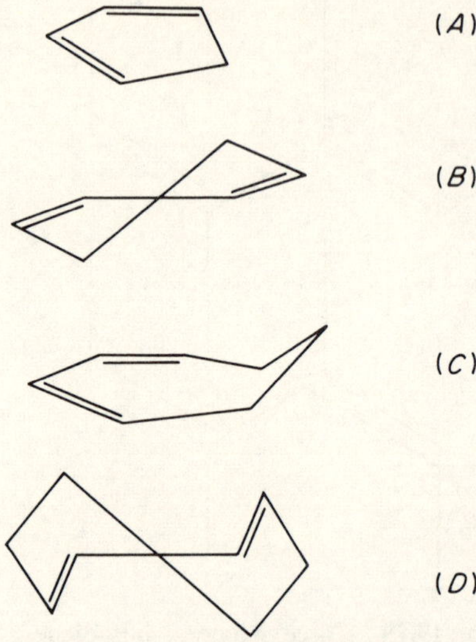

Figure 10.20 The preferred conformations of cyclopentadiene, cyclohexa-1,3-diene, cyclohepta-1,3-diene and cycloocta-1,3-diene, as determined by electron diffraction experiments[127–131]

0, 18, 0 and 38°. These conformations and dihedral angles are in fact quite consistent with the ideas of valence-angle strain, torsional strain, and Van der Waals interaction normally invoked when rationalizing the preferred geometries of cyclic molecules.

10.9.2 *Polyenes*

Two linear trienes, 1,3,5-*cis*- and 1,3,5-*trans*-hexatriene (see Figure 10.21) have been studied.[132,133] The *trans* derivative has been shown to have a flat zig-zag structure whilst steric interaction between hydrogen atoms in the *cis* isomer introduces a small angle of twist of 10° about the central double bond. No barrier information was obtained for these molecules, but some effects arising from a limited torsional motion about the single carbon–carbon bonds were observed.

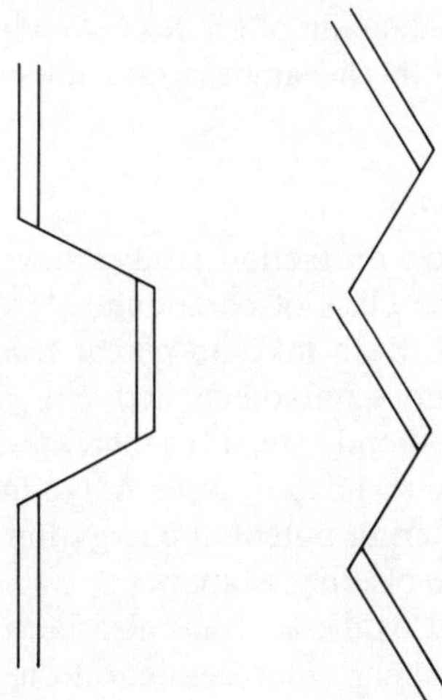

Figure 10.21 Two isomeric hexatrienes studied by electron diffraction[132,133]

Conformations found by electron diffraction experiments for the cyclic polyenes, cycloheptatriene,[134] and cyclooctatetraene.[121] are shown in Figure 10.22. In neither case is the conjugated system planar, as this would require the introduction of considerable angle strain, and the final structures appear

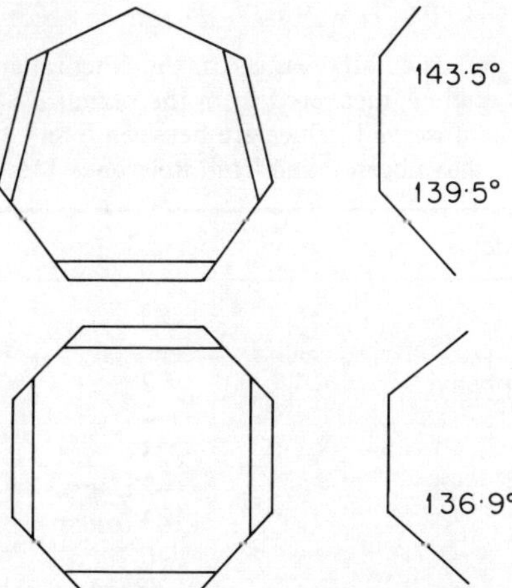

Figure 10.22 Electron diffraction results for two cyclic polyenes, cycloheptatriene and cyclooctatetraene. Both molecules are non-planar[134,121]

to be determined by a balancing of the loss of stability produced by breakdown of conjugation with the angle strain introduced by attempting to maintain π overlap.

10.9.3 *Ene-ones and di-ones*

Fairly detailed electron diffraction studies have been published for the simplest members of this class of compounds,[122] acrolein, CH_2CHCHO, and glyoxal, $HCOCOH$. Both take up planar *trans* conformations, and it was found that in the series butadiene, acrolein, glyoxal, the length of the central carbon–carbon bond steadily increases from 1.463 ± 0.003 Å, through 1.478 ± 0.005 Å, to 1.525 ± 0.003 Å. No information was obtained, however, about the hindering potential energy function, nor was there any evidence for the presence of other isomers.

It is worth noting that oxalic acid has also been investigated recently[135] and shown to have a planar *trans* configuration, but in this case such a structure is fairly inevitable on account of the degree of hydrogen bonding it allows.

10.9.4 *Biphenyls*

These compounds are rather similar to dienes in that their conformations in the gas phase are determined by balancing conjugative and steric effects. If conjugation alone mattered, biphenyls should be planar, but the size of the *ortho* substituents usually causes the rings to have an angle of orientation midway between 0° and 90° in order to allow some reduction of steric strain.

Table 10.4   Observed and calculated values for the dihedral angle ϕ in various derivatives of biphenyl. This angle is measured from the planar *cis* conformation and error limits appropriate to the observed values are between 5 and 10 degrees. These results have been taken from Reference 141

Molecule	ϕ (calc.)	ϕ (obs.)	Reference
Biphenyl	23	42	137
3,3′-Dibromobiphenyl	—	45	140
3,5,3′,5′-Tetrabromobiphenyl	22	45	140
4,4′-Difluorobiphenyl	—	44	139
2-Fluorobiphenyl	33	49	139
2,2′-Difluorobiphenyl	42(*cis*)	60	139
	143(*trans*)	—	
Decafluorobiphenyl	46	70	141
2,2′-Dichlorobiphenyl	72(*cis*)	74	138
	120(*trans*)	—	
2,2′-Dibromobiphenyl	82(*cis*)	75	138
	112(*trans*)	—	
2,2′-Diiodobiphenyl	93	79	138

Thus, although the parent compound is planar in the crystalline state,[136] the free molecule is non-planar in the vapour,[137] with an angle of rotation from planarity of some 42°.

Several halogenated biphenyls have also been studied by the electron diffraction technique,[138–140] and results for these are listed in Table 10.4 which has been taken from a recent publication by Almenningen and co-workers.[141] It is evident from this collection of data that as the bulk of the *ortho* substituent increases, the dihedral angle between the rings also increases, and values for these angles calculated theoretically by balancing the effects of steric repulsion and loss of conjugative stability (see Table 10.4) show similar trends to the experimental data even if good agreement is not always obtained.

It is of interest that although two non-equivalent minima are predicted for the 2,2′-dihalo-compounds (the numbering system adopted is shown in Figure 10.23) corresponding to *cis* and *trans* forms, only the *cis* form has so far been detected by diffraction experiments.

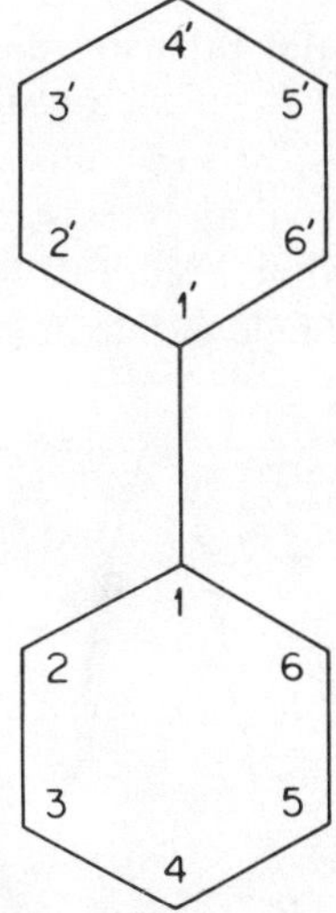

Figure 10.23   The numbering system adopted for biphenyl and its
derivatives

Although in most analyses, an average dihedral angle and the amplitude of torsion about the central bond is all that has been established for derivatives of biphenyl, there is one good example of an investigation in which the potential energy has been treated quantitatively. This is a study of perfluorobiphenyl by Almenningen and coworkers[141] in which these authors employed modern computational methods to study the slow torsional oscillations of the benzene rings, and their preferred dihedral angle. As usual this torsional vibration was assumed separable from the faster motions of the molecular frame, and the angular probability distribution function $w(\phi)$ was assigned three possible analytical forms

$$w_1(\phi) = N\exp\left[-P_1(R_\alpha - P_2)^2\right] \tag{10.63}$$

in which R_α is the shortest distance between *ortho* fluorine atoms

$$w_2(\phi) = N[\exp(-P_1(\phi - P_2)^2) + \exp(-P_1(\pi - \phi - P_2)^2)] \tag{10.64}$$

and

$$w_3(\phi) = N\exp\left[-P_1(\sin^3\phi + P_2\cos^3\phi)\right] \tag{10.65}$$

the angle ϕ in all of these expressions being measured from the planar configuration. The intensity contributions of all rotation-dependent distances in the molecule were then calculated by applying equation (10.35), the framework amplitudes being assumed independent of ϕ, and integration being carried out over the range 0–90°.

When fitting the experimental intensity data by the least-squares procedure, the parameters P_1 and P_2 in each trial function were varied and it was found that all three forms for $w(\phi)$ allowed a similar quality of fit to the experimental observations. Final forms for these angular probability distribution functions are shown in Figure 10.24, which has been reproduced from the original article. The angle of maximum probability is clearly close

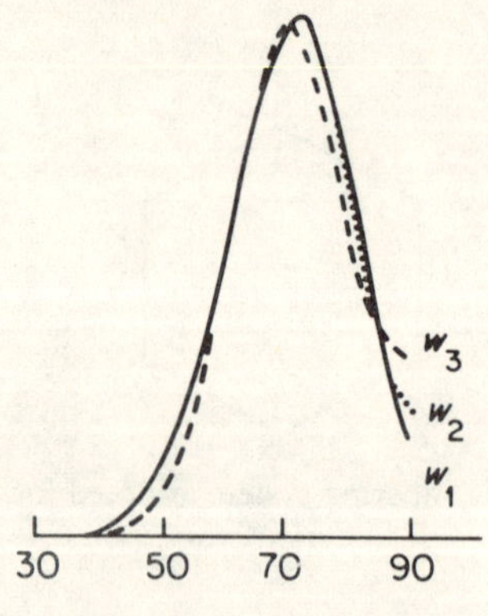

Figure 10.24 Plots of three angular probability distribution functions, $w_1(\phi)$, $w_2(\phi)$ and $w_3(\phi)$, determined for decafluorobiphenyl by Almenningen and coworkers.[141] The preferred dihedral angle is close to 70° in all cases

to 70°, and the barrier to internal rotation may be roughly estimated from this result by applying the formula

$$\Delta V = RT \ln \left[w(\phi_{\text{max}})/w(90°) \right] \tag{10.66}$$

which is based on the classical relationship (equation 10.39) between $w(\phi)$ and $V(\phi)$. It was concluded that this barrier must be in the range 0·4–2·0 kcal/mole.

10.9.5 *Diboron tetrachloride*

B_2Cl_4 may be included in the present discussion as it also consists of two planar rotors coupled together. Conjugative effects are not expected to be important, however, and in this respect diboron tetrachloride differs from the molecules already discussed. A thorough electron diffraction study of B_2Cl_4 at several temperatures has recently been published,[142] and in this investigation the barrier height V_0 was estimated, as well as the molecular geometry.

The experimental radial distribution curve, A, corresponding to a temperature of $-22°C$, is reproduced in Figure 10.25, and this is compared with

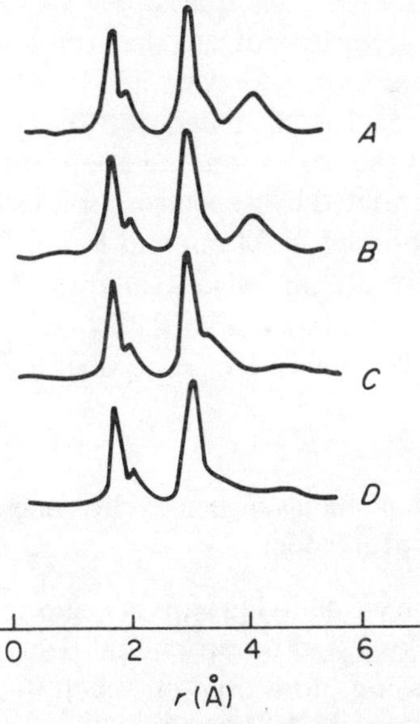

Figure 10.25 The experimental radial distribution curve, A, for B_2Cl_4, and theoretical curves calculated for B, a staggered D_{2d}, C, a planar D_{2h}, and D, a freely-rotating model, by Ryan and Hedberg[142]

a series of theoretical functions calculated on the assumption of B, a completely staggered, C, a planar, and D, a freely-rotating model. The staggered conformation is clearly in best agreement with the experimental data, the broad peak at 4 Å indicating one type of long Cl..Cl distance only, and a considerable amount of libration about the central B—B bond.

By assuming separation of this motion from frame vibrations, it was possible to calculate a theoretical radial distribution curve by applying equation (10.37), assuming $w(\phi)$ to be related classically to one or other of the two trial potential energy functions

$$V(\phi) = V_0\phi^2 \tag{10.67}$$

$$V(\phi) = V_0(1 - \cos 2\phi)/2 \tag{10.68}$$

In this calculation frame vibrational amplitudes were assumed constant, and both functions were used to fit data obtained at the various temperatures. For both choices of $V(\phi)$, and for each temperature, a value of V_0 and a corresponding standard deviation were calculated, and from these results two average values for V_0 and their standard deviations were obtained. The average corresponding to the cosine form was 1.85 ± 0.035 kcal/mole, and that corresponding to the quadratic 1.45 ± 0.020 kcal/mole, but although these results seem to indicate that the quadratic potential is the more correct, the cosine form was concluded to be the more realistic, as each individual value of V_0 calculated for this function had a smaller standard deviation than the corresponding quadratic result. This conclusion is to some extent substantiated by a spectroscopic estimate for V_0 in B_2Cl_4 of 1.7 ± 0.6 kcal/mole published by Mann and Fano.[143]

Electron diffraction results are also available,[144] for the related system dinitrogen tetroxide, N_2O_4, but unlike B_2Cl_4 this molecule is planar, with a very long central N—N bond of 1.75 Å. In this case some kind of π overlap may well be involved.

10.10 Electron diffraction studies of non-cyclic compounds with two or more torsional degrees of freedom

If the internal rotation of methyl groups is neglected, all of the compounds so far discussed have possessed one torsional degree of freedom only. The study of systems possessing more than one such freedom is naturally more difficult, and in most cases the information derived has been limited to a determination of that geometrical arrangement of the rotating groups which gives a best fit to the experimental data. Detailed analysis of the potential energy function restricting internal rotation has rarely been attempted.

Results published for the following classes of compound indicate the problems involved and the kind of information which has been obtained.

10.10.1 *The n-alkanes*

In a study of n-butane Bonham and Bartell[42] found that the *trans* isomer is more stable than the *gauche* by about 630 cal/mole, and this information was applied during a subsequent electron diffraction investigation[145] of the higher n-alkanes, n-pentane, n-hexane and n-heptane.

If the internal rotation of methyl groups is ignored these molecules have two, three and four torsional degrees of freedom, and gas samples should contain several rotational isomers, each corresponding to a particular sequence of *gauche* and *trans* configurations along the chain. The stability of the *trans* isomer of n-butane suggests that species having a large number of *trans* arrangements should predominate.

When analysing the experimental radial distribution curve obtained for each of these compounds, Bartell and Kohl[145] assumed a series of values for the free-energy difference, ΔG^0, between a single *gauche* and *trans* conformation, and then calculated a distribution of isomers for each value, from the equation

$$n_i/n_j = (m_i/m_j)\exp\left[-(V_i - V_j)\Delta G^0/RT\right] \tag{10.69}$$

Table 10.5 Isomer distributions for n-pentane, n-hexane and n-heptane, at 14°C. These results have been taken from Reference 145 and correspond to a free-energy difference of 610 cal/mole between a single *gauche* and *trans* configuration

	Isomer	Multiplicity	Percentage
n-Pentane	*TT*	1	38·4
	TG	4	52·7
	GG	2	9·0
n-Hexane	*TTT*	1	24·5
	TTG	4	33·6
	TGT	2	16·8
	TGG	4	11·6
	GTG	4	11·6
	GGG	2	2·0
n-Heptane	*TTTT*	1	15·7
	TTTG	4	21·5
	TTGT	4	21·5
	TTGG	4	7·4
	GTTG	4	7·4
	TGTG	8	14·8
	TGGT	2	3·7
	TGGG	4	2·8
	GTGG	8	5·1
	GGGG	2	0·4

in which n_i and n_j are the numbers of conformers i and j, m_i and m_j are their multiplicities, and V_i and V_j are the numbers of *gauche* arrangements in each. In these calculations a temperature of 14°C was assumed and the optimum ΔG^0 for each molecule was selected according to the quality of the fit to the experimental radial distribution curve which each distribution allowed. Final results for n-pentane, n-hexane and n-heptane were 625 cal/mole, 520 cal/mole and 625 cal/mole, and when the corresponding value for n-butane is added to this list, and an average taken, this latter turns out to be 610 cal/mole. This average free-energy difference was assumed when calculating the isomer distributions listed in Table 10.5, and it is clear from the figures quoted that rotamers with large numbers of *trans* configurations and high multiplicities predominate.

10.10.2 *Compounds with two equivalent rotors*

Dimethyl ether, $(CH_3)_2O$, is a typical member of an important series of compounds all of which have two equivalent rotors of three-fold symmetry attached to a central atom. Such a molecule has several possibilities for its equilibrium conformation (see Figure 10.26) and, in the case of the ether,

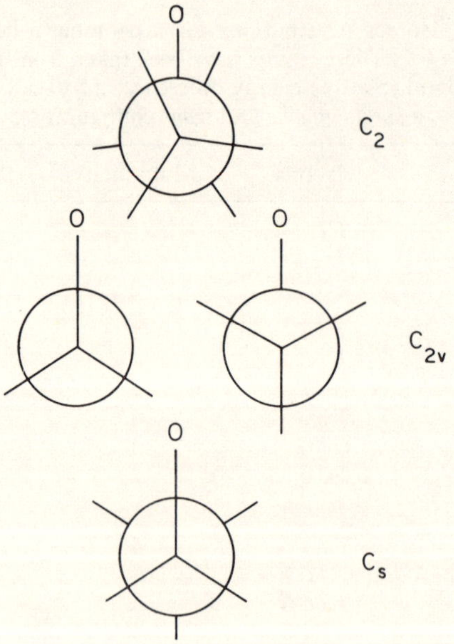

Figure 10.26 Four possible conformations for an AX_3OAX_3 molecule shown in projection. In the first of the two C_{2v} structures both AX_3 groups are staggered with respect to the opposing A—O bond, whilst in the second they are eclipsed

an electron diffraction experiment[146] has indicated a C_{2v} geometry with both methyl groups staggered relative to the opposite carbon–oxygen bond. This is demonstrated in Figure 10.27 which has been taken from the original article, and which shows the experimental radial distribution curve obtained for the ether, and several theoretical versions calculated for different trial models. Peaks corresponding to the rotation-dependent non-bonded

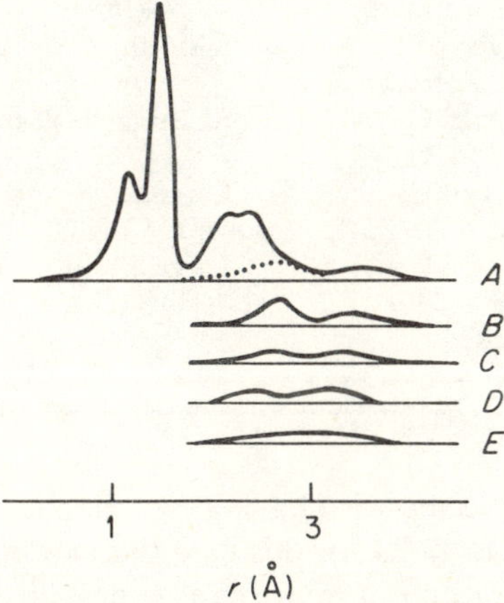

Figure 10.27 The experimental radial distribution curve, A, for $(CH_3)_2O$, and theoretical curves calculated for B, a C_{2v} staggered, C, a C_2 twisted, D, an eclipsed C_{2v}, and E, a freely-rotating model, by Kimura and Kubo[146]

C..H distances are compared, and it is evident that all but the C_{2v} model may be discarded. This conformation seems fairly well defined for, according to one spectroscopic result,[147] the barrier to internal rotation in dimethyl ether is around 2·7 kcal/mole.

Conformational information is also available[148–154] for a number of other molecules of this type, and electron diffraction data for a selection of examples are summarized in Table 10.6. This list of results shows that for some of these systems a twisted C_2 conformation has been proposed in preference to the more symmetrical C_{2v} structure. In some instances it is possible that this angle of twist is indeed a feature of the molecule's equilibrium geometry, but it must be added that in many cases small deviations from a higher symmetry, obtained in this way, are spurious, and indicate an insufficiently detailed treatment of the molecule's twisting motion when analysing diffraction data. Indeed in many determinations, this analysis is extremely difficult because of the relatively low contribution of the rotation-dependent distances to the total scattering.

Table 10.6   Electron diffraction results for some AX_3BAX_3 molecules

Molecule	Model	Remarks	Reference
CH_3OCH_3	Staggered C_{2v}	—	146
CF_3SCF_3	Staggered C_{2v}	—	149
CF_3SeCF_3	(assumed)		
SiH_3OSiH_3	Staggered C_{2v} (uncertain)	Large SiOSi angle $[144 \cdot 1(0 \cdot 9)°]$	150
SiH_3SSiH_3	Staggered C_{2v} (uncertain)	SiSSi angle $97 \cdot 4(0 \cdot 7)°$	151
SiH_3SeSiH_3	Twisted C_2 $[\phi \sim 30(5)°]$	SiSeSi angle $96 \cdot 6(0 \cdot 7)°$	152
SiF_3OSiF_3	Twisted C_2 $[\phi \sim 35(1 \cdot 5)°]$	Large SiOSi angle $[155 \cdot 7(2 \cdot 0)°]$	153
$SiCl_3OSiCl_3$	Twisted C_2 $[\phi \sim 29(1 \cdot 5)°]$	Large SiOSi angle $[146(4)°]$	154
ClO_3OClO_3	Twisted C_2 $[\phi \sim 15 \cdot 4(3 \cdot 7)°]$	—	148

Another type of molecule, also having two identical rotors, is that exemplified by acetone, $(CH_3)_2CO$. In this case the two methyl groups are connected to a trigonal carbon atom forming a double bond, and, from what has been said about vinyl and carbonyl derivatives, it might reasonably be expected that a fully-staggered C_{2v} geometry would be preferred, as this would ensure eclipsing of one hydrogen atom from each methyl group and the central double bond. An electron diffraction study of acetone,[155] however, and the related series of compounds, hexafluoroacetone, $(CF_3)_2CO$, hexafluoropropylimine, $(CF_3)_2C.NH$, and hexafluoroisobutene, $(CF_3)_2C.CH_2$, has shown that in these molecules the attached rotors prefer a twisted C_2 arrangement. This has been claimed to be a real effect, and for the series described, the twist angles were estimated to be $33 \cdot 0$, $36 \cdot 6$, $36 \cdot 9$ and $35 \cdot 0°$ respectively.

Another molecule which can be included in the same category as acetone, is the di-t-butyl nitroxide free radical, as this has been found[156] to have a planar configuration around nitrogen with the two t-butyl groups twisted some $27°$ away from the C_{2v} conformation. It is worth noting, however, that the related fluorine derivative, $(CF_3)_2NO$, appears to have a different structure, as it has been shown[157] to have the N—O bond bent $22°$ out of the CNC plane, and to have a totally staggered arrangement of the two CF_3 groups.

10.10.3 *Compound with three equivalent rotors*

In view of the behaviour of systems with two attached rotors it might be expected that the AX_3 groups in $B(AX_3)_3$ and $CB(AX_3)_3$ molecules would take up positions in which they are completely (or almost completely)

staggered with respect to the opposing A—B bonds. Early electron diffraction studies of such examples as $N(CF_3)_3$,[158] $P(CF_3)_3$,[159] $As(CF_3)_3$,[159] and $Sb(CF_3)_3$[159] certainly confirmed this view, but were not sufficiently exact to provide a rigorous demonstration.

A recent electron diffraction study of tris(trifluoromethyl)methane, $(CF_3)_3CH$,[160] has been more informative, however, as in this analysis Stølevik and Thom have fitted accurate data both by the usual semi-rigid model approach, and by a more sophisticated method involving a dynamic model. They showed that the best semi-rigid structure is one in which each rotor is twisted some 18° away from the staggered conformation, and that if a dynamic model is assumed, a similar result is obtained, the twist angle being only slightly different. In this second treatment the torsional motion was separated from the fast skeletal vibrations and the energy of any conformation was expressed as a sum of two contributions, one from non-bonded interactions, and one from the barrier restricting internal rotation. The angular probability function was taken as a function of all three torsional angles, and expressed classically in terms of the potential energy. The experimental data were then fitted by summing curves, assuming various values for the barrier height V_0, and constant values for the frame amplitudes of vibration, and the parameters describing the non-bonded energy contribution. In this way V_0 was estimated to be close to zero and the equilibrium conformation was found to correspond to a 15° angle of twist from the fully-staggered position.

Reports also appear in the literature of electron diffraction studies of other examples of this type of system of a more complicated nature. For example, triethylphosphite, $P(OC_2H_5)_3$, and trivinylphosphite, $P(OC_2H_3)_3$, have both been assigned C_{3v} models with a *trans* arrangement of the P—O and C—C bonds,[161,162] and investigations of $P(N(CH_3)_2)_3$[163] and $ClSi(N(CH_3)_2)_3$[164] have suggested that in these molecules the dimethylamino groups are planar, and twisted some 20° away from the position in which the normal to each group lies in the same plane as the overall three-fold axis.

Propeller models of this type have also been proposed for the aromatic compounds, triphenylmethane, $(C_6H_5)_3CH$,[165] the triphenylmethyl radical, $(C_6H_5)_3C\cdot$,[166] and triphenylamine, $(C_6H_5)_3N$.[167] The first of these has a definitely pyramidal configuration at the central carbon atom, but the other two are nearly planar, and may indeed be so at equilibrium. In all three cases the best fit between theory and experiment was achieved by twisting the benzene rings, in-phase, about 45° from the position in which each ring normally lies in a plane containing the three-fold axis.

10.10.4 *Compounds with four equivalent rotors*

Early studies of molecules of this type include work on tetranitromethane, $C(NO_2)_4$,[168] tetraphenylsilane, $Si(C_6H_5)_4$,[169] and tetramethoxysilane, $Si(OCH_3)_4$.[169] The last of these was described as having a fully-staggered structure, whilst the first two were assigned conformations in which each planar group eclipsed one of the other bonds to the central atom.

More recently information has also become available for tetrakis(trimethylsilyl)silane, $Si(Si(CH_3)_3)_4$,[170] and nickel tetrakis(trifluorophosphine), $Ni(PF_3)_4$.[171] The second of these has been shown to execute essentially free rotation, the experimental radial distribution curve being fitted by a sum of many semi-rigid models (see Figure 10.28) whilst the first is certainly more rigid and appears to take up a conformation in which each rotor is twisted in-phase away from the position of maximum staggering.

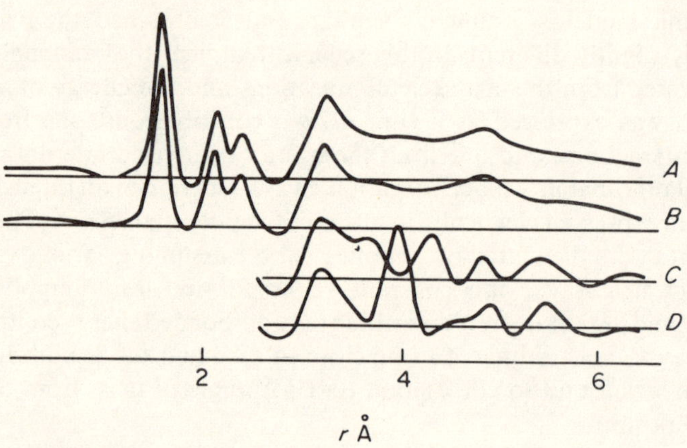

Figure 10.28　The experimental radial distribution curve, *A*, for $Ni(PF_3)_4$, and theoretical curves calculated for *B*, a freely-rotating, and *C* and *D*, two semi-rigid models, by Almenningen and coworkers[173]

10.11 Electron diffraction studies of cyclic molecules

Rings containing fewer than seven atoms have no independent torsional degrees of freedom, the various dihedral angles being dependent on the sizes of particular bond lengths and valence angles. Nonetheless, the conformations preferred by these systems are partly determined by torsional energy and for this reason the present section includes results for molecules containing four-, five- and six-membered rings, in addition to data for larger systems. The small rings are considered first.

10.11.1 Four-membered rings

Electron diffraction data are available for a number of simple four-membered ring compounds including cyclobutane, C_4H_8,[172,173] perfluorocyclobutane, C_4F_8,[174] biscyclobutane, C_8H_{14},[175] cyclobutene, C_4H_6,[176] and perfluorocyclobutene, C_4F_6.[174] Not unexpectedly, these last two examples have been found to have planar frameworks, but studies of the parent hydrocarbon, the corresponding fluorine derivative and biscyclobutane, have shown that the saturated hydrocarbon ring is puckered, with a dihedral angle of roughly 20°. Indeed, in the case of biscyclobutane, this property made interpretation of the intensity data particularly difficult, as it allows both the *trans* and *gauche* isomers to have three possible structures corresponding to equatorial–equatorial (ee), equatorial–axial (ea) and axial–axial (aa), coupling of the two rings.

Apart from conformational information, these investigations also produced accurate molecular dimensions, and it is of interest that for cyclobutane the carbon–carbon bond length was found to be 1.548 ± 0.003 Å, whereas in the totally fluorinated derivative this value increased to 1.566 ± 0.008 Å. These results may be compared with corresponding estimates for cyclopropane $(1.510 \pm 0.002$ Å)$,$[177] cyclopentane $(1.546 \pm 0.0012$ Å)$,$[178] and cyclohexane $(1.520 \pm 0.003$ Å)$,$[179] and with the average carbon–carbon bond length in the n-alkanes $(1.533 \pm 0.002$ Å)$.$[145]

10.11.2 *Five-membered rings*

The parent hydrocarbon cyclopentane, C_5H_{10}, has been the subject of a recent electron diffraction study,[178] and has been shown to have a non-planar skeleton. The exact equilibrium geometry was found difficult to establish, however, as the C_2 half-chair, and C_s envelope models (see Figure 10.29)

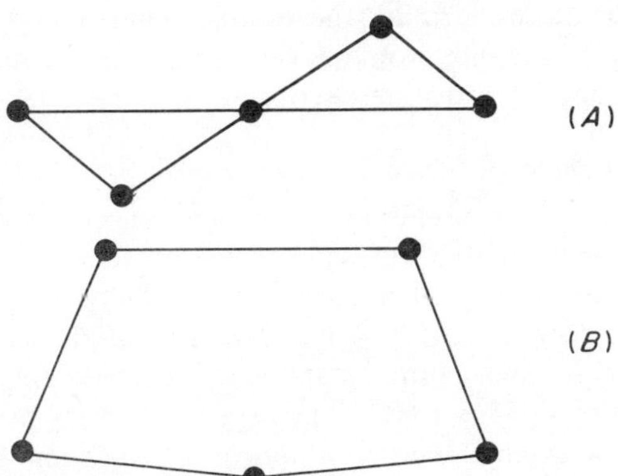

Figure 10.29 Two non-planar structures for a five-membered ring: (*A*) the half-chair, and (*B*) the envelope model

both gave equally good fits to the experimental data, as did a whole series of other puckered structures of intermediate character. These trial geometries were all obtained by varying the phase angle ϕ in the expression

$$z_j = \sqrt{(2/5)}q \cos\left[2(2\pi j/5 + \phi)\right] \qquad (10.70)$$

in which z_j is the perpendicular displacement of the jth atom from its position as part of a regular, planar five-membered ring, and q is the amplitude of puckering. Energy calculations, taking into account strain contributions from bond length, valence angle, and torsional sources, as well as non-bonded interactions, indicated a similar degree of stability for all of these conformations, and it was concluded that cyclopentane must exhibit free pseudo-rotation, a type of intramolecular motion in which ring-puckering moves in a circle without the development of angular momentum. This interpretation allowed a satisfactory fit to the experimental data to be achieved and yielded an average value for q of 0.427 ± 0.015 Å.

Free pseudorotation is not, however, a property restricted to cyclopentane. The heterocyclic molecule tetrahydrofuran, C_4H_8O, appears to behave in this way too,[180,181] for in a recent series of electron diffraction studies,[180–183] Seip and coworkers have shown that although tetrahydrothiophene, C_4H_8S,[182] and tetrahydroselenophene, C_4H_8Se,[183] both have C_2 structures, no definite conformation can be assigned to the oxygen analogue. As in the case of cyclopentane, the best interpretation of the data collected, was to assume free pseudorotation, an explanation again supported by the results of calculations of conformational energy.

Other examples of five-membered ring molecules which have also been investigated in detail, are cyclopentene[184] and perfluorocyclopentene.[185] These have quite definite non-planar geometries, and it has been established that the two atoms forming the double bond and their neighbours are coplanar, whilst the fifth atom lies out of this plane defining dihedral angles of $29 \pm 2.5°$ and $22 \pm 0.5°$, respectively.

10.11.3 *Six-membered rings*

A recent reinvestigation of cyclohexane[179] has confirmed the chair structure well-known for this molecule and provided ring measurements only slightly different from earlier estimates.[186] The new result for the carbon–carbon bond length is 1.520 ± 0.003 Å, compared with the older value of 1.528 ± 0.005 Å, and both distances are considerably smaller than the bond length of 1.551 ± 0.0023 Å, recently found for perfluorocyclohexane, C_6F_{12}.[187] The expansive effect of fluorine substitution is again evident.

Several sets of results[188–190] have also been published for cyclohexene, C_6H_{10}, which takes up a half-chair conformation similar to that shown in Figure 10.20 for cyclohexadiene. In this molecule the single bond was found to vary in length from 1.504 ± 0.006 Å when adjacent to the double bond, to 1.55 ± 0.04 Å when opposite to it.

10.11.4 *Cyclooctane*

Cyclooctane, C_8H_{16}, has two torsional degrees of freedom, and by choosing suitable values for these, it may be assigned several symmetrical conformations. In a recent electron diffraction investigation[191] it was found, however, that although the inner peaks of the experimental radial distribution curve enabled unambiguous calculation of a ring bond length and angle of 1·54 Å and 116·6°, respectively, it was exceedingly difficult to reproduce the outer part of this function on the assumption of any single symmetrical structure or combination of these. It was concluded therefore that the cyclooctane ring is extremely flexible, a computation involving a continuous series of structures being required.

10.11.5 Cis,cis-*cyclodeca-1,6-diene*

In contrast to cyclooctane, this molecule, $C_{10}H_{16}$, has been found[192] to be fairly rigid. Two symmetrical conformations involving staggering about single bonds are shown in Figure 10.30, structure (*A*) being the chair and (*B*) the boat form. Calculation showed that the chair structure gave the

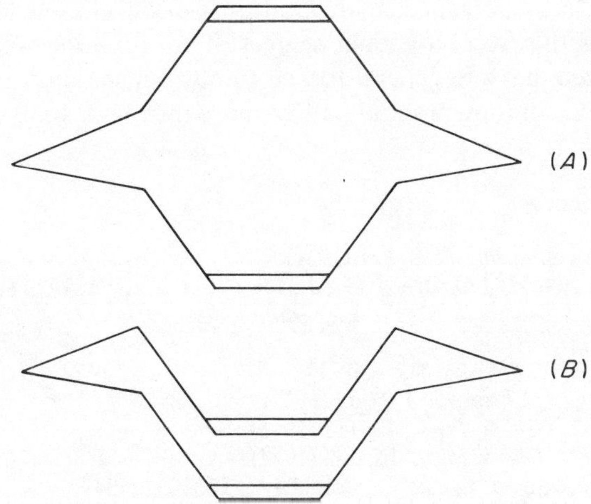

Figure 10.30   The chair (*A*) and boat (*B*) models for
cis, cis-cyclodeca-1,6-diene. The chair structure is preferred[195]

best fit to the experimental data, the derived parameters being exceptional only in the large values ($128·2 \pm 0·3°$) assigned to the CCC angles adjacent to the double bonds.

10.11.6 *Cyclotetradeca-1,8-diyne*

This molecule, $C_{14}H_{20}$, is probably the largest ring system so far investigated by the electron diffraction technique. Like cyclooctane it was found[191] to be extremely flexible, no single structure or combination of structures being discovered which could explain the experimental data satisfactorily.

10.12 Conclusions

The results discussed in the last six sections indicate the effectiveness of the electron diffraction technique as a means of solving conformational problems but they also reveal its limitations. Thus, whereas the internuclear distances of a particular isomer can be calculated to within a few hundredths, or even thousandths of an Angstrom unit, rotamer proportions are rarely determined to better than five or ten per cent, and barrier heights are usually subject to uncertainties of 30 per cent or more. Although these error limits can be reduced a little, by the assumption of values for vibrational amplitudes estimated from spectroscopic data, experimental developments appear necessary if the hindering potential energy function is ever to be studied with anything approaching sensitivity.

Even in its present form, however, the technique is capable of providing much new information, and apart from accurate re-examination of simple systems, it is to be expected that future work will concentrate on larger molecules with several torsional degrees of freedom. Past treatments of this type of system have in general lacked rigour but the increasing availability of high-speed computers should allow more thorough analyses in the future.

10.13 References

1. R. Wierl, *Ann. Physik*, **8**, 521 (1931).
2. L. Pauling and L. O. Brockway, *J. Am. Chem. Soc.*, **59**, 1223 (1937).
3. C. Finbak, *Avhandl. Norske Videnskaps-Akad. Oslo I. Mat. Naturv. Kl.*, No. **13**, 17 (1937).
4. P. Debye, *Z. Physik.*, **40**, 66 (1939).
5. P. Debye, *J. Chem. Phys.*, **9**, 55 (1941).
6. J. Karle, *J. Chem. Phys.*, **13**, 155 (1945).
7. J. Karle, *J. Chem. Phys.*, **15**, 202 (1947).
8. I. L. Karle and J. Karle, *J. Chem. Phys.*, **17**, 1052 (1949).
9. J. Karle and H. Hauptman, *J. Chem. Phys.*, **18**, 875 (1950).
10. J. Karle and I. L. Karle, *J. Chem. Phys.*, **18**, 957 (1950).
11. I. L. Karle and J. Karle, *J. Chem. Phys.*, **18**, 963 (1950).
12. J. Karle, *J. Chem. Phys.*, **22**, 1246 (1954).
13. R. W. James, *Z. Physik*, **33**, 737 (1932).
14. V. Schomaker and R. Glauber, *Nature*, **170**, 290 (1952).
15. R. Glauber and V. Schomaker, *Phys. Rev.*, **89**, 667 (1953).
16. O. Bastiansen, L. Hedberg and K. Hedberg, *J. Chem. Phys.*, **27**, 1311 (1957).
17. O. Bastiansen, O. Hassel and F. Risberg, *Acta Chem. Scand.*, **9**, 232 (1955).
18. W. Zeil, J. Haase and L. Wegmann, *Z. Instrumentenk.*, **74**, 84 (1966).
19. B. Beagley, A. H. Clark and T. G. Hewitt, *J. Chem. Soc(A)*, 658 (1968).
20. B. Beagley, D. W. J. Cruickshank, T. G. Hewitt and A. Haaland, *Trans. Faraday Soc.*, **63**, 836 (1967).
21. B. Andersen, H. M. Seip, T. G. Strand and R. Stølevik, *Acta Chem. Scand.*, **23**, 3224 (1969).
22. V. P. Spiridonov, N. G. Rambidi and N. V. Alekseev, *J. Struct. Chem. (USSR) (Engl. Transl.)*, **4**, 717 (1963).

23. Y. Morino and E. Hirota, *J. Chem. Phys.*, **23**, 737 (1955).
24. N. G. Rambidi, V. P. Spiridonov and N. V. Alekseev, *J. Struct. Chem. (USSR) (Engl. Transl.)*, **3**, 331 (1962).
25. L. S. Bartell, *J. Chem. Phys.*, **23**, 1219 (1955).
26. L. S. Bartell, L. Brockway and R. Schwendeman, *J. Chem. Phys.*, **23**, 1854 (1955).
27. S. J. Cyvin, *Molecular Vibrations and Mean Square Amplitudes*, Universitetsforlaget, Oslo (Amsterdam: Elsevier, 1968).
28. A. Almenningen, O. Bastiansen, A. Haaland and H. M. Seip, *Angew. Chem.*, **4**, 819 (1965).
29. H. M. Seip, T. G. Strand and R. Stølevik, *Chem. Phys. Letters*, **3**, 617 (1969).
30. S. Mizushima, *Structure of Molecules and Internal Rotation* (New York: Academic Press, 1954).
31. D. A. Swick and I. L. Karle, *J. Chem. Phys.*, **23**, 1499 (1955).
32. D. A. Swick, I. L. Karle and J. Karle, *J. Chem. Phys.*, **22**, 1242 (1954).
33. Y. Morino and E. Hirota, *J. Chem. Phys.*, **28**, 185 (1958).
34. K. Kveseth, H. M. Seip and R. Stølevik, *Acta Chem. Scand.*, **25**, 2975 (1971).
35. D. R. Lide, *J. Chem. Phys.*, **29**, 1426 (1958).
36. B. Kirtman, *J. Chem. Phys.*, **41**, 775 (1964).
37. A. Almenningen, A. Haaland and J. E. Nilsson, *Acta Chem. Scand.*, **22**, 972 (1968).
38. S. H. Bauer and J. Y. Beach, *J. Am. Chem. Soc.*, **64**, 1142 (1942).
39. B. Beagley, J. J. Monaghan and T. G. Hewitt, *J. Mol. Structure*, **8**, 401 (1971).
40. B. Beagley (Private Communication, 1971).
41. H. A. Skinner and L. E. Sutton, *Trans. Faraday Soc.*, **36**, 1209 (1940).
42. R. A. Bonham and L. S. Bartell, *J. Am. Chem. Soc.*, **81**, 3491 (1959).
43. K. Kuchitsu, *Bull. Chem. Soc. Japan*, **32**, 748 (1959).
44. Y. Morino and K. Kuchitsu, *J. Chem. Phys.*, **28**, 175 (1958).
45. J. Ainsworth and J. Karle, *J. Chem. Phys.*, **20**, 425 (1952).
46. K. Kuratani, T. Miyazawa and S. Mizushima, *J. Chem. Phys.*, **21**, 1411 (1952).
47. E. Komaki, I. Ichishima, K. Kuratani, T. Miyazawa, T. Shimanouchi and S. Mizushima, *Bull. Chem. Soc., Japan*, **28**, 330 (1955).
48. K. S. Pitzer, *J. Chem. Phys.*, **8**, 711 (1940).
49. K. Kuchitsu, *Bull. Chem. Soc. Japan*, **30**, 399 (1957).
50. A. Almenningen, O. Bastiansen and F. Harshbarger, *Acta Chem. Scand.*, **11**, 1059 (1957).
51. M. Yamaha, *Bull. Chem. Soc. Japan*, **29**, 865 (1956).
52. M. Igarashi and M. Yamaha, *Bull. Chem. Soc. Japan*, **29**, 871 (1956).
53. M. Yamaha, *Bull. Chem. Soc. Japan*, **29**, 876 (1956).
54. A. Almenningen, O. Bastiansen, L. Fernholt and K. Hedberg, *Acta Chem. Scand.*, **25**, 1946 (1971).
55. A. Almenningen, O. Bastiansen and L. Fernholt (Private Communication, 1971).
56. M. Iwasaki, S. Nagase and R. Kojima, *Bull. Chem. Soc. Japan*, **30**, 230 (1957).
57. M. Iwasaki, *Bull. Chem. Soc. Japan*, **31**, 1070 (1958).
58. M. Iwasaki, *Bull. Chem. Soc. Japan*, **32**, 194 (1959).
59. M. Iwasaki, *Bull. Chem. Soc. Japan*, **32**, 205 (1959).
60. W. D. Gwinn and K. S. Pitzer, *J. Chem. Phys.*, **16**, 303 (1948).
61. K. Kuratani and S. Mizushima, *J. Chem. Phys.*, **22**, 1403 (1954).
62. J. R. Thomas and W. D. Gwinn, *J. Am. Chem. Soc.*, **71**, 2785 (1949).
63. Y. Morino, T. Iijima and Y. Murata, *Bull. Chem. Soc. Japan*, **33**, 46 (1960).
64. B. Beagley (Private Communication, 1971).
65. A. Yamaguchi, I. Ichishima, T. Shimanouchi and S. Mizushima, *J. Chem. Phys.*, **31**, 843 (1959).
66. E. R. Nixon, *J. Phys. Chem.*, **60**, 1054 (1956).

67. M. J. Cardillo and S. H. Bauer, *Inorg. Chem.*, **8**, 2086 (1969).
68. C. Glidewell, D. W. H. Rankin, A. G. Robiette and G. M. Sheldrick, *J. Chem. Soc(A)*, 318 (1970).
69. L. S. Bartell and H. K. Higginbotham, *Inorg. Chem.*, **4**, 1346 (1965).
70. A. McAdam and B. Beagley (Private Communication, 1971).
71. A. McAdam, B. Beagley and T. G. Hewitt, *Trans. Faraday Soc.*, **66**, 2732 (1970).
72. R. W. Rudolphi, R. C. Taylor and R. W. Parry, *J. Am. Chem. Soc.*, **88**, 3729 (1966).
73. S. G. Frankiss and F. A. Miller, *Spectrochim. Acta*, 1235 (1965).
74. S. G. Frankiss, F. A. Miller, H. Stammreich and T. T. Sans, *Spectrochim. Acta*, **23A**, 543 (1967).
75. L. V. Vilkov and L. S. Khaikin, *Dokl. Akad. Nauk. SSSR*, **168**, 810 (1966).
76. G. C. Holywell, D. W. H. Rankin, B. Beagley and J. M. Freeman, *J. Chem. Soc(A)*, 785 (1971).
77. P. A. Giguere and V. Schomaker, *J. Am. Chem. Soc.*, **65**, 2025 (1943).
78. M. Winnewisser and J. Haase, *Z. Naturforsch.*, **23A**, 56 (1968).
79. E. Hirota, *J. Chem. Phys.*, **28**, 839 (1958).
80. M. K. Wilson and R. M. Badger, *J. Chem. Phys.*, **17**, 1232 (1949).
81. E. Hirota, *Bull. Chem. Soc. Japan*, **31**, 130 (1958).
82. B. Beagley, G. H. Eckersley, D. P. Brown and D. Tomlinson, *Trans. Faraday Soc.*, **65**, 2300 (1969).
83. K. T. McAloon and B. Beagley (Private Communication, 1971).
84. P. D'Antonio, C. George, A. H. Lowrey and J. Karle, *J. Chem. Phys.*, **55**, 1071 (1971).
85. R. H. Jackson, *J. Chem. Soc.*, 4585 (1962).
86. R. L. Kuczkowski, *J. Am. Chem. Soc.*, **86**, 3617 (1964).
87. T. Ukaji and R. A. Bonham, *J. Am. Chem. Soc.*, **84**, 3631 (1962).
88. E. A. Seibold and L. E. Sutton, *J. Chem. Phys.*, **23**, 1967 (1955).
89. R. K. Bohn and A. Haaland, *J. Organometallic Chem.*, **5**, 470 (1966).
90. A. Haaland and J. E. Nilsson, *Acta Chem. Scand.*, **22**, 2653 (1968).
91. I. A. Ronova, D. A. Bochvar, A. L. Chistjakov, Yu. T. Struchkov and N. V. Alekseev, *J. Organometallic Chem.*, **18**, 337 (1969).
92. A. Almenningen, A. Haaland and T. Motzfeldt, *Selected Topics in Structure Chemistry* (Oslo: Universitets Forlaget, 1967) p. 105.
93. A. Almenningen, O. Bastiansen and A. Haaland, *J. Chem. Phys.*, **40**, 3434 (1964).
94. A. Haaland, *Acta Chem. Scand.*, **22**, 3030 (1968).
95. A. Haaland, *Acta Chem. Scand.*, **19**, 41 (1965).
96. A. Almenningen, A. Haaland and T. Motzfeldt, *J. Organometallic Chem.*, **7**, 97 (1967).
97. D. R. Herschbach and L. K. Krisher, *J. Chem. Phys.*, **28**, 728 (1958).
98. R. W. Kilb, C. C. Lin and E. B. Wilson, Jr., *J. Chem. Phys.*, **26**, 1695 (1957).
99. L. Pierce and L. C. Krisher, *J. Chem. Phys.*, **31**, 875 (1959).
100. S. Kondo, E. Hirota and Y. Morino, *J. Mol. Spectr.*, **28**, 471 (1968).
101. S. S. Butcher and E. B. Wilson, Jr., *J. Chem. Phys.*, **40**, 1671 (1964).
102. H. Murata, *J. Chem. Phys.*, **21**, 181 (1953).
103. T. Shimanouchi, Y. Abe and K. Kuchitsu, *J. Mol. Structure*, **2**, 82 (1968).
104. A. De Meijere and W. Luttke, *Tetrahedron*, **25**, 2047 (1969).
105. N. L. Owen and H. M. Seip, *Chem. Phys. Letters*, **5**, 162 (1970).
106. S. Samdal and H. M. Seip (Private Communication, 1971).
107. J. Derissen, *J. Mol. Structure*, **7**, 81 (1971).
108. T. Shimanouchi, Y. Abe and M. Mikami, *Spectrochim. Acta*, **24A**, 1037 (1968).
109. M. Abe, K. Kuchitsu and T. Shimanouchi, *J. Mol. Structure*, **4**, 245 (1969).
110. J. P. Guillory and L. S. Bartell, *J. Chem. Phys.*, **43**, 654 (1965).

111. L. S. Bartell and J. P. Guillory, *J. Chem. Phys.*, **43**, 647 (1965).

112. L. S. Bartell, J. P. Guillory and A. T. Parks, *J. Phys. Chem.*, **69**, 3043 (1965).

113. L. Pauling, *Nature of the Chemical Bond*, 3rd Ed. (Ithaca: Cornell University Press, 1960) p. 137.

114. A. D. Walsh, *Nature*, **159**, 165, 712 (1947).

115. W. J. Adams and L. S. Bartell, *J. Mol. Structure*, **8**, 199 (1971).

116. L. V. Vilkov, V. S. Mastryukov and P. A. Akishin, *J. Struct. Chem. (USSR) (Eng. Transl.)*, **5**, 834 (1964).

117. L. V. Vilkov and V. S. Mastryukov, *Dokl. Akad. Nauk SSSR*, **162**, 1306 (1965).

118. L. V. Vilkov and V. S. Mastryukov, *J. Struct. Chem. (USSR) (Engl. Transl.)*, **8**, 534 (1967).

119. L. V. Vilkov and N. I. Sadova, *Dokl. Akad. Nauk SSSR*, **162**, 565 (1965).

120. L. V. Vilkov and T. P. Timasheva, *Dokl. Akad. Nauk SSSR*, **161**, 351 (1965).

121. W. Haugen and M. Traetteberg, *Acta Chem. Scand.*, **20**, 1726 (1966).

122. K. Kuchitsu, T. Fukuyama and Y. Morino, *J. Mol. Structure*, **1**, 463 (1968).

123. C. F. Aten, L. Hedberg and K. Hedberg, *J. Am. Chem. Soc.*, **90**, 2463 (1968).

124. M. Traetteberg, *Acta Chem. Scand.*, **24**, 2295 (1970).

125. W. G. Fateley, R. K. Harris, F. A. Miller and R. E. Witkowski, *Spectrochim. Acta*, **21**, 231 (1965).

126. F. A. Miller, W. G. Fateley and R. E. Witkowski, *Spectrochim. Acta*, **23A**, 891 (1967).

127. V. Schomaker and L. Pauling, *J. Am. Chem. Soc.*, **61**, 1769 (1939).

128. M. Traetteberg, *Acta Chem. Scand.*, **22**, 2305 (1968).

129. H. Oberhammer and S. H. Bauer, *J. Am. Chem. Soc.*, **91**, 10 (1969).

130. J. F. Chiang and S. H. Bauer, *J. Am. Chem. Soc.*, **88**, 420 (1966).

131. M. Traetteberg, *Acta Chem. Scand.*, **24**, 2285 (1970).

132. M. Traetteberg, *Acta Chem. Scand.*, **22**, 2294 (1968).

133. M. Traetteberg, *Acta Chem. Scand.*, **22**, 628 (1968).

134. M. Traetteberg, *J. Am. Chem. Soc.*, **86**, 4265 (1964).

135. Z. Nahlovska, B. Nahlovsky and T. G. Strand, *Acta Chem. Scand.*, **24**, 2617 (1970).

136. E. B. Robertson, *Nature*, **192**, 1026 (1961).

137. A. Almenningen and O. Bastiansen, *Kgl. Norske Videnskab. Selskabs Skrifter*, No. **4** (1958).

138. O. Bastiansen, *Acta Chem. Scand.*, **4**, 926 (1950).

139. O. Bastiansen and L. Smedvik, *Acta Chem. Scand.*, **8**, 1593 (1954).

140. O. Bastiansen and A. Skancke, *Acta Chem. Scand.*, **21**, 587 (1967).

141. A. Almenningen, A. O. Hartmenn and H. M. Seip, *Acta Chem. Scand.*, **22**, 1013 (1968).

142. R. R. Ryan and K. Hedberg, *J. Chem. Phys.*, **50**, 4986 (1969); **41**, 2214 (1964).

143. D. E. Mann and L. Fano, *J. Chem. Phys.*, **26**, 1665 (1957).

144. D. W. Smith and K. Hedberg, *J. Chem. Phys.*, **25**, 1282 (1956).

145. L. S. Bartell and D. A. Kohl, *J. Chem. Phys.*, **39**, 3095 (1963).

146. K. Kimura and M. Kubo, *J. Chem. Phys.*, **30**, 151 (1959).

147. R. J. Myers, *J. Chem. Phys.*, **30**, 1096 (1959).

148. B. Beagley, *Trans. Faraday Soc.*, **61**, 1821 (1965).

149. H. J. M. Bowen, *Trans. Faraday Soc.*, **50**, 452 (1954).

150. A. Almenningen, O. Bastiansen, V. Ewing, K. Hedberg and M. Traetteberg, *Acta Chem. Scand.*, **17**, 2455 (1963).

151. A. Almenningen, K. Hedberg and R. Seip, *Acta Chem. Scand.*, **17**, 2264 (1963).

152. A. Almenningen, L. Fernholt and H. M. Seip, *Acta Chem. Scand.*, **22**, 51 (1968).

153. W. Airey, C. Glidewell, D. W. H. Rankin, A. G. Robiette, G. M. Sheldrick and D. W. J. Cruickshank, *Trans. Faraday Soc.*, **66**, 551 (1970).

154. W. Airey, C. Glidewell, A. G. Robiette and G. M. Sheldrick, *J. Mol. Structure*, **8**, 413 (1971).

155. R. Hilderbrandt, A. L. Andreassen and S. H. Bauer, *J. Phys. Chem.*, **74**, 1586 (1970).

156. B. Andersen and P. Andersen, *Acta Chem. Scand.*, **20**, 2728 (1966).

157. C. Glidewell, D. W. H. Rankin, A. G. Robiette, G. M. Sheldrick and S. M. Williamson, *J. Chem. Soc(A)*, 478 (1971).

158. R. L. Livingston and G. Vaughan, *J. Am. Chem. Soc.*, **78**, 4866 (1956).

159. H. J. Bowen, *Trans. Faraday Soc.*, **50**, 463 (1954).

160. R. Stølevik and E. Thom, *Acta Chem. Scand.*, **25**, 3205 (1971).

161. L. V. Vilkov, P. A. Akishin and G. E. Salova, *J. Struct. Chem. (USSR) (Eng. Transl.)*, **6**, 339 (1965).

162. L. S. Khaikin and L. V. Vilkov, *J. Struct. Chem. (USSR) (Eng. Transl.)*, **10**, 614 (1969).

163. L. V. Vilkov, L. S. Khaikin and V. V. Evdokimov, *J. Struct. Chem. (USSR) (Eng. Transl.)*, **10**, 978 (1969).

164. L. V. Vilkov and N. A. Tarasenko, *Chem. Comm.*, 1177 (1969).

165. P. Andersen, *Acta Chem. Scand.*, **19**, 622 (1965).

166. P. Andersen, *Acta Chem. Scand.*, **19**, 629 (1965).

167. Y. Sasaki, K. Kimura and M. Kubo, *J. Chem. Phys.*, **31**, 477 (1959).

168. A. J. Stosick, *J. Am. Chem. Soc.*, **61**, 1127 (1939).

169. M. Yokoi, *Bull. Chem. Soc. Japan*, **30**, 100 (1957).

170. L. S. Bartell, F. B. Clippard and T. L. Boates, *Inorg. Chem.*, **9**, 2436 (1970).

171. A. Almenningen, B. Andersen and E. E. Astrup, *Acta Chem. Scand.*, **24**, 1579 (1970).

172. J. D. Dunitz and V. Schomaker, *J. Chem. Phys.*, **20**, 1703 (1952).

173. A. Almenningen, O. Bastiansen and P. N. Skancke, *Acta Chem. Scand.*, **15**, 711 (1961).

174. C. H. Chang, R. F. Porter and S. H. Bauer, *J. Mol. Structure*, **7**, 89 (1971).

175. A. de Meijere, *Acta Chem. Scand.*, **20**, 1093 (1965).

176. E. Goldish, K. Hedberg and V. Schomaker, *J. Am. Chem. Soc.*, **78**, 2714 (1956).

177. O. Bastiansen, F. N. Fritsch and K. Hedberg, *Acta Cryst.*, **17**, 538 (1964).

178. W. J. Adams, H. J. Geise and L. S. Bartell, *J. Am. Chem. Soc.*, **92**, 5013 (1970).

179. H. R. Buys and H. J. Geise, *Tetrahedron Letters*, **34**, 2991 (1970).

180. H. M. Seip, *Acta Chem. Scand.*, **23**, 2741 (1969).

181. A. Almenningen, H. M. Siep and T. Willadsen, *Acta Chem. Scand.*, **23**, 2748 (1969).

182. Z. Nahlovska, B. Nahlovsky and H. M. Seip, *Acta Chem. Scand.*, **23**, 3534 (1969).

183. Z. Nahlovska, B. Nahlovsky and H. M. Seip, *Acta Chem. Scand.*, **24**, 1903 (1970).

184. M. I. Davis and J. W. Muecke, *J. Phys. Chem.*, **74**, 1104 (1970).

185. C. H. Chang and S. H. Bauer, *J. Phys. Chem.*, **75**, 1685 (1971).

186. M. Davis and O. Hassel, *Acta Chem. Scand.*, **17**, 1181 (1963).

187. K. E. Hjortaas and K. O. Stromme, *Acta Chem. Scand.*, **22**, 2965 (1968).

188. J. F. Chiang and S. H. Bauer, *J. Am. Chem. Soc.*, **91**, 1898 (1969).

189. V. A. Naumov, V. G. Dashevskii and N. M. Zaripov, *J. Struct. Chem. (USSR) (Eng. Transl.)*, **11**, 736 (1970).

190. H. J. Geise and H. R. Buys, *Rec. Trav. Chim.*, **89**, 1147 (1970).

191. A. Almenningen, O. Bastiansen and N. Jensen, *Acta Chem. Scand.*, **20**, 2689 (1966).

192. A. Almenningen, G. C. Jacobsen and H. M. Seip, *Acta Chem. Scand.*, **23**, 1495 (1969).

11 Ab initio *calculations of barrier heights*

A. Veillard

11.1 Introduction

Calculations of the height of rotational barriers aim either to reproduce or to predict the barrier height, with some possible explanation of the origin of the barrier. They have been carried out at three levels of increasing sophistication, corresponding respectively to empirical, semi-empirical quantum-mechanical and *ab initio* quantum-mechanical methods. Empirical treatments are based on classical molecular mechanics or electrostatic models, or they may try to relate the barrier height to some specific molecular property.[1] Through the use of adjustable parameters which may be derived for instance from spectroscopic data, they are intended to obtain quick and practical information for complex systems. This is usually done at the expense of accuracy and without a clear understanding of the physical nature of the problem, since they may be based on a number of unproven assumptions. Although they may achieve some limited success, no single empirical relationship seems capable of correlating the large amount of experimental data.

Semi-empirical quantum-mechanical treatments are based on quantum mechanics, and hence they give a better physical insight while still leading to relatively quick and practical information through empirical parameterization. Although the semi-empirical methods have met with a limited success for some classes of molecules as long as a great deal of similarity between members is preserved,[2,3] even the most refined may lead to erroneous results for some rotational barriers.[4,5] Attempts to compute the barriers with a high reliability have been met through the use of non-empirical or *ab initio* quantum-mechanical treatments which may also provide an interpretation of the barrier within the quantum-mechanical framework. *Ab initio* calculations of rotational barriers have been made possible in the last few years through the advent of fast computers and the achievement of sophisticated programs. The increasing development of conformational analysis has stimulated a strong interest in the calculation of barrier heights.

This review has three major aims : to describe *ab initio* quantum-mechanical methods used to compute barriers, to present a compendium of the barriers so far computed and to describe the theories given for the origin of rotational barriers in the light of *ab initio* calculations.

11.2 Outline of the theoretical and computational methods

11.2.1 *Approximate solutions of the Schrödinger equation—the LCAO–MO– SCF method*

With the neglect of relativistic effects and the Born–Oppenheimer approximation of fixed nuclei,[6] the exact wave-function Ψ and energy E for a system of n electrons are the solutions of the Schrödinger equation

$$\mathscr{H}\Psi = E\Psi$$

where the Hamiltonian $\mathscr{H}$ includes the kinetic operators for the electrons and the electron–nucleus, electron–electron and nucleus–nucleus potentials:

$$\mathscr{H} = \sum_{v}\left(-\frac{1}{2}\nabla_v^2 - \sum_{N}\frac{Z_N}{r_{Nv}} \right) + \sum_{\mu < v}\frac{1}{r_{\mu v}} + \sum_{M < N}\frac{Z_M Z_N}{r_{MN}}$$

The subscripts μ and v refer to the electrons, M and N to the nuclei. The last term is a constant for a given geometrical configuration. Atomic units are used in this equation. The total wave-function Ψ is a function of the $4n$ coordinates of the n electrons, namely the space coordinates x_v, y_v, z_v and the spin coordinate σ_v.

Since the Schrödinger equation cannot be solved exactly for systems with more than one electron, approximate solutions are needed. Usually the total wave-function is built from molecular orbitals (MO). Each molecular orbital φ_i is a one-electron function which depends only on the coordinates x_v, y_v, z_v and σ_v of the electron v. Such a molecular orbital is also called a spin orbital since it depends on the spin of the electron. The simplest wave-function built from molecular orbitals is a Hartree product

$$\Psi(1, 2 \ldots n) = \varphi_1(1)\varphi_2(2) \ldots \varphi_n(n)$$

$\Psi(1, 2 \ldots n)$ is now an approximate wave-function and stands for $\Psi(x_1, y_1, z_1, \sigma_1, \ldots x_n, y_n, z_n, \sigma_n)$. In the same way $\varphi_1(1)$ stands for $\varphi_1 \cdot (x_1, y_1, z_1, \sigma_1)$. In $\varphi_1(1)$ the subscript refers to the molecular orbital and the number between brackets refers to the electron. However, the Hartree product does not satisfy the antisymmetry requirement with respect to the exchange of two electrons (Pauli principle). This requirement is satisfied if the wave-function is an antisymmetrized product of spin orbitals or Slater determinant. For a two-electron system, this antisymmetrized product is written as

$$\Psi(1, 2) = \frac{1}{\sqrt{2}}[\varphi_1(1) \quad \varphi_2(2) - \varphi_1(2) \quad \varphi_2(1)]$$

$$= \frac{1}{\sqrt{2}}\begin{vmatrix} \varphi_1(1) & \varphi_1(2) \\ \varphi_2(1) & \varphi_2(2) \end{vmatrix}$$

An approximate wave-function for a system of n electrons will also be expressed as a Slater determinant

$$\Psi(1, 2 \ldots n) = \frac{1}{\sqrt{(n!)}} \begin{vmatrix} \varphi_1(1) \ldots \varphi_1(n) \\ \varphi_2(1) \ldots \varphi_2(n) \\ \cdots \cdots \cdots \cdots \\ \varphi_n(1) \ldots \varphi_n(n) \end{vmatrix}$$

Given this expression for the approximate wave-function, one has to find the best molecular orbitals φ_i. This is achieved through the variation principle by minimizing the energy with respect to the orbitals φ_i. Since the Hamiltonian is spin-independent, the spin orbitals are written as a simple product of a spin function α or β and a function of the space coordinates $\Phi(x_v, y_v, z_v)$ which is called a space orbital. For a closed-shell system of $2n$ electrons with paired spins, the best space orbitals are the solutions of the Hartree–Fock or Self-Consistent Field (SCF) equations[7]

$$F\Phi = \varepsilon\Phi$$

with F called the Fock operator and ε the orbital energy.

The Hartree–Fock equations are mathematically complicated non-linear integro-differential equations which can be solved directly only for the atomic problem. Instead of using numerical integration, molecular systems may be treated by expanding the unknown molecular orbitals in terms of a fixed basis set χ:

$$\Phi_i = \sum_p C_{ip}\chi_p$$

This is the LCAO (Linear Combination of Atomic Orbitals)–MO method. The corresponding LCAO MO–SCF equations for the closed-shell case are the Roothaan equations[8]

$$FC = \varepsilon SC$$

where F is the Fock matrix, C the matrix of the expansion coefficients and S the overlap matrix defined by

$$S_{pq} = \langle \chi_p | \chi_q \rangle$$

Since the Fock matrix F depends upon the coefficients C, an iterative procedure is needed to solve the Roothaan equations. Nearly all calculations of barrier heights have been done with the LCAO–MO–SCF method.

The SCF theory can be extended in different ways to open-shell systems (systems with unpaired electrons). The Restricted Hartree–Fock (RHF) theory is an extension of the closed-shell treatment: in the open-shell wave-function some space orbitals are combined with both spin functions α and β to form two degenerate spin orbitals. The corresponding SCF equations have been given by Roothaan.[9] It is also possible to develop Unrestricted Hartree–Fock (UHF) SCF methods with an open-shell wave-function of the form[10,11]

$$\Psi = \frac{1}{\sqrt{(n!)}} \| \Phi_1(1)\Phi_2(2)\ldots\Phi_\alpha(\alpha)\overline{\Phi}'_1(\alpha+1)\overline{\Phi}'_2(\alpha+2)\ldots\overline{\Phi}'_\beta(\alpha+\beta)\|$$

The notation $\|\Phi_1(1)\ldots\|$ stands for a Slater determinant, Φ_1 and $\overline{\Phi}'_1$ stand for space orbitals associated with the spin functions α and β respectively and the space orbitals Φ_1 and Φ'_1, Φ_2 and $\Phi'_2 \ldots$ are different. A disadvantage of this method is that the corresponding state has no definite symmetry or multiplicity.

Molecular orbitals which are solutions of the SCF equations are called the canonical SCF orbitals. These are delocalized orbitals extending over the whole molecule. The approximate wave-function can be constructed as well from orbitals which are localized in the region of a particular bond or atom, at least to a good approximation. These localized orbitals are linear combinations of the occupied molecular orbitals and are obtained by subjecting the canonical SCF orbitals to a unitary transformation, subject to some additional criterion.[12,13]

11.2.2 *Choice of the expansion basis set—Slater and Gaussian functions*

The determination of a molecular SCF function is tedious and expensive. In practice, *ab initio* calculations are performed with the aid of electronic digital computers, using fully automatic programs. Frequently several LCAO–MO–SCF wave-functions are available for the same molecule and they differ in the basis set used in the LCAO expansion. There is one numerical solution to the Hartree–Fock equations and the use of a complete basis set for the LCAO expansion would lead to an analytical wave-function which is equivalent to the numerical solution. In practice the basis set used for the expansion is a finite one with a limited number of basis functions. Therefore one obtains only an approximation to the molecular Hartree–Fock function. From calculations on first-row atoms, it has become clear that, if the basis set is sufficiently large and well chosen, an expansion SCF function is virtually the same as the numerical Hartree–Fock function.[14]

The quality of a wave-function is inextricably associated with the choice of the basis set. As the number of functions is increased, the wave-function and its associated energy will get closer to the Hartree–Fock limit, in which the molecular orbitals approach the exact solutions of the Hartree–Fock integro-differential equations. However, this limit is presently accessible only for diatomic and small polyatomic molecules.[15]

A great variety of mathematical functions can be used in the expansion of the molecular orbitals. In practice, nearly all calculations for molecules of chemical interest use either Slater functions or Gaussian functions centred on the nuclei.

Slater functions are defined as

$$N r^{n-1} e^{-\zeta r} Y_{lm}(\theta, \varphi)$$

with N a normalization factor, Y_{lm} the spherical harmonic, r, θ and φ the polar coordinates with respect to the nucleus, n the principal quantum number and ζ the orbital exponent. They are well suited for the expansion of molecular orbitals since they show the correct behaviour both near the nucleus and at large distance (this means that a good fit to the atomic Hartree–Fock function either near the nucleus or at large distance is obtained with only one Slater function). Consequently a fairly accurate molecular wave-function will be achieved with only a small number of Slater functions.[16] It has been customary to classify the basis sets of Slater functions as single-zeta basis sets (or minimal basis sets) and double-zeta basis sets. In the former, one Slater function is used for each atomic sub-shell $1s$, $2s$, $2p_x$, $2p_y$, $2p_z$ .... In the latter, two Slater functions are used for each atomic sub-shell. Near Hartree–Fock wave-functions require at most two, three or four Slater functions per atomic sub-shell. Orbital exponents are usually optimized either for the free atom[17] or for the molecule (since the optimized value for the free atom may not be the best one for the atom combined in the molecule). The usual way of determining the optimum orbital exponent for the molecule is to repeat the molecular calculation many times with different values of the exponent and hence to determine those exponents which do actually minimize the energy. This time-consuming procedure appears important either when an accurate wave-function is required[18] or when a minimal basis set is used.[19]

Apart from linear molecules, the use of Slater functions leads to electron repulsion integrals which are extremely difficult to evaluate. For this reason, molecular calculations for molecules of any geometry have to rely mostly on the use of Gaussian functions.[20,21]

(Another expansion technique is based on the use of Gaussian lobe functions (GLF) of the form $\exp(-ar^2)$ without the use of spherical harmonic angular functions. The angular dependence is achieved by centring Gaussian functions at different points in space determined in part by the symmetry of the orbital to be expanded.[22,23])

Gaussian functions are defined as

$$N r^{n-1} e^{-\zeta r^2} Y_{lm}(\theta, \varphi)$$

The electron repulsion integrals are then relatively easy to evaluate. However, a disadvantage of the Gaussian functions is their incorrect behaviour both near the nucleus and at large distances. Consequently many more Gaussian functions are needed to obtain a wave-function and an energy comparable to that obtained with a given number of Slater functions. For the first-row atoms, Gaussian basis sets range from $5s, 2p$ (namely five s functions and two p_x, two p_y and two p_z functions) to $11s, 7p$. Gaussian basis sets $6s, 3p$ and $10s, 6p$ are respectively comparable to the single-zeta and double-zeta Slater basis sets.[24] The total number of electron repulsion integrals to be computed rises greatly since it varies as the fourth power of the number of basis functions; but the computing time needed to evaluate each integral is much smaller for Gaussian functions than for Slater functions. However, the total number of electron repulsion integrals to be stored would far exceed the storage possibilities of present computers when large Gaussian basis sets are used. One way to overcome this difficulty is to replace the Gaussian functions (sometimes called the primitive functions) with some appropriate linear combinations of Gaussian functions. These linear combinations, called the contracted functions,[25] are usually taken from the atomic wave-functions. The quality of a wave-function will depend not only on the number of Gaussian basis functions but also on the size of the contracted basis set, namely the number of contracted functions. The size of the contracted basis set ranges from a minimal set (one contracted function for each atomic sub-shell) to sets with two or three contracted functions for each atomic sub-shell.

The large number of Gaussian basis functions usually precludes any orbital exponent optimization in the molecule. Orbital exponents are generally optimized for the free atom or derived from the exponents of Slater orbitals through a least-square fitting.[24] For calculations using a very limited basis set like the $5s, 2p$ set, exponent optimization in the molecule is indeed feasible and should bring a notable improvement, especially with respect to the exponents of the valence shell orbitals. In a similar way, the choice of the linear combinations to be used as contracted orbitals is sometimes based on model molecular calculations for related molecules. For instance the contracted functions in a series of calculations for hydrocarbon molecules may be derived from the wave-function obtained for the methane molecule.[26]

Accurate molecular wave-functions necessitate the inclusion of additional basis functions with respect to atomic calculations to account for the polarization of the electronic cloud in the molecule. These 'polarization' functions[27] describe the distortion of atomic orbitals to form molecular orbitals and are provided by functions with suitable angular characteristics, like p functions on the hydrogen or lithium atoms and d functions on the first-row atoms. The corresponding orbital exponents are sometimes optimized in the molecule, but more often chosen on the basis of a maximum in the electron radial density.[28-30]

11.2.3 *The bond-orbital approach*

An interpretive disadvantage of the LCAO–MO–SCF procedure is that the canonical molecular orbitals do not in general correspond to the chemist's intuitive picture of molecules in terms of localized chemical bonds, lone pairs and inner shells. As mentioned previously, this can be overcome by subjecting the canonical orbitals to a unitary transformation which yields a set of localized orbitals. The bond-orbital approach is an attempt to build these localized orbitals directly without using the SCF procedure.[31] It is assumed that the wave-function can be written as an antisymmetrized product of doubly occupied orbitals representing the atomic inner shells and the localized bonds A—B. The method has been applied almost exclusively to hydrocarbons. For the methane molecule, the four C—H bond orbitals are written as

$$b_i = N(t_i + \lambda h_i)$$

where N is a normalization factor, t_i is a tetrahedral hybrid directed towards the ith hydrogen and h_i is a 1s orbital centred at this same hydrogen nucleus. The method is convenient as long as the basis set is a minimal one of 1s, 2s and 2p orbitals centred at the carbon atom and 1s orbitals centred at the hydrogen. Either Slater functions or contracted Gaussian functions can be used to represent these orbitals. The energy is minimized with respect to the parameter λ. For other hydrocarbons like ethane, it is possible to abandon the assumption of a tetrahedral hybridization with the introduction of an additional parameter in each type of bond orbital[32]

$$b(C - H_i) = N(2s + \lambda 2p_i + \delta h_i)$$

$$b(C_i - C_j) = N[2s_i + 2s_j + \gamma(2p_i + 2p_j)]$$

where $2p_i$ is a normalized linear combination of the $2p_x$, $2p_y$ and $2p_z$ orbitals with its positive lobe pointing towards the hydrogen atom H_i or the carbon atom C_j. The method usually proceeds with the assumption of transferable bond orbitals. For instance, the parameters in the C—H bond orbital are found by minimizing the energy of the methane molecule and are carried over without change to ethane and other saturated hydrocarbons.

In addition to a clear picture of the molecular structure, another advantage of the bond-orbital approach is that it is not necessary to solve the SCF equations. It is not even necessary to compute the two-electron integrals as long as only one-electron properties like the dipole moment are to be evaluated. However, these advantages appear rather limited. Calculation of barrier heights necessitates the computation of the energy, and hence the evaluation of the two-electron integrals. Avoiding the SCF equations brings little benefit in an *ab initio* calculation, since most of the computation time is devoted not to solving these equations but to the calculation of the two-electron integrals. From comparison of SCF and bond-orbital wave-functions, the assumption of transferability of the C—H bond orbitals has been found unjustified.[32]

11.2.4 *Correlated wave-functions and energies*

In the Hartree–Fock method, the wave-function of a closed-shell system of electrons is represented as one Slater determinant. The molecular orbitals are the solutions of one-electron equations with each electron moving in the 'average field' of all the other electrons. This is an independent particle model which neglects the correlation between the motions of electrons. In reality, each electron is surrounded by a Coulomb 'hole'. As a consequence of the $1/r_{\mu\nu}$ term in the Hamiltonian, the probability of finding two electrons with the same space coordinates should be zero. This is completely neglected in the Hartree scheme of a wave-function given as an orbital product. It is partly taken into account in the Hartree–Fock scheme as some correlation is introduced between the electrons with parallel spins through the Pauli principle, leaving the main correlation error arising from electrons with anti-parallel spins.

A variety of methods has been used to obtain correlated wave-functions. One popular approach is the Configuration Interaction (CI) method.[33] It usually requires as a first step the computation of a Hartree–Fock wave-function represented by one Slater determinant built over 'occupied' orbitals. These occupied orbitals satisfy the variational principle. A by-product of this calculation is a set of $m-n$ 'virtual' orbitals, when the SCF equations for a closed-shell system of $2n$ electrons are solved with m basis functions. The CI wave-function is expanded as a superposition of Slater determinants Φ_k

$$\Psi = \sum_k C_k \Phi_k$$

built from both sets of occupied and virtual orbitals. A leading term represents the Hartree–Fock approximation and the other terms represent the configurations obtained by promoting electrons from the occupied orbitals to the virtual ones. The expansion coefficients C_k are determined by minimizing the energy associated with this wave-function. A drawback of the CI method is that the configurational expansion is only slowly convergent. A different approach which has been used for the calculation of barrier heights is the group-function method.[34] The wave-function is written as an anti-symmetrized product of pair functions

$$\Psi(1, 2 \ldots n) = \mathscr{A}[\Phi_A(1, 2)\Phi_B(3, 4) \ldots]$$

where $\mathscr{A}$ is an antisymmetrizer and Φ_A a pair function written in the form of a linear combination of Slater determinant $\varphi_{\mu A}$

$$\Phi_A(1, 2) = \sum_\mu C_{\mu A}\varphi_{\mu A}(1, 2)$$

Each Slater determinant is built from an orthonormal set of orbitals $r_1, r_2 \ldots s_1, s_2 \ldots$ in such a way that the orbitals r_i appear only in the functions $\varphi_{\mu R}$ of group R, the orbitals s_i only in the $\varphi_{\mu S}$ of group S, etc.

11.3 Discussion of the factors influencing the computed barrier

Nearly all the calculations of barrier heights have been done within the LCAO–MO–SCF scheme. If the basis set of atomic orbitals is a complete one, the computed energy value for a given conformation is the Hartree–Fock energy. This Hartree–Fock energy is above the experimental energy and the difference includes (excluding the zero-point energy[35]):

(a) The relativistic energy since the Hamiltonian in the Schrödinger equation is a non-relativistic one. For the lighter elements, this relativistic term amounts to 0·1 % or less of the total energy. However, it increases with the atomic number and for the silicon atom its value is of the order of 300 kcal/mole.[36]

(b) The correlation energy, defined as the difference between the Hartree–Fock energy and the non-relativistic experimental energy. The correlation energy generally amounts to 0·5 % or less of the total energy[37,38] (the experimental value for the total energy of the ethane molecule is −79·84 a.u.[39] and the correlation energy is of the order of 0·58 a.u. or 361 kcal/mole[40]).

Since the basis set is an incomplete one, the computed energy value is higher than the Hartree–Fock energy value. The difference will depend on the size of the basis set and will range between 1 a.u. (623 kcal/mole) and 0·01 a.u. (about 6 kcal/mole). Within this model, calculation of a rotation barrier as the difference between the computed energies for the ground state and the transition state involves the following approximations (Figure 11.1).

(a) The relativistic energy is assumed to be the same for both states. This assumption appears reasonable since the relativistic term can be viewed as a sum of the atomic contributions and is then expected to be fairly constant for different conformations. Lehn pointed out[41] that another indication of the invariance of the relativistic term is found in the relativistic treatment of the H_2 molecule by Kolos and Wolniewicz.[42] The kinetic energy and the relativistic energy show a similar dependence on H—H bond length, with a change of about 600 kcal/mole in kinetic energy corresponding to a variation of only 0·01 kcal/mole in relativistic energy. Kinetic energy changes being in general of the order of 100 kcal/mole or less from one formation of a molecule to another, the relativistic term may be considered as invariant.

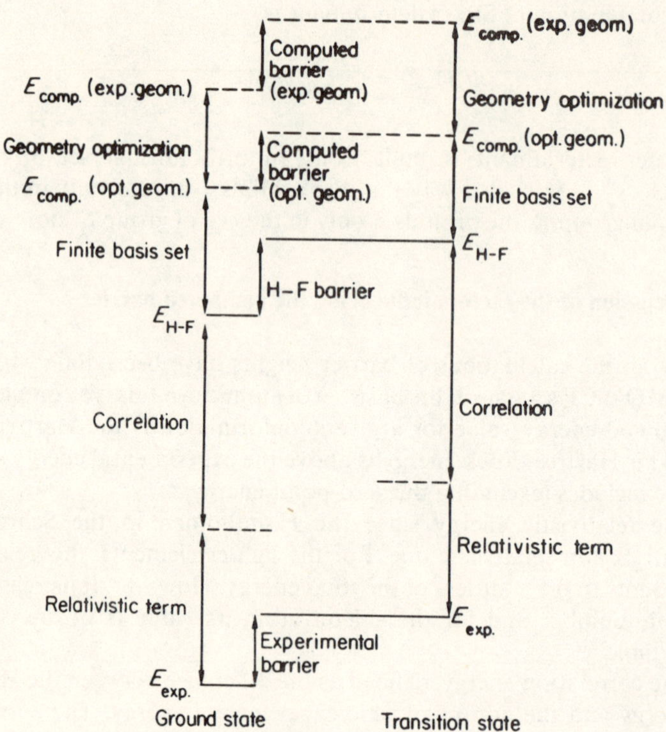

Figure 11.1 Approximations in the computation of a barrier height

(b) The correlation energy is assumed to remain the same for both states. Justification of this invariance requirement stems principally from the fact that calculations approaching the Hartree–Fock limit satisfactorily reproduce the energy barriers (cf. § 11.4). A qualitative *a priori* justification may be found in the usual division of correlation energy between intra-pair and inter-pair correlation.[43] The intra-pair correlation energy is considered to remain roughly constant for isoelectronic closed-shell systems with the inter-pair correlation energy depending mostly on the pair distances. Since the average pair distances remain approximately constant in a rotation process, the inter-pair correlation energy and therefore the total correlation energy are expected to remain approximately the same for different conformations.

(c) The error introduced by the use of a limited basis set is assumed to be the same for both conformations. This corresponds to the frequent assumption that satisfactory barrier heights will be obtained with 'medium quality' basis sets, for instance with a minimal basis set of Slater orbitals.[39,44] In other words, the basis set is assumed to be equally good for both conformations. This assumption seemed to be corroborated for some time by the satisfactory results for the computed barrier of ethane (between 2·5 and 3·5 kcal/mole) whatever the basis set used.[19,31,32,45–48] This good agreement for the ethane molecule is probably explained on the following basis. Firstly, the high symmetry in the ethane molecule may result in a large cancellation of errors.[1] Secondly, the rotation process in ethane may be considered as a 'second-order' process, since it involves a change of orientation only between bonds originating from different atoms, as opposed to a 'primary' process like the inversion of ammonia. A primary process brings about a change of orientation between bonds originating from the same atom, implying a change in hybridization and a first-order effect on the electron density. It was later realized that the barriers and conformations computed with a limited basis set are not always reliable; for instance the results obtained for the hydrogen peroxide molecule have been found very sensitive to the basis set used.[19,40,46,47,49–55] In this respect, the polarization functions appear to be rather important by allowing for a greater flexibility in the molecular wavefunction. The inclusion of polarization functions in the basis set is crucial for the calculation of the *trans* barrier in the hydrogen peroxide molecule.[40,54,55] From some preliminary results, it would seem that a large basis set containing only s and p functions leads to less satisfactory results than a smaller s, p set to which polarization functions have been added.[55]

(d) Barrier heights are generally computed with the same geometrical parameters (bond lengths and bond angles) assumed for both conformations (except for the rotation parameter). These parameters are usually taken from the experimental geometry for the conformation of lowest energy (conformational ground state). A really complete *ab initio* calculation would require that all geometrical parameters be optimized for each conformation. In practice, this is nearly impossible except for very simple molecules. If the calculation is of good quality, the geometry of the conformational ground state should be close to the experimental geometry since the Hartree–Fock method is known to lead to fairly accurate geometries.[56] Geometry optimization would then appear as more important for the transition state than for the ground state. On this basis, one would expect that optimizing the geometry would lower the computed barrier, since the experimental geometry of the ground state should be closer to the optimized geometry for the ground state than to the optimized geometry for the transition state. It will be seen in the discussion of the computed barriers (see § 11.4) that this is generally true, but not always. Anyway, most of the barriers have been computed with the experimental bond lengths and bond angles for both conformations.

Figure 11.1 summarizes the various approximations in the calculation of a barrier height through the LCAO–MO–SCF method with a finite basis set. Each calculation represents a compromise between the size of the system, the chemical interest of the problem, the refinements in the theoretical treatment (size of the basis set, exponent and geometry optimization) and the amount of computer time available (or its economic and financial cost). For a limited number of small, well-chosen molecules, near Hartree–Fock calculations with geometry optimization are feasible. For medium-size molecules with two to four atoms of the first row, the use of a medium-size basis set together with the experimental geometry or partial geometry optimization will lead to less accurate but still reliable results. For larger systems with more than four first-row atoms, *ab initio* calculations are still feasible with a suitably chosen basis set. For instance, small Gaussian basis sets calibrated for hydrocarbons have been relatively successful in the calculation of hydrocarbon barrier heights.[26,57] The same small basis set was later used for a study of the inversion process in the cyclopentane and cyclohexane rings, again with nearly quantitative agreement.[58]

11.4 Survey of the computed barrier heights

11.4.1 *Near Hartree–Fock calculations with geometry optimization*

So far, near Hartree–Fock calculations with at least partial geometry optimization have been reported for three molecules only: ethane, hydrogen peroxide and formamide. The first *ab initio* calculation of a barrier height was presented by Pitzer and Lipscomb for the ethane molecule[39] in 1963, giving a moderately good agreement with experiment despite the crude level of the theoretical treatment. This work opened the way for the calculation of barrier heights in hydrocarbons and further refinements in the theoretical treatment have led to only a slight improvement in the barrier height. A different situation has been found for the hydrogen peroxide molecule. Despite the fact that it is the simplest molecule with internal rotation, early calculations were unable to reproduce the shape of the potential curve for internal rotation. This could be achieved only at a rather sophisticated level in the calculation, with the inclusion of polarization functions and geometry optimization. Formamide, which is the simplest molecule containing the N—C—O linkage characteristic of polypeptides, is a more complicated problem since the rotation around the C—N bond is now coupled with nitrogen inversion. We shall review successively the various results obtained for the three above-mentioned molecules.

11.4.1.1 *Ethane* The first calculation by Pitzer and Lipscomb for this molecule used a minimal basis set of Slater functions with the experimental geometry for both conformations and yielded a computed barrier of 3·3 kcal/mole as compared with an experimental value of 2·93 kcal/mole.[39] This computed value was raised to 3·5 kcal/mole with the use of orbital exponents optimized for the methane molecule instead of those obtained from Slater's rule.[48] Various calculations with a Gaussian basis set lacking polarization functions and with the experimental geometry yielded a computed barrier in the range 2·5–3·6 kcal/mole (see Table 11.5) and indicated that the barrier is not too sensitive to a particular choice of the basis set.[31,32,45–47] A refined calculation was achieved with the use of a large Gaussian basis set including polarization functions on both carbon and hydrogen atoms and separate optimization of the C—C bond length and the HCH bond angle for the staggered and eclipsed conformations.[40] It was then apparent that most previous calculations had benefited from a slight cancellation of errors by neglecting simultaneously the polarization functions and the geometry optimization. In the above calculation, introduction of polarization functions raised the barrier (computed with the experimental geometry) from 3·42 kcal/mole to 3·65 kcal/mole, while the geometry optimization decreased it to 3·07 kcal/mole, in very close agreement with the experimental determination of 2·928 kcal/mole from the infrared torsional spectrum.[59] The optimized values for the C—C bond length and the HCH bond angle in the staggered conformation are in good agreement with the experimental data and the eclipsed conformation is found to have a more 'open' structure with smaller HCH angles and a longer C—C bond than the staggered conformation. The corresponding variations were computed as 0·019 Å for the C—C bond length and 0·34° for the HCH angle. Rather similar results regarding the optimized geometrical parameters were obtained by Stevens using a Slater minimal basis set with optimized exponents[19] and by Lathan and coworkers using both a minimal (for geometry optimization) and an extended set.[60]

The very good agreement between the above computed value and the experimental barrier for ethane implies that the change in correlation energy between the two conformations is practically negligible. Attempts to go beyond the Hartree–Fock scheme have been made in two ways for the ethane molecule. The group-function method has been used in two independent calculations and both predicted the eclipsed conformation to be more stable than the staggered conformation.[61,62] A configuration interaction calculation has been reported by Levy and Moireau.[63] The improvement in the wave-function with respect to the one-determinant approximation resulted in a decrease of the barrier of about 0·13–0·17 kcal/mole. This is nearly exactly the difference (0·14 kcal/mole) between the best LCAO–MO–SCF value[40] and the experimental value. However, although this calculation includes part of the correlation energy, the computed energy is still above the Hartree–Fock limit.

11.4.1.2 *Hydrogen peroxide* This molecule is non-planar in its ground state, the equilibrium dihedral angle being in the range 111–120°,[64,65] with two barriers of 7·0 and 1·1 kcal/mole corresponding to the *cis* and *trans* conformations respectively.[64] Calculations with a minimal set of Slater functions failed to reproduce the *trans* barrier, even with optimization of the geometries and orbital exponents.[19,49,50] Calculation with a large

Gaussian basis set but without any polarization function equally predicted the molecule to be *trans*-planar in the equilibrium conformation.[47] However, two calculations with a smaller Gaussian set, but including *p* polarization functions on the hydrogens, produced a slight well of about 0·1 kcal/mole in the energy curve for a dihedral angle close to 150°.[46,47] Newton and coworkers have reported a calculation using a minimal basis of Slater functions, each one mimicked by three Gaussian functions with standard exponents and geometry optimization, leading to a dihedral angle of 125°; however, no barriers to rotation are quoted.[53] Most of the above calculations yield a barrier to the *cis* form which is too high (10–15 kcal/mole). Hillier and coworkers reinvestigated the use of both a minimal basis and a more flexible basis in which the form of the valence orbitals was allowed to vary (they used a least-square fit of Slater functions or atomic orbitals), with again no barrier to the *trans* form.[52] Only the use of three Gaussian *p* functions on each hydrogen gave satisfactory values for the dihedral angle and the *cis* and *trans* barriers.[52] The effect of both polarization functions and geometry optimization was first investigated by Veillard.[40] With a large Gaussian set and one *p* function on each hydrogen, the *cis* and *trans* barriers are 14·7 and 0·2 kcal/mole respectively, with a dihedral angle of 136°. Geometry optimization decreases the *cis* barrier to 10·9 kcal/mole and raises the *trans* barrier to 0·6 kcal/mole with an equilibrium of 123°. This has been recently reinvestigated by Dunning and Winter,[54] who have achieved very accurate values of the barriers (8·3 and 1·1 kcal/mole) and dihedral angle (114°). The effect of polarization functions and geometry optimization when a medium-size Gaussian set is used has been reported recently with nearly quantitative agreement.[55] Again, the *cis* conformation is found to have a more 'open' structure than the *trans* one,[40,54] the change in O—O bond length being 0·025 Å and the change in the OOH angle of the order of 5–6°. The general agreement regarding the calculation of barrier heights in hydrogen peroxide seems to be that the inclusion of polarization functions in the basis set is crucial for the calculation of the *trans* barrier, with the optimization of geometry also of considerable importance.[40,54,55]

11.4.1.3 *Formamide*   Only one *ab initio* calculation is reported for the barrier hindering internal rotation around the amide carbon–nitrogen bond in formamide.[66] The large number of geometrical parameters precludes a full geometry optimization and only the magnitude of the changes in the carbon–nitrogen and carbon–oxygen bond lengths accompanying the internal rotation were established. The calculated barriers are in good agreement with experiments and appear rather insensitive to geometry optimization (due probably to the high value of the barrier). The carbon–nitrogen bond length increases by 0·055 Å upon a 90° rotation, due to the loss of double-bond character during the twisting motion. Likewise a smaller shortening of the carbon–oxygen double bond seems to occur.

No definite conclusion was reached regarding the planar or non-planar conformation of formamide.

Although no definite conclusions may be reached from the few near Hartree–Fock calculations with geometry optimization, some trends and qualitative indications are apparent. Good agreement between calculated and experimental barrier heights may be fortuitous when small basis sets are used with the experimental geometry. This will be the case when the inclusion of polarization functions and the geometry optimization have an opposite effect on the barrier height, as happens for the ethane molecule. However, it is then possible to extrapolate the use of small sets to larger molecules, as has been done for hydrocarbons. On the other hand, both the inclusion of polarization functions and the geometry optimization will be of prime importance when they have a similar effect on the barrier height, as happens for the *cis* and *trans* conformations of hydrogen peroxide. The assumption that geometry optimization will have a larger effect on the transition-state energy, hence decreasing the barrier, is not supported by the results for the *trans* barrier of hydrogen peroxide.[40] Finally, the practical limitations of geometry optimization are clearly apparent even for as small a molecule as formamide. All the above-mentioned optimizations have been performed by brute force techniques through a numerical minimization of the energy. A promising approach consists of determining analytically the forces acting on the nuclei, and then relaxing the geometry until the forces are zero.[67]

11.4.2 *Calculations with a limited basis set*

These will range from calculations which are still fairly accurate ($9s, 5p$ or $10s, 6p$ Gaussian basis sets for the first-row atoms with a flexible contraction) to calculations which are considered as approximate (sub-minimal Gaussian basis sets, $6s, 3p$ or $5s, 2p$ for the first-row atoms). A compromise corresponds to the use of a minimal Slater basis set (represented either with Slater functions or through their Gaussian expansion) or of a Gaussian basis set of similar accuracy ($7s, 3p$ or $8s, 4p$ for the first-row atoms). The spirit of the calculation will be different depending upon whether it attempts to reproduce some experimental value or to predict an unknown barrier height. If the calculation is intended for its predictive value, the method will usually be tested for similar and experimentally known barriers.

Saturated hydrocarbon molecules other than ethane were first considered by Hoyland. His calculations were carried out first through the bond-orbital approach,[31,32] and then by the SCF procedure.[26,57] The energy curve for the propane molecule could be accounted for by the following expression[26]

$$E(\alpha_1, \alpha_2) = \tfrac{1}{2}V_0(2 - \cos 3\alpha_1 - \cos 3\alpha_2) - \tfrac{1}{2}V_1[1 - \cos 3(\alpha_1 + \alpha_2)]$$

in which α_1 and α_2 are the angles representing the rotation of the two methyl groups, V_0 is the rotational barrier of a methyl group due to the methylene and the term involving V_1 represents the top–top coupling. An excellent agreement was found for the values of V_0 and V_1 from the microwave data and the SCF calculations.[26] For the butane molecule, an excellent agreement with the experimental data was obtained for the energy difference and the angle of rotation between the *trans* and *gauche* conformations together with the barriers hindering conversion from the *trans* to the *gauche* conformations and between the two *gauche* conformations.[57] A systematic survey of internal rotation in hydrocarbons has been carried out by Radom and Pople, with a flexible rotor model in which each CCC angle is separately optimized for each conformation considered, in order to relieve steric strain.[44] Standard values were used for the bond lengths and the other bond angles. In addition, the form of the potential function required to describe the rotation is also discussed. Comparison is made of the potential barrier between closely related molecules (for instance ethane and 1-butyne). The overall agreement with the experimental data is good and many barrier heights are predicted. Barrier heights for substituted propanes have also been reported.[68] Rotational barriers in propene and fluoropropene have been investigated by several authors,[69–71] with the low barrier for *cis*-fluoropropene accounted for.

Several studies have been devoted to the rotational barriers of alkyl cations. The rotation barrier of the $C_2H_6^+$ ion (corresponding to the removal of one electron from the $3a_{1g}$ orbital) was found nearly identical to the one in ethane.[72] In the ethyl cation $C_2H_5^+$ with the CH_2^+ group held in planar geometry, the six-fold rotational barrier is nearly zero,[60,73,74] while, when the CH_2^+ groups is made tetrahedral, the three-fold barrier is 2·8 kcal/mole (close to the value for ethane). The 1-fluoroethyl cation has a very low barrier to rotation (0·62 kcal/mole) with hydrogen eclipsing fluorine in the most stable conformation.[75] The most stable conformer of the 2-fluoroethyl cation also has hydrogen eclipsing fluorine, but the barrier is now 10·53 kcal/mole.[75] Two-fold barriers in the n-propyl or isobutyl cations are found to be higher than the six-fold barriers in the ethyl or neopentyl cations.[76] Incorporation of the methyl groups into ring systems leads to a barrier height of 17 kcal/mole in the cyclopropylcarbinyl and 4 kcal/mole in the cyclobutylcarbinyl.[76] The barrier in a number of substituted 1-propyl cations has been found highly dependent on the substituent, the dominant effect producing the barrier being a preferential stabilization of the perpendicular conformation through interaction of the CH_2X group with the $2p(C^+)$ orbital.[68] The effect of geometry optimization has also been considered for the 1-propyl cation.[77]

A calculation of the rotational barrier in ethylene has been reported by several authors.[44,78,79] Since the N and Z electronic states of ethylene are separated by a large energy gap in the planar configuration but become degenerate in the D_{2d} symmetry, a CI (configuration interaction) calculation, which mixes the N and Z states at each angle of rotation, is needed to correctly predict the barrier height.[78] Potential curves for the rotation in excited states and in the positive and negative ions of ethylene are also given.[78] A more recent calculation with optimization of the C—C distance at each rotational angle gives a rotational barrier of ethylene in its ground state of 63·7 kcal/mole, in excellent agreement with the generally accepted experimental value of 65 kcal/mole.[80] Theoretical calculations of the barrier height in allene have been reported by several authors.[78,81,82]

Calculations of barrier heights have been reported for compounds with a C=O or N=O double bond adjacent to a methyl group. In the acetaldehyde molecule, the equilibrium conformation is found to be the one with one methyl hydrogen eclipsing the C=O double bond,[83–85] in agreement with experiment and with the results for the propene molecule[69–71] (with a C=C double bond in place of a C=O double bond). A similar situation holds in the acetyl radical,[86] with a barrier smaller by about 1 kcal/mole than the one in acetaldehyde.[87] The nitrosomethane CH_3—N=O has also been found to have the oxygen eclipsed to a methyl hydrogen,[88] with a barrier height close to the one in acetaldehyde.

The rotation barrier around a C—N bond has been investigated in the methylamine molecule[46,47] and in the methylnitroxide radical CH_3NHO.[84] With the assumption of a planar conformation at the nitrogen, the methylnitroxide radical is found to have one C—H bond perpendicular to the CNO plane. The relative stabilities of the possible conformers in substituted methylamines have been given by Radom and coworkers (no barrier heights are reported).[89]

Rotation around a C—O bond has been studied in methanol[46,47] and in formic acid.[90] A barrier height of 31 kcal/mole has been found for the rotation around the C=O bond in protonated formaldehyde H_2COH^+, due to the character of the double bond (for the in-plane motion of the OH bond, the barrier height is only 17 kcal/mole).[91] A similar situation holds for the protonated acetaldehyde.[91] In the protonated formic acid with a CO bond character between that of a single and a double bond, rotation around the bond is much easier (13·5 kcal/mole). The relative stabilities (but no barrier heights) of the possible conformers in substituted methanols have been given by Radom and coworkers.[89]

Theoretical work has also been devoted to the rotation about a C—X bond, with X a second-row atom. Barrier heights have been computed for the methylsilane[72] and methylphosphine[92] molecules. For hydrogen methyl sulphoxide, HMSO, rotation about the C—S bond leads to a minimum of energy for a conformation midway between the staggered and eclipsed conformations.[93] Hydrogen methyl sulphinyl anion HMSO$^-$ is found to have its minimum energy conformation with the carbanion lone pair situated in the plane of the bisector of the oxygen–sulphur lone-pair angle.[93] Similar energy curves as a function of the rotation angle about the C—S bond were found for the HMSO$^-$ and DMSO$^-$ (methyl sulphinyl carbanion) anions.[93] The energy surface corresponding to the rotation about the C—S bond and to the inversion of the carbanion angle was used to predict the stereochemistry of proton exchange alpha to a sulphinyl group.[93] This work was extended to the hydrogen methyl sulphonyl carbanion HMSO$_2^-$.[94] A three-dimensional energy surface was also computed for the thiomethoxide ion HMS$^-$ for rotation about the C—S bond and inversion of the HCH angle, leading to an energy minimum for the conformation of HMS$^-$ which maximizes *gauche* interactions between adjacent electron pairs.[95] For dimethyl sulphide dicarbanion DMS^{2-}, the most stable structure also contains the maximum number of *gauche* interactions between adjacent electron pairs.[95] The *d*-orbitals of sulphur do not appear to be necessary to explain the above structures.[93–95]

Work on the rotation about an X—Y bond, with X and Y two atoms of the first or second row other than carbon, has been reported for a few molecules. The borazane molecule BH$_3$NH$_3$ which is isoelectronic to ethane has a rotation barrier very close to the one in ethane.[72,96] Barrier heights and conformations in the hydrazine molecule were investigated by several authors.[46,97,98] Barrier heights for rotation around a N—O bond have been reported for hydroxylamine[46,97] and nitrous acid.[90] The diphosphine molecule is found to have a *gauche* conformation like the hydrazine molecule, with much smaller barrier heights.[99] Experimental results regarding the barrier height in hydrogen persulphide have been discussed and complemented on the basis of theoretical calculations.[100,101]

11.4.3 *Shape of the barrier*

For a three-fold barrier like the one in ethane, the potential energy function may be written as a Fourier series

$$V(\alpha) = \tfrac{1}{2}V_3(1 - \cos 3\alpha) + \tfrac{1}{2}V_6(1 - \cos 6\alpha) + \cdots$$

where α is the angle of internal rotation. It is usually assumed that the higher terms of this expansion are much smaller than the first one. Whereas V_3 is the value of the barrier height, V_6 modifies the shape of the barrier. Little theoretical work has been devoted to the shape of the barriers (the calculation of V_6 requires computation of the energy for four different angles of rotation, while only two are needed for the calculation of V_3). The calculated values of V_6 for three-fold barriers are summarized in Table 11.1. Calculations for the propane molecule seem to indicate that the computed value of V_6 may be rather sensitive to the geometry used.[44] In most cases they are also on the threshold of significance as determined by the computational accuracy.

Table 11.1 Calculated values of V_6 for three-fold barriers

	V_6 (kcal/mole)	Reference
Ethane	-0.012	45, 96
	-0.012	44
	-0.007	31
	-0.010	102
Propane	-0.004	44
1-Butyne	-0.003	44
Propene	0.004	44
1,2-Butadiene	0.004	44
2-Methylpropene	0.014	44
cis-2-Butene	0.020	44
trans-2-Butene	0.002	44
Borazane	0.006	96

In a molecule with two equivalent methyl groups (e.g. propane), the potential function may be expressed as a function of the two angles of rotation α_1 and α_2

$$V(\alpha_1, \alpha_2) = \tfrac{1}{2}V_3(1 - \cos 3\alpha_1) + \tfrac{1}{2}V_3(1 - \cos 3\alpha_2) + \tfrac{1}{2}V_6(1 - \cos 6\alpha_1)$$
$$+ \tfrac{1}{2}V_6(1 - \cos 6\alpha_2) + \tfrac{1}{4}V_3'(1 - \cos 3\alpha_1)(1 - \cos 3\alpha_2)$$
$$+ \tfrac{1}{4}V_3'' \sin 3\alpha_1 \sin 3\alpha_2 + \cdots$$

Computed values of V_3' and V_3'', which represent the interaction of the two rotors, have been reported by Hoyland[26] and by Radom and Pople.[44]

If the rotor and the framework are asymmetric, a general expansion must be used:

$$V(\alpha) = \tfrac{1}{2}V_1(1 - \cos \alpha) + \tfrac{1}{2}V_2(1 - \cos 2\alpha) + \tfrac{1}{2}V_3(1 - \cos 3\alpha) + \cdots$$

Values of the computed parameters V_i have been reported for 1,3-butadiene and n-butane,[44] hydrazine,[98] hydrogen peroxide[40] and hydrogen persulphide.[101]

11.5 The origin of rotation barriers from *ab initio* calculations

In the earlier work on rotation barriers, various hypotheses were put forward concerning the origin of these barriers.[1] These include classical electrostatic and repulsive potential models, together with the application of quantum-mechanical theorems, like the virial theorem or the Hellmann–Feynman theorem, to relatively simple wave-functions. *Ab initio* calculations have been directed rather to the determination of barrier heights and conformations than to the elucidation of the origin of these barriers. However, these calculations have produced a large amount of data from which it has been possible to test the various hypotheses regarding the origin of the barriers. We do not intend in this chapter to review the theories of the origin of the barriers, but only to report on whatever progress has been made in the elucidation of their origin through *ab initio* calculations. So far, these attempts have led to rather negative conclusions, that is, many effects have been found to be not responsible for the barrier, but it has not been possible to pinpoint a simple explanation of what causes the barrier. The fact that quantum-mechanical calculations have been able to satisfactorily reproduce the barrier heights indicates that the origin of the barrier is to be found in the interactions between electrons and nuclei. However, this kind of explanation has no predictive value and does not provide any framework to rationalize the experimental or computational results.

The good agreement with experiment obtained for barrier heights computed from Hartree–Fock wave-functions indicates that the origin of the barrier is to be found in the Hartree–Fock approximation. If Van der Waals forces were the dominant factor, it would not be possible to account for it, since a theoretical treatment of the dispersion forces requires the inclusion of correlation effects. Levy and Moireau have attempted to account for the lack of any correlation effect on the barrier height of ethane.[63] According to the Brillouin theorem, the correction to the rotation barrier, obtained through the inclusion of one-electron excitations in the CI wave-function, is nearly zero when the molecular orbitals are either the SCF orbitals or a set of localized orbitals resulting from the preceding ones. The dipolar terms associated with the two-electron excitations and corresponding to the dispersion forces are found to yield only very small contributions to the barrier. The maximum variation of the interaction between two CH bonds from *cis* to *trans* conformation is 0.3 kcal/mole, with a total contribution which is much smaller because of a large cancellation due to the three-fold symmetry. With a set of bond orbitals, the direct intermethyl correlation effects (including both the dispersion and overlap terms) are very small and the other terms completely balance each other. A further conclusion was that the correlation effect in molecules with a smaller symmetry is not likely to exceed 0.3 kcal/mole.

There are two main approaches to the origin of rotation barriers: by investigating the energetical origin of the barrier or by scrutinizing the electronic changes accompanying the rotation process.

11.5.1 *Energy component analysis of the rotation barriers*

The total energy of a molecule is a sum of four terms, the nuclear repulsion V_{nn}, the nucleus–electron attraction V_{ne}, the electron–electron repulsion V_{ee} and the electronic kinetic energy T

$$E_{tot} = V_{nn} + V_{ne} + V_{ee} + T$$

It would seem natural to study the variation of each of these terms during the rotation process. However, in order to describe the origin of the barrier in terms of energy components, one should retain a division of the total energy which is relatively independent of small changes in the wave-function, those due to the differences in the degree of completeness of the basis set or to the uncertainties in the experimental specification of the molecular geometry. The rotation barrier itself has been found remarkably stable under these two perturbations. It is hoped that, for a suitable division of the energy into two components X and Y, the comparative changes ΔX and ΔY during the rotation process, although probably less stable under the above mentioned perturbations than the change in total energy (i.e. the rotation barrier), will be stable enough to provide a qualitative picture of the rotation barrier.

The division of the energy into its four components does not exhibit the required stability under the above-mentioned perturbations. This has been emphasized by Epstein and Lipscomb[103] for the ethane molecule from the results of calculations by Pitzer and Lipscomb[39] and Stevens.[19] These two calculations give a calculated barrier within 10^{-5} a.u. of one another, yet the contributions to the barrier of the nuclear repulsion, nuclear attraction and electron repulsion all have different signs in the two calculations.

A first two-component decomposition of the total energy was put forward by Fink and Allen[47] as

$$E_{\text{tot}} = (V_{\text{nn}} + V_{\text{ne}} + T) + V_{\text{ee}}$$

With several wave-functions for the ethane and methanol molecules, the one-electron and nuclear energy $V_{\text{nn}} + V_{\text{ne}} + T$ remained out of phase with the two-electron energy V_{ee}. However, it was found that these 'invariants' showed no invariance between the two above-mentioned calculations for ethane.[103]

Allen later introduced a division into an attractive term $V_{\text{a}} = V_{\text{ne}}$ and a repulsive term $V_{\text{r}} = V_{\text{nn}} + V_{\text{ee}} + T$, with the assumption that it is only the components' combination that retains an adequate stability under the basis set and geometry perturbations.[104] A noticeable feature of this partitioning is the possibility of attractive interactions as the major element of the barrier, although for a long time it has been tacitly assumed that barriers are produced by repulsive forces. Thus, depending on which variation ΔV_{a} or ΔV_{r} is the larger, the barrier may be considered as either attractive dominant (Figure 11.2) or repulsive dominant (Figure 11.3). In these figures the plots have been adjusted so that the lowest calculated points of V_{a} and V_{r} match the zero of energy. It turns out that the variations ΔV_{a} and ΔV_{r} are of opposite phase when plotted as a function of the rotation angle.[104]

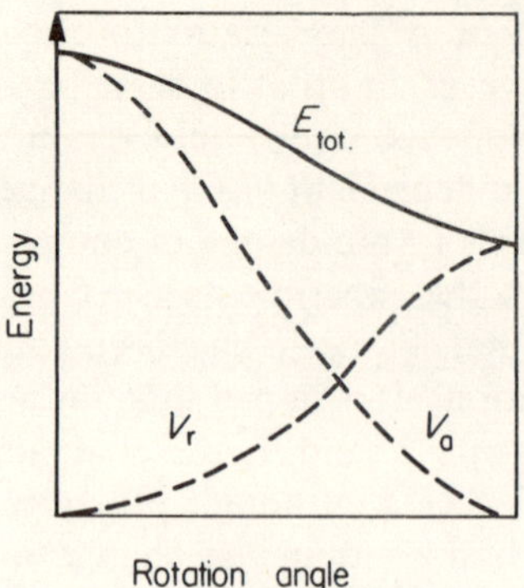

Figure 11.2 Attractive-dominant
barrier

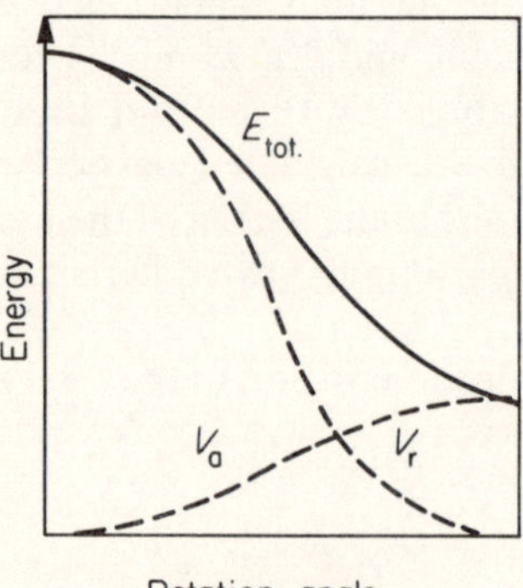

Figure 11.3 Repulsive-dominant
barrier

With this energy partitioning, the barriers in ethane, methanol, methylamine, ethyl fluoride, hydrazine (both *cis* and *trans*) and propene were found
to be repulsive dominant, while the barriers in hydroxylamine, acetaldehyde
and nitrosomethane appeared to be attractive dominant.[69,85,88,104] Examination of different calculations for the same compound will tell us how much
this image is insensitive to the size of the basis set and to small changes in
the geometry. The attractive and repulsive energy differences in ethane
from various SCF calculations are compared in Table 11.2. Calculations
with the experimental geometry assumed for both conformations agree
that the barrier in ethane should be repulsive dominant, despite the differences
in the basis sets used. However, the barrier is found to be attractive dominant
in the calculation by Stevens using different geometries for each conformation.[19] It is apparent that geometry optimization results in rather drastic
changes of the energy components, for instance the nuclear repulsion energy

Table 11.2   Attractive and repulsive energy differences for ethane SCF calculations

Wave-function	$\Delta V_{\mathrm{a}}^{a}$	$\Delta V_{\mathrm{r}}^{a}$	Nature of the barrier	Reference
Fink–Allen	+0·0313	−0·0353	Repulsive	47
Pedersen	+0·0390	−0·0436	Repulsive	46
Pedersen	+0·0415	−0·0470	Repulsive	46
Pitzer–Lipscomb	+0·0490	−0·0542	Repulsive	39
Stevens[b]	−0·2011	+0·1959	Attractive	19

$^{a}\,\Delta V = V_{\mathrm{staggered}} - V_{\mathrm{eclipsed}}.$
^{b} Optimized geometry.

turns out to be smaller for the eclipsed conformation than for the staggered one. Even the use of experimental geometry does not always ensure the stability of this energy partitioning with respect to the basis set. The attractive and repulsive energy differences for the *cis* barrier of hydrogen peroxide are summarized in Table 11.3. With the experimental geometry used for the different conformations, the *cis* barrier is found to be attractive dominant in four calculations and repulsive dominant in two calculations. Again, the same calculation leads to opposite conclusions regarding the nature of the barrier depending on the use of the experimental geometry or the optimized geometry.[40] In the hydrogen persulphide molecule, both the *cis* and *trans* barriers are found to be repulsive dominant in one calculation,[100] while the *cis* barrier appears as repulsive dominant and the *trans* barrier as attractive dominant in a different calculation.[101] Since the same geometrical parameters were used in both calculations, the difference regarding the nature of the barrier should be traced to the differences in the basis set. We are thus led to the rather unfortunate conclusion that the energy division into attractive and repulsive terms does not exhibit the required stability with respect to differences in the basis set and to geometry optimiza-

Table 11.3   Attractive and repulsive energy differences for the *cis* barrier of hydrogen peroxide

Wave-function	$\Delta V_{\mathrm{a}}^{a}$	$\Delta V_{\mathrm{r}}^{a}$	Nature of the barrier	Reference
Fink–Allen	−0·1606	+0·1388	Attractive	47
Pedersen	−0·0439	+0·0185	Attractive	46
Palke–Pitzer	−0·0724	+0·0516	Attractive	50
Hillier and coworkers	−0·0639	+0·0500	Attractive	52
Davidson–Allen	+0·0724	−0·0945	Repulsive	105
Veillard	+0·068	−0·091	Repulsive	40
Veillard[b]	−1·112	+1·094	Attractive	40

$^{a}\,\Delta V = V_{\mathrm{minimum\ energy}} - V_{cis}.$
^{b} Optimized geometry.

tion. A similar conclusion regarding the effect of geometry optimization on the nature of the barrier has been reached for the inversion process of phosphine.[106]

A different energy partitioning has been used recently by Clementi and von Niessen for the ethane barrier.[107] It is based on a break-down of the total molecular energy into one-, two-, three- and four-centre energy contributions:

$$E = \sum_A E_A + \sum_{A,B} E_{AB} + \sum_{A,B,C} E_{ABC} + \sum_{A,B,C,D} E_{ABCD}$$

For instance the summation on the two-centre bond energies $\sum_{A,B} E_{AB}$ includes the interaction between one carbon atom with the second carbon atom and with any of the six hydrogen atoms, and the interactions between one hydrogen atom and a second hydrogen atom. A convenient feature of this analysis is the possibility of grouping the many terms into a small number of sets. For instance, the ethane molecule may be considered as being formed from two methyl groups $(CH_3)-(CH_3)$ and the energy may be written as a sum of three terms corresponding to the two methyl groups and to their mutual interaction respectively. It is then found that the barrier is again the result of heavy cancellation of large terms with opposite signs. For instance, the energy term associated with each methyl group favours the eclipsed conformation, but the methyl–methyl interaction more than compensates this effect. However, results for other molecules are needed in order to judge the usefulness of this analysis and of its stability with respect to the basis set and to small changes in geometry.

Pitzer has given an energy analysis for the ethane molecule in terms of localized orbitals corresponding to the CH and CC bonds and the carbon inner-shells.[108] Although the larger changes in the energy components all involve the CH orbitals, he could not find any simple explanation for the barrier.

Although the SCF total energy is not equal to the sum of the individual molecular orbital energies, it has been shown that in a certain number of situations these two quantities vary in a parallel way.[109] It was then of interest to investigate the possibility of barrier height predictions from the change in the sums of molecular orbital energies. It turned out that even the shape of the energy versus angle curve may be incorrect if the total energy is replaced with the sum of the molecular orbital energies.[47] A related point is the proposal by Lowe that a simple explanation for the ethane barrier may be found in the difference of the energies associated with the e orbitals in the staggered and eclipsed conformations.[110] Although the orbital energies reported by Pitzer and Lipscomb would seem to support Lowe's arguments, there is nothing similar in the calculation by Stevens.[103]

11.5.2 *Analysis of rotational barriers in terms of wave-function and electron density*

Bond-orbital wave-functions for ethane and their influence on the computed barrier were examined by Sovers and coworkers.[102] The wave-function was either a Slater determinant or a Hartree product. Reasonable values for the barrier are obtained within a large range of the C—H ionicity parameter, provided that the wave-function is antisymmetrized. The barrier obtained from the Hartree product model has the wrong sign for a reasonable ionicity. Through antisymmetrization, the wave-function satisfies the Pauli exclusion principle with the constraint that the orbitals be orthogonal. Each bond orbital must therefore mix to a small extent with all of the others, and the extent of this mixing changes during internal rotation because the relative orientation of the bond orbital changes. It is concluded that the dominant contribution to the barrier is the overlap repulsion between closed-shell localized C—H bond orbitals which are not orthogonal to each other. A further conclusion was the relatively minor importance of changes in bond orbitals other than those required by orthogonality, as indicated by the similarity of barriers calculated with a variety of LCAO–MO functions.

Changes in the charge density distribution during the rotation process have been analysed by Jorgensen and Allen for the ethane and acetaldehyde molecules.[111] A difference plot of the eclipsed ethane electron density minus the staggered ethane electron density shows that there is more charge behind the hydrogen atoms in the eclipsed conformation. This would result from an increased repulsion between the eclipsing hydrogens which forces charges behind them. This increased repulsion seems confirmed by the accompanying decrease in the H—H overlap population (Table 11.4). Simultaneously a decrease in electron density behind the carbon atoms is observed by rotating to the eclipsed conformation, indicative of a slightly weaker C—C bond in eclipsed ethane. This is supported by the corresponding reduction of the C—C overlap population (Table 11.4). According to the authors, these charge shifts support the repulsive-dominant nature of the ethane barrier which they have derived from the energy partitioning.[104] In contrast, the attractive-dominant character of the acetaldehyde barrier would result from a charge build-up around the oxygen atom which draws electron density from one methyl hydrogen in the stable conformation. A weak covalent bond is thus formed between the oxygen and the methyl hydrogen that eclipses it. So far, there is no indication of how much this analysis of the changes in charge density is insensitive to changes in the basis set and the geometry. Similar trends in the electron population analysis are obtained for the ethane molecule from different calculations, as indicated in Table 11.4. The conclusion that the lowest energy configuration has the

Table 11.4 Gross atomic populations and overlap populations for the staggered (S) and eclipsed (E) conformations of ethane

Wave-function	C		H		C—C		H_1–H_7[a]	H_1–H_8	H_1–H_4	H_1–H_5
	(S)	(E)	(S)	(E)	(S)	(E)	(S)	(S)	(E)	(E)
Allen[111]	6·816	6·817	0·7278	0·7276	0·493	0·481	0·0010	−0·0017	−0·0058	0·0008
Pitzer[39]	6·3384	6·3408	0·8871	0·8863	0·7652	0·7578	0·017	−0·016	−0·047	0·011
Pitzer[48]	6·0096	6·0123	0·9967	0·9958	0·7422	0·7353	0·0063	−0·0070	−0·0209	0·0044
Hoyland[32]	6·5190	6·5207	0·8269	0·8264	0·6294	0·6186	0·0020	−0·0029	−0·0100	0·0016

[a] The numbering is that of Reference 39.

largest overlap population for the bond about which the rotation occurs seems fairly general. In hydrogen peroxide the O—O overlap population increases for the near equilibrium conformation with respect to the *cis* and *trans* conformation.[40] Since the H—H antibonding character increases from the *trans* to the *cis* conformation with the O—H bonding character decreasing at the same time, it has been pointed out that the *cis* barrier appears as the result of three unfavourable situations at the level of the O—O, O—H and H—H bondings. Conversely, the *trans* barrier results from a balance between a decreased O—O bonding, an increased O—H bonding and a decreased H—H bonding. These results are not affected by the geometry optimization.[40] Kaufman has predicted a close correspondence between the molecular total overlap population (sum of the overlap populations) and the energy of the various conformations, with the molecular total overlap population being a maximum for the most stable conformation.[112] This appears to be true for the ethane,[39] butane[57] and propene[69] molecules and for the *cis* barrier of hydrogen peroxide and hydrogen persulphide but not for the *trans* barriers of these two compounds.[40,100] The relatively high *trans* barrier of hydrogen persulphide with respect to the one in hydrogen peroxide has been traced, together with the near orthogonality of the dihedral angle, to some hyperconjugation between the S—H bonds and the sulphur lone pairs.[100,101] The pronounced dependence of the rotational barrier in 1-propyl cations $CH_2X—CH_2—CH_2^+$ upon the substituent X has been interpreted on the basis of the gross population of the empty $2p$ orbital at the positive carbon atom, which is strongly dependent on the substituent X through an interaction with the CH_2X group.[68] Changes in the electron density upon rotation have also been analysed either through the density difference maps for the propene molecule[69] or through a population analysis for the butane molecule.[57]

11.6 Conclusion

Ab initio calculations of barrier heights have in a few years met with considerable success. The question now is not so much whether they will yield satisfactory barrier heights but rather *why* do they yield such satisfactory results? In fact, calculation may be used at the present time as confidently as any experimental method and, given the computer program to perform it, little human work is needed. However, computational limitations, both practical and economic, prevent, at the moment, the application to large-size molecules or even to medium-size molecules if a high accuracy is required. Progress in this direction will depend not only on the advent of faster computers and on the development of more efficient programs, but also on the various possibilities of making *ab initio* calculations more economical for large-size molecules (e.g. the transferability of localized orbitals or the use of a variable criterion of accuracy). Given the present possibilities, there are many systems of interest which have not been investigated theoretically even when the calculation appears quite feasible. For instance, no calculation has yet been reported for as simple a molecule as 1,2-disubstituted ethane, which raises several interesting questions in connection with the role of steric repulsion and electrostatic interactions in the rotation barrier. There are numerous classes of compounds where the calculation may have a useful predictive value, for instance in organometallic chemistry. Little is known about the barrier to internal rotation of the ligand rings in the sandwich complexes like the ferrocene molecule. Such systems are on the verge of the present computational possibilities, but may become routine studies in a few years.

11.7 Preface to Table 11.5

The barrier for a given internal rotation is given in Table 11.5 on the basis of the atoms forming the bond about which torsion occurs (C—C bonds first, C—X and X—Y bonds next, with X and Y not a carbon atom). The bond about which rotation occurs is always explicitly shown in the formula given in the first column of the table. The method used to obtain the barrier is indicated in the fourth column. Unless otherwise stated (BO bond-orbital method, CI configuration interaction, GF group-function method), the LCAO–MO–SCF method is used either with Slater-type functions (STF) or with Gaussian-type functions (GTF). The Gaussian or Slater basis set is given in parentheses. For instance, the notation (11, 7, 1/6, 1) for the ethane molecule indicates a $11s$, $7p$, $1d$ set for the carbon atoms and a $6s$, $1p$ set for the hydrogen atoms. Square brackets are similarly used for the contracted basis set whenever the Gaussian orbitals have been contracted. The use of optimized exponents (opt. exp.) and the process of geometry optimization (g. opt.) which may be partial (p.g. opt.), are also indicated.

Table 11.5 Barriers to internal rotation: *ab initio* LCAO–MO–SCF results

Formula	Name	Barrier (kcal/mole)	Basis set	Reference
H_3C-CH_3	Ethane	3·3	STF (2, 1/1)	39
		3·5	STF (2, 1/1), opt. exp.	48
		3·62	GTF (10, 6/5)[4, 2/2]	45
		2·51	GTF (10, 5/5)[3, 1/1]	47
		2·88	GTF (5, 2/2)[2, 1/1]	46
		3·45	GTF (5, 2/2, 1)[2, 1/1, 1]	46
		3·04	GTF (7, 3/3)[2, 1/1]	32
		2·40	GTF (7, 3/3)[2, 1/1], BO	32
		3·3	STF (2, 1/1), opt. exp., g. opt.	19
		3·14–3·30	STF (2, 1/1), CI	63
		2·8	GTF (8, 4/4)[3, 2/2], g. opt.	60
		3·32	STF (2, 1/1)	44
		2·72	GTF (7, 3/3)[3, 1/1]	75
		3·14	GTF (9, 5/4)	107
		3·07	GTF (11, 7, 1/6, 1)[5, 3, 1/3, 1], g. opt.	40
		2·93	Experiment	Ref. in 40
H_3C-CH_2F	Ethyl fluoride	2·59	GTF	104
		2·60	GTF (7, 3/3)[3, 1/1]	75
		3·33	Experiment	Ref. in 1
$H_3C-CH_2CH_3$	Propane	3·48	GTF (7, 3/3)[2, 1/1]	26
		3·45	STF (2, 1/1), p. g. opt.	44
		3·57	Experiment	Ref. in 44
$H_3C-CH_2CH_2CH_3$	Butane	2·94	GTF (7, 3/3)[2, 1/1]	57
		3·40	STF (2, 1/1), p. g. opt.	44
		3·25–3·40	Experiment	Ref. in 44
$CH_3CH_2-CH_2CH_3$	Butane	3·62 (*trans → gauche*) 6·83 (*gauche → cis*) 0·76 (*trans–gauche* energy difference)	GTF (7, 3/3)[2, 1/1]	57

Formula	Name	Barrier (kcal/mole)	Basis set	Reference
CH_3CH_2—CH_2H_3 —contd.	Butane	3·58 (*trans* → *gauche*)	STF (2, 1/1), p. g. opt.	44
		5·72 (*gauche* → *cis*)		
		1·13 (*trans–gauche* energy difference)		
		3·6–4·2 (*trans* → *gauche*)	Experiment	Ref. in 44
		5·3–6·7 (*gauche* → *cis*)		
		0·77 (*trans–gauche* energy difference)		
HCC—CH_2CH_3	1-Butyne	3·46	STF (2, 1/1), p. g. opt.	44
H_3C—$CCHCH_3$	2-Butyne	0·005	STF (2, 1/1)	44
H_3C—$CHCH_2$	Propene	1·48	GTF (5, 2/3)	69
		1·25	GTF (10, 5/5)[3, 1/1]	71
		3·6	GTF (4, 3/2)[2, 1/1]	70
		1·55	STF (2, 1/1), p. g. opt.	44
		0·46	GTF (5, 3/3)[2, 1/1], BO	31
		1·98	Experiment	Ref. in 71
H_2CCCH—CH_3	1,2-Butadiene	1·35	STF (2, 1/1), p. g. opt.	44
		1·59	Experiment	Ref. in 44
H_3C—CHCHF	*cis*-Fluoropropene	1·58	GTF (5, 2, 3)	113
		1·07	GTF (10, 5/5)[3, 1/1]	71
		1·06	Experiment	Ref. in 71
H_3C—CHCHF	*trans*-Fluoropropene	1·53	GTF (5, 2/3)	113
		1·34	GTF (10, 5/5)[3, 1/1]	71
		2·20	Experiment	Ref. in 71
H_3C—$CFCH_2$	2-Fluoropropene	1·20	GTF (5, 2/3)	113
		2·45	Experiment	Ref. in 71
H_3C—$C(CH_3)CH_2$	2-Methylpropene	1·70	STF (2, 1/1), p. g. opt.	44
		2·12–2·35	Experiment	Ref. in 44
H_3C—$CHCHCH_3$	*cis*-2-Butene	0·42	STF (2, 1/1), p. g. opt.	44
		0·45	Experiment	Ref. in 44

$H_3C-CHCHCH_3$	*trans*-2-Butene	1·54	STF (2, 1/1), p. g. opt	44
		1·95	Experiment	Ref. in 44
$H_3C-CH(CH_3)_2$	Isobutane	3·88	STF (2, 1/1), p. g. opt.	44
		3·6–3·9	Experiment	Ref. in 44
$H_3C-CH_2CH_2F$	1-Fluoropropane	3·46	STF (2, 1/1)	68
$H_3C-CH_2CH_2OH$	1-Propanol	3·49	STF (2, 1/1)	68
$H_3C-CH_2CH_2CN$	Propyl cyanide	3·64	STF (2, 1/1)	68
$H_2CCH-CHCH_2$	1,3-Butadiene	6·73 (*trans* → *cis*)	STF (2, 1/1), p. g. opt.	44
		2·05 (*cis–trans* energy difference)		
		5·0 (*trans* → *cis*)	Experiment	Ref. in 44
		2·3 (*cis–trans* energy difference)		
$H_3C-CH_2CHCH_2$ (skew)	1-Butene	3·46	STF (2, 1/1), p. g. opt.	44
		3·16	Experiment	Ref. in 44
$H_3C-CH_2CHCH_2$ (*cis*)	1-Butene	4·96	STF (2, 1/1), p. g. opt.	44
		3·99	Experiment	Ref. in 44
$H_3CCH_2-CHCH_2$	1-Butene	1·62 (skew → *cis*)	STF (2, 1/1), p. g. opt.	44
		1·17 (*cis*–skew energy difference)		
		1·74 (skew → *cis*)	Experiment	Ref. in 44
		0·15 (*cis*–skew energy difference)		
H_3C-CH_2	Ethyl radical	0·62	GTF (8, 4/4)[3, 2/2], g. opt.	60
$H_3C-CH_2^+$ (CH$_2^+$ planar)	Ethyl cation	0·29	GTF (8, 4/4)[3, 2/2], g. opt.	60
		0·00	GTF (10, 5/5)[3, 1/1]	73
		0·22	STF (2, 1/1), g. opt.	73
		0·011	GTF (10, 6/5)[4, 2/2]	74
$H_3C-CH_2^+$ (CH$_2^+$ tetrahedral)	Ethyl cation	2·8	GTF (10, 5/5)[3, 1/1]	73
H_3C-CHF^+	1-Fluoroethyl cation	0·62	GTF (7, 3/3)[3, 1/1]	75
$H_2FC-CH_2^+$	2-Fluoroethyl cation	10·53	GTF (7, 3/3)[3, 1/1]	75

Formula	Name	Barrier (kcal/mole)	Basis set	Reference
$CH_2^+-CH_2CH_3$ (CH_2^+ planar)	n-Propyl cation	2·26	GTF (10, 5/5)[3, 1/1]	76
		2·52	STF (2, 1/1)	76
$CH_2^+-CH_2CH_3$ (CH_2^+ tetrahedral)	n-Propyl cation	6·04	STF (2, 1/1)	76
$CH_2^+-CH_2CH_2CH_3$	n-Butyl cation	3·73	STF (2, 1/1)	68
$CH_2^+-CH_2CH_2F$		2·11	STF (2, 1/1)	68
$CH_2^+-CH_2CH_2OH$		0·91	STF (2, 1/1)	68
$CH_2^+-CH_2CH_2CN$		0·87	STF (2, 1/1)	68
$CH_2^+-CH(CH_3)_2$	Isobutyl cation	2·68	STF (2, 1/1)	76
$CH_2^+-C(CH_3)_3$	Neopentyl cation	0·0	STF (2, 1/1)	76
$CH_2^+-CH\triangleleft$	Cyclopropylcarbinyl cation	17·54	STF (2, 1/1)	76
$CH_2^+-(CH_3)C\triangleleft$	1-Methylcyclopropylcarbinyl cation	16·00	STF (2, 1/1)	76
$CH_2^+-CH\pentagon$	Cyclobutylcarbinyl cation	4·08	STF (2, 1/1)	76
$CH_2^+-(CH_3)C\pentagon$	1-Methylcyclobutylcarbinyl cation	2·32	STF (2, 1/1)	76
$H_3C-CH_2CH_2^+$	n-Propyl cation	0·5	GTF (8, 4/4)[3, 2/2], g. opt.	77
$H_3C^+\cdots\substack{CH_2\\ \| \\ CH_2}$	Corner-protonated cyclo-propane	0·01	GTF (8, 4/4)[3, 2/2], g. opt.	77
$H_2C=CH_2$	Ethylene	83·7	GTF (10, 5/5)[3, 1/1], limited CI	78
		83·2	STF (2, 1/1)	79
		138·6	STF (2, 1/1)	44
		63·7	GTF(p. g. opt., CI)	80
		65·0	Experiment	Ref. in 44
$H_2C=C=CH_2$	Allene	82·1	GTF (10, 5/5)[3, 1/1]	78
		74·6	GTF (7, 3/3)[2, 1/1]	81
		72·0	GTF (5, 2/2)	82

Formula	Name	Value	Method	Ref.
		91·9	STF (2, 1/1)	44
$H_2C=CC=CH_2$	Butatriene	73·9	STF (2, 1/1)	44
$H_2C=CHCCH$	Vinylacetylene	137·7	STF (2, 1/1), p. g. opt.	44
H_3C-CHO	Acetaldehyde	0·82	GTF	83
		1·48	GTF (9, 5/4)[3, 1/2]	84
		1·09	GTF (10, 5/5)[4, 2/2]	85
		1·16	Experiment	Ref. in 85
$H_3C-\dot{C}O$	Acetyl radical	0·38	GTF (10, 6/5)[4, 2/2]	87
H_3C-NO	Nitrosomethane	1·05	GTF [3, 1/1]	88
		1·10	Experiment	Ref. in 88
H_3C-NH_2	Methylamine	2·42	GTF (10, 5/4)[3, 1/1]	47
		2·02	GTF (5, 2/2)	46
		1·98	Experiment	Ref. in 46
$H_3C-NHO\cdot$	Methylnitroxide radical	0·93	GTF (9, 5/4)[3, 1/2]	84
$OHC-NH_2$	Formamide	19–21	GTF (11, 7, 1/6, 1)[5, 3, 1/2, 1], g. opt.	66
		17–21	Experiment	Ref. in 66
H_3C-OH	Methyl alcohol	1·06	GTF (10, 5/4)[4, 2/2]	47
		1·59	GTF (5, 2/2)	46
		1·07	Experiment	Ref. in 46
$HOC-OH$	Formic acid	14·2	GTF (5, 2/2)	114
		9·4 (*cis–trans* energy difference)		
		10·9	Experiment	Ref. in 114
		2·0 (*cis–trans* energy difference)		
H_2C-OH^+	Protonated formaldehyde	31·4	GTF (3, 2/2)	91
		25·0	GTF (4, 3/3)	91
$H_3CHC-OH^+$	Protonated acetaldehyde	27·9	GTF (3, 2/2)	91
$(HO-CH-OH)^+$	Protonated formic acid	13·5	GTF (3, 2/2)	91

Formula	Name	Barrier (kcal/mole)	Basis set	Reference
H_3C-SiH_3	Methylsilane	1·44	GTF (12, 9, 1/10, 6, 1/5, 1)[6, 4, 1/4, 2, 1/2, 1]	72
		1·71	Experiment	Ref. in 72
H_3C-PH_2	Methylphosphine	1·83	GTF (9, 5, 1/5, 2/3)	92
		1·96	Experiment	Ref. in 92
H_3C-SHO	Hydrogen methyl sulphoxide	2·68	GTF (5, 2, 1/3, 1/1)	93
H_2C^--SHO	Hydrogen methylsulphinyl anion	7·6	GTF (5, 2, 1/3, 1/1)	93
$H_2C^--SOCH_3$	Methylsulphinyl anion	7·6	GTF (5, 2, 1/3, 1/1)	93
$H_2C^--SHO_2$	Hydrogen methylsulphonyl anion	5·0	GTF (5, 2, 1/3, 1/1)	94
H_2C^--SH	Thiomethoxide ion	18·8	GTF (5, 2, 1/3, 1/1)	95
H_3B-NH_3	Borazane	3·29	GTF (9, 5, 1/4)[4, 2, 1/2]	96
		3·06	GTF (11, 7, 1/6, 1)[5, 3, 1/3, 1]	72
H_2N-NH_2	Hydrazine	11·5 (*cis* barrier) 4·7 (*trans* barrier)	GTF (9, 3/3)	98
		11·05 (*cis* barrier) 6·21 (*trans* barrier)	GTF (5, 2/2)	46
		11·88 (*cis* barrier) 3·70 (*trans* barrier)	GTF (10, 5/5)[3, 1/1]	97
H_2N-OH	Hydroxylamine	11·95 1·16 (*cis–trans* energy difference)	GTF (10, 5/5)[3, 1/1]	97
		9·90 2·53 (*cis–trans* energy difference)	GTF (5, 2/2)	46
$HO-OH$	Hydrogen peroxide	11·8 (*trans*) 2·2 (*cis*)	STF (2, 1/1)	49
		13·6 (*trans*) 0·0 (*cis*)	GTF (11, 6/4)[3, 1/1]	47

		Barrier	Method	Ref.
		13·2 (*trans*)	GTF (10, 5/5, 1)[4, 2/2, 1]	47
		0·1 (*cis*)		
		13·0 (*trans*)	STF (2, 1/1)	50
		0·0 (*cis*)		
		15·9 (*trans*)	GTF (5, 2/2)	46
		0·0 (*cis*)		
		0·1 (*cis*)	GTF (5, 2/2, 1)	46
		9·4 (*trans*)	STF (2, 1/1), opt. exp., g. opt.	19
		0·0 (*cis*)		
		10·9 (*trans*)	GTF (11, 7, 1/6, 1)[5, 3, 1/2, 1], g. opt.	40
		0·6 (*cis*)		
		8·85 (*trans*)	GTF (8, 4/4, 3)[2, 1/1, 3]	52
		0·54 (*cis*)		
		8·3 (*trans*)	GTF (9, 5, 1/4, 1)[4, 3, 1/2, 1], g. opt.	54
		1·1 (*cis*)		
		0·63 (*cis*)	GTF (7, 3, 1/4, 1)[4, 2, 1/2, 1], g. opt.	55
		13·87 (*trans*)	GTF (10, 5, 1/5, 1)[4, 2, 1/2, 1], g. opt.	105
		0·24 (*cis*)		
		12·8 (*trans*)	STF (2, 1/1)	51
		0·0 (*cis*)		
		12·7 (*trans*)	STF (2, 1/1), GF	51
		0·0 (*cis*)		
		7·0 (*trans*)	Experiment	Ref. in 40
		1·1 (*cis*)		
H_2P-PH_2	Diphosphine	2·32 (*cis*)	GTF (8, 4/2)	99
		0·51 (*trans*)		
$HS-SH$	Hydrogen persulphide	7·4 (*cis*)	GTF (10, 6/3)[3, 2/1]	100
		1·9 (*trans*)		
		9·33 (*cis*)	GTF (12, 9, 1/5, 1)[6, 4, 1/2, 1]	101
		5·99 (*trans*)		

11.8 Recent developments

There is a rapid expansion of the field discussed above. Recent developments are summarized here.

Additional calculations have been reported for the rotation barrier in ethane.[115] The rotational barrier in ethyl fluoride has been computed by Allen and Basch and compared to the value for ethane (respectively 2·59 and 2·58 kcal/mole versus experimental values of 3·33 and 2·93 kcal/mole).[116] Both barriers are found to be repulsive dominant. The use of an extended basis set increases the barrier in ethyl fluoride to a value of 3·4 kcal/mole.[117]

Additional calculations have been reported for the rotation barrier of hydrogen peroxide with an extended basis set of Slater functions.[118] The relaxation effect was not considered and the computed *cis* and *trans* barriers are respectively 10·9 and 0·72 kcal/mole. It was concluded that the nuclear relaxation is crucial for a quantitative agreement, but not necessary for a qualitative answer on the existence of the barriers.

A Fourier component analysis of the potential functions has been carried out for a number of saturated molecules and their monomethyl and monofluoro derivatives.[119] The potential functions have been rationalized in terms of three principal effects (preference for a staggered arrangement of bonds, for an arrangement of a lone-pair orbital which is coplanar with an adjacent electron-withdrawing polar bond or orthogonal to an adjacent lone-pair orbital and for an opposed arrangement of dipole moment components perpendicular to the internal rotation axis) respectively reflected in the three-fold, two-fold and one-fold components.

Additional work based on component analysis has assigned the rotation barrier in alkanes to repulsive interactions and the one in π-electron molecules (propene and acetaldehyde) to attractive interactions.[120]

A unified account of rotation barriers in substituted ethyl, propyl and butyl cations has been given in Reference 121.

Ab initio calculations of the rotation barrier in the cations $XCH_2{-}CH_2^+$ and the anions $XCH_2{-}CH_2^-$ (X = F and BH_2) have been used to support estimates of the barrier due to hyperconjugation with an electronegative or electropositive substituent.[122]

Barriers to rotation for substituted benzenes have been determined with a standard geometric model and, in the case of strong steric repulsions, possibility of bond-angle deviation from this model.[123] The conformations of substituted benzenes are found to be determined largely by conjugation effects. In *para*-substituted phenols, the barrier to rotation around the $C{-}O$ bond is lowered by substituents which are π-electron donors and raised by π-electron acceptors.[124]

Theoretical potential energy curves for the rotation around the central single bond and the ethylenic double bond have been reported for acrolein in its ground and excited states.[125]

Vinylcyclopropane and vinylcyclobutane are found most stable in their s-*trans* forms. Limited basis set calculations predict also secondary s-*cis* and *gauche* minima, but extended basis set calculations confirm only the *gauche* minimum for vinylcyclopropane. The effect on the potential curves of π-electron donor and acceptor substituents on the vinyl group has also been investigated.[126]

A study of the rotation curve in the ethylene dicarbanion $^-CH_2{-}CH_2^-$ points to a *gauche* effect in this molecule.[127]

Rotation from *cis*-butadiene to the *trans* isomer has been discussed for the ground and excited states through both SCF and CI calculations,[128] with the assumption of an otherwise rigid model. The energy surface close to the *cis* form appears extremely flat and there is still some doubt as to whether the *cis* conformer corresponds to a potential minimum or a potential maximum. Similar results have been achieved with a minimal basis set and a perturbative treatment of electron correlation.[129]

A rotation barrier for *cis–trans* isomerization of glyoxal has been reported.[130]

The potential for internal rotation around the C—O bond in fluoromethanol has been reported and provides evidence for the destabilization of a conformation which places a polar bond between two electron pairs.[131]

While hydrazine, N_2H_4, has an energy barrier for the *trans* position, both the *gauche* and *trans* forms of N_2F_4 are found stable (with the *trans* configuration more stable by 1·5 kcal/mole).[132] The *gauche* and *trans* conformations of diphosphine P_2H_4 are both stable with a comparable energy and the potential between these two conformations is relatively flat. Only the *trans* conformation of P_2F_4 is stable.[133]

For the monomethylphosphine molecule, the staggered conformation is found to be the most stable with a computed barrier of 1·83 kcal/mole (experimental 1·96 kcal/mole).[134]

It has been shown that the binding energy of a linearly hydrogen-bonded dimer of formamide is insensitive to rotation around the hydrogen bond.[135]

A change in the sign of the rotational barrier for the methyl group has been predicted during the $CH_3NC \rightarrow CH_3CN$ isomerization.[136]

11.9 References

1. J. P. Lowe, in *Progress in Physical Organic Chemistry*, Vol. 6, Ed. A. Streitwieser and R. Taft (New York: Interscience, 1968).
2. R. Hoffmann, *J. Chem. Phys.*, **39**, 1397 (1963).
3. J. A. Pople and G. A. Segal, *J. Chem. Phys.*, **43**, S 136 (1965).
4. W. C. Herndon, J. Feuer and L. H. Hall, *Tetrahedron Letters*, **22**, 2625 (1968).
5. M. S. Gordon, *J. Am. Chem. Soc.*, **91**, 3122 (1969).
6. F. L. Pilar, *Elementary Quantum Chemistry* (New York: McGraw-Hill, 1968) p. 414.

7. F. L. Pilar, *Elementary Quantum Chemistry* (New York: McGraw-Hill, 1968) p. 341.
8. C. C. J. Roothaan, *Rev. Mod. Phys.*, **23**, 69 (1951).
9. C. C. J. Roothaan, *Rev. Mod. Phys.*, **32**, 179 (1960).
10. J. A. Pople and R. K. Nesbet, *J. Chem. Phys.*, **22**, 571 (1954).
11. G. Berthier, *J. Chim. Phys.*, **51**, 363 (1954).
12. S. F. Boys and J. Foster, *Rev. Mod. Phys.*, **32**, 305 (1960).
13. C. Edmiston and K. Ruedenberg, *Rev. Mod. Phys.*, **35**, 457 (1963).
14. E. Clementi, C. C. J. Roothaan and M. Yoshimine, *Phys. Rev.*, **127**, 1618 (1962).
15. M. Krauss, *Compendium of* ab initio *Calculations of Molecular Energies and Properties*, National Bureau of Standards, Technical Note 438, Washington D.C. (1967).
16. J. W. Richardson, *J. Chem. Phys.*, **35**, 1829 (1961).
17. E. Clementi, Atomic energy tables, Supplement to *IBM J. of Res. Develop.*, **9**, 1 (1965).
18. W. Huo, *J. Chem. Phys.*, **43**, 624 (1965).
19. R. M. Stevens, *J. Chem. Phys.*, **52**, 1397 (1970).
20. S. F. Boys, *Proc. Roy. Soc.* (*London*), **A201**, 125 (1950).
21. I. Shavitt, in *Methods in Computational Physics*, Vol. 2 (New York: Academic Press, 1963) p. 1.
22. H. Preuss, *Z. Naturforsch.*, **11**, 823 (1956).
23. J. L. Whitten, *J. Chem. Phys.*, **44**, 359 (1966).
24. B. Roos and P. Siegbahn, *Theoret. Chim. Acta*, **17**, 209 (1970).
25. E. Clementi and D. R. Davis, *J. Computational Phys.*, **2**, 223 (1967).
26. J. R. Hoyland, *J. Chem. Phys.*, **49**, 1908 (1968).
27. R. S. Mulliken, *J. Chem. Phys.*, **36**, 3428 (1962).
28. B. Roos and P. Siegbahn, *Theoret. Chim. Acta*, **17**, 199 (1970).
29. S. Rothenberg and H. F. Schaefer, *J. Chem. Phys.*, **54**, 2764 (1971).
30. Th. H. Dunning, *J. Chem. Phys.*, to be published.
31. J. R. Hoyland, *J. Am. Chem. Soc.*, **90**, 2227 (1968).
32. J. R. Hoyland, *J. Chem. Phys.*, **50**, 473 (1969).
33. A. D. McLean, A. Weiss and M. Yoshimine, *Rev. Mod. Phys.*, **32**, 211 (1960).
34. M. Klessinger and R. McWeeny, *J. Chem. Phys.*, **42**, 3343 (1965).
35. G. Herzberg, *Molecular Spectra and Molecular Structure, I—Spectra of Diatomic Molecules* (Princeton: Van Nostrand, 1950) p. 76.
36. E. Clementi, *J. Chem. Phys.*, **38**, 2248 (1963).
37. C. Hollister and O. Sinanoglu, *J. Am. Chem. Soc.*, **88**, 13 (1966).
38. C. D. Ritchie and H. F. King, *J. Chem. Phys.*, **47**, 564 (1967).
39. R. M. Pitzer and W. N. Lipscomb, *J. Chem. Phys.*, **39**, 1995 (1963).
40. A. Veillard, *Theoret. Chim. Acta*, **18**, 21 (1970).
41. J. M. Lehn, in *Conformational Analysis* (Proceedings of the International Symposium on Conformational Analysis, Brussels, 1969) (New York: Academic Press, 1971) p. 129.
42. W. Kolos and L. Wolniewicz, *J. Chem. Phys.*, **41**, 3663 (1964).
43. O. Sinanoglu, *J. Phys. Chem.*, **66**, 2283 (1962).
44. L. Radom and J. A. Pople, *J. Am. Chem. Soc.*, **92**, 4786 (1970).
45. E. Clementi and D. R. Davis, *J. Chem. Phys.*, **45**, 2593 (1966).
46. L. Pedersen and K. Morokuma, *J. Chem. Phys.*, **46**, 3941 (1967).
47. W. H. Fink and L. C. Allen, *J. Chem. Phys.*, **46**, 2261, 2276 (1967).
48. R. M. Pitzer, *J. Chem. Phys.*, **47**, 965 (1967).
49. U. Kaldor and I. Shavitt, *J. Chem. Phys.*, **44**, 1823 (1966).
50. W. E. Palke and R. M. Pitzer, *J. Chem. Phys.*, **46**, 3948 (1967).

51. P. F. Franchini and C. Vergani, *Theoret. Chim. Acta*, **13**, 46 (1969).
52. I. H. Hillier, V. R. Saunders and J. F. Wyatt, *Trans. Faraday Soc.*, **66**, 2665 (1971).
53. M. D. Newton, W. A. Lathan, W. J. Hehre and J. A. Pople, *J. Chem. Phys.*, **52**, 4064 (1970).
54. T. H. Dunning and N. W. Winter, *Chem. Phys. Letters*, **11**, 194 (1971).
55. J. P. Ranck and H. Johansen, *Theoret. Chim. Acta*, **24**, 334 (1972).
56. W. A. Lathan, W. J. Hehre and J. A. Pople, *J. Am. Chem. Soc.*, **93**, 808 (1971).
57. J. R. Hoyland, *J. Chem. Phys.*, **49**, 2563 (1968).
58. J. R. Hoyland, *J. Chem. Phys.*, **50**, 2775 (1969).
59. S. Weiss and G. E. Leroi, *J. Chem. Phys.*, **48**, 962 (1968).
60. W. A. Lathan, W. J. Hehre and J. A. Pople, *J. Am. Chem. Soc.*, **93**, 808 (1971).
61. M-Cl. Moireau (Unpublished Results).
62. M. Klessinger, *Symposia of the Faraday Society*, No. 2 (London: The Faraday Society, 1968) p. 73.
63. B. Levy and M-Cl. Moireau, *J. Chem. Phys.*, **54**, 3316 (1971).
64. R. H. Hunt, R. A. Leacock, C. W. Peters and K. T. Hecht, *J. Chem. Phys.*, **42**, 1931 (1965).
65. W. C. Oelfke and W. Gordy, *J. Chem. Phys.*, **51**, 5336 (1969).
66. D. H. Christensen, R. N. Kortzeborn, B. Bak and J. J. Led, *J. Chem. Phys.*, **53**, 3912 (1970).
67. P. Pulay, *Mol. Phys.*, **17**, 197 (1969).
68. L. Radom, J. A. Pople, V. Buss and P. V. R. Schleyer, *J. Am. Chem. Soc.*, **92**, 6987 (1970).
69. M. L. Unland, J. R. Van Wazer and J. H. Letcher, *J. Am. Chem. Soc.*, **91**, 1045 (1969).
70. E. Zeeck, *Theoret. Chim. Acta*, **16**, 155 (1970).
71. E. Scarzafava and L. C. Allen, *J. Am. Chem. Soc.*, **93**, 311 (1971).
72. A. Veillard, *Chem. Phys. Letters.*, **3**, 128 (1969).
73. J. E. Williams, V. Buss, L. C. Allen, P. V. R. Schleyer, W. A. Lathan, W. J. Hehre and J. A. Pople, *J. Am. Chem. Soc.*, **92**, 2141 (1970).
74. L. J. Massa, S. Ehrenson and M. Wolfsberg, *Intern. J. Quant. Chem.*, **4**, 625 (1970).
75. D. T. Clark and D. M. J. Lilley, *Chem. Commun.*, 603 (1970).
76. L. Radom, J. A. Pople, V. Buss and P. V. R. Schleyer, *J. Am. Chem. Soc.*, **92**, 6380 (1970).
77. L. Radom, J. A. Pople, V. Buss and P. V. R. Schleyer, *J. Am. Chem. Soc.*, **93**, 1813 (1971).
78. R. J. Buenker, *J. Chem. Phys.*, **48**, 1368 (1968).
79. U. Kaldor and I. Shavitt, *J. Chem. Phys.*, **48**, 191 (1968).
80. R. J. Buenker, S. D. Peyerimhoff and H. L. Hsu, *Chem. Phys. Letters*, **11**, 65 (1971).
81. J-M. André, M-Cl. André and G. Leroy, *Chem. Phys. Letters*, **3**, 695 (1969).
82. L. J. Schaad, L. A. Burnelle and K. P. Dressler, *Theoret. Chim. Acta*, **15**, 91 (1969).
83. M. E. Schwartz, G. V. Pfeiffer and S. R. Rothenberg (Unpublished Results, quoted in Reference 86).
84. P. Millié, These de Doctorat d'Etat, Paris (1970).
85. R. B. Davidson and L. C. Allen, *J. Chem. Phys.*, **54**, 2828 (1971).
86. H. Veillard and B. Rees, *Chem. Phys. Letters*, **8**, 267 (1970).
87. J. R. De La Vega, Y. Fang and E. F. Hayes, *Intern. J. Quant. Chem.*, S, 113 (1969).
88. P. A. Kollman and L. C. Allen, *Chem. Phys. Letters*, **5**, 75 (1970).
89. L. Radom, W. J. Hehre and J. A. Pople, *J. Am. Chem. Soc.*, **93**, 289 (1971).
90. M-E. Schwartz, E. F. Hayes and S. R. Rothenberg (Unpublished Results, quoted in Reference 87).
91. P. Ros, *J. Chem. Phys.*, **49**, 4902 (1968).

92. I. Absar and J. R. Van Wazer, *Chem. Commun.*, 611 (1971).
93. A. Rauk, S. Wolfe and I. G. Csizmadia, *Can. J. Chem.*, **47**, 113 (1969).
94. S. Wolfe, A. Rauk and I. G. Csizmadia, *J. Am. Chem. Soc.*, **91**, 1567 (1969).
95. S. Wolfe, A. Rauk, L. M. Tel and I. G. Csizmadia, *Chem. Commun.*, 96 (1970).
96. M-Cl. Moireau and A. Veillard, *Theoret. Chim. Acta*, **11**, 344 (1968).
97. W. H. Fink, D. C. Pan and L. C. Allen, *J. Chem. Phys.*, **47**, 895 (1967).
98. A. Veillard, *Theoret. Chim. Acta*, **5**, 413 (1966).
99. J-B. Robert, H. Marsmann and J. R. Van Wazer, *Chem. Commun.*, 356 (1970).
100. M-E. Schwartz, *J. Chem. Phys.*, **51**, 4182 (1969).
101. A. Veillard and J. Demuynck, *Chem. Phys. Letters*, **4**, 476 (1970).
102. O. J. Sovers, C. W. Kern, R. M. Pitzer and M. Karplus, *J. Chem. Phys.*, **49**, 2592 (1968).
103. I. R. Epstein and W. N. Lipscomb, *J. Am. Chem. Soc.*, **92**, 6094 (1970).
104. L. C. Allen, *Chem. Phys. Letters*, **2**, 597 (1968).
105. R. B. Davidson and L. C. Allen, *J. Chem. Phys.*, **55**, 519 (1971).
106. J-M. Lehn and B. Munsch, *Mol. Phys.*, **23**, 91 (1972).
107. E. Clementi and W. von Niessen, *J. Chem. Phys.*, **54**, 521 (1971).
108. R. M. Pitzer, *J. Chem. Phys.*, **41**, 2216 (1964).
109. S. D. Peyerimhoff, R. J. Buenker and L. C. Allen, *J. Chem. Phys.*, **45**, 734 (1966).
110. J. P. Lowe, *J. Am. Chem. Soc.*, **92**, 3799 (1970).
111. W. L. Jorgensen and L. C. Allen, *J. Am. Chem. Soc.*, **93**, 567 (1971).
112. J. J. Kaufman, *Intern. J. Quant. Chem.*, **S1**, 485 (1967).
113. M. L. Unland, J. H. Letcher and J. R. Van Wazer, quoted in Reference 71.
114. A. C. Hopkinson, K. Yates and I. G. Csizmadia, *J. Chem. Phys.*, **52**, 1784 (1970).
115. E. Clementi and H. Popkie, *J. Chem. Phys.*, **57**, 4870 (1972).
116. L. C. Allen and H. Basch, *J. Am. Chem. Soc.*, **93**, 6373 (1971).
117. W. E. Palke, *Chem. Phys. Letters*, **15**, 244 (1972).
118. C. Guidotti, U. Lamanna, M. Maestro and R. Moccia, *Theoret. Chim. Acta*, **27**, 55 (1972).
119. L. Radom, W. J. Hehre and J. A. Pople, *J. Am. Chem. Soc.*, **94**, 2371 (1972).
120. A. Liberles, B. O'Leary, J. E. Eilers and D. R. Whitman, *J. Am. Chem. Soc.*, **94**, 6894 (1972).
121. L. Radom, J. A. Pople and P. von R. Schleyer, *J. Am. Chem. Soc.*, **94**, 5935 (1972).
122. R. Hoffmann, L. Radom, J. A. Pople, P. von R. Schleyer, W. J. Hehre and L. Salem, *J. Am. Chem. Soc.*, **94**, 6221 (1972).
123. W. J. Hehre, L. Radom and J. A. Pople, *J. Am. Chem. Soc.*, **94**, 1496 (1972).
124. L. Radom, W. J. Hehre, J. A. Pople, G. L. Carlson and W. G. Fateley, *Chem. Commun*, 308 (1972).
125. A. Devaquet, *J. Am. Chem. Soc.*, **94**, 5160 (1972).
126. W. J. Hehre, *J. Am. Chem. Soc.*, **94**, 6592 (1972).
127. S. Wolfe, L. M. Tel, J. H. Liang and I. G. Csizmadia, *J. Am. Chem. Soc.*, **94**, 1361 (1972).
128. B. Dumbacher, *Theoret. Chim. Acta*, **23**, 346 (1972).
129. U. Pincelli, B. Cadioli and B. Levy, *Chem. Phys. Letters*, **13**, 249 (1972).
130. T. K. Ha, *J. Mol. Struct.*, **12**, 171 (1972).
131. S. Wolfe, A. Rauk, L. M. Tel and I. G. Csizmadia, *J. Chem. Soc. (B)*, 136 (1971).
132. E. L. Wagner, *Theoret. Chim. Acta*, **23**, 115 (1971).
133. E. L. Wagner, *Theoret. Chim. Acta*, **23**, 127 (1971).
134. I. Absar and J. R. Van Wazer, *J. Chem. Phys.*, **56**, 1284 (1972).
135. H. Berthod and A. Pullman, *Chem. Phys. Letters*, **14**, 217 (1972).
136. D. H. Liskow, C. F. Bender and H. F. Schaefer, *J. Chem. Phys.*, **57**, 4509 (1972).

12 Ring inversion in some six-membered heterocyclic compounds

Miss V. M. Gittins, E. Wyn-Jones and R. F. M. White

12.1 Introduction

Several experimental methods[1-14] have been used to study the conformational behaviour of many six-membered heterocyclic molecules including tetrahydropyrans, piperidines, morpholines, diazines and thiazoles. In this chapter we have specifically chosen 1,3-dioxan and cyclic sulphites to illustrate how the conformational analysis in heterocyclic compounds has been studied using many of the techniques described in the preceding chapters.

12.2 General concepts of ring inversion

12.2.1 Cyclohexanes

The conformational behaviour of saturated six-membered rings is often discussed in relation to the corresponding conformational behaviour of cyclohexane. This method of discussion will also be used in this chapter, and for this purpose a brief review of the structure and conformational properties of the cyclohexanes is given below. Since Sasche's hypothesis[15] that there are two forms of cyclohexane, the chair and boat forms shown in Figure 12.1, which are free from angle strain, physicochemical evidence,

Figure 12.1   The two forms of cyclohexane; (a) chair form
and (b) boat form

including X-ray diffraction,[16] Raman spectroscopy,[17] electron diffraction[18] and calculation of thermodynamic properties,[19] has shown that the most stable conformer is the chair isomer which has the dimensions shown in Figure 12.2. The stability of the chair form is due to the fact that all the bond angles are tetrahedral, all dihedral angles (i.e. angles of torsion) are 60° and thus all 1,3-diaxial H—H distances are greater than about 2·5 Å.

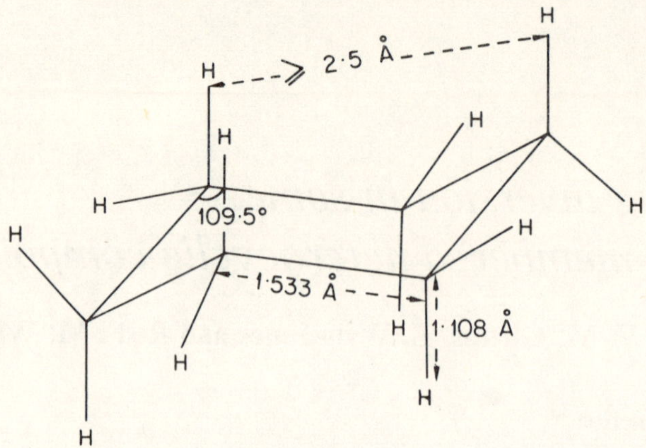

Figure 12.2   Dimensions of the chair form of cyclohexane

By means of internal rotation about the C—C bonds of chair form (I)
'ring inversion' into chair form (II) is possible. At room temperature, the
magnitude of the free-energy barrier (~ 10 kcal/mole) opposing this ring
inversion is such that the chair form (I) rapidly inverts into the chair form (II)

$$\text{(I)} \rightleftharpoons \text{(II)}$$

due to thermal excitations of the molecules. The ring-inversion process for
cyclohexane is now well understood and Hendrickson[20] has shown that the
three main factors contributing to the potential energy of a particular
conformation are: torsional energy arising from the eclipsing of C—H
bonds as the dihedral angles change from the 60° of the staggered conformation, bond-angle strain due to distortion of valency angles away from the
tetrahedral angle, and non-bonded interactions resulting from repulsive
and attractive forces between non-bonded atoms or groups. The reaction
path followed by the molecule during the ring-inversion process is shown in
Figure 12.3. From the stable chair form (I), the molecule passes through
the 'cyclohexene'-like half-chair form (III), in which four carbon atoms lie
in the same plane. This form is normally associated with the transition
state for chair-to-chair inversion. Further internal rotation takes the molecule

(III)   (IV)   (V)

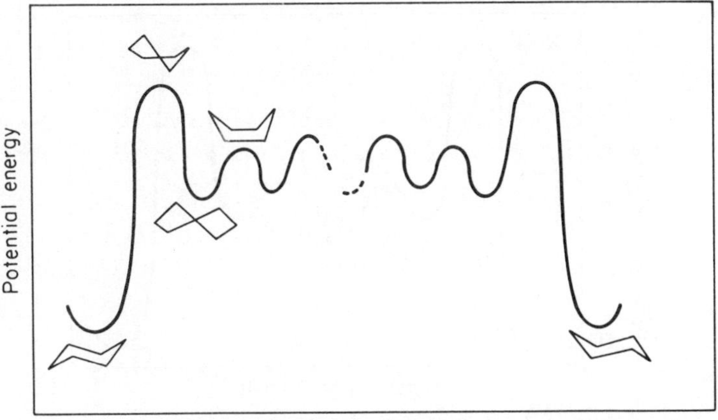

Reaction coordinate

Figure 12.3   Ring-inversion   process   for
cyclohexane

into the so-called 'flexible' forms which are the boat (IV) and twist-boat (V) conformations. According to Hendrickson's calculations the energies of these forms are intermediate between those of the chair forms (I) and (II) and the half-chair form (III); the relative energies are summarized in Figure 12.4. From the 'flexible' boat forms, the reaction pathway further proceeds through a second half-chair form before the molecule is finally converted into the stable chair form (II).

In the case of monosubstituted cyclohexanes such as methylcyclohexane, there are two energetically different chair forms, (VI) and (VII), which are

(VI)   (VII)

in dynamic equilibrium with each other. The reaction pathway for the ring-inversion process associated with the equatorial (VI) to axial (VII) isomerism is shown in Figure 12.5 and is similar in shape to that for the parent molecule (Figure 12.3) except that minima corresponding to the chair conformers differ in energy; in addition, each of the intermediate states (III)–(V) may have two or more non-equivalent conformations due to the possible relative orientation of the methyl groups. In Table 12.1, a summary of the experimentally determined conformational energies for some cyclohexanes is given.

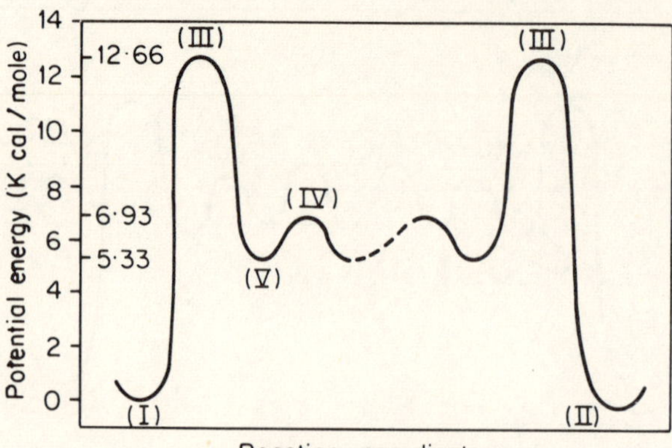

Figure 12.4   Relative energies for the cyclohexane ring-inversion process

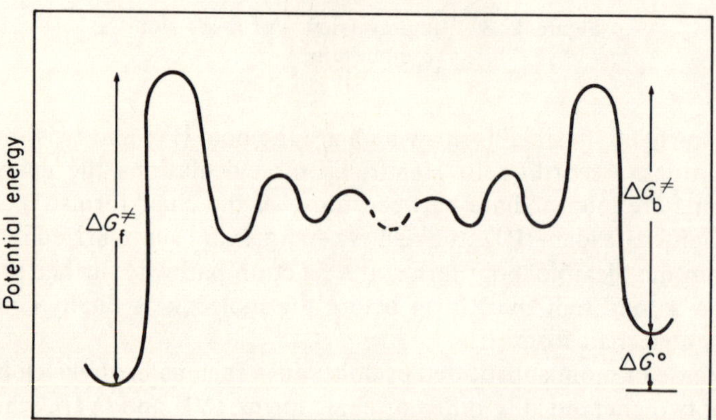

Figure 12.5   Ring-inversion process for methylcyclohexane

Table 12.1   Conformational energies of some substituted cyclohexanes

Substituent	Reference	Method	Solvent	Preferred conformation	ΔG^0 (kcal/mole)
Methyl	21	NMR	CCl_4	Me-e	1·7
Ethyl	22	NMR	2-D-2-PrOH	Et-e	1·75
Isopropyl	23			i-Pr-e	2·1
Hydroxy	24	NMR	DMSO	OH-e	0·54
Methoxy	25	NMR	CCl_4	OMe-e	0·6
Fluoro	26	NMR	Cyclohexane	F-e	0·15
Chloro	26	NMR	Cyclohexane	Cl-e	0·43
Bromo	26	NMR	CH_3CN	Br-e	0·37
Iodo	27	NMR	CS_2	I-e	0·43
Cyano	28	Equilibration	t-BuOH	CN-e	0·17
Amino	29	NMR	Various	NH_2-e	1·1–1·2

12.3 Conformational analysis of 1,3-dioxans

12.3.1 *Introduction*

Early dipole moment measurements[30] suggest that 1,3-dioxan also exists almost exclusively in the chair form (VIII). This was supported by further dipole moment[31] and NMR[32] measurements, and a recent X-ray

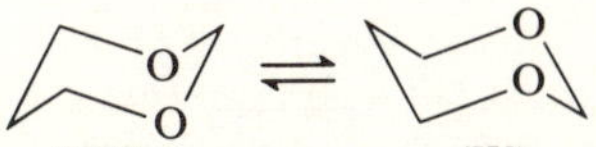

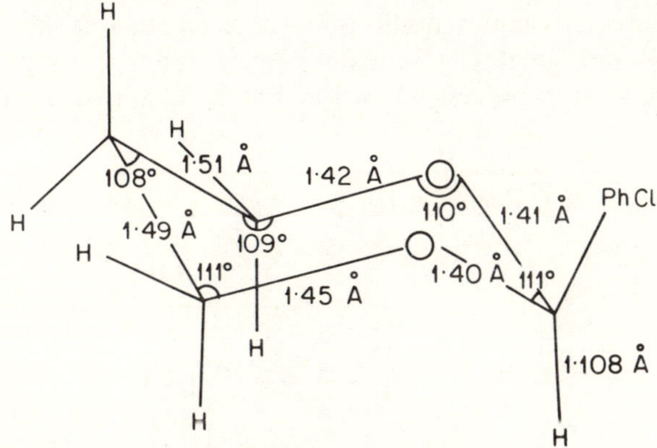

study of 2-*p*-chlorophenyl-1,3-dioxan[33] has confirmed that the molecule exists in a chair form which, relative to cyclohexane (Figure 12.2), is slightly puckered in the O—C—O region and slightly flattened in the C—C—C region (Figure 12.6). One result of this puckering and also the shorter C—O bond lengths is that non-bonded distances between substituted atoms and groups can be quite different from the corresponding distances in cyclohexane. For example Riddell and Robinson[34] have calculated the distance

Figure 12.6   Dimensions of the chair form of 2-*p*-chloro-
phenyl-1,3-dioxan

between the proton of an axial methyl group at C-2 and the synaxial protons at C-4 and C-6 to be 1·94 Å, compared with 2·29 Å in methylcyclohexane, as shown in Figure 12.7.

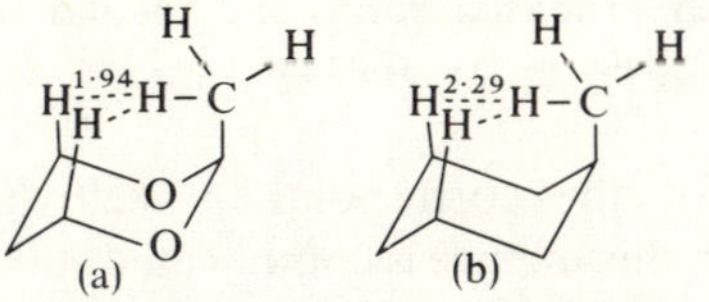

Figure 12.7   Distance between the proton of an axial
methyl group at C-2 and the synaxial protons at C-4 and
C-6 of (a) 2-methyl-1,3-dioxan and (b) methylcyclohexane

The process of ring inversion in 1,3-dioxan is assumed to be similar to that in cyclohexane, thus the reaction path followed by the former compound will be similar in shape to that shown in Figure 12.3. However, the structure of the 1,3-dioxan molecule is such that energetically different conformers for the half-chair, (X)–(XII), boat, (XIII) and (XIV), and twist-boat, (XV) and (XVI), forms are possible.

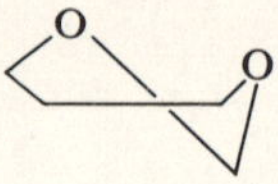

(X) 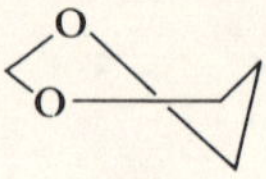   (XI) 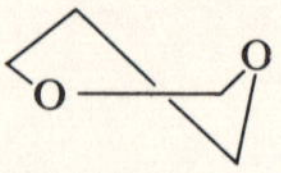   (XII)

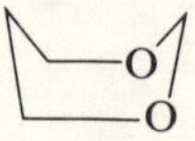

(XIII) 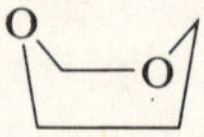   (XIV) 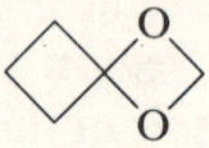   (XV) 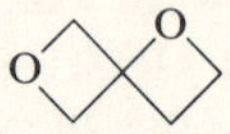   (XVI)

However, in the case of a monosubstituted 1,3-dioxan, as in the corresponding cyclohexane, there are only two energetically different chair forms which are in dynamic equilibrium with each other as shown in Figure 12.8 for 2-, 4- and 5-methyl-1,3-dioxans. Again, the reaction path followed by the ring-inversion process, shown in Figure 12.8, can be compared to

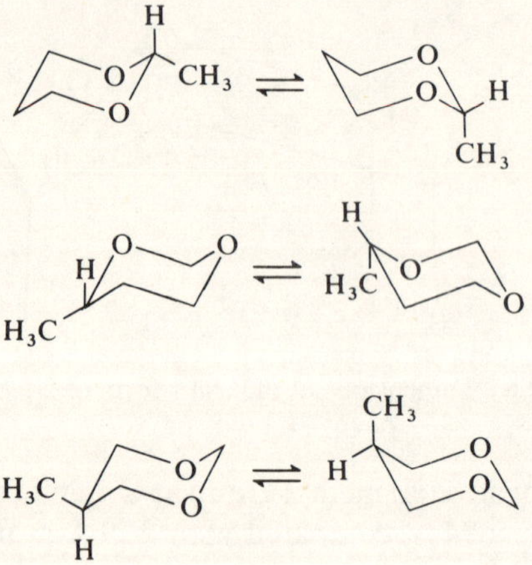

Figure 12.8   The chair forms of 2-, 4- and 5-methyl-1,3-dioxans

the reaction path for the corresponding methylcyclohexane shown in Figure 12.5. However, this picture is complicated still further by the existence of two possible non-equivalent conformations for each of the intermediate states corresponding to (X)–(XVI), depending on the relative position of the substituted methyl group. For 2-methyl-1,3-dioxan, this leads to the possibility of six energetically different half-chair forms (XVII)–(XXII).

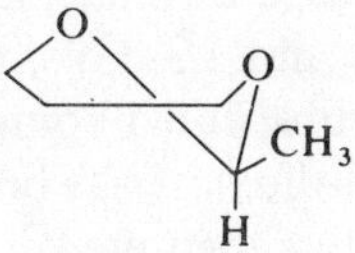

(XVII) 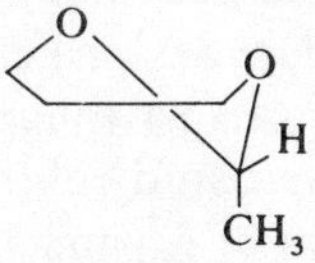   (XVIII)

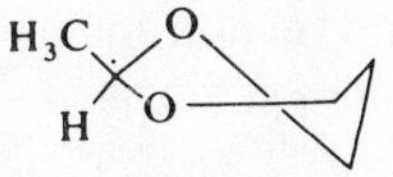

(XIX) 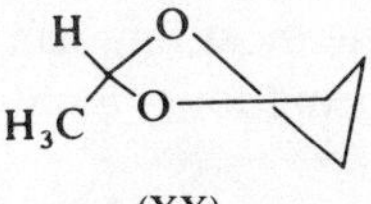   (XX)

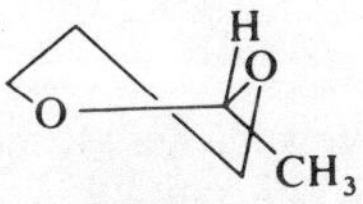

(XXI) 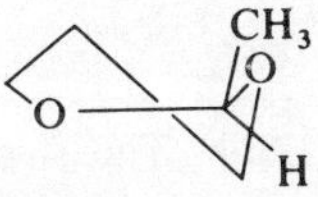   (XXII)

We have chosen to discuss the conformational behaviour of 1,3-dioxans because they are ideal subjects for study by a variety of methods for the following reasons:

(a) Being acetals or ketals, 1,3-dioxans are readily synthesized from 1,3-diols and carbonyl compounds in the presence of an acid catalyst as illustrated in Figure 12.9. In practice, equimolar amounts of the diol and

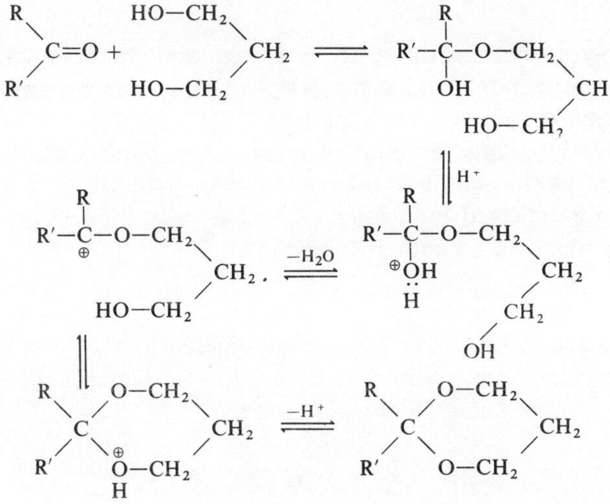

Figure 12.9 Synthesis of a substituted 1,3-dioxan; R = alkyl, R′ = alkyl or H

carbonyl compound are refluxed, in the presence of the catalyst, in a suitable azeotropic solvent until the theoretical amount of water is eliminated. *p*-Toluene sulphonic acid has often been used as the catalyst. Using appropriately substituted aldehydes or ketones and 1,3-diols, most of which are available commercially, a vast range of substituted compounds have been prepared and an extensive study of the 1,3-dioxans has been possible.

(b) The stereoisomers of asymmetrically substituted 1,3-dioxans (Figure 12.10) can be equilibrated easily through reversible ring opening effected by an anhydrous acid catalyst (see Figure 12.5). This is why chemical equilibration is one of the most important and reliable methods used to study the energetics of conformational equilibria in 1,3-dioxans.

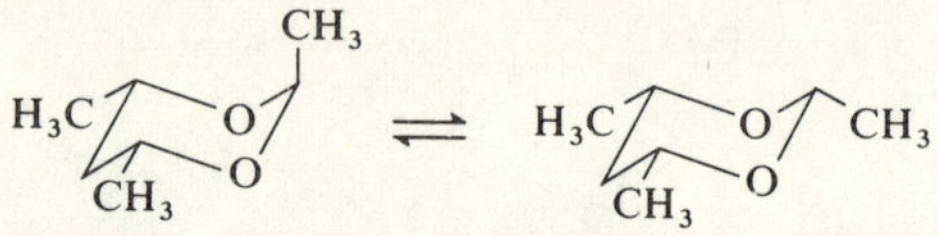

Figure 12.10   The stereoisomers of an asymmetrically substituted 1,3-dioxan

(c) The NMR method is well suited to the study of the 1,3-dioxan system since, compared to cyclohexane, replacement of two methylene groups by oxygen reduces the number of coupling nuclei and also prevents, in many cases, mutual spin–spin coupling by groups on either side of the hetero-atoms. Well-defined spectra, which often facilitate first-order analysis, are generally observed owing to the large chemical shifts between resonances of methylenic protons in the 2-(acetal), 4(6)-(ether) and 5-(hydrocarbon) positions.

(d) Energy barriers in the range 6–10 kcal/mole between chair isomers make the compounds suitable for NMR (8–10 kcal/mole) and also ultra-sonic relaxation studies (3–8 kcal/mole).

Consequently a large amount of quantitative data is available on the energetics of conformational changes in these molecules; this wealth of information is reflected by Tables 12.2–12.4 which report conformational energies for well over a hundred 1,3-dioxans.

12.3.2 *Equilibrium studies*

12.3.2.1 *Steric effects*   The free-energy difference, ΔG^0, (see Figure 12.5) between the chair isomers shown in Figure 12.8 is related to the equilibrium constant, K, for the chair–chair isomerism by

$$K = \frac{n_a}{n_e} = \exp\left(-\Delta G^0/RT\right) \tag{12.1}$$

where *n* refers to the populations and a and e refer to the axial and equatorial isomers respectively. The populations of the chair conformers, and hence ΔG^0, have been measured directly using NMR and equilibration methods as described in the preceding chapters. Table 12.3 lists ΔG^0 values and also the methods used to determine this quantity. As shown in Figure 12.7 the geometry of the 1,3-dioxan ring results in a substituent on C-2 being closer to the synaxial protons on C-4 and C-6 than it would be in the corresponding cyclohexane, thus the free-energy difference between an axial and an equatorial methyl group on C-2 of 1,3-dioxan will be greater than the conformational energy of methylcyclohexane. The conformational energy of 4(6)-methyl-1,3-dioxan is increased (over methylcyclohexane) in the same way but to a lesser extent. The shorter C—O bonds and puckering in the C-2 region would be expected to have but little effect on the free-energy difference between an axial and an equatorial methyl group at C-5. Conversely the replacement of the synaxial hydrogens of methylcyclohexane by electron pairs in the 1,3-positions, and the flattening of the ring in the C—C—C region of the 1,3-dioxan molecule, reduces the repulsive interactions of an axial methyl group in 5-methyl-1,3-dioxan. This latter statement is based on the assumption that the steric requirements of the oxygen axial lone-pairs are smaller than those of the proton, or alternatively, that the methyl–proton synaxial interaction is more repulsive than the methyl lone-pair synaxial interaction. The above facts predict that the free-energy differences between an axial and an equatorial methyl substituent should decrease in the order: 2-methyl-1,3-dioxan > 4(6)-methyl-1,3-dioxan > methylcyclohexane > 5-methyl-1,3-dioxan. Reference to Table 12.3 shows that the average ΔG^0 values for 2-, 4- and 5-methyl-dioxans, which have been obtained by a variety of methods, are 3·8, 2·9 and 0·9 kcal/mole respectively whilst the ΔG^0 value for methylcyclohexane is 1·7 kcal/mole. A similar pattern is followed by other alkyl substituents in positions 2 and 5. The conformational energies of all 5-alkyl-1,3-dioxans are considerably less than in the corresponding cyclohexanes and the difference is quite marked in 5-t-butyl-1,3-dioxan which has a ΔG^0 value of 1·5 kcal/mole compared to ~ 4 kcal/mole for t-butylcyclohexane. It is possible that other factors may be contributing to the conformational energy in this case. For example because of the flattening of the ring in the C-5 region there will be an increase in the synaxial interactions between the 5-t-butyl group in the equatorial position and the axial hydrogen at C-4 and C-6 (Figure 12.11).

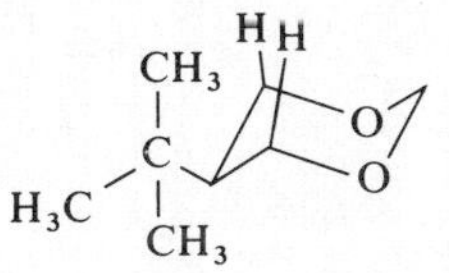

Figure 12.11   5-t-Butyl-1,3-dioxan

Table 12.2 Activation parameters associated with chair–chair equilibria in symmetrically substituted 1,3-dioxans

Compound	Ref.	Temp. (°K)	Solvent	$\Delta G^{\neq}$ (kcal/mole)	$\Delta H^{\neq}$ (kcal/mole)	$\Delta S^{\neq}$ (cal/deg/mole)
1,3-Dioxan	35	200	Me_2CO	9·0	10·2 ± 1·0	6 ± 5
	36	193	Me_2CO	9·7 ± 0·2		
	37	203	Me_2CO–d_6	9·9 ± 0·2	10·3	2 ± 6
	38	191	CS_2	9·7		
	39	205·4	$CFCl_3$	10·0 ± 0·1	11·0 ± 1·0	12 ± 5
	39	208·3	Me_2CO	10·1 ± 0·1	12·0 ± 1·1	17 ± 5
2,2-Dimethyl-1,3-dioxan	35	200	F_2CCl_2	7·8	$(6·3)^a$	$(-7)^a$
	36	<185	Me_2CO	<8·0		
	37	203	Me_2CO	8·0 ± 0·4	10·2	11 ± 9
4,4-Dimethyl-1,3-dioxan	37	203	Me_2O	8·6 ± 0·4	11·2	13 ± 10
4,4,6,6-Tetramethyl-1,3-dioxan	37	<137	Me_2O	<7·0		
5,5-Dimethyl-1,3-dioxan	35	200	Me_2CO	10·5	12·4 ± 0·8	9 ± 4
	36	223	Me_2CO	11·2 ± 0·25	12·8 ± 2·0	
	37	203	Me_2CO–d_6	11·2 ± 0·2	13·0	9 ± 5
				11·0 ± 0·2	12·4	7 ± 6
	38	215·7	CS_2	10·6		
	38	213·9	CS_2	10·5		
	38	217·0	CS_2	10·5		
	39	226·2	$CFCl_3$	10·9 ± 0·1	10·7 ± 1·7	7 ± 8
	40	253	Me_2CO–d_6	10·8	13·0	8·6
4,5,5-*trans*-6-Tetramethyl-1,3-dioxan	38	195·5	CS_2	10·3		
	38	189·0	CS_2	10·2		

5,5-Diethyl-1,3-dioxan	38	211·2	CS_2	10·1		
	38	210·3	CS_2	10·2		
5,5-Diisopropyl-1,3-dioxan	38	~168	CS_2	~8		
2,2,5,5-Tetramethyl-1,3-dioxan	35	200	Me_2CO	8·2	9·1 ± 0·9	4 ± 5
	37	203	Me_2O	8·9 ± 0·2	10·1	6 ± 9
	39	183·1	$CFCl_3$	9·0 ± 0·1		
2,2-Diethyl-5,5-dimethyl-1,3-dioxan	41	171·0	$Me_2CO–d_6$ / $CFCl_3$	6·8		
2,2-Dimethoxy-5,5-dimethyl-1,3-dioxan	37	175	CS_2	8·6		
2,2-Diphenyl-5,5-dimethyl-1,3-dioxan	37	199	CS_2	9·6		
2,2-Diisopropyl-5,5-dimethyl-1,3-dioxan	38	203	CS_2	9·9		
7,7-Dimethyl-5,9-dioxaspiro[3,5]nonane	42	199·5	CF_2Cl_2	9·4 ± 0·1		
	43	196	$CFCl_3$	9·3	8·9 ± 0·2	−2 ± 1
8,8-Dimethyl-6,10-dioxaspiro[4,5]decane	42	192·5	CF_2Cl_2	9·0 ± 0·1		
	43	189	$CFCl_3$	8·9	7·4 ± 0·2	−8·0 ± 1·2
	43	191	$CFCl_3$ / $Me_2CO–d_6$	9·0	7·1 ± 0·5	−10·0 ± 2·6
	38	195·5	CS_2	9·5		
	38	193·4	CS_2	9·5		

Table 12.2 Continued

Compound	Ref.	Temp. (°K)	Solvent	$\Delta G^{\neq}$ (kcal/mole)	$\Delta H^{\neq}$ (kcal/mole)	$\Delta S^{\neq}$ (cal/deg/mole)
9,9-Dimethyl-7,11-dioxaspiro[5,5]undecane	39	183·1	$CFCl_3$	$9·0 \pm 0·1$		
	42	177	CF_2Cl_2	$8·3 \pm 0·1$		
	43	178	$CFCl_3$	8·4	$5·1 \pm 0·2$	$-18·8 \pm 1·0$
	38	183	CS_2	8·9		
	37	183	Me_2CO–d_6	9·0		
3-t-Butyl-9,9-dimethyl-7,11-dioxaspira-[5,5]undecane	39	185·2	$CFCl_3$	$8·9 \pm 0·2$		
5,5′-Dimethyl-2,2′-adamanto-1,3-dioxan	38	185·7	CS_2	9·1		
	38	186·5	CS_2	9·0		
3,3,12,12-Tetramethyl-1,5,10,14-tetraoxa-dispiro[5,4,5]hexadecane	39	190·4	$CFCl_3$	$8·7 \pm 0·1$		
10,10-Dimethyl-8,12-dioxaspiro[6,5]-dodecane	42	171·5	CF_2Cl_2	$8·1 \pm 0·1$		
	38	180·3	CS_2	8·8		
		179·7	CS_2	8·7		
11,11-Dimethyl-9,13-dioxaspiro[7,5]-tridecane	42	173·5	CF_2Cl_2	8·2		
	38	180·5	CS_2	8·8		
	38	182·5	CS_2	8·8		
5,10-Dioxadispiro[2,4,3]undecane	44	197	CF_2Cl_2	$8·9 \pm 0·3$		
5,11-Dioxadispiro[2,4,4]dodecane	44	194	CH_2Cl_2	$8·8 \pm 0·3$		

5,7-Dioxaspiro[2,5]octane	44	206	CH_2Cl_2	9.5 ± 0.3		
	38	202.9	CS_2	9.6		
	38	197.3	CS_2	9.6		
6,8-Dioxaspiro[3,5]nonane	38	210.6	CS_2	10.1		
	38	208.2	CS_2	10.2		
7,9-Dioxaspiro[4,5]decane	38	216.2	CS_2	10.7		
	38	218.2	CS_2	10.7		
8,10-Dioxaspiro[5,5]undecane	38	208.3	CS_2	10.1		
	39	225	$CFCl_3$	10.9 ± 0.1	11.4 ± 1.4	9 ± 6
2,4,8,10-Tetraoxaspiro[5,5]undecane	39	200.1	$CFCl_3$	9.7 ± 0.1	9.2 ± 0.6	5 ± 3
3-t-Butyl-2,4,8,10-tetraoxaspiro[5,5]-undecane	39	199.1	$CFCl_3$	9.5 ± 0.1	7.8 ± 1.4	-1 ± 7
3,3,9,9-Tetramethyl-2,4,8,10-tetraoxaspiro[5,5]undecane	39	168	$CFCl_3$	8.4 ± 0.2		
3,3-Dimethyl-9-t-butyl-2,4,8,10-tetra-oxaspiro[5,5]undecane	39	168.5	$CFCl_3$	8.5 ± 0.1		

[a] Degree of uncertainty due to restricted temperature range.

Table 12.3 Conformational energies associated with chair–chair inversion in asymmetrically substituted 1,3-dioxans

Substituent	Ref.	Method	Temp. (°K)	Solvent	Preferred conformation	ΔG^* (kcal/mole)	ΔH^{*a} (kcal/mole)	ΔS^{*a} (cal/deg/mole)	ΔG^0 (kcal/mole)	ΔH^0 (kcal/mole)	ΔS^0 (cal/deg/mole)
2-Methyl	45	Ultrasonics			Me-e		4.40 ± 0.9	-8.7 ± 1.7			
	46	Equilibration		Et_2O	Me-e				3.55		
	47	Equilibration	298	Et_2O	Me-e				$3.0(3.4)^b$		
	48	Heats of formation			Me-e				4.07 ± 0.46		
	49	Appearance potential by electron impact		C_6H_6	Me-e				3.8		
	50	Equilibration	298	Et_2O	Me-e				3.98 ± 0.09		
2-Ethyl	45	Ultrasonics			Et-e		4.5 ± 0.9	-8.3 ± 1.7			
	47	Equilibration	298	Et_2O	Et-e				$2.8(3.2)^b$		
	50	Equilibration	298	Et_2O	Et-e				4.04 ± 0.04		
2-n-Propyl	45	Ultrasonics			n-Pr-e		5.0 ± 1.0	-5.1 ± 1.0			
2-Isopropyl	45	Ultrasonics			i-Pr-e		4.8 ± 0.95	-6.3 ± 1.3			
	50	Equilibration	298	Et_2O	i-Pr-e				4.17 ± 0.05		
2-n-Butyl	45	Ultrasonics			n-Bu-e		3.5 ± 0.7	-6.3 ± 1.3			
2-t-Butyl	47	Equilibration	298	Et_2O	t-Bu-e				2.9		
2-Methoxy	50	Equilibration	298	Et_2O	OMe-a				0.41 ± 0.016		
				Et_2O	OMe-a				0.50 ± 0.014 $(0.60)^c$		
				Et_2O	OMe-e				0.05 ± 0.003		
	50	Dipole moments	298	C_6H_6	OMe-a				0.62 ± 0.15		
	45	Ultrasonics			OMe-a		8.6 ± 1.7				
					OMe-a		4.1 ± 0.8	-5.5 ± 1.0			
	51	Equilibration		Et_2O	OMe-a				0.35 ± 0.01		
2-Ethoxy	51	Equilibration		Et_2O	OEt-a				0.35 ± 0.01		
2-Phenyl	45	Ultrasonics		1,4-dioxan	Ph-e		3.7 ± 0.75	-3.2 ± 0.6			
	50	Equilibration	298	Et_2O	Ph-e				3.12 ± 0.02		
				THF	Ph-e				3.11 ± 0.02		
	52	Equilibration		C_6H_{14}	Ph-e				2.85 ± 0.04		
				CH_3CN	Ph-e				3.40 ± 0.09		

2-p-Fluorophenyl	50	Equilibration	298	Et$_2$O	p-FPh-e			3.13 ± 0.03		
				THF	p-FPh-e			3.15 ± 0.02		
2-p-Trifluorophenyl	52	Equilibration		C$_6$H$_{14}$	p-CF$_3$Ph-e			3.08 ± 0.04		
				CH$_3$CN	p-CF$_3$Ph-e			3.29 ± 0.04		
	50	Equilibration	298	Et$_2$O	p-CF$_3$Ph-e			3.16 ± 0.02		
				THF	p-CF$_3$Ph-e			3.20 ± 0.02		
2-Phenylethynyl	53	Equilibration		CCl$_4$	C$_6$H$_5$C≡C-a			0.32 ± 0.09		
				Et$_2$O				0.00 ± 0.02		
				CH$_3$CN	C$_6$H$_5$C≡C-e			0.59 ± 0.07		
2-Ethynyl	53	Equilibration		CCl$_4$	HC≡C-a			0.21 ± 0.08		
				Et$_2$O	HC≡C-a			0.06 ± 0.09		
				CH$_3$CN	HC≡C-e			0.94 ± 0.03		
2-Vinyl	54	Equilibration						3–4		
2-p-Bromophenyl	50	Equilibration	298	THF	p-BrPh-e			3.18 ± 0.04		
2-Chloromethyl	45	Ultrasonics			MeCl-e	4.7 ± 0.92	−5.3 ± 1.1			
2-Chloromethyl-5,5-dimethyl	45	Ultrasonics			MeCl-e	5.3 ± 1.1	−4.6 ± 0.9			
2-Bromomethyl	55	Ultrasonics			MeBr-e	5.0	−4.7			
2-Bromomethyl-5,5-dimethyl	55	Ultrasonics			MeBr-e	5.1	−5.8			
2-Methyl-2-ethyl	56	NMR		CFCl$_3$	Et-e, Me-a				0.225 ± 0.009	−0.76 ± 0.04
	57	Equilibration	298		Et-e, Me-a			0.281	0.05 ± 0.03	0.77 ± 0.11
	58	NMR	306.5	CCl$_4$	Et-e, Me-a			0.41		
	59	Equilibration			Et-e, Me-a			0.32		
2-Methyl-2-n-propyl	56	NMR		CFCl$_3$	n-Pr-e, Me-a				0.227 ± 0.014	−0.75 ± 0.06
	58	NMR	306.5	CCl$_4$	n-Pr-e, Me-a			0.3		
2-Methyl-2-isopropyl	58	NMR	306.5	CCl$_4$	i-Pr-e, Me-a			0.61		
	56	NMR		CFCl$_3$	i-Pr-e, Me-a				0.863 ± 0.034	0.26 ± 0.15
	57	Equilibration	303		i-Pr-e, Me-a			0.551	0.09 ± 0.02	2.13 ± 0.05
	59	Equilibration			i-Pr-e, Me-a			0.63		
2-Methyl-2-phenyl	59	Equilibration			Me-e, Ph-a			2.42		
2-n-Butyl-2-methyl	58	NMR	306.5	CCl$_4$	n-Bu-e, Me-a			0.44		

Table 12.3 Continued

Substituent	Ref.	Method	Temp. (°K)	Solvent	Preferred conformation	$\Delta G^{\neq}$ (kcal/mole)	$\Delta H^{\neq a}$ (kcal/mole)	$\Delta S^{\neq a}$ (cal/deg/mole)	ΔG^{0} (kcal/mole)	ΔH^{0} (kcal/mole)	ΔS^{0} (cal/deg/mole)
2-Isobutyl-2-methyl	58	NMR	306.5	CCl$_4$	i-Bu-e, Me-a				0.40		
	57	Equilibration	298		i-Bu-e, Me-a				0.458	0.17 ± 0.09	0.99 ± 0.28
2-s-Butyl-2-methyl	58	NMR	306.5	CCl$_4$	s-Bu-e, Me-a				0.48		
2-t-Butyl-2-methyl	56	NMR		CFCl$_3$	t-Bu-e, Me-a				>2.9		
cis-2,4-Dimethyl	50	Equilibration	298	Et$_2$O	Me-e				⩽9.6		
5-Methyl	34	Equilibration	303	CHCl$_3$	Me-e				0.89	0.86 ± 0.09	−0.1 ± 0.3
					Me-e				0.87	0.98 ± 0.17	0.4 ± 0.6
	46	Equilibration	298	Et$_2$O	Me-e				0.97 ± 0.01		
	60, 61	Heats of combustion			Me-e				0.8		
	62	NMR		CFCl$_3$	Me-e				0.8 ± 0.2		
	63	NMR		C$_6$H$_6$ and CDCl$_3$	Me-e				1.05		
5-Ethyl	13, 34	Equilibration	305	CHCl$_3$	Et-e				0.81	0.74 ± 0.08	−0.2 ± 0.3
					Et-e				0.75		
	46	Equilibration	298	Et$_2$O	Et-e				0.73 ± 0.01		
					Et-e				0.67 ± 0.01		
5-Isopropyl	13, 34	Equilibration	303	CHCl$_3$	i-Pr-e				1.10	1.13 ± 0.10	0.1 ± 0.3
					i-Pr-e				1.05		
	46	Equilibration	298	Et$_2$O	i-Pr-e				0.98 ± 0.01		
5-t-Butyl	13, 34	Equilibration	303	CHCl$_3$	t-Bu-e				1.7	1.9 ± 0.5	0.5 ± 1.7
	46	Equilibration	298	Et$_2$O	t-Bu-e				1.36 ± 0.01		
				Et$_2$O	t-Bu-e				1.40 ± 0.01		
				Et$_2$O	t-Bu-e				1.38 ± 0.01		
				Et$_2$O	t-Bu-e				1.43 ± 0.02		
				Et$_2$O	t-Bu-e				1.46 ± 0.01		
	64	Equilibration	298	C$_6$H$_{12}$	t-Bu-e				1.37 ± 0.01		
				CCl$_4$	t-Bu-e				1.53 ± 0.03		
				C$_6$H$_6$	t-Bu-e				1.52 ± 0.01		
				Et$_2$O	t-Bu-e				1.47 ± 0.01		
				Neat	t-Bu-e				1.54 ± 0.01		
				Et(OMe)$_2$	t-Bu-e				1.57 ± 0.01		
				C$_6$H$_5$NO$_2$	t-Bu-e				1.66 ± 0.01		
				MeCN	t-Bu-e				1.73 ± 0.01		
				CHCl$_3$	t-Bu-e				1.78 ± 0.01		
				MeOH	t-Bu-e				1.62 ± 0.02		
				t-BuOH	t-Bu-e				1.59 ± 0.02		
				AcOH	t-Bu-e				~1.85		
				HCOOH	t-Bu-e				~1.97		

5-Phenyl	46	Equilibration	298	Et_2O	Ph-e	1.03 ± 0.02
5-Hydroxy	65	IR	353	C_6H_{12}	OH-a	0.915
	53, 61	Equilibration		C_6H_{12}	OH-a	0.89
				C_6H_{12}	OH-a	0.86
				C_6H_{12}	OH-a	0.81
				$CHCl_3$	OH-a	0.91
				CH_3CN	OH-a	0.04
				Et_2O	OH-e	0.41
	65	IR	353	i-Pr-OH	OH-a	0.51
	61	Equilibration		i-Pr-OH	OH-e	0.71
	65	IR	353	t-Bu-CH	OH-a	0.50
			353	$Et(OMe)_2$	OH-a	0.27
	61	Equilibration		$Et(OMe)_2$	OH-e	0.51
5-Methoxy	66	Equilibration	308	C_6H_6-HCl	OMe-e	0.46 ± 0.02
				CCl_4	OMe-e	0.71 ± 0.01
	65	IR	301	CCl_4	OMe-e	0.89
			323	Et_2O	OMe-e	0.83
			301	$CHCl_3$	OMe-e	0.18
			301	MeOH	OMe-e	0.03
			298	MeCN	OMe-a	0.01
	67	Equilibration		C_6H_6	OMe-e	0.58
	68	Equilibration		$PhCH_3$	OMe-e	0.71
				Ph-t-Bu	OMe-e	0.83
				$C_6H_3Me_3$	OMe-e	0.87
				$CDCl_3$	OMe-e	0.19
				CH_3CCl_3	OMe-e	0.58
				THF	OMe-e	0.65
				$PhNO_2$	OMe-e	0.2
				Me_2CO	OMe-e	0.34
5-Ethoxy	66	Equilibration	308	C_6H_6-HCl	OEt-e	0.65 ± 0.02
				CCl_4	OEt-e	0.99 ± 0.02
	53	Equilibration		C_6H_6	OEt-e	0.82
				CCl_4	OEt-e	1.09
				$CHCl_3$	OEt-e	0.51
				CH_3CN	OEt-e	0.19
	68	Equilibration		$PhCH_3$	OEt-e	0.94
				Ph-t-Bu	OEt-e	1.01
				$C_6H_3Me_3$	OEt-e	1.12
				$CDCl_3$	OEt-e	0.51
				CH_3CCl_3	OEt-e	0.82
				THF	OEt-e	0.86
				CH_3COCH_3	OEt-e	0.56
				$PhNO_2$	OEt-e	0.38

Table 12.3 Continued

Substituent	Ref.	Method	Temp. (°K)	Solvent	Preferred conformation	ΔG^{*} (kcal/mole)	ΔH^{*a} (kcal/mole)	ΔS^{*a} (cal/deg/mole)	ΔG^{0} (kcal/mole)	ΔH^{0} (kcal/mole)	ΔS^{0} (cal/deg/mole)
5-Isopropoxy	66	Equilibration	308	C_6H_6-HCl	i-OPr-e				1.04 ± 0.02		
				CCl_4	i-OPr-e				1.34 ± 0.02		
5-Acetoxy	65	IR	298	Et_2O					0		
5-Acetate	53	Equilibration		Et_2O					0.0		
5-Fluoro	67	Equilibration		$CHCl_3$	F-a				0.87		
				CCl_4	F-a				0.36		
	65	NMR	298	Et_2O	F-a				0.62		
				MeOH	F-a				0.605		
				C_6H_6	F-a				0.83		
				MeCN	F-a				1.225		
5-Chloro	65	NMR	298	Et_2O	Cl-e				1.20		
	67	Equilibration		CCl_4	Cl-e				1.40		
				C_6H_6	Cl-e				0.89		
				$CHCl_3$	Cl-e				0.94		
				CH_3CN	Cl-e				0.25		
5-Bromo	65	NMR	298	Et_2O	Br-e				1.44		
	67	Equilibration		CCl_4	Br-e				1.71		
				C_6H_6	Br-e				1.17		
				$CHCl_3$	Br-e				1.35		
				CH_3CN	Br-e				0.68		
5-Nitro	65	Dipole moments	303	CCl_4	NO_2-a				0.38		
				$CHCl_3$	NO_2-a				0.63		
				CH_2Cl_2	NO_2-a				0.81		
				$CDCl_3$	NO_2-a				0.89		
	53	Equilibration		MeCN	NO_2-a				1.0		
5-Cyano	65	NMR	303	Et_2O	CN-e				0.21		
				MeCN	CN-a				0.55		
5-Carbomethoxy	65	NMR	303	Et_2O	CO_2Me-e				0.82		
				MeCN	CO_2Me-e				0.22		

5-Hydroxymethyl	65	NMR	300	CCl$_4$	CH$_2$OH-a		0.27		
			303	Et(OMe)$_2$	CH$_2$OH-e		0.01		
5-Methoxymethyl	65	NMR	303	Et$_2$O	CH$_2$OMe-e		0.05		
5-Methyl-5-ethyl	56	NMR	214	CFCl$_3$	Me-e, Et-a[d]	10.4[e]			
	69	NMR	300		Me-e, Et-a[d]		0.077 ± 0.002	0.09 ± 0.004	0.05
							0.124	0.231	0.36
							0.122	0.117	−0.01
							0.205	0.336	0.44
5-Methyl-5-n-propyl	56	NMR	216	CFCl$_3$	n-Pr-e, Me-a	10.52[e]			
	69	NMR	300		n-Pr-e, Me-a		0.039 ± 0.001	0.069 ± 0.002	0.10
							0.055	0.140	0.28
							0.254	0.184	0.09
							0.254	0.451	0.66
5-Methyl-5-isopropyl	69	NMR	300	CS$_2$	i-Pr-e, Me-a		0.190 ± 0.002	0.364 ± 0.004	0.58 ± 0.01
5-Methyl-5-n-butyl	69	NMR	300		n-Bu-e, Me-a		0.020 ± 0.001	0.05 ± 0.002	0.10
							0.025	0.077	0.18
							0.115	0.174	0.28
							0.115	0.291	0.58
5-Methyl-5-s-butyl	69	NMR	300	CS$_2$	s-Bu-e, Me-a		0.178 ± 0.001	0.346 ± 0.003	0.56 ± 0.01
5-Methyl-5-isobutyl	69	NMR	300	CS$_2$	i-Bu-e, Me-a		0.033 ± 0.002	0.132 ± 0.006	0.33 ± 0.02
5-Methyl-5-n-pentyl	69	NMR	300		n-Pe-e, Me-a		0.024 ± 0.001	0.067 ± 0.002	0.14
							0.028	0.096	0.14
							0.087	0.171	0.33
							0.087	0.258	0.57
5-Methyl-5-cyclo-pentyl	69	NMR	300	CS$_2$	cyclo-Pe-e, Me-a		0.117 ± 0.001	0.270 ± 0.001	0.51 ± 0.02
5-Methyl-5-cyclo-hexyl	69	NMR	300	CS$_2$	cyclo-Hex-e, Me-a		0.190 ± 0.001	0.362 ± 0.004	0.57 ± 0.01
5-Methyl-5-phenyl	69	NMR	300	CS$_2$	Ph-e, Me-a		0.316 ± 0.003	0.396 ± 0.008	0.27 ± 0.03
					Ph-e, Me-a		0.495 ± 0.008	0.168 ± 0.015	−1.09 ± 0.05
5-Methyl-5-amino	58	NMR	300	Me$_2$CO	Me-e, NH$_2$-a		0.156	0.339	0.61
5-Methyl-5-acetyl	70	NMR	300	CS$_2$	Ac-e, Me-a		0.906	0.517	1.3
							1.327	0.527	2.6

Table 12.3 Continued

Substituent	Ref.	Method	Temp. (°K)	Solvent	Preferred conformation	$\Delta G^{\ddagger}$ (kcal/mole)	$\Delta H^{\ddagger a}$ (kcal/mole)	$\Delta S^{\ddagger a}$ (cal/deg/mole)	ΔG^0 (kcal/mole)	ΔH^0 (kcal/mole)	ΔS^0 (cal/deg/mole)
5-Methyl-5-vinyl	71	NMR	200	CS_2	Vinyl-e, Me-a				0.255	0.335	0.44
									0.257	0.335	0.44
									0.264	0.358	0.45
				Me_2CO-d_6	Vinyl-e, Me-a				0.230	0.290	0.32
									0.207	0.208	0
									0.188	0.297	0.54
				$CDPyC^f$	Vinyl-e, Me-a				0.278	0.380	−3.1
5-Methyl-5-ethynyl	71	NMR	200	Me_2CO-d_6	Acetylene-e, Me-a				0.229	0.224	−2.3
				$CDPyC^f$	Acetylene-e, Me-a				0.284	0.238	−2.4
									0.284	0.876	−5.8
										0.888	−5.3
										0.860	−5.2
5-Methyl-5-difluoro-methyl	72	NMR	300	CS_2	CF_2H-e, Me-a				1.870		
				Freon-12	CF_2H-e, Me-a				1.690		
5-Methyl-5-fluoro-methyl	72	NMR	300	Freon-12	CFH_2-e, Me-a				1.397	0.490	−3.0
5-Ethyl-5-isopropyl	69	NMR	300	CS_2	i-Pr-e, Et-a				0.172 ± 0.001	0.349 ± 0.003	0.59 ± 0.01
5-Ethyl-5-phenyl	69	NMR	300	CS_2	Ph-e, Et-a				0.380 ± 0.003	0.452 ± 0.011	0.24 ± 0.04
					Ph-e, Et-a				0.399 ± 0.004	0.423 ± 0.007	0.08 ± 0.01
5-Ethyl-5-n-butyl	38	NMR	210.2	CS_2	n-Bu-e, Et-a	10.1^e					
			209.8	CS_2	n-Bu-e, Et-a	10.2^e					
5-n-Propyl-5-phenyl	69	NMR	300	CS_2	Ph-e, n-Pr-a				0.289 ± 0.001	0.397 ± 0.007	0.36 ± 0.02
					Ph-e, n-Pr-a				0.279 ± 0.003	0.40 ± 0.005	0.40 ± 0.02
5-spiro-2′-Tetra-hydropyran	66	Equilibration			O-a				0.37		
		Dipole moments			O-a				~0.24		
4e, 5e -or 4a, 5e -or 4e, 5a-dimethyl	48	Heats of formation							0.69 ± 0.29		
4-Methyl	48	Heats of formation			Me-e				2.72 ± 0.52		
	46	Equilibration	298	Et_2O	Me-e				2.92 ± 0.50		
					Me-e				2.92 ± 0.02		
					Me-e				2.77 ± 0.08		
					Me-e				2.73 ± 0.02		
					Me-e				2.87 ± 0.04		

Compound	Ref.	Method	T (K)	Solvent	Assignment					
	73, 74	Ultrasonics			Me-e	9.2	5.03	3.44	2.5	
	49	Appearance potential by electron impact								
4-Phenyl	73, 74	Ultrasonics			Ph-e	4.5	−8.85		5.3	
4-Methyl-6-trifluoro-methyl	72	NMR	300	CS$_2$	Me-e, CF$_3$-a			0.585		
					Me-e, CF$_3$-a			0.695		
					Me-e, CF$_3$-a			0.588		
					Me-e, CF$_3$-a			0.599		
					Me-e, CF$_3$-a			0.743		
				C$_6$H$_{12}$	Me-e, CF$_3$-a			0.782		
4-Methyl-6-difluoro-methyl	72	NMR	300	CS$_2$	Me-e, CF$_2$H-a			0.294		
					Me-e, CF$_2$H-a			0.299		
4-Methyl-6-fluoro-methyl	72	NMR	300	CS$_2$	CFH$_2$-e, Me-a			0.145		
4,4,5-Trimethyl $4e, 4a, 5a \rightleftharpoons 4a, 4e, 5e$	62	NMR	306.5	CFCl$_3$ and CCl$_4$	Me-e			1.24		
cis-2,4,4,5-Tetramethyl $4e, 4a, 5a \rightleftharpoons 4a, 4e, 5e$	75	Equilibration	298		Me-e			1.45 ± 0.006	1.33	0.41
cis-4,5-trans-6-Tri-methyl $4e, 5a, 6a \rightleftharpoons 4a, 5e, 6e$	62	NMR	306.5	CFCl$_3$ and CCl$_4$	Me-e			0.58 ± 0.13		
cis-2,4,5-trans-6-Tetramethyl $4e, 5a, 6a \rightleftharpoons 4a, 5e, 6e$	75	Equilibration	298		Me-e			0.607 ± 0.003		
2,4,5,5-Tetramethyl $4a, 5a, 5e \rightleftharpoons 4e, 5a, 5e$	37	Equilibration	298		Me-e			3.10 ± 0.1		
cis-4,5-trans-6-Tri-methyl-2-oxo	63	NMR		C$_6$H$_6$ and CDCl$_3$	Me-e			0.2		
2,9-Dimethyl-cis-1,3-dioxadecalane	76	Equilibration	298	CCl$_4$	O-inside			1.278	0.515	2.6
2,4-Dimethyl-cis-1,3-dioxadecalane	76	Equilibration	298	CCl$_4$	O-inside			0.511	0.093	1.4

[a] Energy barriers refer to less stable → more stable transition.
[b] Figures in parentheses corrected for the assumption that the two conformations of the *trans* isomer are equally populated.
[c] Value corrected for conformational heterogeneity.
[d] Based on above ΔG^0 values for 5-Me-a and 5-Et-a.
[e] Energy barrier for more stable → less stable transition.
[f] CS$_2$:Py:CDCl$_3$ = 4:1:1.

Table 12.4 Conformational energies associated with chair–twist-boat inversion in 1,3-dioxans

Compound	Ref.	Method	Temp. (°K)	Solvent	$\Delta G^{\neq a}$ (kcal/mole)	$\Delta H^{\neq a}$ (kcal/mole)	$\Delta S^{\neq a}$ (cal/deg/mole)	ΔG^{0a} (kcal/mole)	ΔH^{0} (kcal/mole)	ΔS^{0} (cal/deg/mole)
1,3-Dioxan	77	Thermochemical						5·88		
	78	NMR		CS$_2$ and Me$_2$CO					5·7 $\pm$ 1·9	
				CS$_2$ and C$_6$H$_5$NO$_2$					6·0 $\pm$ 0·3	
				CS$_2$ and Me$_2$CO					6·3 $\pm$ 0·3	
2,2-Dimethyl-1,3-dioxan	48	Thermochemical						5·82 $\pm$ 0·46		
	45	Ultrasonics	300		5·2 $\pm$ 1·0	4·53 $\pm$ 0·9	$-2\cdot0 \pm 0\cdot4$	1·7[b]	5·67[b]	13[b]
2,2,4-Trimethyl-1,3-dioxan	45, 79	Ultrasonics	300		5·4 $\pm$ 1·1	4·52 $\pm$ 0·9	$-3\cdot0 \pm 0\cdot6$			
2,2,4,6-Tetramethyl-1,3-dioxan	50	Equilibration	300					$\geqslant$5·5	>8·3	
	49	Appearance potential by electron impact							8·5	
	48	Thermochemical						5·7 $\pm$ 0·6	7·1 $\pm$ 0·8	4·8 $\pm$ 1·0
	77	Thermochemical						5·6 $\pm$ 0·6	6·8 $\pm$ 0·7	3·9 $\pm$ 0·5
2-Ethyl-2-methyl-1,3-dioxan	45	Ultrasonics	300		3·9 $\pm$ 0·8	3·1 $\pm$ 0·6	$-2\cdot8 \pm 0\cdot6$			
2-Ethyl-2,4-dimethyl-1,3-dioxan	45	Ultrasonics	300		5·6 $\pm$ 1·1	2·9 $\pm$ 0·6	$-9\cdot0 \pm 2\cdot0$			
2,2-Diethyl-1,3-dioxan	45, 79	Ultrasonics	300		5·6 $\pm$ 1·1	2·5 $\pm$ 0·5	$-10\cdot3 \pm 2\cdot0$			
2,2-Diethyl-4-methyl-1,3-dioxan	45	Ultrasonics	300		5·5 $\pm$ 1·1	2·2 $\pm$ 0·4	$-11\cdot0 \pm 2\cdot0$			
6,10-Dioxaspiro(4,5)decane	45, 79	Ultrasonics	300		5·4 $\pm$ 1·1	4·64 $\pm$ 0·9	$-2\cdot7 \pm 0\cdot6$			
7-Methyl-6,10-dioxaspiro(4,5)decane	45, 79	Ultrasonics	300		5·3 $\pm$ 1·1	4·6 $\pm$ 0·9	$-2\cdot3 \pm 0\cdot5$			
5,5-Dimethyl-1,3-dioxan	45	Ultrasonics	300		6·3 $\pm$ 1·3	2·7 $\pm$ 0·6	$-12\cdot0 \pm 2\cdot4$	3·4[b]	9·7[b]	21[b]
								4·4[b]	10·1[b]	19[b]
								4·0[b]	10·3[b]	21[b]
								4·0[b]	9·7[b]	19[b]
								2·3[b]	8·0[b]	19[b]
								4·1[b]	10·3[b]	20·6[b]
8,8-Dimethyl-6,10-dioxaspiro(4,5)decane	45	Ultrasonics	300		6·0 $\pm$ 1·2	2·8 $\pm$ 0·6	$-10\cdot6 \pm 2$	3·8[b]	4·6[b]	2·6[b]
								4·1[b]	4·3[b]	0·6[b]

2,2,5,5-Tetramethyl-1,3-dioxan	45	Ultrasonics	300		$6\cdot1 \pm 1\cdot2$	$2\cdot5 \pm 0\cdot5$	$-12\cdot0 \pm 2\cdot4$	$1\cdot8^b$ $2\cdot2^b$	$6\cdot6^b$ $7\cdot6^b$	16^b 18^b
4-t-Butyl-6-methyl-1,3-dioxan	78	NMR	300	CS_2 and Me_2CO				1·83	$2\cdot7 \pm 1\cdot8$	$2\cdot9 \pm 3$
4-t-Butyl-6-isopropyl-1 3-dioxan	78	NMR	300	CS_2 and $C_6H_5NO_2$				1·71	$2\cdot35 \pm 0\cdot2$	$4\cdot4 \pm 1$
4-t-Butyl-6-cyclohexyl-1,3-dioxan	78	NMR	300	CS_2 and Me_2CO				1·09	$2\cdot53 \pm 0\cdot2$	$4\cdot8 \pm 0\cdot1$
$trans$-4-Trifluoromethy-6-t-butyl-1,3-dioxan	72	NMR							3·72	6·1

[a] Energy barriers refer to less stable → more stable transition.
[b] Values obtained by combining ultrasonic chair–twist-boat activation parameters with all the independent NMR chair–chair activation parameters quoted in Table 12.2.

12.3.2.2 *Anomeric effect*   Another important phenomenon which has emerged from the study of 1,3-dioxans is the preference for the axial position over the equatorial position of the 2-alkoxy group in 1,3-dioxans. This is due to the anomeric effect which has been defined[51] as the greater preference of an electron-withdrawing group for the axial position when it is located adjacent to a heteroatom in a ring than when it is located elsewhere. The anomeric effect has also been explained[80] in terms of dipole–dipole interactions (Figure 12.12) and by Eliel[81] in terms of the 'rabbit-ear' effect.

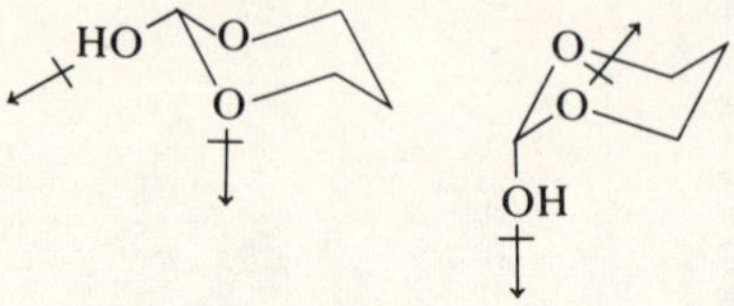

Figure 12.12   Dipole–dipole interactions in 2-alkoxy-1,3-dioxans

Generally speaking, the anomeric effect predicts that in a fragment of the type shown in Figure 12.13, where X and Y are oxygen or nitrogen, those conformations will be preferred in which the number of 1,3-interactions

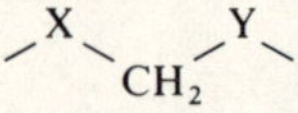

Figure 12.13

between lone-pair electrons is a minimum. Figure 12.14 shows the axial (a) and equatorial (e) conformers of 2-methoxy-1,3-dioxan. All the unshared electron pairs of the oxygen atoms are included and those involved in 1,3-interactions are shaded for ease of identification. It can be seen that in the axial conformer (a) there is only one such interaction whereas in the equatorial conformers (e₁) and (e₂) there are three and two such interactions respectively. Thus the anomeric effect predicts that the axial conformer of 2-methoxy-1,3-dioxan will be the most stable. Experimental results show (Table 12.3) that the free-energy difference between the axial and equatorial

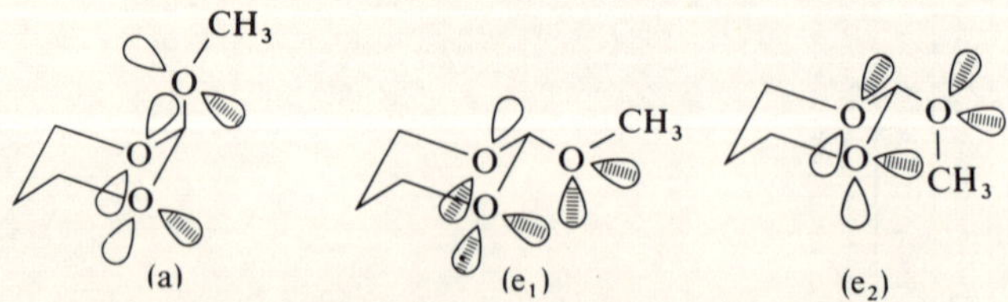

Figure 12.14   Axial (a) and equatorial (e) conformers of
2-methoxy-1,3-dioxan

conformers of this compound is 0·5 kcal/mole in favour of the axial conformer. The extent of the anomeric effect appears to be influenced quite considerably by the solvent used and in diethyl ether the methoxy group actually prefers the equatorial position to the extent of 0·05 kcal/mole.

The conformational equilibrium of the 5-methoxy group is also strongly solvent dependent and the conformational energy, ΔG^0, has been correlated[65] to the dielectric constant of the solvent. In the highly polar solvent acetonitrile, ΔG^0 is ~ 0; this result has been attributed to an attractive Van der Waals interaction between the axial methoxy group and the ring oxygens. In a more recent study[67] of 5-heterosubstituted compounds, dipole repulsion is evidently important as shown by the strong solvent dependence of ΔG^0. In carbon tetrachloride, which has a low dielectric constant ($\varepsilon = 2·29$), the equilibrium shown in Figure 12.15 is apparently controlled by dipole repulsion for all molecules except the fluoro compound.

The preference of fluorine for the axial position is enhanced in acetonitrile, a solvent of high dielectric constant ($\varepsilon = 37·5$). In this solvent the cyano group also prefers the axial position to a slight extent, and in the 5-bromo and 5-chloro compounds, there is a distinct shift in the equilibrium (Figure 12.15) towards the axial conformer.

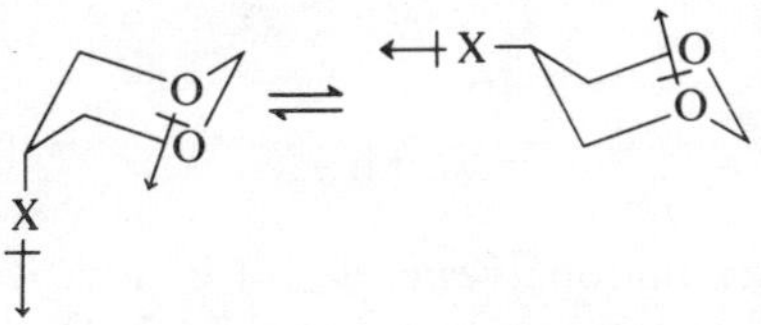

Figure 12.15   Dipole–dipole interactions in 5-substituted 1,3-dioxans; X = F, Cl, Br or CN

Comparison of ΔG^0 values of 5-fluoro-1,3-dioxan and fluorocyclohexane shows the remarkable preference of fluorine for the axial position in the former compound. This indicates an attractive fluorine–oxygen interaction in the axial conformer of 5-fluoro-1,3-dioxan. A theoretical treatment of the effect of the medium on the conformational energy difference between the axial and equatorial conformers was applied to these molecules by Abraham and coworkers.[67] This energy difference was calculated in terms of dipole and quadrupole effects (see Chapter 13) and excellent agreement with the observed conformational energies was obtained.

Recent NMR investigations[72] on 4-fluoromethylated-1,3-dioxans have shown that an anomeric effect is pronounced in these molecules. A study of 4-trifluoromethyl-6-methyl-1,3-dioxan has shown that the CF_3 group prefers the axial position at equilibrium whereas on purely steric grounds the equatorial positions would have been most likely. Similar experiments on the 6-methylated compounds predict that the 4-CHF_2 group also prefers the axial position at equilibrium, but to a lesser extent, whereas the 4-CH_2F group prefers the equatorial position. The existence of an anomeric effect can again be explained in terms of dipole–dipole interactions when the fluoromethyl group (CF_3 in Figure 12.16) is in the equatorial position. In the case of the 4-CH_2F substituent, presumably mutual dipole interactions in the equatorial conformer are not sufficient to overcome the steric effect.

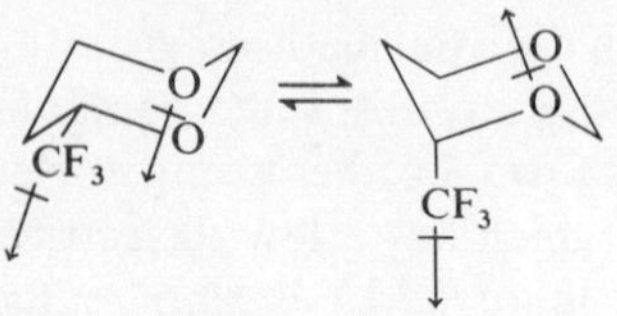

Figure 12.16

Similar experiments on 5-fluoromethylated-5-methyl-1,3-dioxans[82] have shown that the conformational energy of compound (XXIII) is ~ 1.87 kcal/mole (in CS_2) in favour of the Me-equatorial conformer (Figure 12.17).

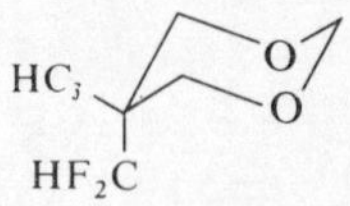

(XXIII)

Assuming that conformational energies are additive [ΔG^0(5-Me) = 0.9 kcal/mole] the conformational energy of 5-CF_2H can be estimated at ~ 1.0 kcal/mole. Similar experiments predict that the conformational energy of the 5-CFH_2 group is ~ 0.5 kcal/mole.

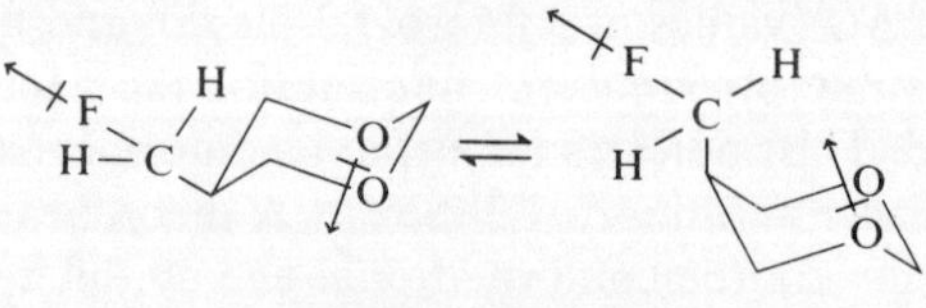

Figure 12.17

12.3.2.3 *Internal rotation in exocyclic groups* The study of conformational energies of 5-methoxy-[66] and 2-halomethyl-1,3-dioxans[83] has been extended to the determination of energy differences and activation energies between rotational isomers formed by internal rotation about the exocyclic bonds in these molecules. Katritzky and coworkers[66] determined the conformational energy difference between the rotamers of *cis*-5-alkoxy-2-phenyl-1,3-dioxan (XXIV)–(XXVI). This molecule is conformationally biased towards the Ph-equatorial isomer since ΔG^0 for 2-Ph is ~ 3.2 kcal/mole (Table 12.3). Figure 12.18 shows the three rotamers (XXIV)–(XXVI) of 5-alkoxy-2-phenyl-1,3-dioxan, one with the alkyl group 'inside' (XXIV) and two equivalent forms with the alkyl group 'outside' (XXV) and (XXVI). Dipole moment measurements were carried out on 5-methoxy-, 5-ethoxy-

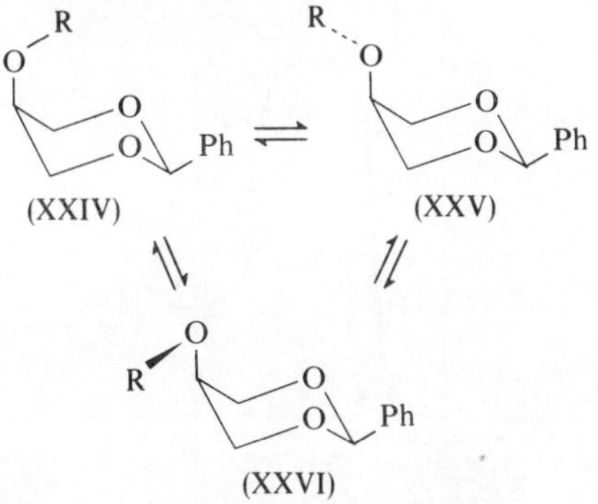

Figure 12.18   Rotamers of 5-alkoxy-2-phenyl-1,3-dioxan

and 5-isopropoxy-2-methyl-1,3-dioxans, from which equilibrium constants, corresponding to ΔG^0 (5-methoxy) = 1·18 kcal/mole and ΔG^0 (5-ethoxy) = 1·23 kcal/mole, were calculated.

Figure 12.19 shows the rotation isomers of the chloromethyl group in *cis*-2-chloromethyl-4-methyl-1,3-dioxans. The large substituents at C-2 and C-4 produce such large synaxial interactions in the diaxial conformer that the ring is essentially locked in the diequatorial conformation. Construction of molecular models of the three isomers shows that the conformation of least interaction is (XXVII); rotamers (XXVIII) and (XXIX) have equal energies. The ultrasonic relaxation spectra which were measured by Eccleston and coworkers[83] showed a single relaxation which was assigned

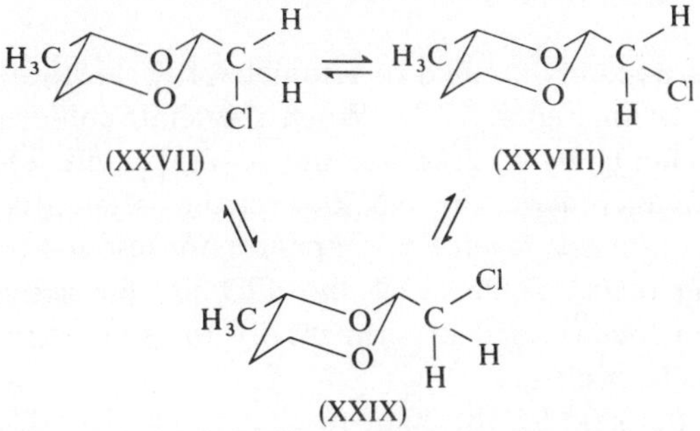

Figure 12.19   Rotamers of the chloromethyl group in
cis-2-chloromethyl-4-methyl-1,3-dioxans

to the rotation of the side-chain about the exocyclic C-2–carbon bond. The barrier to internal rotation, $\Delta H_b^{\neq}$ in Figure 12.20, was found to be 3·0 kcal/mole. The conformational energy ΔH^0 (Figure 12.20) between the rotational isomers was determined from the temperature dependence of carbon–halogen bands in the IR spectrum and the value of 0·57 kcal/mole was obtained.

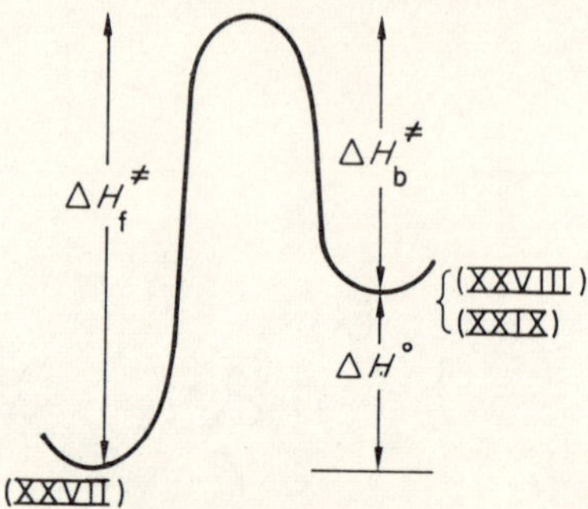

Figure 12.20   Relative energies of the
rotamers in Figure 12.19

12.3.3 *Kinetic studies*

12.3.3.1 *Methods of study*   The ultrasonic and NMR techniques have been used successfully to determine the barriers opposing ring inversion in many substituted 1,3-dioxans. Whilst the ultrasonic technique is applicable to unsymmetrical systems such as those shown in Figure 12.21, the NMR

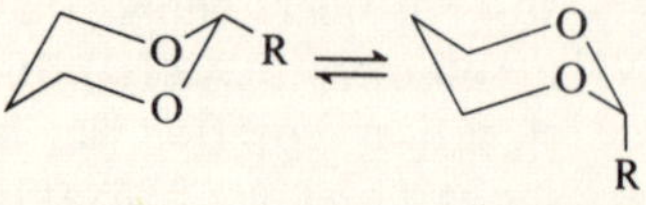

Figure 12.21   Asymmetrically substituted dioxan

technique is more readily applied to symmetrically substituted compounds of the type shown in Figure 12.22. When the chair conformers in Figure 12.21 differ in energy by a large amount ($>2\cdot0$ kcal/mole), i.e. when the equilibrium is anancomeric, the NMR spectrum is essentially that of the abundant isomer and as a result line-shapes do not respond to any exchange processes taking place. Tables 12.2 and 12.3 list the activation energies which have been determined experimentally for symmetrical and asymmetrical systems respectively.

The rate constants (k) for the chair-to-chair isomerism shown in Figures 12.21 and 12.22 are determined by the ultrasonic and NMR techniques in terms of the relaxation time or mean lifetime associated with the conforma-

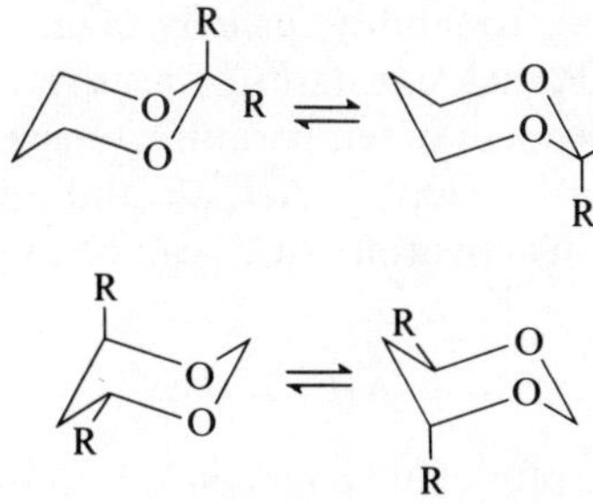

Figure 12.22 Symmetrically substituted dioxans

tional equilibrium. This quantity, τ, is related to the rate constants for the forward (k_f) and reverse (k_b) isomeric changes by

$$\tau = (k_f + k_b)^{-1} \tag{12.2}$$

In the case of the symmetrical compounds studied by NMR techniques, the energies of the ground-state chair forms are equal, thus $k_f = k_b$ and $\tau = 2k_b^{-1}$. The mean lifetime in one site can be related to the chemical shift difference and the coupling between the sites; the reader is referred to Chapter 5 for the actual analysis of the NMR spectra.

If K is the equilibrium constant for the chair-to-chair isomerism then

$$\tau = [k_b(1 + K)]^{-1} \tag{12.3}$$

and in turn is related to the critical ultrasonic frequency, f_c, by equation (12.4). Thus the rate constant for the less stable to more stable transition, k_b, can be expressed in terms of the critical frequency according to equation (12.5).

$$\tau = \frac{1}{2\pi f_c} \tag{12.4}$$

$$k_b = \frac{2\pi f_c}{(1 + K)} \tag{12.5}$$

If the free-energy difference, ΔG^0, between the chair forms is sufficiently large, i.e. $\Delta G^0 > 1\cdot8$ kcal/mole, then $(1 + K) \approx 1$ and $k_b \approx 2\pi f_c$.

It is always assumed that the chair–chair isomerism is governed by first-order kinetics, thus the enthalpy of activation $\Delta H_b^{\neq}$ and the entropy of activation, $\Delta S_b^{\neq}$, for the less stable to more stable transition, can be derived using the Eyring rate equation

$$k_b = \frac{\kappa \mathbf{k} T}{h} \exp\left(-\Delta H_b^{\neq}/RT\right) \exp\left(\Delta S_b^{\neq}/R\right) \tag{12.6}$$

where κ is the transition probability, usually taken as unity and $\mathbf{k}$ and h are the Boltzmann and Planck constants respectively. If it is assumed that $\Delta H_b^{\neq}$ and $\Delta S_b^{\neq}$ are independent of temperature, then a plot of $\ln k_b/T$ versus $1/T$ should be linear with a slope $-\Delta H_b^{\neq}/R$ and an intercept $[\ln \kappa\mathbf{k}/h + \Delta S_b^{\neq}/R]$. The free energy of activation, $\Delta G_b^{\neq}$, can be calculated from equation (12.6) since

$$\Delta G^{\neq} = \Delta H^{\neq} - T\Delta S^{\neq} \tag{12.7}$$

It has been found that the plot of $\ln k_b$ against $1/T$ is also linear in accordance with the Arrhenius equation

$$\ln k_b = -E_a/RT + \ln A \tag{12.8}$$

and the energy of activation, E_a, and the frequency factor, A, have often been quoted. In order to use this plot the assumption is made that both E_a and A are temperature independent. A is in fact temperature dependent but the dominant factor in equation (12.8) is the exponential term and to within the limits of the experimental data, the approximation has proved to be satisfactory. $\Delta H_b^{\neq}$ and $\Delta S_b^{\neq}$ are related to E_a and A respectively by equations (12.9) and (12.10).

$$\Delta H^{\neq} = E_a - RT \tag{12.9}$$

$$\Delta S^{\neq} = R\left[\ln\left(\frac{hA}{\kappa\mathbf{k}T}\right)^{-1}\right] \tag{12.10}$$

It is more meaningful to discuss the relative energies of ground and transition states in terms of $\Delta H^{\neq}$ and $\Delta S^{\neq}$. However, the experimental results show that the derivation of these parameters is subject, in many cases, to systematic errors. In addition, it is difficult to make accurate measurements of the rate constant, k, over a sufficiently large temperature range by the NMR technique, so that the derivation of $\Delta H^{\neq}$ and $\Delta S^{\neq}$ from the plot of the temperature dependence of $\ln (k/T)$, is further subject to errors (see Chapter 5). Although the experimental errors incurred are reasonably small when measuring the ultrasonic absorption of liquids, large errors sometimes arise from the analysis of the ultrasonic relaxation data (see Chapter 9), especially when the relaxation strength is small. However, a consideration[84] of the systematic errors involved in the Eyring rate equation (12.8), has shown that although the derived values of $\Delta H^{\neq}$ and $\Delta S^{\neq}$ are subject to correction terms, the systematic errors do not affect $\Delta G^{\neq}$. Thus much useful information has been obtained from a comparison of the activation free energies of a variety of compounds at a given temperature.

12.3.3.2 *Entropy of activation*   In order to explain the chair-to-chair isomerization in terms of first-order rate theory, a single transition path must be considered. In the case of 1,3-dioxan, there are three possible half-chair forms, (X)–(XII), each of which is asymmetric and therefore its mirror image is non-superimposable, for example (XIa) and (XIb) in Figure 12.23.

(XIa)   (XIb)

Figure 12.23

On the other hand, there are six energetically different half-chair forms of a monosubstituted 1,3-dioxan, e.g. (XVII)–(XXII). If the energies of (XVII)–(XXII) differ considerably, the barrier estimated from the rate equation will probably correspond to the energy difference between the chair form and the half-chair form of lowest energy. However, if (XVII)–(XXII) have similar energies, the estimated barrier is more likely to correspond to the average of the energy differences between the chair and each of the individual half-chair forms.

The relative energies of the half-chair forms also affect the activation entropy. The factors which contribute to the total activation entropy, $\Delta S^{\neq}$, are the vibrational, rotational and translational partition functions, and the entropy of mixing of isomeric species of equal energy. The facts will be discussed individually in relation to the chair (c) (VIII) to half-chair (hc) (XI) equilibrium in 1,3-dioxan (Figure 12.24).

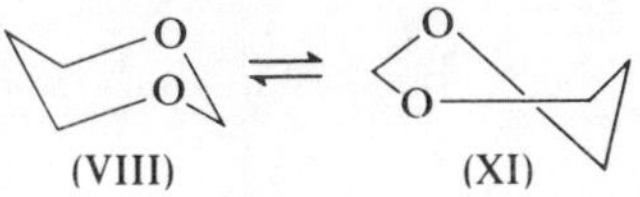

(VIII)   (XI)

Figure 12.24

The translational contribution, $\Delta S_T^{\neq}$, to the total activation entropy is given by

$$\Delta S_T^{\neq} = \tfrac{3}{2} R \ln \left(\frac{M_{hc}}{M_c} \right) \tag{12.11}$$

where M is the molecular weight. The chair and half-chair forms of 1,3-dioxan have the same molecular weight, thus the contribution to the entropy of activation from the translational partition function is zero.

The rotational contribution, $\Delta S_R^{\neq}$, to the activation entropy is given by

$$\Delta S_R^{\neq} = R \ln \left(\frac{Q_c}{Q_{hc}} \right) \tag{12.12}$$

where Q is the rotational partition function given by

$$Q = [8\pi^2 kT/h^2]^{\frac{3}{2}}[\pi(I_x I_y I_z)^{\frac{1}{2}}/\sigma]$$

where I_x, I_y and I_z are the principal moments of inertia and σ is the symmetry number of the species under consideration. The product $(I_x . I_y . I_z)$ is not expected to vary considerably for the chair and half-chair forms,[85] thus equation (12.12) reduces to

$$\Delta S_R^{\neq} = R \ln \left(\frac{\sigma_c}{\sigma_{hc}} \right) \tag{12.13}$$

where σ, for a particular conformation, represents the total number of possible arrangements of equivalent atoms which can be obtained by rotation about a symmetry axis. There are no such axes of symmetry in the chair or the half-chair forms of 1,3-dioxan, consequently each conformer has $\sigma = 1$. (The chair form of 1,3-dioxan has a vertical plane of symmetry and the identity element, and belongs to point group C_s. The half-chair form has only the identity element of symmetry and belongs to point group C_1. Both point groups have symmetry numbers of 1. In the case of substituted 1,3-dioxans, some will have a plane of symmetry, others will be asymmetric. However, in all cases the symmetry number is 1 and there is never a contribution to the activation entropy from $\Delta S_R^{\neq}$ on the basis of the preceding argument.) This in turn means that there is no contribution to the total activation entropy by the rotational partition function.

The vibrational contribution, $\Delta S_V^{\neq}$, to the total activation entropy is given by

$$\Delta S_V^{\neq} = R \sum_{hc} \{[x_{hc} \exp(-x_{hc})]/[1 - \exp(-x_{hc})] - \ln[1 - \exp(-x_{hc})]\}$$

$$- R \sum_c \{[x_c \exp(-x_c)]/[1 - \exp(-x_c)] - \ln[1 - \exp(-x_c)]\} \tag{12.14}$$

and $x_i = h\nu_i/kT$ where ν_i is the vibrational frequency. The significant contribution to the vibrational entropy of a molecule comes from those vibrational modes whose frequencies are less than $\sim 1000\,\text{cm}^{-1}$. However, in considering the difference in vibrational entropy between two conformers it is generally considered that the sum of the vibrational terms for one isomer is very close to the corresponding sum for the other and thus $\Delta S_V^{\neq} = 0$.

The entropy of mixing of a given state, S^M, is given by

$$S^M = -Rx_1 \ln x_1 - Rx_2 \ln x_2 - \cdots \tag{12.15}$$

where x_i is the mole fraction of species i in that state. In the equilibrium (VIII) $\rightleftharpoons$ (XI) the half-chair form (XI) exists as the dl pairs (XIa) and (XIb) (Figures 12.23 and 12.24). Since there is an equal chance of (XIa) and (XIb) being formed from (VIII) during ring inversion, the mole fractions will be equal and in equation (12.15) $x_1 = x_2 = x$. Thus

$$S^M = -R \sum x \ln x$$

$$= -R \ln x \qquad \text{since } \sum x = 1$$

When equal populations of isomers are being considered, $x = 1/w$ where w is the number of isomers of the same energy (i.e. the statistical weight of that energy level), thus

$$S^M = R \ln w$$

and the entropy of mixing of the half-chair state, S^M_{hc}, is given by

$$S^M_{hc} = R \ln 2$$

Since 1,3-dioxan has a plane of symmetry, its mirror image is superimposable and there is only one distinguishable chair form making $S^M_c = 0$. The contribution to the total activation entropy, $\Delta S^{\neq}$, by the entropy of mixing term, $\Delta S^{\neq}_M$, is

$$\Delta S^{\neq}_M = S^M_{hc} - S^M_c$$

$$= R \ln \left(\frac{w_{hc}}{w_c} \right)$$

$$= R \ln 2$$

If we now consider each of the half-chair forms (X)–(XII) in turn, then S^M_{hc} will have a value $R \ln 2$ for (X), (XI) and (XII). However, if the energies of the half-chair forms are very close and the chair–chair inversion proceeds via all three forms, the statistical weight becomes 6 and $S^M_{hc} = R \ln 6$.

It turns out that the only contribution to the total activation entropy, $\Delta S^{\neq}$, for the ring inversion in 1,3-dioxan is from the entropy of mixing term and $\Delta S^{\neq}$ can take a value from $R \ln 2$ to $R \ln 6$.

It must be pointed out that the above method of calculating the total activation entropy is essentially the same as that described by Harris and Sheppard.[86]

Harris and Spragg[87] have calculated the torsional contribution to the enthalpy of formation of the half-chair forms from the chair forms of 1,3-dioxan. These energies, which are calculated in relation to the half-chair of cyclohexane are given in Figure 12.25. Since the torsional strain amounts

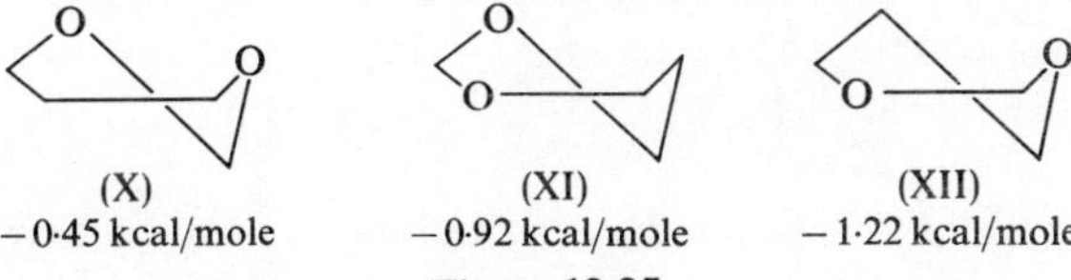

(X)   (XI)   (XII)

-0.45 kcal/mole -0.92 kcal/mole -1.22 kcal/mole

Figure 12.25

to about 60% of the inversion barrier and the differences in the torsional strain energy of (X)–(XII) appear to be very small compared to the actual activation enthalpy, the ring-inversion path probably proceeds via all three intermediates. This in turn means that $\Delta S^{\neq}$ should have a value close to $R \ln 6$. In the case of monosubstituted 1,3-dioxans, e.g. 2-methyl-1,3-dioxan, all six half-chair forms (XVII)–(XXII) differ in energy owing to the relative position of the methyl group in the different half-chair forms. The chair forms of 2- and 5-methyl-1,3-dioxan contain a plane of symmetry, thus for each of the half-chair forms in these compounds there will be an entropy of mixing contribution, $R \ln 2$ and $\Delta S^{\neq}$ should always be positive with a minimum value of $R \ln 2$. On the other hand if we consider 4-methyl-1,3-dioxan the chair form is asymmetric and exists as a *dl* pair of enantiomers. This means that the entropy of mixing of the chair state equals $R \ln 2$. If one of the half-chair forms is preferred during ring inversion $\Delta S^{\neq}$ should be zero.

In 2,2-dimethyl-1,3-dioxan each of the half-chair forms (XXX)–(XXXII) are doubly degenerate. Construction of molecular models of these half-chair forms indicates that in (XXXI) and (XXXII) there are severe 1,3-synaxial interactions which appear to be greater than the corresponding

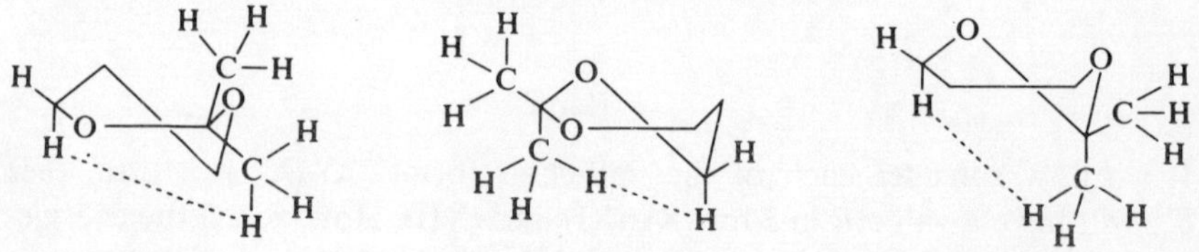

(XXX)   (XXXI)   (XXXII)

interactions in the chair form. Conversely the 1,3-synaxial interactions in (XXX) are less than in the chair form, which points to this half-chair form being favoured as the transition state and $\Delta S^{\neq}$ would then be expected to have a value $R \ln 2$.

12.3.3.3 *Discussion of experimental results*   From the foregoing discussion on activation entropies it follows that $\Delta S^{\neq}$ for 2-methyl- and 2,2-dimethyl-1,3-dioxan would be expected to be lower than for 1,3-dioxan and that all the entropies of activation are positive. Reference to Tables 12.2 and 12.3 shows, however, that the values for 1,3-dioxan and 2,2-dimethyl-1,3-dioxan obtained by Friebolin,[37] and that for 2-methyl-1,3-dioxan obtained by Eccleston[45] (assuming $\Delta S^{0} = 0$), are 2 ± 6, 11 ± 9 and -9 ± 2 e.u. respectively. The value for 4-methyl-1,3-dioxan, predicted as zero, was found[74] to be ~ 5 e.u.

Further examples of even more anomalous activation entropies determined from experiment and which cannot be explained in terms of the above discussion, are found in the tables. For example, the activation entropy values found for 2,2,5,5-tetramethyl-1,3-dioxan[37] (XXXIII) and 9,9-dimethyl-7,11-dioxaspiro [5, 5] undecane[43] (XXXIV) are $\sim 6 \pm 9$ e.u. and -19 ± 1 e.u. respectively. On the basis of the above discussion these

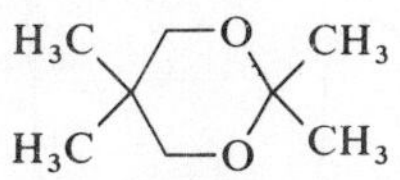

$$H_3C \quad O \quad CH_3 \qquad H_3C \quad O$$
$$H_3C \quad O \quad CH_3 \qquad H_3C \quad O$$

$$\text{(XXXIII)} \qquad\qquad \text{(XXXIV)}$$

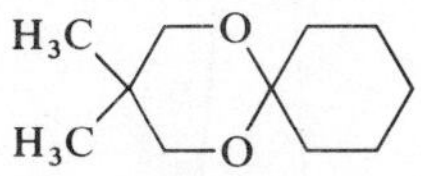

compounds should have similar activation entropies. Although the activation entropies $\Delta S^{\neq}$ quoted in Tables 12.2 and 12.3 are undoubtedly subject to errors, the discrepancies with those predicted using the above arguments are so large that it leads us to suspect that the difference cannot be explained completely in terms of experimental and systematic errors. It is worth noting, however, that in addition to activation entropies the above arguments can also be used to predict the entropy difference ΔS^0 between the stable isomers in a conformational equilibrium. For axial/equatorial chair isomerism, ΔS^0 should always be zero and inspection of Table 12.3 shows that of all the ΔS^0 values quoted from experiment, over 80 % have values in the range ± 1 cal/mole/deg. Summarizing, it appears that the basic concepts and ideas concerning ring inversion in 1,3-dioxans are well understood and in some examples the details are clear also; in other cases, especially concerning activation entropies, there are results which cannot be explained at present.

Further reference to Table 12.2 shows that the free energy of activation, $\Delta G^{\neq}$, for 2,2-dimethyl-1,3-dioxan (XXXV) is ~ 8 kcal/mole compared with ~ 10 kcal/mole for 1,3-dioxan (XXXVI). The conformational energy of a 2-methyl group in 1,3-dioxan is ~ 4 kcal/mole which suggests that the chair form of (XXXV) should be destabilized by approximately this amount relative to the chair form of (XXXVI). From the above activation energies it follows that the half-chair form of (XXXV) is also destabilized relative to the half-chair form of (XXXVI), but to a lesser extent, as shown in Figure 12.26.

The activation free energy for 5,5-dimethyl-1,3-dioxan (XXXVII) is ~ 11 kcal/mole, more than 1 kcal/mole higher than that of the parent compound. The ground-state destabilization of the chair form is expected to be ~ 1 kcal/mole, i.e. the conformational energy of a 5-methyl substituent in 1,3-dioxan. It follows that the half-chair state of (XXXVII) must be destabilized by at least 2 kcal/mole relative to that of 1,3-dioxan and reference to Figure 12.26 shows that the transition states of (XXXV) and (XXXVII) have similar energies.

The ground-state destabilization of (XXXV) and (XXXVII) is due to the severe steric 1,3-interactions between the axial 2-methyl group with the

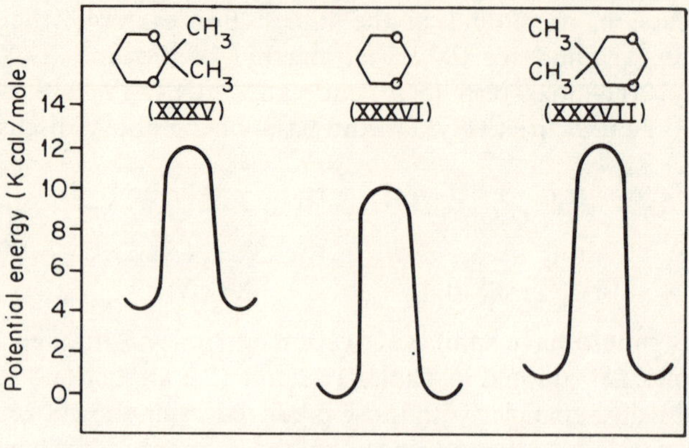

Figure 12.26 Relative transition paths of (XXXV), (XXXVI)
and (XXXVII)

4- and 6-hydrogens and the axial 5-methyl group with the lone-pair electrons
respectively. The destabilization of the transition states of (XXXV) and
(XXXVII), relative to that of (XXXVI), is probably due to a combination of
non-bonded steric interactions and torsional strain. The effect of ground-
state destabilization due to steric overcrowding is illustrated by the regular
decrease in $\Delta G^{\neq}$ (kcal/mole) for the following series of 1,3-dioxans:[37] 5,5-
dimethyl (11·1), parent (9·9), 2,2,5,5-tetramethyl (8·9), 4,4-dimethyl (8·6),
2,2-dimethyl (8·0) and 4,4,6,6-tetramethyl (<7).

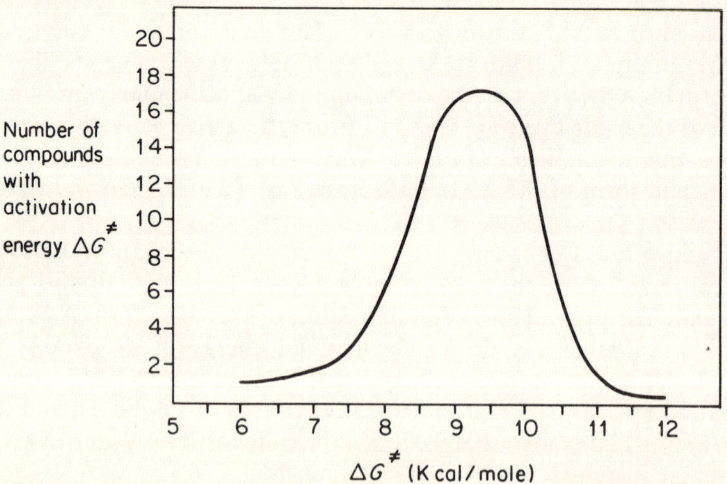

Figure 12.27 Distribution curve of compounds for which $\Delta G^{\neq}$ has been
determined

A further consideration of the activation free energies quoted in Tables 12.2 and 12.3 show that although systematic differences in $\Delta G^{\neq}$ for a series of compounds can be detected, it is also interesting to note that of the forty compounds for which free-energy barriers have been determined, 80% of these have $\Delta G^{\neq}$ values in the range 9.5 ± 1.0 kcal/mole as shown by the distribution curve in Figure 12.27. Table 12.3 also shows, however, that the free-energy difference, ΔG^{0}, between the chair isomers of substituted 1,3-dioxans, varies by $\sim \pm 5$ kcal/mole. This further shows that substituents affect the energy of the chair state more than that of the half-chair state.

12.3.4 *Twist-boat conformation*

Recent experiments using ultrasonic relaxation,[45] NMR[78,88] and thermo-chemical data[48,77] have provided some quantitative information on the relative energies of the twist-boat conformers in 1,3-dioxans. Table 12.4 lists the data which have been obtained and the techniques employed.

There are two different twist-boat conformations of the 1,3-dioxan ring, the 1,4-twist[88] (XXXVIII) and the 2,5-twist[88] (XXXIX). In contrast to the

(XXXVIII)   (XXXIX)

two different positions of substitution (axial and equatorial) of the chair form the twist-boat affords three different substituent positions. These are the pseudo-axial and pseudo-equatorial positions at C-4 and C-6 (in (XXXIX) and the equivalent positions at C-2 and C-5 (in XXXIX) which exhibit a minimum of non-bonded repulsions.

The most recent study[88] of twist-boat conformations of alkylsubstituted 1,3-dioxans has been discussed in terms of the stability of the twist-boat forms in relation to the position of the substituents. The study has shown that the twist-boat conformation, (XXXVIII) or (XXXIX), in which the number of alkyl groups in pseudo-axial positions is a minimum, will be preferred. For example, of the four possible twist-boat conformations (XL)–(XLIII) of *trans*-2,2,4,6-tetramethyl-1,3-dioxan, conformation (XL), in which there are no methyl groups in pseudo-axial positions, would be

(XL)   (XLI)

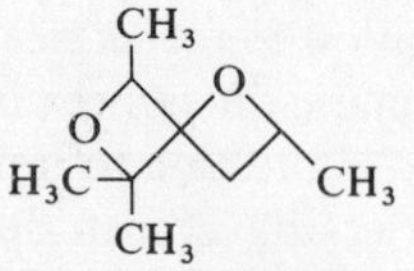

(XLII) 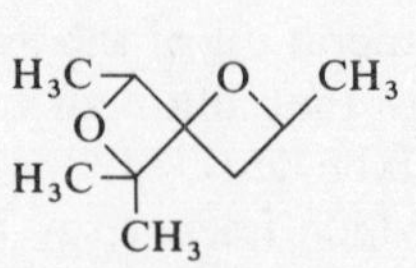   (XLIII)

predicted. This was, in fact, confirmed by NMR measurements to be the abundant conformer. Furthermore, when there is the possibility of more than one conformer with the minimum number of pseudo-axial substituents, the preferred conformation can be deduced from the coupling constants of proton resonances. For example, it was shown that of the two more stable conformers (a) and (b) of 2,2,4,4,6-pentamethyl-(XLIV) and 2,2,4,4,5-penta-methyl(XLV)-1,3-dioxans, the '1,4-twist' conformations (XLIVa) and (XLVa) are preferred to the '2,5-twist' conformations (XLIVb) and (XLVb).

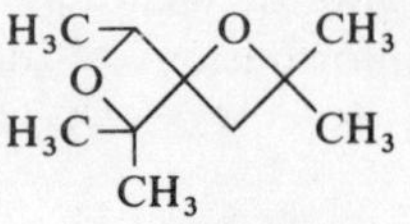

(XLIVa) 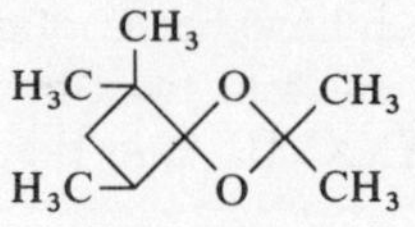   (XLIVb)

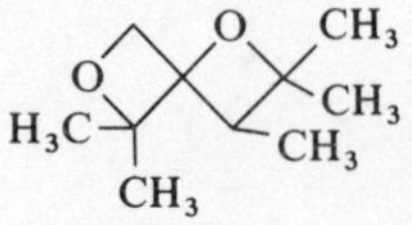

(XLVa) 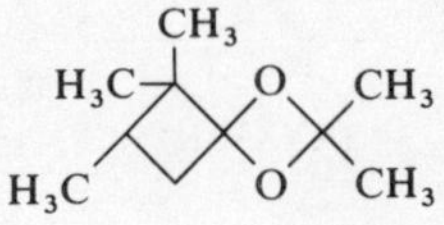   (XLVb)

Anteunis and Swaelens[78] studied a series of *trans*-4-*t*-butyl-6-alkyl-1,3-dioxans (XLVI)–(XLVIII) by the NMR technique and detected the presence of twist-boat conformations. The enthalpy and entropy differences for the chair to twist-boat equilibrium, shown in Figure 12.28, were derived from the

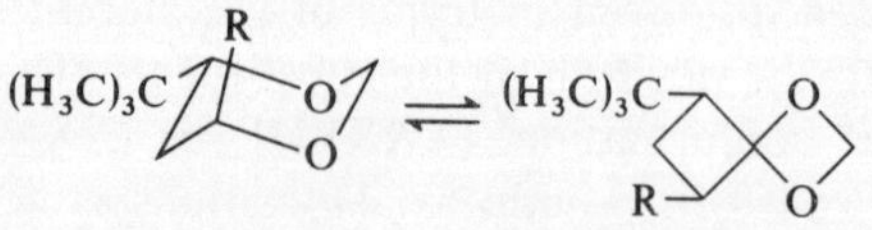

Figure 12.28   Chair to twist-boat equilibrium for a series of *trans*-4-*t*-butyl-6-alkyl-1,3-dioxans; R = ethyl (XLVI), R = isopropyl (XLVII), R = cyclohexyl (XLVIII)

temperature dependence of the coupling constants and the values are in Table 12.4. The enthalpy difference between the twist-boat and chair form of 1,3-dioxan were then estimated as follows. Firstly, it was assumed that the interaction of pseudo-equatorial groups in the twist-boat conformer is the same as the corresponding equatorial interaction in the chair form. Secondly, the chair forms (XLVI–XLVIII) are assumed to be destabilized with respect to that of 1,3-dioxan by an amount equal to the conformational energy of the 4- (or 6-) alkyl substituent. Figure 12.29 illustrates the relative potential energy diagrams for the ring inversion of (XLVII) and 1,3-dioxan. Using the

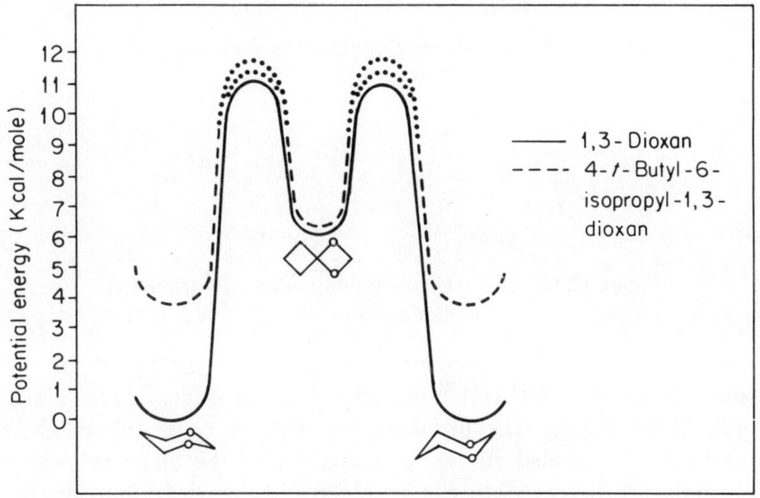

Figure 12.29 Relative potential energy diagrams for 1,3-
dioxan and 4-t-butyl-6-isopropyl-1,3-dioxan

above assumptions and the NMR data for (XLVI)–(XLVIII) the enthalpy difference between the chair and twist-boat forms of 1,3-dioxan was estimated to be 6 kcal/mole, in excellent agreement with the value of 5·88 kcal/mole obtained[77] from thermochemical data.

An ultrasonic relaxation which was consistent with a conformational change was observed for several compounds: 2,2-dimethyl-1,3-dioxan (XLIX), 8,8-dimethyl-6,10-dioxaspiro(4,5)-decane(L), 2,2,5,5-tetramethyl-(LI)

(XLIXa)   (XLIXb)   (L)

and 2,2,4-trimethyl(LII)-1,3-dioxans. The chair–chair equilibria for (XLIX)–(LI), shown in Figure 12.30 for (LI), cannot be perturbed by an ultrasonic wave since there is no enthalpy or volume difference between the conformers. In addition the chair-to-chair equilibrium shown for (LII) in Figure 12.31 does not give rise to a measurable ultrasonic relaxation. Due

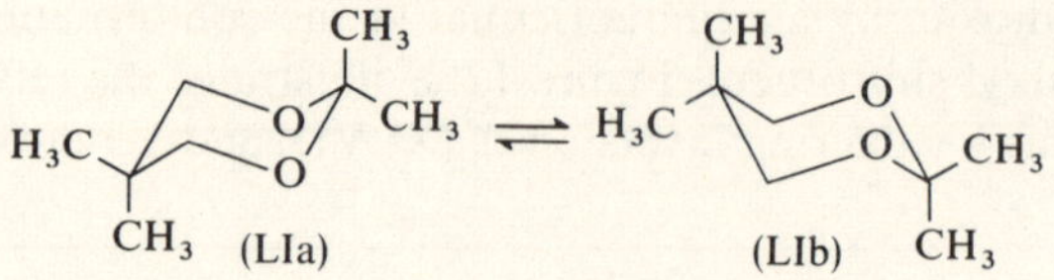

Figure 12.30 Chair–chair equilibrium for 2,2,5,5-tetra-methyl-1,3-dioxan

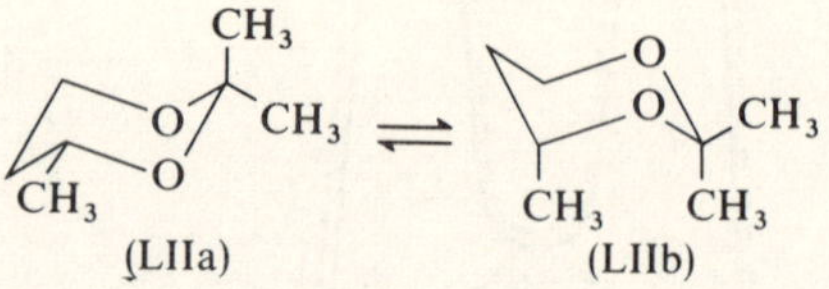

Figure 12.31 Chair–chair equilibrium for 2,2,4-trimethyl-1,3-dioxan

to severe 1,3-diaxial steric interaction the energy of (LIIb) will be ~ 8–11 kcal/mole in excess of (LIIa),[77] which results in a negligible relaxation strength. Consequently, the ultrasonic relaxations observed for (XLIX)–(LII) have been assigned to the perturbation of the chair to twist-boat equilibrium, shown for (LI) in Figure 12.32, and the energy barriers for the twist-boat to chair transitions are in Table 12.4.

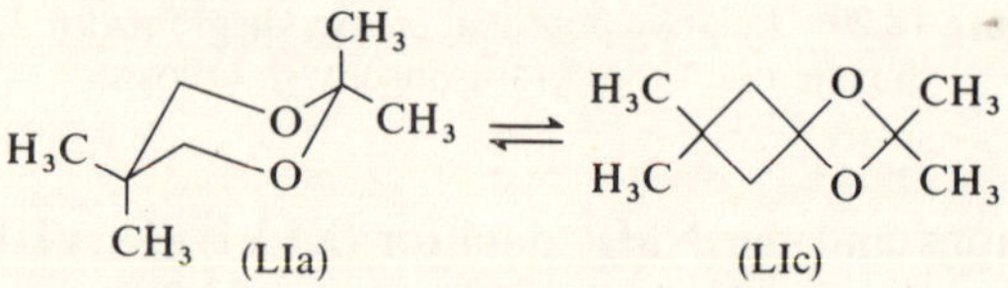

Figure 12.32 Chair to twist-boat equilibrium for 2,2,5,5-tetramethyl-1,3-dioxan

The free-energy barrier, $\Delta G_b^{\neq}$ at $300°K$ for the twist-boat to chair transition of (XLIX) was found[45] to be 5·2 kcal/mole. Figure 12.33 shows the relative potential energy diagrams for the ring inversion of (XLIX) and 1,3-dioxan. The free-energy difference, ΔG^0, between the chair and twist-boat forms of 1,3-dioxan was obtained from thermochemical data by Pihlaja[77] and the energy of the half-chair form was taken from Table 12.2. Although the chair form of (XLIX) is destabilized by ~ 4 kcal/mole relative to the chair form of 1,3-dioxan, the energy of the twist-boat form of (XLIX) is only about 1 kcal/mole higher than that of 1,3-dioxan. This indicates that the severe steric interactions of the chair form of (XLIX) are somewhat alleviated in the twist-boat form. This confirms Eccleston's earlier suggestion[45] that the '2,5-twist' conformer (XLIXa) in which the geminal methyl groups occupy equivalent positions, will be more stable than the '1,4-twist' conformer (XLIXb). This prediction is further supported by the

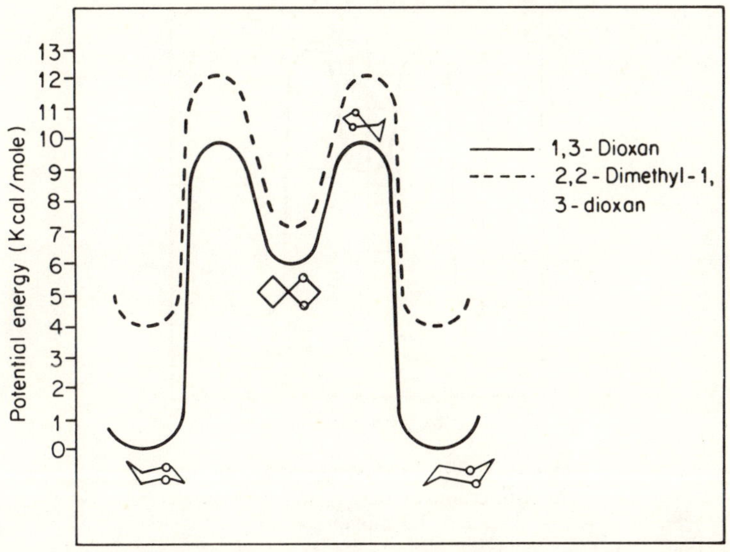

Figure 12.33   Relative potential energy diagrams for the ring inversion
of 1,3-dioxan and 2,2-dimethyl-1,3-dioxan

arguments put forward by Riddell and coworkers.[88] Although the free-
energy barrier opposing the twist-boat to chair inversion in 1,3-dioxan is
~4 kcal/mole which suggests that the rates of conformational change
shown in Figure 12.34 are of the order expected for the equilibrium to show
a relaxation in the MHz region in ultrasonic studies, no such relaxation
was observed. The absence of a relaxation is due to the large energy differences between the chair and twist-boat conformers, which results in a very
small relaxation strength and consequently no observable relaxation

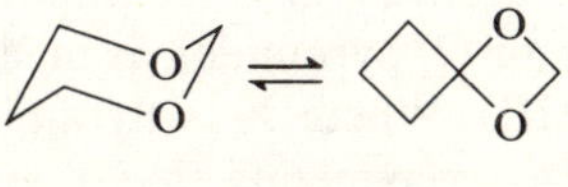

Figure 12.34

By correlation of the free-energy barriers, $\Delta G_b^{\neq}$, for the twist-boat to
chair isomerism, obtained by ultrasonics, with the free-energy barriers,
$\Delta G_f^{\neq}$, for the chair-to-chair isomerism, obtained by NMR, it is possible to
estimate the free-energy difference, $\Delta G_{ch/tw}^0$, between the chair and twist-
boat forms of some substituted 1,3-dioxans as shown in Figure 12.35.
These values are listed in Table 12.4.

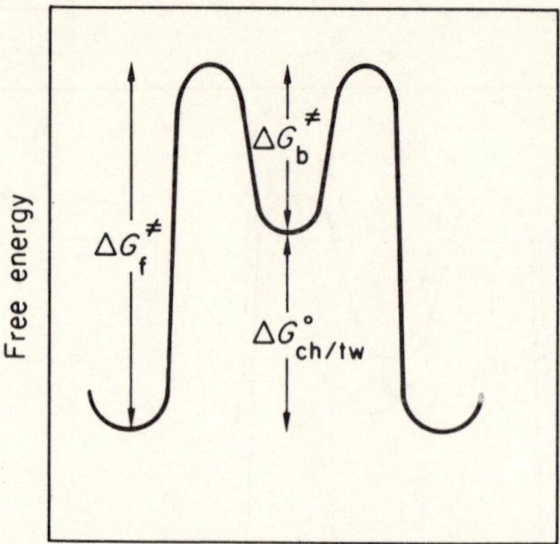

Figure 12.35 Free-energy diagram for the estimation of $\Delta G_{ch/tw}^{0}$ for substituted 1,3-dioxans

12.4 General review of the conformational analyses of cyclic sulphites

In the concluding section of this chapter the conformational properties of six-membered cyclic sulphites will be discussed. Here it is convenient to centre the discussion on the individual physical techniques that have been used to study these compounds, and, after reviewing these, to summarize the present state of understanding of the conformational problems. As a result of this approach the order in which material is presented will bear no relation to the historical order in which results have been published.

In recent years a wide variety of six-membered cyclic esters of inorganic acids have been investigated. These compounds, derivatives of propane-1,3-diols, have in common the ring skeleton (LIII) where, for example, X has been one of sulphur, selenium, phosphorus, carbon, boron, silicon or tin. The stereochemistry of the atom X may be either (1) planar, as in borates (LIV) and carbonates (LV), or (2) pyramidal, as in sulphites (LVI), sulphates (LVII) or phosphates (LVIII). Where X is pyramidal the stereochemistry may be represented as (LIX), recognizing that Y and Z may be either substituent groups or lone-pair electrons. Esters with pyramidal X may be subdivided into two categories, (2a) those with symmetric pyramidal arrangement of groups around X, e.g. sulphates (LVII) and (2b) those with unsymmetric pyramidal arrangements of groups around X, e.g. sulphites (LVI), phosphites (LX) and phosphates (LVIII). For esters derived from propane-1,3-diol itself (trimethylene esters) each of the three ring-carbon atoms carries two hydrogen atoms. Trimethylene esters of types (1) and (2a) have the property that a chair-to-chair inversion leads to a conformation identical with the original chair (LXI), while for esters of type (2b) such an

(LIII)   (LIV)   (LV)   (LVI)

(LVII)   (LVIII)   (LIX)   (LX)

inversion leads to a conformation of the molecule that is different from the original (LXII). Trimethylene esters types (1) and (2a) are then the ester analogues of unsubstituted 1,3-dioxans while those of type (2b), such as

(LXI)

(LXII)

sulphites, are the ester analogues of 2-monosubstituted 1,3-dioxans. Furthermore, the cyclic sulphites obtained from monosubstituted propane-1,3-diols, e.g. (LXIII), will be the analogues of disubstituted 1,3-dioxans, so that in determining the structure of such an ester it will be necessary to consider the orientation (*cis* or *trans*) of the C-substituent relative to the S=O group.

CH_2OH

$tBu-CH$

CH_2OH

(LXIII)   (LXIV)

Although trimethylene sulphite is a liquid at room temperature, Altona and coworkers[85] have carried out an X-ray crystallographic structure determination on a sample of the ester cooled to $-100°C$, where the compound crystallizes in the orthorhombic system. The structure was solved by means of a Patterson synthesis refined by the least-squares method using three-dimensional reflexion data. The six-membered ring was found to have a chair conformation with exocyclic oxygen in the axial position (LXIV). The absence of flattening effects in the ring is ascribed to the small ring bond angles around the sulphur atom.

Most of the other studies have involved the examination of sulphites in the liquid phase, either as pure liquids or as solutions in appropriate solvents. At this juncture it is, then, appropriate to mention evidence for the associated nature of cyclic sulphites. Wood and Miskow[89] have used vapour-phase osmometry to determine M^* the apparent molecular weight for a number of trimethylene sulphites at about 1% by weight concentration in three different solvents. Their results, summarized in Table 12.5 show that the compounds may be appreciably associated in solution. Examination of carbon

Table 12.5 Ratios of apparent molecular weight M^* to formula weight M

Compound[a]	Solvent		
	C_6H_{12}	CCl_4	CH_3CN
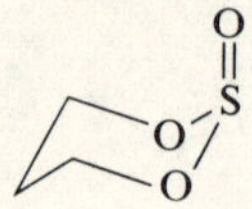	5·35	2·36	1·21
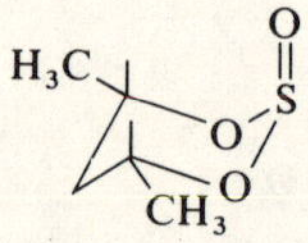	1·99	1·44	1·00
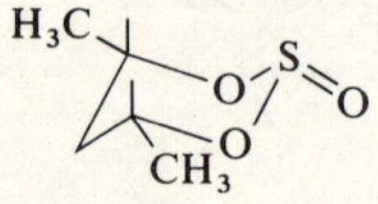	1·42	1·25	1·00
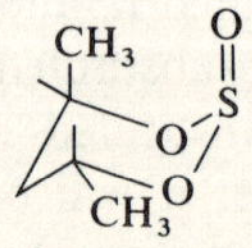	1·85	1·43	1·10
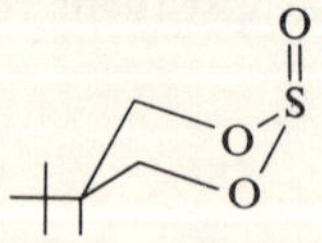	1·10	1·13	1·05
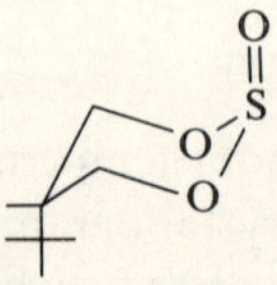	1·08	1·13	1·08

[a] The structure shown does not imply anything about preferred conformation, it merely indicates position of groups relative to one another.

tetrachloride solutions of trimethylene sulphite at various concentrations indicates the presence of dimers in dilute solution (below 10 millimolar). Wood and Miskow point out that their results show that the dimeric structure (LXV) suggested[90] for *meso*-1,3-dimethyl trimethylene sulphite (isomer *B*) is unnecessary and that a monomeric form best describes this compound.

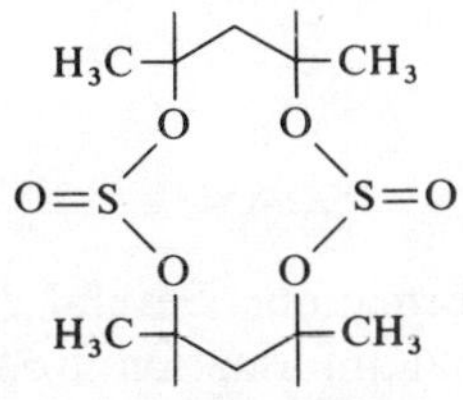

$$
\begin{array}{c}
H_3C-\!\!\!\!\bigsqcup\!\!\!\!-CH_3 \\
O \quad\quad O \\
O{=}S \quad\quad S{=}O \\
O \quad\quad O \\
H_3C-\!\!\!\!\bigsqcup\!\!\!\!-CH_3
\end{array}
$$

(LXV)

In studies of cyclic sulphites, NMR spectroscopy has mainly provided information about structures of isomeric compounds, preferred orientations of groups and the gross conformational behaviour (that is, indicating whether a particular equilibrium is anancomeric or not), while relatively few thermodynamic data have been obtained by use of this technique. In the interpretation of NMR spectra of cyclic esters some use is made of chemical shift data, but it is the observed values of vicinal proton–proton coupling constants that have proved to give most information about both structure and conformation. Important contributions to the calculation of vicinal coupling constants have been made by Karplus[91,92] using the valence bond approach. In a given situation the vicinal proton–proton coupling constant depends on ϕ the dihedral angle between the two H—C—C planes; this dependence is summarized in the well-known Karplus equation. Although this equation has been used to calculate dihedral angles from observed values of coupling constants, Karplus has drawn attention to the need to consider other aspects of molecular environment, such as the effect of electronegative substituents or departure from tetrahedral angles at carbon. In a series of closely related compounds, however, many useful qualitative conclusions may be drawn by judicious use of vicinal coupling constants. The dihedral angle dependence of vicinal coupling constants is such that

$$J_{\phi=\pi/3} < J_{\phi=0} < J_{\phi=\pi}$$

in fixed cyclohexane systems.

J_{aa} (the coupling between two vicinal axial protons) is in the range 10–13 Hz, while J_{ae} (the coupling between an axial and an equatorial proton) is usually in the range 3·5–4·5 Hz, and J_{ee} (the coupling between two vicinal equatorial protons) is about 2–3 Hz. If there is rapid (on the NMR time-scale) rotation about the C—C bond, 'averaged' spectra are obtained for which the observed value of J_{vic} will be a weighted mean of J_{aa}, J_{ae} and J_{ee}.

Booth[93] has argued that rapid inversion between two chair forms would lead to different values of the average vicinal coupling constants J_{ad} and J_{ac} in the moiety (LXVI), while observation of a single value (of about 5·5 Hz) for these two couplings is indicative of important contributions

(LXVI)

from twist or flexible conformations. Several authors have discussed the extraction of conformational information from coupling data. Booth[94] has described a method for use with mobile systems of substituted cyclohexanes which allows rough values of ΔG (equatorial $\rightarrow$ axial) to be calculated.

For six-membered rings having two equivalent chair conformations and a CH_2-CH_2 moiety which gives a spectrum from which two vicinal coupling constants J_{cis} and J_{trans} can be calculated, the 'R-value' method[95,96] can be used for obtaining qualitative information about conformational effects. (Here the terms *cis* and *trans* refer to the relative orientation of hydrogen atoms with respect to the ring, so in (LXVI) a and d are *trans*, b and d are *cis*, a and c are *cis*, b and c are *trans*.) When the population ratio is 1:1 the ratio

$$R = {}^3J_{trans}/{}^3J_{cis} = ({}^3J_{aa} + {}^3J_{ee})/({}^3J_{ae} + {}^3J_{ea})$$

should have a value of 1·9–2·2 for 'perfect' chair forms, while 'puckered' chairs should have $R > 2·5$ and for 'flattened' chair or flexible forms $R < 1·8$, the values of R being nearly independent of the electronegativities of the atoms in the ring. Buys[97] has applied the method to six-membered rings *not* existing as an equilibrium of two equivalent conformers, extending the treatment by use of the Karplus equation to relate the ratio R to ψ, the ring torsional angle. Values of ψ, calculated from the observed coupling constants of *trans*-2,3-dichloro-1,4-dioxan, oxathiane, 2-*p*-chlorophenyl-1,3-dioxan, 2-phenyl-1,3-dithiane, trimethylene sulphite and 2-oxo-2-phenoxy-1,3,2-dioxaphosphorinane, show good agreement with the corresponding torsional angles obtained from X-ray data.

Several workers have investigated the NMR spectrum of trimethylene sulphite and their results are summarized in Table 12.6. Hellier and White (unpublished results used as the basis of the discussion presented in Reference 98) obtained parameters from the 60 MHz spectrum using a first-order treatment; this of course gives no information about the signs of the coupling constants. Samitov[99] reported parameters from the 60 MHz spectrum, but

Table 12.6   NMR parameters for trimethylene sulphite

	Author, frequency of RF source, sample details			
	Hellier and White 60 MHz CCl_4[a]	Samitov 60 MHz 30% in CCl_4	Albriktsen 60 MHz Neat liquid	Green and Hellier 100 MHz 10% in CCl_4	Buys 220 MHz 10% in CCl_4
	J (Hz)				
AX	11·5	12	−11·38	−11·68	−10·9
AM	2·5	3	2·73	1·89	2·3
AR	4·5	4·56	4·63	4·85	4·7
AA'	—	—	1·58	−1·60	2·5
XM	2·5	3	2·61	2·64	2·3
XR	11·5	12	12·08	12·77	12·6
XX'	—	—	−0·04	−0·27	—
AX'	—	—	−0·52	−0·55	—
MR	14·5	14	−14·46	−14·18	−14·2
	δ (p.p.m.)[b]				
A	3·78	3·72	3·90	3·794	3·72
X	4·94	4·73	4·85	4·928	4·73
M	1·62	1·49	1·70	1·618	1·49
R	2·52	2·21	2·45	2·541	2·50

[a] These authors found no significant change of NMR parameters in CCl_4 solutions containing different concentrations of trimethylene sulphite.

[b] From tetramethyl silane.

did not indicate details of the analysis used; as he quotes no signs of coupling constants, his may also have been a first-order analysis. More detailed, computer-based analyses have been reported by Albriktsen[100] (60 MHz), Green and Hellier[101] (100 MHz) and Buys[102] (220 MHz). The line-width in the 220 MHz spectrum is rather large (about 1 Hz) so that small long-range couplings may not be detected; the use of higher frequencies may lose certain information about signs of coupling constants either through loss of detail because of the large line-width or because at higher frequencies (and associated higher field strength) spectra approach more closely to first order. The NMR parameters obtained by these different groups of workers are in good agreement, in particular it would appear that the use of vicinal coupling constants (which are important in the discussion of conformations) obtained from the first-order treatment is not likely to lead to erroneous conclusions. The magnitudes of the coupling constants, particularly the large *trans* coupling J_{XR}, indicate a strong conformational preference for the ring.

Each equatorial hydrogen nucleus is more shielded than its geminal axial counterpart; this is the opposite of the relative positions found for cyclohexane derivatives, and is presumably due to intramolecular effects arising from the sulphite group, in particular the electric field effect and the magnetic anisotropy of the S=O bond. Similar 'anomalous' shifts of C-5 methylene protons (e.g. equatorial-H peak about 0·8 p.p.m. to high field of that[103] arising from the geminal axial-H) have been attributed to a lone-pair interaction of the 1,3-oxygen atoms with the equatorial proton at C-5; this kind of interaction must also influence the C-5 methylene proton shifts in the anancomeric trimethylene sulphites.

Calculation[99,104] of the effects due to the S=O group in different orientations indicate that the S=O bond is axial. Because of uncertainties in the values of magnetic anisotropies, these calculations alone would not be compelling evidence for the orientation of the S=O bond; fortunately this conclusion is in agreement with those obtained from other methods.

Returning to deductions made from the vicinal coupling constants, substantiation of the proposed strong conformational preference comes from substituted trimethylene sulphites. Two isomers of 5-*t*-butyl trimethyl-ene sulphite (that is the sulphites formed from 2-*t*-butyl-propane-1,3-diol) have been separated[105,106] and their PMR spectra reported. The spectrum of isomer *A* (LXVII, m.p. 47°) shows a high-field triplet of triplets arising

(LXVII)

from the proton directly bonded to C-5, with couplings of 11·5 Hz (an axial–axial coupling) and 4·5 Hz (an axial–equatorial coupling)[107] showing that this hydrogen atom must be axial and consequently the *t*-butyl group must preferentially be equatorial. The conformational preference of a *t*-butyl group in the cyclic sulphites will probably be smaller than in the cyclo-hexane system (cf. discussion of 5-alkyl-1,3-dioxans); the fact that replacement of an equatorial hydrogen atom by a *t*-butyl group has virtually no effect on the vicinal coupling constants of the remaining hydrogen at C-5 would indicate that the conformational equilibrium in the unsubstituted ester is indeed anancomeric.

Again the more compelling evidence for the axial S=O in isomer A comes from dipole moment and IR studies, though the chemical shifts of axial and equatorial hydrogen at C-4 (or C-6) again give an indication of the *cis* relationship between the *t*-butyl group and S=O.

On the basis of benzene solvent effects on chemical shifts in the PMR spectrum of trimethylene sulphite Wood and Miskow[108] suggested that if it is assumed that the association between the benzene molecule and the S=O group is similar to that between benzene and say C=O, the conclusion reached is that trimethylene sulphite appears to be in the flexible form. There is now sufficient evidence supporting the conclusion that the major conformer is *not* a flexible form that one would prefer to accept the anancomeric chair form of trimethylene sulphite and to use the NMR benzene solvent shifts as a means of investigating the nature of the association between the sulphite and benzene.

Isomer B of 5-*t*-butyl trimethylene sulphite (m.p. 38°) gives an NMR spectrum that is not closely related to those of trimethylene sulphite or isomer A. In particular the multiplet arising from hydrogen at C-5 is a quintet showing a splitting of 5·5 Hz. Isomer B, with the *t*-butyl group *trans* to S=O, exists as an equilibrium mixture of inverting forms. Support for this interpretation of the observed values of the vicinal coupling constants involving the C-5 proton is provided from the spectrum of trimethylene sulphate (LVII) which consists of a triplet and a quintet with $J = 5·5$ Hz. This sulphate is a type (2a) molecule in which the extreme chair forms are identical. Observation of a single geminal coupling constant of 5·5 Hz, together with equivalence of geminal axial and equatorial protons indicates that the molecule is undergoing rapid conformational change with flexible forms being involved in the equilibrium. So far it has only been possible to obtain an upper limit of 8·5 kcal/mole for the barrier to inversion in this molecule by means of the low-temperature NMR spectrum.[107] In the related 5,5-dimethyl trimethylene sulphite the barrier to chair-to-chair inversion has been found[109] to be 8·2–8·4 kcal/mole from low-temperature NMR measurements. The low-temperature spectrum of isomer B of 5-*t*-butyl trimethylene sulphite (to $-99°$C in acetone-d$_6$) is identical with the room-temperature spectrum;[110] this would indicate a low barrier to flexing in the sulphite ring as in the cyclic sulphates.

The importance of dipole moment measurements in the study of cyclic esters has already been indicated. Earlier work making use of dipole moments often resulted in erroneous conclusions being drawn because calculations of dipole moments for different structures and conformers were based on an incorrect moment for the sulphite group, 'derived' by vectorially adding experimentally derived contributions for S—O and S=O bonds. Detailed studies[105,111,112] by van Woerden and his coworkers have provided a firm basis for the calculation of dipole moments for the cyclic sulphites.

In a sulphite group the sulphur lone-pair electrons occupy an approximately sp^3 hybridized orbital, and the asymmetry of its charge distribution contributes significantly to the overall moment. From a detailed analysis of measured moments of a number of cyclic sulphites van Woerden and co-workers were able to demonstrate that the partial moment of the sulphite group is about 2·5 Debye, the direction of the moment deviating significantly from the S=O bond direction.

For sulphites in a chair conformation, the isomer with axial S=O would have a smaller molecular dipole moment than the corresponding isomer with S=O equatorial. This is well illustrated[113] by the moments of the isomeric sulphites derived from *trans*-2-hydroxymethylcyclohexanol (LXVIII and LXIX). In both isomers the *trans* 4,5-ring-junction gives a strong bias

(LXVIII)   (LXIX)

to the sulphite ring in favour of the chair conformation. The NMR spectra of the separate compounds show vicinal couplings, $J_{AX} \sim 11$ Hz and $J_{BX} \sim 4$ Hz, in both isomers. The observed moments are (LXVIII) $\mu = 3\cdot85$ D (CCl$_4$ solvent) and (LXIX) $\mu = 5\cdot15$ D (CCl$_4$ solvent). Observation of a moment $\mu = 3\cdot46$ (CCl$_4$ solvent) for trimethylene sulphite again indicates an axial S=O.

Solvent effects on the measured dipole moments have been reported, in these a more polar solvent will favour a more polar conformation. In practice the differences in interaction energy between the dipolar molecule and different solvent molecules appear to be moderate and it is only in cases such as that of isomer *B* of 5-*t*-butyl trimethylene sulphite that significant changes of moment are observed on changing solvent. It appears that in anancomeric systems the change in conformation population resulting from a change in the dielectric constant of the medium may not be large enough to affect the dipole moment significantly.

Generally there is satisfactory agreement between the experimentally measured dipole moments reported by different workers. Comparison of measured and calculated moments leads to structural and conformational information, particularly for anancomeric systems. Detailed calculations of the moments of flexible conformations are at present lacking, but may in the future throw further light on possible preferences amongst flexible forms.

Wood and Miskow[110] have used their data on the moments of the three isomeric 4,6-dimethyl trimethylene sulphites to propose a revision of an earlier[114] conformational assignment for these compounds. Of the two *meso* isomers, isomer *A* (LXX) with a dipole moment (CCl$_4$ solvent) of 3·51 D can be assigned the structure and conformation in which S=O is axial and the two methyl groups are equatorial. The second *meso* isomer (*B*) has a large moment of 5·31 D which is indicative of a chair conformation with equatorial S=O (LXXI). The racemic isomer has a moment of 3·93 D, similar to that of 5-*t*-butyl trimethylene sulphite isomer *B* (3·76), indicating that the racemic compound is either a distorted chair or a flexible form (LXXII).

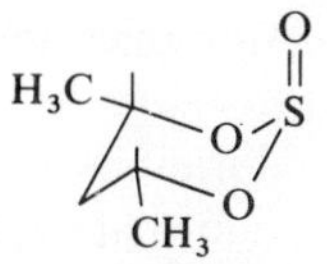

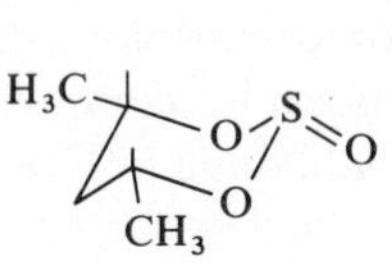

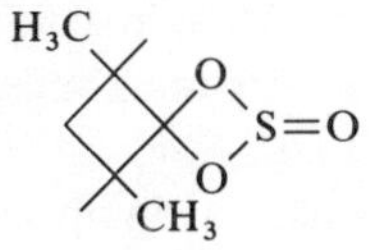

In addition to their work using dipole moments, van Woerden and his coworkers have made considerable use of infrared spectroscopy in the study of cyclic sulphites.[105,111] Considering solid state spectra, practically all derivatives of trimethylene sulphite have an intense absorption band in the 1180–1195 cm^{-1} range which has been assigned to the S=O stretching frequency. On changing to the dissolved state this band, in some cases, is shifted to a slightly higher frequency appearing just beyond 1200 cm^{-1}. A notable exception is isomer *B* of 5-*t*-butyl trimethylene sulphite which in the solid state has an S=O band at about 1230 cm^{-1}, which on going to the dissolved state undergoes changes and shows a striking solvent dependence.

The fact, that the 1180–1195 cm^{-1} band found in the solid-state spectra of some sulphites persists when the compound is dissolved, provides evidence that the preferred orientation of the S=O bond found in the solid state is maintained in solution, an important conclusion in view of the X-ray work described earlier.[85] From detailed consideration of dipole moments and infrared spectra van Woerden and coworkers have assigned the 1180–1195 cm^{-1} band to axial S=O and the 1230–1235 cm^{-1} band to equatorial S=O. Investigations of 4,6-substituted trimethylene sulphites established[115,116] that the higher infrared frequency at about 1230 cm^{-1} is associated with either an equatorial S=O bond, or one in a ring which is so distorted that a true axial S=O does not exist.

The moderate to weak bands observed in the 1230–1240 cm^{-1} region of the spectra of trimethylene sulphite and many of its derivatives have been taken as evidence for the presence of conformers with S=O in other than an axial orientation. Green and Hellier[101] have examined the concentration and solvent variation of the 1234 cm^{-1} band in trimethylene sulphite and conclude that an exclusive and direct assignment of this band to an equatorial

S=O or S=O in flexible conformations is an oversimplification. Spectra for other six-membered carbocyclic compounds also show bands in the region 1220–1260 cm^{-1}, probably due to CH_2 twist vibrations.

The physical methods described so far give more information about the structural problems rather than about thermodynamic parameters. The method of ultrasonic relaxation, on the other hand, gives values of thermodynamic functions but indicates little or nothing about the species involved in the conformational equilibrium. Wyn-Jones has commented[117] on the scope and limitations of ultrasonic relaxation methods in conformational analysis. Care is needed in comparing ultrasonic data with that obtained by other methods, as the reliable data available from ultrasonic experiments are the energies and entropies of activation for the change from the *less* stable to the *more* stable conformation, whereas most other techniques lead to values of thermodynamic quantities for the change from *more* to *less* stable conformer.

Studies of ultrasonic absorption in cyclic sulphates and sulphites in different solvents and at different concentrations have been reported.[118–121] Trimethylene sulphate showed no ultrasonic absorption (in a (2a) type molecule no ultrasonic absorption would be expected for the chair-to-chair inversion). Although, in principle, an absorption should occur for chair → twist or chair → boat changes, the lack of an observed relaxation indicates that the relaxation times for these changes were outside the experimentally attainable frequency range.

The absorption observed for trimethylene sulphite occurs at a characteristic frequency that is sensibly the same for the pure liquid as for the solution in chloroform. This eliminates viscosity relaxation, vibration relaxation and equilibrium between single and associated molecules as sources of the observed relaxation, as these would all be solvent dependent. It is concluded that the absorption must be due to an equilibrium between different conformers. As the chair → twist or chair → boat changes in trimethylene sulphate are outside the available frequency range, it was considered likely that they would also be unobservable for the closely related trimethylene sulphite, the observed absorption for sulphites was, therefore, assigned to a chair → chair inversion in which the S=O bond is changed from an axial to an equatorial position. Values of thermodynamic parameters obtained by this technique are given in Table 12.7. The values of ΔG^0 for 5,5-dimethyl trimethylene sulphite and for trimethylene sulphite imply that at room temperature over 90% of the molecules are in the preferred conformation.

Limited experiments on chemical equilibration have been reported.[113,122] These give values of about 1·5–2·5 kcal/mole for the free-energy difference between S=O axial and equatorial. A preference of this magnitude is sufficient to force a conformational equilibrium to be anancomeric with well over 90% of the molecules having the preferred conformation.

Table 12.7   Thermodynamic parameters obtained from ultrasonic absorption measurements on cyclic sulphites[121]

Compound	ΔH^0 (e–a) (kcal/mole)	ΔS^0 (e–a) (cal/deg/mole)	$\Delta H^{\neq}$ (e–a) (kcal/mole)	$\Delta S^{\neq}$ (e–a) (cal/deg/mole)
Trimethylene sulphite (TMS)	1·3	−2·50	5·50	−4·00
4-Methyl TMS	≤1·2	—	4·80	—
5,5-Dimethyl TMS	1·5	−1·50	4·30	−7·01
5,5-Diethyl TMS	≤1·3	—	4·20	—
5-Methyl-5-ethyl TMS[a]	≤1·4	—	4·30	—
4,6-Dimethyl TMS (*trans, meso*)	≤1·2	—	5·81	—
4,5,6-Trimethyl TMS[a]	≤1·2	—	3·60	—

[a] Equilibrium mixtures of stereoisomers.

In discussing conformational behaviour some care is necessary when using qualitative descriptions of certain energy differences. A free-energy difference of 2 kcal/mole is sufficient for a conformational equilibrium to be anancomeric with almost 100% of the molecules in the preferred form, in this instance 2 kcal/mole is 'large'. On the other hand a barrier of 10 kcal/mole allows rapid (on say the NMR time-scale) inversion, in this context 10 kcal/mole is 'small'. It is possible that some of the controversy about the behaviour of cyclic esters arises from different workers meaning different things when talking in qualitative terms about the energies involved in the consideration of conformational inversion.

At present there is good evidence for the structure and preferred conformation for many cyclic sulphites and sulphates and there are good values for some thermodynamic quantities. What is lacking is good experimental evidence giving definite information about the minor constituents (their relative amounts and conformations) contributing to these conformational equilibria. In view of the evidence for certain esters (e.g. 5-*t*-butyl trimethylene sulphite, isomer *B*) undergoing flexing equilibrium, together with the reported chair → chair barriers of 8·2–8·4 kcal/mole for 5,5-dimethyl trimethylene sulphate[109] and 12·3–12·6 kcal/mole for 5,5-dimethyl-1,3,2-dioxathiane[123] (LXXIII) it would appear unlikely that there is any large barrier to inversion or flexing in the molecules of the cyclic sulphites.

(LXXIII)

12.5 References

1. E. L. Eliel, N. L. Allinger, S. J. Angyal and G. A. Morrison, *Conformational Analysis* (New York: Interscience, 1965).
2. G. Binsch, *Topics in Stereochemistry*, **3**, 97 (1968).
3. C. Romers, C. Altona, H. R. Buys and E. Havinga, *Topics in Stereochemistry*, **4**, 39 (1969).
4. E. Wyn-Jones and R. A. Pethrick, *Topics in Stereochemistry*, **5**, 205 (1970).
5. J. B. Lambert, *Topics in Stereochemistry*, **6**, 19 (1971).
6. W. A. Thomas, *Ann. Rep. NMR Spectr.*, **1**, 44 (1968); *Ann Rep. NMR Spectr.*, **3**, 92 (1970).
7. K. Jones and E. F. Mooney, *Ann. Rep. NMR Spectr.*, **3**, 261 (1970).
8. I. O. Sutherland, *Ann. Rep. NMR Spectr.*, **4**, 71 (1971).
9. H. Booth, *Progr. NMR Spectr.*, **5**, 149 (1969).
10. J. E. Anderson, *Quart. Rev.*, **19**, 426 (1965).
11. F. G. Riddell, *Quart. Rev.*, **21**, 364 (1967).
12. R. A. Pethrick and E. Wyn-Jones, *Quart. Rev.*, **23**, 301 (1969).
13. E. L. Eliel, *Acc. Chem. Res.*, **3**, 1 (1970).
14. A. V. Bogatskii and N. L. Garkovik, *Russ. Chem. Rev. (English Transl.)*, **37**, 264 (1968).
15. H. Sasche, *Chem. Ber.*, **18**, 2269 (1885); *Z. Physik. Chem.*, **10**, 203 (1892).
16. R. G. Dickinson and C. Bilicke, *J. Am. Chem. Soc.*, **50**, 764 (1928).
17. K. W. F. Kohlrausch, A. W. Reitz and W. Stockmair, *Z. Physik. Chem.*, **B32**, 229 (1936).
18. O. Hassel, *Tidsskr. Kjemi, Bergvesen. Met.*, **3**, 32 (1943).
19. C. W. Beckett, K. S. Pitzer and R. Spitzer, *J. Am. Chem. Soc.*, **69**, 2488 (1947).
20. J. B. Hendrickson, *J. Am. Chem. Soc.*, **83**, 4537 (1961).
21. J. Reisse, J. C. Celotti, D. Zimmermann and G. Chiurdoglu, *Tetrahedron Letters*, 2145 (1964).
22. A. H. Lewin and S. Winstein, *J. Am. Chem. Soc.*, **84**, 2464 (1962).
23. E. L. Eliel, *Angew. Chem.*, **4**, 761 (1965).
24. J. J. Uebel and H. W. Goodwin, *J. Org. Chem.*, **31**, 2040 (1966).
25. E. L. Eliel and M. H. Gianni, *Tetrahedron Letters*, 97 (1962).
26. E. L. Eliel and R. J. L. Martin, *J. Am. Chem. Soc.*, **90**, 689 (1968).
27. A. J. Berlin and F. R. Jensen, *Chem. Ind. (London)*, 998 (1960).
28. M. Tichý, F. Šipoš and J. Sicher, *Collection Czech. Chem. Comm.*, **31**, 2889 (1966).
29. E. L. Eliel, E. W. Della and T. H. Williams, *Tetrahedron Letters*, 831 (1963).
30. M. M. Otto, *J. Am. Chem. Soc.*, **59**, 1590 (1937).
31. B. A. Arbuzov, *Bull. Soc. Chim. France*, 1311 (1960).
32. N. Baggett, B. Dobinson, A. B. Foster, J. Homer and L. F. Thomas, *Chem. Ind. (London)*, 106 (1961).
33. A. J. De Kok and C. Romers, *Rec. Trav. Chim.*, **89**, 313 (1970).
34. F. G. Riddell and M. J. T. Robinson, *Tetrahedron*, **23**, 3417 (1967).
35. J. E. Anderson and J. C. D. Brand, *Trans. Faraday Soc.*, **62**, 39 (1966).
36. H. Friebolin, S. Kabuss, W. Maier and A. Luttringhaus, *Tetrahedron Letters*, **16**, 683 (1962).
37. H. Friebolin, H. G. Schmid, S. Kabuss and W. Faisst, *Org. Mag. Res.*, **1**, 67 (1969).
38. E. Coene and M. Anteunis, *Bull. Soc. Chim. Belges*, **79**, 37 (1970).
39. A. Greenberg and P. Laszlo, *Tetrahedron Letters*, **30**, 2641 (1970).
40. H. G. Schmid, H. Friebolin, S. Kabuss and R. Mecke, *Spectrochim. Acta*, **22**, 623 (1966).
41. P. Delaney (Private Communication).

42. J. E. Anderson (Private Communication).
43. V. I. P. Jones and J. A. Ladd, *Trans. Faraday Soc.*, **66**, 2948 (1970).
44. J. E. Anderson (Private Communication).
45. G. Eccleston, Ph.D. Thesis, Salford (1970); G. Eccleston and E. Wyn-Jones, *J. Chem. Soc. (B)*, 2469 (1971).
46. E. L. Eliel and M. C. Knoeber, Sr., *J. Am. Chem. Soc.*, **90:30**, 3444 (1968).
47. E. L. Eliel and M. C. Knoeber, Sr., *J. Am. Chem. Soc.*, **88**, 5347 (1966).
48. K. Pihlaja and S. Luoma, *Acta Chem. Scand.*, **22**, 2401 (1968).
49. K. Pihlaja and J. Jalonen, submitted to *Org. Mass. Spectr.*
50. F. W. Nader and E. L. Eliel, *J. Am. Chem. Soc.*, **92:10**, 3050 (1970).
51. E. L. Eliel and C. A. Giza, *J. Org. Chem.*, **33**, 3754 (1968).
52. E. L. Eliel and W. F. Bailey (Unpublished Observations).
53. E. L. Eliel, in the press.
54. E. L. Eliel and M. C. Knoeber (Unpublished Observations).
55. G. Eccleston, B. Walsh, E. Wyn-Jones and H. Morris, *Trans. Faraday Soc.*, **67**, 587 (1971).
56. V. I. P. Jones and J. A. Ladd, *J. Chem. Soc. (B)*, 567 (1971).
57. K. Pihlaja and A. Tenhosaari, *Suomen Kemistilehti (B)*, **43**, 175 (1970).
58. K. Pihlaja and P. Ayras, *Suomen Kemistilehti (B)*, **43**, 171 (1970).
59. E. L. Eliel and J. M. McKenna (Unpublished Observations).
60. K. Pihlaja and J. Heikkila, *Acta. Chem. Scand.*, **21**, 2390 (1967).
61. R. M. Enanoza, Ph.D. Dissertation, University of Notre Dame; cf. E. Eliel, *Pure Appl. Chem.*, in the press.
62. K. Pihlaja and P. Ayras, *Suomen Kemistilehti (B)*, **42**, 65 (1969).
63. K. Pihlaja, K. J. Teinonen and P. Ayras, *Suomen Kemistilehti (B)*, **43**, 41 (1970).
64. E. L. Eliel and D. I. C. Raileanu, *Chem. Commun.*, 291 (1970).
65. E. L. Eliel and M. Kaloustian, *Chem. Commun.*, 290 (1970).
66. B. J. Hutchinson, R. A. Y. Jones, A. R. Katritzky, K. A. F. Record and P. J. Brignell, *J. Chem. Soc. (B)*, 1224 (1970).
67. R. J. Abraham, H. D. Banks, E. L. Eliel, O. Hofer and M. K. Kaloustian, *J. Am. Chem. Soc.*, **94**, 1914 (1972).
68. E. L. Eliel and O. Hofer (Unpublished Observations).
69. E. Coene and M. Anteunis, *Bull. Soc. Chim. Belges*, **79**, 25 (1970).
70. E. Coene and M. Anteunis, *Tetrahedron Letters*, **8**, 595 (1970).
71. M. Anteunis, R. Camerlynck and F. Borreman, in preparation.
72. P. Dirinck and M. Anteunis, *Can. J. Chem.*, **50**, 412 (1972).
73. P. C. Hamblin, R. F. M. White and E. Wyn-Jones, *Chem. Commun.*, 1058 (1968).
74. P. C. Hamblin, R. F. M. White and E. Wyn-Jones, *J. Mol. Struct.*, **4**, 275 (1969).
75. K. Pihlaja, *Suomen Kemistilehti (B)*, **42**, 74 (1969).
76. G. Swaelens and M. Anteunis, *Tetrahedron Letters*, **8**, 561 (1970).
77. K. Pihlaja, *Acta Chem. Scand.*, **22**, 716 (1968).
78. M. Anteunis and G. Swaelens, *Org. Mag. Res.*, **2**, 389 (1970).
79. G. Eccleston and E. Wyn-Jones, *Chem. Commun.*, 1511 (1969).
80. J. T. Edward, *Chem. Ind. (London)*, 1102 (1955).
81. E. L. Eliel, *Kemisk Tidskrift*, 22 (1969).
82. M. Anteunis and P. Dirinck, *Can. J. Chem.*, **50**, 423 (1972); (Private Communication).
83. G. Eccleston, B. Walsh, E. Wyn-Jones and H. Morris, *Trans. Faraday Soc.*, **67**, 3223 (1971).
84. A. Allerhand, Fu-Ming Chen and H. S. Gutowsky, *J. Chem. Phys.*, **42**, 3040 (1965).
85. C. Altona, H. J. Geise and C. Romers, *Rec. Trav. Chim.* **85**, 1197 (1966).
86. R. K. Harris and N. Sheppard, *J. Mol. Spectr.*, **22**, 231 (1967).

87. R. K. Harris and R. A. Spragg, *J. Chem. Soc.* (*B*), 684 (1968).
88. K. Pihlaja, G. M. Kellie and F. G. Riddell, *J. Chem. Soc.* (*Perkin II*), 252 (1972).
89. G. Wood and M. H. Miskow, *Tetrahedron Letters*, 1775 (1970).
90. R. E. Lack and L. Tarasoff, *J. Chem. Soc.* (*B*), 1095 (1969).
91. M. Karplus, *J. Chem. Phys.*, **30**, 11 (1959).
92. M. Karplus, *J. Am. Chem. Soc.*, **85**, 2870 (1963).
93. H. Booth, *Tetrahedron Letters*, 1449 (1964).
94. H. Booth, *Tetrahedron*, **20**, 2211 (1964).
95. J. B. Lambert, *J. Am. Chem. Soc.*, **89**, 1836 (1967).
96. J. B. Lambert and R. G. Keske, *Tetrahedron Letters*, 4755 (1967).
97. H. R. Buys, *Rec. Trav. Chim.*, **89**, 1253 (1970).
98. D. G. Hellier, J. G. Tillett, H. F. van Woerden and R. F. M. White, *Chem. Ind.* (*London*), 1956 (1963).
99. Y. Y. Samitov, *Dokl. Akad. Nauk. SSSR.*, **164**, 347 (1965).
100. P. Albriktsen, *Acta Chem. Scand.*, **25**, 478 (1971).
101. C. H. Green and D. G. Hellier, *J. Chem. Soc.* (*Perkin II*), 458 (1972).
102. H. R. Buys, *Rec. Trav. Chim.*, **89**, 1244 (1970).
103. M. Anteunis, D. Tavernier and F. Borremans, *Bull. Soc. Chim. Belges*, **75**, 396 (1966).
104. Y. Y. Samitov and R. M. Aminova, *Zh. Strukt. Khim.*, **5**, 207 (1964).
105. H. F. van Woerden, Thesis, Leiden (1964).
106. H. F. van Woerden and E. Havinga, *Rec. Trav. Chim.*, **86**, 353 (1967).
107. R. F. M. White, *J. Mol. Struct.*, **6**, 75 (1970).
108. G. Wood and M. H. Miskow, *Tetrahedron Letters*, 4433 (1966).
109. G. Wood, J. M. McIntosh and M. H. Miskow, *Tetrahedron Letters*, 4895 (1970).
110. G. Wood and M. H. Miskow, *Tetrahedron Letters*, 1109 (1969).
111. H. F. van Woerden and E. Havinga, *Rec. Trav. Chim.*, **86**, 341 (1967).
112. H. F. van Woerden and E. Havinga, *Rec. Trav. Chim.*, **86**, 353 (1967).
113. H. F. van Woerden and A. T. de Vries-Miedena, *Tetrahedron Letters*, 1687 (1971).
114. P. C. Lauterbur, J. G. Pritchard and R. L. Vollmer, *J. Chem. Soc.*, 5307 (1963).
115. L. Cazaux and P. Maroni, *Tetrahedron Letters*, 3667 (1969).
116. S. Sarel and V. Usieli, *Israel J. Chem.*, **6**, 885 (1968).
117. E. Wyn-Jones, *Tetrahedron Letters*, 907 (1971).
118. R. A. Pethrick, E. Wyn-Jones, P. C. Hamblin and R. F. M. White, *J. Mol. Struct.*, **1**, 333 (1967).
119. P. C. Hamblin, R. F. M. White, G. Eccleston and E. Wyn-Jones, *Can. J. Chem.*, **47**, 2731 (1969).
120. R. A. Pethrick, E. Wyn-Jones, P. C. Hamblin and R. F. M. White, *J. Chem. Soc.* (*A*), 1638 (1969).
121. G. Eccleston, R. A. Pethrick, E. Wyn-Jones, P. C. Hamblin and R. F. M. White, *Trans. Faraday Soc.*, **66**, 310 (1970).
122. G. Wood, J. M. McIntosh and M. H. Miskow, *Can. J. Chem.*, **49**, 1202 (1971).
123. G. Wood and R. M. Srivastava, *Tetrahedron Letters*, 2937 (1971).

13 Medium effects on rotational and conformational equilibria

R. J. Abraham and E. Bretschneider

13.1 Introduction

This chapter will be concerned with detailing the effect of changing the medium on various rotational and conformational equilibria, and in particular, with examining the generally scattered and diverse experimental data in the light of present theories of this phenomenon.

The energy differences between rotational isomers are nearly always fairly small (0–3 kcal/mole) and as the energy of solvation of a polar solute in a polar medium is at least as large as this, and often much larger, it is not surprising that changing the medium may affect a particular rotational equilibrium very considerably. Indeed there are a number of compounds in which the more stable rotamer in polar solvents is not the same as the more stable rotamer in the vapour or even non-polar solvents, and others in which one rotamer is predominant in one medium but much less so in other media. The neglect of such considerations has caused much confusion in the literature, particularly as it is a not uncommon occurrence in such studies to obtain spectra of the distinct rotational isomers without any unambiguous method of assigning them.

For example, there has been considerable controversy surrounding the rotational equilibrium in furfuraldehyde (OO *cis* $\rightleftharpoons$ OO *trans*)[1] due mainly to the fact that there is no unambiguous method of assigning the separate NMR spectra of the two isomers observed at low temperatures. Thus although the value of the energy difference between the isomers was known

precisely (1·0 kcal/mole in dimethyl ether solution),[2] this could have either sign depending on the assignment. At first sight, it would appear that the determination of ΔH of -0.99 kcal/mole in the vapour by microwave spectroscopy[3] in which the assignment is unambiguous, provides conclusive evidence for using the negative value of ΔH in solution, and this has been stated. However, more recent research shows that the agreement in the values of ΔH is entirely fortuitous as the value in dimethyl ether solution is reversed from that in the vapour,[1] i.e. the OO *trans* form is more stable in the vapour, but the OO *cis* form is more stable in polar media (see § 13.5.2 for a full treatment).

In the development of a general treatment for the conformational analysis of acyclic compounds,[4] but applied specifically to 1,1,2-trisubstituted ethanes, a basic assumption was made that rotamers with two *gauche* interactions are highly unfavoured, i.e. in Figure 13.1 rotamer C may be

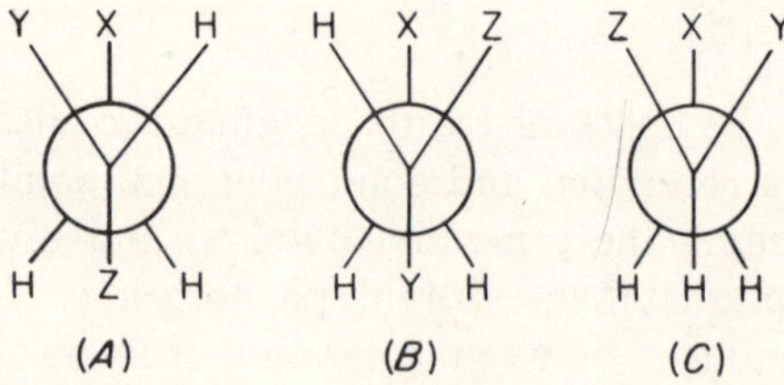

Figure 13.1

disregarded. The strongest evidence produced to support this assumption was that in 1,1,2-trichloroethane (X = Y = Z = Cl), rotamer C is more than 2 kcal/mole higher in energy than the enantiomorphic forms A and B. In fact whilst the energy difference between C and A in 1,1,2-trichloroethane is 2 kcal/mole *in the vapour*, the energy difference in solution is much less than this, varying from about 1·3 kcal/mole in non-polar solvents to zero in very polar solvents (see Table 13.7). Thus the proportion of the higher energy form C cannot be neglected in solution, and in consequence the quantitative validity of the treatment given suffers.

The major problem in these examples and many others has been the lack of any general theory of medium effects of even a semi-quantitative nature. It is the purpose of this chapter to describe and illustrate such a general theory of medium effects. The implications of such a theory are considerable.

Firstly, any successful theoretical treatment provides a framework on which a large mass of experimental data may be coordinated and the inter-actions revealed. Thus, any explanation of the solvent dependence of con-formational equilibria should provide information on the nature of the solvent–solute interactions responsible for the equilibria. Also, a successful calculation of the variation in rotamer energies with the medium would also give, by extrapolation, the rotamer energies in the vapour phase from experimental measurements in solution. This would be of value for those molecules in which gas-phase measurements are not practical, as of course it is the values of the vapour-state energy differences which are of use in any theoretical calculation of rotamer energies. Finally, any general theory provides a means whereby exceptional cases may be recognized. These can be either solutes or solvents. It will be shown later that a number of common solvents are not, so far, amenable to the general treatment given, largely due to specific, in particular hydrogen-bonding, interactions. Also, because we are concerned in this chapter only with the variation of rotamer energies with the medium, we will exclude in general all studies performed in only one medium. In particular, the large amount of data on conformational isomerism in aqueous media will not be considered here. In fact this is a very special case for which the general theory given does not apply.

A general theoretical treatment of medium effects can of course be applied to the medium dependence of any chemical equilibrium, provided only that true thermodynamic equilibrium is reached. For example, the medium dependence of the *cis* to *trans* isomerization of olefins is in principle covered by the same theoretical treatment. However, in this book we are concerned solely with rotational and conformational equilibria and therefore this chapter is limited to these types of equilibrium.

Solvent effects on rotational and conformational equilibria have not been reviewed as such, but many examples of the influence of solvent on rotational equilibria are to be found in Mizushima's classic text,[5] and in the reviews of Sheppard[6] and, more recently, Lowe.[7] A comprehensive coverage of solvent effects in NMR has also been given.[8] The (very small) effect of solvent on conformational energies in monosubstituted cyclohexanes has been well detailed by Hirsch,[9] and the technique and results of the measurement of conformational equilibria are well documented.[10–12]

In this chapter, after some general principles and definitions, we will first consider briefly the various techniques for measuring rotamer and conformer energies, with particular emphasis on the unavoidable systematic assumptions necessarily made in deducing rotamer energies from the measured parameters. This is of some importance, as virtually all techniques suffer from some systematic errors. The derivation of a general theoretical treatment is then given, again with particular emphasis on the limitations of the treatment. The final section deals with the application of the theory to the numerous measurements available. In some compounds the data is sufficiently accurate to allow a quantitative test of the theory and these are considered first. In other cases only qualitative considerations are meaningful but the data can still be usefully considered in the light of the theoretical predictions. Finally exceptional cases, particularly solvents, will be discussed.

13.2 General principles and methods of measurement

Consider a molecule equilibrating between two states A and B. If the populations of the two states in any solvent (s) are n_A and n_B, we have

$$A \rightleftharpoons B$$

$$K = n_B/n_A = \exp(-\Delta G^s/RT) \tag{13.1}$$

and

$$n_A + n_B = 1$$

where ΔG^s is the free-energy difference for the equilibrium in solvent s. The nomenclature we shall use always retains superscripts for the medium (v = vapour, l = pure liquid, s = solvent) and subscripts for the states.

In any other solvent, or in the vapour, an identical equation holds save that ΔG^s becomes ΔG^v. We are only concerned here with the *differences* between different solvents or between the vapour and any given solvent, i.e. we wish to consider merely

$$\delta\Delta G = \Delta G^v - \Delta G^s$$

In this case, considerable simplification of the basic equation occurs, as follows.

(a) It may be assumed that the entropy difference between the states A and B is not affected by the medium, i.e. that $\delta\Delta S$ equals zero. Of course a very common assumption made in studies of rotational and conformational equilibria is that ΔS itself is zero in all solvents. It is not necessary for our purposes to make this more drastic assumption. It is, however, notoriously difficult to obtain accurate values of ΔS and certainly many estimates of ΔS in such equilibria are within the often large experimental error, not significantly different from zero. Thus our assumption that $\delta\Delta S = 0$ would appear generally valid.

(b) Again, because we are considering only differences between solvents, the contributions to the free energy due to the zero-point energy, contributions of higher vibrational states and $P\,dV$ terms would all be expected to cancel.[12]

As a consequence of these two considerations, we may replace $\delta\Delta G$ by $\delta\Delta E$, the potential energy difference between the states, and therefore we only need to calculate the variation of $E_A - E_B$ with solvent.

The two states of the molecule considered may be any equilibrating states of the molecule, most commonly the stable rotamers or conformers. However, the medium dependence of the barrier to interconversion could be treated in an exactly analogous manner, though now of course, ΔG becomes ΔG^*, the free energy of activation of the process $A \rightarrow B$, and similarly ΔE becomes ΔE^*, the energy of activation.

The calculations of the variation of ΔE with solvent are given in the next section. It is convenient to consider first the various experimental methods of obtaining ΔE and more importantly the systematic errors which may occur.

It is not our intention in this chapter to discuss the experimental techniques used to obtain rotamer energies as these are given elsewhere in this book.

However, it is a common factor of virtually all the experimental techniques, which have been used to measure rotamer energies, that given good experimental techniques the systematic errors in the derived energy differences are much larger than the experimental errors. This is particularly important when values of the energy differences in different media are being compared, as these differences are usually quite small. Thus, it is of some importance to consider the various assumptions made in the different investigations and how they may affect the final values.

Basically there are two types of experiment which may be used to obtain rotamer energies. They may be termed the static and the dynamic methods.

13.2.1 Static methods

In the static experiments the time-scale of the experimental technique is such that interconversion between the rotamers occurs during a single measurement, and what is measured is a single quantity, M, say. The value of this parameter is then the weighted average of the values in the various rotamers. For example in the equilibrium, $A \rightleftharpoons B$, if M_A and M_B are the values of M for the states A and B then

$$M_{\text{observed}} = n_A M_A + n_B M_B \tag{13.2}$$

The determination of ΔG from equations (13.1) and (13.2) may be performed in various ways. If values of M_A and M_B are known, or can be derived in some manner, then the value of ΔG is given immediately from one measurement of M. This method thus critically depends on the reliability of the values of M_A and M_B obtained, and also of course the error in the final value of ΔG will depend on the difference between M_A and M_B. That is, in order to achieve meaningful values of ΔG by this method, the approximations in the values of M_A and M_B and the experimental error in M_{obs} must together be $\ll |M_A - M_B|$. It can easily be shown from equations (13.1) and (13.2) that *for the most favourable case* of $\Delta G \sim 0$, if the errors in M_{obs} and $M_A - M_B$ are both about 1/10 of the value of $M_A - M_B$, then the error in ΔG is about 0·3 kcal/mole.

The two commonest examples of this method are the measurement of molecular dipole moments and room-temperature NMR coupling constants and chemical shifts.

Measurements of dipole moments The evaluation of ΔG from one measurement of the averaged molecular dipole moment of the equilibrium is of long standing. The determination of the molecular dipole moment from the observed value of the dielectric constant via the molar polarization (of the solute at infinite dilution for solutions) using the Debye or Onsager equations is only applicable to gases or polar solutes in non-polar solvents. The major source of uncertainty in the method is the allowance to be used for the atomic polarization (see Reference 13 for full details) However, the evaluation of ΔG from the observed dipole moment from equations (13.1) and (13.2) is straightforward, and the major problem is to estimate reliable values of μ_A and μ_B. (Note that equation (13.2) becomes $\mu_{obs}^2 = n_A\mu_A^2 + n_B\mu_B^2$ for this method.)

The evaluation of ΔG is particularly difficult for acyclic compounds. For example, in order to obtain the value of ΔG from the observed dipole moment of 1,2-dichloroethane a value of the dipole moment of the *gauche* isomer must be estimated (clearly $\mu_t = 0$ neglecting molecular vibrations). This is not an easy task, particularly as the exact geometry of the *gauche* isomer is still uncertain. For this reason this method has not been used extensively for acyclic compounds. The method has been used more successfully with cyclic compounds in which appropriate model compounds and thus estimates of μ_A and μ_B are much easier to obtain.

A typical example is provided by the work of Bender and coworkers on 1,2-dihalocyclohexanes.[14] These compounds were measured in dilute benzene and carbon tetrachloride solutions and the dipole moments obtained from the Debye equation. In the case of the *trans*-1,2-dihalocyclohexanes, the measured dipole moment is the average of the diaxial and diequatorial conformers. The assumption was made that $\mu_{aa} = 0$ and μ_{ee} equals the observed moment of the corresponding *cis*-1,2-dihalo compound, which can only exist in the equatorial–axial conformer. Although one of these assumptions is now known to be incorrect ($\mu_{aa} \neq 0$), the values of ΔG obtained compare very well with other measurements of this equilibrium (cf. § 13.5.1). Later workers[15] have used the *t*-butyl derivatives of these cyclohexanes which exist in only one conformation to provide direct values of μ_{aa} and μ_{ee} and this is a fairly general technique, provided the *t*-butyl derivatives do exist in the same conformation as the unknown conformers A and B.

NMR measurements This method is in principle straightforward. Merely one measurement of a nuclear chemical shift or internuclear coupling constant, in any solvent, coupled with the estimation of the chemical shifts or couplings for the individual conformers A and B *in the same solvent*, will give with equations (13.1) and (13.2) a value of ΔG. Furthermore the accuracy of measurement of nuclear chemical shifts and coupling constants is in most cases very high, thus the experimental errors in ΔG are low. It is however, the systematic errors due to the other uncertainties mentioned which restrict this method. This is particularly so for acyclic compounds. The problem of estimating proton chemical shifts of the individual rotamers of any acyclic compound in any given solvent has proved too great for any use to be made of such measurements, despite being able to measure them to one part in 100 very easily. The problem is compounded by the intrinsic solvent dependence of proton chemical shifts[8] which is comparable to the difference between rotamers ($\delta_A - \delta_B$). Similar considerations may prevail for other nuclear chemical shifts, though little systematic work has been attempted.

The measurement of internuclear coupling constants shows more promise, though the same comments apply. The commonest and most useful internuclear coupling for this purpose is the vicinal proton–proton coupling ($^3J_{HH}$), which has been shown to be virtually independent of solvent and may also differ considerably between the rotational isomers. For example in $CHCl_2CHCl_2$ the *gauche* and *trans* couplings are about 2 and 11 Hz.[16,17] Again however, the major difficulty is the estimation of the values of the couplings in the individual rotamers. If these couplings can be estimated to ±0.5 Hz and the difference $J_A - J_B$ is approximately 10 Hz, then from the last section it can be shown that for $\Delta G \sim 0$ the error in ΔG is about ±0.2 kcal/mole and for a $\Delta G \sim 1$ kcal/mole the error will be twice as large. Thus the method may be used semi-quantitatively in general. If however the couplings can be measured by any method (e.g. by low-temperature NMR) then of course this method is capable of providing very precise and rapid measurements of ΔG.

Any other nuclear coupling may in principle also be used in this method, (for example $^3J_{HF}$ and $^3J_{FF}$ couplings) but it is always essential to consider both the intrinsic variations of the couplings of the isomers with solvent and the expected difference between J_A and J_B. The available evidence to date indicates that the differences between the $^3J_{HF}$ couplings in different rotamers may be large (e.g. for $CHBr_2CFBr_2$ $J_g = 1.2$ Hz, $J_t = 22.2$ Hz)[18] but they have an intrinsic solvent dependence. Thus this coupling can be used, with care.

However, the $^3F_{FF}$ couplings are unique in that in many compounds the *gauche* and *trans* couplings are comparable (e.g. for $CF_2BrCFBr_2$ 16.2 and 18.6 Hz),[19] and they also show a solvent dependence. Thus the problem of estimating the couplings in the rotamers to the required accuracy becomes insuperable and the method does not give reliable values of ΔG.

In complete contrast the measurement of chemical shifts and coupling constants has been used very widely and with great success in acyclic compounds and is the commonest method of determining axial–equatorial energy differences in monosubstituted cyclohexanes (cf. Reference 9). In the commonest example of this technique, the chemical shift of the C_1 proton in a monosubstituted cyclohexane is measured in any solvent and the chemical shifts of the axial and equatorial conformers obtained by direct measurement of the corresponding *cis-* and *trans*-4-*t*-butyl derivatives in the same solvent. The method has been used both to obtain the axial–equatorial ΔG values for a range of substituent groups and also in a general study of solvent effects.[20] There are small systematic errors involved in the use of the *t*-butyl derivatives as model compounds. For example the $-\Delta G$ value for chlorocyclohexane in CS_2 solution was obtained by this method as 0·39 kcal/mole which may be compared with the value from the absolute method of low-temperature NMR (§ 13.2.2.1) of 0·57 kcal/mole.[21]

In an alternative method the peak-width of the C_1 proton is used. This is roughly equivalent to the coupling-constant method, and is only to be preferred in compounds (such as monosubstituted cyclohexanes) giving NMR spectra which are too complex to be fully analysed.

The method is not so accurate as the chemical shift method, due to experimental uncertainties in the measurement of the peak-width, but may be used without measuring the 4-*t*-butyl model compounds in every solvent due to the invariance of $^3J_{HH}$ couplings with solvent.

Variable temperature methods If the values of M_A and M_B cannot be estimated with any reasonable degree of accuracy, then further experimental measurements will be required in order to obtain the value of ΔG. The commonest procedure used is to measure M as a function of the temperature. The observed plot of M versus temperature when calculated using equations (13.1) and (13.2) is a four-parameter curve in the unknowns ΔH, ΔS, M_A and M_B. Very commonly ΔS is assumed zero to reduce the problem to a three-parameter fit.

The basic assumption underlying all such treatments is that M_A, M_B and ΔH are independent of temperature, or more precisely that the observed temperature dependence of M due to changes in M_A, M_B and ΔH with temperature is negligible compared to the variation due to the changing proportions of the conformers with temperature according to equations (13.1) and (13.2).

We shall see that it is absolutely essential for this condition to be tested in some manner before carrying out this procedure.

A further critical condition for the success of this method is that it is necessary to obtain a pronounced variation of M with temperature. Specifically it is advantageous to obtain values of M over such a wide temperature range that a pronounced curve of M versus temperature is obtained. Any measurement which merely gives a straight line of M versus temperature can be fitted by only two parameters. The fit with three parameters becomes very uncertain.

Variable temperature dipole moment measurement This method has been used frequently to obtain values of rotamer energies of acyclic compounds for which the single temperature method is inappropriate, and was one of the first techniques to be used systematically in the study of rotational isomerism.[5]

As a typical example, the favourable case of 1,2-dichloroethane may be considered.[5] The dipole moment in the vapour was obtained from the molar polarization, and in this case the value of $P_E + P_A$ was obtained from the dielectric constant of the solid which is entirely in the *trans* form with therefore no permanent dipole moment. The dipole moment of the gas varied from 1·12 D at 305·0°K to 1·54 D at 543·7°K. Assuming the dipole moment of the *trans* isomer as zero the application of equations (13.1) and (13.2) gave $\Delta E^v = 1·2\,\text{kcal/mole}$ and $\mu_g = 2·55\,\text{D}$.

The authors corrected these results to allow for (a) the different partition functions of the *trans* and *gauche* forms, i.e. included a non-zero entropy difference, and (b) the possible non-zero dipole moment of the *trans* rotamer due to torsional vibrations. These refinements did not appreciably change the results.

The method assumes the temperature independence of μ_t and μ_g, which is not easy to test experimentally. The dipole moments of ethanes with degenerate rotamers, e.g. ethyl chloride, do not change with temperature, which provides some support. This is not unequivocal because torsional vibrations in these molecules should not give rise to any significant temperature dependence of the dipole moment whereas they could in principle do so in less symmetric compounds such as *gauche* 1,2-dichloroethane. The method also assumes the temperature independence of $P_E + P_A$, which is also not strictly valid.

Nevertheless this method was of considerable value in the early investigations of rotational isomerism and provided much of the bases on which later developments built. Its major limitation as a technique to study solvent effects is that it is limited to vapours and solutions in non-polar solvents.

Variable temperature NMR   This method has been developed to a very considerable degree of sophistication and is in principle a powerful and general technique. However, considerable controversy has arisen over the application of this method, due, as we shall see, mainly to the neglect of the fundamental conditions mentioned earlier, so that we shall consider the method in some detail.

In principle one merely measures any NMR parameter, usually a chemical shift or coupling constant, and then uses equations (13.1) and (13.2) to fit the experimental curve. As NMR couplings and chemical shifts can be measured accurately virtually in any solvent, in principle the method is a very general one.

The formalism used to obtain the best fit to the experimental curve was developed and applied to a variety of substituents by Gutowsky and co-workers.[22] Subsequently a number of their results were shown to be widely in error, and a critical discussion of the method was given by Govil and Bernstein[18] and we shall follow the latter authors simplified, but equivalent formalism.

If we consider the equilibrium, $A \rightleftharpoons B$, then from equations (13.1) and (13.2) the value of the observed quantity M at any temperature, M_T, is given by

$$M_T = (M_A + KM_B)/(1 + K)$$

i.e.

$$K = (M_T - M_A)/(M_B - M_T) \tag{13.3}$$

In the original treatment and most subsequent work ΔS is taken as zero (where there are two equivalent forms, for example in any 1,2-disubstituted ethane, the factor 2 is included explicitly), thus for equilibrium $A \rightleftharpoons B$ we have

$$n_A = 1/[1 + \exp(-\Delta H/RT)]$$

and

$$n_B = \exp(-\Delta H/RT)/[1 + \exp(-\Delta H/RT)] \tag{13.4}$$

A function ϕ is formed given by

$$\phi = \sum_T (M_T - n_A M_A - n_B M_B)^2 \tag{13.5}$$

and this function is then minimized computationally by varying the values of ΔH, M_A and M_B. The sum over T includes all temperatures at which there are measurements M_T. In practice a value of ΔH is fed in and from this n_A and n_B calculated at each temperature by equation (13.4). These when combined with the experimental measurements of M_T and equation (13.2) (for each temperature) allow the best-fit values of M_A and M_B to be obtained, and finally these from equation (13.5) give a value of ϕ. ΔH is varied until the minimum value of ϕ is obtained, which defines the best values of M_A and M_B.

The basic assumptions in the method are that $\Delta S = 0$ and that the temperature dependence of the measured quantity M is not significantly affected by the temperature variation of ΔH, M_A and M_B. Also there is the question of the accuracy of the 'best-fit' parameters so obtained.

Govil and Bernstein were able to check these assumptions in the case of $CHBr_2CFBr_2$ by measuring the average NMR parameters at high temperature, using the above treatment to obtain the three unknowns and then comparing these values with those obtained by direct measurement of the spectrum at such low temperatures that the individual spectra of the rotamers were obtained. This gave directly the values of the couplings and chemical shifts of the rotamers and also of course the equilibrium constant at these temperatures from the relative intensity of the peaks.

They found that both the ^{1}H and ^{19}F chemical shifts of each rotamer changed with temperature. This is to be expected as nuclear chemical shifts are invariably temperature-dependent. More significantly the evidence suggested that the temperature dependence of the chemical shifts was very different for the two isomers. For example the ^{19}F shift (at 56·4 MHz from $CFCl_3$ solvent) over 27°C varied from 13·7 Hz (*gauche*) to 24·5 Hz (*trans*). Thus the use of the temperature dependence of nuclear chemical shifts to determine rotamer populations would appear very doubtful unless firm evidence is obtained that the intrinsic temperature dependence of the shifts in the rotamers has been eliminated.

However, the coupling constants ($^3J_{HF}$) in the rotamers appeared to be substantially independent of temperature over the small range of temperature available (178–171°K), and these values were used to determine the equilibrium constants at all temperatures. The authors obtained values of ΔH and ΔS of -0.20 kcal/mole and -0.14 e.u. from this method, with values of J_g and J_t of 1·15 and 22·2 Hz. In contrast, use of the high-temperature values of J_t alone with the above analysis gave values of $\Delta H = -0.80$ kcal/mole, $J_g = 5.6$ Hz and $J_t = 17.2$ Hz with $\Delta S = 0$ (assumed).

The two methods gave very different results. In this case the authors were able to show that the assumptions of $\Delta S = 0$ and the temperature independence of J_g, J_t and ΔH were not the major cause of the discrepancy. The real reason was that the observed temperature dependence of the coupling at high temperatures was not sufficient to allow a precise evaluation of all three unknowns. The average coupling constant only varied from 6·50 Hz at 218°K to 7·18 Hz at 343°K. With an experimental error of ± 0.025 Hz, it was shown that the range of acceptable values (those giving calculated curves within this error of the observed results) was $\Delta H = -0.1$ to -1.3 kcal/mole, $J_g = 1$–8 Hz and $J_t = 15$–25 Hz.

This is a special case in that $\Delta H \approx 0$ and therefore the variation of the coupling with temperature will be small. However, it is also a favourable example in that the variation in the rotamer couplings with temperature is small. In other cases this variation is not small and cannot be ignored. For example, Gutowsky and coworkers[22] analysed the temperature dependence of the $^3J_{FF}$ coupling in $CF_2ClCFCl_2$, which varied from 9·41 Hz at 239°K to 8·59 Hz at 470°K, in the normal way, and obtained $\Delta E(E_t - E_g) = +2.8$ kcal/mole, $J_g = -21.2$ Hz and $J_t = +18.9$ Hz. They neglected to check for any temperature variation of the coupling, and it was later found that such couplings are intrinsically temperature-dependent. In CF_3CFCl_2 the temperature dependence of the FF coupling is -0.409 Hz/100°C,[18] which can only be due to an intrinsic temperature dependence. This intrinsic temperature dependence is comparable to the observed temperature dependence in $CF_2ClCFCl_2$, and thus the results obtained on the basis of the above analysis are quite meaningless.

There are also other compounds in which the variation in ΔH with temperature is substantial and cannot be neglected. These are precisely those compounds which show a large solvent dependence, and discussion of this will be deferred to § 13.4.1. However, for any given solute this variation can be minimized by the use of non-polar solvents.

In conclusion the erroneous results obtained, and the subsequent controversy surrounding the application of this method, are due entirely to the neglect of the basic assumptions involved rather than to any inherent defect in the method.

It is essential in any application of this technique (a) to provide some estimate of the intrinsic temperature dependence of the parameters involved, and (b) to obtain a sufficient variation in the measured quantity with temperature to provide well-defined values of the three unknown parameters involved. When these conditions are met this method is capable of accurate results.

Variable solvent NMR   The only examples of the method of varying the solvents are from the measurement of NMR coupling constants and thus we can consider these together. Of course there is no reason why any other appropriate physical technique cannot be used in conjunction with this method.

In this method the solvent is changed in order to vary the proportions of the rotamers. The advantage which this has over varying the temperature is that the effects on the rotamer populations can be much larger. For example the two $^3J_{HH}$ couplings in CH_2BrCH_2Cl only change from 5·99 and 9·14 Hz at $-10°C$ in the pure liquid, to 6·02 and 8·77 Hz at $137°C$.[24] The variation in the couplings in going from cyclohexane solution (5·54 and 10·63 Hz) to acetonitrile solution (6·23 and 7·63 Hz, see Table 13.8) is many times larger. There is unfortunately a compensating disadvantage in this method in that whereas the variation in rotamer populations with temperature is known precisely from equation (13.2), this is not the case for the solvent. If a theoretical treatment capable of predicting the variation of ΔE with solvent was available, then the solvent method could be used in a precisely analogous manner as the temperature variation method, to obtain rotamer energies and couplings. This was the original motivation behind the theoretical development to be given in the next section, which provides a method of calculating the energy difference in any solvent from the energy difference in the vapour ΔE^v, various calculable or measurable parameters of the solute, and the solvent dielectric constant.

Thus the variation in the observed couplings with solvent dielectric constant becomes a three-parameter fit in the parameters ΔE^v and the couplings in the distinct rotamers. Apart from the quantitative validity of the model used to evaluate the solvent dependence of the rotamer energies, precisely analogous assumptions and conditions apply to this method as to the temperature variation method. Thus the observed solvent dependence of the NMR couplings in the compound studied must be much larger than any change due to the intrinsic solvent dependence of the rotamer couplings.[23] Also the observed change must be sufficient to provide well-defined values of the three unknown parameters.

In practice the solvent and temperature methods are complementary as when there is a large solvent dependence of the rotamer energies then ΔH in equation (13.2) will be intrinsically temperature-dependent due to the change in the dielectric constant of the solvent with temperature. This is a considerable factor for many common solvents but for non-polar solvents the variation in the dielectric constant with temperature is small and the subsequent correction to ΔH may usually be ignored.

Of course, the solvent and temperature variations may be combined in one investigation and there are a number of examples of this method. Here the advantage to be obtained is that there is considerably more experimental information to be fitted. However, the limitations and conditions detailed for both methods above still apply to these more complete investigations thus they do not need to be considered separately.

13.2.2 Dynamic methods

Low-temperature NMR   This method can be used at any temperature, depending on the interconversion process being studied. However, for the rotational isomerism processes considered here low-temperature NMR is necessarily involved.

This is the first example of a dynamic process. In these techniques the time-scale of the experiment is such that interconversion between the rotamers is slow compared to the measuring frequency and thus two or more signals are obtained, one from each rotamer.

An estimate of the time-scale required is given by the simple formula for the coalescence point of two equal signals, for which the lifetime in any one state (τ) is given by[25]

$$\tau = \sqrt{2}/\pi\delta \tag{13.6}$$

where δ is the separation (in Hz) of the signals of the distinct rotamers.

For proton resonance a typical value of δ is approximately 30 Hz, thus for separate signals to be observed the lifetime of a state must be greater than $1\cdot5 \times 10^{-2}$ s. The lowest practical temperature, due to solubility problems etc., is about $-100°C$, and these two conditions combined give a minimum value of the energy of activation of the interconversion process of ~ 8–9 kcal/mole. For any given compound the value of δ (in Hz) increases with the applied magnetic field and thus the coalescence point can be raised by operating at higher fields. Also the values of δ are much larger for other nuclei and this also raises the coalescence point for the same equilibrium. Thus in the rotational isomerism of $CHBr_2CFBr_2$, the coalescence point for the ^{19}F nuclei ($\delta \sim 120$ Hz) was $238°K$ but for the protons ($\delta \sim 2$ Hz) it was $198°K$.[23]

However, for those equilibria which can be 'frozen out' at low temperatures the fact that the NMR signals are directly proportional to the concentration of the nuclei means that this method gives immediately and unambiguously the value of ΔG for the equilibrium. This is the only method of those detailed here in which the systematic errors due to assumptions in the derivation of ΔG are negligible. The accuracy of the value of ΔG obtained is limited solely by the accuracy of measurement of the relative intensities of the NMR signals. This is most accurate for approximately equal amounts of the rotamers but decreases sharply for very unequal amounts. Indeed, due to the solubility (and therefore signal-to-noise) problems at the low temperatures needed the accurate measurement of any rotamer present in less than 1% proportion is difficult. This gives a practical upper limit of $\Delta G \sim 2$ kcal/mole.

If a sufficient range of temperature below the coalescence point can be achieved then of course ΔH and ΔS for the equilibrium can be found, but in most cases these are much less accurate than ΔG. A better method of obtaining these parameters is by the combination of these low-temperature and variable temperature NMR methods, as mentioned previously.

Variable temperature infrared The NMR method above is restricted to rotamers with relatively high barriers to interconversion and is thus applicable only to multisubstituted ethanes. The application of equation (13.6) to infrared spectra, in which a typical value of δ would be $10\,cm^{-1}$ (3×10^{11} Hz), shows that the lifetime (τ) of each rotamer must be greater than about 10^{-12} s for separate signals to be observed. This corresponds to an energy of activation of only about 1 kcal/mole, at room temperature.

Thus all the simple ethanes give IR spectra which are the superposition of the IR spectra of the distinct rotational isomers, and indeed this observation was one of the key factors in the original determination of the configurations of substituted ethanes. Furthermore, as virtually all the original determinations of the rotamer energy differences used this technique, it is of some interest to consider the assumptions involved.

Unlike the equivalent NMR technique, the intensities of IR absorption bands are functions of the number of molecules in the cell *and* the integrated absorption coefficient (α) which is a molecular property. Thus the intensity of each band is given by $A = \alpha Cl$ where A is the band area, C the concentration and l the cell length.

The expression for the equilibrium constant K in equation (13.1) is thus

$$K = C_B/C_A = A_B\alpha_A/A_A\alpha_B \tag{13.7}$$

K cannot be calculated from this expression as the absorption coefficients are usually unknown. The assumption is usually made that the absorption coefficients are constant over the temperature range involved and thus ΔH is readily calculated from equations (13.1) and (13.7) to give

$$\ln A_B/A_A = -\frac{\Delta H}{RT} + \text{const.} \tag{13.8}$$

The plot of $\ln A_B/A_A$ versus $1/T$ gives a line of slope $-\Delta H/R$, but in this case neither ΔS nor ΔG is obtained.

One practical difficulty with this method is to ensure that the bands used for the intensity measurements are not contaminated by bands of the other isomer. For this reason it is normal practice to measure a number of pairs of bands (one from each rotamer) to obtain ΔH.

A second practical problem is that of the assignment of the bands to the individual rotamers, which except in special cases (e.g. where one of the rotamers has a centre of symmetry) can be considerable. Thus the value of ΔH may be found without the sign being known.

The major assumption in the method concerns the absorption coefficients. Theoretically, the integrated absorption coefficient of an infrared band is temperature independent in the gas phase, but need not be so in solution, varying with the density and refractive index of the medium.

Recently Hartman and coworkers[26] outlined a simple method of avoiding this assumption. If C_T is the total concentration of the solute, then $C_A + C_B = C_T$ at all temperatures and the substitution of $A = \alpha Cl$ gives

$$\frac{A_A}{\alpha_A l} + \frac{A_B}{\alpha_B l} = C_T$$

or

$$A_A = -\frac{\alpha_A}{\alpha_B} A_B + \alpha_A l C_T$$

Thus the plot of the intensities A at different temperatures will give a straight line of slope $-\alpha_A/\alpha_B$ if the ratio of the absorption coefficients is temperature independent. Now the free energy and entropy can also be obtained as the hitherto undetermined ratio is known. These authors illustrated this use of this method, but to date few other examples have been given.

The other assumption common to this method and all the methods involving variable temperature measurements is that ΔH is temperature independent. Whilst this is theoretically justified for the vapour state, we shall show in the next section that this is not the case for many rotational isomers in solution for which the temperature variation of ΔH is considerable.

All determinations of ΔH (ΔE) from variable temperature measurements use the Van't Hoff equation (13.8), which is usually written

$$\frac{\mathrm{d}\ln K}{\mathrm{d}(1/T)} = -\frac{\Delta H}{R} \tag{13.9}$$

We may define the value of ΔH obtained from this equation as ΔH^0. If however, ΔH is intrinsically temperature dependent then the differentiation of equation (13.8) leads to the more complex expression

$$\frac{\mathrm{d}\ln K}{\mathrm{d}(1/T)} = -\frac{1}{R}\left(\Delta H - T\frac{\mathrm{d}H}{\mathrm{d}T}\right)$$

Thus the relationship between the true value of ΔH at any temperature and the apparent constant obtained from equation (13.9) is simply

$$\Delta H(t) = \Delta H^0 + T(\mathrm{d}H/\mathrm{d}T) \tag{13.10}$$

We shall show that the correction factor $T(\mathrm{d}H/\mathrm{d}T)$ arises for solutions predominantly from the variation in the solvent dielectric constant with temperature, and is therefore negligible for non-polar solvents but a considerable factor for other solvents. In particular the determinations of ΔH for the halogenated ethanes (the pure liquids) need to be corrected appreciably by this factor.

Variable solvent infrared The infrared technique may be extended to the use of different solvents in a similar manner to the NMR technique, though of course the spectra of the individual rotamers are observed, not one average. For this reason it is possible to obtain the energy difference in different solvents directly from this technique, as follows.[27]

Consider the case of equilibrium, $A \rightleftharpoons B$, in which the intensities of two bands, one due to each rotamer, have been measured in two solvents (s^1 and s^2). Then we have, using the same nomenclature as in the preceding section

$$A_A^1 = \alpha_A C_A^1 l$$

$$A_B^1 = \alpha_B C_B^1 l$$

$$A_A^2 = \alpha_A C_A^2 l$$

$$A_B^2 = \alpha_B C_B^2 l$$

where $C_T^1 = C_A^1 + C_B^1$ and $C_T^2 = C_A^2 + C_B^2$ are the total concentrations of solute in the two solvents. If we make the simplifying assumption that $C_T^1 = C_T^2$ (this is not necessary but convenient) then eliminating the unknowns α_A, α_B, C_A^2 and C_B^2 gives the equilibrium constant in solvent (1) as

$$K_1 = \frac{C_B^1}{C_A^1} = \frac{A_B^1(A_A^2 - A_A^1)}{A_A^1(A_B^1 - A_B^2)}$$

Thus the equilibrium constant in solvent (1) is obtained directly from measurements of the intensity of two bands in different solvents.

Again this method has similar assumptions to previous techniques. In particular, the apparent extinction coefficients α_A and α_B are assumed to be independent of solvent. Also, of course, there has to be sufficient variation of the intensities with solvent to make the equation meaningful. A practical problem in this method is that the choice of solvents is often limited due to the solvents' own strong absorption in the infrared. Finally, the bands chosen must not be contaminated with any from the other rotamer.

However, the method has been used successfully in a number of determinations of rotational equilibria, though in view of the assumptions necessarily made in the method most authors suggest that most significance should be given to the relative energy differences obtained in different solvents, and that the absolute values should be treated with caution.

13.2.3 *Other methods*

There are a number of other methods of obtaining rotamer energy differences, none of which are of very general applicability, and these are conveniently grouped together.

Microwave spectroscopy (see Chapter 6)   The determination of rotamer energy differences by microwave spectroscopy is exactly analogous to the variable temperature infrared method. It has certain advantages over this method, in that the possibility of overlapping bands is much smaller and in the molecules studied so far, the assignment of the bands to the various rotamers is known absolutely from the resulting moments of inertia. The major and severe limitation of this method is that it is restricted to the gas phase.

Electron diffraction (see Chapter 10)   The electron diffraction method also suffers from this limitation. In this technique an average electron diffraction pattern is obtained which is the sum of the patterns from the discrete rotamers. As each of these is dependent on the precise geometry of that rotamer (which is often not known to the required accuracy) the results obtained should be viewed with caution.

Chemical methods (see Chapter 2)   There are also a number of chemical equilibration and rate methods. One general method involves the equilibration between two stable non-interconverting species which can be made to occur with a catalyst. Removal of the catalytic reagent and subsequent determination of the proportions of the equilibrated species gives the required equilibrium constant.

Obviously this method requires a large energy barrier for the direct interconversion of the species and therefore is not applicable to rotamer energy differences. It is, however, used in conformational studies where the equilibrated species may be, for example, the *cis* and *trans* isomers of a cyclic compound. These methods have been well documented and reviewed previously.[10]

13.3 Theory of solvent effects

It is instructive to consider first the various possible approaches to the calculation of relative rotamer energies in solution. In principle this calculation, like any other involving molecular energies, could be performed by quantum-mechanical methods. In this approach the total energy of each rotamer when surrounded by solvent molecules would have to be calculated. In practice this leads to very considerable problems. The molecular system to be considered for even small solute molecules would be quite large, as they are usually approximately six solvent molecules surrounding each solute molecule. The position and orientation of each solvent molecule has to be defined. The accuracy of the calculation has to be very high, as relative rotamer energies can be measured to about 0·2 kcal/mole. To date such considerations have precluded any attempt at this approach. The nearest analogy is the calculation of complexes in solution and the calculation of molecular properties, such as NMR coupling constants, in such complexes, which have been performed using the semi-empirical CNDO programme.[28,29]

Also Johnston and Barfield, using the same programme, have calculated the solvent effect on NMR coupling constants using various models for the solvent structure, including a cubic close-packed cluster model[30] and the reaction field model.[31]

The alternative approach is to use classical calculations in which the dominant interactions present in solution are identified and then the energy of the interactions for the different rotamers can be calculated. The advantage of this approach is that only a very small part of the molecular energy is calculated, as the assumption is usually made that all other interactions present in solution are constant for the different rotamers. Note that as the calculations are concerned solely with calculating *differences* in rotamer energies the much more questionable assumption that all other inter- actions are negligible is unnecessary. As the dependence of rotamer energies on the solvent can be seen to be primarily a polar effect, historically this provided the basis for the two major methods of calculation. (An interesting attempt to identify other interactions was the demonstration by Ouellette and Williams that internal solvent pressure affected the *trans* to *gauche* equilibria in some sila-alkanes.[32] However, the energies involved were so small that they could only be measured in the absence of polar effects.)

These methods may be termed the direct dipole–dipole calculation and the reaction-field approach.

The first method has been well summarized in Reference 10, Chapter 3, and thus we shall only briefly consider it. It was used to estimate the large solvent effects observed in the α-halocyclohexanones. This stems from the well-known formula for the history of interaction of two dipoles m_1 and m_2 a distance $\mathbf{r}$ apart in a medium of dielectric constant ε

$$W = \{(\overline{m}_1 \cdot \overline{m}_2)/\mathbf{r}^3 - 3(\overline{m}_1 \cdot \bar{\mathbf{r}})(\overline{m}_2 \cdot \bar{\mathbf{r}})/\mathbf{r}^5\}/\varepsilon$$

$$= \frac{m_1 m_2}{\varepsilon \mathbf{r}^3} f(\theta, \phi) \tag{13.11}$$

where θ and ϕ define the relative orientation of the dipoles and the vector $\mathbf{r}$.

This equation is completely valid for two point dipoles in any medium. The difficulty lies in extending this to systems of two or more intramolecular dipoles surrounded by solvent. In this case there is no adequate method of defining ε, the 'effective dielectric constant', and the value used is not that of the solvent, nor that of the usually hydrocarbon bulk of the solute, but some intermediate value adjusted for each case. Thus the significance of the agreement between the calculated and observed values is uncertain.

A more serious failure of the theory is that when applied to other systems it gives qualitatively incorrect answers. It predicts that for any system the dipolar interaction merely decreases to zero for media of very high dielectric constant ($\varepsilon \rightarrow \infty$). In 1,2-dichloroethane $E_g - E_t$ is $+1.2$ kcal/mole in the vapour, 0.4 in the pure liquid and ~ 0 in acetonitrile solution (see Table 13.4). On the basis of the above theory, the polar interaction between the C—Cl dipoles will be negligible in acetonitrile solution and therefore the steric interactions involving the chlorine atoms, which are the only interactions remaining, must be negligible. This is contrary to most normal expectations. Although the theory is no longer used, the underlying idea is widely used in qualitative discussions. This is that any rotational or conformational equilibrium of a solute with polar substituents in a very polar solvent may be considered as if the energy difference observed was not due to the polar character of the groups, i.e. as if the polar interactions between the groups were zero. This is only valid for intermolecular interactions in which the dipoles can be completely solvated by the solvent. We shall find a number of molecules in which the rotamer energy difference actually changes sign on changing the solvent. This behaviour can be simply explained on the reaction-field model.

13.3.1 The reaction-field model

The general theory of dielectrics is well known and stems from the work of Debye,[33] Onsager[34] and Bottcher.[35] We are concerned merely with that part which is relevant to the calculation of rotamer energies. The rotamer energy difference will vary with solvent due to the different electric fields of the rotamers in the solvent. These electric fields are due to the polarization of the surrounding dielectric by the solute and the calculation of their energy follows directly from the theory of dielectrics.

In this, firstly the potential of any system of charges, such as the solute molecule, is represented as due to a charge plus a dipole plus a quadrupole etc. situated at the origin. In the cases we are considering the molecules are uncharged, thus the first term is the dipole term. The solvation energy of a non-polarizable molecule of dipole moment (m) in a solvent of dielectric constant ε is the difference between the energy of the molecule in the vapour (E^{v}) and in the solvent (E^{s}) and is given by

$$E^{\mathrm{v}} - E^{\mathrm{s}} = \left(\frac{\varepsilon - 1}{2\varepsilon + 1}\right)\left(\frac{m^2}{a^3}\right) \tag{13.12}$$

where a is the radius of the solute cavity, assumed spherical.

For the equilibrium between two rotamers A and B equation (13.12) is valid for both rotamers, and the resulting equations can be combined to give

$$\Delta E^{\mathrm{v}} - \Delta E^{\mathrm{s}} = \frac{(m_{\mathrm{A}}^2 - m_{\mathrm{B}}^2)}{a^3}\left(\frac{\varepsilon - 1}{2\varepsilon + 1}\right)$$

which can be rewritten as

$$\Delta E^{\mathrm{v}} - \Delta E^{\mathrm{s}} = (k_{\mathrm{A}} - k_{\mathrm{B}})x \tag{13.13}$$

where $k_{\mathrm{A,B}}$ equals $m_{\mathrm{A,B}}^2/a^3$ and x equals $(\varepsilon - 1)/(2\varepsilon + 1)$.

This is the basis for the long-established rule that the rotamer of higher dipole moment is more favoured in media of high dielectric constant.

The next approximation to consider is the polarization of the solute molecule by its own reaction field. These calculations are also well known and give

$$\Delta E^{\mathrm{v}} - \Delta E^{\mathrm{s}} = (k_{\mathrm{A}} - k_{\mathrm{B}})x/(1 - lx) \tag{13.14}$$

where $l = 2\alpha/a^3$, α being the molecular polarizability of the solute, usually given by $2(n_{\mathrm{d}}^2 - 1)/(n_{\mathrm{d}}^2 + 2)$, n_{d} being the solute refractive index. If the dipolar term was the only important term, then equation (13.14) should predict the observed values of the differences in energy of rotational isomers in going from the gas to the liquid. In fact equation (13.14) predicts changes in energy of at least twice as much as is observed.

Two treatments evolved from this anomaly. Wada[36] calculated directly the energy of the molecular field for the 1,2-dihaloethanes on the basis of two CX dipoles situated along the appropriate bonds and obtained reasonable agreement with experiment. However, this approach lead to a complex formula and has not been generalized to any other system. Abraham and coworkers merely calculated the second, i.e. quadrupole, term in the molecular electric field and showed that this gave a simple and general formula.[24,37] It is this theory we wish to consider in detail.

Consider a quadrupole q at the origin. The potential of this quadrupole at any point $P(r, \theta, \phi)$ is

$$\Phi = q_{zz}r^{-3}(3\cos^2\theta - 1) + q_{xx}r^{-3}(3\sin^2\theta\cos^2\phi - 1)$$

$$+ q_{yy}r^{-3}(3\sin^2\theta\sin^2\phi - 1) + (q_{xz} + q_{zx})r^{-3}3\sin\theta\cos\theta\sin\phi \quad (13.15)$$

$$+ (q_{xy} + q_{yx})r^{-3}3\sin^2\theta\sin\phi\cos\phi + (q_{yz} + q_{zy})r^{-3}\sin\theta\cos\theta\sin\phi$$

where q_{ij} equals $\mu_i \times j$, thus q_{ii} are axial quadrupoles of the form ($\leftarrow\!+\!\rightarrow$) and q_{ij} perpendicular quadrupoles of the form ($\updownarrow\updownarrow$). The potential functions for q_{ij} and q_{ji} are the same, as the charge distribution for ($\updownarrow\updownarrow$) is identical to ($\rightleftarrows$). The energy of interaction of this quadrupole in a spherical cavity of radius a with an external medium of dielectric constant ε is given by[35]

$$W = -\frac{5(\varepsilon - 1)}{\pi(2 + 3\varepsilon)} \int \bar{E} \cdot \bar{E}\, dV \quad (13.16)$$

where the integral extends throughout the medium outside the cavity and $\bar{E} = \mathrm{grad}\,\Phi$ is the electric field of the quadrupole.

When the requisite operations and integration are carried out, equations (13.15) and (13.16) combine to give

$$W = -h \cdot \frac{(\varepsilon - 1)}{(2 + 3\varepsilon)}$$

where

$$h = \frac{3}{2a^5} \sum_{\substack{i,j=x,y,z}}^{i \neq j} [4q_{ii}^2 + 3(q_{ij} + q_{ji})^2 - 4q_{ii}q_{jj}] \quad (13.17)$$

and is the quadrupole analogue of k in equation (13.13). It is convenient to write h as $3q^2/2a^5$ where q^2 is the summation in equation (13.17). Thus q may be considered to be for our purposes the molecular quadrupole moment.

For a system containing dipole and quadrupole terms, it can be shown that cross-terms vanish in the integration, and the resultant energy difference of two rotamers A and B in any solvent s is given by combining equations (13.14) and (13.17) to give the basic equation

$$\Delta E^s = \Delta E^v - kx/(1 - lx) - 3hx/(5 - x) \quad (13.18)$$

where

$$k = k_A - k_B \quad \text{and} \quad h = h_A - h_B.$$

(In the original derivation,[24,37] h was defined as $h_B - h_A$, which merely alters the sign of the quadrupole term.) This equation can be applied directly to any system of interconverting rotamers once the rotamer dipole and quadrupole moments have been calculated and as we shall see gives results in very good agreement with experiment for a variety of molecules.

13.3.2 *Dipole–dipole interactions*

We shall subsequently consider in more detail the assumptions underlying the derivation of this equation. However, one basic assumption is conveniently considered here as it is possible to take account of this to obtain a further development of the theory. One of the basic assumptions of the Onsager reaction-field theory[34] is that the solvent is treated as a continuum of given dielectric constant. Thus the solute reaction field and therefore the solute external electric fields, which are integrated to obtain equation (13.18), follow instantaneously the orientation of the solute dipole.

In very polar media the Onsager theory, on which equation (13.18) is based, breaks down and a number of attempts have been made to extend the treatment to these solvents. For example Looyenga described an amendment to the theory in that the term $(\varepsilon - 1)/(\varepsilon + 2)$ in the Clausius–Mosotti–Debye equation was replaced by $\varepsilon^{\frac{1}{3}} - 1$.[38,39] On this basis x in equation (13.18) would be replaced by $1 - \varepsilon^{-\frac{1}{3}}$. A very recent fundamental extension of the basic theory has been given by Block and Walker who derived the fundamental equations using a continuously varying cavity dielectric constant to take account of dielectric inhomogeneities.[40] On their model x in equation (13.18) would be given by

$$x - 1 = \varepsilon \log^2 \varepsilon / 2[1 - \varepsilon(1 - \log \varepsilon)]$$

This equation has the considerable advantage over all the other treatments that it predicts that as $\varepsilon \to \infty$, x will also $\to \infty$ and the more polar rotamer becomes infinitely stabilized.

The breakdown of equation (13.18) in very polar solvents was noted from NMR studies of rotational isomerism[41] and the deviation was ascribed to dipole–dipole interactions between the solvent and solute dipoles. An alternative approach to that above follows on this basis by the calculation of this dipole–dipole interaction.

The energy of interaction of two dipoles m_1 and m_2 a distance r apart is given immediately by equation (13.11) where now as we are considering dipoles in contact the value for the dielectric constant is unity. The average energy of interaction $\langle W \rangle$ for all solute–solvent orientations is obtained by integrating W over θ and ϕ at constant r, taking into account the thermal motion of the dipoles. This gives[41]

$$\langle W \rangle = \frac{\int W \exp\left(-W/kT\right) \mathrm{d}N_1\,\mathrm{d}N_2}{\int \exp\left(-W/kT\right) \mathrm{d}N_1\,\mathrm{d}N_2} \tag{13.19}$$

The integrals in equation (13.19) cannot be evaluated explicitly except for some limiting cases (e.g. $W/kT \ll 1$), but were solved numerically and shown to agree with a simple exponential function. The calculation was extended to consider an octahedral arrangement of solvent molecules and for this arrangement the average energy can be written as

$$\frac{\langle W \rangle}{kT} = -C[1 - \exp(-C/2)] \tag{13.20}$$

where $C = m_1 m_2 / r^3 kT$.

The dipole moment of the solvent (m_1) was related to its dielectric constant by the Onsager formula, which assuming a solvent refractive index of $2^{\frac{1}{2}}$, is

$$m_1^2 = \frac{9kT}{32\pi N_0} \cdot (\varepsilon - 2)(\varepsilon + 1)/\varepsilon \tag{13.21}$$

Combining equations (13.20) and (13.21) and converting to one mole of solute gave the final equation

$$\langle W \rangle_m = -bf[1 - \exp(-bf/16RT)] \tag{13.22}$$

where

$$b = \frac{3m_2}{r^3} \left\{ \frac{2V_m RT}{\pi} \right\}^{\frac{1}{2}}$$

and

$$f = \{(\varepsilon - 2)(\varepsilon + 1)/\varepsilon\}^{\frac{1}{2}}$$

V_m is the solvent molar volume, m_2 the solute dipole moment and ε the solvent dielectric constant.

This situation predicts that as $\varepsilon \to \infty$ the solvation energy $\langle W \rangle$ also $\to \infty$, and thus is consistent with the most recent alternative approach.[40] In order to achieve a general treatment the simplifying assumption was made that the variation in r (the solvent–solute distance) could be ignored compared to the variation in the solvent dipole moment, which is explicitly accounted for in the f term. In order to evaluate the difference in energy between two rotamers the convenient assumption was also made that the m_2/r^3 term is proportional to k (equation 13.13).

Therefore in this amendment of the original reaction-field theory, equation (13.18) is now modified by an additional term

$$\langle W \rangle = -bkf[1 - \exp(-bkf/16RT)] \tag{13.23}$$

where the parameter b is a constant which has been determined experimentally to be 0·06–0·12.

13.3.3 *Generalized polar interactions*

Although the inclusion of the dipole–dipole term (equation 13.23) gives somewhat better agreement with experiment than the simple theory (equation 13.18), there are a number of difficulties concerning this term. Perhaps the most important limitation is that whilst it predicts that the solvation energy of a dipolar solute molecule becomes infinitely large for solvents of infinite dielectric constant (from equations 13.22 and 13.23 $\langle W \rangle \rightarrow \infty$ as $\varepsilon \rightarrow \infty$), no comparable interaction occurs for the essentially similar case of a very polar solute quadrupole.

This limitation can be overcome by evaluating in a similar manner the interactions between the solute quadrupole and the solvent dipoles. This is also more consistent with the original theoretical treatment in that the solute potential function was considered explicitly up to and including the quadrupole term.

Recently these calculations have been performed,[42] following the same procedure as given in the last section, but using the quadrupole potential (equation 13.15) as the basis of the calculation in addition to the dipole–dipole interactions given by equation (13.11).

As before, cross-terms between the dipole and quadrupole interactions vanish, and the calculations give the extra term to be added to equation (13.18) as identical to equation (13.22) but where now

$$b = \frac{3}{r^3}\left\{\frac{2V_\mathrm{m}RT}{\pi}\right\}^{\frac{1}{2}}\left\{m_2^2 + \frac{3q^2}{2r^2}\right\}^{\frac{1}{2}} \tag{13.24}$$

where q is the solute quadrupole moment.

This term can now be immediately changed to the same form as equation (13.18) by introducing the definitions of k and h.

Also using a reasonable value of V_m (80 ml) and converting to the appropriate units gives finally

$$b = 4{\cdot}35\left(\frac{T}{300}\right)^{\frac{1}{2}}\frac{a^{\frac{3}{2}}}{r^3}(k + ha^2/r^2)^{\frac{1}{2}} \tag{13.25}$$

where T is the temperature (°K) of the experiment.

The only parameter to be determined is r the solute–solvent distance and this was taken as equal to a + constant (a is the solute radius). A value of 1·8 Å for the constant was found to give excellent agreement with the experimental results and will be used henceforth. That is, the additional term given by equations (13.22) and (13.25) now contains no additional parameters over those needed to evaluate equation (13.18), which are the solute molecular dipole and quadrupole moments and the molecular volume and refractive index.

13.3.4 *Assumptions and limitations of the theory*

The three successive developments of the theory, i.e. the simple reaction-field theory (§ 13.3.1), the reaction-field term plus the dipole–dipole interactions, and the reaction-field term plus the dipole–dipole and quadrupole–dipole interactions (§13.3.3), have all been applied to various problems of rotational isomerism and they will be referred to later. If should be emphasized that the developments of the basic theory do give only small changes in the rotamer energy differences except for very polar solvents and thus can be neglected for approximate calculations.

However, before any of these theories can be used, it is necessary to consider in detail the assumptions on which they are based. The basic assumptions in the theory stem firstly from the general assumptions implicit in the theory of dielectrics on which it is based, and secondly from the simplifying approximations made in order to obtain a reasonably general and useful expression. Four assumptions may be considered in detail.

(a) It is implicit in the derivation of equation (13.18) that there cannot be any specific (i.e. chemical) interactions present in the system. The most important will be solvent–solute chemical interactions but any other interactions between solute or solvent molecules will distort the energetics of the system from those predicted by the theory. The outstanding examples of such systems are those in which hydrogen bonding occurs and this of course immediately excludes from consideration by the theory such common solvents as the alcohols and water.

Although hydrogen-bonding solvents would not be expected to behave according to the theory, there are other common solvents which also show anomalous behaviour which is however consistent. The outstanding example of this is benzene. It has been found in almost every example of rotational and conformational isomerism studied that benzene behaves as a much more polar solvent than its bulk dielectric constant of 2·3 would suggest, and a number of specific examples are given subsequently. Furthermore, other comparative studies of solvent behaviour give similar results[43,44] so that the phenomenon is not confined even to rotational isomerism. From an investigation of the free-energy differences between *cis*- and *trans*-2-isopropyl-5-alkoxy-1,3-dioxans, Eliel and Hofer also concluded that toluene behaved like benzene but that mesitylene and *t*-butyl benzene behaved normally, and proposed a new solvent scale to remove the benzene anomaly.[45] An alternative simpler proposal was adopted in studies of the application of the theory to conformational equilibrium in halogenocyclohexanes (see Reference 46 and § 13.5.1) which merely calculated the energy differences in benzene solution using a value of the dielectric constant in equation (13.18) of 7·5. This of course does not mean that it is inferred that benzene has a static dielectric constant of 7·5 nor does it provide an explanation for the anomalous behaviour of benzene as solvent. It simply provided a means by which a diverse amount of experimental data was fitted to the theory. Inspection of the results with benzene as solvent for the various rotational equilibria given later shows that the value of 7·5 should be used with care as the extent of the benzene anomaly varies from solute to solute. However, it does give a reasonable indication of the rotamer energy in a variety of cases.

(b) The solvent is regarded in the theory as a uniform medium with no structure. This gives rise to apparent anomalies when mixtures of solvents are used to study rotational isomerism, or indeed any other molecular property which is dependent on the dielectric constant of the medium, such as NMR chemical shifts[47] and IR frequency shifts.[48] These anomalies have been at times wrongly attributed as failures in the reaction-field model,[47] but in fact can be in principle treated by the theory. (For a quantitative treatment of IR frequency shifts in solvent mixtures see Reference 48 and for a thermodynamic treatment based on the Onsager model see Reference 49). If we consider a polar solute molecule dissolved in a solvent mixture of a polar and non-polar constituent then it is quite clear that simply as a result of the electrostatic forces considered in the theory, there will be preferential solvation of the solute by the more polar constituent of the solvent. Thus the mixture will behave, as far as the solute is concerned, as a medium of higher dielectric constant than would be estimated from its composition, or measured. For precisely similar reasons solvent molecules which contain a small polar group and a large non-polar (e.g. hydrocarbon) residue will behave anomalously, appearing to be more polar towards the solute than would be expected from the bulk dielectric constant.

(c) The solute molecular field is limited in the theory to merely the dipole and quadrupole components. The extent to which this is an approximation may be quantified by considering the next term in the series, the octupole term. By precisely similar reasoning to that involved in the derivation of equation (13.17) but starting from the potential of the general octupole, the solvation energy of an octupole in a spherical cavity of radius a is given by[50]

$$W = -\frac{G(\varepsilon - 1)}{(3 + 4\varepsilon)} \frac{O_{\mathrm{c}}^2}{a^7} \qquad (13.26)$$

where O_{c}, the octupole moment, is given by

$$O_{\mathrm{c}}^2 = \sum_{i,j,k = x,y,z} \{3O_{iii}^2 + 2(2O_{iij} + O_{ijj})^2 + 5(O_{ijk} + O_{jik} + O_{kij})^2$$
$$- 3O_{iii}(2O_{jji} + O_{ijj}) - 6(2O_{iij} + O_{jii})(2O_{kkj} + O_{jkk})\}$$

and O_{ijk} equals $\mu_i \times j \times k$.

The condition for the validity of equation (13.18) is therefore that the octupole energy as given by equation (13.26) should be very small. Note that it is not sufficient merely to include this term with the dipole and quadrupole terms, simply because in those cases where this term is large the higher order terms will also be appreciable. Thus the expansion of the molecular potential into multipoles is not now a converging series and cannot be used. The limiting cases in which this expansion breaks down are those in which the constituent dipoles are far apart. For example in the series $X(CH_2)_nX$, as n is increased from 2 the evaluation of the molecular potential into merely the dipole and quadrupole terms will at some stage become increasingly approximate and the theory will break down. Fortunately however, the interactions between the polar groups will by then be reduced to zero so that the molecule is probably better considered as two essentially independent parts. To date only one interesting example of this type of investigation has been given. Reynolds and Wood[51] observed that for 1,2-dibromoethylbenzene the variation in the rotamer distribution with solvent (as indicated by the $^3J_{HH}$ coupling) was in reasonable accord with the qualitative predictions of the theory. However, for 4-nitro-(1,2-dibromoethyl)benzene the variations in the observed couplings with solvent were very similar to those of the unsubstituted compound suggesting that the remote 4-nitro group is not affecting the solvent dependence of the rotamer equilibrium despite its very polar character. These authors did not evaluate the quadrupole term explicitly, so that their conclusions are qualitative, but they inferred that only the polar bands near the C—C bond whose rotational isomerism is being studied are concerned in the solvation process as it affects the rotational isomerism.

(d) The other simplifying assumption in the derivation of the theory is that of a spherical solute cavity. Expressions have been derived for the dipolar reaction field in an ellipsoidal cavity.[35] However, the assumption of cylindrical symmetry for most molecules is not much better than the assumption of a spherical cavity. Also, as this would introduce more subjective considerations, such as the proper enveloping ellipsoid for any given molecule, this has not been attempted.

13.3.5 *The model used*

The theory can now be used for any rotational equilibrium once the various parameters required in the equations have been evaluated. These are (equation 13.18) the dipole and quadrupole moments of the solute rotamers, the molecular radius (a) and polarizability (l). An important consideration in evaluating parameters for any general theory is that it is necessary to have a reasonably consistent scheme whereby these parameters are not introduced arbitrarily for each new molecule, and a model has been proposed whereby such a scheme could be realized.[24]

The solute polarizability term (l) in all such calculations is usually related to the refractive index (n_d) by the equation

$$l = 2(n_d^2 - 1)/(n_d^2 + 2) \qquad (13.27)$$

The solute refractive index is usually known or it can be estimated from the variety of additive schemes available.

The correct radius of the solute cavity (a) to use has been the subject of considerable debate in the past. We adopt the simple and explicit model of obtaining the radius directly from the solute molar volume (v_m)

$$4\pi a^3/3 = v_m/N \qquad (13.28)$$

where v_m is given directly from experiment (mol. wt./density) and N is Avogadro's number. This means that the molecular radius is unambiguously defined experimentally.

Note that both l and a are by definition the same for all rotamers of a given molecule; the only parameters which vary are the dipole (k) and quadrupole (h) moments.

In order to evaluate the rotamer dipole and quadrupole moments both the charge distribution and the centre of the molecule must be known. The simplest charge distribution to consider is that due to bond dipoles, and in the model used, these bond dipoles are placed along and at the centre of the polar bonds in the molecule. In cyclic compounds it is often possible to obtain model compounds in a fixed conformation in which the dipole moments can be measured directly. These then provide explicit models on which to base the bond dipoles, provided the molecular geometry is known (cf. § 13.5).

However, in acyclic compounds, the dipole moments of the rotamers are not known accurately, as the observed moments are of course averages and it is necessary to base the bond dipoles used on model compounds with equivalent rotamers. In the original derivation of the model[24,37] the ethyl compounds were used as models for this purpose. Thus the bond dipole for the $-CH_2Cl$ group in 1,2-dichloroethane was taken as 2·03 D, the observed moment of ethyl chloride. Similarly the bond dipole of each $C-Cl$ bond in the $CHCl_2$ group was taken as 1·79 D, the value which reproduced the observed moment of CH_3CHCl_2 of 2·06 D (using a $Cl-C-Cl$ angle of 111°), and so on.

Recent comparisons of the dipole moments obtained from the model with those of the CNDO programme[29] and the observed dipole moments (cf. Table 13.3) suggest that a better model for the CH_2X group is the appropriate methyl compound. Thus the dipole moments now used are CH_2X: $\mu(C-X) = 1·87$ D, $C-CHX-C:\mu(C-X) = 2·03$ and $CHX_2:\mu(C-X) = 1·79$ for X = Cl, Br and for fluorine we use only one bond dipole in all cases: $\mu(C-F) = 1·8$ D (cf. Table 13.1). For this reason there are minor differences between the values of k and h given here and those given previously. These are, however, only minor differences (cf. Table 13.3 for a detailed comparison for 1,2-dichloroethane).

It is still necessary to define the molecular geometry and centre of the molecule, as the quadrupole term, unlike the dipole term, is dependent on the origin of the coordinate system. Originally tetrahedral geometry was used but this has now been superseded by a standarized, but not tetrahedral, geometry in which all C—C—X angles are 111°. The full details are given in Table 13.1. This geometry was adopted for calculations on rotamer energies,[52] for which the differences from tetrahedral geometry are considerable, but used in these calculations essentially for convenience. The difference between this geometry and tetrahedral geometry for these calculations is not very significant (cf. Table 13.3). It also has the advantage of being unchanged for cyclohexanes (all C—C—C angles are equal to 111°).

Table 13.1   Parameters used for DIPQUADMOMS[a]

Atom	C	H	F	Cl	Br	I	N	O	S	Me	tBu	Ph
Atomic number	6	1	9	17	35	53	7	8	16	61	62	63
Van der Waals' radius	1·60	1·20	1·35	1·80	1·95	2·15	1·50	1·40	1·85	2·00[b]	3·00[b]	3·30[b]

Bond lengths

C—C 1·53	C—F 1·36	C—Br 1·93	C—tBu 2·04[c]
C—H 1·09	C—Cl 1·77	C—I 2·13	C—Ph 2·93[d]

Bond angles

(1) C—CH$_2$X, C—CHX$_2$

$$C—C—H = H—C—H = F—C—F = 109·47$$
$$C—C—X = 111·00 \quad (X = C, F, Cl, Br, I)$$
$$X—C—X = 111·00 \quad (X = C, Cl, Br, I)$$

thus

$$C—CH_2X \quad H—C—X = 108·72$$
$$C—CHX_2 \quad H—C—X = 107·10 \quad (X = C, Cl, Br, I)$$
$$C—CHF_2 \quad H—C—F = 107·90$$

(2) C—CX$_3$

$$C—C—X = X—C—X = 109·47 \quad (X = H, C, F, Cl, Br, I)$$

Dipole moments

C—F 1·8	CH$_2$X 1·87	C.CHX.C 2·03	CHX$_2$ 1.79 (X = Cl, Br)

[a] The Van der Waals' radii are incorporated in the programme. All the remaining parameters are input.

[b] Treats Me, tBu and Ph as single atoms with appropriate Van der Waals' radii.

[c] C_1-tBu from C_1 to 'effective' centre of tBu.

[d] C_1-Ph from C_1 to centre of Ph ring.

The final problem is to obtain some estimate of the centre of the molecule, in order to calculate the quadrupole term. The quadrupole term is calculated as the solvation energy of a quadrupole placed at the centre of a spherical solute cavity. Thus the centre of the molecule as far as these calculations are concerned is essentially a volume centre. Originally this was estimated from Dreiding models, but a systematic method of obtaining this was evolved as follows. The volume 'centre of gravity' of any rotamer was obtained by weighting each atom according to its Van der Waals volume and then using the normal centre of gravity equation, i.e. for the x coordinate

$$X = \sum_i x_i V_i / \sum V_i \qquad (13.29)$$

where V_i is the Van der Waals volume of the ith atom. The Van der Waals volumes used are obtained directly from the normal Van der Waals radii, also given in Table 13.1.

A computer programme (DIPQUADMOMS) was written to calculate these parameters. Input for the programme consists of firstly the atomic numbers and the bond distances, angles and dihedral angles needed to specify the geometry of the molecule. Using part of the programme MODEL-BUILDER,[53] the atomic coordinates are then obtained. The second part of the programme utilizes these results to calculate the dipole and quadrupole moments, the values of k and h (in equation 13.18) and the values of the solvation energy $(E^v - E^s)$ for a number of different values of the solvent dielectric constant according to equations (13.18), (13.22) and (13.25). The input required for this is the molar volume (ml), the refractive index, the bond dipole moments to be used, and the values of the solvent dielectric constants required, at the temperature considered.

For convenience it is useful to treat large non-polar groups as single entities rather than to itemize each individual atom, and Table 13.1 gives the equivalent parameters used for the methyl, t-butyl and phenyl groups.

This model has been used for all our calculations, but it should be emphasized that the solvation theory is independent of the model used. Any method of calculating molecular dipole and quadrupole moments can be used immediately with the basic equations, and the validity (or otherwise) of the model does not necessarily reflect on the solvent theory itself.

Finally, we note that the basic equations (13.18) (and 13.22 and 13.25 for the polar term) are intrinsically dependent on temperature via the temperature dependence of the dielectric constant of the solvent. The explicit dependence of b (equation 13.24) on $T^{\frac{1}{2}}$ is incorporated in the variation in dielectric constant with temperature.

Immediately the variation of the dielectric constant of the medium with temperature is known, $\Delta E^v - \Delta E^s$ can be expressed as a function of temperature and therefore the correction term $T(\mathrm{d}H/\mathrm{d}T)$ in equation (13.10) can be calculated. This has been performed in some of the subsequent examples.

13.4 The solvent dependence of rotamer energies

We are now in a position to consider the various experimental results in the light of the theory given in the last section. It is convenient to divide this section arbitrarily into sub-sections dealing with the different types of compound which have been systematically studied in various solvents. In particular we will consider first certain 'basic' compounds which have been very extensively studied and for which all the data needed for the calculations is available. We shall consider these in some detail so as to elucidate, for example, the relative size of the various interactions we are considering, the difference between the true value of ΔE^1 and the value obtained assuming temperature independence etc. The remaining compounds will not be discussed in such depth as the general pattern will be similar to those of these basic compounds. The five compounds are the 1,2-dichloro-, 1,2-dibromo-, 1-bromo-2-chloro-, 1,1,2-trichloro- and 1,1,2,2-tetrachloro-ethanes.

The nomenclature used is straightforward for any 1,2-disubstituted, 1,1,2-trisubstituted and 1,1,2,2-tetrasubstituted ethane as there are only two energetically different rotamers the *trans* (t) and *gauche* (g) forms which are illustrated in Figure 13.2. Note that all subsequent energy differences are given as $E_g - E_t$. The nomenclature for unsymmetric and polysubstituted compounds which are not covered by Figure 13.2 is given subsequently. The definition of the dihedral angle is also given in Figure 13.2.

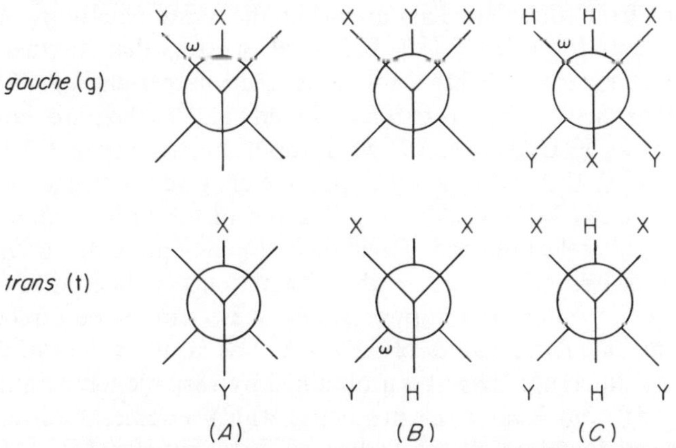

Figure 13.2 Nomenclature for *trans* (t) and *gauche* (g) rotamers of disubstituted (*A*), trisubstituted (*B*) and tetrasubstituted (*C*) ethanes

13.4.1 *Basic compounds—chloro- and bromoethanes*

The five compounds given above have been chosen to illustrate a range of solvent polarity and degree of substitution, and also because all the parameters needed for the calculations are available—in particular the variation of the dielectric constant of the pure liquid with temperature, necessary for the calculation of the dH/dT term (equation 13.10).

All these experimental values are given in Table 13.2 which also gives the values of k and h for these compounds obtained from these parameters and the standardized model detailed in Table 13.1. This together with the symmetry requirements completely defines the geometry of the symmetric rotamers (the *trans*-1,2-, *gauche*-1,1,2- and *trans*-1,1,2,2-ethanes) but not that of the unsymmetric rotamers. The dihedral angles of these rotamers (see Table 13.2) were taken from the results of molecular mechanics calculations which successfully reproduced the observed rotamer energy differences in the vapour phase.[52] Thus, these are essentially vapour phase geometries (see later).

In these and all the subsequent calculations the model used is that of Table 13.1, with the full theory (equations 13.18, 13.22 and 13.25). However, it is of interest to see the consequences of using different models and Table 13.3 gives the results for 1,2-dichloroethane with other possible geometries and dipole moments. These are (a) the standardized model of Table 13.1, (b) as for (a) with $\mu(C—Cl) = 2.03$ D, the original value,[24] (c) as for (a) but with $\mu(C—H) = 0.4$ D and $\mu(C—Cl) = 1.4$ D, the values recommended by Smith[56] and (d) and (e) tetrahedral geometry with $\mu(C—Cl) = 1.87$ and the dihedral angle of the *gauche* form 60° and 70° respectively.

Table 13.3 also gives the calculated dipole moments of the *gauche* isomer, the values of k and h obtained and the calculated values of $\Delta E^v - \Delta E^l$ and $-T(dH/dT)$ from these models, all using the same equations and remaining parameters from Table 13.1. It is vital to stress that the theory only calculates $\delta\Delta E$'s, i.e. the difference between the rotamer energy differences in the various media. Thus from the observed value of the pure liquid dielectric constant at 30°C (Table 13.2) when input into equations (13.18), (13.22) and (13.25) together with the other parameters given in Tables 13.2 and 13.3, the calculated value of $\Delta E^v - \Delta E^l$ given in the fifth column of Table 13.3 is immediately obtained. However, although there are a considerable number of observed values of the rotamer energy differences for dichloroethane in the liquid and vapour states, these cannot be directly compared with the calculated values of $\Delta E^v - \Delta E^l$ as all the values of the liquid-state energy differences have been obtained by variable temperature methods (usually IR and Raman spectroscopy) which assume a constant ΔH (ΔE). These will therefore all give values of the parameter ΔH^0 (ΔE^0) of equation (13.10), and it is necessary to calculate the correction term $T(dH/dT)$ in order to compare the experimental and theoretical values.

Table 13.2 Molecular constants and calculated parameters for haloethanes

Compound		Density[a]	$n_D{}^a$	Molar volume (ml)	l	k (kcal/mole)	h	Temp. (°C)	ε (pure liq.)[b]
CH_2ClCH_2Cl	(g)[c]					3·77	1·81	−10·0	12·83
		1·256	1·4448	78·80	0·5321			30·0	10·08
	(t)					0·00	6·74	50·0	9·04
CH_2ClCH_2Br	(g)[c]					3·61	1·81	−10·0	7·98
		1·739	1·4917	82·47	0·5799			30·0	6·92
	(t)					0·00	6·59	90·0	5·69
CH_2BrCH_2Br	(g)[c]					3·45	1·80	10·0	4·90
		2·180	1·5389	86·19	0·6264			30·0	4·80
	(t)					0·00	6·45	60·0	4·62
$CH_2ClCHCl_2$	(g)					4·46	0·82	−25·0	8·80
		1·4405	1·4706	92·62	0·5586			30·0	6·95
	(t)[d]					0·93	3·73	80·0	5·80
$CHCl_2CHCl_2$	(g)[e]					2·40	1·39	0·0	8·60
		1·5984	1·4944	105·03	0·5826			30·0	7·80
	(t)					0·00	3·74	50·0	7·35

[a] Reference 54.
[b] Reference 55.
[c] $\omega_g = 70°$.
[d] $\omega_t = 50°$.
[e] $\omega_g = 66°$.

Table 13.3　Calculated k and h values and energy differences (kcal/mole) for various models for 1,2-dichloroethane

Description		$\mu(D)$	k	h	$\Delta E^v - \Delta E^l - T(dH/dT)$		$\Delta E^v - \Delta E_0^l$
(a) Standard geom.	(g)	2·86	3·77	1·81	0·77	0·44	1·21
$\mu(C{-}Cl) = 1·87$	(t)		0·00	6·74			
(b) As (a)	(g)	3·10	4·45	2·13	0·91	0·51	1·42
$\mu(C{-}Cl) = 2·03$	(t)		0·00	7·94			
(c) As (a)							
$\mu(C{-}H) = 0·4$	(g)	2·91	3·91	1·71	0·96	0·51	1·46
$\mu(C{-}Cl) = 1·4$	(t)		0·00	6·30			
(d) Tet. geom.	(g)	3·05	4·30	1·31	0·98	0·53	1·51
$\mu(C{-}Cl) = 1·87$	(t)		0·00	6·60			
$\omega_g = 60°$							
(e) As (d)	(g)	2·89	3·85	1·69	0·82	0·46	1·28
$\omega_g = 70°$	(t)		0·00	6·60			

This is simply done by calculating ΔE^l at two other temperatures ($-10°$C and $50°$C using the dielectric constants given in Table 13.2) and replacing $T(dH/dT)$ by $T(\Delta H/\Delta T)$. This term is always negative as the dielectric constant of the medium always decreases with increasing temperature, and is given in column 6 of Table 13.3. The theoretical value of $\Delta E^v - \Delta E_0^l$, which can be directly compared to the observed data is given simply (equation 13.10) as $\Delta E^v - \Delta E^l - T(dH/dT)$, i.e. by the sum of columns 5 and 6, and is given also in Table 13.3.

The results in Table 13.3 are of some interest. The calculated values of the dipole moment of the *gauche* rotamer given in Table 13.3 may be compared with the calculated moment (CNDO)[29] of 3·28 D and the moment obtained by variable temperature dipole moment studies of 2·55 D.[5] (It has been noted that the CNDO dipole moments for chlorocarbons are generally larger than observed.) It should also be noted that the models (a) and (c) are very similar, i.e. the model of one $C{-}Cl$ dipole to represent a CH_2Cl group gives very much the same answer as the more complex model of $C{-}H$ and $C{-}Cl$ dipoles. This is further substantiated by the comparison of the calculated energy differences given in the table. The generally accepted values of ΔE^v and ΔE_0^l for this compound are 1·20 ($\pm 0·05$) and 0·00 ($\pm 0·06$) kcal/mole[7] which can be directly compared with the results of the last column in Table 13.3. The standard model used does give the best answer, but it is of importance to note that all the models give satisfactory answers, and in particular the picture of separate $C{-}H$ and $C{-}Cl$ dipoles is entirely compatible with the theoretical treatment given here. Also we note that the use of tetrahedral geometry makes very little difference, provided the dihedral angle in the *gauche* form is known. This is in complete contrast to the calculation of the actual (vapour state) rotamer energy differences by molecular mechanics, in which the assumption of non-tetrahedral geometry is crucial.[52] It is more convenient for us to use the same geometry for both calculations, but the results in Table 13.3 do prove conclusively that the theory may be used with tetrahedral geometry without any appreciable error.

The results of Table 13.3 do provide some insight into an interesting suggestion originally due to Le Fevre and coworkers[57] and more recently revised by Pachler and Wessels[58] that the molecular geometry of the *gauche* rotamer may alter with solvent. The suggestion of Le Fevre was based on the observation of a non-linear concentration-dependent parameter (the Kerr effect) which could be due to a variety of causes. However, Pachler and Wessels considered in some detail the reasons for the discrepancy between the results of variable temperature NMR studies (of 1-chloro-2-bromoethane) in the gas phase and variable solvent studies. The suggestion is that the *gauche* rotamer dihedral angle may decrease in polar solvents. It is an implicit assumption of all solvent studies that the molecular geometry remains unchanged, thus it is pertinent to consider the suggestion quantitatively. It is easy to visualize that decreasing the dihedral angle of the *gauche* form and thus increasing the total dipole moment of the molecule will further stabilize the molecules and this effect will be most pronounced in polar solvents. However, the dipole term is compensated to a very large extent by the quadrupole term (cf. Table 13.4) and thus will considerably overestimate the gain in energy.

Table 13.4 Observed and calculated rotamer energy differences (kcal/mole) for 1,2-dichloroethane in various solvents

Solvent	ε	Contributions (*gauche–trans*)			ΔE^1	
		Dipole term	Quad. term	Polar term	(calc.)	(obs.)
Vapour	1·0	0·00	0·00	0·00	(1·20)	1·20
Cyclohexane	2·0	0·85	−0·61	0·00	0·97	0·91
Ethylene tetrachloride	2·3	1·00	−0·72	0·00	0·92	0·89
Benzene-d	2·3	1·00	−0·72	0·00	0·92	0·60
p-Xylene	2·3	1·00	−0·72	0·00	0·92	0·70
Carbon disulphide	2·6	1·13	−0·80	0·01	0·87	0·83
Diethyl ether	4·3	1·59	−1·09	0·02	0·68	0·69
Ethyl acetate	6·0	1·83	−1·23	0·03	0·57	0·42
Pure liquid	10.1	2·10	−1·39	0·06	0·43	0·31
Mesityl oxide	15·0	2·24	−1·47	0·09	0·33	0·47
Acetone	20·7	2·33	−1·51	0·13	0·26	0·18
Acetonitrile	36·0	2·43	−1·57	0·22	0·12	0·15

Table 13.3 gives a clear answer. Comparison of examples (d) and (e) of the table shows that the relatively large change in the *gauche* dihedral angle from 70° to 60° only gains 0·16 kcal/mole solvation energy in the pure liquid. For the standard geometry (a), which may be used for both solvation energy and molecular mechanics calculations, the gain in solvation energy of the *gauche* rotamer on going from 70° to 65° is 0·07 kcal/mole for the pure liquid and 0·09 kcal/mole in acetone solution. The steric repulsion energy however, increases by 0·11 kcal/mole. Thus although the possibility that some molecular deformation can occur for these compounds in polar solvents still exists, the energy increments are so small that it will be relatively small. However, in more polar molecules with lower barriers to rotation this effect may well be appreciable.

The results of Table 13.3 shows that the variation in the rotamer energy from the vapour to the pure liquid 1,2-dichloroethane is given satisfactorily by the theory.

A related problem is whether the variation in different solvents is also reproduced. Two experimental problems arise for this case. Almost all the direct estimates of the rotamer energy difference in various solvents have been by variable temperature measurements and have assumed a constant energy difference. These therefore give, in precisely the same manner as for the pure liquid, a value of ΔE_0^s (s = solvent), and the correction term $-T(\mathrm{d}H/\mathrm{d}T)$ needs to be calculated. This will be considered later.

Estimates of the rotamer energy differences by measurements at one temperature removes this problem, but produces other disadvantages. In particular the NMR method does not directly measure the rotamer energy difference but merely one average parameter (usually a coupling constant). Thus both the rotamer energy difference and the values of the couplings in the distinct isomers are parameters obtained from the best fit of the observed data. It is therefore more appropriate to consider these results in the form of observed versus calculated values of the couplings and this will be done subsequently.

The IR method does not suffer from this limitation in that the rotamers give rise to separate bands, but no estimate of the energy difference is possible without some assumptions concerning the integrated absorption coefficient. However, recently two systematic studies of the solvent dependence of the IR frequencies and intensities of a number of haloethanes, including 1,2-dichloroethane, have been performed and it is of interest to utilize these results in order to obtain comparative values of the rotamer energy differences. In these studies Oi and Coetzee[59] recorded the intensity ratios of two pairs of bands, from the *trans* and *gauche* rotamers respectively in various solvents, and El Bermani and coworkers[60] used these values, together with their measurements of the corresponding vapour-state values and converted the data to a logarithmic scale. If the assumption is made that the ratios of the integrated absorption coefficients for the *trans* and *gauche* rotamers are independent of solvent (note that this is not the same as assuming that the individual rotamer absorption coefficients are solvent independent), then this data can be immediately transferred into relative rotamer energies by the equation

$$\Delta E^s - \Delta E^t = RT \log_e 10[\log_{10}(A_t/A_g)^s - \log_{10}(A_t/A_g)^t] \qquad (13.30)$$

This does not give a value of either ΔE^s or ΔE^t. However, as the intensity ratios for the vapour state were measured, then equation (13.30) together with the value of ΔE^v of 1·20 immediately gives values for the energy differences for 1,2-dichloroethane in all the solvents measured. These results can be compared with the corresponding calculated values in Table 13.4, which also gives a breakdown of the calculated solvation energy for this molecule into the three terms of the fundamental equation. These are the dipole term, the quadrupole term (equation 13.18) and the polar term (equations 13.22 and 13.25). These contributions when totalled subtract from the vapour-state energy to give the calculated values.

The comparison between the observed and calculated values of the rotamer energy difference is excellent. Apart from the aromatic solvents benzene and *p*-xylene which are known exceptions, the largest deviation of the calculated and observed energies is 0·15 kcal/mole and the average deviation over the nine solvents 0·07 kcal/mole, which is of the order of the experimental error in the measurements. Thus over a wide range of dielectric constants of the medium (2–36) the theory gives good accord with experiment. It is of particular interest that this direct method gives a value of ΔE^l for the pure liquid of 0·35 kcal/mole, in very good agreement with the calculated value (0·47) and quite different from the value obtained by variable temperature measurements (0·0). This provides very considerable support for the validity of this correction term.

In a recent paper, Tanabe[61] obtains a value for ΔE^l of 0·0 ($\pm$0·10) kcal/mole from similar measurements using a complex minimization procedure based on a virtually identical assumption that 'all intensity parameters have equal values for the different rotamers'. This value is however compared with the previous values of ΔE^l_0 to give satisfactory agreement. As there are 22 force constants and 21 electro-optical parameters to be obtained from the observed spectra the process may not be overdetermined and therefore the value of ΔE^l is subject to some uncertainty. Indeed when only values of the rotamer energy differences are required there would appear to be very little advantage to be gained in using this very complex treatment over the simple method given here.

The other point of interest in Table 13.4 is the comparison of the different terms which contribute to the total solvation energy difference. The dipole term, which for this molecule is only non-zero for the *gauche* isomer, is compensated to a large extent by the quadrupole term, which favours the *trans* rotamer. (Both rotamers have quadrupole moments but that of the *trans* is far larger, cf. Table 13.2). This is particularly noteworthy in non-polar solvents, e.g. in cyclohexane the calculated solvation energy difference is only 0·25 kcal/mole in favour of the *gauche* rotamer compared to the value of 0·85 kcal/mole from the dipole term alone. Although this is a special case in that one rotamer has zero dipole moment, it is generally true that estimates of relative rotamer stabilities based only on the dipole moments of the individual rotamers can be seriously in error. Finally we note that the polar term is always small and only becomes appreciable in very polar solvents, as expected.

It is convenient to consider in detail the comparison between the values of the rotamer energy difference obtained by variable temperature measurements (ΔE_0^s) and the true energy difference (ΔE^s) in any solvent. The difference, i.e. the correction term, $-T(\mathrm{d}H/\mathrm{d}T)$, can be calculated immediately for any solvent simply from the variation of the dielectric constant with temperature. This has been done for dichloroethane in four solvents covering a wide range of dielectric constant, viz n-pentane, chloroform, CH_2Cl_2 and acetonitrile. The calculated value of the correction term in these four solvents was 0·07, 0·34, 0·33 and 0·36 kcal/mole respectively. The correction is negligible for n-pentane, and by inference for all other non-polar solvents. This has the important practical consequence that any variable temperature study in dilute non-polar solvents will not be subject to any significant error due to this term. As expected the correction term increases with solvent polarity but is not in any sense linear, being essentially the same for the last three solvents. This means that any variable temperature measurement of this equilibrium in these solvents which assumes a constant rotamer energy difference will give a value 0·3 kcal/mole less than the true value, with a consequent error in any other parameter determined from the measurements.

We note also that the correction term for the pure liquid (Table 13.3) is significantly larger than that of any other solvent. Large correction terms have been calculated for some of the pure liquids of the other standard compounds considered here (e.g. 1,1,2-trichloroethane = 0·49, Table 13.5). These large correction terms are an immediate consequence of the large variation in the dielectric constant of the pure liquids with temperature, and could well be connected with the rotational isomerism exhibited by these molecules. It is unfortunate that the medium which is experimentally the easiest to study at variable temperatures, i.e. the pure liquid, gives results which are the most seriously in error when the correction term is applied. The results of NMR studies will be considered in more detail for the unsymmetric compounds as the data for these is more accurate and extensive. However, it should be noted that the calculated slope of the coupling constant versus temperature curve for pure liquid 1,2-dichloroethane is *reversed* in sign from that with a constant energy difference, when the correction term is included, and only in the latter case does this agree with experiment.[62]

There are not many results of variable temperature studies in different solvents to compare with those of Table 13.4. Wada[36] collected the data available at that time for 1,2-dichloroethane. In n-hexane and n-heptane ΔE_0^s was 0·92 and 0·98 ($\pm$0·1) kcal/mole in precise agreement with the cyclohexane value in Table 13.4 (0·94) confirming the lack of any correction term. In carbon disulphide the results are 0·36 (IR) and 0·53 (dipole moment) in less good agreement with each other and the result in Table 13.4 (0·83). In methyl alcohol solution ($\varepsilon = 33\cdot6$) the IR method gave $\Delta E_0^s = -0\cdot26$ ($\pm$0·1) kcal/mole and this is in very good agreement with the predictions. (In acetonitrile the values of ΔE^s and $-T(\mathrm{d}H/\mathrm{d}T)$ of 0·15 and 0·36 kcal/mole give a value of ΔE_0^s of $-0\cdot21$). This agreement should be considered with some reserve, as methyl alcohol is not strictly a correct solvent to apply to the theory due to extensive hydrogen bonding. However, the results do in general confirm the theoretical predictions of the correction term.

The results for the remaining compounds considered in this section may be treated in a similar manner to those of 1,2-dichloroethane, although it is not necessary to consider them in such detail. Table 13.5 gives the observed

Table 13.5   Observed and calculated rotamer energy differences (kcal/mole) for haloethanes in the vapour and liquid states

Compound	Calculated values		
	$\Delta E^v - \Delta E^l$	$-T(\mathrm{d}H/\mathrm{d}T)$	$\Delta E^v - \Delta E_0^l$
CH_2ClCH_2Cl	0·77	0·44	1·21
CH_2ClCH_2Br	0·67	0·30	0·97
CH_2BrCH_2Br	0·54	0·12	0·66
$CH_2ClCHCl_2$	1·11	0·49	1·60
$CHCl_2CHCl_2$	0·70	0·23	0·93
	Observed values[a]		
	ΔE^v	ΔE_0^l	$\Delta E^v - \Delta E_0^l$
CH_2ClCH_2Cl	1·20	0·00	1·20
CH_2ClCH_2Br	1·46[b]	0·46[c]	1·00
CH_2BrCH_2Br	1·64[d]	0·74	0·90
$CH_2ClCHCl_2$	3·0; 2·0[e]	0·25	2·75; 1·75
$CHCl_2CHCl_2$	0·0	$-1\cdot1$	1·1

[a] From References 6 and 7 unless stated otherwise.
[b] 1·43 (Reference 62a), 1·49 (Reference 62b).
[c] 0·49 (Reference 63), 0·42 (Reference 62b).
[d] Average of values quoted in Reference 61.
[e] 2·3 (Reference 64), 2·0 (Reference 37), 1·9 (Reference 17).

values of the rotamer energy differences in these compounds in the vapour and liquid states and these are compared with the calculated differences after including the correction term. (For completeness the results for the 1,2-dichloroethane are also included.) Comparison of the observed and calculated values of $\Delta E^v - \Delta E_0^l$ (columns 4 and 7) shows generally good agreement; for the 1,2-dichloro, 1-chloro-2-bromo and 1,1,2,2-tetrachloro the agreement is within the experimental error of the measurements, for the 1,2-dibromo there may be some discrepancy in the two values but there is also some uncertainty about the value of ΔE^v, variously quoted as 1·54 or 1·77 kcal/mole. In the case of the 1,1,2-trichloroethane the values of ΔE^v fall into two groups. The earlier data (mainly IR) gave several values of about 3·0 kcal/mole but more recent determinations suggest a much lower value of about 2·0 kcal/mole. Some of the latter values are extrapolated values from solution measurements[17,37] and are therefore not entirely independent of the theory. However, the value of 3·0 would seriously affect the validity of the treatment and is also possibly on the limit of accurate measurements by this technique (vapour-phase variable temperature IR).

The extensive work of Oi and Coetzee[59] and El Bermani and coworkers[60] on the measurement of the solvent dependence of the relative intensities of the IR bands of the different rotamers has included 1,2-dibromo- and 1,1,2,2-tetrachloroethane and their results can be treated in precisely the same manner as in the case of 1,2-dichloroethane to give relative rotamer energies, i.e. $\delta\Delta E$ values. A comparison of these results with the calculated rotamer energies is given in Table 13.6.

In the case of dibromoethane the measurements included the values for the vapour and pure liquid and thus the energies can be directly related to ΔE^v. This has been done in Table 13.6 and inspection shows generally reasonable agreement between the observed and calculated rotamer energies, apart of course from benzene which gives an amazing value of ΔE^s, equivalent to a solvent polarity equal to that of methyl thiocyanate. The range of the calculated energies is slightly greater than observed and the pure liquid value is also significantly lower than the calculated value, suggesting that the discrepancy noted in Table 13.5 is genuine. However, the difference between the observed values of ΔE^l (0·86) and ΔE_0^l (0·74) exactly equals the calculated value of the correction term (0·12) which is encouraging.

The intensity ratio data for the 1,1,2,2-tetrachloroethane did not include the ratios for either the vapour or pure liquid states and the data are therefore $\delta\Delta E$ values. The observed data in Table 13.6 have been obtained by fixing the scale at one point (at the value for diethyl ether) with the calculated energies. Thus the significance of the agreement between the observed and calculated energies lies in the variation between the different solvents. It can be seen that there is virtually complete agreement between the observed and calculated values.

Table 13.6   Observed and calculated rotamer energy differences (kcal/mole) for
1,2-dibromo- and 1,1,2,2-tetrachloroethane in various solvents

Solvent	ε	Energy difference $\Delta E^s(E_g - E_t)$			
		CH_2BrCH_2Br		$CHCl_2CHCl_2$	
		(obs.)	(calc.)	(obs.)	(calc.)
Vapour	1·0	1·64	(1·64)	—	(0·0)
C_6H_{12}	2·0	—	—	−0·32	−0·25
Pentene	2·1	—	—	−0·36	−0·27
CCl_4	2·2	1·31	1·40	—	—
C_2Cl_4	2·3	1·24	1·38	−0·27	−0·30
Benzene	2·3	0·69	1·38	−0·57	−0·30
p-Xylene	2·3	—	—	−0·52	−0·30
Mesitylene	2·3	—	—	−0·44	−0·30
CS_2	2·6	1·11	1·33	−0·38	−0·35
Et_2O	4·3	—	—	−0·53	−0·53
EtAc	6·0	—	—	−0·68	−0·63
Pure liquid	4·8	0·86	1·10	—	—
Acetone	20·7	—	—	−0·83	−0·91
CH_3SCN	35·0	0·69	0·58	—	—
CH_3CN	36·0	0·66	0·57	−0·81	−1·02
CH_3NO_2	36·7	—	—	−0·91	−1·02

The other series of extensive measurements which have been performed
to investigate the solvent dependence of the rotational isomerism in these
compounds have used the NMR technique. This is the static NMR method
(§ 13.2.1) as the energy barriers to interconversion of the rotamers are too
low in these compounds to observe the separate rotamers by NMR. The
observed parameter in nearly all cases is the observed (average) coupling
constant between the vicinal protons (CH.CH), measured as a function of
solvent and/or temperature.

All the standard compounds have been extensively investigated by this
technique.[17,22,24,37,62] However, there are considerable experimental
difficulties for the symmetric compounds as the coupling can only be
observed via the ^{13}C satellite spectrum. This necessitates the use of relatively
concentrated solutions and even then the spectra are subject to signal-to-
noise limitations. In contrast the couplings in the unsymmetric compounds
(the 1-chloro-2-bromoethane and the 1,1,2-trichloroethane) can be measured
with relative ease and high accuracy and thus the data for these compounds
are both more extensive and accurate. As the same phenomenon is illustrated
in all the compounds we shall only consider the two unsymmetric ones in
detail.

In 1,1,2-trichloroethane there is only one observed averaged CH—CH
coupling constant due to the fact that the unsymmetric rotamer has two

mirror-image forms (cf. Figure 13.1). Thus from the point of view of high-temperature NMR experiments the unsymmetric rotamer may be considered as one averaged rotamer (with of course a statistical weight of 2) and in consequence the coupling in this rotamer obtained by these experiments is the average of the two possible couplings. In contrast in the symmetric (*gauche*) rotamer there is by symmetry only one CH—CH coupling and this is obtained uniquely from the experiments.

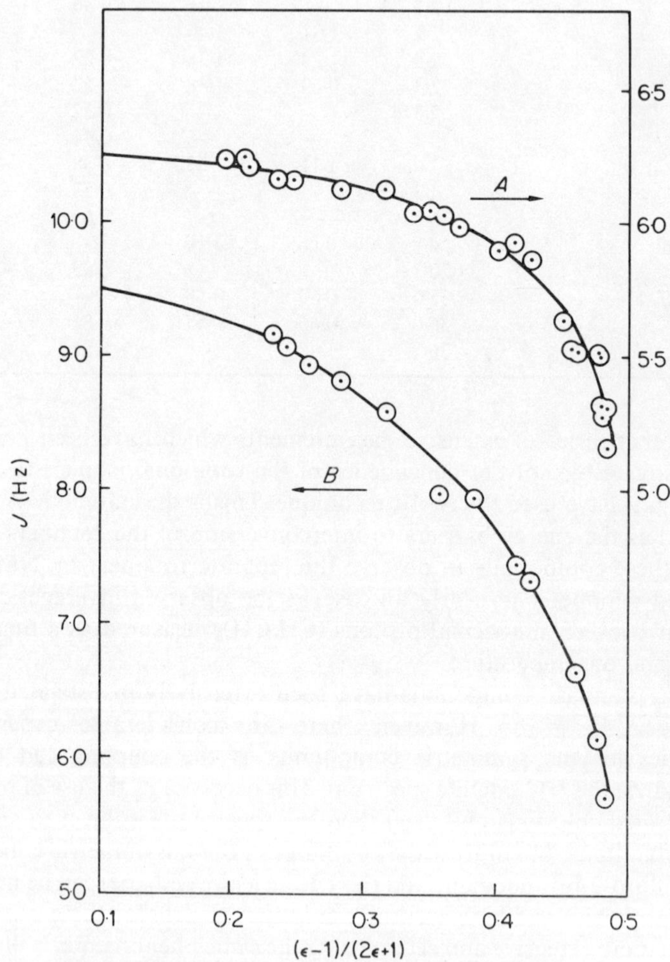

Figure 13.3 The solvent dependence of the vicinal proton coupling constants in 1,1,2-trichloroethane (*A*) and *meso*-2,3-dibromobutane (*B*) with the calculated best-fit curves

The observed coupling has been measured over a variety of solvents and temperatures.[17,37,41] The values of the coupling in a variety of solvents at room temperature from Reference 41 are given in Table 13.7, together with the measured solution dielectric constants. It should be noted here

Table 13.7 Proton coupling constants and calculated rotamer energy differences for 1,1,2-trichloroethane in various solvents[a]

Solvent	ε (soln.)	J (Hz) (obs.)	J (Hz) (calc.)	ΔE^s (kcal/mole)[c]
n-Hexane[b]	1·95	6·24	6·23	1·28
Pentane	2·07	6·25	6·22	1·25
n-Hexane	2·10	6·21	6·22	1·24
Decalin	2·32	6·17	6·20	1·17
CCl_4	2·45	6·17	6·19	1·14
Benzene	2·60	5·94	—	0·63
CS_2	2·95	6·13	6·15	1·03
CCl_2CHCl	3·55	6·13	6·10	0·92
i-Pr_2O	4·10	6·04	6·07	0·84
Et_2O	4·50	6·05	6·04	0·79
$CHCl_3$	4·90	6·03	6·02	0·75
$C_7H_{16}Br$	5·38	5·99	5·99	0·71
Pure liquid	7·15	5·92	5·91	0·58
CH_2Cl_2	8·25	5·93	5·87	0·52
CH_2ClCH_2Cl	9·75	5·87	5·82	0·46
Mesityl oxide	15·0	5·64	5·68	0·32
CH_3COEt	17·3	5·53	5·64	0·27
Acetone	19·2	5·52	5·61	0·24
CH_3CN	32·0	5·52	5·42	0·08
CH_3NO_2	32·6	5·51	5·41	0·07
CH_3CN[b]	34·4	5·52	5·39	0·05
DMF	35·0	5·34	5·38	0·05
CH_3CONMe_2	35·8	5·28	5·37	0·04
Sulfolane	42·2	5·31	5·29	−0·02
DMSO	44·9	5·17	5·26	−0·04

[a] 10% solutions.
[b] 5% solutions.
[c] With $\Delta E^v = 1·9$ kcal/mole.

that due to the relative insensitivity of the NMR method, it is usual practice to work with 5 or 10% solutions, in which case the solution dielectric constant can be very different from that of the pure solvent (cf. Tables 13.6 and 13.7). The coupling varies from 6·24–5·17 Hz with a quoted accuracy of measurement of ±0·03 Hz.

From the data of Table 13.2 and the fundamental equations (13.18), (13.22) and (13.25) the rotamer energy difference in any solution is given directly from the value of ΔE^v and the solution dielectric constant.

The observed coupling (J) is merely the weighted average of the couplings in the distinct rotamers (cf. equation 13.2) and thus the plot of J versus ε is a three-parameter fit in the unknown quantities ΔE^v and the couplings in the two rotamers. Such three-parameter fits are notoriously insensitive to the variation of the different parameters. For example in this case, all solutions with ΔE^v from 1·7 to 2·7 kcal/mole give the same r.m.s. error (to within 0·01 Hz), though of course the associated values of the rotamer couplings vary widely. These various possibilities can be restricted by making use of the known value of $\Delta E_0^1 = 0.25$ kcal/mole for which there is general agreement (Table 13.5). This with the calculated value of $T(dH/dT)$ gives $\Delta E^1 = 0.74$ and thus $\Delta E^v = 1.9$ kcal/mole, which with the observed couplings gives values of the couplings in the two isomers of 6·40 Hz(*trans*) and 2·30 Hz(*gauche*). The r.m.s. error between the observed and calculated couplings for these 24 solutions was 0·06 Hz, roughly the combined errors in the dielectric constants and coupling constant measurements. The observed versus the calculated couplings are given in Figure 13.3 and Table 13.7 and the calculated ΔE values for the different solutions from the results are also given in Table 13.7. (The figure is from Reference 41 in which the curve was calculated on the basis of the earlier theory with in consequence slightly different values of the couplings. The difference in the calculated values on the two schemes is negligible.)

The agreement between the observed and calculated couplings is of interest and worthy of comment. The range of dielectric constants used was 2·0–44·9, i.e. over the total normal range of solvent dielectric constants. The agreement, whilst providing strong support for the general validity of the theory, does not of course define it unambiguously simply because the couplings in the rotamers are not known independently. However, what the agreement does indicate unambiguously is that *the* only property of the solvent which directly affects the proportions of the rotamers is the dielectric constant or any related property. If any other molecular parameter was affecting the rotamer energies then the observed plot of coupling versus dielectric would be discontinuous. This is of course precisely the case for the exceptional solvents such as benzene mentioned previously. In Table 13.7 the quoted value for benzene is the value of the energy difference obtained from these measurements. As usual it is comparable with the values for much more polar solvents. A more comprehensive study of the variation in the coupling constants of this molecule (and of the analogous tribromo- and the 1,1,2,2-tetrachloro- and tetrabromoethanes) with solvent and temperature has been given by Heatley and Allen.[17] Their results were again given a reasonable quantitative explanation on the basis of a treatment analogous to the above.

The results of the NMR measurements on 1-chloro-2-bromoethane may be treated in exactly analogous fashion. In this compound there are two different CH—CH couplings, which are essentially a *cis* (J) and a *trans* (J')

Table 13.8 Proton coupling constants and rotamer energies for
1-chloro-2-bromoethane in various solvents

Solvent	ε (soln.)[a]	Coupling (Hz) (obs.)		(calc.)	ΔE^s (kcal/mole)[c]
		J	J'	J'	
n-Pentane	2·06	5·56	10·88	10·82	1·17
n-Hexane	2·23	5·54	10·63	10·71	1·14
CCl_4	2·43	5·64	10·54	10·60	1·11
Benzene	2·7[b]	5·96	9·12	—	0·70
CS_2	2·81	5·67	10·36	10·40	1·05
$CHClCCl_2$	3·54	5·71	10·23	10·07	0·96
$CDCl_3$	4·74	5·81	9·59	9·66	0·86
CH_2Cl_2	8·63	5·96	8·98	8·85	0·67
Acetone	19·51	6·22	7·79	7·92	0·47
CH_3CN	34·14	6·24	7·63	7·29	0·34
DMF	35·16	6·34	7·13	7·25	0·33
DMSO	44·0	6·40	6·80	6·97	0·27
Pure liquid	6·92	5·91	9·10	9·13	0·74

[a] 5% solutions.
[b] 10% solution.
[c] With $\Delta E^v = 1·40$ kcal/mole.

coupling and these are given in Table 13.8 as a function of the solvent. (This
data is somewhat better than previous published data[24] being from more
accurate measurements of more dilute solutions.[65]) In one case (J) the
coupling changes very little with solvent and this is due to the similar values
of this coupling in the two rotamers. (As in the trichloroethane, the couplings
obtained from such measurements for the *gauche* isomer are averages over
the two mirror-image forms.) In contrast the variation of J', the *trans*
coupling, is large, 10·82–6·97 Hz, and this therefore provides a good test
of the technique. The couplings can be treated as in the case of trichloroethane, i.e. computationally searching for the value of ΔE^v which results in
the lowest r.m.s. error between the observed and calculated couplings.
Again this gives a range of values within the accepted experimental error
and it is preferable to use the measured values of ΔE^v and ΔE_0^1 (together
with the calculated $-T(\mathrm{d}H/\mathrm{d}T)$ term, Table 13.5) to give the final answer.
We use $\Delta E^v = 1·4$ kcal/mole and $\Delta E^1 = 0·74$ kcal/mole and these values
give the best-fit values of the couplings in the *trans* and *gauche* rotamers as
$J = 5·1$ and 7·4 Hz and $J' = 13·3$ and 2·0 Hz. The calculated values of J'
and ΔE for the various solvents are given in Table 13.8, and there is again
good agreement between the observed and calculated values of this coupling.
(Because the variation in J is so small this coupling does not provide a
critical test of the method and the calculated values are not given). Only one
coupling is in error by more than 0·2 Hz in a measured range of 4 Hz (with
of course the exception of benzene) and the r.m.s. error is 0·15 Hz.

It is of interest to compare the values of the couplings obtained here with those obtained from earlier solvent studies using the basic formula (equation 13.18) and including the dipole–dipole term (§ 13.3.2). The comparable couplings were obtained in the first case as 5·1, 7·1, 13·4 and 3·0[24] and in the second case as 5·1, 7·3, 13·3 and 3·3.[66] The results are all in reasonable agreement except for the last value, J' in the *gauche* isomer. The reason for this is that this value is essentially the 'observed' value for the limiting case of a solvent of infinitely high dielectric constant, in which the solute will be all in the *gauche* form. The successive approximations made in the theory are most noticeable in this region of very high dielectric constant and therefore affect the extrapolation to the *gauche* coupling most seriously. In contrast the extrapolation to very low dielectric constants, in which the *trans* isomer predominates, is much less affected by the successive approximations considered and in consequence the *trans* coupling is much more consistent.

In conclusion, this section has shown that the results obtained by the application of the two most general techniques for studying solvent effects, i.e. IR and NMR spectroscopy, can be given a reasonable quantitative explanation by the theory outlined. The large number of substituted ethanes also investigated in this manner, though not to the same depth as for these standard compounds may be conveniently divided into two groups. The first group comprises those compounds which are covered by direct application of the standard parameters of Table 13.1. These include the simple haloethanes, propanes, butanes, etc. and will be considered quantitatively in the next section. The second group which comprises all other substituted ethanes will be discussed in the light of the theory, but in a qualitative manner, in the following section.

13.4.2 *Halogenated alkanes*

The standard parameters described in Table 13.7 combined with the basic equations (13.18), (13.22) and (13.25) may be used essentially unchanged for all substituted ethanes with substituent groups covered by Table 13.7. This includes all the halogens (though iodine is not itemized in the table due mainly to the absence of sufficient available data to test the parameters, but it will be considered later in a specific example) and alkyl substituents which on our model merely act as space-filling groups with zero dipole moment. This may be an oversimplification as we shall see later. Although phenyl substituents could be included, phenylsubstituted ethanes will be dealt with separately, as there is some evidence that other interactions occur with phenyl groups (§ 13.4.4). Although any haloethane could be included, the standard model given in Table 13.1 is strictly only applicable to ethanes in which the halogen atoms are directly bonded to the central C—C bond. For halogen atoms further away from this axis, e.g. CCl_3 group, further consideration would need to be given to the choice of bond moments for the calculation of the molecular dipole and quadrupole moments. (This has been done for the acyclic compounds considered later, cf. § 13.5.) However, the parameters in Table 13.1 can still be directly used for a large variety of substituted ethanes and the results for a selection of such compounds are given in Table 13.9.

Table 13.9 gives the values of $k(k_g - k_t)$ and $h(h_g - h_t)$ which are calculated directly from the parameters of Table 13.1, using the values of the dihedral angles of the *gauche* (non-symmetric) rotamers (ω_g) obtained as previously from steric energy calculations.[52] These are given in the table. The remaining quantities needed for the calculation of k and h (the densities and refractive indices) are taken from the literature (cf. Reference 54 and the original references) and omitted for clarity. The next three columns give the values of $\delta\Delta E^s$ ($\Delta E^v - \Delta E^s$) obtained by inserting the values of k and h into the basic equations with values of the solvent dielectric constant of 2·0, 20·0 and of the pure liquid, if known. The latter can be directly compared with the observed values of ΔE^v, ΔE^l and/or ΔE_0^l given in the remaining columns. In order to simplify the format, where a number of different experimental techniques have been used, the individual results are given in the footnotes and merely the average value recorded in the table. A number of these energy differences have been obtained by the variable solvent method but using the older versions of the theory with slightly different parameters to Table 13.1. In these cases, the results have for consistency been recalculated using the standard parameters and basic equations given here. This only makes small differences to the final values of the energy differences from those quoted previously in the references given, but does remove some minor inconsistencies (see later).

The first four compounds in Table 13.9 complete the possible range of 1,2-dihaloethanes and may be profitably considered along with the more detailed results for the chloro and bromo compounds (Table 13.5).

Inspection of the results shows that for a medium of given dielectric constant the solvation energy difference ($\delta\Delta E$) of all the 1,2-dihaloethanes is remarkably constant. For a medium of dielectric constant 20 the calculated solvation energy difference for 1,2-difluoroethane is only 20% larger than that for 1,2-diiodoethane. This is because (a) the C—X bond dipole moments are very similar for all the haloethanes and (b) there is again considerable balancing between the dipole and quadrupole terms (cf. Table 13.4). For example the actual dipole term for difluoroethane (i.e. the k value) is 50% larger than for diiodoethane reflecting the much smaller molecular volume of the former, but this is considerably balanced by the larger quadrupole term.

The major reason for the differences in the values of $\Delta E^v - \Delta E^l$ in this series is thus simply due to the considerable variation in the dielectric

Table 13.9　Observed and calculated rotamer energy differences (kcal/mole) in haloethanes

Molecule	ε_1	ω_g (deg.)	$k(k_g - k_t)$	$h(h_g - h_t)$	Calculated $\delta\Delta E^v(\Delta E^v - \Delta E^s)$			Observed		
					$\varepsilon = 2.0$	$\varepsilon = 20.0$	Pure liq.	ΔE^v	$\Delta E^l(\Delta E_0^l)$	$\Delta E^v - \Delta E^l(\Delta E^v - \Delta E_0^l)$
CH_2FCH_2F	34.4	65	4.78	−5.33	0.37	1.24	1.42	−0.3	−2.0[a]	1.7
CH_2FCH_2Cl	20.7	64	4.37	−5.15	0.32	1.16	1.17	0.4	−0.65(−1.0)[b]	1.1(1.4)
CH_2FCH_2Br	16.3	64	4.12	−4.86	0.30	1.16	1.11	0.5	−0.3(−0.9)[c]	0.8(1.4)
CH_2ICH_2I	—	70	3.09	−4.11	0.25	1.0	—	2.5	2.1[d]	—
CH_2FCHF_2	9.0	56	4.56	−4.33	0.44	1.38	1.13	1.2	0.1[e]	1.1
CHF_2CHF_2	7.0	62	3.78	−3.49	0.37	1.15	0.87	1.2	0.3(<0.4)[f]	0.9(>0.8)
$CH_2BrCHBr_2$	—	48	3.21	−2.41	0.44	1.57	—	2.0	0.9(0.5)[g]	1.1(1.5)
$CHBr_2CHBr_2$	7.0	66	2.16	−2.12	0.24	0.94	0.69	0.2	−0.4(−0.8)[h]	0.6(1.0)
$CH_3CH_2CH_2Cl$	7.7	70	0.00	−0.66	−0.08	−0.22	−0.19	−0.1	(−0.2)[i]	(0.1)
$CH_3CH_2CH_2Br$	8.1	68	0.00	−0.57	−0.07	−0.20	−0.16	−0.2	(−0.5)[j]	(0.3)
$CHCl_2CFCl_2$	—	64	−0.18	−0.13	−0.06	−0.17	—	−0.8	(−0.4)[k]	(−0.4)
$CFCl_2CFCl_2$	—	64	0.01	−0.08	0.01	0.02	—	<0.3	0.12[l]	0.2
$CH_2ClCHClCH_3$	8.93 (g$_1$)	62	3.57	−3.68	0.34	1.20	0.97	1.9	(0.9)[m]	(1.0)
	(g$_2$)	292	3.35	−3.64	0.29	1.05	0.85	1.2	(0.2)[m]	(1.0)
CH_2ClCMe_2Cl	7.15	66	3.27	−3.14	0.34	1.17	0.89	1.26	(0.0)[n]	(1.3)
meso-$CH_3CHClCHClCH_3$	8.93	66	3.27	−2.19	0.33	1.16	0.95	1.3	0.84(0.64)[o]	0.5(0.7)
meso-$CH_3CHBrCHBrCH_3$	5.93	68	2.95	−3.18	0.27	1.01	0.70	1.6	0.75(0.45)[p]	0.8(1.1)

[a] $\Delta E^v < 0.2$ (IR), −0.6 (NMR); ΔE^l − 2.0 (NMR); References 67, 68.

[b] ΔE^v 0.5 (DM), 0.2 (IR), 0.1 (IR), 0.5 (NMR); ΔE_0^l − 1.0 (IR), −0.5 (IR); ΔE^l − 0.65 (NMR); References 66, 69, 70, 71.

[c] ΔE^v 0.3 (IR), 0.8 (NMR); ΔE_0^l − 0.9 (IR); ΔE^l − 0.3 (NMR); References 66, 70.

[d] CS_2 solution (NMR); Reference 72.

[e] NMR; Reference 68.

[f] ΔE^v 1.2 (IR); $\Delta E_0^l < 0.4$ (IR); ΔE^l 0.3 (NMR); References 73, 74.

[g] ΔE^v 2.0 (NMR), >1.5 (IR); ΔE^l 0.9 (NMR); ΔE_0^l 0.5 (IR), 0.0 (NMR); References 17, 22, 75, 76.

[h] ΔE^v 0.5 (NMR), 0 (IR); ΔE^l − 0.4 (NMR); ΔE_0^l − 0.75, −0.91 (IR); References 6, 17, 75, 77.

[i] ΔE^v −0.05, −0.05, −0.5; ΔE_0^l 0.05, −0.3; cf. References 6, 7.

[j] ΔE^v −0.1, −0.28 (IR); ΔE_0^l −0.5, −0.44 (IR); cf. References 6, 7.

[k] ΔE^v −0.82 (IR); ΔE_0^l −0.42 (IR); −0.40 (NMR); References 22, 78.

[l] ΔE^v <0.3 (IR), 0.25 (±0.25) (ED); ΔE^l (CFCl$_3$ soln.) 0.12 (NMR); References 6, 7, 8, 79.

[m] IR; Reference 80.

[n] IR; Reference 81.

[o] ΔE^v 1.3 (IR), 1.4 (NMR); ΔE_0^l 0.64 (IR); ΔE^l 0.84 (NMR); References 82, 83, 84.

[p] ΔE^v 1.5 (IR), 1.5 (NMR); ΔE_0^l 0.45 (IR); ΔE^l 0.75 (NMR); References 41, 82.

constant of the pure liquid, which varied from 4·8 in dibromoethane (di-iodoethane is a solid) to 34·4 in difluoroethane, and this in turn is a natural consequence of the increased proportion of the *gauche* isomer in all media as the size of the halogen substituent decreases. The difference in stability between the *gauche* and *trans* rotamers is naturally largest for the two extreme cases, the 1,2-difluoro- and 1,2-diiodoethanes and makes the estimation of these energy differences more difficult in these cases. Thus in the difluoroethane the observed value of the CH—CH coupling J'_{HH} was virtually constant to within the experimental error for all solvents from carbon tetrachloride to acetonitrile,[68] due to the great preponderance of the *gauche* isomer (cf. J'_{HH} for chlorobromoethane, Table 13.8). Only in very dilute solutions of non-polar solvents did the coupling change significantly. Also of note was the observation of a large concentration dependence of the coupling in these non-polar solvents which was however considered to be due to the aggregation of this very polar solute in the non-polar solvent and not to molecular deformation as suggested previously for dicyano-ethane. Indeed these results support the absence of any pronounced molec-ular deformation in any solvent. As considered earlier (§ 13.4.2) solvation effects would be expected to decrease the CX—CX dihedral angle in the *gauche* form in polar solvents and this would lead to an *increase* in the value of the measured coupling J'_{HH}. However, no significant variation is observed which can only be explained on the basis of no molecular deforma-tion and about 100 % *gauche* isomer or on a fortuitous balance between the decrease due to the varying rotamer populations and of the increase due to molecular deformation. Also because of the overwhelming preponderance of the *gauche* rotamer in solution the computational search for a best fit between the observed and calculated coupling is virtually meaningless and the value of the energy differences in the liquid was obtained from the NMR studies from an estimate of the probable values of the coupling constants in the *trans* rotamer. (Note that the value of $-0·9$ kcal/mole for ΔE^l often quoted as a measured value from Reference 67 is not a measured value but an estimated value based on the measured value of ΔE^v and using the simple dipole formula, equation 13.12).

The solvation studies for 1,2-diiodoethane,[72] which in this case are the only reported data, were treated in a very similar manner to the difluoro-ethane results, due in this case to an overwhelming preponderance of the *trans* isomer in all solvents. The values of k and h quoted in Table 13.9 are taken directly from this work, and were obtained using a C—I dipole moment of 1·9 D, and tetrahedral angles and ω_g 70°. For 1,2-dichloro-ethane a similar model (d, Table 13.3) gives identical results to the standard-ized parameters we use.

The data for the remaining 1,2-dihaloethanes, i.e. the 1-fluoro-2-chloro- and 1-fluoro-2-bromoethanes is more amenable to analysis and also much more extensive, particularly for the fluorochloro compound. In this case variable temperature IR and Raman studies[70,71] provide estimates of ΔE^v which are in very reasonable agreement (0·2 and 0·1 kcal/mole, see footnote b Table 13.9) and of ΔE_0^l which are not ($-0·5$ and $-1·0$). These vapour energy differences are also lower than the much earlier dipole moment value of 0·5 kcal/mole.[69] A solvent study of the NMR spectrum has been given[68] and the results can be reconsidered using the present model. If we assume that the $-T(\mathrm{d}H/\mathrm{d}T)$ term should be about 0·35 kcal/mole (§ 13.4.1) then Table 13.9 shows that the calculated value of $\Delta E^v - \Delta E_0^l$ should be about 1·5, somewhat larger than the IR results. Furthermore, the value for ΔE^v of 0·2 kcal/mole gives an unreal result with the NMR data, but if the value of ΔE_0^l of $-1·0$ is used with the estimated $-T(\mathrm{d}H/\mathrm{d}T)$ term above (i.e. $\Delta E^l = -0·65$, $\Delta E^v = 0·5$) the NMR data is fitted satisfactorily, and these values are given in the footnote.

Further extensive data is available for this molecule as the ratio of the IR intensities of bands due to the *trans* and *gauche* rotamers has been measured in a variety of solvents.[60] Using the same procedure as previously (cf. Tables 13.6 and equation 13.30) the data given in Reference 60 can be converted into $\delta\Delta E$ values and these are given in Table 13.10 where the

Table 13.10 Rotamer energy differences (kcal/mole) in 1-fluoro-2-chloroethane in various solvents

Solvent	ε	$\Delta E^s(E_g - E_t)$ (obs.) (IR)	$\Delta E^s(E_g - E_t)$ (calc.)
Vapour	1·0	0·90	(0·50)
C_6H_{12}	2·0	0·30	0·18
CS_2	2·6	0·13	0·05
CH_2I_2	5·3	$-0·31$	$-0·25$
CH_3I	7·0	$-0·17$	$-0·35$
MeCO.iBu	13·1	$-0·53$	$-0·54$
MeCO.iPr	15	$-0·53$	$-0·58$
Acetone	20·7	$-0·65$	$-0·67$
Pure liquid	21·1	$-0·70$	$-0·67$

origin is taken for the liquid value and the energies compared with those calculated on the model. It can be seen that there is generally good agreement for all the solvents used but there is a considerable discrepancy in the value of ΔE^v obtained, which by this method would be 0·9 kcal/mole whereas the calculated solvent energies use the much lower value of 0·5 kcal/mole. If the model was in error then this would be seen by a discrepancy in the scales of the solvent energies in Table 13.10. This is not the case, and the discrepancy is between the two IR values of ΔE^v, one from variable temperature measurements and one on the basis of constant integrated intensity ratios for the two isomers. The latter is an assumption which, however, as shown earlier gave excellent results for all the other haloethanes studied. In Table 13.9 average values of ΔE^v and ΔE^l are shown, which are not too different from the separate values.

The results for the bromofluoroethane show a similar anomaly, although the data is not so extensive, in that the NMR solvent studies[66] (which have again been reconsidered in the present model) gave more positive values of both ΔE^v and ΔE^l than the IR data.[70] The discrepancy is most marked for the liquid value (footnote c, Table 13.9), but the values of $\Delta E^v - \Delta E^l$ are in much better agreement and both values are reasonably consistent with the calculated value of $\delta\Delta E$ (Table 13.9).

The remaining simple haloethanes considered in the table, the 1,1,2-tri- and 1,1,2,2-tetrafluoro- and bromoethanes, can be also profitably considered together with the results for the analogous chloro compounds (Tables 13.2 and 13.5). As in the case of the 1,2-dihaloethanes substitution of one halogen for another does not greatly affect the values of $\delta\Delta E^s$ for a medium of given dielectric constant, again due to the partial cancellation of the dipole and quadrupole terms. The values of $\delta\Delta E^s$ for a medium of dielectric constant 20 are for the trifluoro, trichloro and tribromo compounds 1·4, 1·7 and 1·6 kcal/mole respectively, and for the analogous tetrasubstituted compounds 1·13, 0·91 and 0·69 kcal/mole. Thus in the trisubstituted series, the increase in the partial cancellation of the dipole and quadrupole terms more than compensates for the smaller molecular volume in going from bromine to chlorine to fluorine, resulting in a decreased $\delta\Delta E$ value. The tetrasubstituted compounds behave in contrast very similarly to the di-substituted ethanes considered previously.

The only data for the trifluoroethane is from variable solvent NMR studies,[68] however in the case of the tetrafluoroethane both IR and NMR studies have been reported. Klaboe and Nielsen[73] obtained a value of 1·2 kcal/mole for ΔE^v from variable temperature IR measurements but were unable to observe any temperature dependence of the liquid spectrum, and quoted a value (of ΔE_0^l) of <0.4 kcal/mole in consequence. The variable solvent NMR data,[74] reinterpreted on the basis of the present model gives results in good agreement with these. The calculated value of $\Delta E^v - \Delta E^l$ is 0·9 kcal/mole, which assuming a value of the $-T(\mathrm{d}H/\mathrm{d}T)$ term of about 0·3 kcal/mole gives a value of $\Delta E^v - \Delta E_0^l$ of 1·2 kcal/mole. Furthermore, the NMR data interpreted on the basis of the above model with $\Delta E^v = 1.2$ and $\Delta E^l = 0.3$ kcal/mole gives results in substantial agreement with those previously obtained. In Reference 74 the discrepancy between the earlier models which used a C—F dipole of 2·03 D and the IR results was noted. As the values of k and h, and therefore $\delta\Delta E$, are proportional to μ^2 this gave an overestimate of about 20%, which is completely removed by the more accurate value of $\mu(C-F)$. The NMR solvent studies essentially give a value of the rotamer energy differences in the solvents studied, i.e. ΔE^s. The extrapolation to ΔE^v, however, depends on the model used. Thus the data on tetrafluoroethane when reevaluated on the present model gives essentially the same ΔE^l and rotamer coupling constants, but the value of ΔE^v is now decreased from 2·1 to 1·2 kcal/mole, in agreement with the IR data.

The data for the tribromoethane are more extensive and reasonably consistent (footnote g, Table 13.9) with the possible exception of the NMR value of ΔE_0^l obtained from variable temperature studies of the CH—CH coupling constant in the pure liquid.[22] The observed coupling only changed from 6·60 ($\pm$0·06) Hz at 240°K to 6·59 ($\pm$0·05) Hz at 398°K which may be considered within the error limits as either a decrease or an increase of 0·1 Hz. Furthermore the effective couplings in the individual isomers (effective because in the *trans* isomer the coupling is the average between the two mirror-image forms, cf. Figure 13.1), obtained from variable solvent studies, are about 3·0 and 6·8 Hz.[17] This small difference combined with the error limits, means that the value of ΔE_0^l obtained by this method is very approximate.

An extensive investigation of both the solvent and temperature dependence of the coupling combined with the solvent theory given here, gives the more reasonable value of $\Delta E^l = 0·9$ kcal/mole.[17] The NMR and IR data for tetrabromoethane are also in reasonable agreement, the NMR results again being obtained in the same investigation as the tribromoethane.[17] There are discrepancies in that the NMR results are consistently about 0·5 kcal/mole more positive than the IR data. This may be due to the model used by these authors which gave somewhat larger values than our present model (cf. $k = 2·79$ compared to 2·16 in Table 13.9), or this may be related to the anomaly noted previously with the 1,2-dibromoethane. In general, however, the simple di-, tri- and tetrahaloethanes give a reasonably consistent picture on our model, but they suffer as tests of the model in that the values of $\delta\Delta E$ obtained for these compounds are all very comparable. We now wish to consider other substituted ethanes in which very different solvent effects are observed to see whether these can also be satisfactorily explained.

The first examples to consider are propyl chloride and bromide (Table 13.9 and footnotes i and j). In both cases, as a consequence of the standard model used, the dipole moments (and therefore k values) of the *gauche* and *trans* rotamers will be the same. It is important to note that the individual values of k_g (and k_t) are of course appreciable (1·44 and 1·40 for the chloro and bromo compounds) and therefore both rotamers will have considerable solvation energy. The quadrupole terms are a function of both the bond dipole moments and also of the geometry and position of the origin of the coordinate system, and are therefore not the same for the two rotamers.

They are, however, comparable and the result is that the theory predicts that there will be only a very small change of the rotamer energy with the medium with the *trans* form slightly more stabilized.

In propyl chloride the observed data are in good agreement with these predictions in that there is virtually no change in the rotamer energy difference in going from the vapour to the liquid. If anything, there is a slight stabilization of the *gauche* rotamer, but this is within the experimental error of the measurements. The data for propyl bromide are more definite and again there is only a very small change in going from the gas to the liquid. However, there now appears to be a significant, though small, extra stabilization of the *gauche* rotamer of about 0·3 kcal/mole in the liquid. This is the opposite of the theoretical predictions and is very likely due to the over-simplification of considering the alkyl group as merely a space-filling group. The dipole moments of the *trans* and *gauche* rotamers of the propyl halides may well be slightly different and an increased dipole moment of the *gauche* isomer would give the observed result. This would need to be included in a further refinement of the theory, but the major prediction of an essentially negligible solvent dependence of the rotamer energy is confirmed by the experimental results.

The theoretical predictions for the propyl halides may be immediately extended to any other haloalkane with only one halogen atom, or in which the halogen substituents are all attached to the same carbon atom, as again the dipole terms of the rotamers will be identical and the quadrupole terms comparable. Thus all such compounds should exhibit a negligible solvent effect on the rotamer populations. This has been shown to be the case for a number of such compounds.

For example, variable temperature IR studies on the analogous *s*-butyl halides $Me_2CHCH_2X(X = Cl, Br)$ gave values for ΔE^v and ΔE_0^l of 0·23 and 0·37 kcal/mole (X = Cl) and 0·30 and 0·26 kcal/mole (X = Br),[81] in exact agreement with predictions. (The nomenclature follows the definitions of Figure 13.2 by defining the rotamers relative to the CH—CX dihedral angle.)

Similarly Heatley and Allen[17] observed the CH—CH coupling constant in $CHCl_2CHMe_2$ to be totally independent of solvent, whilst varying considerably with temperature, and were able to analyse their results on the basis of a temperature-independent value of ΔE^s of $-0·50$ kcal/mole and reasonable values of the couplings in the two rotamers. Note that for all such compounds the $-T(dH/dT)$ term is negligible and therefore ΔE^l and ΔE_0^l are identical.

A very similar analysis of the NMR coupling constants of a number of 3,3-dimethylbutyl derivatives t-Bu—CH_2CH_2X showed exactly analogous behaviour in that there was no observed solvent dependence of the couplings but a considerable temperature dependence and from this the rotamer energy differences were obtained.[85]

The one series of outstanding exceptions to these predictions are the results of Park and Wyn-Jones on the 2-halo-2-methylbutanes CMe_2XCH_2ME (X = Cl, Br, I).[86] Here the rotamer energies measured by the variable temperature IR method were 1·1 and 1·4 in the vapour, and 0·36, 0·38 and 0·69 kcal/mole in the liquid. Nothing in the theoretical treatment can explain these results, and this may suggest that the observed data is capable of more than one interpretation.

The next examples in Table 13.9, $CHCl_2CFCl_2$ and $CFCl_2CFCl_2$, may be considered together as they exhibit generally similar behaviour. The fluorotetrachloroethane is very analogous to the propyl halides. The equivalence of the C—X bond moments of the $CFCl_2$ group means that the dipole moments of the two rotamers, and therefore the dipole terms, will be very similar. Again note that the individual values may be appreciable; it is merely the difference which is very small. The quadrupole terms are also very similar resulting in a very small calculated solvent dependence of about 0·2 kcal/mole from the vapour to a very polar solvent ($\varepsilon = 20$). In this case the *trans* rotamer is again predicted to be slightly more favoured due to its larger dipole moment. The observed results from IR[78] and NMR[22] data are in excellent agreement and in this case the small solvent effect observed ($\sim$0·4 kcal/mole) is in the same direction as the predicted effect. The agreement also confirms the original assignment of the IR measurements which was made on similar reasoning to the above.

The analogous bromo compound, $CHBr_2CFBr_2$, was made the subject of an extensive NMR investigation[18,87] and the rotamer energy differences (actually free-energy differences) measured directly from the separate NMR signals observed at low temperatures in a number of solvents.[87]

This gave $\Delta E^1(E_g - E_t)$ values of -0.42, -0.40, -0.38 and -0.39 kcal/mole at 180°K in $CFCl_3$, CS_2, Me_2O and acetone solutions. Thus in this case there was no detectable change in the entire range of possible solvents.

For the 1,2-difluorotetrachloroethane, a further effect is obtained from the calculations. Because all the C—X bond moments are so similar, as far as the molecular dipole and quadrupole moments are concerned the molecule has effective C_{3v} symmetry. As a consequence the dipole and quadrupole terms of the *individual* rotamers are all very small. Thus the molecule behaves as far as the dipole and quadrupole fields are concerned to all intents and purposes as a non-polar solute molecule. This is quite different to the case of the propyl halides, or to $CHCl_2CFCl_2$, both of which are polar molecules in which the rotamers have equal molecular fields. This difference would of course be observed most clearly in studies related to the solvation of the individual rotamers, rather than, as in our case, the energy difference between them. For example the absence of molecular dipolar and quadrupolar fields may be one reason for the very low latent heat of vaporization of polyhalogenated ethanes. However, it is also a consequence of the absence of these molecular fields that the rotamer energy difference is completely independent of the medium and this is confirmed by the observed results.

We may generalize the results of these examples as follows. For all fully halogenated ethanes the solvent dependence of the rotamer energies will be negligible. The only possible exception may be with iodine as a substituent as here the differences in the C—X bond moments may lead to observable effects. In any case these will be small. Furthermore for all halogenated ethanes with a CXYZ group (X, Y, Z equals halogen) the effect of the medium on the rotamer energies will be small, usually not greater than the normal experimental errors of about ± 0.1 to 0.2 kcal/mole.

A wide variety of such compounds have been studied by the various techniques catalogued earlier (cf. References 6 and 7 for a comprehensive survey) and inspection of these results shows that these conclusions are fully confirmed. One further useful consequence follows. In many fully halogenated fluoroethanes the low-temperature NMR method, observing the ^{19}F resonance, permits a precise measurement of the energy differences of a variety of such compounds. These values which are usually obtained in a low melting solvent, may be considered with some confidence as the true (vapour-state) energy differences, for example, in steric energy calculations.

The last results in Table 13.9 demonstrate the effect of the replacement of hydrogen by methyl in haloethanes. The successive substitution of hydrogen for methyl on both the rotamer energies and the solvent effects can be seen very clearly by considering the results given for 1,2-dichloropropane (CH$_2$ClCHClMe), 1,2-dichloro-2-methylpropane (CH$_2$ClCMe$_2$Cl) and *meso*-dichlorobutane (CH$_3$CHClCHClCH$_3$) in Table 13.9 with the analogous results for the unsubstituted compound, 1,2-dichloroethane (Tables 13.4 and 13.5).

In the 1,2-dichloropropane there is the additional complication of three non-equivalent rotamers, as shown in Figure 13.4.

The predicted solvent effects ($\delta\Delta E$'s) are thus given in Table 13.9 in terms of $(E_{g_1} - E_t)$ and $(E_{g_2} - E_t)$ although in fact these $\delta\Delta E$'s are virtually identical. That is, the energy difference between the *gauche* rotamers is

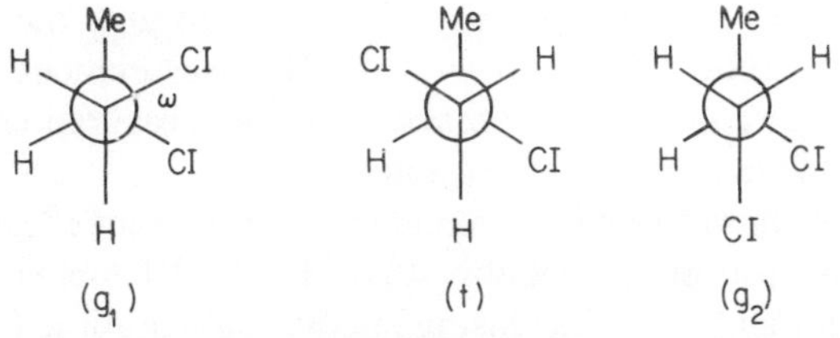

Figure 13.4

almost completely unaffected by the medium, but of course they both gain energy at the expense of the *trans* rotamer in polar solvents. The observed values from variable temperature IR measurements[80] confirm this prediction. Although the energies of the two *gauche* forms differ by 0·7 kcal/mole, the difference is the same in the vapour and liquid. The actual gain in energy of the *gauche* rotamers over the *trans* form (about 1 kcal/mole in the pure liquid) is almost identical to that of the 1,2-dichloroethane, showing the relatively minor effect of methyl substitution.

This is confirmed by the results for the 1,2-dichloro-2-methylpropane, CH_2ClCMe_2Cl, in which now there are, as usual in this section, only two non-equivalent forms, and in which both the solvent dependence of the rotamer energies and also the actual vapour-state energy differences are very similar to those of the 1,2-dichloroethane.

A similar pattern is predicted for *meso*-dichlorobutane, in which again there are only two non-equivalent rotamers, as shown in Figure 13.5 and

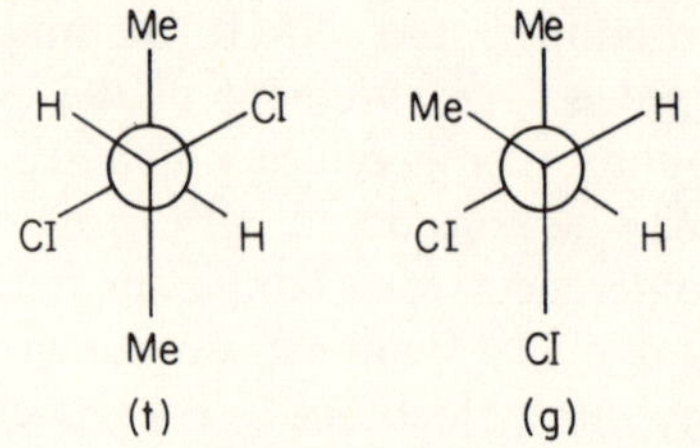

Figure 13.5

the observed results generally agree with this prediction, though the difference between the observed $\delta\Delta E$'s for the two dimethylsubstituted compounds is noticeable, as one is slightly less than the calculated value and the other somewhat greater.

The observed results for the *meso*-dichlorobutane include both variable temperature IR measurements[82] and variable solvent NMR data.[83] The NMR result was obtained via a solvent study of the CH—CH coupling which estimated the likely values of the coupling in the individual isomers and then extrapolated the resulting values of the rotamer energy differences in the different solvents against $(\varepsilon - 1)/(2\varepsilon + 1)$ to give the value of ΔE^v. As this is in very good agreement with the IR measurement, we have used these values of the couplings combined with the measured coupling in the pure liquid[84] to give the ΔE^l value quoted.

The reason for the very small dependence of the solvent effect with increasing methyl substitution is probably that the model increases the C—Cl bond dipole moment with increasing methyl substitution (Table 13.1) in accordance with the observed moments of the haloethanes,[88] and this almost exactly balances the increasing molar volume, resulting in no nett effect. The remarkable similarity between the actual values of ΔE^v in all the compounds is, however, very striking.

A similar phenomenon occurs in the dibromo compounds. For example in 1,2-dibromo-2-methylpropane $CH_2BrCBrMe_2$ the values of ΔE^v and ΔE_0^l are 1·67 and 0·74 kcal/mole,[81] which are almost identical to those for 1,2-dibromoethane. In *meso*-dibromobutane (Table 13.9), for which both IR[82] and NMR[41] investigations have been performed, the corresponding values are 1·5 and 0·45, both somewhat less but giving the same value of $\delta\Delta E$. The NMR solvent data (reproduced in Figure 13.4) has again been recalculated on the present model to give the values of ΔE^v and ΔE^l quoted in the Table. As in previous examples the value of ΔE^l and the rotamer coupling constants are not significantly different from the original values, but the value of ΔE^v is reduced markedly, from 2·1 kcal/mole to 1·6 kcal/mole, which is now in good agreement with the IR data.

In this section we have extended the theory to those compounds for which a quantitative prediction can be made and for which there is experimental evidence to test those predictions. However, there are a number of solvent studies in the literature, many of which do not obtain definite values of the rotamer energy differences, and there are also many studies of compounds for which a quantitative model has not yet been established. These results can, however, be profitably considered in the light of the theory and are given in the next sections.

13.4.3 *Aldehydes and ketones*

Acyclic aldehydes and ketones represent a class of compounds which have been studied by a variety of techniques. Some of the first groups of compounds to come under study were the haloacetones,[89–92] ω-chloro-acetophenones[89] and α-halogenated esters and amides.[89,93] The rotational isomerisms of many of these compounds were investigated by Raman and infrared spectroscopy. For example the spectra of n-methylchloroacetamide $(I, R = Cl, R' = NHCH_3)$[93] demonstrated that the molecule exists in *trans*

(Ia)   (Ib)

and *gauche* forms as shown, and as one would expect, with increasing dielectric constant of the solvent the *trans* isomer (more polar) becomes more stable.

A similar investigation was carried out on chloroacetone (I, $R = Cl$, $R' = CH_3$)[89,90] by following the changes in the intensities of the Raman and infrared stretching frequencies with solvent. As expected, the relative population of the more polar form (*trans*) increases as the polarity of the solvent increases. The results were explained on the basis that the steric repulsions between the chlorine atom and oxygen atom were less than that between the chlorine atom and methyl group which would make the *trans* isomers more stable. In the case of bromoacetone[91] it was found that the *trans* rotamer was more stable in solution whereas in the fluoro and iodo compounds both rotamers were found to be equally stable.[91,92] However, in the vapour state, the less polar form was determined to be the most stable isomer.

Karabatsos and coworkers have done a great deal of work on the conformational analysis of aldehydic-type compounds.[94-98] In their investigations they have studied alkyl- and arylsubstituted aldehydes,[95] mono- and dihaloacetaldehydes[96,97] and several other types of ring-substituted aldehydes.[98]

They were concerned with the problem of trying to elucidate the relative stabilities of rotamers (II) and (III) as a function of Y, Z and R. The relative

$$
\begin{array}{cc}
\text{(II)} & \text{(III)}
\end{array}
$$

stabilities of the various substituted compounds were determined by the solvent dependence of their spin–spin coupling constants.

In the case of monosubstituted acetaldehydes ($Y = H$, $Z = O$) there are three possible rotamers (IVa, IVb, V). When R was methyl, ethyl, n-amyl or isopropyl, rotamer (V) was found to be more stable than (IVa) or (IVb).[95]

$$
\begin{array}{ccc}
\text{(IVa)} & \text{(IVb)} & \text{(V)}
\end{array}
$$

The relative populations of (IVa) and (V) were shown to be solvent dependent (Table 13.11).

The data in Table 13.11 show that as the polarity of the solvent increases (cyclohexane $\rightarrow$ acetonitrile) the population of rotamer (V) decreases (i.e. when R is Me, population of (V) is 67% in cyclohexane and 65% in acetonitrile). In terms of free energy, the ΔG^0 between (IVa) and (V) becomes

Table 13.11 Solvent effect on relative stabilities of aldehyde rotamers[95]

	Percentage of rotamer (IV)	
Aldehyde	Cyclohexane	Acetonitrile
$MeCH_2CHO$	33	35
$t\text{-}BuCH_2CHO$	78	83
$(Et)_2CHCHO$	36	70
$Me(Et)CHCHO$	27	29
$Et(n\text{-}Bu)CHCHO$	38	42
$C_6H_5CH_2CHO$	65	53

more positive on going from cyclohexane to acetonitrile (i.e. when R is t-butyl, ΔG^0 in cyclohexane is $+330$ cal/mole and in acetonitrile $+550$ cal/mole) (Table 13.12). These changes with solvent polarity can be explained by the expected higher dipole moment of (IV) over (V) as shown in (VI) and (VII).

(VI)   (VII)

From the data shown in Table 13.11 it was concluded that non-bonded repulsions between vicinal groups play only a minor role in determining the relative stabilities of the rotamers. Such interactions apparently only become significant when large bulky groups (e.g. t-butyl) are substituted. The data also show that phenylacetalaldehyde, in contrast to the alkyl monosubstituted acetalaldehydes, has (IVa) and (V) energetically equivalent in non polar solvents with (V) instead of (IVa) becoming more stable in

Table 13.12 ΔG^0 (cal/mole) as a function of solvent[95]

Aldehyde	Cyclohexane	Acetonitrile
$MeCH_2CHO$	-880	-820
$t\text{-}BuCH_2CHO$	$+330$	$+550$
$(Et)_2CHCHO$	$+\ 70$	$+180$
$Me(Et)CHCHO$	-180	-120
$Et(n\text{-}Bu)CHCHO$	$+130$	$+230$
$C_6H_5CH_2CHO$	$-\ 50$	-340

polar solvents. For example, in acetonitrile (V) is 350 cal/mole more stable than (IVa). These at first surprising results were explained on the basis that (V) should have a higher dipole moment than (IVa).

In the case of disubstituted acetaldehydes $(R_1 = R_2)$ (VIII, IX) it was shown that when R is methyl, (IX) is more stable then (VIII) by 500 cal/mole; and when R_1 and R_2 are ethyl or *t*-butyl groups (VIII) is more stable than (IX) by 250 and 1100 cal/mole respectively. The fact that (VIII), with the hydrogen eclipsing the carbonyl, is more stable than (IX) with ethyl eclipsing

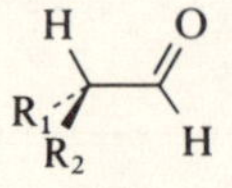

(VIII) 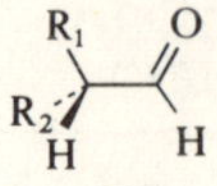   (IX)

the carbonyl, whereas (V) (ethyl eclipsing carbonyl) is more stable than (IVa) (hydrogen eclipsing carbonyl) is explained by 1,3-eclipsing methyl proton non-bonded repulsive interactions. In ethylacetaldehyde the rotamer with the carbonyl eclipsing the ethyl group does not have this unfavourable interaction present.[95]

The authors emphasize that although their data have been interpreted in terms of perfectly eclipsing conformations, small variations from a perfectly eclipsed conformation would not alter the conclusions drawn from their data.

The factors that make (IVa) less stable than (V) are not completely understood. One reasonable explanation is of course the more favourable electrostatic energy of (V) over (IVa), or perhaps that the possibility of hydrogen bonding in (V) increases its relative stability. However, these cannot be the only factors responsible for the greater stability of (V), as shown by phenylacetaldehyde. The authors concluded that these are two interacting groups whose non-bonded distance falls within the attractive part of their Van der Waals curve.

Similar studies on chloro- and bromoacetaldehyde (II, III, IVa, IVb, V, R = Cl or Br) have also been carried out.[96] The data show that as the polarity of solvent increases, there is an increase in the population of the more polar rotamer (V), as shown in (X) and (XI). The free-energy and enthalpy differences also show this effect (Table 13.13). It should be pointed out that

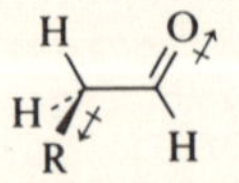

(X) 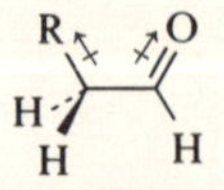   (XI)

because of the large difference in dipole moments between (X) and (XI), in solvents of high polarity, ΔH^0 is calculated to be much more negative than ΔG^0. This is due to the fact that for solvents of high dielectric constant, with increasing temperature (temperature measurements are required for determination of ΔH^0) there is a large decrease in dielectric constant. For this reason, in highly polar solvents, ΔG^0 values reflect better the enthalpy differences between highly polar rotamers whose dipole moments are very different, than do ΔH^0 values. The ΔH^0 values calculated by the temperature dependence of vicinal coupling constants are really only meaningful in solvents of low dielectric constant which have weak solute–solvent interactions. Hence, in *trans*-decalin, $\Delta H^0 \simeq \Delta G^0$ and ΔS^0 must be approximately zero between rotamers.

An example of how ΔG^0 shows the change in rotamer population with solvent polarity is that with chloroacetaldehyde in pentane (least polar solvent used), (V) is only favoured by 70 cal/mole whereas in formamide

Table 13.13 Free energy and enthalpy difference between rotamers of haloacetaldehydes[96]

Solvent	Chloroacetaldehyde ΔG^0 (ΔH^0) (kcal/mole) for (IVa) $\rightleftharpoons$ (V)	Bromoacetaldehyde ΔG^0 (ΔH^0) (kcal/mole) for (IVa) $\rightleftharpoons$ (V)
$CH_3(CH_2)_3CH_3$	-70	—
trans-Decalin	$-300(-300)$	$+40(0)$
Cyclohexane	$-310(-400)$	$+40$
CCl_4	-350	~ 0
$CHCl_3$	-560	-100
CH_2Br_2	-700	-200
CH_2Cl_2	-710	-230
CH_3COCH_3	-1000	-430
CH_3CN	$-1100(-2500)$	-500
$(CH_3)_2NCHO$	$-1250(-2100)$	$-570(-1500)$
H_2NCHO	-1500	-700
C_6H_5Cl	$-650(-2500)$	-180

(most polar solvent used) it is favoured by 1500 cal/mole. The same trend is observed in bromoacetaldehyde but to a small extent. For example in formamide, (V) is only favoured by 700 cal/mole.

The increase in rotamer population of (V) with increased solvent polarity is reasonable in view of its higher dipole moment (XI). It is also reasonable that this increase be more pronounced among the rotamers of chloro-acetaldehyde than among those of bromoacetaldehyde on account of the carbon–chlorine bond being more polar than the carbon–bromine bond.

The data was consistent with a three-fold barrier around the carbon–carbon bond. Among analogous-type monohalo compounds only fluoro-acetyl fluoride has been found to have a two-fold barrier about the sp^2–sp^3

carbon–carbon bond (XII, XIII).[99] Also, it was assumed that the rotamers

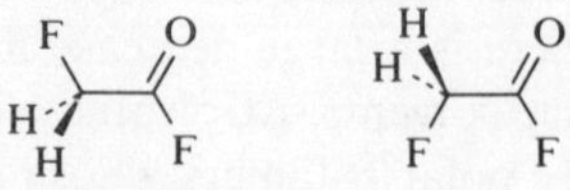

$$\text{(XII)} \qquad\qquad \text{(XIII)}$$

are perfectly eclipsed. However, experimental results from vibrational spectroscopy are incompatible with this assumption.[90] It was pointed out[96] that in cases where assignments were made from NMR studies, any accurate determination of dihedral angles could not be made.

Results from infrared spectroscopy show that the most stable rotamer in chloroacetaldehyde is (IVa) (hydrogen eclipsing carbonyl group)[100] whereas the conclusions drawn from NMR measurements has rotamer (V) being more stable (halogen eclipsing carbonyl). However, it was shown that if the degeneracy factor of two which favours (IV) over (V) were to be removed, it becomes less stable than (V) by 300 kcal/mole.[96]

When the results for chloroacetaldehyde are compared to those of chloroacetone, differences appear. In the liquid state ($\varepsilon \sim 30$) chloroacetone was found to have rotamers (Ia) and (Ib) of comparable stability. If, as stated, it had not been for the non-bonded repulsions in (Ib) (*gauche* rotamer) between the chlorine atom and methyl group, then (Ib) might have been expected to be the more stable isomer. If this is correct then (IV) in chloroacetaldehyde might be expected to be more stable than (V), as the above interaction is absent. However, in highly polar solvents the rotamer ratio (IV):(V) was determined to be smaller rather than larger than the ratio (Ib):(Ia) for chloroacetone. Consequently, if this non-bonded interaction was affecting the rotamer population, the interaction would have to be attractive rather than repulsive.

Also, since in solvents of low polarity, the enthalpy difference for (IV) $\rightleftharpoons$ (V) of bromoacetaldehyde is less negative by 300 cal/mole than the corresponding ΔH^0 for chloroacetaldehyde, this argues against the polarizability of the halogen as being very significant in affecting their rotamer populations.

Some of the minor differences of rotamer stabilities were explained in terms of non-bonded interactions. For example, in the monoacetaldeyhdes and in the ethyl acetates,[101] when bromine is substituted for chlorine, ΔH^0 increases for (II) $\rightleftharpoons$ (III). They rationalized the results on the basis of increased non-bonded repulsions between the halogen and oxygen in rotamer (III) as the size of the halogen increases.

The data show that the ΔH^0 of the haloacetaldehydes are less negative than those of the ethyl acetates[96,101] and haloacetyl halides.[90,99,102] This may be attributed to the differences in the electrostatic energies of (II) and (III). Since for the haloacetaldehydes these differences are larger than the corresponding differences for the haloesters and haloacetyl halides, this would make the ratio (II):(III) smaller for the haloacetaldehydes. This argument has been used to partly explain the differences in rotamer stabilities for chloroacetyl chloride, chloroacetone and N-methylchloroacetamide.

The data on dichloro- and dibromoacetaldehyde gave the expected results, that the relative stability of (XV) increases as the polarity of the

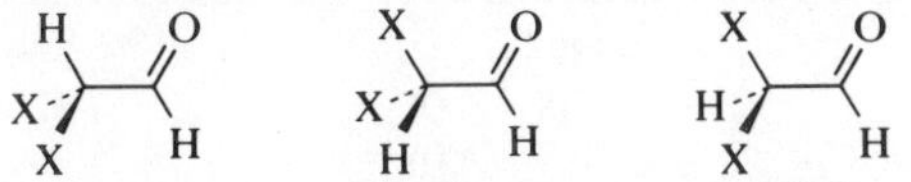

(XIV)   (XVa)   (XVb)

solvent increases, this is, of course, in line with its higher dipole moment (Table 13.14).[97] As with the monohaloacetaldehydes a three-fold barrier around the carbon–carbon bond was found to be most consistent with the

Table 13.14 Solvent dependence of the rotamer population of dichloro- and dibromoacetaldehydes[97]

Solvent	Dichloro(XIV) (%)	Dibromo(XIV) (%)
$CH_3(CH_2)_3CH_3$	44	47
Cyclohexane	47	46
trans-Decalin	45	45
CCl_4	45	43
$CHCl_3$	38	37
CH_2Br_2	33	31
CH_2Cl_2	33	30
CH_3COCH_3	16	17
$(CH_3)_2NCHO$	10	21
CH_3CN	9	15
$(CH_3)_2SO$	7	7
C_6H_6	29	29
$C_6H_5CH_3$	30	30
C_6H_5CN	19	19
$C_6H_5NO_2$	21	20
Neat	27	26

data. This is quite different from the ethyl dihaloacetates which exhibit two-fold barriers to rotation.[19]

Changes in solvent could not explain all the experimental results. For example benzene and toluene have dielectric constants of 2·3 and 2·4 respectively. Consequently, it would be expected in these solvents that (XIV) would be more stable than (XVa), however, the reverse was found to be true. The free-energy difference for (XIV) $\rightleftharpoons$ (XVa) in benzene was evaluated as -120 cal/mole for both the dichloro- and dibromoacetaldehyde (Table 13.15). This result was interpreted in terms of solute–solvent interactions which destabilize (XIV) with respect to (XV).

Table 13.15   Solvent dependence of the free energy and enthalpy differences between rotamers of dichloroacetaldehydes[97]

Solvent	ΔG^0 (ΔH^0) (kcal/mole)	
	Dichloroacetaldehyde	Dibromoacetaldehyde
$CH_3(CH_2)_3CH_3$	$+380$	$+340$
Cyclohexane	$+350(+300)$	$+320$
trans-Decalin	$+300(+300)$	$+290(+500)$
CCl_4	$+300$	$+260$
$CHCl_3$	$+120$	$+100(+500)$
CH_2Br_2	-15	$-75(-100)$
CH_2Cl_2	-14	-80
CH_3COCH_3	-590	-550
$(CH_3)_2NCHO$	$-930(-1400)$	$-370(-500)$
CH_3CN	-1000	$-650(-800)$
$(CH_3)_2SO$	-1200	-1200
C_6H_6	$-120(-450)$	-120
$C_6H_5CH_3$	$-100(-500)$	$-90(-200)$
C_6H_5CN	$-340(-1000)$	-480
$C_6H_5NO_2$	-390	-420
Neat	-190	-210

The results obtained by the NMR method disagreed with those from infrared spectroscopy,[100] which concluded that dichloroacetaldehyde exists in one minimum-energy conformation. It is also interesting that in the dichloro- and dibromoacetaldehydes the most stable conformation is the one in which the carbonyl group and hydrogen are eclipsing one another (in saturated hydrocarbon solvents), whereas in the monohaloacetaldehydes the more stable rotamer is the one in which the halogen eclipses the carbonyl function.

The ΔH^0 values for $(XIV) \rightleftharpoons (XVa)$ were found to be larger than the enthalpy differences between $(II) \rightleftharpoons (III)$ $(R = Cl$ or $Br)$. A plausible explanation is that the dipole moment difference between (XIV) and (XVa) is greater than that between (II) and (III), hence the electrostatic interaction would lead to greater energy differences between (XIV) and (XVa) than between (II) and (III).

From a study of the vicinal coupling constant as a function of both temperature and solvent for methoxyacetaldehyde $(IV, V, R = OCH_3)$, phenoxyacetaldehyde $(IV, V, R = OC_6H_5)$, methyl mercaptoacetaldehyde $(IV, V, R = SCH_3)$, glycidaldehyde (XVI) and cyclopropanecarboxaldehyde (XVII), it was shown that a three-fold barrier to rotation about the carbon–

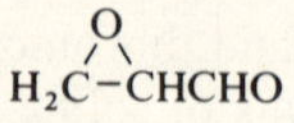

$$\overset{\displaystyle O}{\overset{/\ \backslash}{H_2C-CHCHO}} \qquad \overset{\displaystyle CH_2}{\overset{/\ \backslash}{H_2C-CHCHO}}$$

(XVI) 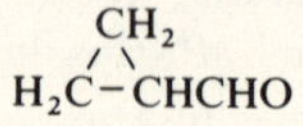 $\qquad\qquad\qquad$ (XVII)

carbon bond agreed best with the data for the first three compounds with the results of the latter two compounds in doubt.[98] However, since cyclopropanecarboxaldehyde[103,104] cyclopropylmethyl ketone[104] and cyclopropanecarboxylic acid chloride[105] exhibit two-fold barriers to rotation in the gas phase, the authors assume a similar barrier for cyclopropanecarboxaldehyde in solution. On the basis of the similarity of the vicinal coupling constants for glycidaldehyde and cyclopropanecarboxaldehyde, they assume a two-fold barrier to rotation about the carbon–carbon bond.

The most stable rotamers for methoxy- and phenoxyacetaldehyde are the ones where the carbonyl function is eclipsed by the carbon–halogen bond. For the remaining compounds the more stable rotamer was found to be the one that has the carbon–hydrogen bond eclipsing the carbonyl group. The relative stabilities of (IVa) and (V) as a function of R are shown below.[98] The order was determined to be valid only in solvents of low dielectric constants, such as carbon tetrachloride. For the groups ahead of bromine, the free enthalpy difference for (IV) $\rightleftharpoons$ (V) is negative, and for those after bromine, ΔH^0 is positive. On the basis of the order below with the more polarizable methylmercapto groups ahead of the less polarizable methoxyl (also note relation of bromine to chlorine), the authors concluded

$$R = CH_3 > CH_2CH_3 \sim C_6H_5O \sim CH_3O > CH(CH_3)_2 > C_6H_5 \sim Cl > Br$$

$$> C(CH_3)_3 > CH_3S$$

Increased stability of (V) $\leftrightarrow$ Decreased stability of (V)

that dipole-induced dipole interactions play insignificant roles in determining the relative stabilities of (IV) and (V). Non-bonded repulsions are probably partly responsible for the position of the bulky *t*-butyl and mercapto groups. Their relative positions, however, imply that they are not the main factor in determining the relative stability of (IV) and (V).

The vibrational spectra of fluoroacetyl halides[106] and chloroacetyl halides have been interpreted on the basis of more than one rotational isomer being present in both the liquid and vapour state.[106,107] The most polar and stable rotamer (*trans*) in both states was determined to be the one with an approximately zero dihedral angle (e.g. fluoroacetyl halides, C—O bond eclipses C—F bond). However, as might be expected the relative stability of the *trans* isomer in comparison with the staggered isomer was determined to be relatively diminished in the vapour state.[106] Similar studies have been reported for dihaloacetyl chlorides[108] and dichloroacetyl halides.[109] Only in these systems the least polar isomer is the one with an approximately zero dihedral angle (e.g. C—O bond eclipses C—H bond).

13.4.4 *Other acyclic compounds*

Several studies have appeared on rotational isomerism of trisubstituted ethanes of the following type, XCH$_2$CHXR.[4,51,110–113] Reynolds and coworkers in particular have been interested in the determination of energy differences in these molecules, where X is a polar and R a non-polar group.[51,110,111] All three rotamers, in the above compounds, have different energies. A semi-quantitative approach was developed for evaluating conformational preferences of the isomers as a function of the dielectric constant of the solvent and the average observed vicinal coupling constant J_{ij}.[51] Snyder, who earlier carried out a similar investigation, concluded that the polarity of the solvent used had only a small effect on the energy differences between rotamers.[4,112,113] However, his conclusions were based on results from only a limited number of solvents. Consequently, Reynolds decided to investigate the rotational isomerism of these compounds in a wider range of solvents to test Snyder's conclusions.

The initial studies dealt with the effect of solvents on 1,2-dibromoethylbenzene and some of its 4-substituted (1,2-dibromoethyl)benzene derivatives.[51] In this way it would be possible to study the effect of a remote dipolar group on rotamer populations.

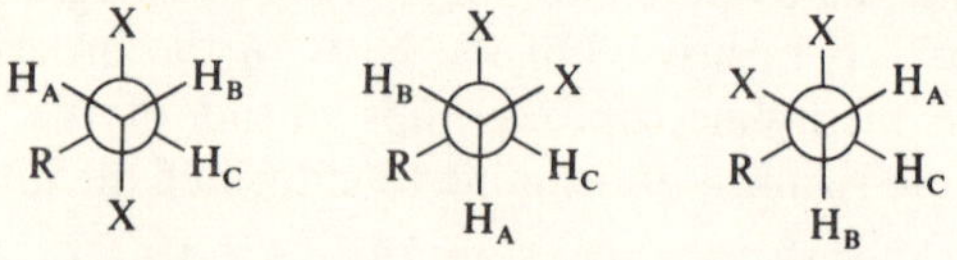

$$
\begin{array}{ccc}
\text{(XVIII)} & \text{(XIX)} & \text{(XX)}
\end{array}
$$

From steric and electrostatic considerations, one would expect rotamer (XX) to be the least stable isomer of 1,2-dibromoethylbenzene, since there are two steric interactions and one dipolar interaction present, whereas in rotamers (XVIII) and (XIX) the number of interactions is less. The relative stability of the remaining two isomers was more difficult to predict. However, by examining the effect of the dielectric constants of solvents on the coupling constants a scheme was devised whereby such distinctions could be determined. It is well known that the more polar rotamers (XIX) and (XX) should be relatively more stable in solutions of high polarity. Experimentally if isomer (XVIII) is the most stable in solutions of low polarity then J_{AC} (*trans* coupling in rotamer XVIII) should show a decrease as the dielectric constant of the medium increases, since this relatively non-polar rotamer will become less populated. If, however, $E_{XVIII} > E_{XIX}$ then $J_{BC} > J_{AC}$ and the coupling constant J_{AC} should decrease while J_{BC} increases as the polarity of the solvent increases. Hence $|J_{AC} - J_{BC}|$ should increase with increasing dielectric constant of the medium. In rotamer (XX) all the couplings are *gauche*, consequently any increase of population with solvent polarity should lead to a decrease in $(J_{AC} + J_{BC})$. The fractional population of (XX) should increase with increasing solvent polarity. However, $(J_{AC} + J_{BC})$ will remain relatively constant if $E_{XX} \gg E_{XVIII}$ since the population of (XX) would be present to only a small extent in all solvents.

Table 13.16 shows that the difference between J_{AC} and J_{BC} is smaller in solvents of high dielectric constant than in solvents of low polarity. However, a large deviation was observed when the solvent used was benzene. From these results the authors concluded that rotamer (XVIII) must be the most stable isomer, since $|J_{AC} - J_{BC}|$ increased with increasing solvent polarity. This is the same conclusion that Snyder came to earlier.[51,113] As expected, the values for $(J_{AC} + J_{BC})$ were found to be relatively constant, indicating a low population of rotamer (XX).

Results for the 4-substituted derivatives (4-chloro, 4-methoxy and 4-nitro) were found to be similar to those of the parent compound.[51] Apparently, the remote dipole does not significantly change the effect of

Table 13.16 Coupling constants for
1,2-dibromoethylbenzene[51]

Solvent	ε	$(J_{AC} + J_{BC})$	$|J_{AC} - J_{BC}|$
Hexane	1·98	15·98	5·92
Cyclohexane	2·07	16·02	6·06
CCl_4	2·30	16·07	6·07
C_6H_6	2·31	15·95	5·11
CS_2	2·59	15·93	6·01
$CHCl_3$	4·26	16·06	5·74
t-Butyl chloride	8·40	15·97	5·67
Acetone	18·1	15·94	5·16
DMSO	42·4	16·03	4·79

solvent upon rotamer populations. Consequently Snyder's conclusion, that the energy differences of substituted ethanes are only weakly related to dielectric constants, appears to be in error.

In the case of *threo*-1,2-dibromopropylbenzenes it was found that the observed average vicinal coupling constant increased with solvent polarity. Conclusions based on their solvent and low-temperature studies gave the

relative order of stabilities as $E_{XXIII} > E_{XXI} > E_{XXII}$. A similar investigation of *erythro*-1,2-dibromopropylbenzene showed no solvent-dependent effects. However, since the coupling constant was large they concluded that the *trans* isomer must be the more stable rotamer.

Reynolds has also investigated a series of dihaloalkanes [for rotamers (XVIII), (XIX) and (XX); X = Cl, Br; R = CH_3, CH_2CH_3, $CH(CH_3)_2$, $C(CH_3)_3$] as a function of solvent polarity.[110] As shown in Table 13.17,

Table 13.17   Coupling constants (Hz) for (a) 1,2-dibromo- and (b) 1,2-dichloro-3,3-dimethylbutanes (0·2M solutions)[110]

Solvent	J_{AB}	J_{AC}	J_{BC}	$\mid J_{AC} - J_{BC} \mid$	$(J_{AC} + J_{BC})$
(a) CCl_4	−11·37	3·17	9·11	5·94	12·28
C_6H_6	−11·50	2·75	9·65	6·90	12·40
CS_2	−11·32	3·12	9·17	6·05	12·29
$CDCl_3$	−11·45	2·73	9·70	6·97	12·43
$(CH_3)_3CCl$	−11·51	2·75	9·64	6·89	12·39
$(CD_3)_2C{=}O$	−11·72	2·42	9·98	7·56	12·40
$(CH_3)_2S{=}O$	−11·83	2·34	10·13	7·79	12·47
(b) CCl_4	−11·88	2·97	9·06	6·09	12·03
C_6H_6	−11·99	2·60	9·56	6·96	12·16
$CDCl_3$	−11·95	2·58	9·61	7·03	12·19
$(CD_3)_2C{=}O$	−12·19	2·34	9·81	7·47	12·15

it was found that $\mid J_{AC} - J_{BC} \mid$ increased with dielectric constant of the surrounding medium. This means that $E_{XVIII} > E_{XIX}$ in both compounds. In both these cases the sum of coupling constants was found to be relatively constant, implying that the fraction of rotamer (XX) present is negligible. Hence, $E_{XX} \gg E_{XVIII} > E_{XIX}$. However, once again note the anomalous result of $\mid J_{AC} - J_{BC} \mid$ in benzene.

The energy levels established here corresponded with those determined by Snyder.[112,113] From the dipole moments of the dibromo compound the fractional population of rotamer (XIX) was calculated to be 0·73 in carbon tetrachloride and 0·83 in benzene with the population of isomer (XX) approximately zero.[114] From their coupling-constant data Dawson and Reynolds[110] determined the corresponding rotamer populations to be 0·75 and 0·81.

For the other compounds studied, the difference in coupling constant decreased with increasing dielectric constants, establishing rotamer (XVIII) to be more stable than rotamer (XIX). Also the sum of the vicinal coupling constants was found to be not constant, but decreasing with increasing solvent polarity, suggesting that the population of rotamer (XX) was not negligible. The relative order of stabilities for this series of compounds was found to be $E_{XX} > E_{XIX} > E_{XVIII}$.

The rotational isomerism of 1,2-disubstituted propanes has been investigated as a function of solvent polarity.[115]

For example, from a study of the coupling constants of 2-bromo-1-phenylpropane with solvent it was deduced that there was a predominance of one rotamer in solution, that with the aromatic and halogen groups *trans*, as would be expected on steric grounds. In the case of 1-methoxy-2-propanol the *gauche* form was predicted to be the more stable in solution. Hydrogen bonding between the glycol hydrogen and ether oxygen would tend to stabilize the *gauche* rotamer.

Dipole moment studies have been useful in determining the major rotamer present in solution.[114] A series of substituted bromoethanes have been investigated to see what effect varying the size of the alkyl substituent would have on the equilibrium and what changes, if any, there would be in the displacement of the equilibrium on changing solvents.[114] Table 13.18 clearly demonstrates that when the solvent is changed from carbon tetrachloride to benzene the equilibrium shifts towards the more polar *gauche* rotamer.

Table 13.18 Dipole moments[114]

Compound	$\mu(CCl_4)$	μ(benzene)
CH_2BrCH_2Br	1·02	1·29
$CH_3CHBrCH_2Br$	1·13	1·43
$(CH_3)_2CHBrCH_2Br$	1·19	1·47
$(CH_3)_2CBrCHBr(CH_3)$	1·23	1·53
$(CH_3)_2CBrCBr(CH_3)_2$	0·87	1·06
$(CH_3)_3CCHBrCH_2Br$	2·59	2·76

The effect of solvent upon the observed coupling constant for 1-phenyl-1,2,2-trihaloethanes was found to be somewhat unusual.[116] The values were found to be relatively constant or to decrease slightly on changing from solvents of low to intermediate polarity, but they increase significantly in solvents of higher dielectric constant. For example in 1-phenyl-1,2,2-tribromoethane, J_{AB} in CCl_4 and $CDCl_3$ is 7·35 and 7·39 respectively,

(XXIV)   (XXV)   (XXVI)

whereas in acetone and dimethyl sulphoxide the coupling constant increases to 7·81 and 8·14 Hz. Other studies have shown either a regular increase or decrease in the vicinal coupling constant with increasing polarity, for the solvents used in their study. It was calculated that rotamer (XXVI) should be more stable than rotamers (XXIV) and (XXV) by 0·33 and 0·23 kcal/mole respectively on going from the gas phase to carbon tetrachloride, and by 0·52 and 0·35 kcal/mole on going from carbon tetrachloride to acetone. Consequently rotamer (XXIV) should decrease with increasing solvent polarity leading to a decrease in the vicinal coupling constant. The odd result was explained by hydrogen bonding which occurs between the protons of 1-phenyl-1,2,2-trihaloethanes and a strong hydrogen bonding solvent, such as acetone and DMSO, and that the hydrogen bonding favours the *trans* isomer.

Recently Roberts and coworkers[117] did a solvent-dependent study of 2,2,3,3-tetrachloro- and tetrabromobutanes. Low-temperature NMR spectra of the two compounds showed a significant solvent dependence. For example when the solvent for 2,2,3,3-tetrachlorobutane was changed from acetone-d_6 to a carbon disulphide–acetone-d_6 (10:1 v/v) mixture, the relative intensity of the peaks assigned to the *gauche* isomer decreased in intensity. This is expected since decreasing the dielectric constant of the solvent will reduce the stability of the rotamer possessing the largest dipole moment, in this case the *gauche* isomer. Table 13.19 gives the free-energy differences.

Table 13.19   Conformational free-energy differences for tetrahalobutanes[117]

	Solvent	Temp.(°C)	ΔG(kcal/mole)
2,2,3,3-Tetrachloro	Acetone-d_6	− 44	170
	CS_2–acetone-d_6 (10:1 v/v)	− 50	630
2,2,3,3-Tetrabromo	C_6H_5Cl–acetone-d_6 (2:1 v/v)	− 25	680
	CS_2	− 3	∼1800

A recent infrared and Raman study of 1,2-diphenylethane, 1,2-di-(*p*-bromophenyl)ethane,   1,2-di-(*p*-chlorophenyl)ethane,   *meso*-2,3-diphenyl-butane and its *pp'*-dibromo derivative suggested an equilibrium of both *gauche* and *trans* rotamers being present in solution, with the latter being the least populated isomer.[118] The *trans*:*gauche* ratio in solution for 1,2-di-(*para*-substituted phenyl)ethanes was determined to be 63:16[119] and 84:16 for *meso*-2,3-di-(*p*-bromophenyl)butane. However, in the case of 1,2-di-phenylethane no *gauche* rotamer could be detected by infrared. This is in contrast to its *pp'*-disubstituted derivative, where both isomers were shown to be present. The result was explained as being due to the fact that the dipole moment of 1,2-diphenylethane is essentially zero and not subject to any significant dielectric energy change when the polarity of the solvent is increased, whereas the polar *gauche* rotamer of the *pp'*-disubstituted derivative should be appreciably stabilized by this effect.

The solvent dependence of the rotational isomers of dihalopropenes has also been investigated.[120–123] One of the earlier studies was concerned with the solvent dependence of 2-bromo-3-chloropropene and 2,3-dichloro-propene by IR, NMR and dipole moments.[120] For example the IR shows doublet splittings for the C=C and C—C stretching bands and =CH_2

symmetrical bonding fundamentals. However, the spectra of related compounds such as 2-bromopropene, allyl chloride, methallyl chloride and *trans*-1,2,3-tribromopropene failed to show these splittings. The relative intensities of these doublets were affected by changing the solvent, suggesting the presence of rotational isomers. The energy difference for 2,3-dichloropropene liquid was estimated to be 914 cal/mole between (XXVII) and (XXVIII), with (XXVIII) being the more stable isomer. However, an IR

(XXVII)   (XXVIII)

and Raman study on 2,3-dichloropropene suggested that the more polar conformation was the most stable form in solution.[122] The same conformation was found to be the most stable in the vapour state from dipole moment measurements. If the polar form is more stable, then the chlorine atoms cannot be in a *trans* position because the dipole moment would have its smallest value for this configuration. They concluded that the most stable conformation has C_1 symmetry with the chlorine atoms intermediate between a *trans* and *cis* position.

The rotational isomerism of *trans*-1,3-dichloropropene was investigated by NMR and shown to be solvent dependent.[123] The solvent dependence of the proton coupling was interpreted on the basis of two discrete rotamers with the *gauche* form favoured by 0·5 kcal/mole in polar solvents. It was pointed out that it is the quadrupole term which differentiates between *gauche* and *trans* forms, since the dipole moments of the two isomers were found to be very similar. The above result could not have been forecast on simple grounds.

Substituted nitriles are known to exist in more than one form.[60,124–129] An early IR investigation established that the β-chloro- and β-bromopropionitriles exist as a mixture of conformers in solution, the *gauche* isomer being more stable.[126] An NMR study of β-bromopropionitrile in a variety of solvents was interpreted on the basis of two rotational isomers.[129]

Table 13.20 shows that the *gauche* rotamer is more stable due to the larger value of the dipole moment associated with this configuration, except for carbon tetrachloride ($L < 0$) indicating greater stability for the *trans* isomer. Probably strong dipole–dipole interactions appearing in carbon tetrachloride solution lower the energy of the *trans* form to make it relatively more stable than the *gauche* rotamer in carbon tetrachloride. The energy difference for the *trans* to *gauche* isomerization was determined to be −0·45 kcal/mole. This is in agreement with an IR study of succinonitrile which shows that

Table 13.20   The coupling parameters (N and L)
at infinite dilution[129]

Solvents	N_∞	L_∞
Methyl formamide	12·66	2·30
Dimethyl formamide	12·67	2·35
Acetone	12·69	2·14
Methyl chloroacetate	12·94	2·00
1,2-Dichloroethane	13·27	1·48
CH_2Cl_2	13·37	1·25
$CHCl_3$	13·85	0·55
C_2HCl_3	14·14	0·12
CCl_4	14·74	−0·76

at below $-44°C$ the *gauche* rotamer is the more stable isomer.[124] However,
β-fluoropropionitrile exists as a mixture of rotamers even at $-165°C$.[128]
A similar IR investigation of β-chloro-, bromo- and iodopropionitrile
have shown the existence of *trans* and *gauche* isomers.[60,108] In Table 13.21
are reported some of the known energy differences for these compounds.

Table 13.21   Enthalpy differences (kcal/mole) for the
trans–gauche equilibrium for the liquid
β-halopropionitriles

Compound	Reference		
	128	127	129
FCH_2CH_2CN	∼0·0	—	—
$ClCH_2CH_2CN$	−0·44	−0·38	—
$BrCH_2CH_2CN$	−0·54	−0·54	−0·54
ICH_2CH_2CN	−0·87	—	—

The ratio of *gauche* to *trans* is expected to be considerably greater in the
liquid state than in the vapour. As mentioned previously for succinonitrile
the *gauche* isomer is the more stable form, although it was estimated that the
trans isomer would be more stable in the vapour phase, it seems likely that
this would be the case for the β-halogenopropionitriles, unless there is
some specific association such as hydrogen bonding as in the case of 2-halo-
genoethanols.[130]

There are a number of cases where specific interactions such as hydrogen
bonding help to stabilize one isomer.[130–136] The IR of ethylene glycol for
example has shown that in both the liquid and vapour states, it is only
the *gauche* isomer which is present, presumably due to intramolecular
hydrogen bonding.[134] A similar result for 2-fluoroethanol was obtained.[132]

However, in this case there was evidence that there were some *trans* molecules present in the vapour state at 60–70°C. Both the *gauche* and *trans* rotamers of 2-nitroethanol populate the vapour and liquid states, with the *gauche* isomer being stabilized by a weak hydrogen bond.[134]

An NMR study of both ethylene glycol and 2-fluoroethanol[135] has shown a solvent dependence indicating that the previous IR results for these compounds may be in error. From the solvent dependence study it was concluded that 2-fluoroethanol exists in an equilibrium in which the population of the *trans* isomer amounts to not more than 5 %. In the case of ethylene glycol up to 20 % *trans* form is believed to be present.

Table 13.22 shows that the *gauche*:*trans* ratio increases with increasing dielectric constant of solution. Only dioxan behaves differently but this is not surprising since dioxan is known to exhibit strong preferential interactions.[137]

Table 13.22 Relative population of the *gauche* rotamer and enthalpy difference for (a) 2-fluoroethanol and (b) ethylene glycol[135]

Solvent	Concentration (Vol. %)	$n_g(N)$	$n_g(L)$	ΔH(kcal/mole)
(a) $CHCl_3$	10	0·95	0·95	1·35
Neat	—	0·95	0·95	1·35
D_2O	10	0·97	0·98	1·8
(b) $(C_2H_5)_3N$	10	0·81	0·77	0·38
Dioxan-d_8	7	0·86	0·84	0·61
CH_3COCH_3	20	0·83	0·81	0·46
CH_3CN	20	0·87	0·85	0·67
Neat	—	0·86	—	0·67
D_2O	20	0·88	0·87	0·75

13.5 The solvent dependence of conformer energies

In this final section we shall consider the application of the theory to cyclic compounds exhibiting conformational isomerism and, as in the case of the acyclic compounds, we shall consider first those compounds for which a quantitative treatment has been given; other classes of compound will then be discussed in the light of the theory but in a qualitative manner. The basic theory given in Table 13.1 defining the molecular geometries and bond dipole moments of haloethanes now needs to be extended to these more complex systems. Fortunately, both the molecular geometries and dipole moments are better known for the simple cyclic compounds we shall consider quantitatively than for the substituted ethanes, and thus the geometries and bond dipole moments required for each class of compound can be obtained from reliable experimental data. It should be explained at this stage that the bond dipole moments used for these calculations are not regarded as molecular invariants. For example, different $C-Cl$ bond moments are used in ethanes, cyclohexanes and 1,3-dioxans. The important point to note is that we are not concerned with a comprehensive calculation of molecular dipole moments, but are merely using those bond dipole moments for each class of compound which reproduce the observed dipole moments of the individual conformers. The first class of cyclic compounds to consider are the halocyclohexanes.

13.5.1 *Halocyclohexanes*

The determination of conformational preferences in halocyclohexanes has been investigated extensively in the last decade,[9] due to the relative ease of obtaining reliable experimental data and the insight which such determinations give on interatomic interactions in the molecule and on reactivities in these systems.

However, in most of these investigations the influence of the solvent has not been considered explicitly, even though this may be considerable. Recently the solvent theory has been applied to these compounds with some success,[46] and we shall follow this treatment.

In cyclohexyl halides, which have been the most extensively studied, $-\Delta G$(ax. → eq.) is remarkably independent of solvent, the range of values being virtually within the experimental error. For example Eliel obtains values for the chloride ranging from 0·34 to 0·51 kcal/mole for sixteen solvents.[20]

In complete contrast $-\Delta G(aa \to ee)$ for the *trans*-1,2-dihalocyclohexanes shows a pronounced dependence on the medium, varying from about $-1·0$ kcal/mole in non-polar media to about $+0·5$ kcal/mole in polar media (see Table 13.23 and references therein). Obviously this is due to the very different dipole moments of the conformers, the diequatorial isomer being favoured in more polar media.

The *cis*-1,3-dihalocyclohexanes should also on this basis show a large solvent effect, but in the few investigations performed so far no evidence of an equilibrium has been found,[144] presumably due to the large repulsive interaction of two axial halogens.

However $-\Delta G(aa \to ee)$ for the *trans*-1,4-dihalocyclohexanes also shows a pronounced solvent dependence, varying by about 0·6 kcal/mole in different solvents (Table 13.24 and references therein), even though in this case the dipole moments and therefore the dipolar interactions of both conformers are very small.

Thus it is clear that the dipole term alone cannot explain these results which are therefore a good test of the general treatment of the theory.

Table 13.23 Energy differences in *trans*-1,2-dihalocyclohexanes

$\Delta E^s(E^s_{ee} - E^s_{aa})$ (kcal/mole)

Solvent	ε	1,2-Dichloro						1,2-Dibromo				
		(calc.)		(obs.)				(calc.)	(obs.)			
Vapour	1.0	(0.70)	0.61[c]	0.72[d]	—	—	—	(1.30)	—	—	—	—
C_6H_{12}	2.0	0.15	—	0.72	0.41[f]	—	—	0.83	1.55[i]	0.85[f]	—	—
CCl_4	2.2	−0.08	—	0.58	0.37	−0.2[g]	0.0[h]	0.77	1.32	0.77	0.65[g]	0.50[h]
CS_2	2.6	−0.11	0.17	0.20	0.24	—	—	0.66	0.98	0.55	0.65	—
$CHCl_3$	4.7	0.40	—	—	0.0	—	—	0.35	—	0.26	—	—
Benzene	7.5[a]	−0.60	−0.43[e]	—	0.0	−0.65	−0.69	0.16	—	0.20	0.10	0.13
Pure liquid	b	−0.72	—	−0.60	0.0	—	—	0.18	0.33[j]	0.26	—	—
Acetone	20.2	−0.86	—	—	−0.42	—	—	−0.09	—	−0.10	−0.10	—
CH_3CN	35.9	−0.93	—	—	−0.47	−1.0	—	−0.15	—	−0.24	−0.30	—
CH_3NO_2	36.5	−0.94	—	—	−0.53	—	—	−0.16	—	−0.32	—	—

[a] See text.
[b] Estimated as 10.0 for the dichloro and equal to 6.9 for the dibromo.[138]
[c] Reference 139.
[d] Reference 140.
[e] For methyl acetate solution (ε = 6.8).
[f] Reference 141.
[g] Reference 142.
[h] See text.
[i] Reference 138.
[j] Reference 143.

Table 13.24   Energy differences in *trans*-1,4-dihalocyclohexanes

Solvent	ε	$\Delta E^s(E^s_{ee} - E^s_{aa})$					
		1,4-Dichloro			1,4-Dibromo		
		(calc.)	(obs.)	(obs.)	(calc.)		(obs.)
Vapour	1·0	0·80	—	—	0·70		—
C_6H_{12}	2·0	0·35	0·23[c]	—	0·26		—
CCl_4	2·2	0·30	0·41	—	0·21	0.2[c]	—
Et_2O	4·3	0·03	−0·16	—	−0·05		—
$CDCl_3/CD_2Cl_2$	11[a]	−0·22	—	−0·16[d]	—		—
C_6H_6	7·5[b]	−0·15	−0·16	−0·16	−0·22		−0·21[d]
Acetone	20·2	−0·29	—	—	−0·37	−0.45	—
CH_3CN	35·9	−0·34	—	—	−0·41		—

[a] 40/60 mixture.
[b] See text.
[c] Reference 145, see text.
[d] Reference 146.

One advantage of halocyclohexanes over the analogous haloethanes is that the dipole moments of a large number of conformationally fixed derivatives have been measured, and these provide useful checks on the model. The model used was relatively simple, based on a fixed cyclohexane geometry and a constant C—X bond moment, as follows.

The molecular geometry was taken direct from the recent determination of the cyclohexane geometry[147] together with that for haloalkanes (Table 13.1). Buys and Geisse obtain C—C—C and C—C—C—C angles of $111 \cdot 05(\pm 0 \cdot 15)°$ and $55 \cdot 9(\pm 0 \cdot 3)°$ for cyclohexane.[147] As the C—C—C angle used in Table 13.1 is identical to that in cyclohexane, the standardized geometry was used virtually unchanged for these systems. The only minor amendment is that in cyclohexane itself the C—C—C and H—C—H angles define the C—C—H angles as $109 \cdot 08°$ and this value was used throughout.

To summarize, all bond lengths were from Table 13.1, all C—C—C and C—C—X angles were $111 \cdot 00$ and all H—C—H angles were tetrahedral. This gave the C—C—C—C dihedral angle as $56 \cdot 00°$, the C—C—H angle as $109 \cdot 08°$ and the H—C—X angle as $105 \cdot 5°$.

The C—Cl and C—Br bond dipole moments were taken as $1 \cdot 9$ D and the remaining parameters needed in equation (13.18) (the molar volume and l) are obtained directly from the experimental values of the density and refractive index.[54] The complete set of parameters used in the calculations is given in Table 13.25.

Dipole moments   Unlike the corresponding ethanes, the dipole moments of the conformationally mobile halocyclohexanes can be estimated reasonably from conformational fixed analogues. Thus in Table 13.25 the dipole moments of the *ee* and *aa* conformers of the *trans*-1,2-dihalocyclohexanes are obtained as the observed moments of the 2*a*, 3*a*- and 2*e*, 3*e*-*trans*-decalins[148] and of *aa*- and *ee*-1,2-dibromo-4-*t*-butylcyclohexanes.[15] The values for the *cis* 1,3-dibromo and *trans*-1,4-dibromo and dichloro compounds are, however, the observed values for these compounds.

Also, the dipole moments of those compounds which are equilibrating between two identical conformations (the *cis*-1,2-, *trans*-1,3- and *cis*-1,4-dihalocyclohexanes) can be directly measured and compared with the calculated values and these are also given in Table 13.25.

Inspection of the table shows remarkably good agreement between the observed and calculated dipole moments. The observed and calculated values are identical for the *cis*-1,2-, *trans*-1,3- and *cis*-1,4-dihalocyclohexanes and the agreement with the *trans*-1,2-dihalo(*ee*) conformer is satisfactory when it is considered that small deformations of the cyclohexane ring may occur in the *t*-butyl and *trans*-decalin derivatives used for the observed value. Note that although the cyclohexane geometry is not tetrahedral, the condition that all the C—C—C and C—C—X angles are equal necessarily

Table 13.25   Molecular constants and calculated parameters for halocyclohexanes

Compound	Density[a]	$n_D{}^a$	Molar volume (ml)	l	Dipole moment (D)		k	h
					(calc.)	(obs.)[d]	(kcal/mole)	
1-Chloro {eq. / ax.}	1·016	1·4626	116·74	0·5504	1·90	2·2	1·12	{2·07 / 1·27}
1-Bromo {eq. / ax.}	1·3264	1·4956	122·93	0·5838	1·90	2·2	1·07	{1·84 / 1·12}
cis-1,2-Dichloro	1·2018	1·4963	127·35	0·5845	3·13	3·12	2·80	1·94
$trans$-1,3-Dichloro	—	1·49[b]	129[b]	0·5782	2·15	—	1·31	3·01
cis-1,4-Dichloro	—	1·49[b]	129[b]	0·5782	2·94	2·89	2·44	2·25
cis-1,2-Dibromo	1·803	1·5514	134·20	0·6385	3·13	3·12	2·66	1·62
$trans$-1,3-Dibromo	—	1·5472[c]	135·5[b]	0·6344	2·15	2·17[c]	1·24	2·96
cis-1,4-Dibromo	1·7714[c]	1·5476[c]	136·59	0·6348	2·94	2·92	2·31	2·18
$trans$-1,2-Dichloro {ee / aa}	1·1842	1·4907	129·24	0·5789	{3·13 / 0·37}	{— / 1·21[f]}	{2·76 / 0·04}	{2·28 / 2·81}
$trans$-1,2-Dibromo {ee / aa}	1·784	1·5495	135·63	0·6367	{3·13 / 0·37}	{3·28[e], 3·3[g] / 1·15[e], 1·19[f]}	{2·63 / 0·04}	{1·86 / 2·90}
cis-1,3-Dichloro {ee / aa}	—	1·49[b]	129[b]	0·5782	{2·15 / 3·78}	{— / —}	{1·31 / 4·03}	{3·97 / 1·45}
cis-1,3-Dibromo {ee / aa}	—	1·5445[c]	135·5[b]	0·6318	{2·15 / 3·78}	{2·19[c] / —}	{1·24 / 3·84}	{3·95 / 1·16}
$trans$-1,4-Dichloro {ee / aa}	—	1·49[b]	129[b]	0·5782	0·0	0·0	0·0	{9·16 / 5·59}
$trans$-1,4-Dibromo {ee / aa}	1·7714[c]	1·5476[c]	136·59	0·6348	0·0	0·0, 0·55[c]	0·0	{8·91 / 5·43}

[a] Reference 54.
[b] Average values, see text.
[c] Reference 138.
[d] Reference 88 unless stated otherwise.
[e] 2a,3a- and 2e,3e-dibromo-$trans$-decalin.[148]
[f] aa-1,2-dihalo-4-t-butylcyclohexanes.[15]
[g] ee-1,2-dibromo-4-t-butylcyclohexane.[15]

leads to the same calculated dipole moments for the *ea-* and *ee*-1,2-dihalo-cyclohexanes, providing some justification for the use of the dipole moment of the former as a model for the latter in obtaining the conformational preference in *trans*-1,2-dihalocyclohexanes.[14] The observed dipole moment for *cis*-1,3-dibromocyclohexane is identical to the calculated moment for the *ee* conformer, in agreement with the calculations of Franzus and Hudson[144] which are based on tetrahedral geometry, and confirm their interpretation that the molecule is entirely in this conformation.

The difference in the observed and calculated dipole moment of the cyclo-hexyl halides is a real effect and analogous to the situation in haloalkanes. The dipole moments of methyl, ethyl, n-propyl and isopropyl chlorides are 1·87, 2·03, 2·06 and 2·2 D respectively,[88] showing an almost constant effect of successive methyl substitution at the α-carbon atom and a negligible effect of β-methyl substitution. The dipole moment of cyclohexyl chloride is identical to that of isopropyl chloride in agreement with this scheme.

The only serious discrepancies in the table are for the *trans*-1,2-diaxial conformers and for the *trans*-1,4-dihalo compounds. The large observed dipole moment of the 4-*t*-butyl-1*a*,2*a*-dibromocyclohexane[15] and the analogous *trans* decalin derivative[148] has been noted previously, without any real explanation for this value.

Even more anomalous is the reported dipole moment of 0·55 D for *trans*-1,4-dibromocyclohexane.[144] The molecule has by symmetry zero dipole moment in both the chair forms; there is therefore no likelihood of any populated conformation with a large dipole moment and no good reason to postulate large amplitude molecular vibrations. The generally good agreement, however, provided considerable justification for the use of the bond dipoles in the calculation of the k and h values for these compounds.

The solvent dependence of the conformational equilibrium The data in Table 13.25 and equation (13.18) give the solvent dependence of the free-energy difference of any equilibrium between any two of the species shown. In Reference 46, only the basic theory (equation 13.18) was used, and these values are given here. The addition of the polar term for these compounds would make very little difference (cf. Table 13.4).

Cyclohexyl halides The first equilibrium to consider is the well-documented one in cyclohexyl chloride and bromide (Figure 13.6). The appropriate values of k, h and l given in Table 13.25 and equation (13.18)

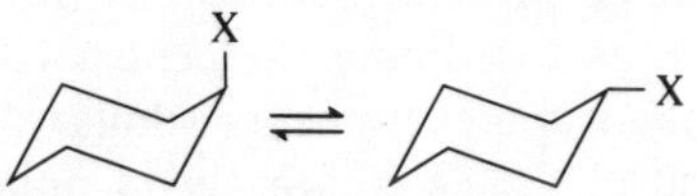

Figure 13.6 Conformational equilibrium for cyclohexyl chloride and bromide; X = Cl, Br

gives directly $\delta\Delta E(\Delta E^v - \Delta E^s)$ in terms of the dielectric constant of the solvent. The dipole moments (and therefore k values) for the axial and equatorial isomers are equal and therefore the dipolar term in equation (13.18) vanishes for this equilibrium. The quadrupole moments are not the same, as they depend on the position of the centre of the molecule and on the distance and orientation of the bond dipole from the centre. However, in this equilibrium the quadrupole terms are small and comparable and thus the predicted influence of the medium on the equilibrium is very small. As the quadrupole term is larger for the equatorial conformer this conformer will be more stabilized in polar media. Equation (13.18) and Table 13.25 predict that the free-energy difference $(-\Delta G)$ of this equilibrium should increase by 0·10, 0·21 and 0·26 kcal/mole on going from the vapour to media of dielectric constant 2·0, 7·5 and 35·9. This change is virtually within the experimental error of the determinations and this is in agreement with observation. For example, the systematic study of the solvent dependence of this equilibrium by Eliel[20] using the chemical-shift method gave $-\Delta G^0$ values ranging from 0·34 to 0·46 kcal/mole (9 solvents) for cyclohexyl chloride, and 0·27 to 0·38 kcal/mole (8 solvents) for cyclohexyl bromide, omitting associated solvents. It may be significant that the $-\Delta G$ values for cyclohexyl chloride and fluoride (which would behave similarly) measured in the vapour state are both at the low end of the reported ranges of values[9] but the only firm conclusion is that the predicted and observed solvent dependence of this equilibrium is extremely small.

trans-1,2-*Dihalocyclohexanes*   The conformational equilibrium in *trans*-1,2-dihalocyclohexanes (Figure 13.7) has been studied by a number of different techniques and investigations.

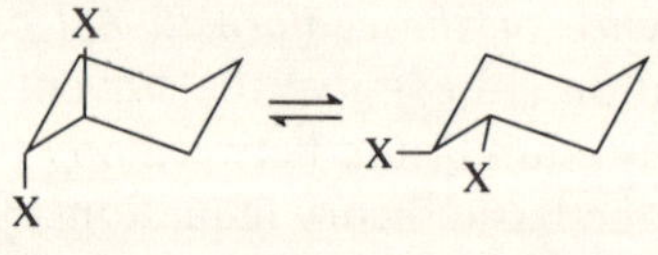

Figure 13.7   Conformational equilibrium for the *trans*-1,2-dihalocyclohexanes; X=Cl, Br

Again the effect of the medium on this equilibrium is given directly from Table 13.25 and equation (13.18). However, mere inspection of Table 13.25 shows that there will be now, in contrast to the monohalocyclohexanes, a large solvent effect, as the dipole moments of the conformers are very different but their quadrupole moments similar. Thus there will be considerable stabilization of the *ee* conformer in polar media. This is indeed observed and the observed and calculated results are compared in Table 13.23.

In this table, each separate column under the observed heading gives the results obtained by different techniques of measurement. These are in order, variable temperature IR measurements, single temperature IR measurements, NMR studies and dipole moment measurements. The values obtained by dipole moment measurements (h in the table) are obtained from the calculated moments for the ee and aa conformers of Table 13.25 with the observed dipole moments. These values are virtually identical to those given by previous workers[14] based on $\mu_{aa} = 0$ and $\mu_{ee} = \mu_{ea}$.

The dielectric constants used are those of the pure solvents at 30°C except that the free-energy differences in benzene solution were calculated for a solvent dielectric constant of 7·5 (cf. § 13.3.4).

For the dichloro compound, two groups of workers[139,140] measured the energy differences by variable temperature IR and these are given together in the table. They agree very well in the value of ΔE^{v} obtained and the average of these values (0·70 kcal/mole) was used together with equation (13.18) to obtain the calculated ΔE^{s} values shown.

In almost every measurement considered here, it is the systematic errors necessarily introduced in order to obtain the results, rather than the experimental errors, which are dominant. This is clearly shown in the very considerable differences in the ΔE values of the dichloro compound shown in the table. However, these errors are minimized when one considers the $\delta\Delta E$ values, i.e. the difference between two solvents measured by the same technique. These values are much more consistent and provide a better test of the theory; for example, $\delta\Delta E(\text{vapour} \rightarrow \text{pure liquid})$: observed = 1·32 kcal/mole (d, Table 13.23), calculated = 1·42; $\delta\Delta E(\text{CCl}_4 \rightarrow \text{CH}_3\text{CN})$: observed = 0·84 ($f$), 0·80 ($g$), calculated = 0·85. On this basis there is excellent agreement between the observed and calculated effects.

For the dibromocyclohexane the experimental results are much more consistent. Indeed the last three columns of Table 13.23 giving the results of single-temperature IR studies, NMR studies and dipole moment measurements agree to within the experimental error and also agree precisely with the calculated values. This agreement allows the prediction of the value of ΔE^{v} of 1·30 kcal/mole, which of course is not calculated by equation (13.18) which only gives $\delta\Delta E$ values. The results of Ul'yanova and Prentin[138] from variable temperature IR studies in non-polar media were the only values which were out of line for the dibromo compound.

The Russian school have also used the method of dilute solutions to obtain ΔE^{v}. This is essentially an extrapolation of ΔE^{s} versus $(\varepsilon - 1)/(2\varepsilon + 1) . a^3$ and should therefore give results similar to ours. However, their treatment gives very different results. Their values of ΔE for the dichloro compound in the non-polar solvents cyclohexane, carbon tetrachloride and carbon disulphide (d, Table 13.23) are, on their basis, consistent with the ΔE^{v} value; on our theory they are inconsistent. For the dibromo compound they obtain a ΔE^v value of 1·8 kcal/mole compared with the extrapolated value of Table 13.23 of 1·3, though this is based on different ΔE^s values.

One reason may be that following Powling and Bernstein[62a] they take a to be the radius of the *solvent* molecule. There is no theoretical justification for this procedure except for the case in which the solute is so small that it fits in the holes of the solvent lattice (cf. lithium in lithium iodide). This is a very rare occurrence in liquids and is certainly not valid for the systems considered here.

Furthermore, the extrapolation to the gas value is much less accurate than one would expect even if the simple extrapolation with $x = (\varepsilon - 1)/(2\varepsilon + 1)$ is used. This is easily demonstrated by considering typical values of x. For values of 1·0 (vapour), 2·0 (C_6H_{12}) and ∞ (e.g. acetone) the values of x are 0, 1/5 and 1/2. Thus the solvation effect of cyclohexane is almost half that of acetone, and the extrapolation of ΔE from the small measured range of non-polar solvents of cyclohexane, carbon tetrachloride and carbon disulphide to the vapour involves large uncertainties. A much simpler method of estimating ΔE^v stems from the relationship (equation 13.18 and Table 13.23), $\delta\Delta E(\text{vapour} \rightarrow CS_2) = \delta\Delta E(CS_2 \rightarrow \text{acetone})$. Thus a measurement of ΔE in carbon disulphide and acetone will provide a good estimate of ΔE^v.

It is of interest to compare the calculation of the solvent effect in the above compounds with that of the corresponding 1,2-dihaloethanes (Table 13.5). The observed effects are very comparable $\delta\Delta E(\text{vapour} \rightarrow \text{pure liquid})$ for $X = Cl, Br$ are 1·2 and 1·0 kcal/mole for the ethanes compared to 1·4 and 1·1 kcal/mole for the cyclohexanes (Table 13.23). Furthermore, the dipole moments of the isomers are very similar. However, the details of the calculations are quite different. The much smaller molecular volume of the ethanes (78·9 and 82·5 ml, Table 13.25) results in a corresponding increase in the dipole term, which is about 50% larger than in the cyclohexanes. This is compensated to a great extent in the ethanes by a large quadrupole term favouring the *trans* isomer. In contrast, the quadrupole term in the cyclohexanes is very similar in the two isomers. Thus, the overall result is much the same but the detailed breakdown of the calculations quite different.

cos-1,3-*Dihalocyclohexanes*   The conformational equilibrium in the *cis*-1,3-dihalocyclohexanes (Figure 13.8) has not been studied systematically to any great extent, and there is no data to compare with the theoretical predictions of the solvent effect upon this equilibrium.

The incorporation of the k, h and l values for these conformers given in Table 13.25 into equation (13.18) allows the prediction of the solvent dependence of this equilibrium. The diaxial conformer becomes more favoured (or more accurately less unfavoured) as the polarity of the medium increases and the equation predicts $\delta\Delta E(\text{vapour} \rightarrow \text{solvent})$ values for the

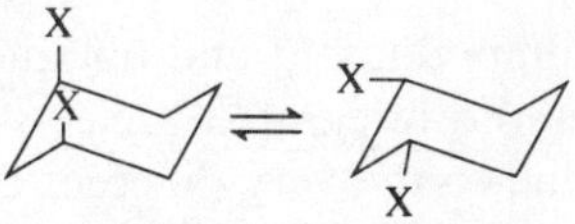

Figure 13.8   Conformational equilibrium for the *cis*-1,3-
dihalocyclohexanes; X = Cl, Br

dichloro compound for solvent dielectric constants of 2·0, 2·6, 7·5, 20·2
and 35·9 of 0·30, 0·42, 0·78, 0·95 and 1·01 kcal/mole respectively. The values
for the dibromo compound are very similar.

The observed dipole moment of *cis*-1,3-dibromocyclohexane is identical
to the calculated moment for the *ee* conformer (Table 13.25), thus on this
basis there is virtually none of the diaxial form present. Franzus and
Hudson[144] reached this conclusion and confirmed it by examining the
NMR spectrum of the compound at $-73°C$ in carbon disulphide. No peaks
corresponding to the *aa* conformer could be observed, even though the
corresponding *trans* isomer showed a definite coalescence point at $-33°C$,
showing that the conformations were frozen out at the lower temperature.
However, small amounts of the *aa* conformer could pass undetected, particularly if the proton resonances were near to those of the major conformer.

The interesting prediction given by the theory is that the repulsive interaction in the *aa* form can be compensated by about 1 kcal/mole of solvation
energy in polar solvents, and obviously the study of this equilibrium in polar
solvents would seem the most profitable means of attack.

trans-1,4-*Dihalocyclohexanes*   The solvent dependence of the conformational equilibrium of *trans*-1,4-dihalocyclohexanes (Figure 13.9) is of some
interest, in as much as both conformers have zero dipole moment and any
solvent dependence must be due to higher-order terms.

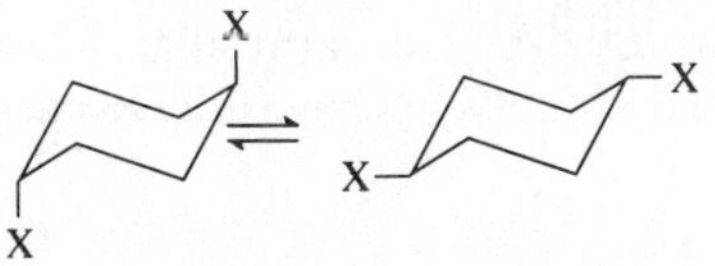

Figure 13.9   Conformational equilibrium for the *trans*-
1,4-dihalocyclohexanes; X = Cl, Br

Inspection of Table 13.25 shows that the diequatorial conformer has a
much larger quadrupole moment than the diaxial conformer and therefore
will be preferentially stabilized in more polar media, and insertion of the
value of *h* and *l* in Table 13.25 into equation (13.18) gives the predicted increase
in energy.

There are only a few experimental observations to test these predictions.
Wood and Woo observed the low-temperature NMR spectrum of the

dichloro and dibromo compounds[146] and obtained the proportions of the two forms directly from the integrated areas. As the assignment of the peaks is unambiguous, due to the very different coupling constants in two conformers, this method is definite. The only limitation is the accuracy with which the peak areas can be measured. Wood and Woo stated that there was no solvent effect in this equilibrium, but this conclusion was based on an unfortunate choice of solvents. They used toluene-d_8 and a 60/40 (v/v) mixture of CD_2Cl_2 and $CDCl_3$. Toluene behaves like benzene in these equilibria, i.e. as if the dielectric constant was about 7·5, and the dielectric constant of the $CD_2Cl_2/CDCl_3$ mixture (solvent dielectric constants of ~ 14 and 6 at $-70°C$) will be very similar.

However, Kozima and Yoshino[145] have observed a pronounced solvent dependence of this equilibrium using Raman spectroscopy. This technique gives $\delta\Delta E$ values (§§ 13.2.2 and 13.4.1) which together with the definite value of ΔE(toluene) obtained by Wood and Woo give the ΔE values in the other solvents studied. These values are given in Table 13.24 together with the calculated values which are obtained directly from equation (13.18) using a value of ΔE^v which reproduces the observed benzene result. The less detailed results for the dibromo compound have been treated similarly.

Inspection of Table 13.24 shows that, within the limited precision of the experimental data, there is good agreement between the observed and calculated results. This is especially pleasing in as much as this demonstrates that the procedure used to obtain the molecular quadrupole moments is a reasonable approximation. In all the other examples so far studied, the dipolar term is dominant and could therefore hide any errors in the quadrupole term.

To summarize, the extension of the solvent-effect model to the conformational equilibria in halocyclohexanes would appear to be in quantitative agreement with the experimental results and no modifications to the theoretical treatment are necessary. Very recent results[149] using the NMR integration technique in a variety of solvents have fully confirmed these conclusions.

13.5.2 *Furfuraldehyde*

The rotational equilibrium between the two planar forms of furfuraldehyde (OO *cis* $\rightleftharpoons$ OO *trans*) has been the subject of numerous investigations and some controversy, since the first demonstration of this equilibrium, by Allen and Bernstein.[150] They found $\Delta H \sim 1$ kcal/mole in the liquid but were unable to observe any temperature dependence of the IR and Raman spectra of the vapour. In agreement with this Karabatsos and Vane, from a study of long range coupling constants, concluded that only the *cis* form was present in solution.[151]

In 1965 two detailed investigations appeared. The microwave spectrum of the vapour gave $\Delta H = -0.99(\pm 0.2)$ kcal/mole and also the torsional potential for the rotation ($V_2 = 8.67$ kcal/mole).[3] Dahlquist and Forsen succeeded in observing the separate isomers by low-temperature NMR in dimethyl ether solution to give $\Delta H = -1.05$ kcal/mole and also an energy of activation (ΔH^*) of 11.5 kcal/mole.[2] Their assignment, based on chemical shift values was the reverse of that of Reference 151.

Subsequent values of the energy of activation were $V_2 = 7.0$ kcal/mole in the vapour from far infrared spectroscopy[152] and $\Delta H^* = 12.1(\pm 2)$ kcal/mole in the liquid from ultrasonic measurements.[153] Also further IR studies[154] showed the *trans* form to be more stable in carbon tetrachloride solution, a conclusion which is also supported by dipole moment measurements.[88]

Recently Roques and coworkers from a comparison of crystal structures and coupling constant data[155] and Martin and coworkers from other NMR studies[156] further supported the assignment of Reference 151, and this is now undoubtedly the correct assignment. However, considerable confusion still exists. For example Arlinger and coworkers use the vapour-state value of ΔH in support of their assignment of the conformers of 2-acetylfuran[157] even though the comparison is with the NMR of furfur-aldehyde in the liquid. Similarly Roques and coworkers dwell at length on the reasons for the preferential thermodynamic stability of the *cis* form[155] when in fact in the vapour it is the *trans* form which is the more stable.

Abraham and Siverns[1] measured the aldehyde-ring proton coupling constants in a number of solvents, using the low-temperature measurements of the couplings in the distinct conformers to obtain the relative conformer energies in these solutions. These and the earlier results were then given a quantitative explanation in terms of the solvent theory, and it is this treatment which is followed here.

The theory was also used to predict the solvent dependence of the barrier to interconversion as well as the conformer energy difference. In order to achieve this the molecular geometry and the dipole moments of the three states involved, i.e. the *cis* and *trans* conformers and the transition state (which was assumed to have the aldehyde group twisted 90° out of the plane of the furan ring) were estimated.

The geometry used (Figure 13.10) was taken from the microwave data of furan, the crystal structure determination of the $C-C-CO_2H$ angle in 2-furoic acid and the known geometry of the aldehyde group.[158]

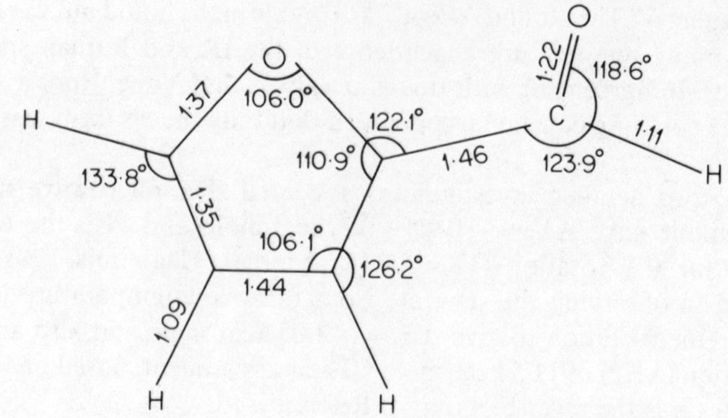

Figure 13.10   The geometry of furfuraldehyde

The bond dipole moments were obtained using the microwave determinations of the dipole moments of the *cis* (3·93 D) and *trans* (3·23 D) conformers[3] and of furan (0·66 D).[88] Using C=O and C—O dipoles of 3·5 and 0·4 D with the above geometry gave values of 3·97 D (*cis*), 3·25 D (*trans*) and 0·48 D (furan), in very reasonable agreement.

With the same geometry the CNDO (INDO) MO programme[29] gave dipole moments for the *cis*, 90° and *trans* forms of 3·67 D (3·70), 2·73 D (2·78) and 2·67 D (2·82). The *cis* and *trans* values were somewhat lower than the experimental ones but compare reasonably in the relative sizes of the dipoles.

The value for the transition state was virtually indentical to that of the *trans* form. A value of the C=O dipole of 2·8 D (the value for an unconjugated carbonyl group) gave the dipole moment of the transition state as 2·93 D, which agreed well with the MO calculations and was used in the calculations. The remaining parameters needed in equation (13.18) are experimentally measured quantities and are given in Table 13.26 together with the resulting values of k and h. These can now be used to evaluate the medium dependence of ΔE and ΔE^*. (As for the halocycloalkanes, the calculations omitted the polar term, which would again be expected to be small.)

Table 13.26   Molecular constants and calculated parameters for furfuraldehyde

Isomer	Density[54]	n_{D}^{54}	Molar volume (ml)	l	Dipole moment (D)	k (kcal/mole)	h (kcal/mole)
cis					3·97	6·92	16·38
trans	1·1598	1·5261	82·84	0·6140	3·25	4·63	13·92
90°					2·93	3·78	9·30

Table 13.27 Observed and calculated energy differences and activation energies (kcal/mole) for furfuraldehyde.

Medium	ε	$\Delta E(E_{cis} - E_{trans})$		$\Delta E(E_{90°} - E_{trans})$		Reference
		(obs.)	(calc.)	(obs.)	(calc.)	
Vapour	1·0	1·0, 1·5, 2·0	(1·5)	8·1, 8·7	(8·5)	2, 3, 150
Carbon tetrachloride	2·2	0·2	0·56	—	—	88
CF_2Cl_2	2·9	0·34	0·29	—	—	1
Benzene	2·2	−0·1	0·54	—	—	88
Dimethyl ether	7·90	−0·21	−0·43	—	—	1
Dimethyl ether (−120°C)	12·0	−0·58	−0·60	—	—	1
		−0·60		10·5	10·4	2
Acetone	19·5	−0·55	−0·72	—	—	1
Acetone (−120°C)	37·4	−0·77	−0·84	—	—	1
Pure Liquid	41·9	−1	−0·85	11	10·6	150, 153
DMSO	45·0	−0·84	−0·85	—	—	1

The results of the various investigations were collected in Reference 1, and are given in Table 13.27. These can be used to test the predictions of the theory, by again inserting the relevant parameters of Table 13.26 into equation (13.18); the calculated energy differences are also given in the table.

Inspection of these data and the corresponding calculated energies shows immediately the reason for the confusion in the literature concerning this equilibrium. The energy difference changes sign with solvent, being approximately zero for a solvent of dielectric constant of about 5 (e.g. chloroform). Thus interpolations even between one solvent and another can result in the wrong assignment. However, the calculated energies are in complete agreement with the observed energies and therefore the theory provides a complete quantitative explanation of the results. The one anomalous solvent, is as usual, benzene, which again gives the result expected for a solvent of dielectric constant 7·0.

The theory also provides a quantitative explanation of the various estimates of the energy of activation, though these are inevitably less accurate than the conformer energy differences. The increase in barrier height is simply due to the increased solvation energy of the planar forms. Note that the *cis* conformer is even further stabilized, due to its larger dipole moment, $\Delta E^*(cis \rightarrow trans)$ changing from 7·0 kcal/mole in the vapour to 12·1 kcal/mole in the pure liquid. Again this is given immediately from equation (13.18). It is not necessary to invoke any other factor to account for the differences in the results.

13.5.3 *5-Substituted*-1,3-*dioxans*

In this section we wish to consider the application of the theory to conformational measurements obtained by chemical means. The chemical equilibrium by acid catalyst of compounds such as 5-substituted 2-isopropyl-1,3-dioxans results in an equilibrium between the *cis* and *trans* isomers (Figure 13.11). The diaxial intermediate *C* may be safely disregarded since

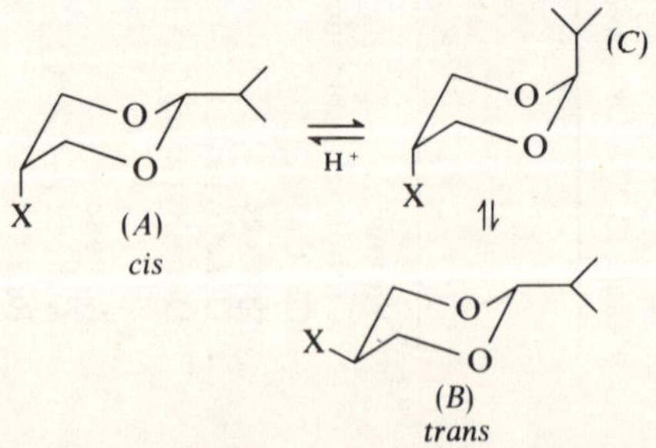

Figure 13.11 The *cis–trans* equilibria in 5-heterosubstituted 2-isopropyl-1,3-dioxans

the ΔG^0 value of a 2-isopropyl group in a 1,3-dioxan is very large (4·2 kcal/mole). Thus the chemical equilibration is in effect an equatorial $\rightleftharpoons$ axial conformational equilibrium between $A(cis)$ and $B(trans)$ and the measurement of this equilibrium in different solvents will produce ΔG values which may be treated exactly as the earlier measurements as a test of the solvent theory.

The observed ΔG values obtained from such measurements are given in Table 13.28 for a variety of 5-substituted 2-isopropyl-1,3-dioxans and do

Table 13.28 ΔG^0 values (kcal/mole) for the *cis-trans* equilibrium (Figure 13.12) in 5-heterosubstituted 2-isopropyl-1,3-dioxans (at 25°)[159]

Solvent	ε	Heterosubstituent				
		F	Cl	Br	CN	OCH$_3$
Carbon tetrachloride	2·24	0·36	$-1·40$	$-1·71$	—	$-0·90$
Diethyl ether	4·33	0·62	$-1·26$	$-1·45$	$-0·21$	$-0·83$
Benzene	2·20	0·83	$-0·89$	$-1·17$	—	$-0·59$
Chloroform	4·81	0·87	$-0·94$	$-1·35$	—	$-0·16$
Acetonitrile	37·5	1·22	$-0·25$	$-0·68$	0·55	0·01

indeed show a pronounced dependence on the solvent. These results are of general interest; in particular the positive value of ΔG for the fluoro compound is noteworthy (for a full discussion see Reference 159). However, we shall only be concerned here with that part of the investigation dealing with the application of the solvent theory to these results.

For this it is necessary to obtain the molecular geometry, bond dipole moments and other parameters needed in the fundamental equations. The 1,3-dioxan ring geometry was taken from electron diffraction measurements on 1,3-dioxans and is shown in Figure 13.12. The substituents were defined in accordance with the standard geometry given previously (Table 13.1 and Reference 52). Thus all C—X—C angles (X = O, F, Cl, Br) were taken as 111·0°.

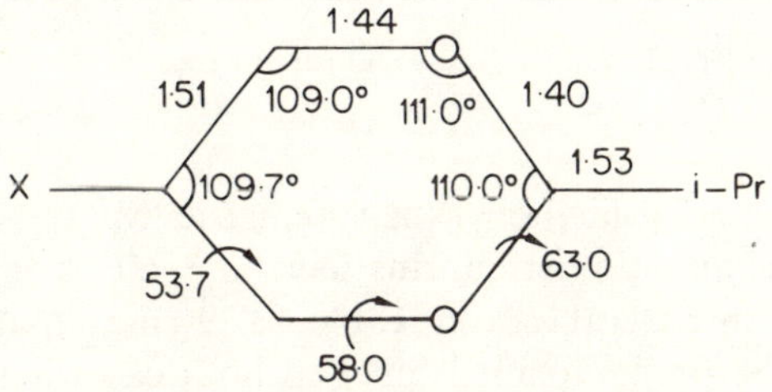

Figure 13.12 Bond lengths, bond angles and dihedral angles ($\curvearrowright$) for 1,3-dioxans

The C—O bond moment was taken from the observed moment (1·30 D) and C—O—C angle (111·5°) of dimethyl ether to give μ(C—O) = 1·16 D. Using this and the defined geometries gave calculated dipole moments for 2-phenyl-1,3-dioxan and 2-*p*-chlorophenyl-1,3-dioxan of 1·93 and 3·46 D (using the observed value of the Ph—Cl dipole of 1·75 D) which compared well with the observed values (2·04 and 3·54 D).

A further check was obtained from the dipole moments of the substituted 2-methoxy-1,3-dioxans though here there is an additional complication of rotational isomerism about the C—O bond of the methoxyl. The calculated and observed values agreed very well and therefore provided a good support for the bond dipole moments used.

For the 5-methoxyl-2-isopropyl-1,3-dioxans studied, the calculated dipole moments for the two rotamers of the *cis* and *trans* compounds are given in Table 13.29. The two rotamers are defined by the H—C—O—Me torsional

Table 13.29 Molecular parameters and calculated and observed energy differences (kcal/mole) for 5-substituted 2-isopropyl-1,3-dioxans

	X = F	X = Cl	X = Br	X = OMe (60°)	X = OMe (180°)
μ(D)(*cis*)[a]	3·09	3·18(3·04)	3·18	2·99(2·85)	2·13
μ(D)(*trans*)[a]	1·07	1·05(0·87)	1·05	1·38(1·30)	2·30
k(*cis–trans*)	2·13	2·18	2·14	1·70	—
h(*cis–trans*)	2·16	1·70	1·47	0·32	—

ΔE^{b}

	X = F		X = Cl		X = Br		X = OMe	
Solvent	(calc.)	(obs.)	(calc.)	(obs.)	(calc.)	(obs.)	(calc.)	(obs.)
Vapour	0·6	—	2·2	—	2·5	—	1·25	—
CCl$_4$	−0·37	−0·36	1·41	1·40	1·75	1·71	0·80	0·90
Et$_2$O	−0·78	−0·62	0·90	1·26	1·26	1·45	0·47	0·83
CHCl$_3$	−0·84	−0·87	0·85	0·94	1·21	1·35	0·43	0·16
CH$_3$CN	−1·43	−1·22	0·29	0·25	0·68	0·68	0·08	−0·01

[a] Calculated values; observed values are shown in parentheses.
[b] E (*cis*) − E (*trans*).

angle, thus for the axial methoxyl the symmetric 'Me-outside' rotamer has $\omega = 180°$ and the enantiomeric forms have $\omega = 60°$. The observed dipole moments shown in parentheses in Table 13.29 show that as expected on steric grounds, the predominant isomer in both cases is the enantiomeric form ($\omega = 60°$). Only this form was considered in the solvation energy calculations.

For the halosubstituted compounds the values of the C—X moments used were C—F 1·4 D, C—Cl and C—Br 1·5 D after Smith. These are quite different from those used for the haloethanes (Table 13.1) and halocyclohexanes (§ 13.5.1), no doubt due to the electronegative oxygen atoms of the ring. However, these values gave calculated dipole moments for the 5-chloro-2-isopropyl-1,3-dioxan of 3·18 D (*cis*) and 1·05 D (*trans*) which are in good agreement with the observed values of 3·04 and 0·87 D and confirm the validity of the bond moments used for these calculations.

The molar values and refractive indices of these solutes were estimated from similar compounds and hence the solvation energies of the isomers were obtained directly by the application of equation (13.18). It is of interest to note that the equation and the bond dipole moment model were used unchanged for the calculations although of course this is a multi-bond moment calculation rather than the two-component systems considered previously. As for the other cyclic molecules considered in this section only equation (13.18) was used. In these cases the application of the polar term would not significantly affect the conclusions.

The solvation energies arrived at by these calculations are given in Table 13.29 together with the observed values. As always the calculations only give $\delta \Delta E$ values and thus it is the agreement between the different solvents which is of significance. In Table 13.29 the comparison between the observed and calculated values is made clearer by equating the observed and calculated energy differences for carbon tetrachloride solution, and from these the calculated values of ΔE^{v} were obtained. This then uniquely defines the calculated values in any other solvent, which from Table 13.29 can be seen to agree in general extremely well with the observed values. The calculated values for the methoxyl substituent, in which case the agreement with the observed is not as good as for the halides, were obtained by taking a value of ΔE^{v}, such that the calculated values have the same mean over all the solvents as the observed values. The less good agreement was ascribed to the conformational (rotameric) inhomogeneity of the OMe substituent.

In general the large solvent effects found in these systems can be given a quantitative explanation on the basis of the theory. This agreement lead to two further useful applications. Significant deviations from the predicted values were suggested as due to specific solvent–solute interactions. Such deviations occurred for the methoxyl compound in chloroform and for all the compounds in benzene solution. Also the agreement lends some significance to the predicted values of ΔE^{v} obtained in Table 13.29. It is noteworthy that it was predicted that in the vapour the *trans*-fluoro compound should be more stable than the *cis* in contrast to all the solution results.

The results of this and the previous two sections do provide good evidence that the theory may be applied, essentially without any change, to solvent effects in conformational isomerism. In the final section we shall detail a number of other systems and consider them in the light of the theory, even though the model has not been quantitatively applied to these systems.

13.5.4 *Cyclic ketones*

Cyclic ketones represent a group of compounds which have been extensively studied by a variety of means. The α-chloro- and bromocyclohexanones were two of the first compounds studied (XXIX, R = H, X = Cl or Br). The equatorial–axial equilibrium (XXIXe $\rightleftharpoons$ XXIXa), as is well known, is

controlled by both steric and polar effects, with only polar effects being solvent dependent.

In the 2-bromo derivative the electrostatic dipole–dipole repulsions would be expected to destabilize conformer (XXIXe), since the negative ends of the dipoles are closer in the equatorial conformer than the axial, with the positive ends approximately equally separated in both conformers.[10] Consequently, electrostatically the axial isomer would be expected to be more stable than the equatorial conformer. However, this interaction should decrease as the polarity of the solvent increases.[160]

The initial study of the composition of the equilibrium mixture of 2-bromocyclohexanone was performed by studying its dipole moment in solvents of varying dielectric constant (Table 13.30).[161] The data was not at first interpreted solely on the basis of an axial–equatorial equilibrium, but around the possible presence of a third isomer, a 'flexible form', whose conformation was intermediate between the other two and could participate in an axial $\rightleftharpoons$ 'flexible form' $\rightleftharpoons$ equatorial equilibrium. Allinger studied the

Table 13.30   Dipole moments of 2-halocyclohexanones[161]

	Solvent			Theoretical	
	Heptane	Benzene	Dioxan	(eq.)	(ax.)
2-Fluoro	3.76^a	4.09^b	4.20^a	4.35^a	3.00^a
	—	3.86	—	4.00	2.70
2-Chloro	3.45	3.78	3.91	4.22	2.30
2-Bromo	3.37	3.50	3.64	4.22	2.30

[a] Reference 162.
[b] Reference 163.

α-bromo and α-chloro compounds by examining their infrared and ultra-violet spectra in several solvents.[164] The results for 2-bromocyclohexanone show that the axial conformer is more stable in both non-polar solvents (74% axial conformer in carbon tetrachloride) and polar solvents but to a lesser extent in the latter (53% axial in dimethyl sulphoxide). These results agreed with previous work, using the equilibration method[165,166] where the axial conformer was shown to be the most stable species in non-polar solvents (78%). The data were interpreted solely in terms of an axial–equatorial equilibrium.[164] In the case of 2-chlorocyclohexanone it was found that the chlorine atom has a greater tendency to occupy the equatorial position than does the bromine atom (24% equatorial in isooctane and 63% equatorial in dioxan). From infrared studies the enthalpy difference was found to be 1·1 kcal/mole in pure liquid,[10,166–168] and 0·8 kcal/mole in carbon disulphide,[10,167,168] the axial isomer being least stable in both cases. These values differ, as would be expected, from those in polar solvents. For example in dioxan the enthalpy difference was shown to be 0·4–0·6 kcal/mole,[10,161,166] and −0·7 kcal/mole to −0·2 kcal/mole in non-polar solvents.[10,161,166]

The data are consistent with the interpretation that the equilibrium is shifted toward the equatorial form in the more polar solvent. This shift in equilibrium is observed to an even greater extent in α-fluorocyclohexanone which exist primarily as the equatorial conformer.[162,169] This is obvious when its dipole moment is compared with the theoretical moment of the equatorial conformer in Table 13.30. Allinger calculated that in heptane the equatorial conformer made up 52% of the mixture, while in dioxan it constituted 85% of the mixture.[160]

NMR gives analogous results for these compounds when compared with the results from other methods.[170] Table 13.31 summarizes the results from the various methods mentioned.

The discrepancies reported above may to some extent be attributed to concentration differences, since not all the measurements were made at infinite dilution. However, the accuracy of the infrared results has been questioned.[173,174]

The differences between the two NMR results may in part be due to a concentration difference or the fact that one of the groups[171] did not correct for solvent effects of the coupling constants.

In Table 13.32 observed and calculated free energies for the α-halocyclo-hexanones are reported.[170]

In the calculation of ΔG^0 account was taken of non-bonded dipole–dipole and induced dipole interactions. The agreement between calculated and observed values for the 2-chloro and 2-bromo derivatives is considered fair in most cases, but the 2-fluorocyclohexanone does not agree very well and is an indication of a deficiency in the model chosen. This is more clearly

Table 13.31   The population of the axial conformer of α-halocyclohexanones[170]

Solvent	Halogen	IR	UV	DM	NMR Ref. 170	NMR Ref. 171
Hydrocarbon	Cl	73[a]	63[a]	76[a]	77	58
	F	—	—	48[b]	43	—
Carbon tetrachloride	Br	74[c,d]	—	—	86	80
	Cl	—	—	—	72	58
Benzene	Br	60[c,d]	—	76[e]	80	62
	Cl	—	—	56[a]	50	35
	F	—	—	23[b]	28	—
p-Dioxan	Br	57[c,d]	—	62[e]	74	54
	Cl	25[a]	37[a]	37[a]	44	33
	F	—	—	15[a]	22	—

[a] Reference 166.
[b] Reference 163.
[c] Reference 172.
[d] Reference 162.
[e] Reference 160.

Table 13.32 ΔG^0 (kcal/mole) of the α-halocyclohexanones[170]

Solvent	F (calc.)	F (obs.)	Cl (calc.)	Cl (obs.)	Br (calc.)	Br (obs.)
C_6H_{12}	1·13	−0·17	1·13	0·74	1·10	1·28
CCl_4	1·02	−0·35	1·07	0·58	1·04	1·11
$CHCl_3$	0·54	−0·85	0·74	−0·12	0·78	0·71
CH_3CN	0·17	−1·17	0·49	−0·38	0·58	0·00

seen in the fact that in acetonitrile the model used predicts for the 2-fluoro compounds 57% axial conformer, whereas the observed value is only 13%.

The studies show that the axial–equatorial ratio increases by changing X from fluorine to chlorine to bromine. The equatorial conformer is more stable when X is fluorine than the axial conformer, and when X is chlorine or bromine the axial conformer is more stable in solvents of low polarity. There are other studies[167,168] which suggest that the equatorial conformer is more stable than the axial conformer in non-polar solvents for 2-chloro-cyclohexanone. However, this result is believed to be in error, and could be due to the fact that the compound is associated at low concentrations.[10,171]

In special cases where the conformer is optically active, optical rotary dispersion (ORD) and circular dichroism (CD)[171,175–177] can be used along with the more standard methods to determine the composition of the

$$(XXXe) \rightleftharpoons (XXXa)$$

(XXXe) (XXXa)

equilibrium mixture. Two compounds which were studied by use of these methods were *trans*-2-chloro- and bromo-5-methylcyclohexanone (XXX, X = Cl or Br).[175-177] In these compounds the methyl group would be expected to prefer to be in the equatorial position while the halogen prefers to be in the axial position. These opposing, but approximately equal, forces will cause the compound to exist as a mixture consisting of both *trans* equatorial and *trans* axial conformers. The equilibrium was investigated in a variety of solvents by ORD, CD, dipole moment, infrared and ultraviolet spectroscopy.

The conformational equilibrium exhibited by the two chair forms (XXXe, XXXa) should be dependent on the polarity of the solvent. A polar solvent would be expected to favour the equatorial conformer at the expense of the axial one while an increased proportion of the axial conformer would be expected in non-polar solvents. Since the axial haloketone rule[171] predicts a positive Cotton effect for conformer (XXXe) and a negative one for conformer (XXXa), a change in the optical spectra would be expected upon changing the solvent. Table 13.33 summarizes and compares the results for 2-bromo- and 2-chloro-5-methylcyclohexanone.

Another series of compounds which have been extensively studied are the 2-halo-4-*t*-butylcyclohexanones (XIX, R = $(CH_3)_3C$)[172,178] Allinger[179] has suggested the possibility that these compounds may exist to a small extent in the boat form.

Table 13.33 Data for conformational isomers of *trans*-2-halo-5-methylcyclohexanone[175-177]

Solvent	% Axial conformer								
	2-bromo				2-chloro				
	DM	IR	UV	ORD	DM	IR	UV	ORD	CD
Heptane	45	39	37	44	—	34	18	18[a]	11[a]
Carbon tetrachloride	—	36	36	38	—	—	—	—	—
Benzene	—	—	—	—	15	24	—	—	—
Dioxan	19	26	18	20	—	22	12	0	4
Ethanol	—	—	16	18	—	—	—	—	—
Methanol	—	—	11	17	—	—	0	1	3

[a] Isooctane used as solvent.

In the case of the *trans*-2-bromo derivative, the only unfavourable inter-action is between the synaxial hydrogen and the bromine to the extent of 0·4 kcal/mole. Allinger calculated the free energy of the cyclohexanone ring in the boat form as 2·8 kcal/mole[179] and stated that the energy of the chair form may be lower than that of the boat form by at least 2·4 kcal/mole. He concluded that the boat form must be present to the extent of 2% at room temperature.

Consequently, the equilibrium between the chair and boat form (if any) would be so one-sided as to be essentially invariant to changes in solvent. Hence, the dipole moment of the mixture would be expected to be essentially independent of solvent polarity (Table 13.34).

Table 13.34   Experimental dipole moments of 4-*t*-butylcyclohexanones[179]

Ketone	Solvent		
	Heptane	Benzene	Dioxan
trans-2-Bromo	3·19	3·20	3·17
cis-2-Bromo	4·20	4·27	4·33
cis-2-Fluoro	4·37	4·35	4·39

The *cis* isomer presents a more complicated picture. In the chair form there are unfavourable electrostatic interactions between the dipoles which can be relieved if the molecule takes up the boat form (XXXI $\rightleftharpoons$ XXXII). The equatorial bromine in (XXXI) has an unfavourable electrostatic energy relative to the axial-like bromine in (XXXII). The free-energy increase on

(XXXI)   (XXXII)

going from the chair to the boat form was calculated to vary from 1·4 kcal/mole in heptane to 1·0 kcal/mole in benzene and 0·8 kcal/mole in dioxane.[180] Consequently, the dipole moment of the *cis*-2-bromo compound should be lower in non-polar solvents (heptane) than in polar solvents (dioxan), since the population of the boat form should increase as the polarity of the solvent decreases. The results bear this out, as shown in Table 13.34. From the data they calculated that in non-polar solvents the boat form should be present to about 10% of the mixture.

On comparing these results with the *cis*-2-fluoro-derivative, it was shown that the electrostatic energy of the boat conformer was lower than that of the chair form by an amount that varied from 0·7 kcal/mole in heptane to 0·2 kcal/mole in dioxan. This means that the fraction of boat form present would only change from 3–1 % under these conditions, so that the observed dipole moment should be invariant with solvent, as was found (Table 13.34).

The conclusions have been questioned by other workers in the field.[181] The authors suggest that, since the geometry of the $-C(=O)-CHBr-CH_2$ moiety is very close to that in the axially-substituted chair form, the values of the dipole moments and coupling constants in the former should equal those in the latter. Therefore they suggest, that if the solvent dependence of the dipole moment of the *cis*-2-bromo derivative is due to a chair–boat equilibrium, the coupling constant should show a significant variation, according to the following relationship which they derived, $\mu^2 \propto J$. The variation in coupling constant was not observed (Table 13.35).

Table 13.35 Dipole moments and coupling constants for 2-halo-4-*t*-butylcyclohexanones[181]

Solvent	2-Bromo		2-Chloro		2-Fluoro	
	μ	J	μ	J	μ	J
Heptane	4·26	—	4·32	—	4·35	—
CCl$_4$	4·30	18·0	4·36	17·7	4·33	18·2
CH$_3$CN	—	17·5	—	17·9	—	18·1

From the data shown above, they concluded that, since there were only small irregular changes of the coupling constant in various solvents, a chair–boat equilibrium is unlikely. They had expected the dipole moments to show a small decrease with increasing polarity (heptane $\rightarrow$ CCl$_4$ $\rightarrow$ CH$_3$CN). The larger value found for benzene was explained as being due to specific solvent–solute interactions.[170] They did not measure the dipole moments in dioxan, as the previous study had done,[180] since, based on their investigations, dipole moments observed in this solvent appear to be higher than predicted from theory. Thus the high values obtained in this solvent[180] may not necessarily be due to conformational effects.

According to Allinger, the bromo derivative, but not the fluoro compound, should occur to a significant extent in the boat form.[180] Hence, the solvent dependence of the dipole moment should decrease in the series bromide $\rightarrow$ chloride $\rightarrow$ fluoride. However, the results in Table 13.35 show practically the same dipole moment for all the compounds. They concluded, on the basis of their data, that a chair–boat equilibrium does not occur to any significant extent.

Heterocyclic ring compounds are also suitable for studies similar to those used for the cyclohexanones.[182,183] An interesting set of compounds that falls into this category is trimethylene sulphite and its derivatives which have been studied by a variety of techniques.[184–196]

One of the first studies on the present compound was a dipole moment investigation which suggested that trimethylene sulphite exists in solution as a mixture of possible chair forms (XXXIII) and (XXXIV).[184] However, NMR studies[185–187] suggest that only a single conformation is present, which is not undergoing inversion at an appreciable rate. Therefore, if one

(XXXIII)   (XXXIV)

considers only chair forms, the $S{=}O$ function must necessarily be in an equatorial or axial position. Dipole moments and infrared spectra[190–192] of a variety of cyclic sulphites indicate that the $S{=}O$ group preferentially occupies the axial position. X-ray results of trimethylene sulphite at $-100°C$ show that the compound exists in the chair conformation with the $S{=}O$ bond in the axial position, with all the dihedral angles close to $60°$.[188,189]

Recently a series of dipole moment studies on a number of these compounds in several solvents was reported.[190,191,193,195] The results are reported in Table 13.36.

Several of these compounds have dipole moments corresponding to a chair form with axial $S{=}O$ compounds (compounds 1,2,4,10,11). The dipole moment for compound (12) is seen to increase on going from cyclohexane to dioxan, the most polar solvent. This is what would be expected for an equilibrium mixture of conformers. However, for most of the other compounds the dipole moments are fairly constant in the different solvents which makes it highly unlikely that any of these compounds exist as equilibrium mixtures of conformations. They stress that, except for compound (12), it is unlikely that any of these compounds exist as an equilibrium mixture of two chair conformations.

Since it is believed that barriers to ring inversion are largely due to the torsional energy for rotation about these bonds, the rings having single bonds with high barriers to rotation would be expected to have a higher barrier to inversion.[195] Consequently, since rotational barriers about single bonds containing unshared electron pairs are believed to be high (10 kcal/mole), the compounds listed in Table 13.36 which have unshared electron pairs on three adjacent atoms would be expected to have a high barrier to inversion.[195]

Table 13.36 Dipole moments of the compounds of the general structure (XXXV)[195]

(XXXV)

Substituents[a]	Dipole moment (D)			
	C_6H_{12}	CCl_4	C_6H_6	Dioxan
(1)	Insoluble	3·34	—	—
(2) $R^2 = CH_3$	—	3·35	3·41	—
(3) $R^1 = CH_3$	—	4·66	4·75	—
(4) $R^2 = R^6 = CH_3$	3·44	3·51	3·60	—
(5) $R^2 = R^5 = CH_3$	3·97	3·93	3·93	—
(6) $R^1 = R^5 = CH_3$	Insoluble	5·31	5·37	—
(7) $R^1 = R^2 = R^6 = CH_3$	3·91	—	3·86	—
(8) $R^1 = R^2 = R^5 = CH_3$	4·53	—	4·71	—
(9) $R^1 = R^2 = R^5 = R^6 = CH_3$	4·22	—	4·30	—
(10) $R^3 = R^4 = CH_3$	3·38	—	—	—
(11) $R^3 = (CH_3)_3C$	3·50	3·54	3·60	3·66
(12) $R^4 = (CH_3)_3C$	3·54	3·63	3·90	4·15

[a] All Rs not specified = H.

13.6 References

1. R. J. Abraham and T. M. Siverns, *Tetrahedron*, **28**, 3015 (1972).
2. K. Dahlquist and S. Forsen, *J. Phys. Chem.*, **69**, 1062 (1965).
3. F. Monnig, H. Dreizler and H. D. Rudolph, *Z. Naturforsch*, **20a**, 1323 (1965).
4. E. I. Snyder, *J. Am. Chem. Soc.*, **88**, 1165 (1966).
5. S. Mizushima, *Structure of Molecules and Internal Rotation* (New York: Academic Press, 1954).
6. N. Sheppard, *Advan. Spectr.*, **1**, 288 (1959).
7. J. P. Lowe, *Progr. Phys. Org. Chem.*, **6**, 1 (1969).
8. P. Laszlo, *Progr. N. M. R. Spectr.*, **3**, 231 (1967).
9. J. A. Hirsch, *Topics in Spectrochemistry*, **1**, 199 (1967).
10. E. L. Eliel, N. L. Allinger, S. J. Angyal and G. A. Morrison, *Conformational Analysis* (New York: Wiley, 1965).
11. E. L. Eliel, *Angew. Chem. Intern. Ed. Enge.*, **4**, 761 (1965).
12. G. Chiurdoglu (Ed), *Conformational Analysis* (New York: Academic Press, 1971).
13. C. P. Smyth, *Dielectric Behaviour and Structure* (New York: McGraw-Hill, 1955).
14. P. Bender, D. L. Flowers and H. L. Goering, *J. Am. Chem. Soc.*, **77**, 3463 (1955).
15. H. J. Hageman and E. Havinga, *Tetrahedron*, **22**, 2271 (1966).
16. R. J. Abraham and M. A. Cooper, *Chem. Commun.*, 588 (1966).

17. F. Heatley and G. Allen, *Mol. Phys.*, **16**, 77 (1969).
18. G. Govil and H. J. Bernstein, *J. Chem. Phys.*, **47**, 2818 (1967).
19. S. L. Manatt and P. D. Elleman, *J. Am. Chem. Soc.*, **84**, 1305 (1962).
20. E. L. Eliel and R. J. L. Martin, *J. Am. Chem. Soc.*, **90**, 689 (1968).
21. F. R. Jensen and B. H. Beck, *J. Am. Chem. Soc.*, **90**, 3251 (1968).
22. H. S. Gutowsky, G. G. Bedford and P. E. McMahon, *J. Chem. Phys.*, **36**, 3353 (1962).
23. S. Ng, *J. Mag. Res.*, **7**, 370 (1972).
24. R. J. Abraham, L. Cavalli and K. G. R. Pachler, *Mol. Phys.*, **11**, 471 (1966).
25. G. Binsch, *Topics in Stereochemistry*, **3**, 97 (1968).
26. K. O. Hartman, G. L. Carlson, R. E. Witkowski and W. G. Fateley, *Spectrochim. Acta*, **24A**, 157 (1968).
27. P. Klaboe, J. J. Lothe and K. Lunde, *Acta Chem. Scand.*, **11**, 1677 (1957).
28. G. E. Maciel, J. W. McIver, N. S. Ostlund and J. A. Pople, *J. Am. Chem. Soc.*, **92**, 1 (1970).
29. J. A. Pople and D. L. Beveridge, *Approximate Molecular Orbital Theory* (New York: McGraw-Hill, 1970).
30. M. D. Johnston and M. Barfield, *J. Chem. Phys.*, **55**, 3483 (1971).
31. M. D. Johnston and M. Barfield, *J. Chem. Phys.*, **54**, 3083 (1971).
32. R. J. Ouellette and S. H. Williams, *J. Am. Chem. Soc.*, **93**, 466 (1971).
33. P. Debye, *Polar Molecules* (New York: Chemical Catalog Co., 1929).
34. L. Onsager, *J. Am. Chem. Soc.*, **58**, 1486 (1936).
35. C. J. F. Bottcher, *Theory of Electric Polarisation* (Amsterdam: Elsevier, 1952).
36. A. Wada, *J. Chem. Phys.*, **22**, 198 (1954).
37. R. J. Abraham and M. A. Cooper, *J. Chem. Soc.* (*B*), 202 (1967).
38. H. Looyenga, *Physica*, **31**, 401 (1965).
39. H. Looyenga, *Mol. Phys.*, **9**, 501 (1965).
40. H. Block and S. M. Walker, *Chem. Phys. Letters*, **19**, 363 (1973).
41. R. J. Abraham, *J. Phys. Chem.*, **73**, 1192 (1969).
42. R. J. Abraham (Unpublished Results).
43. C. Lassau and J. C. Jungers, *Bull. Soc. Chim. France*, 2678 (1968).
44. C. Reichardt, *Angew. Chem. Intern. Ed.* (*Engl.*), **4**, 29 (1965).
45. E. L. Eliel and O. Hofer (Private Communication).
46. R. J. Abraham and T. M. Siverns, *J. Chem. Soc.* (*Perkin II*), 1587 (1972).
47. P. Laszlo and J. I. Musher, *J. Chem. Phys.*, **41**, 3906 (1964).
48. J. Midwinter and P. Suppan, *Spectrochim. Acta*, **25A**, 953 (1969).
49. R. W. Haskell, *J. Phys. Chem.*, **73**, 2916 (1969).
50. R. J. Abraham (Unpublished Results).
51. W. F. Reynolds and D. J. Wood, *Can. J. Chem.*, **47**, 1295 (1969).
52. R. J. Abraham and K. Parry, *J. Chem. Soc.* (*B*), 539 (1970).
53. MODELBUILDER Quantum Chemistry Program Exchange Prog. No. 135.
54. *Handbook of Chemistry and Physics*, 45th ed. (Cleveland, Ohio: Chemical Rubber Publishing Co., 1965).
55. H. H. Landolt and R. Bornstein, *Tables of Physical Constants* (Berlin: Lange and Springer, 1959).
56. J. W. Smith, *Electric Dipole Moments* (London: Butterworth, 1955).
57. R. J. W. Le Fèvre, G. L. D. Ritchie and P. J. Stiles, *Chem. Commun.*, 846 (1966).
58. K. G. R. Pachler and P. L. Wessels (Private Communication).
59. N. Oi and J. F. Coetzee, *J. Am. Chem. Soc.*, **91**, 2478 (1969).
60. M. F. El Bermani, A. J. Woodward and N. Jonathan, *J. Am. Chem. Soc.*, **92**, 6750 (1970).
61. K. Tanabe, *Spectrochim. Acta*, **28A**, 407 (1972).

62. R. J. Abraham, K. G. R. Pachler and P. L. Wessels, *Z. Physik. Chem.* (*Frankfurt*), **58**, 257 (1968).

62a. J. Powling and H. J. Bernstein, *J. Am. Chem. Soc.*, **73**, 1815 (1951).

62b. K. Kuratani, T. Miyazawa and S. Mizushima, *J. Chem. Phys.*, **21**, 1411 (1953).

63. J. K. Wilmshurst and H. J. Bernstein, *Can. J. Chem.*, **35**, 734 (1957).

64. R. H. Harrison and K. A. Kobe, *J. Chem. Phys.*, **26**, 1411 (1957).

65. R. J. Abraham (Unpublished Results).

66. R. J. Abraham and G. Gatti, *J. Chem. Soc.* (*B*), 961 (1969).

67. P. Klaboe and J. R. Nielsen, *J. Chem. Phys.*, **33**, 1764 (1960).

68. R. J. Abraham and R. H. Kemp, *J. Chem. Soc.* (*B*), 1240 (1971).

69. A. D. Giacomo and C. P. Smyth, *J. Am. Chem. Soc.*, **77**, 1361 (1955).

70. M. F. El Bermani and N. Jonathan, *J. Chem. Phys.*, **49**, 340 (1968).

71. P. A. Bazhulin and C. P. Osipova, *Opt. Spectr.* (*USSR*) (*English Transl.*), **6**, 406 (1959).

72. K. G. R. Pachler and P. L. Wessels, *J. Mol. Struct.*, **3**, 207 (1969).

73. P. Klaboe and J. R. Nielsen, *J. Chem. Phys.*, **32**, 899 (1960).

74. L. Cavalli and R. J. Abraham, *Mol. Phys.*, **19**, 265 (1970).

75. I. Mizagawa, *J. Chem. Soc. Japan, Pure Chem. Sect.*, **75**, 1173, 1162 (1954).

76. F. E. Malherbe and H. J. Bernstein, *J. Am. Chem. Soc.*, **74**, 1859 (1952).

77. R. E. Kagarise, *J. Chem. Phys.*, **24**, 300 (1956).

78. R. E. Kagarise, *J. Chem. Phys.*, **29**, 680 (1958).

79. R. A. Newmark and C. H. Sederholm, *J. Chem. Phys.*, **43**, 602 (1965).

80. S. Mizushima, *Pure Appl. Chem.*, **7**, 1 (1963).

81. E. Wyn-Jones and W. J. Orville-Thomas, *Trans. Faraday Soc.*, **64**, 2907 (1968).

82. P. J. D. Park and E. Wyn-Jones, *J. Chem. Soc.* (*A*), 422 (1969).

83. S. Kondo, E. Tagami, K. Iimura and M. Takeda, *Bull. Chem. Soc., Japan*, **41**, 790 (1968).

84. A. A. Bothner-By and C. Naar-Colin, *J. Am. Chem. Soc.*, **84**, 743 (1962).

85. G. M. Whitesides, J. P. Sevenair and R. W. Goetz, *J. Am. Chem. Soc.*, **89**, 1135 (1967).

86. P. J. D. Park and E. Wyn-Jones, *J. Chem. Soc.* (*A*), 2944 (1968).

87. G. Govil and H. J. Bernstein, *J. Chem. Phys.*, **48**, 285 (1968).

88. A. L. McClellan, *Tables of Experimental Dipole Moments* (London: W. H. Freeman and Co., 1963).

89. L. J. Bellamy and R. L. Williams, *J. Chem. Soc.*, 4294 (1957).

90. S. Mizushima, T. Shimanouchi, T. Miyazawa, I. Ichishima, K. Kuratani, I. Nakagawa and N. Shido, *J. Chem. Phys.*, **21**, 815 (1953).

91. G. A. Crowder and B. R. Cook, *J. Chem. Phys.*, **47**, 367 (1967).

92. B. R. Cook and G. A. Crowder, *J. Chem. Phys.*, **47**, 1700 (1967).

93. S. Mizushima, T. Shimanouchi, I. Ichishima, T. Miyazawa, I. Nakagawa and T. Araki, *J. Am. Chem. Soc.*, **78**, 2038 (1956).

94. G. J. Karabatsos and R. A. Taller, *Tetrahedron*, **24**, 3923 (1968).

95. G. J. Karabatsos and N. Hsi, *J. Am. Chem. Soc.*, **87**, 2864 (1965).

96. G. J. Karabatsos and D. J. Fenglio, *J. Am. Chem. Soc.*, **91**, 1124 (1969).

97. G. J. Karabatsos, D. J. Fenglio and S. S. Lande, *J. Am. Chem. Soc.*, **91**, 3572 (1969).

98. G. J. Karabatsos and D. J. Fenglio, *J. Am. Chem. Soc.*, **91**, 3577 (1969).

99. E. Saegebarth and E. B. Wilson, *J. Chem. Phys.*, **46**, 3088 (1967).

100. L. J. Bellamy and R. L. Williams, *J. Chem. Soc.*, 3465 (1958).

101. T. L. Brown, *Spectrochim. Acta*, **18**, 1815 (1962).

102. I. Nakagawa, I. Ichishima, K. Kuratani, T. Miyazawa, T. Shimanouchi and S. Mizushima, *J. Chem. Phys.*, **20**, 1720 (1952).

103. L. S. Bartell, B. L. Caroll and J. P. Guillory, *Tetrahedron Letters*, **No. 13**, 704 (1964).
104. L. S. Bartell, B. L. Caroll and J. P. Guillory, *J. Chem. Phys.*, **43**, 647 (1965).
105. L. S. Bartell, J. P. Guillory and A. T. Parks, *J. Phys. Chem.*, **69**, 3043 (1965).
106. A. Y. Khan and N. Jonathan, *J. Chem. Phys.*, **52**, 147 (1970).
107. A. Y. Khan and N. Jonathan, *J. Chem. Phys.*, **50**, 1801 (1969).
108. A. J. Woodward and N. Jonathan, *J. Mol. Struct.*, **35**, 127 (1970).
109. A. J. Woodward and N. Jonathan, *J. Phys. Chem.*, **74**, 798 (1970).
110. D. A. Dawson and W. F. Reynolds, *Can. J. Chem.*, **49**, 3438 (1971).
111. G. K. Hamer, W. F. Reynolds and D. J. Wood, *Can. J. Chem.*, **49**, 1755 (1971).
112. E. I. Snyder, *J. Am. Chem. Soc.*, **88**, 1155 (1966).
113. M. Buza and E. I. Snyder, *J. Am. Chem. Soc.*, **88**, 1161 (1966).
114. C. Altona and H. J. Hageman, *Rec. Trav. Chim.*, **87**, 279 (1968).
115. H. Finegold, *J. Chem. Phys.*, **41**, 1808 (1964).
116. W. F. Reynolds and D. J. Wood, *Can. J. Chem.*, **49**, 1209 (1971).
117. B. L. Hawkins, W. Brensen, S. Borcic and J. D. Roberts, *J. Am. Chem. Soc.*, **93**, 4471 (1971).
118. K. K. Chiu, H. H. Huang and L. H. L. Chia, *J. Chem. Soc.* (*Perkin II*), 286 (1972).
119. K. K. Chiu, N. H. Huang and P. K. K. Linn, *J. Chem. Soc.* (*B*), 304 (1970).
120. E. B. Whipple, *J. Chem. Phys.*, **35**, 1039 (1961).
121. G. A. Crowder, *J. Mol. Struct.*, **23**, 103 (1967).
122. G. A. Crowder, *J. Mol. Struct.*, **20**, 43 (1966).
123. R. J. Abraham and K. Parry, *J. Chem. Soc.* (*B*), 724 (1967).
124. G. J. Janz and W. F. Fitzgerald, *J. Chem. Phys.*, **23**, 1973 (1955).
125. V. S. Watts, G. S. Reddy and J. H. Goldstein, *J. Mol. Spectr.*, **11**, 325 (1963).
126. E. Wyn-Jones and W. J. Orville-Thomas, *J. Chem. Soc.*, 5853 (1964).
127. E. Wyn-Jones and W. J. Orville-Thomas, *J. Chem. Soc.*, 101 (1966).
128. M. F. El Bermani and N. Jonathan, *J. Chem. Soc.* (*A*), 1711 (1968).
129. K. K. Deb and R. J. Abraham, *J. Mol. Spectr.*, **23**, 393 (1967).
130. P. J. Krueger and H. D. Mettee, *Can. J. Chem.*, **42**, 326 (1964).
131. P. Buckley and P. A. Giguére, *Can. J. Chem.*, **45**, 397 (1967).
132. P. Buckley, P. A. Giguére and D. Yamarota, *Can. J. Chem.*, **46**, 2917 (1968).
133. P. Buckley, P. A. Giguére and M. Schneider, *Can. J. Chem.*, **47**, 901 (1969).
134. P. A. Giguére and T. Kawamuia, *Can. J. Chem.*, **49**, 3815 (1971).
135. K. G. R. Pachler and P. L. Wessels, *J. Mol. Struct.*, **6**, 471 (1970).
136. E. Wyn-Jones and W. J. Orville-Thomas, *J. Mol. Struct.*, **1**, 79 (1967).
137. J. Crossley and C. P. Smyth, *J. Am. Chem. Soc.*, **91**, 2482 (1969).
138. O. D. Ul'yanova and Y. A. Prentin, *Russian J. Phys. Chem.* (*English Transl.*), **41**, 1447 (1967).
139. K. Kozima and K. Sabashita, *Bull. Chem. Soc. Japan*, **31**, 796 (1958).
140. O. D. Ul'yanova, M. K. Astrovskii and Y. A. Prentin, *Russian J. Phys. Chem.* (*English Transl.*), **44**, 562 (1970).
141. P. Klaboe, J. J. Lothe and K. Lunds, *Acta Chem. Scand.*, **11**, 1677 (1957).
142. R. V. Lemieux and J. W. Loan, *Can. J. Chem.*, **42**, 893 (1964).
143. Y. A. Prentin, G. M. Kuz'yants and O. P. Ul'yanova, *Russian J. Phys. Chem.* (*English Transl.*), **38**, 708 (1964).
144. B. Franzus and B. E. Hudson, *J. Org. Chem.*, **28**, 2238 (1963).
145. K. Kozima and T. Yoshino, *J. Am. Chem. Soc.*, **75**, 166 (1953).
146. G. Wood and E. P. Woo, *Can. J. Chem.*, **45**, 2477 (1967).
147. H. R. Buys and H. J. Geise, *Tetrahedron Letters*, 2991 (1970).
148. C. Altona, H. R. Buys, H. J. Hageman and E. Havinga, *Tetrahedron*, **23**, 2265 (1967).

149. R. J. Abraham and Z. L. Rossetti, *Tetrahedron Letters*, 4965 (1972).

150. G. Allen and H. J. Bernstein, *Can. J. Chem.*, **33**, 1055 (1955).

151. G. J. Karabatsos and F. M. Vane, *J. Am. Chem. Soc.*, **85**, 3886 (1963).

152. F. A. Miller, W. G. Fateley and R. E. Witkowski, *Spectrochim. Acta*, **23A**, 891 (1967).

153. R. A. Pethrick and E. Wyn-Jones, *J. Chem. Soc.* (*A*), 713 (1969).

154. D. J. Chadwick, J. Chambers, G. D. Meakins and R. L. Snowden, *Chem. Commun.*, 624 (1971).

155. B. Roques, S. Combrission, C. Riche and C. Pascard-Billy, *Tetrahedron*, 26, 3555 (1970).

156. M. L. Martin, J. C. Roze, G. J. Martin and P. Fourmari, *Tetrahedron Letters*, 3407 (1970).

157. L. Arlinger, K. Dahlquist and S. Forser, *Acta Chem. Scand.*, **24**, 662 (1970).

158. Chemical Society Tables of Interatomic Distances, *Special Publications*, **No. 11**, 18, M162, 179, 815.

159. R. J. Abraham, H. D. Banks, E. L. Eliel, O. Hofer and M. K. Kaloustian, *J. Am. Chem. Soc.*, **94**, 1913 (1972).

160. N. L. Allinger, J. Allinger and N. A. Lebel, *J. Am. Chem. Soc.*, **82**, 2926 (1960).

161. W. D. Kumler and H. C. Huitric, *J. Am. Chem. Soc.*, **78**, 3369 (1956).

162. A. S. Kende, *Tetrahedron*, **No. 14**, 13 (1959).

163. N. L. Allinger and H. M. Blatter, *J. Org. Chem.*, **27**, 1523 (1962).

164. J. Allinger and N. L. Allinger, *Tetrahedron*, 64 (1958).

165. N. L. Allinger and J. Allinger, *J. Am. Chem. Soc.*, **80**, 5476 (1958).

166. N. Allinger, J. Allinger, L. A. Frieberg, R. F. Czaja and N. A. Lebel, *J. Am. Chem. Soc.*, **82**, 5876 (1960).

167. K. Kozima and E. Hirano, *J. Am. Chem. Soc.*, **83**, 4300 (1961).

168. K. Kozima and Y. Yamanouchi, *J. Am. Chem. Soc.*, **81**, 4159 (1959).

169. J. Cantacuzene and R. Jantzen, *Tetrahedron Letters*, 2429 (1970).

170. Y. Pan and J. B. Strothers, *Can. J. Chem.*, **45**, 2943 (1967).

171. C. Djerassi, *Optical Rotatory Dispersion* (New York: McGraw-Hill, 1960) p. 125.

172. E. G. Cummins and J. E. Payer, *J. Chem. Soc.*, 3847 (1957).

173. J. B. Bervelt, R. Ettinger, P. A. Peters, J. Reise and G. Chiurdoglu, *Can. J. Chem.*, **45**, 81 (1967).

174. J. Reisse, P. A. Peters, R. Ottinger, J. B. Bervelt and G. Chiurdoglu, *Tetrahedron Letters*, 2511 (1966).

175. A. Moscowitz, K. Wellman and C. Djerassi, *J. Am. Chem. Soc.*, **85**, 3515 (1963).

176. N. L. Allinger, J. Allinger, L. E. Geller and C. Djerassi, *J. Org. Chem.*, **25**, 6 (1960).

177. J. Allinger, N. L. Allinger, L. E. Geller and C. Djerassi, *J. Org. Chem.*, **26**, 3521 (1961).

178. N. L. Allinger, J. Allinger, L. W. Chow and G. L. Wang, *J. Org. Chem.*, **32**, 522 (1967).

179. N. L. Allinger, H. M. Blatter, L. A. Freiberg and F. M. Karbowski, *J. Am. Chem. Soc.*, **88**, 2999 (1966).

180. N. L. Allinger, J. G. D. Carpenter and M. A. De Rooge, *J. Org. Chem.*, **30**, 1423 (1965).

181. H. R. Buys, M. T. Giesen and E. Havinga, *Rec. Trav. Chim.*, **89**, 114 (1970).

182. C. Romers, C. Altona, H. R. Buys and E. Havinga, *Topics in Stereochemistry*, Vol. 4, Ed. E. L. Eliel and N. L. Allinger (New York: Interscience, 1969) p. 39.

183. E. L. Eliel, *Accts. Chem. Res.*, 3, 3 (1970).

184. B. A. Arbovzov, *Bull. Soc. Chim. France*, 1311 (1960).

185. D. G. Hellier, J. G. Tillet, H. F. van Woerden and R. F. M. White, *Chem. Ind.* (*London*), 1956 (1963).

186. C. H. Green and D. G. Hellier, *J. Chem. Soc.* (*Perkin II*), 458 (1972).
187. P. Albrikstan, *Acta Chem. Scand.*, **25**, 478 (1971).
188. C. Altona, H. J. Geise and C. Romers, *Rec. Trav. Chim.*, **85**, 1197 (1966).
189. J. W. L. van Oxyen, R. C. D. E. Hasekamp, G. C. Vehoor and C. Romers, *Acta Cryst.*, **B24**, 1471 (1968).
190. H. F. van Woerden and E. Havinga, *Rec. Trav. Chim.*, **86**, 341 (1967).
191. H. F. van Woerden and E. Havinga, *Rec. Trav. Chim.*, **86**, 353 (1967).
192. G. Wood, G. W. Buchanan and M. H. Miskow, *Can. J. Chem.*, **50**, 521 (1972).
193. G. Wood and H. M. Miskow, *Tetrahedron Letters*, 1109 (1969).
194. G. Wood and H. M. Miskow, *Tetrahedron Letters*, 1775 (1970).
195. G. Wood, J. M. McIntosh and M. H. Miskow, *Can. J. Chem.*, **44**, 1202 (1971).
196. G. Wood and M. Miskow, *Tetrahedron Letters*, 4433 (1966).

Author Index

Abe, K., 103, 111
Abe, Y., 91, 105, 106, 110, 359, 360, 382
Abraham, R. J., 84, 85, 91, 109, 110, 119,
    154, 312, 322, 441, 442, 449, 477, 479,
    481, 482, 487, 492, 501, 502, 504, 505,
    506, 507, 508, 509, 512, 515, 519, 520,
    521, 523, 524, 525, 526, 527, 528, 529,
    530, 531, 537, 550, 551, 552, 554, 559,
    564, 565, 567, 568, 569, 580, 581, 582,
    583
Absar, I., 401, 402, 418, 421, 423, 424
Achyra, R. V., 22, 25, 27, 28
Adams, W. J., 361, 377, 383, 384
Addhelm, M., 107, 113
Ahmad, M., 94, 110
Ainsworth, J., 348, 349, 381
Airey, W., 373, 374, 383, 384
Akishin, P. A., 362, 375, 383, 384
Akita, K., 100, 111
Albriktsen, P., 471, 480, 578, 584
Alekseev, N. V., 331, 332, 356, 380, 381,
    382
Alexander, S., 127, 155
Aliev, S. S., 319, 320, 322
Allen, G., 83, 108, 240, 242, 248, 253, 255,
    263, 268, 269, 270, 271, 272, 275, 278,
    279, 282, 487, 519, 520, 521, 523, 524,
    528, 532, 533, 565, 568, 580, 583
Allen, H. C., Jr., 172, 187, 212
Allen, L. C., 395, 397, 398, 400, 401, 405,
    406, 407, 408, 409, 410, 413, 414, 415,
    417, 418, 419, 420, 422, 423, 424
Allerhand, A., 138, 155, 454, 479
Allinger, J., 572, 573, 574, 575, 583, 584
Allinger, N. L., 6, 7, 12, 18, 21, 22, 23, 27,
    425, 478, 483, 498, 572, 573, 574, 575,
    576, 577, 580, 583, 584

Allkins, J. R., 83, 109
Almenningen, A., 338, 348, 349, 350, 351,
    356, 357, 366, 367, 368, 373, 374, 376,
    378, 379, 381, 383, 384
Altona, C., 425, 456, 467, 475, 478, 479,
    548, 549, 551, 558, 559, 578, 582, 583,
    584
Amaya, K., 37, 55
Aminova, R. M., 472, 480
Andersen, B., 374, 376, 384
Andersen, P., 374, 384
Anderson, B., 328, 380
Anderson, C. B., 22, 23, 27, 28
Anderson, J. E., 49, 52, 56, 425, 434, 435,
    436, 437, 478, 479
Anderson, R. J., 196, 213
Andrae, J. H., 298, 304, 322
André, J. M., 401, 416, 423
André, M. Cl., 401, 416, 423
Andreassen, A., 374, 384
Andresen, H. G., 233, 245, 247, 249, 250,
    253, 270, 271, 272, 282
Anet, F. A. L., 94, 110, 152, 155
Angyal, S. J., 6, 7, 12, 18, 425, 478, 483,
    498, 572, 573, 574, 580
Anteunis, M., 434, 435, 436, 437, 443,
    444, 445, 446, 447, 449, 450, 461, 462,
    472, 478, 479, 480
Antony, A. A., 38, 55
Araki, T., 537, 582
Arbousow, B. A., 429, 478, 578, 584
Arlinger, L., 565, 583
Armitage, B. J., 22, 27,
Astrouskii, M. K., 555, 561, 583
Astrup, E. E., 376, 384
Aten, C. F., 362, 383
Ayras, P., 439, 440, 443, 445, 479

Badding, V. G., 23, 28
Badger, R. M., 75, 105, 108, 112, 353, 383
Baeyer, A., 5, 17
Baggett, N., 429, 478
Bailey, J., 315, 316, 322
Bailey, W. F., 438, 439, 478
Bak, B., 398, 417, 423
Baker, A. W., 106, 112
Baker, J. G., 197, 213
Balch, A. L., 150, 155
Baldock, R. A., 95, 106, 110
Bambenek, M. A., 150, 155
Banks, H. D., 441, 442, 449, 477, 479, 569, 583
Barabas, A. B., 150, 155
Barfield, M., 498, 580
Barnes, J. D., 107, 112
Bartell, L. S., 334, 337, 348, 349, 352, 360, 361, 371, 376, 377, 381, 382, 383, 384, 545, 582
Barton, D. H. R., 6, 7, 17, 19, 27
Basch, H., 420, 424
Bass, R., 304, 322
Bastiansen, O., 326, 338, 348, 349, 350, 351, 356, 366, 367, 373, 374, 376, 377, 379, 380, 381, 382, 383, 384
Bauer, M. E., 44, 55
Bauer, S. H., 348, 352, 364, 374, 376, 378, 381, 382, 383, 384
Bazhulin, P. A., 528, 530, 581
Beach, J. Y., 348, 381
Beagley, B., 328, 329, 348, 352, 353, 373, 374, 380, 381, 382, 383
Beaudet, R. A., 172, 196, 199, 204, 212, 213, 215
Beck, B. H., 152, 153, 155, 488, 580
Beckett, C. W., 425, 478
Belford, G. G., 101, 111, 117, 118, 154, 489, 491, 521, 528, 532, 534, 580
Bellamy, L. J., 76, 108, 537, 538, 542, 544, 582
Bender, C. F., 421, 424
Bender, P., 486, 559, 561, 580
Benedetti, E., 107, 113
Bent, H. A., 24, 28
Bentrude, W. G., 25, 28
Bergmann, K., 31, 53, 54, 56
Berlin, A. J., 428, 478
Bernstein, H. J., 83, 85, 101, 102, 108, 109, 111, 118, 119, 154, 490, 491, 494, 519, 528, 534, 562, 565, 568, 580, 581, 582

Berti, G., 24, 28
Bertie, J. E., 53, 56
Berther, G., 388, 422
Berthod, H., 421, 424
Bertrand, R. D., 25, 28
Bervelt, J. B., 573, 583, 584
Beveridge, D. L., 498, 508, 514, 566, 580
Beyer, R. T., 313, 322
Bhatia, A. B., 296, 322
Bilicke, C., 425, 478
Binsch, G., 131, 155, 425, 478, 493, 580
Biquard, P., 301, 314, 322
Biros, F. G., 21, 27
Birshtein, T. M., 45, 56
Bischoff, C. A., 3, 17
Blaker, J. W., 172, 187, 212
Blanch, G., 234, 253
Blandamer, M. J., 314, 322
Blank, J. M., 25, 28
Blatter, H. M., 22, 27, 572, 574, 575, 576, 583, 584
Bloch, F., 120, 149, 154, 155
Block, H., 502, 503, 580
Boates, T. L., 376, 384
Bochvar, D. A., 356, 382
Bodat, H., 107, 113
Boerio, F. J., 97, 98, 111
Bogatskii, A. V., 425, 478
Boggs, J. E., 205, 215
Bohn, R. K., 354, 382
Bolton, J. R., 147, 154, 155
Bolton, K., 202, 214
Bonham, R. A., 348, 349, 353, 371, 381, 382
Boobyer, G. J., 79, 108
Boone, D. W., 205, 215
Booth, G. E., 22, 27
Booth, H., 24, 25, 28, 425, 470, 478, 480
Borcic, S., 550, 582
Borgers, T. R., 203, 204, 214
Born, M., 62, 108
Bornstein, R., 513, 581
Borreman, F., 444, 472, 479, 480
Bothner-By, A. A., 528, 536, 581
Bothorel, P., 86, 109
Botskor, I., 206, 215
Bottcher, C. J. F., 499, 501, 507, 580
Boutin, H., 263, 267, 282
Bovey, F. A., 17, 18
Bowen, H. J. M., 373, 374, 375, 383, 384
Bowles, A. J., 106, 112
Boys, S. F., 388, 389, 422

Bradley, C. A., 75, 108
Bradley, R. H., 268, 282
Bragin, J., 268, 282
Brand, J. C. D., 434, 435, 478
Breig, E. L., 197, 213
Brensen, W., 550, 582
Bret, G., 83, 109
Brett, T. J., 21, 23, 27, 28
Brevdo, V. I., 84, 109
Brewster, A. L., 17, 18
Brier, P. N., 268, 269, 282
Bright-Wilson, E., 16, 18
Brignell, P. J., 23, 27, 441, 442, 444, 450, 479
Britt, C. O., 205, 215
Brockway, L. O., 326, 337, 380, 381
Brown, D. P., 353, 382
Brown, F. B., 270, 282
Brown, J. K., 87, 109
Brown, K., 22, 27
Brown, T. L., 542, 582
Buchanan, G. W., 578, 584
Bucker, H. P., 101, 102, 111
Buckley, P., 90, 109, 552, 582
Buenker, R. J., 400, 401, 408, 416, 423, 424
Bunn, C. W., 92, 110
Burkhard, D. G., 184, 213
Burnelle, L. A., 401, 416, 423
Burnitt, D. L., 105, 106, 112
Bus, W. C., 21, 27
Buss, V., 400, 411, 415, 416, 423
Butcher, S. S., 197, 213, 358, 382
Buys, H. R., 377, 378, 384, 425, 470, 471, 478, 480, 557, 558, 559, 577, 578, 583, 584
Buza, M., 545, 546, 547, 548, 582
Bwahulin, P. A., 102, 111

Cabana, A., 79, 108
Cadioli, B., 107, 113, 421, 424
Cahill, P., 197, 213
Camerlynck, R., 444, 479
Campbell, N. C. G., 25, 28
Cantacuzene, J., 573, 583
Caraculacu, A., 105, 112
Cardillo, M. J., 352, 382
Carles Lorjou, M., 107, 113
Carlson, G. L., 93, 101, 105, 110, 111, 420, 424, 495, 580
Caroll, B. L., 545, 582
Carpenter, J. G. D., 576, 577, 584
Carr, H. Y., 136, 155

Carter, V. B., 100, 111
Cavalli, L., 84, 85, 109, 119, 154, 492, 501, 502, 507, 508, 512, 521, 525, 526, 528, 531, 580, 581
Cazaux, L., 475, 480
Cazzoli, G., 202, 214
Cecili, P., 107, 113
Celotti, J. C., 428, 478
Chadwick, D., 205, 215
Chadwick, D. J., 565, 583
Chambers, J., 565, 583
Chan, S. I., 203, 204, 214
Chandra, S., 52, 53, 56
Chang, C. H., 376, 378, 384
Chang, K. Y., 154, 155
Chelkowski, A., 37, 55
Chen, Fu-Ming, 119, 154, 454, 479
Chen, J. H., 313, 322
Chia, L. H. L., 88, 104, 107, 109, 112, 550, 582
Chiang, J. F., 364, 378, 383, 384
Chistjakov, A. L., 356, 382
Chitoku, K., 37, 38, 52, 55, 56
Chiu, K. K., 88, 107, 109, 112, 550, 582
Chiurdoglu, G., 21, 27, 428, 478, 483, 484, 573, 580, 583, 584
Chow, L. W., 575, 584
Christensen, D. H., 398, 417, 423
Christie, G. H., 4, 17
Ciampelli, F., 97, 111
Claasen, H. H., 269, 282
Clague, A. D. H., 270, 282
Clark, A. E., 312, 313, 322
Clark, A. H., 325, 328
Clark, D. T., 400, 413, 415, 423
Cleeton, C. E., 200, 214
Clementi, E., 388, 389, 390, 393, 395, 397, 398, 403, 408, 413, 420, 422, 424
Clemett, C. J., 53, 56
Clippard, F. B., 376, 384
Coene, E., 434, 435, 436, 437, 443, 444, 478, 479
Coetzee, J. F., 516, 520, 581
Cohen, A. D., 118, 154
Coke, J. L., 21, 22, 27
Cole, K. S., 46, 56
Cole, R. H., 46, 56
Combini, M., 100, 111
Combression, S., 565, 583
Cook, B. R., 537, 538, 582
Cook, R. L., 157, 185, 187, 197, 210, 211, 212, 213
Cook, R. R., 91, 110
Cookson, R. C., 19, 27

Cooper, M. A., 312, 322, 487, 501, 502, 508, 519, 520, 521, 523, 580
Cornell, S. W., 100, 111
Corradini, P., 92, 110
Cortilli, G., 97, 111
Costain, C. C., 159, 193, 195, 202, 203, 212, 214, 245, 247, 253
Cotter, F. H., 94, 110
Cotterill, W. D., 22, 27
Cox, A. P., 197, 214
Cramer, H., 244, 253
Craven, S. M., 268, 282
Crawford, B. L., 75, 108, 184, 213
Crook, K. R., 102, 111, 298, 312, 313, 321, 322, 323
Cross, P. C., 76, 95, 108, 110, 172, 187, 212, 218, 219, 230, 252
Crossley, J., 38, 39, 55, 553, 583
Crowder, G. A., 91, 103, 107, 110, 112, 113, 537, 538, 550, 551, 582
Cruickshank, D. W. J., 328, 329, 373, 374, 380, 383
Csizmadia, I. G., 401, 402, 417, 418, 421, 423, 424
Cummins, E. G., 574, 575, 583
Cumper, C. W. N., 53, 54, 56
Cunliffe, A. V., 228, 232, 237, 240, 241, 242, 243, 244, 248, 252, 253, 258, 259, 269, 270, 271, 272, 278, 279, 282
Cunliffe-Jones, D. B., 106, 112
Curby, R. J., 22, 27
Curl, R. F., Jr., 188, 197, 213
Cyvin, S. J., 337, 381
Czaja, R. F., 573, 574, 583

Daasch, L. W., 86, 102, 109, 111
Dahlquist, K., 481, 565, 568, 580, 583
Daily, B. P., 204, 214, 269, 282
Dale, J., 272, 277, 282
Danti, A., 269, 270, 272, 282
D'Antonio, P., 353, 382
Danusso, F., 95, 110
Darling, B. T., 237, 253
Dasgupta, S., 50, 51, 56
Dashevskii, V. G., 378, 384
Davidson, D. W., 46, 56
Davidson, R. B., 401, 406, 407, 417, 419, 423
Davies, M., 30, 46, 49, 54, 56, 95, 110
Davis, D. R., 390, 395, 397, 398, 403, 413, 422
Davis, M., 378, 384

Davis, M. I., 378, 384
Dawson, D. A., 545, 546, 548, 582
Dean, L. B., 16, 18
Deb, K. K., 551, 552, 582
Debye, P., 29, 30, 54, 301, 322, 326, 380, 486, 499, 580
Decius, J. C., 76, 95, 108, 110, 218, 219, 230, 252
DeGroot, M. S., 94, 110, 315, 317, 318, 322
De Kok, A. J., 429, 478
Delaney, P., 435, 478
Della, E. W., 428, 478
Demaison, J., 206, 215
De Meijere, A., 359, 376, 382, 384
De Meo, A. R., 218, 220, 229, 230, 233, 234, 245, 247, 249, 250, 252, 270, 272, 283
Dempster, A. B., 89, 107, 109, 113
Demuynch, J., 402, 403, 407, 411, 419, 424
Dennison, D. M., 184, 213, 237, 253
Derissen, J., 360, 382
De Rooge, M. A., 576, 577, 584
Desantis, P., 97, 111
Desper, C. R., 100, 111
Devaquet, A., 420, 424
de Vries-Miedena, A. T., 474, 476, 480
Dew, G., 240, 242, 248, 253, 269, 270, 271, 272, 278, 279, 282
Di Carlo, E. N., 49, 56
Dickinson, R. G., 425, 478
Dirinck, P., 444, 445, 447, 449, 450, 479
Djerassi, C., 573, 574, 583, 584
Dobinson, B., 429, 478
Dornte, R. W., 32, 55
Dorris, K. L., 205, 215
Dostrovsky, I., 6, 17
Dowling, J. M., 202, 214
Dreizler, H., 157, 197, 211, 212, 214, 215, 482, 565, 566, 580
Dressler, K. P., 401, 416, 423
Dum Bacher, B., 421, 424
Dunitz, J. D., 376, 384
Dunning, Th. H., 391, 395, 398, 419, 422, 423
Durig, J. R., 106, 112, 268, 282
Dzhassati, S., 106, 112

Eccleston, G., 321, 323, 438, 439, 446, 447, 450, 451, 458, 461, 464, 476, 477, 479, 480
Eckersley, G. H., 353, 382

Edmiston, C., 388, 422
Edmonds, P. D., 299, 322
Edward, J. T., 448, 479
Egan, C. J., 21, 27
Eggers, F., 302, 322
Ehrenson, S., 400, 415, 423
Ekers, J. E., 420, 424
El-Bermani, M. F., 87, 102, 109, 516,
   520, 528, 530, 531, 551, 552, 581, 582
Eliel, E. L., 6, 7, 12, 18, 20, 21, 22, 23, 25,
   27, 28, 425, 428, 438, 439, 440, 441,
   442, 443, 444, 446, 448, 449, 477, 478,
   479, 483, 488, 498, 505, 554, 560, 569,
   572, 573, 574, 578, 580, 581, 583, 584
Eliezer, I., 43, 44, 55
Elleman, P. D., 487, 543, 580
Ellestad, O. H., 107, 112
Emptage, M. R., 197, 213
Emsley, J. W., 63, 108, 117, 154
Enanoza, R. M., 440, 441, 479
Engerholm, G. G., 203, 205, 215
Epstein, I. R., 405, 408, 424
Esbitt, A. S., 198, 214
Evdokimov, V. V., 375, 384
Evering, B. L., 21, 27
Ewig, C. S., 190, 214
Ewing, V., 373, 374, 383
Eyring, H., 16, 18, 31, 32, 43, 52, 55
Ezumi, K., 103, 112

Fahrenfort, J., 80, 81, 108
Fairheller, W. R., 93, 94, 106, 110
Faisst, W., 434, 435, 436, 445, 458, 459,
   460, 478
Fanconi, B. M., 107, 112
Fang, Y., 401, 417, 423
Fano, L., 370, 383
Fateley, W. G., 93, 94, 101, 105, 110, 111,
   218, 226, 229, 230, 231, 233, 238, 245,
   248, 249, 250, 252, 253, 268, 269, 270,
   272, 275, 276, 282, 283, 363, 383, 420,
   424, 495, 565, 580, 583
Favero, P. G., 202, 214
Feeney, J., 63, 108, 117, 154
Fenglio, D. J., 538, 540, 541, 542, 543,
   544, 545, 582
Ferguson, E. E., 80, 108
Fernandez, J., 203, 204, 214
Fernholt, L., 351, 373, 374, 381, 383
Feuer, J., 385, 421
Fewster, S., 255, 263, 275, 282
Finbak, C., 326, 380

Finegold, H., 548, 582
Fink, W. H., 395, 397, 398, 401, 402, 405,
   407, 408, 413, 417, 418, 419, 422, 424
Finley, K. T., 25, 28
Fish, K., 42, 55
Fitzgerald, W. E., 87, 102, 109, 551, 552,
   582
Flanagan, C., 196, 213
Flory, P. J., 44, 45, 51, 55, 56
Flowers, D. L., 486, 559, 561, 580
Flygare, W. H., 90, 109, 157, 160, 161, 212
Fong, F. K., 48, 49, 54, 56
Fonken, G. J., 22, 27
Forbes, W. F., 94, 110, 150, 155
Forsen, S., 139, 140, 154, 155, 481, 565,
   568, 580, 583
Foster, A. B., 429, 478
Foster, D. F., 106, 112
Foster, M. J., 314, 322
Fourmari, P., 565, 583
Fraenkel, G. K., 149, 150, 151, 154, 155
Franchini, P. F., 395, 419, 423
Frankiss, S. G., 352, 382
Franzus, B., 554, 559, 563, 583
Freed, J. H., 149, 155
Freeman, J. M., 353, 382
Freiberg, L. A., 21, 22, 27, 573, 574, 575,
   576, 583, 584
Friebolin, H., 434, 435, 436, 445, 458, 459,
   460, 478
Fritsch, F. N., 317, 384
Fujiama, T., 268, 282
Fujishiro, R., 37, 55
Fukushima, K., 245, 253
Fukuyama, T., 362, 366, 383

Gale, L. H., 23, 28
Garg, S. K., 50, 53, 56
Garkovik, N. L., 425, 478
Gatti, G., 526, 528, 531, 581
Gaufres, R., 107, 113
Gazzard, I. J., 89, 109
Gebbie, H. A., 81, 84, 108
Geise, H. J., 377, 378, 384, 456, 467, 475,
   479, 557, 578, 583, 584
Geller, L. E., 574, 575, 584
George, C., 353, 382
George, W. O., 106, 112
Geske, D. H., 144, 150, 151, 155
Ghatak, A., 52, 56
Giacomo, A. D., 102, 111, 279, 283, 528,
   530, 581

Gianni, M. H., 428, 478
Gibson, J. S., 204, 215
Giesen, M. T., 577, 584
Giglio, E., 97, 111
Giguere, P. A., 90, 107, 109, 112, 552, 553, 582, 583
Gilbert, E. C., 21, 27
Giza, C. A., 23, 28, 438, 448, 478, 479
Glarum, S. H., 47, 56
Glauber, R., 326, 380
Glidewell, C., 352, 373, 374, 382, 383, 384
Goering, H. L., 486, 559, 561, 580
Goetz, R. W., 533, 581
Gold, L. P., 197, 213
Golding, D. R., 267, 269, 282
Goldish, E., 376, 384
Goldstein, J. H., 551, 582
Gonze, H., 21, 27
Gooberman, G. L., 304, 322
Goodwin, H. W., 22, 27, 428, 478
Goodwin, T. H., 277, 283
Gordan, R. G., 64, 66, 68, 69, 108
Gordon, M. S., 385, 421
Gordy, W., 157, 161, 185, 187, 197, 210, 211, 212, 213, 397, 423
Goubeau, J., 103, 107, 112, 113
Gourset Leroy, A., 107, 113
Govil, G., 118, 119, 154, 487, 490, 491, 494, 534, 580, 581
Grant, D. M., 25, 28
Green, C. H., 471, 475, 480, 578, 584
Greenberg, A., 434, 435, 436, 437, 478
Guidotti, C., 420, 424
Guillory, J. P., 360, 382, 383, 545, 582
Gunthard, H. H., 245, 253
Gussoni, M., 96, 97, 111
Gutowsky, H. S., 101, 111, 117, 118, 119, 120, 138, 154, 155, 234, 253, 269, 282, 454, 479, 489, 491, 521, 528, 532, 534, 580
Gwinn, W. D., 33, 55, 196, 197, 203, 204, 205, 213, 214, 215, 220, 221, 222, 223, 232, 246, 251, 252, 352, 381

Ha, T. K., 421, 424
Haaland, A., 328, 329, 338, 348, 349, 350, 351, 354, 355, 356, 357, 380, 381, 382
Haase, J., 353, 382
Hageman, H. J., 486, 548, 549, 557, 558, 559, 580, 582, 583
Hainer, R. M., 172, 187, 212

Hall, D. N., 301, 315, 320, 322, 323
Hall, L. H., 385, 421
Hamblin, P. C., 313, 321, 322, 323, 445, 458, 476, 477, 479, 480
Hameed, S., 278, 279, 283
Hamer, G. K., 545, 546, 548, 582
Hamo, K., 107, 112
Hanack, M., 7, 12, 18
Hannon, M. J., 97, 98, 111
Hannum, S. E., 106, 112
Haq, M. Z., 152, 155
Harring, H. G., 89, 109
Harrington, H. W., 203, 204, 214
Harris, D. O., 176, 190, 203, 204, 205, 207, 209, 212, 214, 215
Harris, R. K., 238, 253, 363, 383, 457, 479, 480
Harrison, R. H., 519, 581
Harshbarger, F., 350, 381
Hartman, K. O., 93, 105, 110, 495, 580
Hartmann, A. O., 366, 367, 368, 383
Hasan, A., 52, 56
Hasekamp, R. C. D. E., 578, 584
Haskell, R. W., 506, 581
Hassel, O., 6, 17, 326, 378, 380, 384, 425, 478
Haubenstock, H., 22, 25, 27, 28
Haugen, W., 362, 365, 383
Hauptman, H., 326, 343, 380
Havinga, E., 425, 472, 473, 475, 478, 480, 486, 557, 558, 559, 577, 578, 580, 583, 584
Hawkes, G. E., 24, 28
Hawkins, B. L., 550, 582
Haworth, W. A., 7, 18
Hayashi, M., 103, 107, 112, 181, 185, 186, 197, 213, 234, 253
Hayes, E. F., 401, 402, 417, 423
Hayes, W. P., 94, 110
Hayman, H. J. G., 43, 44, 55
Heasell, E. L., 304, 322
Heatley, F., 278, 279, 283, 487, 519, 520, 521, 523, 524, 528, 532, 533, 580
Hecht, K. T., 184, 213, 237, 245, 253, 397, 423
Hedberg, K., 326, 351, 362, 369, 370, 373, 374, 376, 377, 380, 381, 383, 384
Hedberg, L., 326, 362, 380, 383
Hehre, W. J., 395, 397, 398, 400, 401, 413, 415, 420, 421, 423, 424
Heikkila, J., 440, 479

Heitkemp, N. D., 270, 282
Hellier, D. G., 470, 471, 475, 480, 578, 584
Hendrickson, J. B., 426, 427, 478
Hennelly, E. J., 37, 55
Herndon, W. C., 385, 421
Herschbach, D. R., 159, 181, 184, 185, 186, 187, 188, 190, 194, 195, 196, 197, 198, 212, 213, 219, 220, 221, 222, 223, 229, 232, 234, 235, 245, 247, 252, 269, 282, 357, 382
Herzberg, G., 393, 422
Hester, R. E., 94, 106, 110, 275, 283
Heston, W. M., Jr., 37, 55
Hewitt, T. G., 328, 329, 348, 352, 380, 381, 382
Hiddon, N. J., 314, 322
Higasi, K., 33, 37, 38, 52, 53, 55, 56
Higginbotham, H. K., 352, 382
Higgins, J. S., 268, 269, 282
Higgs, P., 98, 111
Hilderbrandt, R., 374, 384
Hill, N. E., 30, 46, 54, 56
Hill, R. R., 25, 28
Hillier, I. H., 395, 398, 407, 419, 423
Hiraishi, J., 101, 111
Hiraishi, M., 107, 113
Hirano, E., 573, 574, 583
Hironaka, Y., 160, 212
Hirota, E., 157, 160, 176, 197, 199, 202, 207, 209, 212, 214, 215, 269, 282, 332, 342, 346, 347, 353, 355, 358, 381, 382
Hirota, K., 160, 212
Hirsch, J. A., 19, 27, 483, 487, 546, 554, 560, 580
Hisatsune, I. C., 105, 106, 112
Hjortaas, K. E., 378, 384
Hofer, O., 441, 442, 449, 477, 479, 505, 569, 581, 583
Hoffman, R. A., 139, 140, 154, 155
Hoffmann, R., 385, 420, 421, 424
Hollister, C., 393, 422
Holm, C. H., 120, 154
Holness, N. J., 24, 25, 28
Holywell, G. C., 353, 382
Homer, J., 429, 478
Hopkinson, A. C., 417, 424
Horowitz, E., 95, 110
Houeix, A., 103, 112
House, H. O., 22, 27
Howard, J. B., 219, 237, 252, 253

Hoyland, J. R., 196, 214, 249, 253, 390, 391, 392, 395, 396, 397, 399, 400, 403, 410, 411, 413, 414, 422, 423
Hsi, N., 538, 539, 540, 582
Hsu, H. L., 407, 416, 423
Hu, S. E., 21, 22, 27
Huang, H. H., 88, 104, 107, 109, 112, 550, 582
Huang, N. H., 550, 582
Hubbard, J. C., 303, 322
Hudson, A., 152, 154, 155
Hudson, B. E., 554, 559, 563, 583
Huettig, H., Jr., 35, 55
Hughes, E. D., 6, 17
Huitric, H. C., 572, 573, 583
Hunt, R. A., 237, 245, 253, 397, 423
Hunziker, H., 245, 253
Huo, W., 389, 422
Hussain, H. A., 152, 155
Hutchins, R. O., 23, 28
Hutchinson, B. J., 23, 27, 441, 442, 444, 450, 479
Hüttner, W., 161, 212

Ibers, J. A., 203, 214
Ichishima, I., 88, 91, 103, 105, 109, 110, 112, 350, 352, 381, 537, 538, 542, 582
Igarashi, M., 351, 381
Iijima, T., 352, 381
Imura, K., 528, 536, 581
Ingold, C. K., 6, 17
Itoh, T., 197, 213
Iwasaki, M., 351, 352, 381
Iwatoni, K., 103, 112

Jackson, R. H., 353, 382
Jacobs, G. D., 196, 213, 268, 282
Jacobsen, G. C., 379, 384
Jaeschke, A., 197, 214
Jalonen, J., 438, 445, 446, 479
James, R. W., 326, 332, 380
Jannink, G., 97, 111
Jantzen, R., 573, 583
Janz, G. J., 87, 102, 109, 551, 552, 582
Jeffrey, G. A., 92, 110
Jenkins, J. E. F., 91, 110
Jensen, F. R., 22, 23, 27, 28, 428, 478, 488, 580
Jensen, H., 379, 384
Jernigan, R. L., 44, 55
Jewitt, J. G., 25, 28

Jindal, S. P., 25, 28
Johansen, H., 395, 398, 419, 423
Johnson, C. R., 23, 28
Johnson, C. S., 127, 130, 155
Johnston, M. D., 498, 580
Jonas, J., 119, 154
Jonathan, N., 87, 91, 102, 105, 109, 110, 112, 516, 520, 528, 530, 531, 545, 551, 552, 581, 582
Jones, K., 425, 478
Jones, L., 105, 112
Jones, P. W., 278, 279, 283
Jones, R. A. Y., 23, 27, 441, 442, 444, 450, 479
Jones, R. N., 77, 108
Jones, V. I. P., 91, 105, 110, 118, 154, 435, 436, 439, 440, 443, 459, 479
Jorgensen, W. L., 409, 410, 424
Jungers, J. C., 505, 580

Kabuss, S., 434, 435, 436, 445, 458, 459, 460, 478
Kaercher, A., 172, 187, 212
Kagarise, R. E., 80, 86, 101, 102, 108, 109, 111, 528, 534, 581
Kaldor, U., 395, 397, 400, 416, 418, 422, 423
Kalman, O. F., 49, 56
Kaloustian, M. K., 23, 27, 441, 442, 443, 449, 477, 479, 569, 583
Kal'Yanov, B. I., 317, 322
Kaneda, Y., 160, 212
Kaplan, F., 94, 110
Kaplan, J. I., 127, 155
Karabatsos, G. J., 538, 539, 540, 541, 542, 543, 544, 545, 565, 582, 583
Karbowski, F. M., 575, 576, 584
Karle, I. L., 326, 332, 342, 344, 346, 347, 380, 381
Karle, J., 326, 332, 342, 343, 345, 348, 349, 353, 380, 382
Karplus, M., 403, 409, 424, 469, 480
Karpovitch, J., 300, 315, 320, 322, 323
Kasai, P. H., 195, 197, 213
Kassell, K. L., 221, 252
Kastha, G. S., 107, 112
Kasuya, T., 196, 213
Katayama, M., 35, 55
Kato, Y., 83, 109
Katon, J. E., 93, 94, 106, 110
Katritzky, A. R., 22, 23, 27, 95, 106, 110, 441, 442, 444, 450, 479

Kaufman, J. J., 411, 424
Kawamuia, T., 552, 553, 583
Keller, R. A., 203, 215
Kellie, G. M., 461, 465, 480
Kemp, J. D., 4, 7, 17
Kemp, R. H., 528, 529, 530, 581
Kende, A. S., 572, 574, 583
Kenner, J., 4, 17
Kenney, C. N., 157, 161, 211
Kern, C. W., 403, 409, 424
Ketelaar, J. A. A., 43, 55
Khabibullaev, P. K., 319, 322
Khaikin, L. S., 353, 375, 382, 384
Khan, A. Y., 91, 105, 110, 112, 545, 582
Kilb, R. W., 181, 183, 207, 212, 213, 223, 234, 252, 253, 357, 382
Kilp, H., 53, 56
Kilpatrick, J. E., 220, 232, 251, 252
Kim, H., 203, 204, 205, 214, 215
Kimura, K., 37, 55, 373, 374, 375, 383, 384
King, G. W., 172, 187, 212
King, H. F., 393, 411, 422
Kirchoff, W. H., 193, 214
Kirtman, B., 348, 380
Kivelson, D., 176, 178, 212
Kiviat, F. E., 270, 283
Klaboe, P., 85, 87, 101, 107, 109, 111, 112, 278, 279, 283, 496, 528, 529, 531, 555, 580, 581, 583
Klages, G., 54, 56
Klemperer, J., 269, 282
Klemperer, W., 229, 230, 233, 252
Klessinger, M., 392, 397, 413, 422, 423
Klug, D. D., 52, 56
Knoeber, M. C., 23, 27, 438, 439, 440, 444, 478, 479
Knopp, J. V., 193, 195, 207, 214, 215
Kobayashi, M., 97, 99, 100, 111
Kobe, K. A., 519, 581
Koenig, J. L., 97, 98, 100, 111
Koga, Y., 52, 56
Kohl, D. A., 371, 377, 383
Kohler, J. S., 184, 213
Kohlrausch, K. W. F., 57, 85, 108, 425, 478
Koide, T., 103, 112
Kojima, R., 351, 352, 381
Kojima, T., 197, 213
Kollman, P. A., 401, 406, 417, 423
Kolos, W., 393, 422
Komaki, C., 103, 112, 350, 381
Kondo, S., 160, 212, 358, 382, 528, 536, 581

Kortzeboran, R. N., 398, 417, 423
Koster, D. F., 270, 282
Kozima, K., 555, 556, 561, 564, 573, 574, 583
Kranbuehl, D. E., 52, 56
Kraus, G., 54, 56
Krauss, M., 388, 422
Krebs, K., 319, 322
Kreevoy, M. M., 16, 18
Krimm, S., 97, 111
Krisher, L. C., 196, 213, 357, 382
Krueger, P. J., 90, 102, 103, 109, 111, 552, 582
Kubo, M., 373, 374, 375, 383, 384
Kubota, T., 103, 112
Kuchitsu, K., 348, 349, 359, 360, 362, 366, 381, 382, 383
Kuczkowski, R. L., 353, 382
Kuhnle, J. A., 152, 153, 155
Kumar, K., 107, 112
Kumler, W. D., 572, 573, 583
Kuratani, K., 35, 55, 88, 91, 101, 103, 105, 109, 110, 111, 112, 350, 352, 381, 519, 537, 538, 542, 581, 582
Kuz'Yants, G. M., 555, 583
Kwart, H., 25, 28
Kyazimova, A. R., 106, 112
Kyeseth, K., 348, 380

Lack, R. E., 469, 480
Ladd, J. A., 80, 91, 105, 108, 110, 118, 154, 435, 436, 439, 440, 443, 459, 479
Lamanna, U., 420, 424
Lamb, J., 94, 110, 298, 301, 304, 310, 311, 312, 313, 315, 317, 318, 319, 320, 322, 323
Lambert, J. B., 425, 470, 478, 480
Lande, S. S., 538, 543, 544, 582
Landolt, H. H., 513, 581
Lane, G., 228, 252, 268, 278, 279, 282
Lasettre, E. N., 16, 18
Lassau, C., 505, 580
Laszlo, P., 434, 435, 436, 437, 478, 483, 487, 506, 528, 580, 581
Lathan, W. A., 395, 397, 398, 400, 413, 415, 423
Lau, K. K., 268, 282
Laurie, V. W., 159, 178, 193, 196, 197, 202, 203, 204, 205, 212, 213, 214, 215
Lauterbur, P. C., 475, 480
Leacock, R. A., 237, 245, 253, 397, 423
Lebel, N. A., 572, 573, 574, 583

Lecomte, J., 80, 81, 108
Led, J. J., 398, 417, 423
Lee, D., 96, 110
Lee, H. K., 23, 28
Lees, R. M., 197, 213
Lefevre, R. J. W., 515, 581
Legon, A. C., 205, 215
Lehn, J. M., 203, 214, 393, 408, 422, 424
Lemleu, R. V., 555, 583
Leonard, W. J., 44, 55
Leroi, G. E., 261, 268, 269, 397, 423
Leroy, G., 401, 416, 423
Letcher, J. H., 400, 401, 406, 411, 414, 423
Letcher, S. V., 313, 322
Levine, H. B., 269, 282
Levy, B., 397, 404, 421, 423, 424
Lewin, A. H., 428, 478
Li, J. C. M., 234, 253
Liang, C. D., 22, 27
Liang, J. H., 421, 424
Liberles, A., 420, 424
Lide, D. R., Jr., 157, 160, 184, 192, 193, 196, 202, 212, 213, 214, 348, 381
Lilley, D. M. J., 400, 413, 415, 423
Lim, P. K. K., 88, 109, 550, 582
Lin, C. C., 157, 183, 184, 186, 187, 189, 192, 197, 207, 212, 213, 215, 219, 220, 222, 223, 224, 225, 232, 240, 252, 357, 382
Lippincott, E. R., 83, 109
Lipscomb, W. N., 393, 394, 396, 405, 407, 408, 410, 411, 413, 422, 424
Liquori, A. M., 97, 111
Liskow, D. H., 421, 424
Lister, D. G., 202, 214
Litovitz, T. A., 298, 313, 322
Livingston, R. L., 375, 384
Loan, J. W., 555, 583
London, L. H., 218, 220, 229, 230, 233, 234, 245, 247, 249, 250, 252, 270, 272, 283
Long, D. A., 75, 108
Looney, C. E., 120, 154
Looyenga, H., 502, 580
Lothe, J. J., 496, 555, 580, 583
Lovell, W. S., 53, 56
Lowe, J. P., 85, 109, 119, 154, 385, 395, 403, 408, 413, 421, 422, 424, 483, 514, 519, 528, 535, 580
Lowrey, A. H., 353, 382
Lucas, R., 301, 322

Lucier, J. J., 89, 109
Luckhurst, G. R., 154, 155
Lukach, C. A., 25, 28
Lunde, K., 496, 555, 580, 583
Luntz, A. C., 203, 204, 205, 214, 215
Luoma, S., 438, 444, 446, 479
Luongo, J. P., 97, 111
Luttke, W., 359, 382
Luttringhaus, A., 434, 478
Lynden-Bell, R. M., 127, 155

McAdam, A., 352, 382
McAloon, K. T., 353, 382
McAlpine, K. B., 33, 55
McCall, D. W., 120, 154
McCants, D., 23, 28
Macchia, B., 24, 28
Macchia, F., 24, 28
McClellan, A. L., 536, 558, 559, 565, 566, 568, 582
McConnell, H. M., 125, 154
McCrum, N. G., 45, 56
McGlug, F. J., 83, 109
McGraw, G. E., 105, 106, 112
Maciel, G. E., 498, 580
McIntosh, J. M., 473, 476, 480, 578, 579, 584
McIver, J. W., 498, 580
McKenna, J., 26, 28
McKenna, J. M., 26, 28, 439, 479
McKinney, T. M., 151, 155
McKown, G. L., 204, 215
McLean, A. D., 392, 422
McMahon, P. E., 101, 111, 117, 118, 154, 489, 491, 521, 528, 532, 534, 580
McWeeny, R., 392, 422
Maeda, S., 80, 108
Maelder, W. W., 95, 110
Maestro, M., 420, 424
Magee, J. L., 43, 55
Maier, W., 434, 478
Maki, A. H., 146, 155
Malherbe, F. E., 528, 581
Manatt, S. L., 487, 543, 580
Manderkern, L., 100, 111
Mann, D. E., 196, 214, 269, 282, 370, 383
Manning, M. F., 201, 214
Margenau, H., 231, 252
Margolis, E. I., 21, 27
Mark, J. E., 52, 56, 92, 110
Maroni, P., 475, 480

Marsmann, H., 402, 419, 424
Marstokk, K. M., 172, 176, 199, 212
Martin, G., 103, 112
Martin, G. J., 565, 583
Martin, M. L., 565, 583
Martin, R. J. L., 428, 478, 488, 554, 560, 580
Masiko, Y., 34, 35, 55
Mason, E. A., 16, 18
Massa, L. J., 400, 415, 423
Masschelein, W., 21, 27
Mastryukov, V. S., 362, 383
Mateos, J. L., 25, 28
Matsumura, C., 196, 214
Matsuua, H., 98, 107, 111, 113
Mayer, J. E., 6, 17
Meakin, P., 176, 207, 209, 212
Meakins, G. D., 565, 583
Meakins, R. J., 49, 56
Mecke, R., 434, 478
Meinzer, R., 119, 154
Melikhova, L. P., 104, 112
Meloy, G. K., 94, 110
Merlino, S., 24, 28
Mertes, M. P., 23, 28
Mettee, H. D., 90, 102, 103, 109, 111, 552, 582
Miazawa, T., 350, 352, 381
Mibrahaski, S., 97, 111
Michel, S., 90, 109
Midwinter, J., 506, 581
Mihachlan, R. D., 91, 110
Mikami, M., 360, 382
Millen, D. J., 157, 202, 205, 212, 214, 215
Miller, F. A., 94, 110, 218, 226, 229, 230, 231, 233, 238, 245, 248, 249, 250, 252, 253, 268, 269, 270, 272, 275, 276, 282, 283, 352, 363, 382, 383, 565, 583
Miller, R. C., 47, 56
Millie, P., 401, 417, 423
Mills, I. M., 75, 96, 108, 110
Minasso, B., 86, 102, 109
Minden, H. T., 269, 282
Miskow, M. H., 468, 469, 473, 474, 476, 480, 578, 579, 584
Mislow, K., 23, 28
Miyake, A., 105, 112
Miyazawa, T., 88, 91, 98, 103, 105, 107, 109, 110, 111, 112, 113, 228, 245, 252, 253, 519, 537, 538, 542, 581, 582
Mizagawa, I., 528, 581

Mizushima, S., 7, 14, 16, 18, 33, 34, 35, 55, 57, 74, 84, 85, 88, 91, 101, 103, 105, 108, 109, 110, 111, 112, 339, 350, 352, 381, 483, 489, 514, 519, 528, 536, 537, 538, 542, 580, 581, 582
Moccia, R., 420, 424
Mohr, E., 5, 10, 17
Moireau, M. Cl., 397, 402, 403, 404, 418, 423, 424
Møllendal, H., 172, 176, 199, 212
Moller, K. D., 218, 220, 229, 230, 233, 234, 245, 247, 249, 250, 252, 253, 270, 271, 272, 282, 283
Monaghan, J. J., 348, 381
Monnig, F., 482, 565, 566, 580
Mooney, E. F., 425, 478
Moore, G., 105, 112
Moran, A. N., 321, 323
Mori, N., 25, 28
Morino, Y., 34, 35, 36, 55, 84, 85, 101, 109, 111, 157, 159, 160, 196, 212, 214, 269, 282, 332, 342, 346, 347, 348, 349, 352, 355, 358, 362, 366, 381, 382, 383
Morokuma, K., 395, 397, 398, 401, 402, 407, 413, 417, 418, 419, 422
Morris, H., 439, 450, 451, 479
Morrison, G. A., 6, 7, 12, 18, 425, 478, 483, 498, 572, 573, 574, 580
Mortensen, E. M., 43, 55
Moscowitz, A., 574, 575, 584
Motzfeldt, T., 356, 357, 382
Mourning, M. C., 22, 27
Muccini, U., 24, 28
Muecke, J. W., 378, 384
Muenter, J. S., 160, 212
Muir, D. M., 25, 28
Muirakami, M., 103, 112
Mukharji, D. K., 107, 112
Muller, H., 37, 38, 55
Mulliken, R. S., 390, 399, 422
Munsch, B., 203, 214, 408, 424
Murata, H., 103, 107, 112
Murata, Y., 352, 358, 381, 382
Murphy, G. M., 231, 232, 252
Musher, J. C., 506, 581
Myers, R. J., 195, 197, 213, 214, 373, 383

Naar-Colin, C., 528, 536, 581
Naberukhin, Yu. I., 83, 109
Nader, F. W., 438, 439, 440, 446, 479
Nagase, S., 351, 352, 381

Nahlovska, Z., 366, 378, 383, 384
Nahlovsky, B., 366, 378, 383, 384
Nakagawa, I., 88, 91, 105, 109, 110, 112, 537, 538, 542, 582
Nakamura, M., 52, 56
Nandy, S. K., 107, 112
Natta, G., 92, 95, 110
Naumov, V. A., 378, 384
Naylor, R. W., Jr., 193, 214
Neckel, A., 36, 37, 55
Nelson, R., 193, 195, 197, 213
Nesbet, R. K., 388, 422
Newmark, R. A., 117, 118, 154, 528, 581
Newton, M. D., 395, 398, 423
Nielsen, H. H., 184, 213
Nielsen, J. R., 85, 87, 101, 102, 109, 111, 269, 278, 279, 282, 283, 528, 529, 531, 581
Nilsen, W. G., 83, 109
Nilsson, J. E., 348, 354, 355, 380, 382
Nixon, E. R., 352, 381
Nolan, F. J., 172, 187, 212
Nordlander, J. E., 25, 28
North, A. M., 315, 322
Nozdrev, V. F., 317, 322
Nybury, S. C., 92, 110
Nyquist, R. A., 91, 95, 110

Oberhammer, H., 364, 383
Oda, T., 103, 113
Oelfke, W. C., 397, 423
Oesper, P. F., 42, 55
Ohno, K., 107, 112
Oi, N., 516, 520, 581
Oka, T., 159, 161, 212
Okada, T., 100, 111
O'Leary, B., 420, 424
Onishi, T., 160, 212
Onsager, L., 30, 37, 54, 486, 499, 502, 580
Orville-Thomas, W. J., 15, 18, 86, 90, 102, 103, 104, 105, 109, 111, 112, 313, 322, 528, 533, 537, 551, 552, 581, 582, 583
Oshima, I., 103, 112
Osipova, L. P., 102, 111, 528, 530, 581
Ostlund, N. S., 498, 580
Ottinger, R., 573, 583, 584
Otto, M. M., 429, 478
Ouellette, R. J., 22, 27, 498, 580
Overend, F., 96, 110

Owen, N. L., 94, 105, 106, 110, 112, 157, 188, 189, 190, 193, 197, 202, 213, 214, 275, 283, 359, 382

Pachler, K. G. R., 84, 85, 109, 119, 154, 492, 501, 502, 507, 508, 512, 515, 521, 525, 526, 528, 529, 552, 553, 580, 581, 583
Padmanaban, R. A., 298, 313, 322
Pal, A., 52, 56
Palke, W. E., 395, 397, 407, 419, 420, 422, 424
Pan, D. C., 402, 418, 424
Pan, Y., 573, 574, 577, 583
Pankova, M., 24, 28
Pardoe, G. W. F., 53, 56
Parish, J. H., 25, 28
Park, P. J. D., 86, 87, 102, 103, 104, 111, 112, 313, 322, 528, 534, 536, 537, 581
Parks, A. T., 360, 383, 545, 582
Parpiev, K., 319, 322
Parry, K., 509, 512, 515, 527, 550, 551, 569, 581, 582
Parry, R. W., 352, 382
Pascard-Billy, C., 565, 583
Pasquier, B., 107, 113
Pasto, D. J., 21, 24, 27, 28
Patel, D. J., 17, 18
Pauling, L., 16, 18, 326, 361, 364, 380, 383
Payer, J. E., 574, 575, 583
Pedersen, L., 395, 397, 398, 401, 402, 407, 413, 417, 418, 419, 422, 423
Pedinoff, M. E., 320, 323
Penn, R. E., 197, 213
Pentin, Y. A., 85, 101, 103, 104, 106, 109, 112, 555, 558, 561, 583
Peraldo, M. L., 100, 111
Perez, C., 25, 28
Peters, C. W., 237, 245, 253, 397, 423
Peters, P. A., 573, 583, 584
Pethrick, R. A., 92, 110, 313, 317, 318, 321, 322, 323, 425, 476, 477, 478, 480, 565, 568, 583
Peticolas, W., 45, 55
Petit, M. G., 204, 215
Petrauskas, A. A., 313, 322
Peyerimhoff, S. D., 401, 408, 416, 423, 424
Pfeiffer, G. V., 401, 417, 423

Phillips, W. D., 120, 154
Piekara, A., 37, 55
Pierce, L., 181, 185, 186, 193, 194, 195, 196, 197, 213, 214, 234, 245, 253, 357, 382
Piercy, J. E., 298, 312, 313, 314, 320, 321, 322, 323
Pihlaja, K., 438, 439, 440, 443, 444, 445, 446, 461, 463, 464, 465, 479, 480
Pilar, F. L., 386, 387, 421, 422
Pilato, L. A., 23, 28
Pincelli, U., 107, 113, 421, 424
Piseri, L., 98, 111
Pitha, J., 77, 108
Pitzer, K. S., 4, 7, 17, 89, 109, 210, 215, 220, 221, 222, 223, 228, 232, 234, 236, 237, 246, 251, 252, 253, 267, 282, 350, 352, 381
Pitzer, R. M., 393, 394, 395, 396, 397, 403, 405, 407, 408, 409, 410, 411, 413, 419, 422, 424
Placzek, G., 75, 108
Plyler, E. K., 269, 282
Polo, S. R., 210, 215
Popkie, H., 420, 424
Pople, J. A., 91, 110, 385, 388, 394, 395, 397, 398, 400, 401, 402, 403, 411, 413, 414, 415, 416, 420, 421, 422, 423, 424, 498, 508, 514, 566, 580
Porri, L., 92, 110
Porter, R. F., 376, 384
Porto, S. P. S., 83, 109
Powell, F. X., 160, 212
Powles, J. G., 47, 56, 118, 154
Powling, J., 85, 101, 102, 109, 519, 562, 581
Prakash, J., 52, 53, 56
Preuss, H., 389, 422
Price, A. H., 30, 46, 54, 56
Price, K., 89, 107, 109, 113
Pringle, W. C., Jr., 205, 215
Pritchard, J. G., 475, 480
Prouder, T., 51, 56
Pruettiangkura, P., 107, 113
Przybylska, M., 277, 283
Ptitsyn, O. B., 45, 56
Pujenkamp, H., 103, 112
Pullman, A., 421, 424
Purcell, E. M., 135, 155
Purcell, W. P., 42, 55

Quade, C. R., 193, 195, 207, 214, 215, 232, 240, 252, 253
Queneudec, M., 103, 112

Rabi, I. I., 128, 155
Radom, L., 394, 400, 401, 402, 403, 411, 413, 414, 415, 416, 420, 422, 423, 424
Ragle, J. L., 150, 155
Raileanu, D. I. C., 23, 27, 440, 479
Raju, V., 317, 322
Raman, C. V., 61, 108
Rambidi, N. G., 331, 332, 380, 381
Ramsay, D. A., 78, 108
Ranck, J. P., 395, 398, 419, 423
Rank, D. H., 101, 111
Rankin, D. W. H., 352, 353, 373, 374, 382, 383
Rao, R. D., 21, 27
Rao, S., 317, 322
Rasmussen, J. J., 43, 55
Rasmusson, G. H., 22, 27
Rassat, A., 151, 155
Rauk, A., 401, 402, 418, 421, 423, 424
Rea, D. G., 83, 109
Read, B. E., 45, 56, 100, 111
Record, K. A. F., 23, 27, 441, 442, 444, 450, 479
Redfield, A. G., 149, 155
Reddy, G. S., 551, 582
Ree, T., 43, 55
Rees, B., 401, 423
Reese, M. C., 22, 27
Reeves, L. W., 133, 155
Reichardt, C., 505, 580
Reilly, G. R., 120, 154
Reisse, J., 428, 478, 573, 583, 584
Reitz, A. W., 425, 478
Rerick, M. N., 23, 28
Reynolds, W. F., 507, 545, 546, 547, 548, 549, 581, 582
Rhodes, I., 234, 252
Richardson, J. W., 389, 422
Riche, C., 565, 583
Richer, J. C., 21, 27
Rickbourn, B., 22, 27
Riddell, F. G., 23, 27, 425, 429, 440, 461, 465, 478, 480
Rieger, P. H., 151, 155
Rigden, J. S., 197, 213
Ripamonti, A., 97, 111

Risberg, F., 326, 380
Ritchie, C. D., 393, 411, 422
Ritchie, G. L. D., 515, 581
Riveros, J. M., 175, 199, 207, 210, 212
Ro, R. S., 20, 25, 27, 28
Robert, J. B., 402, 419, 424
Roberti, D. M., 49, 56
Roberts, J. D., 550, 582
Robertson, E. D., 367, 383
Robertson, J. Monteath, 277, 283
Robiette, A. G., 352, 373, 374, 382, 383, 384
Robinson, M. J. T., 22, 23, 26, 27, 429, 440, 478
Roebuck, A. K., 21, 27
Rolfe, R. E., 152, 155
Romers, C., 425, 429, 456, 467, 475, 478, 479, 578, 584
Ronn, A. M., 196, 214
Ronova, I. A., 356, 382
Roos, B., 390, 391, 422
Roothaan, C. C. J., 387, 388, 422
Roques, B., 565, 583
Ros, P., 401, 417, 423
Rossetti, Z. L., 564, 583
Rossiter, R. F., 53, 54, 56
Rothenberg, S., 391, 417, 422
Rothenberg, S. R., 401, 402, 423
Rothschild, W. G., 204, 214
Roulph, C., 107, 113
Roussy, G., 206, 215
Roy, S. B., 107, 112
Rudolph, H. D., 157, 160, 197, 206, 212, 214, 215, 482, 565, 566, 580
Rudolphi, R. W., 352, 382
Ruedenberg, K., 388, 422
Russell, G. A., 144, 154, 155
Russell, H., 267, 269, 282
Russell, J. W., 203, 204, 214
Ryan, R. R., 369, 383

Sabashita, K., 555, 561, 583
Sachse, H., 5, 10, 17, 425, 478
Sadova, N. I., 362, 383
Saegebarth, E., 196, 209, 213, 215, 542, 582
Sage, G., 229, 230, 233, 252, 269, 282
Sage, M. L., 195, 197, 213
Saika, A., 120, 154
Saito, S., 160, 212

Sakurai, Y., 160, 212
Salem, L., 420, 424
Sales, K. D., 152, 155
Salova, G. E., 375, 384
Salovey, R., 97, 111
Samdal, S., 359, 382
Samitov, Y. Y., 470, 471, 472, 480
Sandorfy, C., 79, 108
Sarel, S., 475, 480
Sasaki, Y., 375, 384
Sauki, S., 107, 113
Saunders, J. E., 89, 109
Saunders, V. R., 395, 398, 407, 419, 423
Saunders, W. H., 25, 28
Scarzafava, E., 400, 401, 414, 415, 423
Schaad, L. J., 401, 416, 423
Schachtschneider, J. H., 96, 110
Schaefer, H..F., 391, 421, 422, 424
Scharpen, L. H., 204, 205, 215
Schatz, P. N., 80, 108
Schaufele, R., 98, 111
Schiff, L. I., 239, 242, 253
Schleyer, P. V. R., 400, 411, 415, 416, 420, 423, 424
Schmid, H. G., 434, 435, 436, 445, 458, 459, 460, 478
Schneider, B., 105, 112
Schneider, M., 107, 112, 552, 582
Schomaker, V., 326, 353, 364, 376, 380, 382, 383, 384
Schowen, R. L., 23, 28
Schroeter, S. H., 20, 21, 27
Schug, J. C., 118, 154
Schwartz, M. E., 401, 402, 407, 411, 417, 419, 423, 424
Schwartz, S. E., 83, 109
Schwendeman, R. H., 196, 213, 268, 282, 337, 381
Scott, D. W., 103, 112
Sears, W. F., 301, 322
Sederholm, C. H., 117, 118, 154, 528, 581
Segal, G. A., 385, 421
Seibold, E. A., 354, 382
Seiler, H., 197, 214
Seip, H. M., 328, 338, 348, 349, 350, 351, 359, 366, 367, 368, 373, 374, 378, 379, 380, 381, 382, 383, 384
Sepp, D. T., 22, 23, 27, 28
Serbali, C., 86, 102, 109
Seshagiri Rao, M. G., 298, 312, 313, 322
Sevenair, J. P., 533, 581

Shah-Malak, F., 24, 28
Shavitt, I., 389, 395, 397, 400, 416, 418, 422, 423
Shaw, K. N., 133, 155
Shawlow, A. L., 157, 161, 172, 187, 199, 211
Sheldrick, G. M., 352, 373, 374, 382, 383, 384
Shelton, R., 94, 110
Sheppard, N., 57, 87, 88, 89, 101, 102, 104, 105, 107, 108, 109, 111, 112, 113, 118, 154, 457, 479, 483, 519, 528, 535, 580
Sheridan, J., 157, 188, 189, 190, 193, 197, 202, 212, 213, 214
Sherwood, J., 320, 323
Shido, N., 88, 109, 537, 538, 542, 582
Shiengthong, S., 22, 27
Shimanouchi, T., 35, 55, 74, 75, 85, 88, 89, 91, 98, 101, 103, 105, 106, 108, 109, 110, 111, 112, 268, 282, 350, 352, 359, 360, 381, 382, 537, 538, 542, 582
Shindo, Y., 100, 111
Shiner, V. J., Jr., 25, 28
Shiro, Y., 103, 112
Shoemaker, R. L., 161, 212
Shulgin, A. T., 106, 112
Sicher, J., 24, 28, 428, 478
Sidran, M., 172, 187, 212
Siegbahn, P., 390, 391, 422
Siegel, S., 196, 213
Siegl, W. O., 23, 28
Simon, I., 81, 108
Sinanoglu, O., 393, 394, 422
Sipos, F., 428, 478
Siverns, T. M., 481, 482, 505, 554, 559, 565, 567, 568, 580, 581
Skancke, A., 366, 367, 383
Skancke, P. N., 376, 384
Skinner, H. A., 348, 381
Skinner, J. G., 83, 109
Slichter, C. P., 120, 154
Slie, W. M., 298, 322
Smedvik, L., 366, 367, 383
Smith, D. C., 282
Smith, D. R., 218, 220, 229, 230, 233, 234, 245, 247, 249, 250, 252, 270, 272, 283
Smith, D. W., 370, 383
Smith, J. W., 512, 581
Smith, R. P., 16, 18, 43, 52, 55
Smith, W. V., 157, 161, 211

Smyth, C. P., 30, 31, 32, 33, 34, 35, 37, 38, 40, 41, 42, 46, 47, 48, 49, 50, 51, 52, 53, 54, 55, 56, 102, 111, 279, 283, 486, 528, 530, 553, 580, 581, 583
Snowden, R. L., 565, 583
Snyder, E. I., 482, 545, 546, 547, 548, 580, 582
Snyder, R. G., 96, 97, 110, 111
Sourisseau, C., 107, 113
Southam, R. M., 25, 28
Sovers, O. J., 403, 409, 424
Spiridonov, V. P., 331, 332, 380, 381
Spragg, R. A., 457, 480
Srivastava, R. M., 477, 480
Staughan, B. P., 94, 110
Stearn, A. E., 35, 55
Stein, R. S., 100, 111
Stejskal, E. O., 234, 253
Stephens, R. D., 154, 155
Stevens, R. M., 389, 395, 397, 405, 406, 407, 413, 419, 422
Stewart, G. H., 16, 18
Stiefvater, O. L., 176, 199, 209, 210, 212
Stiles, P. J., 515, 581
Stockmair, W., 425, 478
Stockmayer, W. H., 44, 55
Stokr, J., 105, 112
Stølevik, R., 328, 338, 348, 375, 380, 381, 384
Stosick, A. J., 375, 384
Strand, T. G., 328, 338, 366, 380, 381, 383
Strandberg, M. W. P., 203, 214
Strange, J. H., 118, 154
Strauss, H. L., 203, 204, 214
Strom, E. T., 154, 155
Stromme, K. O., 378, 384
Strothers, J. B., 573, 574, 577, 583
Struchkov Yu, T., 356, 382
Subrahmanyam, S. V., 315, 320, 321, 322, 323
Sudi, F., 25, 28
Sugden, T. M., 157, 161, 211, 212
Sullivan, P. D., 147, 150, 154, 155
Suppan, P., 506, 581
Sutcliffe, L. H., 63, 108, 117, 154
Sutherland, G. B. B. M., 203, 214
Sutherland, I. O., 116, 154, 425, 478
Sutton, C., 52, 56
Sutton, L. E., 348, 354, 381, 382
Suzuki, I., 105, 112
Swaelens, G., 445, 446, 447, 461, 462, 479

Swalen, J. D., 157, 183, 184, 186, 187, 189, 193, 195, 196, 203, 212, 213, 214, 219, 220, 222, 223, 224, 225, 232, 245, 247, 252, 253
Swick, D. A., 342, 344, 345, 346, 347, 381
Symons, M. C. R., 314, 322
Szasz, G. J., 86, 102, 109, 111
Szkrybalo, W., 21, 22, 27
Szymanski, H. A., 83, 108

Tabokoro, H., 97, 111
Tabuchi, D., 315, 322
Tadokora, T., 99, 100, 111
Tagami, E., 528, 536, 581
Takeda, M., 528, 536, 581
Takeshita, T., 25, 28
Takuma, H., 83, 109
Talaty, E. R., 154, 155
Taller, R. A., 538, 582
Tamaru, K., 160, 212
Tamies, M., 38, 55
Tanabe, K., 107, 112, 113, 517, 519, 581
Tannaka, M., 315, 322
Tannenbaum, E., 197, 214
Tarasenko, N. A., 375, 384
Tarasoff, L., 469, 480
Tasumi, M., 89, 97, 98, 109, 111
Tatetevskii, V. M., 85, 101, 103, 109
Tavernier, D., 472, 480
Taylor, R. C., 352, 382
Teinonen, K. J., 440, 445, 479
Tel, L. M., 418, 421, 424
Tenhosaari, A., 439, 440, 479
Thom, E., 375, 384
Thomas, E. C., 205, 215
Thomas, J. R., 33, 55, 352, 381
Thomas, L. F., 429, 478
Thomas, T. H., 104, 105, 112, 313, 322
Thomas, W. A., 425, 478
Thorbjørnsrud, J., 107, 112
Thornburrow, P. R., 24, 28
Tichy, M., 24, 28, 428, 478
Tillett, J. G., 470, 480, 578, 584
Timasheva, T. P., 362, 383
Timmons, C. J., 94, 110
Tolles, W. M., 197, 214
Tolman, C. A., 203, 215
Tomlinson, D., 353, 382
Tonelli, A. E., 17, 18
Topping, G., 202, 214
Torchia, D. A., 17, 18

Torgrimsen, T., 107, 112
Torii, T., 97, 111
Townes, C. H., 157, 161, 172, 187, 199, 211
Traetteberg, M., 362, 363, 364, 365, 373, 374, 383
Trambarulo, R. F., 157, 161, 211
Trunel, P., 41, 55
Tryan, M., 95, 110
Tsumura, K., 99, 111
Tsunetake, H. J., 75, 108
Turner, J. J., 101, 111, 118, 154
Tushaus, L. A., 22, 27
Tyler, J. K., 202, 214
Tyulin, V. I., 106, 112

Uebel, J. J., 22, 27, 428, 478
Ukajii, T., 353, 382
Ukita, M., 97, 111
Ul'Yanova, O. D., 104, 112, 555, 558, 561, 583
Unland, M. L., 400, 401, 406, 411, 414, 423
Urey, H. C., 75, 108
Usieli, V., 475, 480
Utley, J. H. P., 24, 28, 152, 155

Van Catledge, F. A., 21, 27
Van Meurs, N., 43, 55
Van Oyen, J. W. L., 578, 584
Van Waser, J. R., 400, 401, 402, 406, 411, 414, 418, 419, 421, 423, 424
Van Woerden, H. F., 470, 472, 473, 474, 475, 476, 480, 578, 584
Vane, F. M., 565, 583
Varma, R., 197, 214
Vaughan, G., 375, 384
Vaughan, W. E., 30, 46, 51, 52, 54, 56
Vega De La, J. R., 401, 417, 423
Vehoor, G. C., 578, 584
Veillard, A., 385, 393, 395, 397, 398, 399, 400, 401, 402, 403, 407, 411, 413, 418, 419, 422, 423, 424
Verdier, P. H., 198, 214
Verma, A. L., 107, 113
Vernai, C., 395, 419, 423
Vilkov, L. V., 353, 362, 375, 382, 383, 384
Vincent-Geisse, J., 80, 81, 108
Visser, W. M., 80, 81, 108
Volk, H., 36, 37, 55
Volkenstein, M. V., 45, 56, 84, 109

Vollmer, R. L., 475, 480
Von Niesson, W., 408, 413, 424

Wada, A., 35, 36, 52, 55, 84, 109, 500, 519, 580
Wagner, E. L., 421, 424
Walker, S., 38, 39, 55
Walker, S. M., 315, 321, 322, 323, 502, 503, 580
Walsh, A. D., 361, 383
Walsh, B., 439, 450, 451, 478, 479
Wang, G. L., 575, 584
Wangsness, R. K., 149, 155
Waring, A. J., 22, 27
Watanabe, I., 34, 35, 55, 85, 101, 109, 111
Watts, V. S., 551, 582
Wegmann, L., 326, 380
Weiner, D., 83, 109
Weiss, M. T., 203, 214
Weiss, S., 261, 268, 269, 282, 396, 423
Wellman, K., 574, 575, 584
Wells, A. J., 77, 78, 108
Wendling, P., 197, 214
Wessels, P. L., 84, 109, 515, 521, 528, 529, 552, 553, 581, 583
Westheimer, F. H., 6, 17
Whipple, E. B., 550, 582
White, R. F. M., 313, 321, 322, 323, 445, 458, 470, 471, 472, 473, 476, 477, 479, 480, 578, 584
Whitesides, G. M., 533, 581
Whiting, M. C., 24, 25, 28
Whitman, D. R., 420, 424
Whitten, J. L., 389, 422
Wierl, R., 325, 326, 380
Willadsen, T., 378, 384
Williams, G., 45, 56, 188, 189, 190, 193, 197, 213
Williams, J. E., 400, 415, 423
Williams, J. W., 31, 54
Williams, N. H., 200, 214
Williams, R. L., 537, 538, 542, 544, 582
Williams, S. H., 498, 580
Williams, T. H., 428, 478
Willis, J. N., 89, 109
Wilmhurst, J. K., 102, 111, 519, 581
Wilson, E. B., 176, 211, 212, 215
Wilson, E. B., Jr., 32, 55, 76, 77, 78, 95, 108, 110, 157, 175, 176, 183, 184, 192, 193, 195, 196, 198, 199, 207, 209, 210, 212, 213, 214, 215, 218, 219, 230, 252, 253, 337, 358, 382, 542, 582

Wilson, M. K., 353, 357, 382
Windle, J. J., 152, 153, 155
Winnewisser, M., 353, 382
Winstein, S., 24, 25, 28, 428, 478
Winter, N. W., 395, 398, 419, 423
Witenhofer, D. E., 100, 111
Witkowski, R. E., 93, 94, 105, 110, 238, 253, 275, 276, 283, 363, 383, 495, 565, 580, 583
Wodarczyk, F. J., 199, 209, 214
Wolfe, S., 401, 402, 418, 421, 423, 424
Wolfsberg, M., 400, 415, 423
Wolkenstein, H., 75, 108
Wollrab, J. E., 157, 161, 185, 187, 197, 202, 210, 211, 212, 213, 214
Wolniewicz, L., 393, 422
Woo, E. P., 556, 563, 583
Wood, D. J., 507, 545, 546, 547, 548, 549, 581, 582
Wood, G., 468, 469, 473, 474, 476, 477, 480, 556, 563, 578, 579, 583, 584
Wood, J. L., 240, 241, 243, 253, 269, 270, 272, 282
Wood, J. W. M., 53, 56
Woods, R. C., 184, 187, 196, 197, 213, 214
Woodward, A. J., 516, 520, 530, 545, 551, 552, 581, 582
Wyatt, J. F., 395, 398, 407, 419, 423
Wyn-Jones, E., 15, 18, 86, 87, 90, 92, 102, 103, 104, 105, 109, 110, 111, 112, 298, 312, 313, 317, 318, 321, 322, 323, 425, 439, 445, 446, 450, 451, 458, 476, 477, 478, 479, 480, 528, 533, 534, 536, 537, 551, 552, 565, 568, 581, 582, 583

Yamaguchi, A., 352, 381
Yamaguchi, S., 35, 55, 85, 101, 109, 111
Yamaha, M., 351, 381
Yamanouchi, Y., 573, 574, 583
Yamarota, D., 552, 582
Yasufuka, K., 97, 111
Yasumi, M., 103, 111
Yates, K., 417, 424
Yee, K. C., 25, 28
Yip, S., 263, 267, 282
Yokoi, M., 375, 384
Yoshimine, M., 388, 422
Yoshino, T., 556, 564, 583,
Yost, D. M., 267, 269, 282
Young, J. M., 313, 322
Young, M. C., 154, 155

Zaripov, N. M., 378, 384
Zbinden, R., 95, 110
Zeeck, E., 400, 401, 414, 423
Zeil, W., 326, 380
Zelinsky, N. D., 21, 27
Zerbi, G., 95, 96, 97, 98, 110, 111
Zimmermann, D., 428, 478
Zinn, J., 203, 204, 214
Zumwalt, L. R., 75, 108

Subject Index

Ab initio calculations, 211, 385
Absorption, 285
  cells, 164
  per wavelength, 291
Adiabatic compressibility, 287–8
Alternating line width effect, 147
Angular frequency, 30
  momentum, 169–71, 184–5, 187, 192, 194
  probability distribution, 340, 343, 346, 355, 368
  probability function, 375
Anharmonicity, 333–4
Anomeric effect, 23, 448
Asymmetric rotating groups, 206–7
  rotor, 170, 172, 174, 176, 184, 259
  top, 170, 236, 240, 258
Atomic background, 325, 334
Attenuated total reflection, 80

Backward wave oscillators (BWO), 161–4
Band contours, 75
  intensity, 76
  shapes, 64, 107
Barrier shape, 355, 402
  to internal rotation (calculation of), 217
Basis set, 388
Beat frequency, 165
Bloch equations, 123, 149
Boltzmann equation, 33–4, 175, 198
Bond orbital, 391, 409
Bootstrap method, 186
Boundary conditions, 220

Carr–Purcell sequence, 136
Centrifugal distortion, 176, 178

Chair/boat interconversion, 320
  relaxation, 320
Clausius–Morotti–Dekyl equation, 502
Cole–Cole arc plots, 46, 52–3
Cole–Davidson equation, 46
Collision absorption, 53
Commercial spectrometers, 166
Concentration dependence, 294
Configuration interaction, 392
Conformational analysis, 2, 7
  behaviour, 211
  energies, 428, 434–47
  equilibria, 19–20, 23, 25–6
Conformations, 2, 7
Conjugation, 363, 366
Constitutional formula, 1
Coriolis forces, 176, 178
Correlation, 392
  energy, 393
  function, 66, 69, 150
  theory, 63, 107
Cosine potential, 341, 343, 347, 355
Coupled asymmetric rotors, 278
Coupling constant, 487, 491, 498
  C—H, 318
Coupling term, 194
Crystal detector, 162
  diode, 164–5
CX_3 tops and C_{3v} symmetry, 245

Damping constant, 341
Davidson–Cole equations,
  skewed arcs (see Cole–Davidson)
Debye behaviour, 46
  equation, 45, 486
Degeneracy, 170
Degree of freedom, 208

Delay line, 304
Density matrix, 127
Depolarization ratio, 63
Dielectric constant, 30, 37
  complex, 45
  optical, 30
  static, 29
Dielectric loss, 30
  relaxation, 30
    time, 37
Diffraction, 306
Diffusion effects, 139
Dipole moment, 29, 36, 159–60, 172, 199–200, 207, 486, 488
Directional cosines, 180, 184, 189
Dispersion methods, 79
Distribution coefficients, 38
Double resonance techniques, 139
Dynamic equilibrium, 6

Echo, 305
Elastic scattering, 266
Electric dipole, 29
Electric field gradient, 160
Electron diffraction, 497
  instruments, 326
  technique, 325
Electron scattering, 329
  spin resonance, 141
Electronegativity, 273
Electrostatic attractive force, 86
Ellipsoid of inertia, 168
Energy barrier, 8
  calculations, 378
  difference, 71, 84, 197, 210
  transfer, 161
Enthalpy difference, 58, 85
Entropy of activation, 454
Equilibrium studies, 432
Exchange matrix, 129
Expansivity, 287

Fabry–Perot etalon, 80
Far infrared absorption, 53
First Born approximation, 326
Five membered rings, 377
Fourier coefficients, 208–10
  component, 208
    analysis, 420
Fourier series, 178, 181, 187, 194, 402
  transform, 81, 335

Four-membered rings, 376
Frame vibrational amplitude, 340, 370, 375
  vibrations, 340, 346
Framework amplitude of vibration, 347, 355, 368
  fixed axis system, 195, 207
Free energy difference, 432
  internal rotation, 181, 192
    quantum number, 192
Free energy rotator, 193
  rotation, 339, 343, 376
    quantum number, 182
Freezing point measurements, 35

Gauche isomers, 236
Gaussian functions, 388
Geometry optimization, 395–6, 406
Generalized mean-square amplitudes of vibration, 329
Gunn microwave oscillator, 167

Halogenoethanes, 279
Hamiltonian, 128, 170–2, 183–5, 187, 191–5, 202
Harmonic approximation, 250
  oscillator, 182, 185
    approximation, 238, 251
    energy levels, 252
    limit, 247, 250
    perturbed, 201
    quantum number, 182
Harmonic vibration, 332
Hartree–Fock, 387
  energy, 393
Heat capacity, 287
Hermite polynomials, 239
High barrier, 181
Hot bands, 229
Hydrogen bonding, 39, 315, 351, 366

Inelastic scattering, 266
Inertia moments of, 158, 168–70, 176, 180, 183–4, 189, 207–8
  reduced, 221
Inertial defect, 158–9, 176–7
  tensor, 168
Intensity of an infrared band, 75
Infrared birefringence, 100
  dichroism, 100
  and Raman activity, 58
  spectroscopy, 260, 494, 496

Interaction between internal and overall
    rotation, 219
  angular momentum, 220
Interferogram, 262
Interferometer, 82
  Michelson, 81, 260–1
Interferometry, 81
Internal axis method (IAM), 183–4, 187,
    195, 219
  field, 34
  and overall rotation separation of,
    220, 222
  rotation, 157, 199, 255
Internal rotation barrier, 196, 211
  methods chemical, 12
    diffraction, 13
    physical, 13
    relaxation, 15
    spectroscopic, 14
    theoretical, 15
  mode, 217
  splittings, 178
Intramolecular relaxation times, 45
Inversion, 49, 178, 199–203
  splittings, 201
Isothermal compressibility, 287
Isotopic effects, 74
Isotropic hyperfine interactions, 142

j factors, 150

'kappa' a parameter which indicates the
    asymmetry of a molecule, 170
Kinetic energy, 221
  for J symmetric tops, 246
  methods for conformational analysis,
    24
Kinematic studies, 452
Klystrons, 161–4
Kramers–Heisenberg formula, 79

Larmor precession, 121
LCAO–MO–SCF method, 386
Least squares refinement, 333
  variation, 326
Linear molecules, 169
Line-shape analysis, 120
Line-shape classical theory, 121
  strength, 199
Liouville operator, 130
Lorentz curve, 78

Loss tangent, 30
Low barrier, 181, 191

McConnell equation, 142
Macroscopic relaxation time, 47
Magnetic quantum number, 165
  susceptibility, 161
Mathieu equation, 180, 185, 198, 223,
    229, 257, 270
  functions, 181, 187, 225
    tables, 223, 233–4, 252
Mechanical coupling, 193
Microphotometer, 328
Microwave double resonance, 161
  line splitting, 183
  spectrometer, 163, 165, 167
  spectroscopy, 157, 160–1, 497
Modification function, 335
Modified intensity curve, 335
Molecular intensity, 333, 335
  large amplitude motion, 199
  polarizability, 29
  quadrupole moments, 161
  relaxation times, 37, 47
  rotation, 178
Molecules with more than one top, 245
Multiple scattering, 332

Neutron incoherent inelastic scattering
  (N.I.I.S.), 260, 263–5, 267
Nielsen transformation, 220–2
NMR parameters, 471
Normal coordinate analysis, 332, 347
  calculations, 245
Normal modes, 59
  vibrations, 221, 230
Nuclear magnetic resonance, 115
  spin functions, 183
Nucleophilic substitution reactions, 22

Oblate top, 169–70
Octupole, 506
Onsager equation, 486, 503, 506
  model, 84

P branches, 172
Partition functions, 72
Permittivity, 29
Phase sensitive detector, 162, 165
Phenyl tops, 275

Photometer measurement, 325
Piezoelectric transducers, 298
Planck's constant, 169
Polarizability, 60
  tensor, 69
Polarization, 30
  functions, 390, 395
Potential barrier, 7, 38, 178–80, 182, 186, 196–9, 202–3, 208, 210–11
Potential energy, 8, 57
  curve, 57, 176
Potential function, 218
Pressure broadening, 199
  or density changes in acoustic relaxation processes, 317
Principal axis method (PAM), 183–4, 194, 219, 247
Principal internal axis system, 220
Principal moments of inertia, 231–3
Prolate tops, 169
Pseudo rotation, 203, 378
Puckering motion, 203

Q branches, 172
Quadratic potential, 343, 370
Quadrupole, 501–2, 504, 507, 510, 518, 560
  coupling constant, 160
  moments, 160
Quantum mechanical theory, 127
  tunnelling, 199–200, 207
Quantum number, 169–70, 172
Quartz crystal, 166

R branches, 172
'Rabbit ear', 448
Radial distribution curve, 335
r_a distances, 334
Raman effect, 61
  line intensity, 34
  scattering, 61, 68
  spectroscopy, 82, 260
Ramsey method, 78
Rayleigh scattering, 61
Reduced barrier, 180, 188–9, 198
  mass, 202, 210
  moments of inertia, 194, 232
Refractive index, 30
Relative intensity, 198
  measurements, 197
  method, 198
  studies, 210

Relativistic energy, 393
Relaxation,
  frequency, 290, 294
    effect of pressure change, 298
  matrix, 149
  processes, 31
  strength, 289
  time, 38, 289–90
  width, 291
Relaxing specific heat, 295, 297
r_g distance, 333
Rigid rotor, 174, 176
Ring puckering, 199, 203
Rotation-inversion phenomena, 202
  barrier, 32, 409
    analysis, 404
    origin, 403
Rotational constants, 160, 171, 173–5, 178, 185–6, 189, 191, 199
  isomerism, 57, 206–7, 209, 211, 255
  isomers, 33, 178, 206–7, 210, 260
  spectra, 168–9, 174, 176
  transitions, 165

Saturation of absorption lines, 199
  effect, 37
Scattering factors, 326
Schlieren technique, 301
Schrödinger wave equation, 180, 192, 257, 343, 355
Sector, 326–7
Selection rules, 161, 169–70, 172, 174, 183, 192, 230, 248, 250
Self consistent field (SCF) equations, 387
Several tops, 245
Shear viscosity, 286
Shrinkage effect, 333, 339, 345
Simple harmonic equation, 198
'Sing around' technique, 305
Single top with N-fold symmetry, 218
Slater and Gaussian functions, 388
Solute-solvent interaction, 39
Solvent basicity, 39
  interactions, 85
Sound absorption coefficient, 286, 290
Spin-echo technique, 136
  isochromats, 136
  quantum number, 160, 174
Stark effect, 160, 202
  electrode, 162, 165
  field, 165–6

generator, 162
lobes, 199
modulated spectrometer, 161–2
modulation, 165
Stark shift, 160
voltage, 165
Steric effects, 363, 366, 432
repulsive forces, 86
Stokes–Navier equation, 286
Strain theory, 5
Structural relaxation, 286
Substituted cyclohexanes, 428
Symmetric top, 169–70, 178, 184, 255, 267

Teller–Redlich product rule, 74
Thermal conductivity, 286
relaxation, 286, 289
Three CX_3 tops, 250
top molecules, 196
Time-of-flight neutron spectrum, 264
Töpler technique, 301
Top–top interaction, 194–5, 247
Torsion and other vibrations mixing, 218, 244
Torsional equation, 223
frequencies, 217
kinetic energy, 219
motion, 340
selection rules for, 230
—rotational transitions, 202
strain, 364
vibration, 255, 368
wave equation, 223
Trans isomers, 236
Transition moment, 172, 200
Tunnelling, 228
Twist-boat conformation, 461
Twisting vibration, 343

Two CX_3 tops, 245
dissimilar tops, 251
tops, 249
top molecules, 195
with C_{2y} and C_s symmetries, 245

Uncertainty principle, 146
Unified theory of exchange, 131
Unimolecular equilibria, 294
reactions, 292
Unstable species, 160
Uphill curve, 329, 332
data, 334

Valence-angle strain, 364
van der Waals forces, 404
interaction, 364
Velocity, 303–5, 319
dispersion, 285, 290, 311, 319
Vibration–rotation interaction
constants, 159, 175–7
Vibrational analysis, 74
bands, 67, 72
frequency assignment, 74
quantum number, 175
specific heat relaxation, 286
Viscothermal, 286
absorption, 290
Volume change, 289, 312
accompanying the relaxation process, 298
Volume viscosity, 286

Wave guides, 164–5, 174
Wavelength–electron beam, 327
Wave-meters, 166
Weight matrix, 338

Zeeman splitting, 161